Routledge R

THE DEVELOPMENT OF TRANSPORTATION IN MODERN ENGLAND

THE DEVELOPMENT OF
TRANSPORTATION
IN MODERN ENGLAND

W. T. JACKMAN

With a new Introduction by
W. H. CHALONER

Routledge
Taylor & Francis Group

First published in 1962 by Frank Cass and Company Limited

This edition first published in 2018 by Routledge
2 Park Square, Milton Park, Abingdon, Oxon, OX14 4RN
and by Routledge
52 Vanderbilt Avenue, New York, NY 10017, USA

Routledge is an imprint of the Taylor & Francis Group, an informa business

© 1962 by Taylor and Francis

Publisher's Note
The publisher has gone to great lengths to ensure the quality of this reprint but points out that some imperfections in the original copies may be apparent.

Disclaimer
The publisher has made every effort to trace copyright holders and welcomes correspondence from those they have been unable to contact.
A Library of Congress record exists under ISBN: 27023241

ISBN 13: 978-0-367-17595-5 (hbk)
ISBN 13: 978-0-367-17596-2 (pbk)
ISBN 13: 978-0-429-05753-3 (ebk)

THE DEVELOPMENT OF TRANSPORTATION
IN MODERN ENGLAND

THE DEVELOPMENT OF
TRANSPORTATION
IN MODERN ENGLAND

W. T. JACKMAN

With a new Introduction by
W. H. CHALONER
University of Manchester

FRANK CASS & CO. LTD.
1962

First published by Cambridge University Press, 1916

Second Edition (Revised) *1962*

© 1962

Published by FRANK CASS & CO. LTD.

10 WOBURN WALK, LONDON, W.C.1

This book has been printed in Great Britain by
offset litho at Taylor Garnett Evans & Co. Ltd,
Watford, Herts, and bound by them

CONTENTS

MAPS

.

The attention of readers is drawn to the two maps of English
river navigations in T. S. Willan, *River Navigation in England,
1600–1750* (1936) and to the maps in Charles Hadfield's *British
Canals* (1st edn. 1950, 2nd edn. 1959).

NEW INTRODUCTION

*An Introductory Guide to recent published
work on the History of British Transport up
to 1850*

BY

W. H. CHALONER

INTRODUCTORY GUIDE

INTRODUCTORY GUIDE

GENERAL

THE late W. T. Jackman's *Development of Transportation in Modern England* was recognized on publication as a standard work of profound scholarship and enormous industry. It was based on research extending over many years in the Public Record Office, the British Museum, the Library of the London School of Economics and Political Science and the Goldsmiths' Library, as well as a number of the great municipal libraries in the provinces.

Although most of the narrative is concerned with the period 1500–1850, the author interpreted the word "modern" in a wide sense, and dealt with Roman roads in Britain, and with mediaeval transport. The book contains sections on roads, river navigations, canals, railways, steam navigation and postal history, besides nearly a hundred pages of valuable appendices. It has never been superseded.

Since May 1953 *The Journal of Transport History* has been published twice yearly by Leicester University Press, and its files contain many important articles on numerous aspects of British transport history. Particularly valuable are the various contributions on the sources of such history, the reviews and lists of recent publications, and H. J. Dyos's list of transport theses in the university libraries of the British Isles (vol. IV, May 1960, pp. 161–73). The *Economic History Review*, founded in 1927, has from time to time published articles on aspects of British transport history, a selection from which appears below. The references to transport, communications, roads and waterways in *An Historical Geography of England before 1800* (ed. H. C. Darby, 1st edn., 1936, 2nd rev. edn., 1948) are most useful.

ROADS

Considering the importance of the subject comparatively little has been written on roads. It is true that Ivan D. Margary's two volumes on *Roman Roads in Britain* (1955–7) have superseded much of the matter previously published on this subject, but for the mediaeval period in England there are only a few outstanding articles: Sir Frank Stenton, "The road system of mediaeval England" (*Economic History Review*, vol. VII, Nov. 1936, pp. 1–21); G. B. Grundy, "The evidence of ancient Saxon land charters on the ancient road-system of Britain" (*Archaeological Journal*, vol. LXXIV (1917), pp. 79–105; G. B. Grundy "The ancient highways and tracks of Wiltshire, Berkshire, and Hampshire, . . ." (ibid., vol. LXXXV (1918), pp. 69–194; G. B. Grundy, "The ancient highways and tracks of Worcestershire and the Middle Severn Basin" (ibid., vol. XCI, 1935, pp. 66–96, 241–68 [Parts I and II] and vol. XCII, 1936, pp. 98–141 [Part III]; G. B. Grundy, "The ancient highways of Dorset, Somerset and South-West England" (ibid., vol. XCIV, 1938, pp. 257–90); H. J. Hewitt, *Mediaeval Cheshire* (1929), chap. IV on transport and communications; C. G. Crump's exposure of the myth of the Pilgrim's Way across Hampshire, Surrey and Kent (*History*, n.s., vol. XXI, 1936, pp. 22–33); and J. F. Willard's two articles ("Inland transportation in England during the fourteenth century", *Speculum*, vol. I, Oct. 1926, pp. 361–74, and "The use of carts in the fourteenth century", (*History*, n.s., vol. XVII, Oct. 1932, pp. 246–50). In the second of his two articles Professor Willard shows that the cart was a far more important means of transport in mediaeval England than is generally supposed; the proportion of pack-horse traffic has been exaggerated by historians. These findings have been confirmed by O. Coleman's edition of *The Brokage Books of Southampton, 1443–1444* (1960). For Scotland, R. P. Hardie's *The Roads of Mediaeval Lauderdale* (1942) should be noted.

Neither is the modern period much more adequately served. R. J. Forbes ("Roads to c. 1900", *A History of Technology*, ed. Singer, Holmyard, Hall and Williams, vol. IV, 1958, pp. 520–47) deals almost entirely with the technical aspects of roadmaking. F. G. Emmison has printed "The earliest turnpike bill (Biggleswade to Baldock road), 1622" (*Bulletin of the Institute of Historical Research*, vol. XII, 1935, pp. 108–22), and G. H. Tupling has given us his admirable paper on "The turnpike trusts of Lancashire" (*Memoirs and Proceedings of the Manchester Literary and Philosophical Society*, Session 1952–3, vol. 94, pp. 1–23), while J. L. Hobbs has published "The turnpike roads of North Lonsdale" (*Trans. Cumberland and Westmorland Antiquarian and Archaeological Society*, n.s., vol. LV, 1956, pp. 250–92), and J. D. Marshall has a section on road history in his *Furness and the Industrial Revolution* (1958, pp. 82–4). Nevertheless, satisfactory studies of individual turnpike trusts remain surprisingly rare, when one considers the compact nature of the original records, i.e. the minute books and Acts of Parliament, relating to them. Of those which have been published, F. A. Bailey's two articles on "The minutes of the trustees of the turnpike roads from Liverpool to Prescot, St. Helens, Warrington and Ashton-in-Makerfield, 1726–89" (*Trans. Historic Society of Lancashire and Cheshire*, vol. LXXXVIII, 1936, pp. 159–200 and vol. LXXXIX, 1937, pp. 31–90), and the two articles by A. E. and E. M. Dodd on "The old road from Ashbourne to Leek" (*Trans. North Staffs. Field Club*, vol. 83, 1948, pp. 29–57, vol. 84, 1949–50, pp. 46–95), based on the minute books of the Ashbourne to Congleton trust from 1762 to 1835, with one gap, constitute perhaps the best and certainly the most exhaustive treatments of trusts in print. Other studies are: (*a*) W. Buckingham, *A Turnpike Key, or an Account of the Proceedings of the Exeter Turnpike Trustees from 12th June 1753 to 1st November 1884* (privately published, Exeter, 1885.[1]) This trust is interesting and unusual in that it controlled about 150 miles of main and cross roads radiating out of Exeter in several directions, and not simply a stretch between two towns. (*b*) P. L. Payne, "The Bermondsey, Rotherhithe and Deptford turnpike trust, 1776–1810" (*Journal of Transport History*, vol. II, May 1956, pp. 132–43). (*c*) F. H. Maud, *The Hockerill Highway: the Story of the Origin and Growth of a Stretch of the Norwich Road* (1957). This trust controlled the road from Harlow Bush Common in Essex to Stump Cross in Hertfordshire from 1744 to 1870; Mr. Maud's study is based on the original records of the trust, but contains far too much topographical detail and lacks a modern map of the road. (*d*) A. Cossons, *The Turnpike Roads of Nottinghamshire* (1934). (*e*) A. W. Goodfellow, "Sheffield turnpikes in the eighteenth century" (*Trans. Hunter Archaeological Society*, vol. V, 1943, pp. 71–90) and R. E. Leader, "Our old roads" (ibid., vol. II, 1924, pp. 7–29) form a guide to the complicated history of the turnpike and other roads out of Sheffield. (*f*) W. B. Crump has written of the roads leading out of Halifax ("Ancient highways of the parish of Halifax", *Papers, Reports, etc. read before the Halifax Antiquarian Society*, 1924–28), in nine articles, and of the roads out of Huddersfield (*Huddersfield Highways through the Ages*, Tolson Memorial Museum, 1949). These essays are accompanied by excellent maps, as is the same author's article on "Saltways from the Cheshire wiches" (*Trans. Lancashire and Cheshire Antiquarian Society*, vol. LIV for 1939, pp. 84–142). (*g*) F. T. S. Houghton, "Salt-ways" (*Trans. Birmingham Archaeological Journal*, vol. LIV, 1929–30, pp. 1–17). (*h*) A. Cossons has dealt with the history of Wiltshire roads in vol. IV of *A History of Wiltshire* (Victoria History of the Counties of England), 1959, pp. 254–71.

There is now a new and superior edition of Celia Fiennes, *Through England on a Side Saddle in the Time of William and Mary* (1888), much used by Jackman, under the title of *The Journeys of Celia Fiennes* (edited with an introduction by Christopher Morris, 1947). For conditions on English and Welsh roads in the 1780s and 1790s, *The Torrington Diaries* (ed. C. B. Andrews, 4 vols., 1934–8) should be consulted. The

---

[1] I was able to consult a copy by courtesy of the Exeter City Librarian. Buckingham was clerk to the trustees for nearly twenty-five years, and his scarce work, which escaped the vigilance of Jackman, is not in the British Museum.

routes of the Scots cattle trade have been admirably chronicled by
A. R. B. Haldane in *The Drove Roads of Scotland* (1952), and J. B.
Salmond's *Wade in Scotland* (1934) deals with the road- and bridge-
building activities of the British Army after the Jacobite Rebellion of
1715. H. Hamilton's *The Industrial Revolution in Scotland* (1932) contains
a section (pp. 226–30) on the progress made in the improvement of Scots
roads in the eighteenth and early nineteenth centuries. There is a useful
section on road history in A. H. Dodd's *The Industrial Revolution in
North Wales* (1933), pp. 89–101, accompanied by a clear map of the turnpike roads in the area. See also R. T. Pritchard's two articles on Caernar-
vonshire turnpikes (*Trans. Caernarvon Hist. Soc.*, vol. xvii, 1956 and
vol. xix, 1958). For Irish road history, see T. W. Freeman, *Pre-Famine
Ireland* (1957), pp. 109–16. and E. R. R. Green, *The Lagan Valley,
1800–1850* (1949), pp. 39–51.

The road engineers of the eighteenth century have received unequal
treatment by their biographers. There are now two excellent books on
Telford: Sir Alexander Gibb's *The Story of Telford: the rise of civil engin-
eering* (1936), which concentrates on the engineering side of his activities,
while L. T. C. Rolt's *Thomas Telford* (1958) is a general biography.
There is, however, no modern study of John ("Blind Jack") Metcalfe of
Knaresborough, and *John Loudon McAdam: chapters in the history of
highways* (1936) by "Roy Devereux" (pseudonym of Mrs. Rose McAdam
Pember-Devereux) does less than justice to the great repairer of roads.
R. H. Spiro has, however, published a most useful article on "John
Loudon McAdam and the Metropolis Turnpike Trust" (*Journal of Trans-
port History*, vol. ii, Nov. 1956, pp. 207–13), and it is to be hoped that his
thesis on McAdam will eventually be published.

G. C. Dickinson ("Stage-coach services in the West Riding of Yorkshire between 1830 and 1840", (*Journal of Transport History*, vol. iv,
May 1959, pp. 1–12) and H. W. Hart ("Some notes on coach travel,
1750–1848", ibid., vol. iv, May 1960, pp. 146–60) treat in a factual way
of subjects which have often been the object of romantic and subjective
speculation. Local historians would do well to apply their methods to
the history of coach travel in their areas. G. A. Sekon (pseudonym of
George Augustus Nokes), in his *Locomotion in Victorian London* (1938)
deals with all types of vehicular traffic in the metropolis.

BRIDGES

E. Jervoise has published a series of handbooks on bridges: *The Ancient
Bridges of the North of England* (1931), *The Ancient Bridges of the South
of England* (1930), *The Ancient Bridges of Wales and Monmouth* (1936),
and *The Ancient Bridges of Mid and Eastern England* (1932).

Patricia Carson, "The building of the first bridge at Westminster, 1736–1750" (*Journal of Transport History*, vol. III, Nov. 1957, pp. 111–22) reveals how much can be discovered about the building and finance of an important bridge.

FERRIES

H. R. Davies's detailed *Review of the Records of the Conway and Menai Ferries* (1942) explores thoroughly one aspect of local transport history which has in general been neglected.

POSTAL HISTORY

In the sphere of postal history Howard Robinson's *The British Post Office: a history* (1948) provides a full-scale treatment from Tudor times to the present day, based on printed sources[1]. T. Todd's *William Dockwra and the Rest of the Undertakers: the Story of the London Penny Post, 1680–82* (1952) deals with the political as well as the postal aspects of this pioneer venture. K. L. Ellis's *The British Post Office in the Eighteenth Century: a study in administrative history* (1958) pays particular attention to the internal organization of the Post Office and its place in the scheme of government. John Palmer's work as Surveyor and Comptroller General of the Mails (1786–92) comes in for heavy criticism, and it is now clear that his reputation has been unduly inflated in the past (see C. R. Clear, *John Palmer (of Bath) Mail Coach Pioneer* (1955).) E. Vale's *The Mail Coach Men of the late Eighteenth Century* (1960) is based largely on the letter-books for 1794–96 of Thomas Hasker, Post Office Superintendent of Mail Coaches, and throws a good deal of light on this very curiously organized undertaking. E. Trory's *A Postal History of Brighton, 1673–1783* (1953) and its *Supplement* (1954) should also be noted.

RIVER NAVIGATION

Although Jackman certainly did not overlook the importance of river navigations he devoted only one chapter of fifty-four pages to their history between 1500 and 1750, with the result that until 1936 insufficient attention was paid to this sector of inland transport by economic historians; and it still tends to be ignored by the writers of textbooks on English economic history. The publication of T. S. Willan's *River Navigation in England, 1600–1750* (1936) left no excuse for neglecting these precursors

---

[1] For some important criticisms of this book and a later one by the same author (*Britain's Post Office*, 1953), see *English Historical Review*, vol. LXIV, 1949, p. 516 and vol. LXVIII, 1953, p. 641.

of canal transport. The same author has also published separate studies of *The Navigation of the River Weaver in the Eighteenth Century* (Chetham Society, 3rd ser., vol. III, 1951) and *The Navigation of the Great Ouse between St. Ives and Bedford in the Seventeenth Century* (vol. XXIV in the *Publications* of the Bedfordshire Historical Record Society, 1946). W. H. B. Court has given a general account of river transport in the West Midlands region in *The Rise of the Midland Industries, 1600–1838* (1938), which is filled out by T. S. Willan's two articles on "The river navigation and trade of the Severn Valley" (*Economic History Review*, vol. VIII, Nov. 1937, pp. 68–79) and "Chester and the Navigation of River Dee, 1600–1750" (*Journal of the Chester and North Wales . . .Archaeological . . . Society*, vol. XXXII, 1938, pp. 64–7). C. N. Parkinson's *The Rise of the Port of Liverpool* (1952) also gives details concerning efforts to improve the Dee.

On the other river navigation systems the following should be consulted:

Aire and Calder:
   (i)   T. S. Willan, "Yorkshire river navigation, 1600–1750" (*Geography*, vol. XXII, 1937, pp. 189–99).
   (ii) G. Ramsden, "Two notes on the history of the Aire and Calder Navigation" (*Thoresby Miscellany*, vol. XII, pt. 4, 1954, pp. 383–95).

Avon:
   T. S. Willan, "Salisbury and the navigation of the Avon" (*Wiltshire Archaeological and Natural History Magazine*, vol. XLVII, 1937, pp. 592–4.)

Clyde:
   H. Hamilton, *The Industrial Revolution in Scotland* (1932), pp. 241–4.

Derwent (Derbyshire):
   F. Williamson, "George Sorocold of Derby" (*Journal of the Derbyshire Archaeological and Natural History Society*, vol. LVII, new ser., vol. X, 1936, pp. 43–63).

Don:
   (i)   A. W. Goodfellow, "Sheffield's waterway to the sea" (*Trans. Hunter Archaeological Society*, vol. V, 1943, pp. 243–55).
   (ii) G. G. Hopkinson, "The development of inland navigation in South Yorkshire and North Derbyshire, 1697–1850" (ibid., vol. VII, 1956, pp. 229–51).

Douglas:
   A. P. Wadsworth and Julia de L. Mann, *The Cotton Trade and Industrial Lancashire, 1600–1780* (1931), pp. 211–19.

Lagan:

E. R. R. Green, *The Lagan Valley, 1800–1850,* pp. 33–9.

Lea:

(i) Anon., *Papers relating to the Navigation on the River Lea* (1734)[1].

(ii) E. M. Hunt, *The History of Ware* (1946), pp. 16–25.

(iii) P. Mathias, *The Brewing Industry in England, 1700–1830* (1959), pp. 442–7.

Little Ouse:

R. H. Clark, "The staunches and navigation of the Little Ouse River" (*Trans. Newcomen Society,* vol. xxx, for 1955–6 and 1956–7, pp. 207–19).

Shannon:

V. T. H. Delany, "The development of the River Shannon Navigation" (*Journal of Transport History,* vol. iii, Nov. 1958, pp. 185–94).

Thames:

(i) F. S. Thacker, *The Thames Highway: General History* (1914).

(ii) F. S. Thacker, *The Thames Highway: Locks and Weirs* (1920).

(iii) T. S. Willan, "The Navigation of the Thames and Kennet, 1600–1750" (*Berkshire Archaeological Journal,* vol. xl, 1936, pp. 146–56).

(iv) E. C. R. Hadfield, "The Thames Navigation and the canals, 1770–1830" (*Economic History Review,* vol. xiv, no. 2, 1944, pp. 172–9).

(v) F. S. Thacker, *Kennet Country* (1932).

(vi) I. G. Philip, "River navigation at Oxford during the Civil War and Commonwealth" (*Oxoniensia,* vol. ii, 1937, pp. 152–65).

Trent:

(i) A. C. Wood, "The history of the trade and transport on the river Trent" (*Trans. Thoroton Society,* vol. liv, 1950, pp. 1–44).

(ii) J. D. Chambers, *The Vale of Trent, 1670–1800* (*Economic History Review,* Supplement no. 3, 1957, pp. 10–11).

Witham and Yare:

T. S. Willan, "River navigation and trade from the Witham to the Yare, 1600–1750" (*Norfolk Archaeology*), (Norwich and Norfolk Archaeological Society, vol. xxvi, 1938, pp. 296–309).

Wye:

I. Cohen, "The non-tidal Wye and its navigation" (*Trans. of the Woolhope Naturalists' Field Club,* vol. xxxv, pp. 83–101).

A. W. Skempton has compiled a useful biographical list of "The engineers of the English river navigations, 1620–1760" (*Trans. Newcomen*

---

[1] This was apparently overlooked by Jackman.

Society, vol. xxix, 1953–4 and 1954–5, pp. 24–54). Unfortunately A. E. Richardson's study of the greatest eighteenth-century expert on the engineering problems of river navigation (*Robert Mylne, Architect and Engineer*, 1955) contains very little about this side of his work.

On the drainage of the Fens, important in its dependence on river improvement, H. C. Darby's *The Draining of the Fens* (1956), L. E. Harris's *Vermuyden and the Fens* (1953) and J. Korthals-Altes's *Sir Cornelius Vermuyden* (n.d.—1925) should be consulted.

CANALS

J. W. F. (now Sir Francis) Hill has assembled the available evidence concerning the Fossdyke, a canal eleven miles long which unites Brayford pool at Lincoln to the Trent at Torksey, only noted incidentally by Jackman. The Fossdyke is very probably of Roman origin, and was certainly in existence in early mediaeval times. It promoted the growth of Lincoln by joining the city to the system of waterways formed by the Trent, the Humber and the Yorkshire Ouse (J. W. F. Hill, *Mediaeval Lincoln*, 1948, esp. pp. 13–14, 306–14; and *Tudor and Stuart Lincoln*, 1956).

W. B. Stephens has set forth a good deal of new evidence regarding the sixteenth-century Exeter canal ("The Exeter Lighter Canal, 1566–1698", *Journal of Transport History*, vol. III, no. 1, May 1957, pp. 1–11). The canal runs from Exeter to Trenchard's Sluice, just above Topsham Quay. This article and W. T. MacCaffrey's *Exeter, 1540–1640* (1958) largely supersede the account of the canal given by Jackman.

The most important advance in British canal history has been the detailed demonstration by T. C. Barker, that the Newry Navigation in Northern Ireland, constructed between about 1730 and 1742, and the Sankey Navigation (1755–7) formed connecting links between the age of river navigations and the canal era of 1759–1835. Thomas Steers (d. 1750), Liverpool's first dock engineer, supervised the construction of the Newry Navigation and his successor Henry Berry was the engineer in charge of the excavation of the Sankey Navigation. Although the Sankey Navigation began as a river improvement project, Berry and his financial backer Nicholas Ashton decided to use the Parliamentary powers granted by the Sankey Navigation Act of 1755 to cut a deadwater canal instead (T. C. Barker, "The beginnings of the canal age in the British Isles" in *Studies in the Industrial Revolution presented to T. S. Ashton*, ed. L. S. Pressnell, 1960, pp. 1–22, and "The Sankey Navigation", *Trans. Historic Society of Lancashire and Cheshire*, vol. c, 1948, pp. 121–55[1].

Charles Hadfield's *British Canals: an illustrated history* (1959, 2nd edn.,

---

[1] See also T. C. Barker, "Lancashire coal, Cheshire salt and the rise of Liverpool" (*Trans. Historic Society of Lancashire and Cheshire*, vol. CIII, 1951, pp. 83–101).

revised and enlarged) is now the best general account of canal history in these islands. Mr. Hadfield has, however, published two regional volumes containing histories of individual canals (*The Canals of Southern England*, 1955[1], and *The Canals of South Wales and the Border*, 1960[2]). The volume on South Wales is particularly valuable as showing the great extent to which tramroads, often of considerable length and complexity, acted as feeders to canals, and the same may be said of B. Baxter's "Early railways in Derbyshire" (*Trans. Newcomen Society*, vol. xxvi (for 1947–9), 1953, pp. 185–97). Mr. Hadfield hopes to extend the series to cover the whole of Britain. L. T. C. Rolt's *The Waterways of England* (London, 1950) is another general historical account and Eric de Mare's *The Canals of England* (1950) is especially valuable for its photographs.

Hugh Malet's *The Canal Duke* (1961), a biography of Francis, third duke of Bridgewater, fills a long-felt gap. The results of recent research are to be found in Herbert Clegg's "The third Duke of Bridgewater's canal works in Manchester" (*Trans. Lancashire and Cheshire Antiquarian Society*, vol. 65 for 1955, pp. 91–103), V. I. Tomlinson's "Salford activities connected with the Bridgewater Navigation" (ibid., vol. 66 for 1956, 1957, pp. 51–86) and Frank Mullineux's *The Duke of Bridgewater's Canal* (1959). All recent research emphasizes the fact that John Gilbert (1720–95) the Duke's steward, and brother of Thomas Gilbert, M.P., the Poor Law reformer, was at least as important as Brindley in the planning and direction of the Duke's coal and canal concerns[3]. Details of the abortive Macclesfield canal scheme of 1765–6, important as illustrating the desire of the Duke of Bridgewater to exercise control over local canal development, are given in W. H. Chaloner, "Charles Roe of Macclesfield (1715–81): an eighteenth century industrialist" (*Trans. Lancashire and Cheshire Antiquarian Society*, vol. LXII (1950–1), pp. 144–56 (see Jackman, pp. 365–6, 404)). A. Temple Patterson's article on "The making of the Leicestershire canals, 1766–1814" (*Trans. Leicester Archaeological Society*, vol. XXVII, 1951, pp. 1–35) sets an example which could usefully be followed for some other counties. J. R. Harris has dealt with "Liverpool canal controversies, 1769–1772" (*Journal of Transport History*, vol. II, no. 3, May 1956, pp. 158–74), controversies which marked the early years of the Leeds and Liverpool canal, and Charles Hadfield's' 'The Grand Junction Canal"

---

[1] See also Charles Hadfield, "Canals", in *A History of Wiltshire*, vol. IV, 1959, pp. 272–9 (*Victoria History of the Counties of England*).

[2] See also Harold Pollins, "The Swansea Canal" (*Journal of Transport History*, vol. I, no. 3, May 1954, pp. 135–54).

[3] W. H. Chaloner, "Francis Egerton, Third Duke of Bridgewater (1736–1803); a bibliographical note" (*Explorations in Entrepreneurial History*, vol. V, no. 3, March 1953, pp. 181–5). Bernard Falk's *The Bridgewater Millions* (1942) contains a popular account of the Duke's life.

(ibid., vol. IV, no. 2, Nov. 1959, pp. 96–112) is an admirable short study of the "greatest canal company south of Birmingham". G. G. Hopkinson's "The inland navigations of the Derbyshire and Nottinghamshire coalfield, 1777–1856" (*Journal, Derbyshire Arch. and Nat. Hist. Soc.*, vol. LXXIX, 1959, pp. 22–41) contains much information from manuscript sources). E. A. Wilson's short analysis of the proprietors of the Ellesmere and Chester Canal Company in 1822 county by county offers an interesting approach to the problem of the financing of the canals (*Journal of Transport History*, vol. III, no. 1, May 1957, pp. 52–4), a problem which is also touched upon in G. H. Evans, *British Corporation Finance 1775–1850* (1936).

For Scotland there are (a) the section on canals in H. Hamilton's *The Industrial Revolution in Scotland* (1932), pp. 230–41, and (b) George Thomson's "James Watt and the Monkland Canal" (*Scottish Historical Review*, vol. XXIX, no. 108, Oct. 1950, pp. 121–33). For Ireland, see T. W. Freeman's *Pre-Famine Ireland* (1957), pp. 116–24.

As noted above[1] in the section on roads there have been two studies of Telford and one of J. L. McAdam. In addition, short studies of "William Jessop (1745–1814), civil engineer" by Jack Simmons (*Parish and Empire*, 1952, pp. 146–54), of "James Green as canal engineer" by Charles Hadfield (*Journal of Transport History*, vol. I, no. 1, May 1953, pp. 44–56), and of "John Rennie, 1761–1821" by C. T. G. Boucher (*Proc. Manchester Association of Engineers*, Session 1956–7, pp. 1–23), have recently been published.

A. L. Thomas, *Geographical Aspects of the Development of Transport and Communications affecting the Pottery Industry of North Staffordshire during the Eighteenth Century* (Wm. Salt Archæological Society's vol. for 1934, Stafford), should also be noted.

RAILWAYS

Research has been most enthusiastically, if at times most uncritically, pursued in the sphere of railway history. R. A. Peddie's *Railway Literature,1556–1830: a handlist* (1931) is an invaluable starting point, and so are the same author's handlist of the histories of individual railway companies ("Railway history"), in *Library World*, vol. XLV, no. 515, Dec. 1942, pp. 74–6) and W. J. Skillern's continuation of it ("Railway history: a supplement to Peddie", ibid., vol. LVI, March 1955, no. 657, pp. 139–43).

For the early history of railways in Britain C. E. Lee's *The Evolution of Railways* (2nd edn., revised and enlarged, 1943) is indispensable, disposing as it does of a number of hoary legends, such as that which derives the word "tram" from the surname of Benjamin Outram, the engineer. R. S. Smith's "Huntington Beaumont, adventurer in coal mines" (*Renais-*

[1] Page xviii.

sance and Modern Studies, Nottingham, vol. I, 1957, pp. 115–53) shows that the celebrated Wollaton waggonway for coal was probably constructed in 1605 and not in 1598; see also, however, the same author's "England's first rails: a reconsideration" (*Renaissance and Modern Studies*, vol. IV, 1960, pp. 119–34).

Detailed studies of British railway and locomotive history are fairly numerous and among the best are:

(i)   H. G. Lewin, *Early British Railways: a short history of their origin and development, 1801–1844* (1925) and its sequel, by the same author, *The Railway Mania and Its Aftermath, 1845–1852* (1936).

(ii) C. F. Dendy Marshall, *A History of British Railways down to the year 1830* (1938).

(iii) C. Hamilton Ellis, *British Railway History: an outline from the accession of William IV down to the nationalisation of railways, 1830–1876* (1954)[1].

(iv) H. Hamilton, *The Industrial Revolution in Scotland* (1932), pp. 224–53.

(v)   a. J. C. Conroy, *A History of Railways in Ireland* (1928).[2]
      b. T. W. Freeman, *Pre-Famine Ireland* (1957), pp. 124–7.
      c. E. R. R. Green, *The Lagan Valley, 1800–1850* (1949), pp. 51–6.

(vi) R. W. Kidner, *The Early History of the Locomotive, 1804–1876* (1956).

(vii) E. L. Ahrons, *The British Steam Railway Locomotive 1825–1925* (1927).

(viii) C. F. Dendy Marshall, *A History of Railway Locomotives down to the end of the year 1831* (1953).

(ix) E. A. Forward, *Handbook of the Collection illustrating Land Transport: III: Railway Locomotives and Rolling Stock: Part I, Historical Review* (Science Museum, London, 1931, repr. 1947).

(x)   H. Parris, "Railway policy in Peel's administration, 1841–1846" (*Bull. of the Institute of Historical Research*, vol. XXXIII, Nov. 1960, pp. 180–94).

Exact information about the dates of opening of particular lines in three counties may be found in M. D. Greville's two articles, "Chronological list of the railways of Cheshire, 1837–1939" (*Trans. Historic Society of Lancashire and Cheshire*, vol. CVI, 1954, pp. 135–44) and "Chronological list of the railways of Lancashire, 1828–1939" (ibid., vol. CV, 1953, pp. 107–202), and C. R. Clinker's "Railways" in *A History of Wiltshire*, vol. IV, 1959, pp. 280–93 (*Victoria History of the Counties of England*).

[1] See, however, the review of this work in the *Journal of the Stephenson Locomotive Society*, vol. XXXI, Feb. 1955, no. 357, pp. 46–7.

[2] See also R. D. C. Black, *Economic Thought and the Irish Question, 1817–70* (1960), pp. 189–201, for early Irish railway policy, and corrections to Conroy.

A volume of general interest is G. Royde Smith, *The History of Brad-shaw: a centenary review of the origin and growth of the most famous guide in the world* (1939).

For the general background of railway finance the following are important as showing that the British landed interest was by no means as hostile to railways and railway investment as is popularly supposed:

(i) David Spring, "The British landed estate in the age of coal and iron: 1830–1880" (*Journal of Economic History*, vol. xi, no. 1, 1951, pp. 3–24).

(ii) H. Pollins, "A note on railway constructional costs, 1825–1850" (*Economica*, Nov. 1952, pp. 395–407).

The legend that the borough of Northampton refused to allow the London and Birmingham Railway to come near the town has been effectively exploded by Miss Joan Wake in her lively pamphlet, *Northampton vindicated: or why the main line missed the town: an episode in early railway history* (1935).

Biographical studies have been few: R. S. Lambert's *The Railway King, 1800–71: a study of George Hudson and the business morals of his time* (1934); H. W. Dickinson and A. Titley, *Richard Trevithick: the Engineer and the Man* (1934); L. T. C. Rolt, *Isambard Kingdom Brunel: a biography* (1957) and the same author's *George and Robert Stephenson: the Railway Revolution* (1960) are the only satisfactory full-scale studies of the outstanding personalities of the early railway age. There is, for example, no modern reassessment of Captain Mark Huish, although a short study of Carr Glyn has recently appeared: anon., "George Carr Glyn and the railways" (*Three Banks Review*, no. 46, June 1960, pp. 34–47).

Harold Pollins's article on "The marketing of railway shares in the first half of the nineteenth century" (*Economic History Review*, 2nd ser., vol. vii, no. 2, Dec. 1954, pp. 230–9) modifies traditional views about both railway and canal promotion, such as those expressed, for example, by G. H. Evans (*British Corporation Finance, 1775–1850*, Baltimore, 1936, Chap. II). S. A. Broadbridge's article on "The early capital market: the Lancashire and Yorkshire Railway" (*Economic History Review*, 2nd ser., vol. viii, no. 2, Dec. 1955, pp. 200–12) is also relevant in this connexion.

Mr. Pollins's two later articles on "Railway contractors and the finance of railway development in Britain" (*Journal of Transport History*, vol. iii, nos. 1 and 2, May–Nov. 1957, pp. 41–51, 103–10) break new ground in an attempt to obtain a more rounded picture of the railway contractor, "an elusive figure in the history of British Railways". On the subject of railway navvies the articles by R. A. Lewis ("Edwin Chadwick and the railway labourers", *Economic History Review*, 2nd ser., vol. iii, no. 1, 1950, pp. 107–18) and Jack Simmons ("The building of the Woodhead Tunnel" in *Parish and Empire*, 1952, pp. 155–65) should be consulted.

Only two of the new communities created by the railway companies in the mid-nineteenth century have been studied in a detailed fashion— Crewe and Swindon. For Crewe, there is W. H. Chaloner's *The Social and Economic Development of Crewe, 1780–1923* (1950) and for Swindon H. B. Wells, "Swindon in the 19th and 20th centuries" in *Studies in the History of Swindon*, 1950, pp. 90–157; D. E. C. Eversley, "Engineering and railway works" in *A History of Wiltshire*, vol. IV (*Victoria History of the Counties of England*), 1959, pp. 207–19, and the same author's "The Great Western Railway and the Great Depression" (*University of Birmingham Historical Journal*, vol. V, no. 2, 1957, pp. 167–90). Wolverton, Darlington, Earlestown, Ashford, Eastleigh and Horwich remain as yet unchronicled.

On the private sector of locomotive building less work has been done, but J. G. H. Warren's *A Century of Locomotive Building by Robert Stephenson and Co., 1823–1923* (1923) is a fitting memorial to the first century of a world-famous firm. The anonymous *History of the North British Locomotive Co., Ltd.* (1953), is a much slighter work.

There have naturally been numerous histories of individual railway lines and companies, and as two lists of these have been referred to above it is proposed to mention only the most important:

Early Lines

W. G. Rimmer's "Middleton Colliery near Leeds (1770–1830)" in the *Yorkshire Bulletin of Economic and Social Research* (vol. VII, March 1955, pp. 41–57), is notable as being the history of the first railway for which an Act of Parliament was passed (in 1758). C. E. Lee has published "Early railways in Surrey" (*Trans. Newcomen Society*, vol. XXI for 1940–1, 1943, pp. 49–79), which deals with the early history of the Surrey Iron Railway of 1801 and its continuation, the Croydon, Merstham and Godstone Railway Company, and "Tyneside tramroads of Northumberland: some notes on the engineering background of George Stephenson" (*Trans. Newcomen Society*, vol. XXVI, for 1947–8 and 1948–9, pp. 199–229). The same author's *The Swansea and Mumbles Railway* (2nd revised edn. 1954) deals with the earliest passenger line in the world, which was incorporated by Act of Parliament in 1804; the transport of passengers on it began on 25 March, 1807. C. R. Clinker's *The Hay Railway* (1960) is the first history of an early railway company to be written largely from the original records of the line, constructed under the powers of an Act of Parliament of 1811. E. A. Forward has translated and reviewed portions of the "Report on Railways in England, 1826–27" by two Prussian engineers, Carl von Oeynhausen and Heinrich von Dechen (*Trans. Newcomen Society*, vol. XXIX for 1953–4 and

1954–5, 1959, pp. 1–12). About seventy per cent of the descriptive part of the report is covered. Miss Lois Basnett's "The history of the Bolton and Leigh Railway, based on the Hulton Papers (1824–8)" explores the early years of the first public railway in Lancashire, opened throughout a year before the line from Liverpool to Manchester (*Trans. Lancashire and Cheshire Antiquarian Society*, vol. LXII, 1950–1, pp. 157–76. Other early lines have been chronicled by R. B. Fellows (*History of the Canterbury and Whitstable Railway*, 1930 (this line was opened in 1830), and by C. R. Clinker "The Leicester and Swannington Railway" (*Trans. Leicestershire Archaeological Society*, vol. XXX, 1954, pp. 59–113).

The centenary celebrations of the opening of the Liverpool and Manchester railway in 1830 called forth a number of notable publications dealing mainly with the early history of this line: G. S. Veitch, *The Struggle for the Liverpool and Manchester Railway* (1930); C. F. Dendy Marshall, *The Centenary History of the Liverpool and Manchester Railway* (1930), and anon., *The Centenary of the Liverpool and Manchester Railway, 1830–1930: a list of printed and illustrated material in the (Liverpool) Reference Library* (1930). In contrast, the later history of this line has been comparatively neglected, but Harold Pollins has examined "The finances of the Liverpool and Manchester Railway" (*Economic History Review*, 2nd ser., vol. V, no. 1, 1952, pp. 90–7) and cleared up the controversies about them which raged in the mid-1830s. G. O. Holt's *Short History of the Liverpool and Manchester Railway* (Railway and Canal Historical Society, 1955) devotes two-thirds of the text to the period 1830–45. Other notable company histories covering the period 1825–50 are E. T. Mac-Dermot's *History of the Great Western Railway*, vol. I (in two parts), 1927[1]; J. S. MacLean, *The First Railway across Britain: the Newcastle and Carlisle Railway, 1825–1862* (1948), a line which began as a canal project; Jack Simmons, *The Maryport and Carlisle Railway* (1947); D. S. M. Barrie, *The Taff Vale Railway* (2nd edn. 1950); George Dow, *The First Railway across the Border* (1946), issued to commemorate the opening of the North British Railway in 1846; George Dow, *The First Railway between Manchester and Sheffield* (1945); George Dow, *Great Central . . . 1813–1863*, vol. I (1960); D. G. Parkes, *The Hull and Barnsley Railway* (2nd edn., 1948) and J. I. C. Boyd, *The Festiniog Railway: a history of the Narrow Gauge Railway connecting the slate quarries of Blaenau Festiniog with Portmadoc, North Wales* (2 vols., 1956–9). The pamphlet by J. Melville and J. L. Hobbs on *Early Railway History in Furness* (Cumberland and Westmorland Antiquarian and Archaeological Society Tract Series, no. XIII, 1951) should be read in conjunction with S. Pollard and J. D. Marshall, "The Furness Railway and the Growth of Barrow" (*Journal*

---

[1] See errata list in *Journal of Transport History*, vol. I, no. 3, May 1954, pp. 185–6.

of Transport History, vol. I, no. 2, Nov. 1953, pp. 109–26).

G. Royde Smith's *Old Euston: an account of the beginning of the London and Birmingham Railway and the building of Euston Station* (1938) is a pleasant and factual introduction to the history of the main constituent of the London and North Western Railway of 1846. The route from London to Ireland is covered by V. S. Haram, *Centenary of the Irish Mail, 1848–1948* (1948). For Ireland, see E. M. Patterson, *The Belfast and County Down Railway* (1958); Railway Executive, Northern Counties Committee: *Centenary of the opening of the Belfast and Ballymena Railway* (1948) and D. S. M. Barrie, *The Dundalk, Newry and Greenore Railway and the Holyhead–Greenore Steamship Service* (1957). See also O. S. Nock, *The Great Northern Railway* (1958) and C. H. Ellis, *The North British Railway* (2nd ed., 1959).

Steam Navigation

The chief new work on this subject has been done by H. P. Spratt, whose article on "The prenatal history of the steamboat" (*Trans. Newcomen Society*, vol. XXX, for 1955–6, 1956–7, pp. 13–23) and book on *The Birth of the Steamboat* (1958) are well documented. Additional information on early experiments with steamboats on the Sankey and Bridgewater canals will be found in T. C. Barker and J. R. Harris, *A Merseyside Town in the Industrial Revolution: St. Helens, 1750–1900* (1954, pp. 193–5), and in F. Mullineux, *The Duke of Bridgewater's Canal* (1959). A. C. Wardle's accurate and valuable paper on "Early steamships on the Mersey, 1815–1820" (*Trans. Historic Society of Lancashire and Cheshire*, vol. XCII, 1940, pp. 85–7) should also be consulted. For other references to early steam navigation, see W. H. Chaloner and W. O. Henderson, "Aaron Manby, builder of the first iron steamship" (*Trans. Newcomen Society*, vol. XXIX, for 1953–4 and 1954–5, pp. 77–91). For the early history of Anglo-Irish steam navigation, see H. S. Irvine, "Some aspects of passenger traffic between Britain and Ireland, 1820–1850" (*Journal of Transport History*, Nov. 1960, pp. 224–41). J. R. T. Hughes and S. Reiter, "The first 1,945 British steamships" (*Journal, American Statistical Assoc.*, June 1958, pp. 360–81) modify Sir John Clapham's statements on the history of British steam navigation in the mid-nineteenth century.

W. H. CHALONER

Manchester, February 1962

PREFACE

IN offering this work as a modest contribution to our knowledge of the economic development of England from the standpoint of transportation, the author must say, in the first place, that he has endeavoured to adhere rigidly to the subject in hand, without making deviations into collateral fields. It is impossible to study at first hand and from original sources such a comprehensive subject as this without being impressed by its intimate and vital connexion with the other phases of the national evolution; and it has required much self-restraint to keep from branching out farther into a discussion of the relation of transportation to the progress of agriculture, the growth of markets, the advance of industry, the increase of wealth, and many other economic and social factors which have affected the welfare of different classes of the people and of the nation as a whole. In the collecting of material for these volumes during almost a decade, a wealth of information has been accumulated upon the other great aspects of the nation's expansion; and it is with much regret that this account of the facilities of conveyance and communication could not be amplified so as to trace more minutely the outworking of cause and effect along all these lines. Contrary to the wishes of some who are interested in this subject, whole chapters, for the writing of which the material is already in hand and outlined, have had to be omitted, and to these friends I tender my sincere apologies for failure to do what would have given me so much satisfaction. The voluminous amounts of material have demanded judicious selection and unbiased sifting of evidence; and, in the process of elimination, the hopes of some who have looked forward to the appearance of this work will be disappointed when they find meagre. if any, mention of certain aspects of the subject, which I would gladly have elaborated had space permitted. Such excision has been necessitated in order to keep within reasonable limits.

A few words are required in regard to the definition of the field covered. In its application to England, the term *modern* has been taken to mean the period beginning with about the close of the fifteenth century and ending with approximately the middle of the nineteenth century. But, in order to get a good background for our study, the first chapter has been devoted to an outline of the conditions between the time of the Roman occupation of Britain and the accession of the Tudor monarchs. It is not intended that this should be anything else than merely a general view of the ante-modern period. Although, with the amount of research involved, a much more complete discussion of this early period could have been given, it has been deemed more consonant with our present purpose to refrain from a consideration *en détail* of the conditions of the mediaeval epoch. Lest any should inquire why the record has not been brought down more closely to the present time, perhaps it may suffice to say, first, that the history of road transportation has been so profoundly affected by the recent introduction of the bicycle, the automobile, the motor truck and the motor omnibus that we are yet too near to these innovations to adequately realize their influence; and, second, that, as far as the railways are concerned, the outlines of the various systems were practically finished by 1850, while the economic problems in the period since that time have been considered with discrimination by such men as Mr Acworth and Mr Knoop. It will be noted that the history of the canals has been sketched to date, in order to give a proper basis for judging of the merits of the existing agitation for the resuscitation of the inland waterways.

It may, perhaps, be said by some that too much attention has been devoted to certain aspects of the historical development of the means of conveyance. My answer to this possible criticism is that he is the best economic historian who can most faithfully depict the past until it lives again in the mind of the reader and can then most judiciously interpret the action and reaction of those forces which have shaped the life of the nation as manifested in its external material forms.

It is my privilege to mention the almost universal spontaneity with which those who are in a position to give assistance have rendered this favour when application has been made to them. But I regret to say that some who could have given invaluable aid, without any inconvenience to themselves or injury to the interests they represent, were antagonistic to granting any such service. For instance, the clerk of the most important navigation company whose headquarters are at the midland metropolis, when permission was requested to examine some of the freight bills before the year 1832, refused to allow any such privilege, even when he was assured that all information taken from these sources would be open for his inspection and that none of it would be used without his authorization. Similar treatment was received from the superintendent of another canal and from the manager of one of the largest firms of carriers and forwarders, both of whom have their offices in the midland centre of the waterway system. Until a finer sense of responsibility and honour comes to such public servants, the cause of truth is impeded by the barriers raised against the advance of research. Happily, there are few persons in important places who are not willing, within the measure of their power, to furnish information so long as all interests are safeguarded.

My obligation to Professor Edwin F. Gay, of Harvard University, under whom this study was begun, and to Professor William Z. Ripley, of the same institution, whose balanced judgment has elicited my admiration, must secure recognition in this public manner. To the late President Matthew H. Buckham, of the University of Vermont, the patron and exemplar of the finest culture and the most scholarly attainments, a man whose friendship during his ripest years has been highly prized, and whose unfailing interest in me and my work has added a stimulus and rendered an aid of the influence of which he was sympathetically unconscious, it is my great privilege to bear this encomium and the tribute of grateful remembrance. The courtesy of Mr Hubert Hall, of His Majesty's Public Record Office, London, of Mr G. F. Barwick, of the British Museum, London, and of the librarians of the London School of Economics and the Goldsmiths' Library, University of London, in facilitating my work in the British archives is acknowledged with gratitude, as is also the kindness of the officials in the civic libraries of Birmingham, Liverpool, Manchester and Bristol, who opened their treasures to me. To one who shall remain unnamed here, whose helpfulness was so graciously bestowed and the memory of whom is a constant incitement to the best achievement, I owe more than can ever be expressed. I cannot but make mention of the considerate and courteous way in which the publishers have rendered their services, at a time and under circumstances which have been very trying; and it is a great pleasure to express my appreciation of their cordial co-operation throughout all our relations. But greatest of all is the assistance I have received from my wife, whose companionship in life's deepest interests has lightened the burden of difficult years and has been the incessant encouragement of my highest endeavours.

W. T. J.

New Year's Day, 1916
Toronto

CHAPTER I

INTRODUCTION

SINCE any discussion of the subject of transportation in modern England would be inadequate without some brief survey of earlier modes of conveyance and communication, we will endeavour to briefly present the conditions before the modern period, which we may roughly date from the time of the accession of the Tudors.

In the early history of Britain, the conquest by the Romans stands out as a very important event, from many points of view; and from no other standpoint is it more important than from the linking together of all parts of the empire, by the construction of great military highways. In order that their conquest might be complete, the Roman legions had to penetrate into distant and even secluded sections, where the early Britons had become ensconced to withstand the advance of their enemy; and even the fens had to be rendered passable to Roman arms by building suitable roads.

Considering the fact that such works of engineering skill had all to be done by patient manual labour, through a wild and meagrely cultivated island, we can easily see what a vast expenditure of time and effort must have been involved in it. In some places, great quantities of timber had to be cut down; in others, the ground had to be embanked and drained before it was firm enough to advance upon. The deeper rivers were crossed by timber bridges, often laid on piers of such solid masonry that they were used in mediaeval times as the foundations of several successive bridges of better construction. The shallower streams were crossed by paved fords, whose courses were marked by large wooden posts. The enduring quality of the works, as well as their great extent, left its impress upon the face of England to such a degree, that these roads are the wonder of the engineering world to-day.

The number and precise direction of these Roman roads have been matters of dispute among many who are competent to judge, but there seems to be almost unanimous agreement in regard to the general direction of the main arteries, which may be represented as in the accompanying outline[1]. It may be noted here that London seems to have been the centre of these early roads, as it was the centre of the road system in the later centuries, and as it became also the centre of the railroad system, when the latter came into existence[2].

The Roman roads varied somewhat in the mode of construction according to their importance and the nature of the materials obtainable. In the main, they consisted of layers of concrete made by pouring slaked lime upon small clean stones, mixed often with broken tiles. Two furrows were first made, at the proper distance apart, the earth between them was then dug up for a foot or two in depth, and the bottom rammed and beaten down firmly. Upon this the first stratum of material was laid and the lime poured over it; then larger stones were placed upon that, and the interstices filled in with mortar; after which sometimes came another layer, similar to the bottom one. The whole was often three feet thick, or more, and was rounded in the centre to prevent water lodging upon it. The usual width was only about fifteen feet, although sometimes twenty-four or twenty-five feet. As a rule, these works were raised somewhat above the surface of the ground, even where no question of keeping them clear of floods could arise. There is reason to think that they were built almost entirely by the soldiers, mostly at public expense, and under the direction of skilled engineers from Italy. They were intended primarily to serve a military, rather than a commercial purpose. The whole system was in charge of the *curator viarum*, and was in every way as essential to the safety of the country as the railways are now[3]. After the conquest had been completed, they built public roads to facilitate the peaceful development of the country and the communication with the capital; and even many private roads, or *viae vicinales*, were put through, which were probably mere lanes leading to solitary estates.

But we must not assume that the Britons had no roads before their

---

[1] On the whole subject of Roman roads and the different views regarding them, see Codrington, *Roman Roads of Britain*; Holinshed, *Description of Britaine*; Bergier, *Histoire des grands chemins de l'Empire Romain*; Horsley, *Britannia Romana*; Scarth, *Roman Britain*; Vigilo (pseud. for H. L. Long), *Observations upon Certain Roman Roads and Towns in the South of Britain*; H., A., *The Roman Roads in Britain, with the Ancient and Modern Names attached to each Station on or near the Route*; Paley, 'The Roman Roads of Britain,' in *Nineteenth Century*, Nov. 1898; Forbes and Burmester, *Our Roman Highways*. See also the manuscripts in the British Museum written and bequeathed by Rev. John Skinner, especially Brit. Mus., Add. MSS. 33,694 giving maps and plans to illustrate this subject.

[2] This is not true, however, in the same sense, of the canal system, which was developed in the latter half of the eighteenth century and the first third of the nineteenth century.

[3] W. B. Paley in *Nineteenth Century*, XLIV, p. 858.

Roman Roads in Britain (*according to the present state of knowledge*)

conquerors came[1]. On the contrary, that they had some passable roads, is directly to be inferred from their possession of wheel carriages. That they had chariots of war with which they resisted the Roman invasion is indubitable; and it is very likely that they had other wheeled vehicles for purposes of peace. Their acquaintance with the great mechanical power of the wheel, and their use of horses, which they harnessed to their chariots, lead us to infer that the Britons had some roads of considerable value; but they seem to have been principally near the sea-coast. The country was also crossed, here and there, by roads which were much better than mere trackways, and occasional examples of these roads have been found in the nineteenth century in various degrees of preservation[2].

But, unlike the Britons, the Romans made their roads almost perfectly straight, probably in order to give the greatest facility for the conveyance of troops from place to place. They had ample means, abundant numbers of labourers, and competent artificers to put their roads through the most difficult positions. They raised causeways through marshes, threw bridges over rivers, and lowered hills or cut their way through them.

In England, one great mark of the Roman origin of roads is their retention of the Roman name of street, as we see in the names of the four principal remaining roads, namely, Watling Street, Icknield Street, Ermine Street, and the Fosse Way. Another mark of Roman origin, of course, is their peculiar construction, which we have already referred to.

A recent writer, in speaking of the Roman roads, says: "It is not too much to say that the internal communications of Britain, when the last of the Roman legions sailed from Richborough, never to return, were superior to what they were little more than a century ago[3];" and others have made declarations with similar import[3]. But these

---

[1] Holinshed, *Description of Britaine*, 1, p. 189 et seq.

[2] Willmore, *History of Walsall*, says that the old Chester road near Walsall was one of the chief British thoroughfares. See also Hoare, *History of Modern Wiltshire*, 1, p. 166.

[3] W. B. Paley in *Nineteenth Century*, xliv, p. 853. See also Forbes and Burmester, *Our Roman Highways*, p. 118, in which it is said: "It is evident from the foregoing chapters that the facilities for travelling in Britain must have been far greater during the Roman occupation than at the beginning of the eighteenth century....." Much of the paragraph from which this quotation is taken seems like pure clap-trap. Its essential teaching accords well with the late Prof. J. Thorold Rogers' statement as to the roads being good in the period of the Middle Ages (v. *History of Agriculture and Prices*, 1, pp. 654, 660); in fact, to one who has gone into this work thoroughly, it appears to be almost a verbatim recital of the opinions of the latter. But that much credence is to be placed in such a statement is extremely doubtful. In all probability, the great Roman roads were better than most of the

statements can scarcely be substantiated by facts. Doubtless, some of the Roman roads were better constructed than most of the roads in England during the third quarter of the eighteenth century; but the road system as a whole was by no means so good nor so complete as at the latter period. By the last quarter of the eighteenth century, England had a network of roads of such completeness that there was scarcely a hamlet of any importance which was not in communication with the great market towns[1].

For centuries after the Romans left Britain, their roads, while still the main highways of internal communication, were inevitably exposed to the natural agencies of decay or disintegration. Gradually, portions would be washed away or broken up; vegetation would force its way through these crevices in the roads; refuse would be heaped upon them by the wind and rain, until they would be buried under it. In other places the roads would sink into the wet, spongy soil upon which they had been laid, and disappear from view, and this, notwithstanding the fact that the monarch required each locality to maintain its roads and bridges so as to facilitate the movements of the army and of the government officials who were charged with the administration of affairs. In addition to these ancient roads, new highways would doubtless be opened up by the Saxons and Danes to aid them in effecting their settlements and in pursuing their trade, which they did with much activity; but we have only insignificant hints as to the roads of this period, and we must leave to later investigators to show the system of management that prevailed from the withdrawal of the Romans until after the establishment of Norman rule.

With the beginning of the manorial records, we have something like authentic information. All landed proprietors were obliged, in theory, to watch over the good condition of the highways adjacent to their own land; but the lord of the manor imposed upon his tenants this necessity for repairing the roads, and the Court Leet settled all such obligations between lord and tenant. Even the religious houses, which had dispensation from every service and rent toward the former proprietor of the soil, had to satisfy the *trinoda necessitas*, or triple

roads of England at the beginning of the eighteenth century; but the network of roads at the latter period was doubtless much more elaborate than that of the time of the Romans. Settlement and occupation of the land had gone on in the intervening centuries, and in order thereto it was necessary that communication should be opened up so that these settlements might secure access to markets. It seems, therefore, incredible that the facilities for travel were greater under the Roman régime than they were fourteen hundred years later.

[1] This is plainly seen by reference to the Road Books of the latter period, such as those of Taylor, Badeslade, Bowles, Cary, Patterson, etc.

obligation, which, among other duties, consisted in repairing roads and bridges.

But in speaking of these roads, we must discard all modern conceptions of a road bounded by fences or hedges; for to the people of England of the twelfth and succeeding centuries, the highway meant nothing more than a legal and customary right of way over others lands, from place to place; and, of course, by perpetual usage, there came into existence a beaten path[1]. The highway, however, was the legal right of passage, and not the well-worn track; and if the one track became too bad to be used, the traveller could turn out into the adjacent land, whether cultivated or in grass. As the roads "were used almost exclusively for foot traffic of man or beast," and as this liberty of changing the path would not need to be exercised much except during the winter half of the year, we can readily understand that it would not be such a serious detriment to the farming interests as a similar procedure would be to-day. The English roads were, in fact, originally tracks struck out by travellers or by the drivers of pack-horses, making their way as best they could from place to place. They made long circuits to reach fords where they could cross the streams; they chose high ground to escape the bogs of the plain or valley; and they deviated from the straight course at all obstructions.

It is sometimes stated that Edward I took up road improvements as a definite policy. In reality, in the thirteenth year of his reign (1285), an Act was passed, but it was rather as a police measure than for improvement of the roads. This law, contained in the important Statute of Winchester[2], provided "that highways leading from one market town to another shall be enlarged, where as woods, hedges or dykes be, so that there be neither dyke, tree, nor bush, whereby a man may lurk to do hurt, within two hundred foot on the one side and two hundred foot on the other side of the way." The former part of the wording of this measure would point to the widening of the highways for commercial purposes; but, as a matter of fact, the other aspect is the more dominant, that is, the prevention of ambuscades of highway robbers. We may say with all truthfulness that the highways engaged very little of the time and attention of the Parliament.

The keeping of these roads in repair was not considered as merely worldly, but rather as pious and meritorious work before God, of the same sort as visiting the sick or caring for the poor; men saw in this work a true charity for certain unfortunate people, namely, travellers. To encourage the faithful to take part in constructing or repairing

---

[1] Webb, *English Local Government: The Story of the King's Highway*, p. 6.

[2] Act 13 Edward I, c. 5.

roads and bridges, the Bishops would often grant indulgence for a certain time to those who would perform such public service[1]. People were more ready to take up pious works when they received such "salutary encouragement of fuller indulgences[2]."

There were also gilds, animated by the religious spirit, which repaired roads and bridges. The Gild of the Holy Cross in Birmingham, founded in the time of Richard II, furnished a notable example of such activity[3]; and also the craft gilds or livery companies of the borough of Leominster. In the case of the latter borough, we have instances in 1685, 1650, and other years, of the presentments of the gilds for not fulfilling their duty; and if they had this duty at that period we may reasonably infer that they had the same duty in earlier days[4]. The gilds of Rochester, Bristol, Ludlow, and other places spent considerable sums in the same kind of charity[5].

As a general thing, it would appear as if the repair of the roads was largely a matter of chance. In some cases, the only real good was done by the voluntary work of the gilds or through the gifts and legacies of charitably disposed persons[6]. In others, the roads were maintained in the immediate vicinity of a large estate because of the good will or devotion of those to whom the adjoining land belonged[7].

---

[1] See Gibbons, *Ely Episcopal Records*, pp. 398, 400, 401, etc., in which the Registers of Bishop Fordham and other Bishops show many cases of such indulgences. See also Worth, *History of Plymouth*, p. 385, showing that Bishop Stafford in 1411 granted indulgence for repair of the road from Plymouth to Smapolemille.

[2] Jusserand, *English Wayfaring Life in the Middle Ages*, p. 42.

[3] Lucy Toulmin Smith, *English Gilds* (E.E.T.S.), p. 249.

[4] Blacklock, *The Suppressed Benedictine Minster*, pp. 383–4, gives records from the Rolls of the Court Leet showing these facts.

[5] Jusserand, op. cit., p. 45.

[6] Camden Society, *Letters and Papers of the Verney Family*, p. 27. It would seem, however, that charitable contributions for this purpose were comparatively unimportant in the early part of the time we are now considering.

[7] It was the usual custom of the manor to bring all matters relating to the repair of the highways, bridges, etc., before the Court Leet to have them decided by that court. Toward the end of the Middle Ages, we find in the Court Leet records of the manors a great many presentments of persons for offences in connexion with the highways; and the almost wearisome repetition of these offences in all such court records would seem to indicate that conditions were much the same on most manors, and that non-compliance with the customary regulations for preservation of the highways was very common. For example, persons are presented before the court for making dunghills in the public road, for leaving waggons, timber, stone, etc., to block up the highway, for refusal to cleanse the watercourses, for allowing bad causeways, for setting their fences out too far on the highway, for non-repair of their part of the road, etc.

To show the nature of some of these offences, and the alertness, in some cases,

The administration of the law was often uncertain; for frequently, in accordance with the law, sheriffs ordered the levying of taxes on those who did not repair the roads, but those who were fined protested that the roads were good enough and so escaped the payment of the penalty.

Under such conditions of uncertainty and vacillation, the roads of the kingdom would have been entirely impassable, and religious zeal would have been no more sufficient to keep them in condition than the Bishops' indulgences, if the nobility and the clergy, that is, the whole of the landed proprietors, had not had an immediate and daily interest in working to provide passable roads. The lands given by the King to the nobles were scattered about the kingdom, and not collected or joined in compact holdings. This would result in benefit to the roads in the different parts of England. The proprietors of these lands then had to go from one estate and country house to another, and thus

of the manorial jury in guarding the public interests, we take the following from Hardy's *Records of Doncaster*:

Vol. II, p. 5 (1456). The jury present six persons "for making dung hills in the high and public roads in Doncaster," and one person "for putting a swine-trough in the public road."

Vol. II, p. 5 (1456). The jury present four men "for blocking up a water-course leading from St George's Street."

Vol. II, p. 18 (1506). "John Rawson is ordered to repair a water-course called Freredyke."

Vol. II, p. 22 (1507). John Humberston has made a dungheap, and keeps his waggon in the high road, to the nuisance of all who pass along the road.

Vol. II, p. 23 (1507). Five men "have failed to scour their water-courses." Three others "have made dunghills in the highroad." "Edw. Cooke, mayor, has suffered his timber to lie in the highroad."

That occasionally the Court Baron had matters relating to the highways brought before it, is shown by the following entries from vol. IV of the *Publications of the Selden Society*:

p. 98. "Also they say that Warin of C. hath caused a purpresture by placing his dungheap on the king's highway to the nuisance of the country. Therefore command is given that it be at once removed...."

"Also they say that Adam of T. hath stopped a water-course at Hamland and the path which leadeth to the church or the mill, both of which are from beyond memory. Therefore command is given that the water be brought back into its old course and that for the future the paths be used."

"Also they say that James Day hath ploughed with his plough and appropriated to himself three furrows from the King's highway which leadeth to Raunbury, in such a place, to the prejudice of the King and the nuisance of the country. Therefore he be in mercy, and command is given that this be put to rights forthwith."

p. 130. "William of Thame was summoned to answer Robert Carter why he unjustly detains from the common highway 12*d.*, which he promised, to (its) damage 6*d.*"

See similar examples, pp. 131, 132, etc.

it was to their advantage, as well as to that of the King, who had royal domains throughout the kingdom, to have roads that would satisfy the demands of the time[1]. In the same way, the monks, the great cultivators, were much interested in the maintenance of the roads, on account of the great distances between their scattered agricultural properties[2]. Besides, the care of the roads was more incumbent on the clergy than on any other class, because it was considered as a work of Christian beneficence, well pleasing to God.

All these motives combined were enough to provide highways that were considered sufficient for the current needs, for people were contented with little. Most of the journeys that were made were performed on foot or on horseback[3]; and the conveyance of goods was effected by

[1] Jusserand, *English Wayfaring Life*, p. 82.

[2] Blomfield, in his *History of Bicester*, II, p. 40, says that after the establishment of the Priory at Bicester the public highways were largely repaired by the monks.

[3] It seems impossible to agree with certain writers who endeavour to maintain that the roads of the country generally, during the Middle Ages, were in good condition. For instance, Blomfield, *History of Bicester*, II, p. 40, says that this seems almost certain from the fact that long journeys were then undertaken both on horseback and in carts in the course of a single day, which would have been impossible if the roads had been mere trackways with an oft-recurring series of ruts and holes. Wylie, *England under Henry IV*, I, p. 95, after speaking of the King riding post at a great speed from Windsor to the capital, says: "The speed with which journeys were then performed fully bears out the assumption of Rogers, *Agriculture and Prices*, I, p. 664, that the high-roads of England were, in that age, far from being so desperately impassable as has been often assumed." Healey, in his *History of Part of Somerset*, p. 264, says: "From the records of the Manor of Porlock (1422) we find that the bailiff had ridden to Exeter and Topsham to buy wine for the Lady of the Manor, taking four days for the journey. In 1424 he went with a waggon and two men to Dunster, to fetch a pipe of wine to be had there. This proves that the highway in those days was available for a heavily laden waggon, and was not a mere path for pack-horses."

Now it is easily conceivable that there were some portions of roads that were kept in fairly good repair; it was to the interest of landowners that this should be so; but to make such possible facts the basis of positive and comprehensive statements like the above, seems unwarranted. Possibly some of the great trunk routes of the Romans were still in sufficiently good condition in the early fifteenth century to render travel, on portions of them, comparatively easy. We have noted above the activity which some manorial courts exercised in looking after the highways; and no one can examine the records of the Court Leet in the case of many manors without being impressed by the great number of presentments because of lack of attention to the maintenance of the roads. The presence of some good stretches of road, capable of accommodating a waggon, will easily agree with the incident mentioned above by Healey.

Forbes and Burmester, *Our Roman Highways*, p. 176, refer to the necessity, after the Norman Conquest, for "the existence of passable roads," in order that the King and his court, the great lords, the monks, and others, might go, either from choice or necessity, from one estate to another often at considerable distance apart. It

pack-horses, and also by means of heavy, lumbering carts which stood the hardest jolts.

When we come to speak of the mediaeval towns, we find that at that time the imposition of tolls played an important part in the upkeep of the streets[1]. Murage tolls, that is, tolls intended primarily for the maintenance of the walls of the towns, but later used also to help in repairing the streets, were levied on persons coming to the towns with wares for sale. How early these tolls were established we do not know; but we know that they were in existence before 1189, for in that year King Richard freed the burgesses of Northampton from the payment of toll throughout England[1]. In 1224, Henry III

is perfectly true that these classes were interested in good roads, but this would by no means ensure the existence of such roads. At a later time, under the system of statute labour, all classes were interested in good roads, but each man wanted to do as little as he could and let the other man do as much as he would; in other words, they were not sufficiently interested in good roads to make their desire for them effective. The same conditions doubtless prevailed at the earlier time also. As to the fact that the fairs were largely attended during mediaeval times, this does not prove that the roads were good; for much of the carrying was done by pack-mules or pack-horses, which could go anywhere where there was sufficient room for a bridle-path.

In reference to the speed sometimes attained, we do not see that the instances given prove anything in support of the assumption of good roads. A saddle-horse could often go at fairly good speed even where the usual track was deep and miry, by merely going a little to the side. But that journeys on horseback (which were the usual kind) were not usually effected at great speed is evidenced by the fact that members of Parliament and those persons who had to come to the aid of the King in time of war and on other occasions, were allowed for travelling twenty miles a day— see *Rot. Parl.*, vi, p. 525; also exceptions given in *Rot. Parl.*, vi, pp. 525–6. For persons walking, this would be an easy rate, even where there was no track at all; and for persons riding on horseback, it would seem to indicate that there must have been great delays on the route. Price, in his *Leeds and its Neighbourhood*, p. 114, shows that a journey at that time was very costly. Then he refers to the fact that, on at least three occasions in the fourteenth century, Parliament had to adjourn because, owing to the state of the roads, not a sufficient number of members were present to go on with the business. These things do not support any assumption that the roads in general, in mediaeval times, were in fairly good condition. There probably were all over the kingdom quite passable bridle-paths, but we must not mistake these for good roads.

[1] In the charter given to the borough of Northampton by King Richard in 1189, it is expressly stated: "And this we have granted to them that all the burgesses of Northampton be quit of toll and lastage through all England and by the ports of the sea...." Tolls were therefore earlier than 1189. *Northampton Borough Records*, i, p. 27.

In the *Records of Gloucester*, edited by Stevenson, p. 5, we find that the burgesses of Gloucester, in 1194, were granted by Richard I, along with other privileges, "the same customs and liberties throughout all our land of toll and of all other things as the citizens of London and those of Winchester had at any time in the

granted permission to Northampton to levy certain tolls, in aid of enclosing the town, for a term of three years[1]. In 1251 and 1301, still more extensive grants were made for this purpose[1]. It was not long before these murage tolls were associated with pavage tolls and pontage tolls. For instance, in Leicester, where previously the wall had been maintained at the expense of the inhabitants, the Mayor, bailiffs and burgesses in 1286 received, by royal patent, a grant of five years' murage and pavage, a toll which was to be spent on walls and road-making[2]. In 1322, Earl Thomas of Lancaster obtained a grant of seven years' murage and pavage for the bailiffs and good men of Leicester. In 1285, Edward I granted the burgesses and inhabitants of Coventry letters patent authorizing them to take a toll for three years upon all saleable commodities that should be brought into their city. This toll was for defraying the expenses of paving the town[3]. According to their charter, granted by the King "to the Mayor, bailiff and good men of the vil of Preston," in 1314, a great many articles were liable to pay tolls when brought for sale within the town; and the revenues from these tolls were to be used for paving the streets, and were to continue for five years[4].

It would take too long to follow the course of these local tolls throughout England. To a large number of the towns such grants were made; and frequently the tolls were revived or continued by subsequent grants from the King[5].

reign of King Henry II...." Evidently, therefore, tolls had been in existence at least by 1154, when the latter King began his reign.

[1] *Northampton Borough Records*, I, pp. 37, 41, 58, where these grants are given in full.

[2] *Cal. Pat. Rolls*, pp. 221, 424.

[3] Poole, *History of Coventry*, p. 343. Paving must have gone on slowly, for in 1305 another patent was granted them to renew and continue such toll on the same articles and commodities. It was not till 1423 that it was ordered that every man should repair the pavement in front of his own house. See also Leonard W. Clarke, *The History of Birmingham* (in MS., 7 vols.), II, p. 58, showing that in 1319 at the instance of the inhabitants of Birmingham, the Earl of Pembroke obtained a license for three years to take toll for paving the streets. All vendible commodities brought there to be sold were to contribute—corn, $\frac{1}{4}d$. per quarter; other things in proportion. From the small amount of trade that was carried on there, this was insufficient for the purpose, and fourteen years afterwards a patent was obtained for a further term of three years. (*Patent Roll*, 7 Ed. III, Part 1, m. 7, etc.)

[4] See Fishwick, *History of the Parish of Preston*, p. 25, for the list in full.

[5] To remedy the bad condition of the streets of Evesham, that town, in 1328, received from Edward III its first grant, empowering it to exact tolls on all provisions, wares and merchandise vended in the market-place during three years, to aid in paving the town. This, apparently, was insufficient, and a second grant was made to continue these tolls for two years more. A third grant of similar tolls

But, in addition to these local tolls, there were in some instances traverse tolls (called also "passage" or "through" tolls) charged on beasts and burdens passing through or entering the town[1]. This was a thoroughly established practice at Northampton as early as 1274[2]. The origin of this toll is probably as follows: Right through the centre of Northampton, north and south, and east and west, ran two main roads, the north and south road being the most important highway between London and the north. These roads had to be kept in good repair throughout the liberties of the borough, and paved when within the town walls. This was a constant source of expense. Moreover, as a town on the royal domain, Northampton had to pay to the Crown or its assigns an annual fee-farm rental, which was very heavy in the earlier days, according to the value of money at that time. It was not unusual for the Crown, in such cases, to grant to royal demesne towns certain special privileges as a sort of set-off to the rental. At Northampton, this took the form of a traverse or passage toll.

for four years was made by Henry IV in the second year of his reign. May, *History of Evesham*, p. 370.

In 1329, Liverpool was given a similar grant by the Crown for three years for local improvement—Picton, *Liverpool Municipal Records*, I, p. 10. These grants became frequent, five having been issued under Edward III.

Note similar case of mending the highway from London to Brentford, by a grant from Richard II in 1380, as given in Davis, *Memorials of the Hamlet of Knightsbridge*, p. 24.

Southampton was given a grant of tolls in aid of paving in 1384. Davies, *History of Southampton*, p. 119.

In 1431 an Act of Parliament was passed for paving and repairing the principal streets of Northampton. Markham and Cox, *Northampton Borough Records*, II, p. 266.

In 1354 Edward III gave a grant of tolls to London, of 1*d.* per cart on all vehicles, and a farthing per horse on all laden beasts coming to or going from the city, except the carriages used to transport provisions for great men. This, too, was to aid the repair of the streets, which were often in bad condition.

Gloucester, in 1384–5, received a grant of tolls in aid of the paving of the town. The tolls were to be levied for seven years on goods coming to the town for sale, which included nearly everything that could be of any service. These tolls included, among others:

Of each load of corn	¼*d.*
„ horse, mare, ox, or cow	1*d.*
„ hide of a horse, mare, ox, or cow, fresh salted or tanned	1*d.*
„ cart carrying salt meat	2*d.*
Of five salted pigs	1*d.*
Of six hams	½*d.*
Of a fresh salmon	½*d.*
Of a dozen shad	1*d.*

For full list, see *Records of Gloucester*, edited by Stevenson, pp. 50–52.

[1] Markham and Cox, *Northampton Borough Records*, II, pp. 197–208.

[2] Ibid., II, pp. 198–9.

Where traverse tolls had been established, it was customary to fix upon toll-gathering points at some distance from the town. This had two objects: to prevent confusion with the local traffic of freemen, which would probably have resulted had the tolls been collected within or at the town gates; and to prevent the evasion of the toll by turning aside on tracks through the open country, so as to pass around the town instead of through it. The tolls were either collected by the bailiff or his agent, or else leased or farmed to a collector; but, in any case, the collectors at these distant points gave tokens to those who paid, which were delivered up at the borough gates[1].

By means of these local and passage tolls, some towns obtained considerable sums for the maintenance of their streets and roads.

Despite all the regulations which were made for keeping the streets in good condition, we find in the records of the boroughs a great many presentments for the usual offences against convenience and decency. Some were presented for piling timber, stone, etc., on the streets[2]. The regulations against obstructing the streets with great heaps of filth, which were dangerous to the community, were often persistently disregarded[3]. The water-channels in the streets were too frequently overloaded, and rendered worse than useless by misdemeanours and neglect of selfish citizens. The butchers, fishmongers, washerwomen, and others, were continually guilty of throwing the refuse from their work into these channels[4]. Pigs and pig-sties were kept within the town walls, despite ordinances to the contrary; and these animals were often allowed to go unringed around the towns, committing all kinds of nuisances that were destructive to the streets[5].

In the later mediaeval times, a change was made in the case of many boroughs, when paving Acts were passed for these places. The important provision of each of these Acts was that the mayor of the town, when he saw the necessity, could compel every owner of a messuage or tenement abutting on any high road or street to newly pave and keep in repair the road or street in front of his own tenement

[1] These traverse tolls were subject to abuse, by collecting them at distances remote from the town to which they belonged—Markham and Cox, *Northampton Borough Records*, ii, p. 200. At a later period, perhaps in the time of Elizabeth, the habit of collecting these tolls at distant points ceased, and they were gathered at the town gates or within the town.

[2] May, *History of Evesham.*

[3] Morris, *Chester in the Plantagenet and Tudor Periods*, p. 260 et seq.

[4] Ibid., pp. 261–2; also *Records of Southampton*, etc.

[5] In some towns a herdsman was appointed to take the cows, pigs, and other animals out of the town to the common pasture, and tend them during the day, bringing them back in the evening.

or landholding, as far as the middle of the road or street[1]. But in many cases, the number of presentments for failure to comply with this law, as shown by the town records, is a clear evidence that there was much attempted evasion of the responsibilities it imposed.

When such a regulation was put in force, it was frequently followed in a few years by the appointment of a town paviour, whose work was to examine the pavement of the town, and to perform the work himself, where necessary[2]. The town paviour remained as an established institution in such places for two or three centuries.

In some instances, there seems to have been very strong opposition to the law that compelled householders to maintain the pavement in front of their own property. In 1473, the bailiffs and stewards of the town of Gloucester presented a petition to Parliament, showing

---

[1] An Order of 17 Richard II, renewed in 6 Henry IV, was made for the town of Chester, directing each householder to cleanse the pavement before his own door and to keep it clean. But little attention was paid to such directions. See Morris, *Chester in the Plantagenet and Tudor Periods*, p. 260. The repair of the pavements of the town appears, however, to have been undertaken by the Muringers (or those who had charge of the maintenance of the walls) till 1566–7, an assessment being made upon the inhabitants. Thus, in 18 Richard II (1395), a grant of murage was made, on the petition of the mayor and commonalty stating that the walls and pavements of the city were ruinous, and that they had, from the earliest times, been supported and repaired by a murage. In 1463, the householders in the four streets were assessed in a 3*d.* lay. By 1532, the duty of repairing seems to have devolved upon the individual householder, for in that year William Robynson was fined 4*d.* for neglecting the pavement at his door. Morris, *Chester*, p. 264.

An Ordinance for paving the town of Northampton was made in 1422. For the full text of this Ordinance, see Markham, *The Liber Custumarum*, pp. 95–98.

In 1477, the following Paving Act was passed for Southampton, and it was ordered that owners of property, on notice from the mayor, sheriff and bailiffs, should be compelled to pave at their own charges before their doors as far as the middle of the street, and on their failing to do so within a quarter of a year after such notice, it became lawful for the mayor, etc., to pave and levy costs by distress. Tenants, on being distrained for what the landlords should pay, might stop the amount out of their rents, or recover by action of debt in the court of the town. Davies, *History of Southampton*, p. 119, gives the text of the Act.

[2] In the records of Southampton for Dec. 1482, pursuant to the Paving Act of 1477, we find it ordered that "A paviour be ordeyned to dwell in a house of the towne, price of 13*s.* 4*d.*, rent free, and to have yerely a gowne to this intent that he shall with a sargent of the same towne doo serche the pavement of the said towne, and also to pave in all places nedefull withyn the said towne and doo all thyng that longeth to that office, takyng his wages for his labour as it is used for a Tese: provided alway that the stone and all maner thyng to the said pavement belongyng be ordeyned by hym or theym afore wose house the pavement shall bee noyouse or nedefull of reparacon." Davies, *History of Southampton*, p. 120.

The Muringers' accounts of Chester contain a statement of payments made to John Paver and others for new paving. Morris, *Chester*, pp. 264–6.

that the town was "feebly paved and full perilous and jeopardous" for passengers; and praying that persons who owned property on the four principal streets be *compelled* to repair and maintain the pavement in front of their property, out to the middle of the street[1].

From what we are able to learn as to the roads in the country and in the towns, it may be stated that they were in accord with the comparatively meagre demands of agriculture and industry, which were then on a small scale. The population was mainly agricultural; the community tended to be self-sufficing, and to have but little business relations with outside communities; and the roads were of such a character as merely to satisfy the demands of the great majority of the people[2]. Toward the end of the Middle Ages, with the breaking down of the manorial system and the consequent decline of tillage, we can understand why the roads and bridges were not so well maintained. Further, we have seen that the activity of the Courts Leet throughout England had had much to do with the preservation of the roads; and with the decay of their authority we find a strongly predisposing condition towards laxity in the management of the roads. About the same time the decline of the fairs and the cessation of the pilgrimages caused considerable reduction in the amount of travel, which was accentuated by the break with Rome and the stoppage of this riding to and fro. All this, accompanying the long-continued and expensive wars to which England was subjected, points with decided emphasis to a decline of the means of communication before the opening of the sixteenth century.

Bridges.

The Romans were noted bridge-builders, and taught the world the construction of stone bridges by means of the semicircular arch; but all their bridges in England, apparently, were built of wood, and though not of artistic merit, they were very durable.

The art of bridge-building seems to have been cultivated in Britain from an early period, with success. The noted Gothic triangular bridge at Croyland, in Lincolnshire, was probably built in the latter half of the ninth century. In 998, the first bridge was built across

---

[1] *Records of Gloucester*, edited by Stevenson, pp. 15–16.

[2] For purposes of husbandry, carts and waggons or waynes, drawn by oxen, were in common use, as we see by the chartularies of the monasteries; and some of the villain services were performed by the tenants in wheel carriages. Chalmers, *Caledonia*, IV, pp. 729–30. But the use of the cumbersome carts and waggons was probably confined almost entirely to work upon the land, and the amount of wheel-carriage employment upon the roads was insignificant.

the Thames, a wooden structure, which after many vicissitudes was supplanted by a stone arch bridge, begun in 1176 and completed in 1209. At each end of this bridge there was a tower erected for purposes of defence; and on one of the piers a chapel was built. Other houses were added to the bridge in later years. This bridge was so substantially built that it lasted for over six hundred years.

But it is of the rebuilding, repair and maintenance of bridges that we would speak more particularly here; for after a bridge had been erected, and it had fallen partly or wholly to ruin, it was by no means easy to ascertain upon whom rested the burden of restoring it. And even after it was known that a certain person or corporate body was responsible for this work, there was still to be considered the question as to the means to be used in obtaining the necessary funds therefor. The consideration of these things for each of the important bridges, while it is an interesting study, must be abandoned here for a more general outline of the chief facts with reference to this early period.

In some cases, the construction or repair of bridges was carried out at the cost of particular individuals, and in this the clergy were perhaps most prominent. Walter Skirlaw, Bishop of Durham, rebuilt, at his own expense, the old stone bridge which, up to the time of Henry IV, had spanned the river Wear at Shincliffe, near Durham. He also bridged the Gaunless at Auckland and the Tees at Yarm[1]. Boyle tells us that Elvet Bridge in the county of Durham was first built by Bishop Pudsey[2]. Doubtless, were our local records more complete, we should find that many other enterprises of this kind were carried out through the efforts of patriotic and self-sacrificing benefactors. The building of a bridge always ranked very high among works of piety. It was also considered a pious act to assist in maintaining bridges, and therefore, closely akin to the foregoing are those instances of bequests which were left for the maintenance or repair of bridges. These, however, were usually not large, nor were they abundant[3].

---

[1] Leland, *Itinerary*, I, p. 60. (Hearne's ed., 9 vols., Oxford, 1710.) See also Jusserand, *English Wayfaring Life*, p. 60.

[2] Boyle, *The County of Durham*, p. 390.

[3] In 1439, William Neel, in his will, bequeathed "for the repair of the great bridge, 3s. 4d." *Market Harborough Records*, p. 173.

Thomas del Bothe had built a chapel on the old bridge at Salford, for the repose of the founder's soul. His will (in 1368) directs the gift to the bridge of £30. Axon, *Annals of Manchester*, p. 19.

From time to time, various bequests were made for the repair of the Trent bridges at Nottingham. A house and portion of land were given in trust by Robert

In other cases indulgences were granted to those who, by contributions of labour or money, would assist in the erection or repair of a particular bridge or bridges. This was another phase of the reward which attended the performance of all "pious and meritorious" work. The Bishops' registers throughout England amply testify to the many cases of indulgences granted for a variety of causes, among them the repairing of bridges[1].

On many bridges, chapels were erected where passers-by could turn aside and have the aid of a priest or hermit in the performance of the duties of religious worship. Religious devotion being ended, the traveller had the opportunity of contributing toward the support of the chapel and of the bridge[2]. At times, the priest was supported,

---

Poole and others on May 23, 1501.—Orange, *History of Nottingham*, p. 644. Thomas Willoughby, an alderman of Nottingham, willed to "Hethbeth" bridge four of his best pieces of timber.—Deering, *Nottinghamia Vetus et Nova*, p. 315. It is recorded that a chapel dedicated to St James stood upon the High Bridge, and that it possessed lands, in 1535, of the value of £2. 6s. 2d. per year.—See *Inquisitiones ad quod Damnum*, p. 127.

The town of Burton-on-Trent had various bequests given to it for charitable purposes, some of which were for repairing of bridges; for instance, note the charity of Mrs Almond in Molyneux, *Burton-on-Trent*, p. 86. Note also ibid., p. 57.

For other cases, see Hudson and Tingey, *Records of Norwich*, I, p. 259; Clark, *Wood's City of Oxford*, I, p. 411. See also description and cost of Maud Heath's causeway in Daniell, *History of Chippenham*, pp. 120–6.

[1] See, for example, Gibbons, *Ely Episcopal Records*, pp. 4, 40, 400, 403, and also occasional instances on pp. 408–20.

The following release from penance, granted by the Bishop of Durham in 1394 to all who would contribute toward the rebuilding of the Chollerford bridge across the Tyne, is interesting: "To all Christian people, to whom these presents shall come, Walter, Bishop of Durham, health in our Lord everlasting. Whereas the bridge of Chollerford, as we hear, is decayed by the inundation of the waters, by which there used to be a frequent passage, and now wants repair, whereby the inhabitants in the neighbourhood are in great danger. We therefore confiding in the mercy of Almighty God, and the sufferings of his Holy Mother, and all the Saints, do release unto all our parishioners, and those in other dioceses where this indulgence shall be received, 13 days of their enjoined penance, upon condition they lend a helping hand to the repairing of the said bridge, or contribute their pious charity thereto. These presents after three years nothing availing. Given at Chester the 8th Kalends of August, of our pontificate the 7th year." Wallis, *Antiquities of Northumberland*, pp. 69–70.

[2] In the Records of St Nicholas Hospital, Salisbury, p. xlvi, we learn that for three hundred years (1244–1545) the salaries of the priests that officiated there, and the repairs of the bridge itself, were paid by the Masters of St Nicholas, who recouped themselves with the offerings made by pious wayfarers who turned aside from the road to this wayside chapel. See also Atkinson and Clark, *Cambridge*, p. 62. A fine treatment of this subject is given in the Brit. Mus., Add. MSS. 30,302, 'Remarks upon Wayside Chapels,' by J. C. Buckler and C. A. Buckler, architects. This is a complete transcript of the printed work of 1843, with many additions.

at least in part, from the products of adjoining land which had been bequeathed for that purpose; but in other cases it took most of the chapel offerings for his support, so that there was but little left to maintain the bridge.

When such votive offerings were insufficient, recourse was had to the occasional or frequent requests for charity, and to other voluntary contributions. In order to give these requests some weight with the people at large, they were signed by influential persons well known to the community, or were issued by the abbot of an adjoining monastery, sometimes associated with many private gentlemen[1], or in some cases by the municipal authorities of the town which was interested in the maintenance of the bridge[2].

---

[1] See the document issued by the abbot of the Burton-on-Trent monastery, associated with many private gentlemen, "to all true Christian people," asking for aid to repair the Burton-on-Trent bridge, in Molyneux. *Burton-on-Trent*, pp. 58–59.

[2] We give in full such a document, taken from the *Records of Nottingham*, edited by Stevenson, II, p. 264:

"Appointment of Collector of Alms for
Hethbeth Bridge. 1467. Oct. 10.

To all the faithful in Christ as well men of religion as of the world seeing these present letters, John Hunt, Mayor of the town of Nottingham, Thomas Babington, Recorder of the same town, Thomas Thurland, Thomas Alestre, John Squire, Robert Stable, Robert English and Thomas Lockton, Keepers of the Peace of the town aforesaid, John Spencer, John Clerk, Robert Smith, John Sergeant, and John Painter and Richard Knight Wardens of the Bridges of Hethbeth, and John Cook and John Draper, Sheriffs of the said town, and also John Ody and William Way, Chamberlains of the aforesaid town, and the other co-burgesses and the whole Community there, greeting in Him that is the true salvation of all. Forasmuch as it is pious and meritorious, as far as human strength can, to augment and to arouse human exertions to pious alms, exchanging temporal things for eternal and worldly things for heavenly, so that according to the word of the Lord (Matt. vi, 20) every faithful man may lay up his treasures in heaven, where neither thieves break through, nor moth destroy, nor rust doth corrupt; and whereas, amongst the divers alms that should be forwarded, it is meritorious to mend dangerous roads and perilous bridges, and especially the bridge of Hethbeth over the Trent, which has nothing whereby it may be sustained, except only by gifts of charity, and in order that every Christian may give more speedily and freely alms and charitable gifts to the fabric and sustentation of the same bridge, there is found there a fit priest in the chapel built upon the said bridge daily celebrating divine service in honour of our Saviour and of the Blessed James the Apostle and of All Hallows, for all the helpers and benefactors of the bridge aforesaid. Furthermore, know ye that we, of the common consent and assent of the whole Community of the town aforesaid, have made, ordained and constituted our well-beloved in Christ William Thornes and William Chase our certain and true attornies jointly and severally to seek and receive alms and charitable gifts for the reparation, sustentation, and mending of the bridge aforesaid, especially beseeching and asking that, if the same William and William, or any other bearer of the same, come amongst you in order to ask for such alms, you will give to them, or one of them, gifts such as you are able,

Occasionally, the gilds showed their activity by the maintenance of bridges as well as roads: a fact which is well exemplified by the history of the Gild of the Holy Cross at Birmingham, which was in very close association with the town government. Along with its other charitable features, the gild laid out considerable sums in the repair of the roads and bridges. These works are set forth in the report made by the Commissioners of 37 Henry VIII, in 1547, which says: "Also theare be mainteigned, wt parte of the premisses, and kept in good Reparaciouns, two greate stone bridges, and divers foule and daungerous high wayes; the charge whereof the towne of hitsellfe ys not hable to mainteign; So that the Lacke thereof wilbe a great noysaunce to the Kinges maties Subjectes passing to and ffrom the marches of Wales, and an utter Ruyne to the same towne—being one of the fayrest and most profittubble townes to the Kinges highness in all the Shyre[1]."

These and other precarious sources[2], however, were not sufficient for the work to be done, and in all directions tolls and taxes were sanctioned, in the form of pontage and pavage, to cover the cost of

and this more favourably in consideration of our supplication. And in order that this our present writing may the more readily appear authentic to you, the seal of the office of the Mayoralty of the town aforesaid is affixed to these presents.

Given at Nottingham, the tenth day of October, in the seventh year of the reign of King Edward the Fourth."

[1] Bunce, *History of Birmingham*, I, p. 30. Whether the craft gilds of Leominster were at this time responsible for repair of the bridges, as they were at a later date, we cannot say; but it is altogether probable that they were; for it is not likely that the responsibility would be transferred in such a case as this. See Court Leet cases in 1635 and 1650 of the gilds having been presented for non-repair of bridges, as given in Blacklock, *The Suppressed Benedictine Minster*, pp. 383–4.

What is true of the Gild of the Holy Cross and of the Leominster craft gilds, we may safely infer regarding some of the other gilds. We know this to be true of the gild of Leicester, for in Miss Bateson's *Records of the Borough of Leicester*, I, p. 34 (about 1225), the alderman of the gild was disbursing money for the work on a bridge.

[2] At Leicester, the profits on the sale of wax on Holy Cross Day went to the bridges. See *Records of the Borough of Leicester*, I, p. 349. In the same town, in the fourteenth century, collections for the bridges were being made in the parish churches. The churchwardens' accounts of the parish of Luccombe in Somersetshire show us similar small disbursements for repair of bridges. See Healey, *History of Part of Somerset*, p. 177. Again, in Leominster, anyone who was received and sworn into the company as one of the capital burgesses, was required to "be at the Lawday, and faire dayes in a gowne, except it raine or snowe, or be foule weather, upon pain of xs. to be levied by distress as in former orders is lymitted, and to be bestowed upon and towards the reparation of highwayes and bridges within this borough." See Minute Book of the Corporation, Jan. 23, 1634. But we need not multiply these small sources of funds.

repairs. These were levied both in ordinary and extraordinary cases: not only to accumulate funds to cover the ordinary costs of maintenance and repairs, but also for the rebuilding of bridges that had been destroyed by an army or swept away by a flood, both of which kinds of accident were frequent at this early period. In some cases the privilege of collecting this pontage was conferred in the charter given to the town by the King[1]; in other cases the right to take these tolls was granted by royal letters patent for a few years only, although frequently the privilege was renewed again and' again[2]. Pontage, of course, was the toll taken on goods passing either over or under a bridge. Where this was granted to an individual, the latter was expected to collect the tolls and keep the bridge in good condition: but where it was granted to a municipality, the latter appointed bridgewardens who were to look after the collection and expenditure of the money for the bridge and to render a faithful account of their trust[3]. That some of

[1] *Charters of Ludlow*, pp. 34–37. The charter granted by Edward IV, in 1461, to the town of Ludlow, gives a full list of the tolls and customs which the town might take for the repair and fortification of the bridges, gates and walls of the town. These tolls were on things which were brought for sale to the town.

[2] Brit. Mus., Add. MSS. 15,662, p. 225. In 1347, Edward III granted a license to the town of Newcastle-upon-Tyne, to levy customs for the repair of the bridge across the river there. These duties, on all goods brought to the town to be sold, are set forth in detail. They were to be collected for fifteen years, but no longer.

[3] Stevenson, *Records of Nottingham*, II, pp. 220, 244, 364; also III, pp. 2, 4, 6–10, 12–14, etc., where the form of the Bridgewardens' accounts, with summary of receipts and expenditures, may be found.

The bridge over the Aln at Alnwick was decayed, and Edward III granted the tolls of this bridge for three years to the men of Alnwick, to enable them to repair it and to pave the town. For full text of the document granting this pontage, see *Patent Roll*, 51 Edward III, m. 19; Tate, *History of Alnwick*, I, pp. 150–1.

Henry IV, to help make a bridge over the Foss river in the city of York, gave to the Mayor and Commonalty of the city, for five years, the right to take stipulated tolls from those crossing the bridge. As soon as the five years were past, the customs and tolls were to cease, and to be applied for the making of the bridge. *Patent Roll*, 4 Henry IV, Part 1, m. 22.

About ten miles of the great road between London and Southampton was foul with mud near Hartford bridge. It was thereupon ordered that, for the next three years, a tax of $\frac{1}{2}d.$ should be laid upon every horse, cow, load of corn or fish, and every barrel of herrings brought into Basingstoke for sale; $1d.$ for every six sheep; $\frac{1}{4}d.$ for every pig; and so on, at the rate of $\frac{1}{2}d.$ for every 5s. of value; which was equivalent to a tax of nearly one and one-fourth per cent. *Patent Roll*, 8 Henry IV, Part 1, m. 26; cf. Rymer's *Foedera*, VIII, p. 634.

The grant of Richard II, in 1381, authorizing the town of Newport Pagnell to take tolls for properly repairing their bridges over the Ouse and Lovat, was made in the following words: "The King to his beloved Thomas Cowe Robert Bowes, John Taillour of Newport Paynel and Simon Swet Greeting Know ye that in aid of mending and repairing of the bridges of Northbrigge and Southbrigge in the

the bridges were ruinous was due to the fact that the grantees of the tolls for their amending used the money for their own purposes and allowed the public to suffer[1].

It may not be amiss to outline the early history of one bridge, as a type of many others in their essential features; and for this purpose we shall take the Boston bridge. It is first mentioned in 1805, when a petition was granted for pontage to be taken by John de Brittany, Earl of Richmond, for repairing the bridge across the river at St Botolph's. He was allowed to levy certain tolls upon saleable articles passing over and under the said bridge, for three years from Mar. 18, 1305[2]. On May 22 of the same year, a similar privilege was granted to William de Ros, of Hamlake, for five years, for the same purpose[3]. The tolls to be levied were exactly the same in the two cases, and it was not intended that there should be any conflict between the grants. The intention seems to have been that William de Ros should take the duties upon the goods passing from the west to the east, his land being on the west side; and that the Earl of Richmond should receive the customs payable upon goods passing from the east, where his land was, to the west. Each was to repair the side of the bridge joining upon his estate. It would seem that no material

town of Newport Paynell which are broken and injured to the grievous damage and danger of the men crossing by the said bridges and for the mending and repairing whereof the men of the said town are from year to year as we understand heavily charged we have by our special grace granted to you that from the day of making these presents up to the end of three years next following fully to be completed you may take by the hands of those whom you trust and for whom you are willing to answer of things saleable passing over and under the said bridges the customs under-written namely:

for every horse load of corn	1 farthing
,, cart load of corn	½ penny
,, horse, mare, ox, cow	1 farthing
,, horse load of cloth	½ penny
,, whole cloth	1 farthing
,, cwt. of linen cloth, canvas cloths of Ireland, Galewath and Worstede	½ penny

etc.   etc.

And therefore we command you that you take the said customs until the end of three years and apply them towards the repair and mending of the said bridges, as is aforesaid, but when the term of the said three years is completed the said customs shall entirely cease and be discontinued. In testimony, etc., to last during the said three years. Witness the King at Northampton the 20th day of November." *Patent Roll*, 4 Richard II, Part 1, m. 4.

[1] See, for example, Atkinson and Clark's *Cambridge*, p. 62; *Rotuli Hundredorum*, I, p. 49; Brit. Mus., Add. MSS. 15,662, p. 225.

[2] *Patent Rolls*, 33 Edward I.

[3] Ibid.

difficulty arose, for several other grants of tolls with concurrent operation occur afterwards.

Another grant of pontage for repairing the bridge and the pavement of the town was made to the Earl of Richmond in 1308, to continue five years[1]; and one to William de Ros, in 1313, for five years[2]. The Earl of Richmond had also a grant of pontage in 1313 for five years[3]

In 1319, pontage and pavage duties and tolls were granted to William de Ros, of Hamlake, for five years[4]; and another grant of the same kind was made to him for three years in 1328, giving him tolls on merchandise passing over or under the bridge, "in aid of the repair of his half of the bridge[5]." At this time the bridge is spoken of as being broken up, and dangerous to pass over. This grant was renewed for three years in 1331[6]. The Earl of Richmond had a grant of pontage tolls in 1337, for five years, when the bridge was declared to be "ruinous and broken up[7]." The Earl of Richmond had another grant of pontage for five years in 1358, in aid of newly erecting and supporting the bridge, and paving the town[8]. This new bridge seems to have been of firm construction, for nothing is recorded about its needing repairs till 1500.

The constant succession of grants of pontage which we have here outlined might be repeated in endless detail for many other bridges, with the same result that the structures were continually "ruinous" and in need of further repairs. The records of presentments before Courts Leet for failure to repair bridges[9], and the accounts of Borough Chamberlains in cases where the towns had to repair their bridges[10], are sufficient evidence that bridges patched up from time to time were as constantly falling into decay. Only as stone bridges were substituted for wooden ones did the continual outlay for repairs greatly diminish.

Enriched by so many offerings and by the revenue from so many tolls, while protected by the *trinoda necessitas*, and by the common interests of the landed proprietors, the bridges should have remained continually in good repair. But there is another side to the subject which must be mentioned in order that we may get a complete view of the conditions under which bridge repair was carried out. It was not always easy to ascertain who should bear this responsibility; and, of course, those who felt or knew no duty resting upon them for this

[1] *Patent Rolls*, Dec. 4, 1308.   [2] Ibid., May 15, 1313.
[3] Ibid., Feb. 25, 1313.   [4] Ibid., June 5, 1319.
[5] Ibid., Mar. 6, 1328.   [6] Ibid., July 8, 1331.
[7] Ibid., Mar. 25, 1337.   [8] Ibid., year 1358.

[9] *Ingoldmells Court Rolls*, pp. 140, 155, 211, 281, etc.; also *Rolls of the Court Leet of Leominster*, in Blacklock, *The Suppressed Benedictine Minster*, pp. 383–4.

[10] Chamberlain's accounts of Leominster, given in Blacklock, *The Suppressed Benedictine Minster*, p. 384 et seq.

work were dilatory to venture upon such improvements at their own expense. In some cases the repair of a bridge was the duty of a private landlord, usually one whose land adjoined the bridge; or in other cases a monastery or other ecclesiastical body was the property owner. In a few cases the bridges were in the hands of the King[1], and his sheriff had to look after the repairs. Many bridges were under the control of the towns that were immediately interested in them[2].

For several bridges the responsibility was divided. The Leen bridge across the Trent at Nottingham was kept up and repaired by the town of Nottingham and the several hundreds of the county, in certain stipulated proportions[3]. Rochester bridge consisted of nine piers, and each pier was to be repaired by a different person from those who repaired the others[4]. With such divided authority, we can easily see the difficulty connected with the work of maintenance. But the tendency in all cases, especially in those the control of which was doubtful, seemed to be to make the whole county responsible for the maintenance of all public bridges[5].

Rivers and Navigation.

Not only were the rivers more useful, in all probability, than the roads in facilitating internal trade, but they were also safer for travel, for the roads were often infested with highwaymen[6]. Brooks that are now shallow and narrow were recorded in Domesday as navigable for vessels. The trade back and forth with the Continent was carried inland by the small vessels pushing their way, in some cases, far into the interior on the rivers and streams that connected with the sea. Hence towns like York and Doncaster are chronicled in Domesday as ports. It was important for towns, therefore, to secure grants giving them control of the waterways on which they were situated[7]; and the towns

---

[1] Such was the control of the two bridges across the river at Cambridge. See Atkinson and Clark, *Cambridge*, p. 62.

[2] For example, the bridge at Alnwick, East bridge at Oxford, etc.

[3] Deering, *Nottinghamia Vetus et Nova*, p. 167, gives these proportions.

[4] Denne, *History of Rochester*, pp. 41–56. (See also Appendix 1.)

[5] Possibly this arose from the fact that streams and rivers were often the boundaries of parishes, and the bridges crossing these rivers could not be said to be in either parish. Hence, neither parish, as a rule, was liable for repair, and so the larger division (the county) would naturally assume this burden.

[6] Travellers from London to Leeds and York, we are told, used to go by land to Nottingham, and thence sail down the Trent to the Humber, on reaching which they could make use of the tide to carry them up the Ouse toward their destination. Price, *Leeds and its Neighbourhood*, p. 115.

[7] This is well exemplified by the case of the city of York. Drake, in his excellent *History and Antiquities of York*, pp. 199–202, gives the early law governing this

were constantly complaining to Parliament about weirs and other obstructions placed in the rivers.

In 1351, Edward III resolved on a strong measure. He passed an Act for the removal of all obstructions placed in rivers since the time of Edward I[1]. But the manufacturing interests continued to assert themselves and weirs and mills were soon constructed[2]. Parliament, therefore, in 1371, attached the great penalty of one hundred marks (over £66) to this offence[3]. Still the Commons complained,

city, in regard to its conservatorship. The Lord Mayor of the city of York always bore the title of "conservator" or "overseer" of the river. The city appointed a water-bailiff, whose duties are given on the aforementioned pages.

"That the free and open navigation of the river from the Humber up to the city, was a great encouragement to trade is most certain. Free and open it must have been anciently, and a strong flow of tide run up it; else such ships as Malmesbury speaks on, which then did navigate the German and Irish Seas, could never get up to unlade their burdens, and lie in the heart of the city. In the Danish invasions, their fleets, sometimes consisting of 500 or 600 sail, came very high up the Ouse, before they landed. Anno 1066, a vast fleet of ships, with 60,000 land forces on board, came up the Humber and Ouse as far as Rickal (within six miles of York), where they moored their vessels." Drake, p. 229.

The tide up the river Ouse in those days must have been very strong, for even small vessels with such cargoes of men, horses, armour, etc., must have drawn deep water. The trade of the city was correspondingly great, and was encouraged by successive Kings and Parliaments.

The *Victoria History of Lincolnshire*, II, pp. 383–5, shows how the rivers were the early means of conveying the agricultural and manufactured products (corn, wool, cloth, etc.) to the markets; and also shows what a good water system the county had for this purpose.

Cambridge, at the head of a waterway communicating with the sea, was in direct communication with the Continent, by means of its river. Thus it early became a distributing centre and the seat of one of the largest fairs of Europe. The importance of its river can scarcely be exaggerated. The same is true of Norwich.

[1] Act 25 Edward III, stat. 3, c. 4. Brit. Mus., Add. Charters 45,960, gives the Latin text of the commission appointed under Edward III (1352) to inquire into the state of the rivers Trent and Don.

[2] The benefits conferred on the navigation of a river by mill-weirs may be thus stated: When the river ran in its natural channel, it passed through alternate series of sharp, shallow streams, and long, deep pools. In summer, many of these rapids were too shallow to float a barge; and it was just at these shallow places that mills were generally constructed, because the descent in the level of the water in the river furnished also the fall necessary for working a mill-wheel. The mill-weir, which kept back the water and forced it over the mill-fall, of course deepened the water for some distance above. Also, when a barge was approaching from below, the miller would open his weir and let a rush of water through, sufficient to tide the barge over the shallows. This rush of water was called a "shoot" or "flash." For the benefit of this flash, the bargemaster paid the miller a fee, the original of our modern payments at locks.

[3] Act 45 Edward III, c. 2.

but now only of the obstructions erected before the reign of Edward I[1], which shows that the measures of the Government had proved effectual. Commissions were accordingly issued in 1398 to Justices of the Peace to have the freeholder remove such weirs within half a year after he was duly notified thereof, upon penalty of one hundred marks. These measures were enforced by Henry IV in Acts of 1399 and 1402, and by Henry V in an Act of 1413[2]. But in 1423, complaints were made that the law was ineffectually executed in Kent, Surrey, and Essex, and fresh Commissions were issued for its enforcement[3].

Another subject that demanded the attention of Parliament was the inundation of land adjoining rivers, where the banks of the rivers were not high enough to keep out the sea-water. Great damage had been done in this way in several parts of the kingdom; and in order to provide a remedy, the Act of 1427 authorized the Chancellor of England to make provision for several Commissions of Sewers to be sent into various parts of the realm, to survey the banks, walls, bridges, etc., of such rivers near the sea-coast, and to inquire about the defects in these and how such defects had occurred. If necessary, in the fulfilment of the statute, they were to distrain upon tenants of lands for the repair of these things; and those who were negligent, or who refused to do their share of the work, were to be compelled to do so by legal procedure[4]. Similar Commissions were granted in subsequent years for the same purpose, a résumé of which is given in the Act of 1472 which continued the authority to the Chancellor for fifteen years[5]. These Commissions were to have power to execute the statutes and ordinances pertaining to such river navigations; but the fact that each Act recites the inefficacy of former Acts would lead us to infer that the statutes were not very rigidly enforced.

The conflict of interests between the commercial and the manufacturing classes, as to keeping rivers clear of weirs and other obstructions, became more equal toward the middle of the fifteenth century, and in 1464 Edward IV, who eagerly courted the favour of the manufacturing classes, refused a petition of the Commons to enforce the statutes of 1351 and 1371, in the case of the Severn river and its tributaries. But

---

[1] Act 21 Rich. II, c. 19.

[2] Acts 1 Henry IV, c. 12; 4 Henry IV, c. 11; 1 Henry V, c. 2.

[3] Act 2 Henry VI, c. 12.

[4] Act 6 Henry VI, c. 5. Dupin, *Voyages dans la Grande-Bretagne*, pt. III, vol. I, p. 64 ff., says that this Act offered a complete system of administration of waterways; but this statement is contrary to fact, for the Act applied only to the conditions in "several parts of the kingdom," where the ocean water flowed up the river channels and spread out over the land. See preamble to the Act.

[5] Act 12 Edw. IV, c. 6.

in 1472, the shipping and mercantile interests again prevailed; appeal was made to the Magna Charta and to subsequent Acts for the removal of obstructions in the rivers, which Acts were confirmed; offenders were ordered to remove or break down these impediments to navigation, according to the tenor of the laws for that purpose; and a fine of one hundred marks was imposed upon all defaulters[1].

While the Parliaments passed statutes for the preservation of the navigation of the great rivers free of artificial obstructions, the forces of nature were incessantly at work tending to overthrow the usefulness of the navigation. The natural result of rains, floods, etc., is to silt up the beds of rivers, and to take the material from some parts of the river and deposit it in other parts, thus rendering the depth of water very far from uniform. Then, too, the boats and barges used on the rivers were usually ballasted, but when they reached a shallow part of the river the ballast was thrown out into the river in order to lighten the boat and permit its passage with its cargo along the route[2]. In this way, the effect of shallows in a river was cumulative. This was the history of many of the rivers of England; but at the period we are now considering there were but few efforts made to free the rivers from these natural impediments.

The river from Norwich to Yarmouth was a very important navigation, for Norwich in early days was a mercantile and trading town, and ships came up by an arm of the sea to an open market, which was held every day in the week. Besides, fairs were held here twice a year, and it was a rendezvous for merchants from all over England and for foreign merchants. At the beginning of the fourteenth century, Yarmouth began to be a rival port to Norwich, and levied duties on vessels and goods passing up to Norwich. But after much litigation, Yarmouth was for a time compelled to desist from exacting such tolls. Norwich, however, was in a precarious position, for her river was becoming narrower and the stream shallower, so that it was often difficult for sea-going vessels to come up to the old inland port. But the people of Norwich were not to be outdone without some struggle for their trading supremacy, and we find from the records of the town that in 1422 a great effort was made to cleanse the river and restore the navigation[3]. Trade, nevertheless, declined; and it was not till

---

[1] Act 12 Edw. IV, c. 7.

[2] We see what action was taken in such cases at Ipswich: "Every ship that shall throw their ballast into the river, shall pay 12*d.* for each ton thrown out, and the informer shall have 12*d.*" Bacon, *Annals of Ipswich*, p. 164 (1492). That this was carried into effect we see from ibid., pp. 195, 200, etc.

[3] From the *Records of Norwich*, edited by Hudson and Tingey, I, p. 277, we learn that at the first assembly on Friday before St James Apostle, 10 Henry V,

early in the nineteenth century that schemes were revived for improvement, so as to admit ocean-borne traffic. Such plans were attended, however, with almost total failure.

Another example of improvement of the bed of a river is that of the river Ley (or Lea), which was used for the conveyance of products to London. By 1430, there was such a great number of "shelfs" in the river, which runs from the town of Ware to the river Thames, that boats could not pass along its course. As there was no statute giving authority for the removal of these impediments, representation of these facts was made to Parliament, which ordained that the Chancellor of England might appoint Commissioners with authority to remove such obstructions. They were given authority to borrow money, if necessary, to accomplish this work, and for three years to take a toll of 4*d.* on every freighted boat passing in this river[1].

But, while there were some rivers which were attended to in such a way as to aid the navigation, most of them were not amended. In these cases the tide was relied upon to carry the boats far inland; and those which were loaded at inland points and were ready to pass outward to the great arms of the sea were, in like manner, often compelled to await the flood tide to carry them over the shallows[2].

Another aspect of river navigation is seen in the early history of the Severn river. In a charter granted to the city of Gloucester between 1163 and 1174, Henry II ordained and commanded that the men of Gloucester, and all those who wished to use the river Severn,

24 July, 1422, "it was ordained and granted that the King's river should be cleansed, from the mills called the Calkemyll to the gates called the Bishopesyates by the men who dwell in the Ward Ultra Aquam sparing no one as their state or position demand. Also that the said River from le Bishopesyates to Thorpe Episcopi should be cleansed by the residue of the City...sparing no one as is aforesaid....And that every person of the said Ward able in body to work or able in goods to pay shall labour at cleansing the river or pay to the Constable of the leet where he resides 4*d.* a day for the wage of a labourer in his place....Everyone is to be at his work at latest at the fifth hour in the morning and remain until the seventh hour after noon. The Community is to pay for all tools and farms of boats. The Mayor for the time being may appoint overseers of the work...." Vol. 1 of the *Records* shows a map of the river leading from Yarmouth to Norwich.

[1] Act 9 Henry VI, c. 9. This is the first instance I have seen of money having been borrowed for such public work, and tolls charged to secure the funds for reimbursement.

[2] The navigation of the river Dee was impeded by sands as early as the reign of Richard II and, more or less, down to the reign of Henry VI, when a quay was formed near Shotwick Castle, about six miles below Chester, from which place the troops were usually embarked for Ireland. Ormerod, *History of Cheshire*, I, p. lxxiii. See also the case of Exeter, as given in Oliver, *History of Exeter*, p. 249.

were to have the right of free passage along this waterway with their coal, wood, timber, and other merchandise, without any disturbance from anyone[1]. Richard I, in 1194, confirmed the "liberties and free customs and acquittances" given by the charter of Henry II, and also confirmed to the burgesses of the city the privileges they had had under the charter of Henry II, regarding the free passage on the Severn[2].

But while the river had been *legally* free to all persons from early times, it was not so in reality. After much annoyance, the people of Tewkesbury in 1429 petitioned the House of Commons against the robberies and injuries that had been done by "rovers of the Forest of Dean," and others, in destroying the goods and ships of the people of Tewkesbury and others[3]. The King had made proclamation against this, but it had not stopped. Consequently, in this year, Parliament gave sheriffs authority to enjoin the trespassers to return the goods taken, or the value of them, with full amends. But the law was apparently a dead letter, for in the next year comes the same complaint. In the Act of this year (1430), it was said that because the river Severn was common to all the King's liege people to be used as a means of conveyance, and because boats carrying goods had been cut to pieces and the goods destroyed, therefore it was enacted that these people must have free passage in the river, without disturbance; and if disturbed, the injured party should have his remedy according to the course of the common law[4]. But even this did not stop the nuisance, nor render the river navigation free, as we shall see in a later chapter[5].

Perhaps the constant use of the waterways, as disclosed by the history of these measures, helps to account for the economic fact of the cheapness of land transportation. According to Rogers' investigations[6], the cost of transporting wine by land was from one and one-half pence to two pence per ton per mile in summer, and twice that amount in winter; while for grain the cost was one penny per ton per mile. By

---

[1] *Records of Gloucester*, edited by Stevenson, pp. 4, 5.

[2] Ibid., pp. 4–5.

[3] See petition in full in Bennett, *History of Tewkesbury*, Appendix No. 3, pp. 325–8.

[4] The Act of 1430 was 9 Henry VI, c. 5. The Act of the preceding year was 8 Henry VI, c. 27.

[5] Beginning of Chapter III. In the later fifteenth or early sixteenth century, Bewdley was becoming a considerable depôt for merchandise directed to Bristol and the south. Worcester, however, began to charge tolls for all boats passing under their bridge across the Severn; and those tolls they had greatly increased. What was still worse, Bewdley merchants were sometimes forced to sell at Worcester at a lower price than they could get nearer the coast. Worcester authorities enforced their claims very violently at times. *Victoria History of Worcestershire*, II, p. 251.

[6] Rogers, *History of Agriculture and Prices*, I, pp. 657, 659–60.

comparison of these charges with those which we give for the later period, it will be recognized how low they were; and Rogers explains this by saying that the horses and carts that did the carrying were hired from the smaller tenants of the manor, or rather the cultivators of small sections, at times when there was no pressing work upon the farm, and, therefore, the expense was low[1]. It would be but natural that the tenants of the manor should be ready to work at a low figure in order to help themselves along in making a living; and they could afford to carry cheap when they had slack time on their own small pieces of land. These rates of land carriage would also be kept down by the low rates of carriage on the rivers[2], many of which, as we have seen, were navigable and much used by small boats[3].

[1] Rogers, *History of Agriculture and Prices*, I, p. 660.

[2] Ibid., I, p. 663, where he gives examples of the cost of water carriage.

[3] It is impossible for us to agree with Prof. Rogers when he goes so far as to say that the "means of communication were kept in far better repair before than they were after the Reformation" (ibid., I, p. 654), and that, because the rate of land carriage was low during the Middle Ages, "consequently communication must have been easy, and probably regular" (ibid., I, p. 660). The fact that land carriage was cheap may indicate nothing more than that there were many who were ready to perform that service at small remuneration; but to say that communication must have been easy is wholly unwarranted. We are not justified in saying that, because the expenses of carriage by land were low, therefore the roads were good (ibid., I, p. 658). After speaking of the amount of travelling that was done in looking after the widely scattered estates, in religious pilgrimages, in attending the great fairs, etc., he says: "Every motive, in short, was present which should suggest the wisdom and utility of good and well-mended roads" (ibid., I, p. 654); but it must be said that the desirability of good roads does not necessarily lead to their being made good. Because of such absurd reasoning, we cannot adopt Rogers' views in regard to the so-called good roads of the Middle Ages, even in the face of the details of journeys which he presents in ibid., II, pp. 610, 612, 613, 614, 635–41. Further, if "the bye-roads were no doubt bad and could not be used except in summer" (ibid., I, p. 654), there must have been the cessation of a vast amount of travel during the rest of the year, for, of course, a great many of those who would be on the roads would be compelled to use the by-roads for some part of their journey. The great estates and the public rendezvous we have spoken of above were not all on the few main roads.

William Harrison, in the *Lancashire and Cheshire Antiquarian Society Transactions*, IX, p. 102 et seq., follows Rogers almost word for word.

In reality, we find no facts from other sources that give any confirmation of Rogers' belief; but, on the contrary, it is inconceivable that the roads of the mediaeval period should have been so good, when, so soon after, they had to be repaired by the expenditure of such great sums as those given in Stow, *Summary of the Chronicles*, pp. 431–86 et seq.

CHAPTER II

ROADS AND ROAD LEGISLATION, 1500–1750

WHILE we do not notice any distinct break from the economic conditions of the Middle Ages, accompanying the accession of the Tudor sovereigns, we can say in general that under the latter greater attention was given to the means by which the interior districts could be brought into closer trade relations with each other and with the coast cities, and through these with the foreign states.

In the early days of the Tudors, the monasteries were still in existence; and while, during the later mediaeval times, many of the smaller monastic houses had become greatly impoverished, the larger ones were still flourishing, and all owned property. As holders of land, they were under obligation to maintain the roads, and it was to their interest to do so, since their properties were considerably scattered and constant communication had to be kept up between these different places and the monastery. It would seem that where such an establishment was founded, the charge of the roads was assigned by custom to the monks[1], and "even the highways were mended out of devotion[2]." In Roger Aske's testimony to the good work of the monasteries, we find him saying: "And such abbeys as were near the danger of sea-banks, were great maintainers of sea-walls and dykes, maintainers and builders of bridges and highways (and) such other things for the commonwealth[3]."

When the monasteries were dissolved by Henry VIII, in 1536–9, their property fell into the hands of a class of rapacious landlords who would be slow to recognize any claim upon their rents for the maintenance of the roads. Besides, there was no longer the need of communication between the various parts of the estates that had been formerly felt. The inevitable result would be the rapid decadence of many highways which had hitherto been in common use; and it seems very probable that these conditions and results had marked

[1] Blomfield, *History of Bicester*, II, p. 40; also Holmes, *Pontefract*, p. 69.
[2] Anderson, *History of Commerce*, II, p. 44.
[3] Quoted in Gasquet, *Henry VIII and the English Monasteries*, pp. 227–8.

influence in causing the passage of the Act of 1555, although, for good reasons, they were not formally put forth in the preamble of the Act.

The mediaeval idea that the maintenance of the roads and bridges was a pious work highly to be commended found expression in the later period also, as is plainly shown by the records of some parishes and of the Episcopal offices. Where the need for repair was urgent, indulgences were sometimes granted, on the authority of the Bishop, to those who were ready to contribute of their labours or their substance for this good work[1]. The accounts also of churchwardens show that the Church found some scope for activity in this direction; but, unfortunately, there are but few of such accounts available, so that it is not advisable to make any comprehensive statements regarding this phase of the efforts of the Church[2].

Closely related to the foregoing, and of vast importance, at least locally, were the gifts of individuals and of corporate bodies, like the towns and gilds, to aid in keeping up the roads. In the borough of Sheffield, which was just rising into prominence during the Tudor period, we have several bequests to aid in establishing good communication to and from the town[3]. Among the donations to the parish of Tiverton, we find that, in 1599, Peter Blundell, a clothier and merchant of that place, by his will, contributed, in addition to his other gifts, one hundred pounds to amend the Tiverton highways[4]. This was a great sum for that time, but similar amounts seem to have been frequently given in other places[5]. In some instances, voluntary

---

[1] Gibbons, *Ely Episcopal Records,* 1445–1587, pp. 408–20.

[2] In 1542–3, the church of the parish of Ashburton paid 5s. 8d. for the "pavyment" in the North Strate (street). In 1543–4, the same church paid 3s. 4d. for mending the road at Holne Bridge. See Churchwardens' accounts, Parish of Ashburton. I do not have access to these accounts before the time of the dissolution of the monasteries, and therefore cannot be certain whether the payments by the church for this purpose were or were not greater after the disestablishment of these religious houses. See also Gasquet, *Parish Life in Mediaeval England,* p. 42.

[3] Leader, *Sheffield Burgery Accounts.*

[4] Dunsford, *Historical Memoirs of Tiverton,* p. 118. For similar gifts see Camden Society, *Wills and Inventories* from the Registers of the Commissary of Bury St Edmunds and the Archdeacon of Sudbury, pp. 96, 112, 154, 252.

[5] By the will of King Henry VII, that monarch left £2000 for the construction of good roads and bridges between Windsor, Richmond, Southwark, Greenwich and Canterbury. These roads were to be substantially ditched on both sides, well gravelled, raised to a good height, and wide enough to admit two carts abreast. See Astle, *The Will of Henry VII,* pp. 20–21.

For other gifts, many of them of large amount, to be devoted to the repair of roads, see Stow, *Summary of the Chronicles,* pp. 434, 437, 440, 441, 442, 443, etc.; also quotations from the will of Thomas Sutton, regarding money left by him for repair of the highways, in Bearcroft, *History of the Charterhouse,* pp. 85–86; also Parsons,

rates were paid by the communities to clear off obligations contracted for the benefit of the communities and for making roads and bridges[1].

Of the work of the gilds, in the repair of the streets and roads, we have already spoken[2]. Where the gild was specially strong in the municipal government of the town, it usually exercised some supervision over all such matters of local importance, and its funds were contributed for the common good[3]. The work effected by the Gild of the Holy Cross, which was in very close association with the town government of Birmingham, is well set forth in the Report of the Commissioners appointed by Henry VIII to investigate the influence and material standing of all bodies partaking of a religious character. To this we have formerly alluded. As an offshoot of the Gild of the Holy Cross, we have another foundation, established in 1525, and known as William Lench's Trust[4]. The founder gave his wife a life interest in the trust, and after her death the trustees were directed to do the same sort of works of charity as the Gild of the Holy Cross, "according as the ordering and will of the bailiffs and commonalty shall appoint." The master of the gild was to be one of the trustees. This trust was distinctly connected with the government of the town by two provisions: first, that its funds should be used "ffor the repairing the ruinous waies and bridges in and about the same towne of Birmingham, where it shall want; and for default of such uses to the poor liveing within the Towne aforesaid, where then shall be most need;" and, second, that the "two said (managing) ffeoffees shall, uppon the Tuesday in Easter weeke, make their just accompts, before the rest of the said ffeoffees, and other the enhabitaunts of the said Towne of Birmingham;......and that uppon Tuesday in Easterweeke, yearely, in the afternoone, uppon the tolling of a bell, such meeting, for the purpose aforesaid, shalbe had, and the like accompte shalbe made and taken & two new officers chosen[5]."

Not only were gifts received from corporate bodies and from private

History of Leeds, I, p. 97; Holloway, *History of Rye*, p. 456; and *Archaeologia*, xx, pp. 457–8.

[1] Gasquet, *Parish Life in Mediaeval England*, p. 42.

[2] Page 6.

[3] See Scott, *Berwick-upon-Tweed*, p. 301. The few presentments given in Scott's work show that about the year 1557 the gilds were closely identified with such forms of improvement. The gild of this place, about the middle of the eighteenth century, subscribed £50 to help a Turnpike Act through Parliament. *Ibid.*, pp. 229–30.

[4] Hutton, *History of Birmingham*, pp. 282–5, describes Lench's Trust. At the present time (1913) this Trust is a large charitable institution in Birmingham, near the Stratford Road Grammar School.

[5] Bunce, *History of Birmingham*, I, p. 30.

persons for the repair and maintenance of roads, but the latter were sometimes actually put in repair by individuals, on their own initiative, at their own expense, and by the labour of their own hands. In the town of Bury St Edmunds we see a good example of private benevolence on the part of John Cowper, whose will, in 1522, directed that his executors should gather and carry "six score loads of small stones," such as he had already gathered, and place them on a certain part of the streets where there was the greatest need[1]. In the later years of the reign of Henry VIII, Huntington Lane, near the city of Chester, had been so bad, that Sir William Hanley and Sir Hugh Cholmondeley had repaired it for two and one-half miles, thus furnishing a convenient access into the city. In order that this good work might not be nullified, an Act was passed[2], authorizing them, or in default of them the mayor of Chester, to make agreement with some one to keep this road repaired; and this person was to have the right to build a house in the highway, to live there, and to have pasture in the highway for five cows[3].

Another method, a similar practice to which was not unknown in the Middle Ages, is reflected in an Act of 1554[4], by which Parliament authorized two roads to be repaired by the owners of adjoining lands, at the orders of Justices of the Peace; and for the charges thereof the latter were to levy rates upon those who were chargeable with the repairs.

Before leaving these special methods of maintaining the roads, an Act of 1523 deserves notice. It was the custom over the open-field area of England, when a road became so bad that it could no longer be used with safety, for the public to drive out to the side of this, and open up a new road. But the Weald of Kent was not under the open-field system; and therefore it was not possible there for the public to drive to the side of the road. To obviate the difficulty, in the above-mentioned year Parliament gave permission to any person in the Weald of Kent to take the ground of the old road, which had become dangerous for travel, and to lay out in place of it a part of his own land, as a new highway, provided he obtained the assent of two Justices of the Peace and twelve other discreet men living within that hundred. Those

---

[1] Camden Society, *Wills and Inventories of Bury St Edmunds*, p. 252. For other instances, see Stow, *Summary of the Chronicles*, pp. 431, 470.

[2] Act 37 Henry VIII, c. 3; see also Brit. Mus., Harl. MSS. 2077, p. 21. See the indenture that was made in 12 Elizabeth for the maintenance of this very road at Chester, as given in Brit. Mus., Harl: MSS. 2046, p. 1.

[3] That this was done much later also, see, for example, *Journal of House of Commons*, xxv, p. 59 (Feb. 10, 1745).

[4] Act 1 Mary, cc. 5, 6.

whose assent was necessary were to see that such new road should be certified in Chancery; and all who had lands adjoining the old way were to have access over that to the new way. Twelve years later, the authority of the previous Act was extended in like manner to the county of Sussex[1].

While, by the foregoing means, the roads in certain localities were amended, it is more particularly of the means applicable to all localities that we would here devote attention. From early days, the obligation to repair the highways lay, according to the common law, upon the parishes or hundreds through which they passed, and they could be indicted at the Quarter Sessions for their neglect; but it was not till 1555 that a special officer was appointed in each parish whose duty it was to call the residents of the parish together to work on the highways. By the legislation of that year[2], entitled the "Statute for Mending of Highways," it was enacted that "the Constables and Churchwardens of every parish shall yearly, upon the Tuesday or Wednesday in Easter week, call together a number of parishioners, and shall then elect and choose two honest persons of the parish to be surveyors and orderers for one year of the works for amendment of the highways in their parish." The Constables and Churchwardens were each year to appoint four days on which the people of the parish were to come together, with the requisite implements and draught animals for repairing the roads, and they were to give knowledge of these days openly in church at the appropriate time. The surveyors were to order and direct the expenditure of labour on the roads during those days. Every householder of the parish except servants who were hired by the year were proportionately liable to contribute labour to the needful work[3]. Fines were to be imposed upon parishioners who neglected to

---

[1] Act 26 Henry VIII, c. 7.

[2] Act 2 & 3 P. & M., c. 8.

[3] The following scale of statute labour was imposed upon the parishioners:

Every person having arable land in tillage or pasture, and every other person keeping a plough in that parish, was to send every day one cart, with oxen, horses, or other necessaries for work, and also two able men. Penalty for default, 10s.

Every other householder, and also every cottager and labourer of that parish, able to work, and not being a hired servant by the year, was to labour, either himself or by a substitute, for the four days in repair of the highways. Penalty for default, 12d. per day.

If the surveyor thought the waggons of the parish would not be required upon any of these days, every person that would have sent such waggon was required to send instead two able men to work for that day—under penalty of 12d. for every man not so sent.

Every person or waggon aforesaid was to bring with them to the repairing of the roads, such shovels, spades, mattocks, and other tools and instruments as they

carry out the provisions of this law; and these fines were to be applied by the churchwardens of the parish to mending the highways. Similarly, a penalty of 20*s.* was imposed upon surveyors who failed to carry out their part of the work.

The work of the surveyor was much more exacting than that of any of the others who had anything to do with the roads. He had to keep his accounts in such order that they would satisfy the Justices of the Peace. At least three times a year he had to go through his district to view all roads, bridges, water-courses and pavements, and then go before the next Justice and show upon oath how he found them. He was expected to see that no nuisances were allowed to continue to the detriment of the highways; all ditches and water-courses were to be kept scoured; all trees and shrubs growing in the highway were to be removed; landowners were to keep the highways adjacent to their lands clear of all obstructions, and to trim their hedges so as to allow the entrance of the sun and wind to dry the roads. In addition to the oversight of this work, he had the unpopular task of going before his fellow parishioners on a Sunday and informing against any of them who had used more than the legal number of draught animals on their waggons or carts or who had these animals arranged in any other than the prescribed manner. When any failed to satisfy the conditions of the law, he was to make presentment of such persons before the Justices of the Peace assembled either in Special or in Quarter Sessions. All this work was to be done without remuneration; and if he refused to carry out the law after the office of surveyor had been thrust upon him, he might be brought to compliance with it by the imposition of a heavy penalty.

This Act was to continue in force for seven years. By it, the repair of the roads was definitely committed to the care of the parishes, and an organized system was established for carrying it into execution. Cases of default or dispute were to be decided by the magistrates, but it would seem as if the fulfilment of their duty was largely optional or capricious, so far as the outworking of it was concerned.

At the expiration of the Act in 1562, it was continued for twenty years, and amended in several particulars[1], giving increased powers to

---

would use in making their own ditches or fences, and such as would be necessary for their said work.

All persons and carriages were to work for eight hours each day at the work given them by the surveyors, unless otherwise excused by the latter.

[1] It was amended in the following particulars:

To get suitable material for amending the ways, the surveyors were authorized to carry away the small broken stones of any quarry within the parish, without

the surveyors, and requiring that six days, instead of four, should be yearly employed in repairing the highways.

By these Acts, statute labour became firmly rooted as the parochial means for keeping the roads in good condition; and it continued in force, in one form or another, until abolished in 1835[1].

In 1562 we have the passage of the first general highway Act of Elizabeth[2], and in 1575 this legislation for amending the highways received a few minor additions[3]. Surveyors were no longer mere clerks or overseers of the works, under the constables and churchwardens, but were made independent, responsible officers of the parish. Provision was made for more effectively keeping open the ditches on land adjoining the highway, so as to give the water on the highway its natural channel to run off; and the liability of a person who had land in several parishes, with reference to the up-keep of the roads, was limited to the repair of the roads in the parish where he resided. There was also an extension of the liability for the performance of statute labour, to a class hitherto not included[4].

license of the owners, to such an amount as they should deem necessary; or in default of such material, the surveyor could go into the land adjoining the way or ways to be repaired and dig for gravel, or to gather stones from such land suitable for road repairing. But regulations were provided under which such material should be taken and the pits filled up; all to cause the least detriment to owners of the land.

Surveyors were authorized to turn springs or water-courses into adjoining ditches, if they prevented the construction of permanent roads.

Ditches were to be scoured; trees or bushes growing in the highways were to be cut down; dykes were to be kept low; all these by the owners of the adjacent lands.

Six days, instead of four, were to be yearly employed on the highways.

Supervisors or surveyors were to present all defaults to the next Justice (under penalty of 40s.) and the Justice was to certify the defaults at the next General Sessions within that county, upon penalty of £5. The Justices of the Peace at their Quarter Sessions were to inquire concerning such defaults or offences, and assess fines; such fines to be used as stated in Act 2 & 3 P. & M., c. 8, for mending the highways.

Justices might present defaults from their own knowledge, and defaulters should be fined.

[1] Act 5 & 6 Will. IV, c. 50.
[2] Act 5 Eliz., c. 13.
[3] Act 18 Eliz., c. 10.
[4] Some of these additions were:

Every person or persons (except residents of the city of London) that should be assessed to the payment of any subsidy to Her Majesty to £5 in goods, or 40s. or more in lands, and not being a party chargeable for amending the highways by any former law, but as a cottager, was to find two able men yearly to labour on the highways during the days appointed by statute.

Every person that should occupy arable land in tillage or pasture lying in

In 1586, it was agreed in Parliament that the Acts of 1555 and
1562[1] had been by "proof and experience tried and found to be very
necessary and profitable for the common wealth of this realm," and
they were decreed to "remain and continue in force and effect for
ever;" or, in other words, this system of statute labour was made
perpetual[2].

Before proceeding to the later history of the road legislation, it
may be well to glance at the condition of the thoroughfares of the
towns during the Tudor period.

From all that we can learn, the streets of London were in a very
ill-kept and often dangerous condition. This was, in all probability,
due to the failure of many to live up to the laws[3], and thus to maintain in repair the roads or streets adjoining their property. That
some parts of London had streets that were good for that time, we
see reflected in the statutes; for the preamble to one of the Acts says
that if certain streets were "sufficiently paved and made after the
manner of the pavement of the street between the said Strand and
Temple Bar, it should not only then be a great comfort to all your
subjects thereabout dwelling but also to all other your liege people[4]...."
But most of the streets would seem to have been but poorly maintained, for we find in the statutes frequent references to the streets
as "noyous and fowle and in many places thereof very jeopardous to
all your liege people...both in winter and summer[5];" "very foule
and full of pittes and sloughs, very pillous and noyous[6];" and one
Act tells us that a certain street was "of late time so well and substantially paved," but it had of late become "so noyous and so full
of sloughs and other incumbrances, that oftentimes many of your
subjects riding through the said street and way be in jeopardy of hurt
and have almost perished[7]." There were, doubtless, many exaggerations

several parishes, was to be chargeable for making of the way⸱ in the parish where
he dwelt, according to the tenor of the statute.

Penalty of 10s., for every default of any person in scouring the ditches, cutting
the trees, or repairing the hedges adjoining the highway.

Any person occupying land adjoining any highway which had a ditch for carrying
off the water from the highway, was to keep any connecting ditch in his own ground
scoured so as to permit free passage of water, upon penalty of 12d. per rod for every
time he caused such offence.

[1] Acts 2 & 3 P. & M., c. 8 and 5 Eliz., c. 13.
[2] Act 29 Eliz., c. 5.
[3] Preamble to Act 24 Henry VIII, c. 11.
[4] Ibid.
[5] Act 24 Henry VIII, c. 11.
[6] Act 32 Henry VIII, c. 17.
[7] Act 25 Henry VIII, c. 8.

in these representations to Parliament; but there must have been much truth also, for in 1582 and 1533[1] owners of lands adjoining the highways in certain sections of London were required to pave these highways, under penalty of 6*d.* per square yard, and to keep them paved. The Justices were given authority to inquire of defaults, and to proceed at their discretion against these offenders.

In 1540, these paving regulations were made general for the whole of London[2]. The Justices at their Quarter Sessions were empowered to investigate delinquency in Middlesex county, and the Clerk of the Peace of that county was to certify the fines into the King's Exchequer, under penalty of £5 for neglect. In London, the mayor and aldermen were to make the inquiry, which if they neglected, a fine of £5 was imposed on them.

But London was not worse than many of the other towns throughout England. In most of the towns whose records we have been able to examine, we find a constant series of complaints against householders who failed to live up to the statutory provisions for the repair and maintenance of the streets. As a rule, the towns which, by Act of Parliament or by royal letters patent, had obtained authority for paving, placed the responsibility therefor upon the property owners, and made it compulsory for each of them to keep in good condition the pavement in front of his tenement out to the middle of the street[3]. But that there was a common disregard of this law by allowing the pavements of the towns to become ruinous, is seen by the many presentments of those who were liable for repair because they had utterly neglected their duty[4]. In other cases, the town assembly passed

---

[1] Acts 24 Henry VIII, c. 11 and 25 Henry VIII, c. 8.

[2] Act 32 Henry VIII, c. 17. The paving was not flat stone slabs, but round cobblestones, with a channel or open gutter running down the middle. In the case of some towns, the middle of the street was raised and paved, and there were two gutters, one at each side of the street, into which were poured from the connecting open drains the various kinds of refuse from the houses.

[3] See provisions of Act 35 Henry VIII, sess. 3, c. 18, An Act for paving Cambridge town; also Brit. Mus., MSS. 5821, p. 93.

[4] From the *Annals of Ipswich*, edited by Bacon, we take the following:

p. 211 (1538). "Every inhabitant shall amend the decays in pathing (i.e., paving) before his own house before Michaelmas next, under forfeiture of 40*s.* for default." This law had been made about 1531 or 1532. (Ibid., p. 217.)

p. 229 (1547). Each inhabitant was to pave the streets before his own door by a certain day, under penalty.

p. 230 (1548). "All inhabitants shall amend the paving in the streets against their own houses and grounds before All Saints day next; every defaulter to pay 6*s.* 8*d.*" This notice seems to have been given each year.

That the paving regulations in Ipswich were followed by good results, we see from its *Annals* in 1576 (p. 311), when the effort was made to preserve the pavement from being cut up. The ordinance of this year reads as follows: "Carts and tombrils of the town, or others carrying muck or other carriage from place to place within the town shall be unshod after Michaelmas next, at such times of carriage, under penalty of forfeiture of 12*d.* for each coming over the paved street otherwise. And all carts that shall come with commodities for transportation, shall pass by the Back streets or lanes, and not upon the paved streets if it may be avoided, and after warning given." It would almost seem as if this paving was more for ornament than for use. Seven years later, a stock of £20 was voted to be disbursed by the town for the providing of paving stones for the streets of this town, and the profits thereof arising were to be employed to the town's use. Apparently, therefore, Ipswich used every means for obtaining good streets.

In Chester, the repair of the pavement appears to have been under the supervision of the Muringers until 1566–7, the expense being defrayed by an assessment upon the inhabitants, and by grants of murage. (Morris, *Chester*, pp. 264–6.) By 1532, the duty of repairs had been imposed upon the individual householder, as we see by the fact that in that year a certain man was fined 4*d.* for neglecting the pavement at his door. But the necessity, as well as the duty, of paving was brought more forcibly to the householder in 1567, when it was ordered that everyone must, "at his own proper costes and charges from tyme to tyme and at all tymes repayre and mayntayne all and every the pavements every man right against his owne dwellinge." (Morris, *Chester*, pp. 266–8.) Two years later, an indenture was signed, Oct. 8, 12 Eliz., by which Thomas Bennett, of Chester, paviour, agreed to keep the streets of the city, and all the lanes that had been accustomed to being paved, in good repair. He was to find everything that was necessary for this work, and was each year to receive therefore 1*d.* from every mansion dwelling house. (Brit. Mus., Harl. MSS. 2150, p. 182.) This work, however, must have been inefficiently carried out, for there were frequent complaints of the dangerous condition of the streets. As in the case of Ipswich above, restrictions were made on the use of iron-bound wheels; see the Ordinance of July 2, 1573, given in full in Morris, *Chester*, p. 268 n., showing what limits were imposed on carts and waggons using iron-bound wheels.

In Liverpool, there had been a law that each householder must repair the pavement in front of his own place. But it does not seem to have been enforced, for in 1583 the late mayor was presented at the Port Moot Inquisition "for not causing the highways to be repaired and amended the last year, according to the statute." (Picton, *Liverpool Municipal Records*, I, p. 92.) In 1592 the same Court ordered the streets of the town to be repaired and amended "where need is, but especiallie Chapel Streete." In 1595, at a convocation in the Common Hall, it was resolved that for repairing the streets "every townes man havinge a team shal serve wth the same half a daye a pece in due order and course, as the necessitie of the work shall require;" and every householder not having a team was to find a satisfactory labourer, and take him to the work on the street. But this system of road work did not have the desired effect, and in 1601 it was agreed to continue the former ordinance that every householder should repair the pavement in front of his own house. (Ibid.)

The *Barnstaple Records*, edited by Chanter and Wainwright, show us that in the time of Elizabeth many persons were fined for allowing their pavements to be out of repair (I, p. 44). In some instances, the highway in front of certain tenements was "muddy and full of ruts," so much so that it was made the occasion of a presentment (I, p. 51). In other cases, persons were presented for "founderous and broken pavements in the streets and highways" (I, pp. 70–71).

ordinance after ordinance to enforce its authority upon those who persistently refused to discharge their obligation in this respect[1]. Some of the towns had a public official known as a paver (or paviour) whose duties seemed to be those of an overseer and director of this work[2]; others, like Leicester, called on the aldermen of the different wards to see that householders attended to this[3]. In Northampton, the assembly of Apr. 19, 1571, ordered that the town chamberlains for the time being should go through the streets in every part of the town once every three months, to see that each man did the paving required of him; and they had to report to the mayor those who had failed in this[4]. But whoever had the oversight of it, it was the recognized duty of the householder to contribute thus to the public good. He might shirk the work, as many did; he might use poor material, or perform the work in other ways very imperfectly; but still the duty was incumbent upon him.

The responsibility was brought to his attention in several ways, as by a public notice from the town authorities that unless the pavement were repaired by a certain time the penalty would be imposed. This was probably done annually in Ipswich[5], and, doubtless, in some other towns more or less regularly. But it was especially emphasized in the case of a proposed or expected visit from the sovereign. At such times the whole town would be astir, and all who were liable for the repair of the streets were incited with the motive of patriotism

For other instances of failure to keep up the pavement, see Brit. Mus., MSS. 16,179, p. 61; Hardy, *Records of Doncaster*, p. 28 et seq.; East, *Records of Portsmouth*, pp. 41, 75, 76, 82, 99, 100, etc.; and the records of almost any other city or town that are available.

[1] See last footnote.

[2] Examples are Chester, Bath, Southampton, etc.

[3] Bateson, *Records of the Borough of Leicester*, III, p. lvi.

[4] Markham and Cox, *Northampton Borough Records*, II, p. 260. From this order, it would seem as if all the streets within the corporation limits were to be paved by the holders of lands or tenements. But there were some parts, probably those which were most thinly built up, that were repaired at the expense of the town. This is evident from a record of the town (Northampton) in 1617: "Whereas the highway leading from the North gate to the hether end of St Seppulchres Churchyarde within this Corporation lyeth very undecent and unfitting for the passage of his Maties said subjects and in the winter time is to the great annoyance and danger of his Maties said subjects that way comeing; for prevention and amendment whereof it is agreed and ordered that every person that hath or holdeth any land about St Sepulchres Churchyarde to pave and mend so much of the same way with pible as by lawe is appointed and the residue thereof to be paved and amended at the charge of the corporation in like manner before the said feast day of All Saintes." (Ibid., p. 267.)

[5] Bacon, *Annals of Ipswich*, pp. 211, 229, 230, etc.

and civic pride, to perform their work faithfully, while at the same time adding an element of beauty to the fronts of the houses[1]. Such towns, through which royalty was to pass, or in which the sovereign was to be entertained, had sometimes to get paviours from considerable distances to come and aid in making the streets presentable for the occasion[2].

This being the general system for paving, it remains merely to mention a few of the other methods which were adopted in particular places or on special occasions. Worthy citizens, under a sense of religious duty, or as a matter of charity, were appealed to and contributed for such work[3]. In Chester, the mayor gave authority to the churchwardens to collect sums of money for repairing the highways and pavements[4]. Some towns undertook to do part of the paving at the general expense[5], and in other cases a special assessment was levied on the several wards to provide funds for the paving[6]. But the streets of the towns suffered in other ways than by insufficient and defective pavements. It was a customary complaint that the streets were not only bad for travelling or carrying purposes, but that they were obstructed in many ways; and this was highly objectionable because they were often very narrow at best. Blocks of timber, piles of wood, heaps of ashes, carts, waggons, and other things were left in the streets, thus blocking up the traffic[7]. Despite ordinances to the contrary,

[1] Bateson, *Records of the Borough of Leicester*, III, p. lvi, showing what was done when Queen Elizabeth was expected in 1575.

[2] The city of Bath was anticipating a visit from Queen Elizabeth in 1602, and the paviour, who occasionally got some meagre remuneration for patching up a street or lane, was not sufficient for the task. The following entries from the city's records show what was done:

"Paid unto Robert Vernam to goe unto Sodburie and unto Cicister to get paviers against the Quenes Cominge, 3*s*. 4*d*.

"Paid unto John Pavior to goe unto Bristoll to get paviers, 1*s*.

"Paid unto Roger Feildes man to goe unto Frome to get paviers, 1*s*.

"Paid to a messenger to goe to Warminster to get paviers, 1*s*.

"Paid to the pavier of Chipnam for his Cominge to Bath to see the worke, 4*d*."
(King and Watts, *Municipal Records of Bath*, pp. 48–49.)

[3] *Bury St Edmunds Wills and Inventories* (published by the Camden Society), p. 252, shows what was done in one case by private exertions and testamentary munificence.

[4] See the document giving such authority, as given in full in Morris, *Chester*, pp. 268–9 n.; also Brit. Mus., MSS. 16,179, p. 61.

[5] Markham and Cox, *Northampton Borough Records*, II, p. 267.

[6] Morris, *Chester*, pp. 264, 270. In 1597, that city raised a large assessment, amounting to £75. 4*s*. 10*d*., for paving.

[7] In Northampton, an order was given in 1568 for the removal of such things, which was revived in 1592. In 1599 it was made more stringent, demanding the removal of all piles of wood, timber, etc., to one of five appointed places, under

house-sweepings and similar refuse of houses, stables and yards, were laid in the streets, forming not only an obstruction but a great danger to the whole community[1]. In some towns, once or twice a week the householders were to sweep up such sullage and refuse on and in front of their own property, even to the middle of the street, and have it removed[2]; in other cases they were to sweep it up and have it ready for a public scavenger, who would cart it away to certain appointed places where it could be disposed of[3]. The regulations were good, but they were little regarded, as may be seen by following out the references here given.

Pigs were allowed sometimes to roam at will through the streets of the towns, and the local records are full of presentments of persons for permitting the offensive and destructive work of these animals[4].

penalty of 10*s.* or else imprisonment. Markham and Cox, *Northampton Borough Records*, II, p. 265. But such orders were continually being disobeyed, and it was not long before a further order was made for the removal of such things within a week, under a penalty of 20*s.* (ibid., II, p. 268). See many similar cases in *Nottingham Borough Records*, edited by Stevenson, IV, p. 161, etc.; Hardy, *Records of Doncaster*, II, pp. 22, 23, et seq.; Morris, *Chester*, p. 260; *Salford Portmote Records*, pp. 32, 41, 53, etc.

[1] Markham and Cox, *Northampton Borough Records*, II, pp. 264, 265; Morris, *Chester*, p. 260; Picton, *Liverpool Municipal Records*, I, pp. 92, 191; Turner, *Records of the City of Oxford*, p. 422; Hearnshaw, *Southampton Court Leet Records*, 1551, paragraphs 21, 54, 60; 1571, paragraphs 12, 29; *Nottingham Borough Records*, IV, p. 161, etc.; Hardy, *Records of Doncaster*, p. 22 et seq.

[2] Morris, *Chester*, p. 260; East, *Records of Portsmouth*, pp. 124, 172, etc.; *Nottingham Borough Records*, IV, p. 276.

[3] *Northampton Borough Records*, II, p. 265; Cowper, *Records of Faversham*, p. 329; Turner, *Records of the City of Oxford*, p. 166; Hearnshaw, *Southampton Court Leet Records*, 1575, paragraph 14; 1577, paragraph 63, etc.

[4] Hearnshaw, *Southampton Court Leet Records*, 1550, paragraphs 14, 19, etc. Presentments were apparently made without respect of persons. In 1550 "Item yt ys presented that Mr maire (mayor) kepith a sowe in his backe-syde, which is brought in and owte contrary to the ordennes of the towne, Wherefore be yt oomanded to hym and all other that theye kepe no hoges wt in the Towne to the anyance of theire neighbour upon payne that every of them that so shall kepe any swyne to forfeyte for every 15 daies he shall so offend 20*d.*" (*Southampton Court Leet Records*, I, p. 7.) See also East, *Records of Portsmouth*, pp. 33, 42, 61, 96, 100, 124, etc.; Morris, *Chester*, p. 263; Tait, *Mediaeval Manchester*, p. 49; Turner, *Records of the City of Oxford*, pp. 422–5; *Salford Portmote Records*, pp. 7, 22, 23, 28, 42, 59, etc.

The town of Ipswich seemed to be especially desirous to rid their streets of these walking nuisances. From what we have formerly learned, Ipswich people took considerable pride in the appearance of their town, and the presence of pigs on their streets was derogatory to this civic pride. In 1536, they passed an order that "No inhabitant shall...suffer his swine to go at large in the town or suburbs thereof. If found there, they shall be put in pound till the owner shall pay 4*d.* for each." (Bacon, *Annals of Ipswich*, p. 209.) In 1543, this punishment was made heavier, for in that year it was ordered that "The Crier shall impound all the swine that he

Cows, sheep, geese and ducks followed this imprudent example. But certain towns appointed each a swineherd, and a cowherd, or one person to perform both services, whose duty was to assemble such animals in the morning, lead or drive them out to the town pasture or the lord's waste, attend to them there during the day, and bring them back at night[1].

Not only were the streets subject to obstructions, but the drainage and sanitation were defective. Where the pavement was neglected, especially with particular kinds of soil, a heavy rain made the streets dangerous or impassable[2]. The gutters or watercourses in the streets were open; with them were frequently connected open drains from the houses, and the former, in turn, poured their contents into larger ditches which were supposed to find access to a stream, river, or the sea. But the system was very defective; gutters and ditches constantly needed scouring, and each one had to do this before his own door[3]. This was rendered difficult by the unsanitary methods of domestic and public life; for, contrary to civic ordinances, butchers used the streets as slaughter-houses[4], and into them threw the offal from their shops[5]; fish dealers poured forth their fishy water, with its

shall find in the streets or lanes, etc., and the owner shall pay 16*d*. a-piece before they shall be delivered. If not redeemed in four days after the proclamation made, they shall be sold, to the use of the town." (Ibid., pp. 221, 223.) In 1557, two other orders were made (ibid., pp. 248, 249), the latter of which states that: "Every inhabitant that keepeth hogs shall mark them with their proper mark, and the Sergeants in their several Wards shall have notice of the said marks, and if any hog shall be taken in the streets, the owner shall pay for every foot 4*d*., and the Sergeants shall pay the one half to the town, and shall retain the other half to themselves; and if any swine shall be taken unmarked the same shall be forfeited to the town, and if any person shall refuse to pay the forfeiture, he shall be committed to prison till payment be made."

[1] Picton, *Liverpool Municipal Records*, I, p. 234. A shepherd and a swineherd were appointed to keep the "sheepe upon the Comon, and to looke to the swyne all the day long from trespassing about the towne." Cf. Tait, *Mediaeval Manchester*, p. 49.

[2] Hearnshaw, *Southampton Court Leet Records*, 1574, paragraph 56; 1575, paragraph 35; 1576, paragraphs 14, 17, etc. Also East, *Records of Portsmouth*, pp. 97, 105–7.

[3] Hearnshaw, *Southampton Court Leet Records*, 1569, paragraph 54; 1575, paragraph 12; 1576, paragraph 6; also Bacon, *Annals of Ipswich*, p. 236; East, *Records of Portsmouth*, pp. 26, 72, 121, etc.; Stevenson, *Nottingham Borough Records*, IV, p. 161, No. 54; Hardy, *Records of Doncaster*, II, pp. 5, 23.

[4] *Southampton Court Leet Records*, 1551, paragraph 38; 1575, paragraph 5; East, *Records of Portsmouth*, pp. 40, 41, 50, etc., which show that pigs were killed and dressed in the streets. See also Morris, *Chester*, p. 261.

[5] *Southampton Court Leet Records*, 1551, paragraph 37; 1574, paragraph 36; 1575, paragraph 27, etc.; Cowper, *Records of Faversham*, p. 329; also Morris, *Chester*, pp. 261–2.

stench, into the streets[1]; and the refuse of house and stable prevented
the water in the gutters from flowing freely[2]. One gets the impression
from reading many records that the towns were, in many cases, in-
describably and incurably filthy, and in a condition to render not only
natural, but inevitable, the plagues and pestilences which we know
infested them.

With reference to facilities for communication and conveyance,
we have noted in our introductory chapter that, during the mediaeval
period, most of the travelling, except what was done in going to and
from the fairs and markets and on the pilgrimages, was done by the
King or his messengers, the Justices, and by the great landlords or
their representatives. Similar conditions prevailed during the Tudor
monarchy. Under the régime of the handicraft system of industry,
when production was carried on with raw materials which were chiefly
supplied by the local community, the necessity for good roads was
not so great. The woad or dye-stuffs for cloth manufacture, and the
wine, spices, etc., for household use, were brought from abroad by those
who traded at the fairs. The goods which were carried from the interior
to the ports of England for shipment to the Continent, and those which
were returned, were conveyed principally on pack-horses. The use of
wheeled vehicles for the carriage of goods in the early sixteenth century
was comparatively unimportant; and the fact that there was no great
economic need for good roads was a sufficient reason why, except
for the principal roads communicating with the important seaports,
the seats of the great fairs, and the fortresses, the highways should be,
in most cases, but little better than bridle-paths.

While the internal trade of England in the sixteenth century was
largely inter-municipal, yet we can note traces, toward the end of the
century, of the extension of that trade over longer distances, so as to
connect the interior with the great trading port towns and with the
metropolis[3]. Norden, in giving a "Description of Hartfordshire" in

---

[1] *Southampton Court Leet Records*, 1574, paragraph 74; 1575, paragraph 39,
etc.

[2] Ibid., 1550, paragraph 72; 1551, paragraph 24; Chanter and Wainwright,
Barnstaple Records, I, p. 44; Hardy, *Records of Doncaster*, III, pp. 5, 18, 23, etc.

Among the Acts and Ordinances agreed upon by the Common Council of the city
of Oxford, June 15, 1582, was one that no person, after a great rain, or any other
time, should sweep any rubbish into the common water-courses in the streets; but
they were to sweep such stuff up out of the water channels, in front of their own
ground, and have it carried away twice a week, under a penalty of 12*d*. every time
of default. Turner, *Records of the City of Oxford*, pp. 422–5, gives the full text of
these "Actes and Ordinaunces."

[3] That most of the trade at that time was municipal or inter-municipal, is shown
by the fact that every county had many market towns, at each of which a weekly

1598, says: "It is much benefited by thorrow-fares to and from London Northwardes, and that maketh the markets to bee the better furnished with such necessaries, as are requisite for Innes, for th' intertainement of travaylers[1]." A little later, after having shown the benefit of this trade in the building up of the town of Hertford, he bemoans the fact that "It (that is, Hertford) hath been most rob-d of her glory, by Wayres (Ware's) advancement, which since the turning of the highway through it hath flourished more and more, and this dayly withered[2]." It is evident, therefore, that these places were on a through route, north and south, between London and the interior, and that there must have been considerable trade along this route when it could cause such a reversal of fortune as that shown here. The town of Northampton was another which was located just at the point where two main roads joined, one running north and south, and the other east and west. This gave it also a strategic position for the imposition of through or passage tolls[3], as well as to derive benefit from the establishment of its large fairs and its increasing trade[4].

By 1599 at least, there was a regularly established carrying trade between Ipswich and London, and two waggons were employed on this route. The town of Ipswich made definite regulations as to the time when each carrier should set forth from that place and when it should return, taking care that one should not get any advantage over the other, and that their business should be conducted in an orderly manner[5]. By 1621, there were still but two carriers by waggon,

---

market was held. For instance, Hertfordshire had eighteen such market towns. (Norden, *Hartfordshire,* p. 3, gives the list.) Northamptonshire had twelve, and according to Norden (*Northamptonshire,* p. 32) no place in the whole county was more than four miles, and few were more than three miles, from a market town, either in the same county or the next adjoining. The same writer, in his *Description of Essex,* pp. 13–14, gives a list of nineteen market towns in that county, at each of which there was a weekly market. When each community was thus provided with a market almost at its door, we are justified in concluding that most of the trading was done at the local market.

[1] Norden, *Description of Hartfordshire,* p. 2.

[2] Norden, *Hartfordshire,* p. 18. This raises another interesting phase of the subject, but at present it is impossible for me to venture to investigate the decay of some towns and the rise and advancement of others through the changes in the local trade routes. See another example given in Holinshed, *Chronicles,* I, pp. 97–98.

[3] *Northampton Borough Records,* II, p. 197.

[4] Ibid., II, p. 188 et seq.

[5] From the *Annals of Ipswich,* we find that "Waggons travelling to London Tuesday and returning Friday, and not coming to this town till the Lord's Day, to the great offence of Almighty God, contrary to the laws of the realm, and the infamy and slander of the government of this town. Its ordered that no waggoner or common carrier of this town, shall return with his waggons or carriage or shall

and the former regulation was renewed, that one should go one week and the other the next week; but both were to give security for carriage according to former orders. One of them, however, refused to submit to this discipline, and the town ordered that all its freemen deliver their commodities for carriage to the other, according to prices set down, on penalty of 8s. 4d. for every offence[1]. The fact that one carrier was able to do the work which two had formerly done, without any increase of cost to the customers, is pretty clear proof that the amount of traffic thus carried was not great.

We have a hint that there was also a great thoroughfare through Tenbury, in Worcestershire, from Wales to the city of London; but no statement is found as to the number of vehicles or amount of carrying that passed along this way[2], so that we are unable to form any estimate as to its importance. Doubtless there were others which might also, in the language of that day, be called great thoroughfares; but we are safe in saying that the whole éxtent of this long distance traffic, during most of the sixteenth century, was comparatively slight[3].

The arrangements for travellers to make their way through the

labour or travel within this town or liberties thereof, on the Sabbath Day...upon pain of forfeiture of 20s. for each offence, to the use of the poor of this town.

"And that Thomas Lane shall set forth with his carriage on Monday, and return on Wednesday, and Richard Lane shall set out on Thursday and return on Wednesday. And he that shall set forth with carriage on the Monday one month, shall set out on the Thursday another month, and so shall continue, *alternis vicib'*, monthly. And hereunto Richard and Thomas Lane do submit." Bacon, *Annals of Ipswich*, pp. 402–3.

[1] Ibid., pp. 476, 478.

[2] *Records of Quarter Sessions*, edited by J. W. Bund for the Worcestershire Historical Society, Part II, see 1615 (133) xxii, 83, p. 212.

[3] If Harrison is to be credited, we learn that there were several thoroughfares radiating from London as a centre:

(a) The way from Walsingham, through Newmarket, Ware and Waltham, to London.

(b) The way from Berwick, through York, Newark, Grantham, Huntingdon, Ware, Wahham, to London.

(c) The way from Carnarvon, through Denbigh, Chester, Lichfield, Coventry, Towcester, Dunstable, St Albans, to London.

(d) The way from Cockermouth, through Keswick, Kendal, Lancaster, Preston, Warrington, Lichfield, Coventry, Dunstable, to London.

(e) The way from Yarmouth, through Ipswich, Colchester, Chelmsford, to London.

(f) The way from Dover, through Canterbury, Gravesend, to London.

(g) The way from St Buryan (in Cornwall), through Truro, Bodmin, Launceston, Honiton, Salisbury, Basingstoke, Staines, to London.

(h) The way from Bristol, through Chippenham, Newbury, Reading, to London.

(i) The way from St David's, through Brecknock, Gloucester, Dorchester, Maidenhead, to London.

kingdom were simple but yet effective. As the greater part of the carriage was performed by pack-horse, so most of the travelling for the longer distances was done on horseback, by posting. The system, doubtless, took its name from the fact that the postmasters throughout the country were required, not only to transmit the mail by horses which they always kept in readiness for that purpose, but also to keep, or to have kept for them, a sufficient number of horses so that a traveller might be accommodated at any time[1]. It would seem that the system was thus adopted probably for two reasons, both of which may have

(*j*)   The way from Dover, through Canterbury, Gravesend, across the Thames to Horndon, through Chelmsford, Thaxted, Linton, to Cambridge.

(*k*)   The way from Canterbury, through London, to Oxford.

(*l*)   The way from London, through Waltham, Ware, to Cambridge.

See Holinshed, *Chronicles*, I, pp. 415–7. Contrast this with Harrison's *Description of England in Shakspere's Youth*, II, pp. 109–17.

This list looks like a pretty complete veining of the kingdom by great roads. But when we remember that each shire had many market towns at which the great majority of the people could meet and trade (Holinshed, I, p. 326), and thus provide themselves with the requisites, it is evident that for them directly there was little need for wider communication. Moreover, the character of the roads precluded the possibility that there could be a great amount of traffic upon them, even though the highways may have existed as here described. It is more probable that these courses represented the general lines of communication and conveyance by post and pack-horse, rather than implying any great extent of trade along these directions.

The number of market towns in the following shires is taken from Holinshed, I, p. 326:

Shires	Number of Market towns	Shires	Number of Market towns
Middlesex	3	Northampton	10
Surrey	6	Buckingham	11
Sussex	18	Oxford	10
Kent	17	Southampton	18
Cambridge	4	Dorset	19
Bedford	9	Norfolk	26
Huntingdon	5	Suffolk	25
Rutland	2	Essex	18
Berkshire	11		

[1] Brit. Mus., G. 6463 (228) gives the articles issued by the Comptroller General of the Posts to his deputy postmasters throughout the kingdom, so that they might know the facts regarding the Queen's Proclamation of that year (1583) concerning the posts between London and Scotland. Among these instructions the following apply here:

Each postmaster was always to have ready for the mail service three good post-horses, suitably equipped, and three good, strong, leather bags to carry the packet in, and three horns to blow by the way. As soon as the mail was brought to him, or within a quarter of an hour after, he was to have it taken with all speed to the next post. Each postmaster, or someone that he appointed under him, was always to

been but vaguely recognized at the time: in the first place, the great majority of those who made long journeys were men who were entrusted with the execution of a public commission of some kind, and it was but natural, therefore, that the public work should be arranged for by public servants acting under instructions from the government; and, in the second place, since the postmasters had to keep horses for carrying the packet from stage to stage, it was only a reasonable extension of their office that they should provide a few extra horses and receive the remuneration that was given by travellers for that service. All persons who were riding post in fulfilment of a commission had to get their horses from the regular postmasters along the route, and the price for this accommodation was usually fixed by the Privy Council[1]; but ordinary travellers probably obtained their horses from private individuals in the towns or else from the postmasters, at a price reached by agreement between the parties at the time. In some cases, the town authorities stipulated the price which these horse-hirers could charge, and even gave assistance to postmasters when they were too poor to keep horses at their own expense; they even determined the maximum distance that any horse could go, and the maximum weight of luggage that he could carry[2]. In other cases, towns

have ready four good post-horses and two horns for those who wanted to ride post. No post or guide was to ride without his horn, which was to be blown as prescribed; and the post was not to ride past the next post.

[1] In Brit. Mus., G. 6463 (232) we have the orders of the Privy Council regarding the posts between London and Scotland, as issued in 1583. These orders, *inter alia*, directed that it would be unlawful for any man riding post by commission to take his horses of any man except of the ordinary and standing posts, or at their appointment; and everyone riding post by commission for Her Majesty's service was to pay $1\frac{1}{2}d$. per mile, but if he were going on urgent business the postmaster might charge him 2*d*. per mile. No man riding post was to ride without a guide, who was to blow his horn as often as he met company, or passed through a town, or at the least three times a mile. The posts were to ride in summer seven miles per hour, and in winter five miles per hour; by which the packet might be carried between London and Berwick in 42 hours in summer and 60 hours in winter.

[2] Blomefield, *History of Norfolk*, III, p. 294, quoting from the *Records of Norwich*, says that about 1568 the order relating to post-horses was first established in that city, by the Duke of Norfolk and the Mayor, who agreed that there should be three postmasters, each of which had £3. 13*s*. and 4*s*. lent him out of the city's treasury, free of interest, and a stipend of £4 per annum, paid by the sheriffs, one-half of which was levied on the innkeepers and tipplers in the city, and the other half on the other inhabitants. No man was to take up any post-horses in the city, unless he was licensed by warrant from the Queen's Majesty, the Privy Council, the Duke of Norfolk, or the Mayor, nor to use any one horse above twelve or fourteen miles together; for which he was to pay 2*d*. each mile outward, and 6*d*. to the guide to go and bring back the horses. The horses were not to carry any cloak-bag, etc., of above ten pounds weight.

appointed certain men, other than postmasters, to keep horses for hire
for the use of travellers, and fixed the prices which they could charge[1].
In Newcastle, in 1593, the borough paid for the keeping of horses to
be used on the by-posts; but whether all the posting arrangements
there were carried out under the same encouragement, we have been
unable to ascertain[2]. When the needs of travellers were greater than
the supply of horses kept by the postmaster, the latter might go to
the fields or stables of his neighbours and take whatever horses he
needed; but under these conditions the full amount paid for a horse
went to the owner of the horse[3].

It is evident from the foregoing that, in the letting of horses for
riding post, private persons entered into competition with the postmasters; and at times the innkeepers or other neighbours were able
to secure to themselves greater returns from the posting traffic than
did the postmasters. But into the details of this competition it is not
necessary for us to enter, since these have already been given to the
public[4].

As to the actual condition of the roads in the greater part of England,
we have not sufficient evidence to make anything but conditional statements. In 1577–8 we learn that the highway in Kent county, from
the market town of Middleton to King's Ferry, "is so decaied that
neither man nor beast is able to pass without great danger[5]." In
1591, the Privy Council notified the Justices in the county of Surrey
that one of the roads in their jurisdiction was so "very farr out of
repaier" that ordinary travellers, as well as Her Majesty's carriages,
were many times forced to take a longer and more inconvenient route;

---

[1] In Southampton, in 1558, certain "horse-hirers" were appointed for the town.
The hire of a horse was fixed at 8*d.* for the first day, afterward at 6*d.*, but longer
journeys were made the subject of special bargains; for instance, they calculated
a week for a journey from Southampton to London or Bristol, and the charge for
a horse was 6*s.*, with 6*d.* extra for every extra day.

[2] Richardson's *Newcastle Reprints*, III, Extracts from the Municipal Accounts
of Newcastle, pp. 23, 29, etc.

[3] *Acts of Privy Council*, 1571–5, p. 181. This privilege became subject to abuse.
It was said that the postmaster sometimes worked off an old grudge against a neighbour by always taking horses from him and allowing other neighbours to be free
from this service; or that some gave the postmaster bribes if he would pass
them by and take horses from others (v. *Cal. S. P. D.*, 1619–23, p. 86; 1631–3,
p. 257; 1635, p. 18). Other complaints against the system of posting were that
travellers abused the horses they had hired by riding them too hard or too far;
that owners of horses often gave travellers poor old plugs that could scarcely creep
along at more than a snail's pace; etc. (v. *Acts of Privy Council*, x, 1577–8, pp. 62,
219; 1580–1, p. 203).

[4] Hemmeon, *History of the British Post Office*, p. 91 et seq.

[5] *Acts of Privy Council*, x (1577–8), pp. 223–4.

and instructions were given them to attend to its immediate repair[1].
In 1587 a letter was sent to Lord Burghley showing that because one
of the great roads of Suffolk was so bad, the people of Norfolk and
Suffolk could not get any fish from the sea-coast[2].   In 1592 and 1610
the two Dukes of Wirtemberg travelled in England; and their journals
tell us that, "on the road (from London to Oxford) we passed through
a villainous, boggy and wild country, and several times missed our
way, because the country thereabouts is very little inhabited, and is
nearly a waste; and there is one spot in particular where the mud is
so deep, that in my opinion it would scarcely be possible to pass with
a coach in winter or in rainy weather[3]." If this was the character
of the roads so near the metropolis, where population was densest
and traffic was greatest, we would naturally expect that the means
for communication in those sections which were more remote from the
national capital would be meagre indeed.   In perfect accord with this
are the facts given by Harrison, who shows that the good intention of
the statute duty was largely defeated by the selfishness and indolence
of parishioners and surveyors[4].

---

[1] *Acts of Privy Council*, xxi (1591), pp. 77–78.

[2] Brit. Mus., Lansd. 55 (Burghley Papers), No. 40, pp. 109, 114.

[3] Rye, *England as seen by Foreigners in the Days of Elizabeth and James I*, p. 31.
See also *Acts of Privy Council*, ix (1575–7), pp. 117, 120, 131, 135. Norden,
Speculi Britanniae (Middlesex), p. 11, after speaking of the fertility of the land
of this county, says: "Yet doth not this so fruitefull soyle yeeld comfort to the
wayfairing man in the wintertime, by reason of the claiesh nature of soyle: which
after it hath tasted the Autume showers, waxeth both dyrtie and deepe." Again,
(p. 15) he says that the road from Gray's Inn to Bernet, an "auncient highway,"
had become so deep and dirty in the winter season, that an agreement was made
between the Bishop of London and the country, under which a new way had been
laid out through the Bishop's parks so as to allow the passage of carriers and travellers
upon the payment of a toll to the Bishop.   In Norden's time this toll gate was
farmed by the Bishop at £40 a year (v. p. 22).   The fact that it became necessary, almost in the metropolis, to pay for the laying out of the new highway through
a private park, would indicate that the old road was in pretty bad condition.

[4] See Harrison's *Description of Britaine*, written in the latter part of the reign
of Elizabeth (Holinshed, *Chronicles*, i, pp. 191–2). He says: "Now to speake
generallie of our common high waies through the English part of the Ile (for the rest
I can saie nothing) you shall understand that in the claie or cledgie soile they are
often verie deepe and troublesome in the winter halfe. Wherefore by authoritie of
parlement an order is taken for their yearlie amendment, whereby all sorts of the
common people doo imploie their travell for six daies in summer upon the same.
And albeit that the intent of the statute is verie profitable for the reparations of
the decaied places, yet the rich doo so cancell their portions, and the poor so loiter
in their labours, that of all the six, scarcelie two good days works are well performed
and accomplished in a parish on these so necessarie affaires. Besides this, such as
have land lieng upon the sides of the waies, doo utterlie neglect to dich and scowre

In the arrangements for post-horses at Southampton, in 1558, the time set for a journey from there to London or Bristol was seven days, and to Salisbury two days[1]. This rate of travelling would be on the average about eleven miles per day. But we have an account of a journey made by Lupold von Wedel through England and Scotland, in 1584 and 1585, which gives us results which are considerably different from the foregoing. This journey, from London to Edinburgh, was performed by riding post, and by making two, three or four changes of horses per day, each relay of horses going ten to twelve miles conveniently. Where no delays were experienced, they rode from twenty-four to fifty miles per day[2]. But on the return journey, from Edinburgh to London, they did not have fresh horses at every stage, and so the distances covered ranged from twelve to twenty-six miles per day, that is, just half the rates going northward[3]. We would naturally expect that the speed at which the mail was carried would

their draines and watercourses, for better avoidance of the winter waters... whereby the streets doo grow to be much more gulled than before, and thereby verie noisome for such as travell by the same. Sometimes also, and that verie often, these daies works are not imploied upon those waies that lead from market to market, but ech surveior amendeth such by-plots & lanes as seeme best for his owne commoditie, and more easie passage unto his fields and pastures. And whereas in some places there is such want of stones, as thereby the inhabitants are driven to seeke them farre off in other soiles: the owners of the lands wherein those stones are to be had, and which hitherto have given monie to have them borne awaie, doo now reape no small commoditie by raising the same to excessive prices, whereby their neighbours are driven to grievous charges, which is another cause wherefore the meaning of that good law is verie much defrauded." Then he goes on to speak of the way in which trees and bushes along the road-sides prevented the roads from drying up and thus becoming firm. We are inclined to think that the "stones" above referred to, must have been used to fill up the holes, sloughs and ruts in the roads; and if so, any pretence at good roads must in some cases, in fact in most cases, have been poorly substantiated.

[1] Davies, *History of Southampton*, p. 274, quoted from the *Southampton Boke of Remembrances*, p. 77 b. In 1577 the journey to London or Bristol was set at eight days. The same writer, quoting from the *Southampton Journal*, of Aug. 1609, mentions that a certain prisoner released from the Bargate received a passport to London, eight days being allowed for the journey.

[2] Lupold von Wedel, *Journey through England and Scotland in 1584 and 1585*, p. 237 et seq. From the nature of the travelling, which was not at all continuous, it is almost impossible to get any accurate figure as an average rate of travel.

[3] The distances covered per day on the return journey were:

On Oct 5,	26 miles	On Oct. 10,	26 miles
,, 6,	22 ,,	,, 11,	21 ,,
,, 7,	12 ,,	,, 12,	12 ,,
,, 8,	26 ,,	,, 13,	24 ,,
,, 9,	28 ,,	,, 14,	20 ,,

be greater than that of an ordinary traveller; yet the former was on an average only about three to five miles an hour, and was sometimes much less than this[1].

The only conclusion, therefore, which we can reach at the present stage of our investigation is that while some portions of roads may have been fairly good for the usual method of travelling by horseback, yet the conditions over most of the kingdom were such as to preclude any extended communication. This was especially the case during the winter and the rainy seasons of the year.

Having now brought this subject almost to the close of the Tudor period, let us return from the foregoing digression, to continue our account of the legislation for the roads in general throughout the kingdom. In the latter part of the reign of Elizabeth, after 1580, eleven different Acts were passed for the repair of highways and for paving the streets of towns, such as Ipswich and Newark-on-Trent. Probably this agitation to have the streets of large towns and cities paved was one result of the introduction of coaches or waggons, the use of which gradually spread among the nobility.

The system in force for repairing the roads during the later sixteenth and most of the seventeenth century was that which was made perpetual in 1586[2]. Under it, all the inhabitants were liable, and the six days' work was done by the people either working themselves or sending others to work in their places. Each parish appointed two surveyors of the highways, who fixed the days when people were to come and work on the roads. The neglect to appoint surveyors rendered the parish liable to indictment before the Quarter Sessions for neglect of duty; if the surveyors were appointed and did not fix the days, they were liable to indictment for neglect; while if the people

---

[1] Hemmeon, *History of the British Post Office*, 98–99. Of course, on special occasions, the rate of posting could be greatly increased. For instance, when Sir Robert Cary in 1603 was sent to Edinburgh to communicate the news of Queen Elizabeth's death to her successor, he performed the whole journey in three days, which would be 130 miles per day (Cary, *Memoirs*, pp. 149–50). In 1605, John Lepton performed his famous journeys between London and York. He rode five different times between these two cities in five days, and these were accomplished in one week, as follows:

Monday,	May 20, left London 2–3 a.m., and got to York	5–6 p.m.
Tuesday,	„ 21, „ York 3 „ „ „ London 6–7 „
Wednesday,	„ 22, „ London 2–3 „ „ „ York 7 „
Thursday,	„ 23, „ York 2–3 „ „ „ London 7–8 „
Friday,	„ 24, „ London 2–3 „ „ „ York 7–8 „

(Stow, *Abridgement of the English Chronicle*, p. 455). This was done under extraordinary circumstances.

[2] Act 29 Eliz., c. 5.

did not attend on the days that were appointed for their work, they might be indicted for non-fulfilment of the work.

During the troubled times of the first two Stuart monarchs we have very little new legislation of any importance. But one measure passed at this time foreshadows much of the legislation of the seventeenth and eighteenth centuries, with its prohibitions or restrictions as to the weight that could be drawn and the way in which it could be drawn. Incidentally, too, it casts some light upon the condition of the roads of that day; for when James I, in 1621[1], forbade the use of any four-wheeled waggon or the carriage of more than a ton weight at a time, because vehicles carrying "excessive burdens so galled the highways, and the very foundations of bridges, that they were public nuisances," he shows us by implication that the highways must have been easily cut up by the stage waggons that were coming into common use by the carriers. The surface of the road must have been very poor, when it was not substantial enough to bear the carriage of more than twenty hundredweight of goods upon a four-wheeled waggon. This proclamation of 1621 was withdrawn, but in 1629 there was another to the same effect issued[2].

Under the Commonwealth, an Ordinance was passed to provide more effectual remedies than the legislation in force for improving the highways[3]. It retained some of the provisions of the preceding statutes, and added a few others. For example, an assessment was to be made by the parishioners upon the property of the parish, according to its yearly value, and also upon every parishioner who paid poor rates; but this assessment was not to exceed 12*d.* in the pound. Where the parish was over-burdened, and the above rate was not sufficient, the Justices might levy upon adjoining parishes in their jurisdiction that did not have to pay that amount for their highway repairs, till the rate amounted to 12*d.* in the pound, and devote that money to aid the over-burdened parish. All money obtained by assessment, fines, and otherwise, was to be employed by the surveyors, in paying for the services of the labourers and carts that were necessary for repairing the highways perfectly. Waggons and carts were not to be drawn on the road or street with more than five horses, upon penalty of seizure of the supernumerary horses and other satisfaction. The important

---

[1] Roberts, *Social History of the People of the Southern Counties*, p. 488. Under James I, one or two private Acts were passed for the repair of roads in particular districts.

[2] Anderson, *History of Commerce*, xix, p. 180; Parnell, *Treatise on Roads* (1834), p. 16.

[3] Brit. Mus., E. 1068 (59) gives this Ordinance in full.

thing for us here to notice is the complete sweeping away of the old system of statute labour, and its displacement by an equivalent money payment, which would be expended upon voluntary labour on the roads. Whether this Ordinance was put into effect or not, we cannot say: if it was, it could have been only for a few years, for with the Restoration all legislation of the Commonwealth was repealed, and it was not until 1835 that the principle underlying this Ordinance was again established, and forced labour on the highway was finally abolished.

As a general rule, the repair and maintenance of the highways were to be effected in three ways: either by some landowner, or by the parish in which the road was situated, or by some district, such as the hundred. In many instances, there was great difficulty in determining upon whose shoulders the liability to repair rested[1]. Usually it was the duty of the parish to attend to the roads, as we see by the fact that in the records of Quarter Sessions the indictments for this cause are largely against the parishes. During the period which precedes the Restoration, the long lists of presentments[2] as to the roads in many parishes furnish us with abundant evidence that the care of the roads often did not weigh very heavily upon the minds of those who were charged therewith, and that in most cases this responsibility devolved upon the inhabitants of the parish[3], or parishes[4], through

[1] See, for example, *West Riding Sessions Rolls*, p. xxxvii.

[2] See, for instance, the majority of the presentments in *Records of Quarter Sessions of Worcestershire*, *West Riding Sessions Rolls*, *North Riding Sessions Rolls*, etc. Note as a particular case, the thirty-nine presentments in the year 1633, in Bund, *Records of Quarter Sessions of Worcestershire*, Part II. See also Cox, *Derbyshire Annals*, II, p. 227 et seq.

[3] A very early instance is found in 1592 in the *Records of Quarter Sessions*, Part II, published by the Worcestershire Historical Society, under 1592 (29), xxxix, 15, p. 6, when the sheriff was directed to distrain on the inhabitants of Great Malvern for the repair of a highway leading from a place called the Red Green...to the bridge adjoining the Old Hills, being the highway between the market town of Upton-upon-Severn and the market town of the city of Worcester.

At the Easter Sessions, 1656, one of the Derbyshire Justices presented a portion of the highway leading to Chesterfield for being out of repair, in the following words (Cox, *Derbyshire Annals*, p. 227 of vol. II):

"The presentment of John Spateman, Esq., one of ye Justices of peace of this County of Derby ye 15th day of April, 1656, upon his owne view.

" I present the Inhabitants of ye Parish of Chesterfield for not repayring of ye highwayes leading betwixt Wingerworth and ye towne of Chesterfield, and being within ye sayd Parish. Jo. Spateman."

See also Cox, *Derbyshire Annals*, II, p. 228 ; *West Riding Sessions Rolls*, pp. 105, 129, 130, etc.

[4] *West Riding Sessions Rolls*, p. 104—"Whereas the highway leading from Leeds to Wikebridge and so to Seacroft and so to Kiddall toward Yorke hath been

which the roads passed. But it is very seldom that we find anything like the modern idea of "through traffic" put forth as a reason why a parish should be relieved of the complete maintenance of a road within its boundaries[1].

When the residents in a parish were given notice, by the Quarter Sessions, of their liability for repairing a certain road, they were bound to come and give their work gratis on the highway according to the tenor of the law; this applied to all, whether husbandmen, cottagers, or other residents, and if they failed they were indicted[2]. Even non-resident occupiers or owners of land were required to work on the roads in the parishes where their land was situated; and if they did not, they were liable to indictment[3]. In this way, all who were responsible for statute labour could be compelled to contribute their share to this work for the up-keep of the roads.

But this system did not always work successfully and smoothly. It sometimes happened that the surveyors did not fulfil their obligations, and the penalty had to be imposed upon them by the Justices, who were empowered to inquire and determine the amount of such offences[4].

heretofore presented by jury to be in great decay for want of amendment so that travellers can very hardly pass to the great hindrance of all her Majesty's subjects that have occasion to travel that way. Therefore the aforesaid jurors by the consent of the Justices here present, do lay a pain that every person occupying a plough tilth within any of the parishes of Leeds, Whitkirk and Berwick (through which parishes the said highway lieth) shall send their draughts (that is, teams) and sufficient labourers according to the statute and repair the same way before the 25th day of August upon pain that every person making default therein shall forfeit 20s."

[1] Note example given in *Records of Quarter Sessions of Worcestershire*, Part II, 1615 (133), XXII, 83, p. 212. Compare also example in note 4 on p. 53.

[2] *Records of Quarter Sessions of Worcestershire*, Part II, 1634 (232), LXIII, 87, p. 557; 1634 (245), LXIII, 108, p. 559; 1600 (45), XIII, 51, p. 29; etc.

[3] *Records of Quarter Sessions of Worcestershire*, Part II, 1634 (230), LXIII, 84, p. 557.

[4] There is given to us a list in 1633 of the fines imposed by the Worcestershire Justices on the different surveyors and owners of teams for not repairing the highways. See *Records of Quarter Sessions*, Part II, 1633 (243), LVIII, 101, p. 525. In the first column the name of the parish is given, and the sum of money (if any) required to repair the roads; in the second, the name of the surveyor, and if he was fined the amount of the fine is set opposite his name; in the third, the names of the owners of teams who should have sent them to work and did not, and the fines they had to pay.

Imprimis		Supervisors		Teams and Owners	
Ripple	(bene)				
Severn Stoke	£10	Thos. Wade 40/-	} £4	John Beste	} 20/-
		Thos. Dalley 40/-		Humphrey Best	
		Edmund Smith	} £4	Jno Best	10/-
		Chas. Bacon		John Smith	10/-
	etc.	etc.		etc.	

At times the surveyor was an innkeeper, and after the men had worked a short time on the road, if they would go to his house and drink he would withdraw them from their work and allow them to spend the rest of the day there in a disorderly manner[1]. Doubtless this is the reason why innkeepers were, at a later time, prohibited from being surveyors[2]. In other cases, personal considerations decided what attitude the surveyor would take toward those who had to work under his direction, and if any offered him personal offence or opposed his wishes, the culprit was charged before the Justices with having failed to do his statute labour[3].

Other fruitful sources of trouble were the misunderstandings and disputes of individuals and of parishes or towns, with reference to the liability for repair of highways. Persons were not always clear as to the amount of these claims which could be made upon them, and as the economic standing of the individual changed, the demands upon him for this public service also changed. Sometimes the son performed the same amount of work on the roads as his father had done, in other cases he performed more; but if he had satisfied the former demands, without knowing of the increase, he was liable for presentment and payment of fine[4]. We have instances also of persons and of

In Ripple, the work was done alright. In Severn Stoke, the work would cost £10. Of this, two of the supervisors had to pay 40s. each, the other two the same, and four owners of teams £2 among them.

See also the complaint against the surveyors of Reading (July 31, 1650) for not executing their office in mending the highways (Guilding, *Records of the Borough of Reading*, IV, p. 370).

[1] *Records of Quarter Sessions of Worcestershire*, Part II, 1633 (230), LVIII, 77, p. 521.

[2] By Act 26 Geo. II, c. 30 (1753).

[3] In the *Records of Quarter Sessions of Worcestershire*, Part II, we find the case of a petitioner who says that although he worked with his team on the highway, he was fined £20 for not working, the reason being that he would not let his team draw a load of wood for the surveyor gratis, 1633 (232), LVIII, 96, p. 522. In another instance, a charge was made against a man because, when he was given notice to attend and work on the highways, he said he would not go himself, but would send his boy-servant and his dog and waggon, 1602 (59), XV, 47, p. 52.

[4] The following protest of an inhabitant of Breadsall (1649), against being fined for neglecting to do his share of the common work on the roads, is taken from Cox, *Derbyshire Annals*, II, p. 226:

"To the right woll his Maties Justices of the peace for this county of Derby.

"The humble petition of Thomas Cheshire Humbly Sheweth

"That whereas your petitioner is tennt of a Cottage in Bredsall of foure poundes per annum uppon the rack, and was never charged to bringe any cart to the mending of the kinges high wayes but only a laborer, neither his father before him when hee enjoyed the same and that both his father and himselfe have beene so carefull to sende such labourers as no exception hath beene taken agst them by the overseers

parishes having been, by Quarter Sessions, adjudged liable for highway repair, but upon further investigation by the Justices they were relieved from this obligation, and those who were legally liable were compelled to bear the burden[1].

When the inhabitants of a parish neglected their work on the roads, or when the statute labour was not sufficient to accomplish the desired result, the Justices in Quarter Sessions, upon petition from the surveyors, would sometimes levy an assessment upon all the parish[2].

of the sayd worke neither was it ever intimated unto your petitioner that more would bee required at his handes so that if hee hath beene any way defective as to the letter of the statute for not bringinge his carte, it was done in ignorance not in contempt beeing very willing henceforwards to doe what shall bee anyway reasonably required at his handes for any publick good worke, wherewth the officers were so well satisfyed that your petitioner was not p'sented at the last sessions when the Constables p'sentments were first delivred into ye Court and yett after by the instigation of some neighbours not well disposed towarde your petitioner (and as your petitioner is informed it was an Ale house plot) your petitioner was p'sented xxs. for not bringing his cart to the common worke without notice.

" Your petitioner therefore most humbly prayeth (the premisses considered) that the sayd penalty may bee taken off, or mitigated with as much favour as your worps lawfully may, and your petitioner will willingly hereafter come so p'pared to the sayd common worke as the Court shall appoynt, and for your favourable p'ceeding for the p'sent (as in duty bound) will dayly pray etc."

    [1] The following is taken from the *West Riding Sessions Rolls,* p. 119:

"Whereas at Pontefract Sessions last an order was made that the town of Pollington (amongst others) should pay 16s. towards the repair of Purston lane near Pontefract, Now forasmuch as it is found by jury that divers towns within the Soake (parish) of Snaith ought rather to repair the same. Then the said town of Pollington, by reason they have often to repair through the same with carriages And for that the same order was repug…to the course of the laws and statutes in such cases provided Therefore the said order hath since been repealed And it is now ordered by this Court that the said sum of 16s. levied of the said town of Pollington and already paid shall upon show of this order be repaid by him that received the same And if he refuse to do so then to carry him before the next Justice of Peace to enter bound with sureties to appear at the next Sessions to answer the same."

    [2] Cox, *Derbyshire Annals,* ii, p. 227, gives us the text of a petition from the surveyor of the roads of two townships of Chesterfield, asking that the Court lay an assessment on these townships, as many neglected to do their share of the work. The request was answered by the Court ordering an assessment. The petition is as follows:

"For ye Right Honorable ye Justices of ye Bench in Sessions

" The humble petition of John Fowler of Stonegravels in ye towneshippe of Newbold and Dunston
Sheweth

"That whereas your petr is Supervisor of ye highwayes for ye said townes wherein hee now dwelleth and but few of ye Inhabitants have helped or paid towards ye mending of ye said highwayes and many doe refuse both to helpe wth their draughts or to give money, yt draughts and labourers might be hyred according to Lawe

It had been, and was still, the custom in exceptional circumstances for the governing officials of a town to levy an assessment upon the inhabitants of the town, which was to be used for putting in repair their own streets[1]. This was the method which was now extended by imposing the assessment over a wider area, so that sometimes the inhabitants of a whole township paid such a levy to aid in repairing the highways within and near to some town in that township[2].

That during all this time when the Acts of 1555 and 1562 were in force, these statutes had "not been found so effectual as is desired[3]," does not prove that the statutes themselves were at fault, else the Parliament of Elizabeth, in 1586, would not have made them permanent. The good intention of, and results from, them were recognized even at that time, when they had been in effect only a short time; and if, in the first half of the next century, they had not proved to be of the utmost satisfaction, the trouble lay, not so much with the law itself, as with the method of its execution and administration. Some of these defects in the carrying out of the law we have just noticed. We have seen how the surveyors were at fault in the fulfilment of their trust; and how the inhabitants were negligent in the performance of their duties. But there were also proceedings against those who had actually injured the highways; for some had enclosed not only the

and custome. And your petr has layd out before what hee had received ye sume of five poundes, and still there is much to bee done, and it will not bee done except your petr should disburse all himselfe for ye finishing thereof. And there is also a bridge called Brearly bridge wch is much out of repair. Your petr humbly prayeth your honours to grant an Order yt an Assessm't may bee made through ye wholl townshipp of Newbold & Dunston, to pay yor petr his disbursemt & to goe on with the perfecting of his worke, namely ye amending ye rest of high wayes wch are not yet done, & alsoe to repaire ye said bridge.

"And your petr shall pray etc."
See similar petition from the inhabitants of Calow, on the same page.

[1] See case of Chester, mentioned on pages 264 and 270 of Morris, *Chester during the Plantagenet and Tudor Periods*. Note also the following:

"Borough of ⎱   At a general meeting of the Mayor, Burgesses, etc., of the town
Pontefract ⎰   and borough, May 17, 1659.
Ordered That an assessm't of £10 be forthwith laid and assessed upon the severall Inhabitants and owners and occupiers of Land within the said Towne & Burroughe by Robert Sutton, Zachary Stable...or any three of them for and towards the repaire of the highwaies belonging to the said Towne & Burroughe: and that the Constables of the said Towne doe collect the said Assessm't & afterwards pay the same into the hands of the p'sent surveyor for the highwaies to be imploied accordingly...." Holmes, *Pontefract*, p. 69. Assessments for the repair of the highways seem to have been the common thing at Pontefract (ibid., pp. 73, 141, 146, 181).

[2] See note 2 on opposite page.
[3] Preamble to Act 14 Car. II, c. 6.

greensward at the sides, but the whole road[1]; some had blocked the highways by erecting various kinds of buildings on them and by piles of refuse[2]; others stopped the water by not scouring out their ditches; and some encroached on the road allowance by planting their hedges out too far on it[3]. From the Sessions' records of the many batches of presentments made at this time, it is apparent that compulsion was necessary to induce the people or the parishes to amend their roads and make them fit for travel[4]. The significance of these facts would seem to be that it was not the legislation, but rather an improved public sentiment as to the need of roads and the necessity for the enforcement of the law, that was delaying road improvement.

We are not surprised that during the reigns of James I and Charles I, and under the Protector, such matters as the public highways received little attention from the ruler or from Parliament, for their interests were almost wholly in the constitutional, military and political problems of the kingdom. But soon after the Restoration, by an Act of 1662 "for enlarging and repairing of the common highways[5]," which was very elaborate in its specifications[6], the former system of statute

---

[1] *Records of Quarter Sessions of Worcestershire*, Part II, 1633 (126), LVII, 48, p. 507.

[2] Ibid., 1634 (256), LXIII, 121, p. 560; 1635 (138), LX, 74, p. 594; 1628 (180), LIII, 74, p. 448, etc.

[3] Ibid., 1633 (182), LIX, 124, p. 515; 1633 (184), LIX, 126, p. 515. See also the presentments given in Cox, *Derbyshire Annals*, II, pp. 228–9.

[4] See the many certificates handed in to Quarter Sessions showing that after indictment the roads had been repaired; and also the many parishes and persons charged with non-repair of highways; e.g., in *Records of Quarter Sessions of Worcestershire*, Part II, pp. 733, 740 in *General Index*, No. 1. See also Cox, *Derbyshire Annals*, II, p. 228.

Whether it was the case in other counties or not, I have not the means of knowing, but in Worcestershire when a parish was indicted, and pleaded that the work was done and that its highways were in repair, it had to produce a certificate of that fact. In 1633 the Court gave the following directions as to these certificates (*Records of Quarter Sessions*, Part II, 1633 (178), LIX, 127, p. 513): "The certificates that are to be brought from towns that their ways are amended must be under the hands of the Minister Churchwardens and Overseers with some other sufficient persons of the neighbourhood and also be delivered into Court by them or some of them by oath made of the truth." Sometimes the certificate, after certifying that the road has been repaired, goes on to make further complaints (see, for instance, the following cases in the *Records of Quarter Sessions of Worcestershire*, Part II, 1633 (182), LIX, 124, p. 515; 1633 (184), LIX, 126, p. 515).

[5] Act 14 Car. II, c. 6.

[6] The chief provisions of this Act were as follows:

(a) Churchwardens and Constables of every parish, town, village or hamlet were annually to elect two or more surveyors of the highways, in Easter week, under penalty of £5.

(b) Surveyors, within twenty days after receiving notice of election, were to

labour was revived, extended, and modified in several particulars. Surveyors were appointed as before, who were to go over their district within twenty days after appointment, and examine the roads, bridges, water-courses, etc., preliminary to the performance of statute labour. They were to consider what repairs were needed, and what amount of money would be required for this purpose, over and above what would be accomplished by the parishioners with their teams; and were to lay upon the householders of the parish one or more assessments, the amount of which was to be applied to repairing and enlarging the

go through their district, and examine the roads, bridges, water-courses, etc., on penalty of £5.

(c) Surveyors were to consider what repairs were needed and what amount of money would be required for this purpose, over and above what would be accomplished by the other laws made for amending the highways; and together with two or more substantial householders of that parish, town, village, or hamlet, were to lay one or more assessments upon those who paid poor rates, and upon every occupier of houses, lands, tithes, mines, saleable underwoods, stock, goods, or other personal estate not being household stuff. The assessment or assessments were not to exceed 6*d.* in the pound in any one year, according to the real value of the property assessed; and £20 in money, goods, or other personal estate should be rated equally with 20*s.* a year in lands.

(d) These assessments must be allowed and signed by the Justice of the Peace before they could become effective.

(e) Except in special cases, those who would not pay the assessment within 20 days after demand by surveyors, should forfeit and pay double as much.

(f) Surveyors were to see that common nuisances in the highways, and unscoured ditches or water-courses adjoining the highways, were reformed, and offenders punished.

(g) Surveyors were to direct the statute labour authorized by former laws. If more work than the required six days was needed to repair the roads, the workmen and owners of teams, carts, etc., hired to do such work, were to be paid for according to the usual rate of the country for such work; or if there were a dispute about this, a neighbouring Justice was to settle the rate of payment.

(h) Where roads were not of the breadth of eight yards, surveyors were to lay out lands adjoining, by order of the Quarter Sessions, to make the road at least that breadth, and to give satisfaction to owners of the land.

(i) If it were necessary, the surveyors, with the consent of the Justices, might dig for gravel and other road material in neighbouring commons, without paying for same; and if a sufficient supply could not be had in the commons, the surveyors might enter and dig in private ground without paying. Damages to the owner were to be assessed by two Justices, and the holes or pits were to be filled up, as required by 5 Eliz., c. 13.

(j) After Sept. 29, 1662, no waggon or cart carrying for hire (except those used in and about husbandry, etc.) was to be drawn on the public roads with above seven horses or their equivalent; and the burden at any one time was not to be more than 20 cwt. between Oct. 1 and May 1, nor 30 cwt. between May 1 and Oct. 1.

(k) All such waggons or carts were to have wheels not less than four inches wide, on penalty of 40*s.* (This clause was repealed by 22 Car. II, c. 12.)

highways as the surveyors and householders should think necessary. It will be noted that the surveyors, whose powers were increased by the Acts of 1562 and 1575[1], were given still greater authority under this Act of 1662; for formerly, whenever assessments were levied, it was done by the Justices of the Peace in Quarter Sessions[2], but under this Act the surveyors were given authority to make this levy, and, with the sanction of the Justices of the Peace, to put it into effect, rendering annual accounts to a parish meeting. The assessment or assessments were not to exceed sixpence in the pound in any one year, according to the real value of the property assessed; and twenty pounds in money, goods, or other personal estate, was to be rated equally with 20s. a year in lands. The distinctive feature of the Act of 1662 was this granting to the surveyors the right to take an assessment to aid in repairing the roads when the six days of statute labour were found to be insufficient[3]. Another provision, which was very useful, was that where roads were not of the width of eight yards, the surveyors, by order of the Quarter Sessions, were to make them at least that breadth, by laying out lands adjoining, for which they were to give satisfaction to the owners[4]. The Act stipulated also that waggons or carts carrying for hire on the public roads should have wheels not less than four inches wide, should not be drawn with more than seven horses (or their equivalent in horses and oxen), and should not carry at any one time more than twenty hundredweight between October 1 and May 1, nor more than thirty hundredweight for the other part of the year[5]. Apparently it was the thought of the framers of the Act that by increasing the width of the rolling surface of the wheel, the roads would be less liable to get cut into ruts and would be more firmly consolidated.

The extent to which this Act was made effective it is almost impossible to determine. That the regulation as to the width of roads was not enforced, is evident from another Act passed thirty years

---

[1] Acts 5 Eliz., c. 13 and 18 Eliz., c. 10.

[2] Except under the Ordinance of the Protector, by authority of which the parishioners were to lay the assessments upon themselves.

[3] This Act of 1662 was to continue in force only three years from Mar. 25, 1662, so far as its power of levying assessments was concerned.

[4] This provision was continued in 1696–7 by another Act, 8 and 9 W. III, c. 16.

[5] This enactment is in accord with the statement in the preamble of the Act, that because former laws had not proven as effectual as was desired, and because of "the extraordinary burdens carried upon waggons and other carriages, divers of the said highways are become very dangerous and almost unpassable." Because of the extraordinary burdens then carried, it was deemed wise to restrict the amount that any waggon might carry in future, making this less in winter than in summer.

later, which made provision for widening the roads to eight yards. The four-inch width of wheels was not made compulsory, for this provision of the Act was repealed in the Act of 1670, and we have numerous complaints for more than a century and a half after this that the narrow wheels were causing the ruin of the roads. The highway rate, which was to be levied in all parishes, unless sufficient proof were given to the Quarter Sessions that it was not needed, was to be continued for but three years from Mar. 25, 1662, after which it was expected that the ordinary statute duty would be sufficient. But we have found no clear case to show that this Act was ever put into effect. If it were, it must have been very rarely, for in 1670 Parliament once more essayed to get an assessment rate for the repairing of the roads[1]. This was to be levied at the discretion of the Quarter Sessions, where need required; it was not annually to exceed 6d. in the pound on lands or £20 of personal estate, and was to be limited to a period of three years. Like its predecessor, this Act failed to accomplish the purpose intended, and it was not till 1691 that the next attempt was made to effectively set in motion the system of highway rates, by re-enacting that clause of the Act of 1670[2].

But a new method was soon to be adopted for road maintenance, under the plausible pretext that those who used the roads should pay for their upkeep. Road tolls were not unknown ĩn the mediaeval times, but it was only in detached instances that they were employed[3]. They were soon, however, to have much more extended use; and as their employment was important till within recent times in England, the statute inaugurating this system may profitably receive some attention.

English road legislation arrived at another milestone when, in 1663[4], the first Turnpike Act was passed, entitled "An Act for repairing the highways within the Counties of Hertford, Cambridge and Huntingdon." After reciting that the ancient highway and post road from London, through York, to Scotland, ran through these counties, and that in many places this road had become ruinous on account of the great number of heavy loads that were drawn upon it[5], the preamble of the Act says that because "the ordinary course appointed by the laws and statutes of this realm is not sufficient for the effectual repairing and amending" of the road, and because the inhabitants of that section through which

---

[1] Act 22 Car. II, c. 12.   [2] 3 & 4 W. & M., c. 12.

[3] v. p. 9 ff.; also Clifford, *History of Private Bill Legislation*, I, pp. 4–5, and II, pp. 3–8.

[4] Act 15 Car. II, c. 1.

[5] This is substantiated by Ogilby's *Britannia*, p. 9, where it is said that after the first twenty or thirty miles out of London the road was generally bad.

this road lay were not able to repair it without some further provision of money, therefore it was considered best to help them to remedy these conditions by the means stipulated in the Act[1]. In each of these three counties, the Justices of the Peace were annually to appoint three surveyors; and within a week after notice of appointment, the surveyors of each county were to meet to consider what repairs were necessary for this highway. To fulfil the purpose of this Act, in each county one toll-gate was to be established on this road, and one toll-gatherer was to be appointed at each place to collect the amounts of toll fixed by the Act[2]. The toll collector at each gate was to be paid a moderate allowance, as approved by the Justices of that county; and all money he received for tolls was to be turned over to the surveyor of that county, under whom it was to be spent in necessary repairs for that road and not elsewhere[3]. The surveyors, in their turn, were to send an annual account to Quarter Sessions of all funds received and spent, and for good services they might be remunerated by the Justices in Quarter Sessions. Penalties were imposed upon those who refused to pay the toll, upon toll collectors who did not duly pay over their receipts to the surveyors, and upon the surveyors who did not render account of their trust.

For the more speedy repairing of the road in these three counties, the surveyors in each county were given authority, with the consent of the Justices, to mortgage the profits of the toll for a period of not more than nine years, to anyone who would advance their present value in money; and if the necessary money could not be borrowed in this way, the Quarter Sessions might levy a rate that would be sufficient to accomplish the purpose when applied by the surveyors.

But although this new method was introduced, the old system of statute labour, as stipulated by the Act of 1662, was still to be in effect, with some slight modifications.

---

[1] The important features of the Act are given in what follows.

[2] The rates of toll were fixed as follows:

For each horse	..	..	..	..	1*d*.
,, coach	..	..	..	..	6*d*.
,, waggon	..	..	..	.. 1*s*.	
,, cart	..	..	..	..	8*d*.
,, score of sheep or lambs			.. 1*s*. 2*d*.		
,, ,, oxen or neat cattle		..			5*d*.
,, ,, hogs	..	..	..		2*d*.

[3] It is regrettable that the plan approved by this first turnpike Act applicable to a considerable length of road—in that case through three counties—was not continued, instead of parcelling out the roads into short divisions under independent bodies of trustees.

The Act of 1663 was to continue in force not longer than eleven years, at the end of which time toll collectors were to pay all money in their hands to the Justices, who should use it as they thought best; but if, before the expiration of that period, the road in any county was sufficiently repaired, tolls were to cease in that county.

It will be noted that, in this statute, no provision was made for the appointment of turnpike trustees; all the functions which they assumed at a later day were now performed by the surveyors of the highway; and it was not till early in the eighteenth century that we have the appointment of trustees, under whose authority and control the improvements of the turnpikes were carried out.

How this first turnpike Act was received is a matter of conjecture. We have but little evidence that it was ever put in force in Huntingdon-shire[1]. In Cambridgeshire, the gate was erected at Caxton, but it was found to be so easy of evasion that no toll was collected. It would appear that only in Hertfordshire was there any substantial result. In that county, several gentlemen, finding that people did not want to lend money on the security of the toll, borrowed £1300 on interest, and expended this amount, together with the amount of the tolls for the first two years, by which means the part of the road which lay in that county was so amended that from a road impassable it came to be "to the satisfaction of all that travel that way[2]." In 1665, the term of the trust, so far as Hertfordshire was concerned, was extended from its original limit of eleven years to twenty-one years, and at the same time power was given to the Cambridgeshire authorities to remove the Caxton gate to Arrington. Both of these Acts were allowed to expire, and the toll-gates were removed; but by 1692 the road had again become "dangerous and impassable" because of the heavy loads carried along it, and another Act was passed reviving for Hertfordshire the former powers for a term of fifteen years[3].

Turning from this special legislation to the statutes that were in force generally throughout England, we are led to infer that there was

---

[1] Ogilby, *Britannia*, p. 9 (1675), tells us that there had been a certain late imposition upon travellers for three years, at Stilton, in Huntingdonshire and a place or two between that and London; but whether he is speaking of a toll actually collected, or of an imposition merely laid by Parliament, is hard to determine, although it seems more probable from his words that it was a toll actually collected during that three years, since the Act had been in force for twelve years. On the other hand, Webb, *The King's Highway*, p. 115, says that the Stilton gate excited so much local opposition that it was never erected; but no authority is given for this except the author's own *ipse dixit*.

[2] Preamble of Act of 1665 (16 & 17 Car. II, c. 10).

[3] Act 4 W. & M., c. 9 (1692).

considerable laxity in their enforcement by the surveyors. Apparently, also, parish residents endeavoured at times to evade their liability for the performance of statute labour on the roads. In consequence of this, in 1670 an Act was passed, demanding that all constables and surveyors of highways should see that all existing laws relating to the roads were put into effect, under penalty of a fine of two pounds for each offence[1]; and that Justices of the Peace should impose stipulated penalties on those who failed to give the required days of statute labour each year. Any person who forcibly resisted the officers in putting into effect the highway Acts was also required to pay the penalty of two pounds. This Act repealed the Act of 1662, concerning the four-inch width of waggon and cart wheels, and required carriers using more than five horses on a waggon to have them arranged to draw in pairs, not singly one behind another[2]. This, it was thought, would tend to prevent the roads from becoming worn down along one track.

In 1691 we have another Act passed, complaining of the impassable condition of the roads, due to some ambiguities in the laws and to insufficient provision for compelling the execution of them[3]. To remedy these conditions, one of the principal changes made was to increase the number and amount of the penalties for failure to comply with the laws; for example, surveyors who failed to put in due execution the existing laws were to pay a fine of five pounds, instead of the previous forty shillings. A change was made in the method of appointment of surveyors, for instead of being appointed directly by the parish, they were now appointed by the Justices of the Peace at special sessions, from a list of suitable men supplied by the parish. To secure more certain and speedy repair of the roads, surveyors were to give notice in church of any defaults or annoyances in the highways, bridges, water-courses and hedges, and if these were not corrected within thirty days, the surveyor was to repair them, and to be repaid by the persons who ought to have done it. In further aid of this, Justices of each county were to hold Petty Sessions every four months, at which surveyors' accounts were to be presented, and the surveyors were to be charged to do their duty.

In this Act, for the first time, we find a reference to the "waggoners and other carriers, by combination among themselves, having raised the prices of carriage of goods in many places to excessive rates, to the

---

[1] Act 22 Car. II, c. 12.

[2] Act 30 Car. II, c. 5 repealed the words "for hire" (applying to waggons and carts carrying for hire) and made the enactments apply to all carriages carrying goods, whether "for hire" or not.

[3] Act 3 W. & M., c. 12.

great injury" of trade; to prevent which the Justices of each county, once a year, were to settle the rates of carriage, and these rates were to be posted up in public places where all might see them. Any carriers taking above that rate were subjected to the penalty of five pounds for each offence[1].

Although the first turnpike Act was passed in 1663, it was not till the session of Parliament in 1695–6 that the second Act of this kind was passed, for the repair of part of the ancient post-road between London and Colchester[2]. As in the former Act, so here, after reciting the dangerous condition of the road, and that the ordinary legal means of repairing it were not sufficient, without some other provision of money, authority was given to the Justices for the appointment of surveyors, collection of tolls[3], etc.; and the surveyors might borrow money on the credit of the tolls, to be repaid, with interest at six per cent., from the yearly tolls. From this time on, it was the great arterial highways, and not the cross roads, which claimed most attention, and on which toll-gates were established. But, as a matter of fact, only a very few turnpike roads were authorized before the early years of the eighteenth century. Probably the chief reasons why more turnpike roads were not sought by those who wished better means of communication were, the opposition to the toll-gates, and the fact that where such a road was established and the taking of tolls authorized, a special Act of Parliament was necessary in each case, and this Act was not to be obtained without the expenditure of a considerable amount of money therefor.

The Act of 1662[4], which, among other things, made provision for the widening of the roads to eight yards from ditch to ditch, or from hedge to hedge where there were no ditches, had expired, and in 1697 another Act was passed which provided that the law for widening the roads might remain still in force[5]. But there were a few modifications

---

[1] Act 8 W. & M., c. 12, sec. 23.

[2] Act 7 & 8 W. III, c. 9.

[3] The tolls in this case were slightly different from those of the Act of 1663. Essex was an agricultural county, and probably for that reason the tolls imposed on animals passing through the toll-gate were higher, to produce more revenue. The tolls on waggons, horses, etc., were the same as those of 1663, but here there were the following differences:

for every score of sheep or lambs			..	1*d*.
,,	,,	calves	.. ..	3*d*.
,,	,,	hogs ..	.. ..	3*d*.
,,	,,	oxen or neat cattle	..	6*d*.

[4] Act 14 Car. II, c. 6.

[5] Act 8 & 9 W. III, c. 16.

made by the latter Act, which we may briefly notice. While, in the former Act, the land taken into the highway from the adjoining land was to be sufficient to make the highway fully eight yards wide, in the Act of 1696–7 the Justices had authority to widen any of their roads "so that the ground to be taken into the said highways do not exceed eight yards in breadth;" in other words, this Act did not stipulate any minimum width of road, but determined the maximum width of land that might be taken into the road from the property of an adjoining landowner. At the same time, in obtaining such land, there was a prohibition against pulling down any house or taking away the ground of any garden, orchard, court or yard. The earlier Act decided that such land should be paid for at a rate not exceeding twenty years' purchase, but the later at a rate not exceeding twenty-five years' purchase; and in both cases the money required for purchasing such land and repairing the road might be obtained by assessment[1]. An entirely new provision was also made in the Act of 1696–7, "for the better convenience of travellers," by giving the Justices authority to order surveyors to put up at cross roads a stone or post, with an inscription in large letters showing the name of the next market town to which each of these roads led. This would seem to be an evidence that the amount of travel had considerably increased in the latter half of the seventeenth century.

We are not surprised at the reiterated complaint that the highways were not improved, notwithstanding the good legislation, when we remember the persistent efforts to thwart the laws. For example, in 1670[2], it was enacted that no waggon, cart or carriage should be drawn on any highway with above five horses at length, and if any person should use a greater number of horses or oxen they should all draw in pairs. To avoid the intention of this Act, the waggoners had fixed an iron or shaft on the side of the waggon, by which none of the horses would go in a line with the wheels; and those which would draw on the side would only make the ruts deeper, and thereby more impair the highways[3]. In 1696, to reinforce the purpose of the preceding law, another Act was passed stipulating that all waggons, carts, or carriages carrying for hire (except those used in husbandry or in His Majesty's service) should not be drawn with more than eight horses, or their

---

[1] The assessment or assessments were not to exceed in any one year the rate of 6*d.* in the pound of the yearly income of any land, houses, etc., nor the rate of 6*d.* in the pound value of personal estates (money, goods, etc.). The rate of this assessment was the same in the two Acts.

[2] Act 22 Car. II, c. 12, sec. 7.

[3] Preamble to Act 7 & 8 W. III, c. 29.

equivalent in horses and oxen; "which said horses, or horses and oxen, shall draw in pairs with a pole between the wheel-horses or in double shafts," and the other horses were to draw in line with the wheel-horses, under penalty of forty shillings for every offence[1]. In 6 Anne, c. 29 it is stated that this proposed remedy had proven impracticable in many parts of the kingdom, and therefore it was repealed; and, instead, it was enacted that no travelling waggon, or cart (with the usual exceptions) should be drawn by more than six horses, oxen or beasts, under penalty of five pounds. But the Justices of the Peace might license a greater number to draw uphill.

But while the waggoners and carriers actively sought means of evading the law, the surveyors and others were constantly negligent in seeing that the law was enforced. No special knowledge of road construction was asked of the surveyors, and most of them had not the opportunity to apply themselves to this. Nor was there much inducement for them to become more efficient in this work, for although they were sometimes permitted to receive what the Quarter Sessions saw fit to give them, their services did not necessarily need to be remunerated. Then, too, the office of surveyor was held by him for only one year, and his public work was arduous in travelling over the section which was under his care, once in three months, while the time occupied in performing his public duties was often a serious detriment to his own work at home. Under these conditions, the surveyor would need to be a very public spirited man if he fulfilled the work of his office according to the tenor of the statutes.

That the money intended for improving the roads was sometimes misapplied, would seem to be evident from the language of the constantly increasing mass of legislation; and a statute of the early part of the reign of George I imposed the heavy penalty of five pounds for misappropriation of such funds[2]. In all probability, this was one of the things referred to in the statutes when the latter repeat

---

[1] Act 7 & 8 W. III, c. 29. Under Act 22 Car. II, c. 12, there were also some abuses, which were set forth in a petition to the House of Commons from the carriers on the western roads (1695–6). They complained of the heavy sums extorted from them by informers and others who were or pretended to be surveyors of the highways, or to be authorized by Justices of the Peace, and who, under the pretence of forfeitures for breaches of 22 Car. II, c. 12 by the carriers using more horses than were allowed by that Act, levied on them considerable sums of money, expressing at the same time their willingness to connive at a breach of the law to any extent, if their demands were complied with. This petition was referred to a Committee of the House of Commons, who examined witnesses and discovered systematic and impudent extortion. Hence in Act 7 & 8 W. III, c. 29, it was directed that forfeitures should in future be paid to the surveyor and to no other person.

[2] Act 1 Geo. I, stat. 2, c. 52, sec. 5.

again and again that there were "some neglects in the execution" of the laws, because of which the highways were not so fully repaired as it was intended they should be.

The history of the English road legislation seems to confirm the view that where one statute was passed, it required to be followed by one or more other statutes to amend it, to supplement it, or to aid in its execution. The inevitable tendency, therefore, was to have a series of related pieces of legislation. This was shown in the early part of the reign of Geo. I[1], by an Act which recited the great evil done to the roads by the heavy loads drawn by six horses, and enacted that carriers' waggons should not be drawn with above five horses at length. This brought the law back to what it was under 22 Car. II, c. 12. The same trend of legislation is seen again in 1718, when a new statute was passed[2], reciting that previous Acts for "better repairing and amending the highways within this kingdom, and for preventing carriers and waggoners from carrying excessive burdens" had proved "wholly ineffectual," and providing that after June 24, 1719, no waggon travelling for hire should be drawn with more than six horses, nor cart with more than three horses, on penalty of forfeiting to the seizor all the horses above that number, with their equipment[3]. And as one great occasion of bad roads was the narrow tires set on the wheels with rose-headed nails, this Act prevented any waggon with a tire less than two and one-half inches wide, fastened with these nails, from being drawn by more than three horses. From this time on, the width of the wheels of waggons used on the roads began to receive more attention.

In regard to this Act of 1718, there was considerable complaint by the carriers. In accordance with what they regarded as its terms, the carriers got their wheels bound with tires two and one-half inches wide, not doubting but that they would be allowed to wear out these tires. But later they were informed that, according to the letter of the law, the tires were to be not less than two and one-half inches wide when worn out; and if this were the case, they would be put to considerable expense several times a year getting new tires[4]. But, further, the Act made it possible for idle persons, who would not work, to maintain themselves by a sort of parasitic existence: they would watch for the carriers' waggons at the regular time they were accustomed to come along the roads, and if the wheels were worn even to the least extent

---

[1] Act 1 Geo. I, stat. 2, c. 11.

[2] Act 5 Geo. I, c. 12.

[3] This applied, whether the horses drew at length, or in pairs, or sideways.

[4] Brit. Mus. 356. m. 1 (66), 'Case of the Carriers and Waggoners who carry Goods to Hire.'

within the limit of two and one-half inches in breadth, they would take off one or more of the carrier's horses and detain them until they had extorted from the carrier a considerable sum of money, and frequently they would follow and meet the waggons for that purpose[1]. In other cases these vagrants would hide themselves near some places where the roads were so bad, or the ascent of the hill was so steep, that the carrier had to hire a horse or two to help draw him through or up the difficult way, where five horses were not enough for the load carried; and then, some time afterward, they would inform against the carrier and force him to pay the 40*s*. penalty each time, of which amount they would claim as much as they could get or keep[2]. These conditions were so aggravating to the carriers, that they presented their case to Parliament, and urged that if not allowed to wear out their tires they might have the privilege of raising the price of carriage of goods; that instead of paying these fines to such idle persons, who lived debauched lives, they should be paid to the landholders adjoining the highways, who would use the money in the repair of the highways; and that, in order to make their occupation pay expenses, they be allowed to travel with six horses to a waggon, since thereby they could draw a weight that would be remunerative[3].

Beginning with the eighteenth century, the Acts passed for establishing turnpike roads became more prominent. Before that time only four such Acts had been passed; but the changes from that time on were such as to accord with the increasing progress of industry which was initiated after the close of the Revolution of 1688. The number of these turnpike and other road Acts will appear from a statistical summary which is elsewhere given[4]; but we may say here that the agitation for better roads, which made considerable advance in the reigns of Anne and George I, continued to gain force as the century wore on. Down to 1702, turnpikes were authorized by "public" Acts, which are found among the other statutes of the regular collections; but from 1702 to 1720 they were all authorized by "private" Acts, of which at most only a few copies of each were printed, and as comparatively few of these are now found our knowledge of this aspect of the legislation is not so complete as we would wish[5]. In nearly every case,

---

[1] See last footnote.

[2] Brit. Mus. 1879. c. 4 (28), 'Case of the Waggoners of England.'

[3] See Brit. Mus. 356. m. 1 (66) and Brit. Mus. 1879. c. 4 (28) mentioned above.

[4] See Appendix 13.

[5] Probably the largest collection of these Private Acts is to be found in the British Museum. See the Acts found in the large volumes press-marked 213. i. 1, 213. i. 2, etc.; also in B. 263 series. From 1720 to 1753 the turnpike Acts were private but were printed and bound with the public general statutes; from 1753 to

however, authority was given for the taking of tolls for twenty-one years, and for borrowing money on the credit of the tolls, so that the repair of the roads might be the earlier effected.

Up to the year 1706, the turnpike Acts gave the Justices of the Peace the supreme authority for the administration of the system, and under them the surveyor was to act in laying out the funds that were collected at the gates. But in that year there was the institution of the first turnpike that was to be administered by a special body of trustees[1], and this was followed by the second and third in 1709[2] and 1710[3]. During the time following 1711 this new method of looking after the turnpikes completely superseded the older, and the Justices were relieved of this burden as a body. It frequently occurred that Justices were appointed as members of such boards of trustees, but in that case they were not acting in their official capacity as magistrates. For over a century and a quarter this new method of administration prevailed, and we may now consider some elements of it in a little more detail.

With the passage of a turnpike Act intended to benefit a particular piece of road, there was named a number of prominent interested men who, in their corporate capacity, were called the turnpike trustees; and it was their duty to see that the Act was put into effect. The portion of road over which they were given jurisdiction was called a turnpike trust, and with the great increase in the number of these Acts there was a vast multiplication of turnpike trusts, each of which was usually but a few miles in length. As a rule, they were not continuous for any great length of road, but portions of the road that were under the turnpike legislation would alternate with other portions that were not. The authority of the trustees, therefore, was confined to their own small piece of road. Upon it they could establish toll-gates and take tolls according to specified rates; and the revenues from these tolls were to be expended by the special surveyors who were appointed by and acted under the trustees. It was not intended that the parochial obligation for statute labour should be by this means abrogated; but merely that additional revenue should be obtained, the expenditure of which would supplement the efforts of the parish in maintaining good

1798 they were Private Acts, but were bound separately; and from 1798 onward, they were grouped as Acts Local and Personal.

[1] This Act of 1706 is one of the Private Acts which have not been printed. See *Journal, House of Commons,* 1707, Mar. 3 and 27, and 1710, Feb. 15 and 24. It is also referred to in preamble of Act 3 Geo. I, c. 15.

[2] Act 8 Anne, c. 15.

[3] Act 9 Anne, c. 7.

roads. When the six days' statute duty had been performed by the parish under its own surveyor, then the special surveyors acting for the turnpike trustees could come and, with the revenues from the tolls, engage teams and men to do as much additional work as they thought requisite. By degrees, however, the special surveyors had transferred to them, for their small portions of road, most of the authority of the parish surveyors, and they were even given the right to require a certain proportion of the statute labour of the parishioners to be performed under their own direction[1]. From ,1716 on, the turnpike surveyor might agree with the parish surveyor to commute this specific share of statute labour into a money equivalent; so that the turnpike surveyor then had the expenditure of funds, part of which were obtained from the tolls and part from the parish rates.

In fulfilment of their trust, the turnpike trustees, when they established toll-gates on the piece of road over which they exercised control, might either appoint their own toll gatherers and their own surveyors, to assume these functions under them; or else they might farm out the tolls at each gate for a definite amount and apply the revenues by letting out contracts for the repairing of the road. The surveyors were expected to account to the trustees for all money received and disbursed; but the trustees in their turn did not need to account to any higher authority. They were supreme in the matter of financing their share of the road; and were not restricted by law as to the amount they could borrow on the credit of the tolls, nor in regard to the way such money should be employed[2]. Of course, while all turnpike Acts were temporary, usually for only twenty-one years, they could be renewed at the expiration of that time; but if the roads were sufficiently repaired before that time the Justices were to have the trustees remove the toll-gates and cease the taking of tolls.

Every improvement has had its opponents, and this is true also of the turnpike roads. In the early years of the establishment of turnpikes, as also in the later, organized bands of men, who were opposed to the payment of tolls on the roads, would collect at nights, would burn or otherwise destroy the toll-gates, and frequently burn down the houses of the toll collectors. So much terror did these men cause, that moneyed men were deterred from lending on the surety of the tolls, when they saw the insecurity of such a source of repayment. This prevailed to such an extent, that, in order to bring offenders to justice,

---

[1] Webb, *Story of the King's Highway*, pp. 117–18.
[2] Act 9 Geo. I, c. 11, shows what authority the Justices had over the roads.

an Act was passed in 1728, by which any person convicted of wilfully breaking down a turnpike gate (or destroying locks, flood-gates, etc., erected to preserve the navigation of rivers made navigable under Acts of Parliament) should be sent to the common gaol or house of correction for three months, and should be publicly whipped at the market cross by the keeper of the gaol or house of correction. If he were convicted of this offence the second time, he was adjudged guilty of felony, and, like other felons, might be transported for seven years[1].

But even this, and another Act passed in 1732[2], did not put a stop to such practices; and in 1785 Parliament increased the severity of the penalty[3], by enacting that persons maliciously destroying turnpike gates or other turnpike equipment, or any locks, flood-gates, or other works erected under authority of Parliament in navigable rivers, should be judged guilty of felony and should suffer death[4].

These turnpike riots, from 1735 to 1750 and after, seem to have been very fierce, and the annals of the time are full of instances of such wholesale destruction. The rioters came at times in such numbers that an armed force was necessary to restrain them, and even this might not be successful. Sometimes the toll-gates were defended by powerful guards of men, who were, nevertheless, unable to drive back their assailants. In other cases, the destroyers in a body took the toll-gates on certain roads, one after another, and totally demolished them. Such encounters were not infrequently attended by some loss of lives[5].

---

[1] Act 1 Geo. II, c. 19.

[2] Act 5 Geo. II, c. 88.

[3] Act 8 Geo. II, c. 20.

[4] In 1754, by Act 27 Geo. II, c. 16, the need for effective measures in such cases was recognized by making the above Acts perpetual.

[5] To show more precisely the nature of some of these riots, we have taken at random a few examples, which will serve to illustrate the general spirit of them all:

Monday, Sept. 22, 1735. "Ledbury turnpike, in Herefordshire, was pulled down by a large body of people, notwithstanding Justice Skip defended it with a good number of armed men, who killed two, and took two others of the rioters. Only two of his party were slightly wounded; but the populace threaten to burn his house and kill him wherever they meet him." (*Gentleman's Magazine*, v, p. 558.)

"The Commissioners of the turnpikes at Ledbury, in Herefordshire, being informed that an attempt would be made to pull them down, about eight in the evening repaired, with their attendants well armed, to that which leads towards Hereford, where a great number of persons provided with guns, axes, etc., advanced against them....Some of the rioters notwithstanding, began to assault the townsmen...." (Ibid., v, p. 618.)

Bristol, Aug. 7, 1749. "On Tuesday the 1st inst., at eight o'clock in the morning, about 400 Somersetshire people cut down a third time the turnpike gates on the

An important step was taken in regard to road legislation in 1741, when an Act was passed for the preservation of the public roads in England[1]. From the time of the passage of Act 13 and 14 Car. II, c. 6, up to this time, the only way tried to limit the weight of loads was by limiting the number of draught animals by which they were drawn. But additional means were now adopted for effecting this object, and the Act of 1741 gave trustees of roads authority to have built, at any or every toll-gate, weighing engines for weighing all carriages and goods passing through the toll-gate, and to take, in addition to the regular toll, a further duty of twenty shillings per hundredweight for all above sixty hundredweight, which extra payment was also to be applied to mending the roads[2]. It also provided that

Ashton road, and burnt the timber; then afterwards destroyed the Dundry turnpike, and thence went to Bedminster, headed by two chiefs on horseback,...; the rest were on foot, armed with rusty swords, pitch-forks, axes, guns, pistols, clubs," etc., etc.

See also Parsons, *History of Leeds*, i, pp. 128–9. The newspapers of the time are full of such incidents of riotous conduct toward the turnpikes.

[1] Act 14 Geo. II, c. 42.

[2] This extra toll was not to apply, however, to carts, waggons, or other carriages employed only about husbandry, nor to private covered carriages of noblemen and gentlemen, nor to waggons employed in the King's service.

There seems to have been much complaint against some of the provisions of this statute. It was said that the law would seriously affect inland traders, landowners, farmers, etc., for the taxing of over-weight would raise the price of carriage. In this way, inland traders would have to pay more for having their goods carried; and because commodities could not be carried to market upon nearly as good terms from inland towns as from towns near the sea-coast, the landowners and farmers would in many instances suffer from want of marketing facilities, for prudent people naturally go to the cheapest market.

Furthermore, the carriers' waggons at that time weighed more than 25 cwt., which was alone sufficient for two horses' strength; and since the waggons were to be drawn by not more than four horses, the inevitable consequence would be that waggons would be set aside and carts would be used. This would only make the roads worse. Besides, the toll for a cart was nearly the same as for a waggon; and if two carts were sent to market instead of one waggon, the toll would be almost double and also the drivers' wages.

Again, the weighing machine might easily be manipulated by the finger of the weigher, so that the carrier might be deceived in his weight, and thereby an extortionate amount of toll might be demanded.

The Act was not to extend to the covered carriages of noblemen, gentlemen, etc. On account of this, complaint was made that to compel poor men, who had to make their living by carrying, to pay toll, while the rich went toll-free, was cruel and inhuman. To continue such a policy as this, would put the poor man and the small trader at a disadvantage, and trade would soon be in the hands of a few rich.

The request was also made that time should be given to all persons alike to wear out their old narrow wheels, and to those who had to make waggons to supply farmers or other persons, who were not carrying goods for hire, from April 15 to September 29, might use carriages with wheels of any breadth; but this provision was repealed the next year, because of the difficulty of convicting and punishing offenders against it[1].

Weighing machines formed a fruitful source of trouble, arousing much opposition and many attempts to evade the law which sanctioned them[2]. Immediately after the law was passed, men with heavy loads, approaching one of these nuisances, would unload part of their goods before driving on to the weighing engine, and then re-load after they had passed. In other cases, to avoid having their loads weighed and paying the extra duty, men would sometimes go out of their way, through narrow lanes and side roads, till they had passed such a place of payment. These things were followed by the passing of a new law in 1748[3], imposing a penalty of twenty pounds upon any person who thus endeavoured to avoid the intention of the statute. In reality, comparatively few of these engines had been erected, for it was merely optional with the road trustees whether they established them or not.

Despite the considerable sums of money spent on the turnpike roads, many of them could not be kept sufficiently repaired, because of the excessive weights allowed to be drawn upon them by the many horses which the law allowed to be used with carts and waggons. In 1751[4], legislation was passed which *required* trustees of the roads to *demand* and take, at all turnpike gates, twenty shillings per hundredweight for every waggon or other carriage drawn by six horses, over and

themselves with the materials of the proper width. (See Newball, *A Concern for Trade*, pp. 13–25.)

Another writer of the time voices almost identically the same opinions as Newball. He showed also the fallacy of limiting the number of horses to four, for he said that six or seven horses of one man might be weaker than the four horses of another man. (v. Phil' Anglus, *The Contrast, etc.*, pp. 6–27.) Both these men wanted the publication of the accounts of the turnpike trusts.

[1] Act 15 Geo. II, c. 2.

[2] The same thing continued into the first quarter of the nineteenth century. They were once or twice abolished, only to be re-established.

[3] Act 21 Geo. II, c. 28. This Act was necessary also to enable trustees to erect weighing engines at other places on the road than the toll-houses. The former limit of weight (60 cwt.) was retained. Additional powers were introduced to enforce the weighing of carriages that were subject to the operation of the Act. Every common waggoner or carrier was to have his name painted on his waggon or cart.

[4] Act 24 Geo. II, c. 43. In this Act turnpike roads are first mentioned as distinguished from other highways; part of the title of the Act reads: "for more effectual preservation of the turnpike roads, and for the disposition of penalties given by Acts of Parliament relating to the highways."

above the tolls or duties already granted[1]. Any person who should
be found to have taken off any horse or horses before coming to the
turnpike gate, with intent to avoid paying the additional toll, was
required, upon conviction, to pay to the informer five pounds. No
waggon, cart, or other carriage was to be driven out of the turnpike
roads to avoid payment of the legal tolls and duties, upon penalty of
forfeiting any one of the horses, except the shaft horse. Under this
Act, it was no longer optional, but *required*, that trustees should erect
weighing engines at one or more turnpike gates, or other convenient
place or places within their district, and should weigh all waggons or
other carriages that were not exempted, and take the regular toll as
well as the additional duty for extra weight. But trustees of roads
beyond thirty miles distance from London were not required to erect
such weighing machines. Even this legislation did not put a stop to
the abuse of the roads in carrying heavy loads. The carrier did not
always know the weight of his load when he started on his journey;
or he might have been well within the limit when he set out with his
load, but have added to it along the way by taking the goods of other
customers until he had more than the legal sixty hundredweight for
passing through the gate. Under these, and similar circumstances,
what was more natural than that the carrier should make a private
agreement with the toll collector to give him a small recompense on
condition that he would allow the waggon to pass without weighing?
By the connivance of the keeper of the weighing machine, many devices
could be resorted to by which the purpose of the law might be evaded,
and the method adopted would prove mutually profitable for both
keeper and carrier. As time passed, new means were found to avoid
the payment of the extra, or "extraordinary," toll as it was called;
and both the weighing machine and the law which required its maintenance proved fruitful sources of deception.

By the middle of the eighteenth century the traffic on the roads
was quite extensive; and it was thought that if the means being used
for repairing the highways were to be most effectual, the heavy burdens
must not be carried on waggons with the prevalent narrow wheels.
Accordingly, by a statute of the year 1753[2], it was enacted that the
wheels of waggons, carts, and other carriages using the turnpike roads[3],
must be *nine inches broad*, under penalty of five pounds or forfeiture
of one of the horses[4]. Trustees were to lessen the *extraordinary* tolls

---

[1] The same exceptions were made as those noted in Act 14 Geo. II, c. 42.
[2] Act 26 Geo. II, c. 30.
[3] Except those exempted under Act 24 Geo. II, c. 43.
[4] An exception was made by prohibiting the forfeiture of the shaft horse.

on carriages with broad wheels; and they might order the width of the wheels of waggons and other vehicles to be measured at any turn-pike gate. Surveyors were required to fill in the ruts in the roads[1], and to widen the roads where necessary, the charges to be paid out of the tolls. All officers found to be negligent in the performance of duty were to be removed; and coupled with this was the express prohibition that no keeper of a public house was to fill any place of trust under the tolls, or to farm the tolls[2].

Against the demand for broad wheels there was considerable com-plaint; for those who had been using the narrow wheels did not want to throw these away, especially when almost as good as new, and replace them by the wide wheel. Then too, there were places where the narrow wheels were the better, as, for instance, in husbandry and on stony roads; and the owners did not want to have to use the broad wheels at some times and then change to narrow wheels at other times. Yet in opposition to these complaints the broad wheels were represented as better for the roads, since, by rolling down a wider surface, they would tend to consolidate the road-bed. To further the adoption and use of the wide wheels, it was deemed advisable to amend the Act of 1753, two years after it was passed[3]. Under this new law, waggons, etc., with wheels nine inches broad, were exempted from payment of toll for three years; while waggons with wheels six inches broad might be drawn by six horses, and carts with wheels six inches broad might be drawn by four horses, and pay reduced tolls. If with this decrease of tolls, trustees should find that their revenues were not sufficient, they were authorized to raise the tolls by one-fourth on all narrow-wheeled vehicles; and the latter were not to be allowed to pass without weighing[4]. Thus we see that while special inducements were given to those who would use wide wheels, there was mild pressure put upon those who used the narrow wheels, to urge them also to comply with the intention of the law to secure the adoption of broad wheels. But the framers of this Act were also convinced that if the number of horses used on each waggon were to be reduced, there would be less occasion of the roads being cut up; and in accordance with this opinion it was enacted that waggons, or other four-wheeled carriages, not being

[1] Some idea may be formed of the miserable state of the roads, due to ignorance or negligence, when a legislative enactment was thought necessary to enforce the levelling and filling up of the ruts.

[2] In extenuation of this prohibition, see the statements made on page 55, in regard to innkeepers as surveyors.

[3] Act 28 Geo. II, c. 17.

[4] If collectors of tolls allowed narrow-wheeled carriages to pass without weighing, they were to be committed to the house of correction for one month at hard labour.

common stage waggons, should be drawn by five horses; but if drawn by more, the owner was to forfeit five pounds for every offence, and the driver was to be committed to the house of correction for one month. In this Act is the first notice of the qualification for trustee. It says that great mischief had arisen from mean (i.e. poor) persons acting as trustees, and the qualification now enforced was the possession of land of the yearly value of £40, or personal estate of £800. Some parts of this Act were altered by 30 Geo. II, c. 28, but always in favour of the nine-inch wheels, for, within one hundred miles of London, the toll on carriages with nine-inch wheels was reduced one-half.

Despite the benefits expected from the nine-inch wheels, new difficulties arose from their introduction. Being calculated, from their additional strength, to bear considerably heavier weights, it was found necessary to limit their size and their width; and therefore all waggons were prohibited having the wheels wider apart than five feet six inches from the middle of the fellies of the wheels on each side. They were also to be drawn by horses in pairs, while carriages with narrower wheels were to be drawn by the team at length.

We have now discussed the character of the legislation under which the roads were to be supported by the public, and some of the difficulties which arose in the enforcement of the laws. But some of the roads were constructed and repaired by private individuals at their own expense; and roads of this kind, when constructed by wealthy landowners, were almost invariably models, both in construction and maintenance. In some cases, large sums of money were expended by individual landowners in such enterprises; and the effects of these public-spirited activities became manifest, at least locally, in the facilitation of travel and in furnishing object lessons of good roads[1].

---

[1] The road into the eastern end of Fordington, in Dorset, being through deep water, by which the lives of people were endangered, and horses injured, an Act was passed, empowering Mrs Lora Pitt to make a new road through Fordington Moor, 1900 feet long and 36 feet wide; which was done at the expense of £1500. She had promised to open and keep up the road for three years to make it a public highway. It was begun in 1746 and finished in the next year. She also built a bridge of three arches over the river Frome, under authority of Act 19 Geo. II, c. 24. See *Journal, House of Commons*, xxv, p. 59; also Hutchins, *History of Dorset*, I, pp. 573–4.

In 1740 the corporation of Nottingham made the south entrance into that town much wider and more convenient. It had formerly been a narrow passage cut out in the rock on which the town stood, where only one coach or waggon could pass at a time; but now, when widened, it was made so that in some places three or four carriages could easily give way to each other. They were animated thereto by the generosity of Lord Middleton, who, the year before, had, at his own expense, levelled part of the sand hills, and thereby much enlarged the entrance to the town from the west. Deering, *Nottinghamia Vetus et Nova*, pp. 267–8.

This could not be said, however, of the roads which were maintained by private persons under the necessity of duty[1].

The Act of 1662[2] provided that if the ordinary statute labour were insufficient for repairing the roads, the surveyors, with the aid of two or more substantial householders, might levy one or more assessments upon the parish, for the purpose of obtaining the necessary funds for their work. The amount of the assessment in any one year was

In the Wolley MSS. (Brit. Mus., Add. MSS. 6692, p. 180), we find a letter from Edmund Evans to Mrs Turnor written from Bonsall, July 10, 1738, showing us some interesting features of the private road enterprise in Derbyshire:

"We have lately been very busy in making a coach or waggon road from Bonsall to Crumford, from which place there is one already made to Mat Bath (wh was done at Mr Pennell's expense, who built the Bath) & another from Crumford to Swanwick made mostly at Mr Turnor's expense for the encouragement of his cole trade. So when ours is completed it will make a through passage from Matlock Bath to Buxton & likewise neigh to the colepitts, to Nottingham or Derby, or where else they have occasion this way. This rough piece of work is not done by any levey, but chiefly by the Miners, who have no wage, but all come to assist, some at the instance of one fd (friend?) and some another who goes along with them & assists the overseer on their respective dayes: the gunpowder they blow away & the ale we allowed 'em is paid out of a collection some of us have made amongst ourselves, only Mrs Hallam (who keeps the Bath) hath sent a guinea & Mr Moore of Wimper ten shillings.

" I have not yet heard any of 'em mention any expectation they had of anything from your Ladyship, neither do I think they will, yr Ladyships late bounties being (I hope) not so soon forgot. And yet (tho' I am under the greatest obligation to be silent) still I beg leave to tell you L— that I think you cd never better bestow a guinea of 'em than now, which wd be enough, & (if pleased to order it) would please 'em more than a gter thing another way or from any other hand."

[1] Usually the repair of the roads was to be effected by the occupiers of lands in the parish where the roads lay. But, according to Burn, particular persons might be subject to the charge of repairing a highway in two cases:

1. When the land, that had formerly been used as a road, was inclosed. When the owner of lands not inclosed, adjoining the highways, finally inclosed his lands on both sides of the road, he was bound to make a good way, and was not excused for making it as good as it had been at the time of the inclosure, if it had then been at all defective. The reason for this was, that before the inclosure, the people were accustomed, when the road was bad, to choose a better road over the adjoining fields; but by inclosure this liberty had been taken away. If after inclosure the way were not sufficient, any passenger might break down the inclosure, and go over the land, and justify it, until a good road were made.

2. A particular person might be bound to repair a highway because of a prescription; that is, if the owner of certain lands had been accustomed to repairing and maintaining a road, his descendant who came into possession of this property by inheritance would be compelled to bear the burden of maintaining the road. The obligation to repair such a way followed by reason of the tenure of the land. (Burn, *Justice of the Peace*, I, p. 511, gives the cases upon which these facts were based.)

[2] Act 14 Car. II, c. 6.

not to exceed sixpence in the pound, according to the real value of the property assessed, or its equivalent in personal estate; and such assessments could be made only with the consent of the Justices in the Court of Quarter Sessions[1]. It was also within the power of the Justices in Quarter Sessions to order an assessment for the repair of roads, even if the parish surveyors or the parishioners failed to petition therefor[1]. Some parishes commonly used this means of improving or keeping up their roads[2]. As a general thing, we find very few assessments for this purpose after the first quarter of the eighteenth century; it would seem that after that time a parish preferred to obtain a turnpike Act under which to repair its roads, rather than levy a direct assessment upon the residents of the parish. The reason for this course is very obvious.

With the foregoing knowledge of the nature of the legislation down

[1] The parish of Ashbourne, in Derbyshire, had refused in 1713 to make a levy for the repair of its highways, and the Quarter Sessions ordered the assessment, as follows:

"Whereas the inhabitants of Ashbourne in this County have made it appear to this Court that they have already done their six days work apiece towards the repairing of their highways (pursuant to the Act of Parliamt in that case made and provided) & it proveing insufficient to amend the same This Court doth order & it is hereby ordered that the sum of sixpence in the pound be raised by assessment for & towards the repairing and amending thereof." (Cox, *Derbyshire Annals*, II, p. 230.)

See the petition of the supervisor of the roads of two townships in Chesterfield, Derbyshire, as given in Cox, *Derbyshire Annals*, II, p. 227.

Sometimes the inhabitants of a parish, not the surveyors, petitioned the Quarter Sessions for an assessment. Note, for example, the petition from the inhabitants of Calow in 1650, as given in Cox, *Derbyshire Annals*, II, p. 227.

The old form of assessment of the highway rate may be well illustrated by an entry from the records of the parish of Twickenham. On May 5, 1673, it was ordered "that for the highways it is agreed with the consent of the whole vestry that:

The laborers of the parish doe pay	2s.
The yeomen not laborers doe pay	4s.
The gentlemen doe pay	6s.
Mr Browne by reason of land doe pay ..	15s.
Mr Knight and others of the better degree ..	15s.
The Lords	20s.

The parish to pay £30, and to be allowed 8s. a day for their worke. The labourers doing a full dayes worke 16d." Cobbett, *Memorials of Twickenham*, pp. 193–4.

[2] Pontefract was an example of this. See Holmes, *Pontefract*, pp. 69, 73, 141, 146, 181, etc. Cf. also Latimer, *Annals of Bristol in the Seventeenth Century*, p. 10.

In the parish of Sefton, in Lancashire, the assessment called the fifteenth was the usual means adopted for raising money for the surveyors of the highways, as well as for several charitable purposes. The tax was not very heavy, for in 1719 four fifteenths and a half amounted to only £4. 3s. 6d.; and in 1749 twenty fifteenths, collected for the highways, amounted to only £19. 0s. 10d. Horley, *Sefton*, p. 110.

to the middle of the eighteenth century, we proceed to inquire how the law was administered, and what effect it had upon the roads of the kingdom.

The condition of the roads did not seem to be taken very seriously until after the rebellion of 1745, when it was seen that the Highlanders could get down nearly to the centre of England before the news could reach the rest of the realm.   But no sooner was the rebellion put down than the Government turned its attention to bringing the Highlands into subordination, and for this the construction of roads was indispensable.   From that time, though slowly, the construction of the great thoroughfares between the north and the south made steady progress[1].

But the extension of the turnpike system met with great opposition, for people regarded it as a restriction upon their freedom of movement from place to place;   and prejudices were so strong, that in some instances the country people and stage-drivers would not use the improved roads after they were made[2].   Petitions were also presented to Parliament against extending the turnpike system[3].   Near London, the agricultural classes did not want the turnpikes continued back into the country, for that would destroy their monopoly of the advantages of their improved means of communication with the capital.   They thought that if the remoter counties should obtain the benefit of better and easier travelling facilities, the greater cheapness of labour there would enable the distant farmers to undersell them in the London market, and thus they would be ruined.   But even this opposition from those who represented "vested interests" was powerless to prevent the advance.

In 1752, the House of Commons appointed a Committee "to inquire into the management and application of all such sums of money as have been collected within ten years last past."   On Mar. 12th of that year, this Committee made a report to the House[4], from which we may gather a few reasons why the roads were not better than they were. The Kensington turnpike received in tolls and compositions for tolls, in 1749, £3383. 1s. 5d., in 1750, £3230. 18s. 2d., and in 1751, £3146. 16s. 8½d. The whole fifteen miles of that road could have been kept in repair for £1500, that is, £100 per mile;   but there was a remaining debt of £3300,

---

[1]   The greater amount of road construction after this date may be gathered from the increase in the number of road Acts which were passed after these years, particularly from 1750 on.   See Appendix 13.

[2]   *Journal, House of Commons*, Report of Committee on Old Stratford road, Apr. 22, 1714.

[3]   Adam Smith, *Wealth of Nations*, Book I, c. XI, Pt. I, p. 148, in Cannan's edition.

[4]   *Journal, House of Commons*, XXVI, pp. 490–3.

to the payment of which the treasurers applied any overplus, when they had a balance in their hands. The annual income at the turnpike on the road from Cranford-bridge to Maidenhead-bridge amounted to nearly £900, and the road was thirteen miles. The business of this trust had usually been transacted by several commissioners, who were small farmers, and who, until the preceding January, had never paid toll, either for themselves or their families; but a late order for making them pay had increased the tolls by between £3 and £4 per week. Their treasurer had lately absconded with £857. 3s. 10d. of the trust money, which had been allowed to remain in his hands, though the trust, during that time, had paid four per cent. for £2500 which was the debt still due on the tolls of the turnpike. The annual receipts at the Puddle-hill gate, on the road from Dunstable to Hockliffe, upon an average of seven years, amounted to £583. 7s. 11d. per annum; but the previous year (May 1751), the old collector having died, a new one was chosen, since which the tolls had risen to 40s. a week in summer and to nearly 20s. a week in winter more than in former years. The road to be mended was three and three-quarter miles, and was still so bad that it would require at least £1000 to put it in good repair. The expenses of management had been lately increased from £46 to £83 per annum, chiefly due to increase in the salaries of collectors, treasurer and surveyor. As to the turnpike leading from Hertford to Basingstoke, the revenue was about £300 per annum; but by paying interest on £1200 at four per cent., paying also £190 a year in salaries, and other bad management, the net amount left to be spent on the road, fourteen miles in length, was only about £60. But the number of officers had been recently lessened, and the expenses of management had been reduced from £190 to £55 per annum; and instead of having, as formerly, a clerk and a treasurer each with a salary of £30 per annum, one person was at this time doing the work of both offices for £10 per annum. In nearly all the cases examined, officers gave no security for the proper fulfilment of their duties; and the Committee recommended that in future none but gentlemen of fortune should be made commissioners of turnpikes[1], and that they should take security of their treasurer for money placed in his hands and for the faithful performance of his duty.

Other instances of misapplication of turnpike trust funds are given

---

[1] In all probability, it was considered that the gentlemen of fortune would be subjected to but little temptation to dishonesty, and that they would be able to devote more time to the proper discharge of their duties in connexion with the roads. Act 28 Geo. II, c. 17 (1755), speaks of the evil of having poor persons acting as trustees and enacted that trustees should henceforth have a certain qualification by the possession of a stipulated amount of property or personalty.

in a Report from the Committee of 1765[1], in which, among other decisions by the members of the Committee, it was agreed that there had been great mismanagement of the public money in the repair of a certain portion of the Kensington road, and they urged that some alteration should be made in the execution of that trust. From the results of these and other similar investigations[2], we are led to infer that there was much dishonesty, that surveyors and toll collectors abused their offices for their own private ends, and that there was much failure on the part of the public officials who had the charge of the roads, to administer effectively and economically the trust reposed in them and the money contributed by the people.

But there were also other reasons why the roads did not profit as much as was intended by the well-meant legislation. A writer in 1754 puts the matter before us very concisely when he says: "It is but too notorious a truth, that as soon as a turnpike Act is obtained, all the parishes through which the road passes consider the Act as a benefit ticket, and an exemption from their usual expenses, and elude the payment of their just quota towards the reparation of the road, by compounding with the trustees for a less sum, or by doing their statute labour in a fraudulent manner; and in both these cases they are generally favoured by the neighbouring Justices and gentlemen, for the ease of their own estates only[3]." In a journey from London to Bath, the writer mentioned above saw a team of three horses in a cart drawing only a bushel of gravel for a load. He said he could "point out a parish also which has compounded with the pike at £15 per annum, for a piece of road that before had annually £60 expended on it." These two methods, of shirking statute labour and of evading the payment of full toll, have always been very prominent; and they are still seen, although probably to a much less extent, where statute labour is adopted for the repair of the roads[4].

Not only those who were chargeable with the maintenance of the roads, but even those who were entrusted with the office of surveyor,

---

[1] Brit. Doc., Reports from Committees, II, 1787–65, pp. 465–8.

[2] See also Report of Committee on the road from Loughborough to Derby, in *Journal, House of Commons,* Feb. 15, 1743; Report of Committee of the House of Commons, in ibid., Mar. 18, 1713.

[3] *Gentleman's Magazine,* XXIV, p. 395.

[4] Ibid. In Brit. Mus., MSS. 12,496, pp. 263–91, 'Orders and Directions, together with a Commission for the better Administration of Justice,' etc., we are informed that one of the great reasons why the roads were decayed was that the statute labour was "so omitted, or idly performed, that there comes little good" from it (p. 290). On farming out the tolls, see Report on the road from Gloucester to Hereford, in *J., H. of C.* (*Journal, House of Commons*), XXV, p. 851.

seemed to be guilty of connivance against the law[1]. By Act of 1670[2], it was enacted that no waggon should travel on the roads with above five horses at length, under penalty of forty shillings for each offence, and that all constables and surveyors of highways should see that all existing laws relating to roads should be put into effect, under penalty of the same amount. This latter provision was probably aimed *inter alia* at officers who extorted from waggoners or carriers great sums of money in return for, or under pretence of, giving them liberty to draw with more horses than the law allowed. There was even the complaint that surveyors seized the teams of some who carried according to the law, as if they were carrying contrary to the law, and induced the drivers to pay them a certain amount of money, on consideration of suppressing proceedings against them[3]. If there was much of this carrying done by eight, nine, or ten horses, when only five horses should have been used, it is no wonder that the roads were not in good repair.

Sometimes materials for road construction had to be drawn considerable distances, when they might have been obtained in an adjoining field had not the meanness of the landowner prevented their being taken from his property[4]. This, of course, necessitated much expenditure of time and labour, all to no purpose.

In the choice of roads that were to be benefited by turnpike Acts, there was no security that the best routes would be selected, for there were so many diverse interests to be served. The existing roads along which pack-horses wended their way, were frequently made to ascend hills and take their course over dreary, dangerous and hilly commons; and when waggon roads were made they often followed these same uneven surfaces, sometimes in order to avoid the valleys, across which roads could not be kept in good order during wet weather, and at other

---

[1] Very often were surveyors and others who had charge of the oversight of the roads warned to be careful that the laws concerning the highways were enforced. See, e.g., Rymer's *Foedera*, xix, pp. 130–1, 697, 'A Proclamation for the Restraint of Excessive Carriages (1629), and 'A Proclamation for Restraint of Excessive Carriages to the Destruction of the Highways' (1635).

[2] Act 22 Car. II, c. 12.

[3] On this whole matter see Brit. Mus. 816. m. 14 (27), 'The Case of Richard Fielder, in Relation to the Petition of the Waggoners,' and Brit. Mus. 816. m. 14 (28), 'The Case of John Littlehales against the Pretended Petition of the Waggoners travelling the Northern Roads of England.'

[4] Report of Committee appointed to inquire into the management and application of sums of money for repairing highways, in *J., H. of C.*, Apr. 22, 1714.

The law provided that surveyors could go into a field adjoining the road, and dig for gravel, etc., for fixing the road, so long as they were careful to not unduly trespass and to fill up the gravel pit when they were done with it. In every case, full compensation was to be given to the landowner.

times because the landowner refused to grant the privilege of making a more even road over part of his estate. Other landowners, it was said, endeavoured to make the turnpikes definitely subservient to their particular advantage, by having them made to this or that country seat[1]. The location of a certain inn along a road would occasionally determine that that immediate portion of the way should be turnpiked, rather than another part or course that would have been more acceptable[2]. Local interests were a strongly determining factor in the location of turnpikes; and instead of the straightest course being selected, the more circuitous road was not infrequently adopted as the line to be improved[3].

The use of narrow-wheeled waggons and carts, upon which heavy loads were carried, must certainly have been a potent factor in preventing the improvement of the roads. These wheels, if set on a smooth stone, we are told, would touch it little more than one-quarter inch[4], probably because of the wearing away of the tire at the sides; and, what was still worse, the large rose-headed nails projecting through the tires acted like a plough, tearing up the surface of the roads faster than they could be mended. Such waggons, in passing along the roads, cut them into ruts and ridges, which were rendered still deeper on account of the water lying in them; and when, as was often the case, the heavy carriages had to keep the same track except when meeting other carriages[5], the evils of the soft roads would tend to be progressively intensified.

And, finally, one reason, upon which we ought, perhaps, to lay considerable emphasis, is that there were no engineers who thought road construction a sufficiently dignified pursuit to worthily engage their time and talent; and not until the time of John Metcalfe was this work definitely taken up as a special occupation. Lack of skill, and lack of knowledge of how roads should be constructed, led to diverse practice; and in some cases so badly was the work done that the road was rendered concave and was lower than the fields on either side, in consequence of which the water flowed from each side into the road, and there lay, softening the road-bed, until by the natural processes it was evaporated or otherwise disappeared[6].

---

[1] *Gentleman's Magazine*, August 1754.

[2] Clark, *General View of the Agriculture of Hereford*, p. 58.

[3] Scott, *Digest of the General Highway and Turnpike Laws*, 1778, p. 317.

[4] *London Magazine*, XXI, p. 609.

[5] Ibid., XXI, p. 609, and XXIV, p. 582.

[6] A Swedish traveller, Kalm, in his *Visit to England*, 1748, p. 881, says: "These high roads had not the character, as with us in Sweden, that the road lay higher than the land around, but here exactly the opposite is the case, viz., so that the road goes in most places deep down in the earth, to a depth of two, four, or six

What is the truth about the roads of England down to the middle of the eighteenth century? Were they good or bad?

Amidst the mass of conflicting testimony, it is extremely difficult to obtain a satisfactory answer to this question. A writer in 1747, in a letter to the *Gentleman's Magazine*[1], says: "In my journey to London, I travelled from Harborough to Northampton, and well was it that I was in a light Berlin, and six good horses, or I might have been over-laid in that turnpike road. But for fear of life and limb, I walked several miles on foot, met twenty waggons tearing their goods to pieces, and the drivers cursing and swearing for being robbed on the highway by a turnpike, screened under an act of parliament." In a note by the editor of that magazine, we have a confirmation of this in the following words: "These complaints we have found experimentally true, in a journey to Derby, and rather than travel the said bad and dangerous road twice, chose to go several miles about into another turnpike road. It is surprising that the adjacent towns, whose interests may be affected, do not raise a subscription on the credit of the Act."

Another writer, in 1752[2], who had travelled much in England, after speaking of how that country had been raised to "so high a pre-eminence over other nations that all foreigners both envy her and admire her," and having defended the English people against the charge of ferocity, which foreign travellers attributed to them, says: "The only solid objection I can make to this amiable recess, secreted, as it were by the hand of nature, from the gross of the European continent, is the wretched state of many public roads." Then he urges that the great public arteries of communication should be kept "open and permeable." In his last journey from London to Falmouth, after the first 47 miles from London he "never set eye of a turnpike for 220 miles."

The same writer, about two years later[3], in speaking of the con-dition of England, says: "were the same persons who made the full tour of England thirty years ago, to make a fresh one now, and a third

---

feet, so that many would believe the road was only some dry stream-course. There is commonly on one side of the road, if not on both sides, on the walls or the high sides, a foot-path for foot passengers.

" That the roads are so deep seems to come from this, that in this country very large waggons (vagnar) are used with many horses in front, on which waggons a very heavy load is laid. Through many years' driving, these waggons seem to have eaten down into the ground, and made the road so deep." See also Coxe, *Historical Tour through Monmouthshire* (reprint of 1904), p. 35.

[1] *Gentleman's Magazine*, XVII, p. 232.
[2] Ibid., XXII, p. 517.
[3] Ibid., XXIV, pp. 347–9.

some years hence, they would fancy themselves in a land of enchant-ment. England is no more like to what England was, than it resembles Borneo or Madagascar." A little further on, he remarks: "In a few years I hope to see all England accessible to travellers, and open to commerce. The North is already, and the West, 'tis to be hoped, will take its turn and come in play soon: for at present 'tis a great tract of terra incognita....It is 172 miles from London to Exeter, further yet to Plymouth, 272 to Falmouth; no turnpike more than 40 miles from London, except...people go round by Bath, or Wells."

But to return to our question, after these diverse statements: we must consider the *relativity* of good and bad roads, and of the judgments upon them. We must also keep in mind the economic state of a dis-trict and the amount of traffic needing good roads. A road which would be good for one section where the roads were generally bad, might be very bad in another section where, as a general thing, the roads were in good condition. Further, a road might appear very good, and fully equal to the necessities, to a man whose range of observation had been very limited; whereas the same road might appear very bad to a traveller who had seen the wider horizon, especially the roads of France.

For the present, we may largely discard the consideration of those portions of England which lie north of the county of York, and also the south-western counties; for these were not the districts in which industry in general was flourishing.   To expect good roads throughout these localities would be unnatural, although we find evidence that even here there were occasional stretches of fair road. Roads that passed through sparsely settled districts, or through places where there was much broken land, or other impediments, were frequently but little repaired, and were usually mere tracks followed by the pack-horses and occasional travellers[1].

[1] One or two out of many entries in Thoresby's *Diary* will well illustrate the state of the roads in the north, about 1680. The following entry is dated Sept. 21, and is to be found in Atkinson, *Ralph Thoresby, the Topographer*, I, p. 129:

"Up by twelve o'clock in order to a journey, and with a guide, were got over *most prodigious high hills and very many of them* by daybreak; thence by Teviotdale, upon the brink of a steep hill for some miles, to Usedale, where, upon the sudden, the precipice grew to that height and steepness, and withal so exceedingly narrow, that we had not one inch of ground to set a foot upon to alight from the horse. Our danger here was most dreadful, and, I think, inconceivable to any that were not present; we were upon the side of a most terrible high hill, in the middle whereof was a track for the horse to go in, which we hoped to find broader, that we might have liberty to turn the horse; but instead of that it became so narrow, that there was an impossibility to get further; for now it began likewise to be a sudden de-clension, and 'the narrow way so cumbered with shrubs, that we might be forced to lie down upon the horses' necks, and have our eyes upon a dreadful precipice,

The portion of England which, by reason of its agricultural and industrial importance, would require the best roads, was from Yorkshire and Lancashire southward and from Staffordshire and Worcestershire eastward. Under the domestic system of industry, the *local* traffic on the roads was much larger than any through traffic; and, therefore, each section was interested, not in long stretches of road, but only in near-by circumjacent portions. The roads which extended past their market town did not much engage their attention; these were left to

such as mine eyes never till then beheld, nor could I have conceived the horror of it by anyone's relation. We had above us a hill so desperately steep, that our aching hearts durst not attempt the scaling of it, it being much steeper than the roofs of many houses; but the hill below was still more ghastly, as steep for a long way as the walls of a house; and the track we had to ride in was now become so narrow that my horse's hinder foot slipped off," etc.

A traveller in 1684 found much the same kind of roads in the north. He says that travelling along the rivers Tyne and Derwent, they met with some dangerous ways, on one of which they expected their horses to fall on them (Brit. Mus., Add. MSS. 34,754, p. 19). In going thence toward Carlisle, along by the Picts' wall, they found the ways "as mountainous, rocky, and dangerous, as those the day before" (ibid., p. 20). At Penrith, he speaks of the "stony wayes." From there they journeyed to Kendall "through such wayes as wee hope wee never shall againe, being no other but clim(b)ing & stony, nothing but Bogs and Myres o'r the tops of those high hills, so as wee were enforc'd to keepe these narrow, loose, stony, base wayes, though never so troublesome & dangerous....On we went for Kendall, desiring much to be releas'd of those difficult & dangerous wayes, which for the space of eight miles travelling a slow marching pace we pass'd over nothing but a most confus'd mixture of Rocks and Boggs."

With regard to the roads in the south-western counties, note what we have said on page 86.

In 1707, Rev. Mr Brome, Rector of Cheriton, in Kent, found the roads in Devonshire so rocky and narrow that it was not possible for the farmers to use waggons; they had to carry their corn on horseback. Mason, *History of Norfolk*, p. 482, quoting from Brome, *Travels over England, Scotland and Wales*.

In the time of William and Mary, Mrs Fiennes travelled through England, and in referring to Cornwall, she says, "Here I entered into Cornwall and soe passed over many very steep stony hills, though here I had some two or three miles of exceeding good way on the downs, and then I came to ye steep precipices—great Rocky hills....Here indeed I met with more inclosed Ground and soe had more Lanes and a deeper Clay Road which by the raine ye night before had made it very dirty and full of water in many places, in the road there are many holes and sloughs where Ever there is Clay Ground, and when by raines they are filled with water its difficult to shun danger;..." Fiennes, *Through England on a Side Saddle*, p. 216.

In 1637, a writer, speaking of Cornwall, says: "the countrie hath no coaches nor any kinde of carte or ought that is moved upon wheels. All carriage is layde upon horses backs either in trusses, or on crookes, or in paniers or beds, which they call pots." His way going to Exeter was "verie ill and most of it causeway," while from Exeter to Honiton it was a "verie stonie and evill way." Brit. Mus., Harl. MSS. 6494, pp. 185–7. See also Brit. Mus., MSS. 15,776, Milles' *Tours in England and Wales*, 1735–43, pp. 102, 108, 109.

be repaired by those who had lands adjoining.   The same may be said even after the turnpikes came into prominence, for the latter were not continuous, but made up of disconnected portions of long roads, the other stretches of which had often been but slightly, if at all, repaired.

In considering the roads in Yorkshire, let us look at that from Leeds to York, which we would expect to be one of the best roads in this northern portion of England, since both these cities were centres of manufacture, and York was the place of export.   In 1654, because "the highway leading from Leeds to Wikebridge and so to Seacroft and so to Kiddall toward Yorke hath been heretofore presented by jury to be in great decay for want of amendment so that *travellers can very hardly pass*, to the great hindrance of all" that had occasion to travel that way; therefore, the West Riding Quarter Sessions imposed a considerable fine upon every person who failed to render the statutory aid for the repair of the highway[1].   But this did not seem to make any permanent improvement, for in the summer of 1680[2] Ralph Thoresby describes the road near York as very bad, "the waters being very great and dangerous;" and in 1708, while journeying to York, he "found the way very deep, and in some places (so) dangerous for the coach" that he walked on foot[3].   In 1712, he was again crossing over the country to see Harwood, about seven miles from Leeds, when he found "some part of the way as rocky as can well be supposed in the most remote parts of the island[4]."   The highways in this neighbourhood must, at times, have been in a sorry state, if many of them at all resembled that between Leeds and York[5].   But it would seem that some

---

[1] *West Riding Sessions Rolls*, p. 104.   The Court ordered "that every person occupying a plough tilth within any of the parishes of Leeds, Whitkirk, and Berwick (through which parishes the said highway lieth) shall send their draughts (teams) and sufficient labourers according to the statute and repair the same way before the 25th day of August upon pain that every person making default therein shall forfeit 20s."

[2] Thoresby's *Diary*, I, p. 50, July 27, 1680.

[3] Ibid., II, p. 5, May 17, 1708.

[4] Atkinson, *Ralph Thoresby, the Topographer*, II, p. 215.

Another hint of the nature of the roads in this locality is given us in Thoresby's *Diary*, I, p. 28, where he says: "From Hull we came by coach to York and thence on horseback to Leeds."   This was the winter of 1678–9.   The note at the bottom of the page says: "The stage-coaches being given over for this winter, I hired one to conduct me safe; though it proved a mortification to us both, that he (i.e., Thoresby's father) was as little able to endure the effeminacy of that way of travelling as I was at present to ride on horseback."   From York to Leeds they rode the manly way, on horses.   Cf. also Atkinson, *Ralph Thoresby, the Topographer*, I, p. 66.

[5] It was on account of the bad roads and the heavy expense of carriage, that there was in 1697 such a strong agitation for making the rivers Aire and Calder navigable.   *J., H. of C.*, Jan. 12 and Feb. 3, 1697.   Whitaker (*Loidis and Elmete*,

slight improvement may have taken place before the middle of the eighteenth century; for in 1740, when the towns of Halifax, Ripponden and Ealand petitioned for an extension of the Calder navigation from Wakefield to Halifax, they stated, contrary to the usual form of expression, that this navigation "would preserve the highways *which are now maintained* at large annual expense[1]." Whatever improvement was made, it was only during the summer that this was noticeable, for in winter the roads were still almost impassable for wheels[2].

We have confirmation of this opinion when we consider the state of the roads between York and London, which were part of the great route between London and the North, and the time occupied in a journey between these points. It is very evident that it took much longer at some times than others; for even in the summer months there were great differences in the condition of the roads. In 1688, although Thoresby had been accustomed to going that distance on

p. 81) says that it is difficult for a modern mind to conceive the impediments which lay in the way of commerce and manufactures. "The roads were sloughs almost impassable by single carts, surmounted at the height of several feet by narrow horse tracks, where travellers who encountered each other sometimes tried to wear out each other's patience rather than either would risque a deviation. Carriage of raw wool and manufactured goods was performed on the backs of single horses at a disadvantage of nearly 200 to 1 compared to carriage by water...On horseback before daybreak, and long after nightfall these hardy sons of trade pursued their object... Sloughs, darkness, and broken causeways certainly presented a field of action no less perilous than hedges and five-barred gates;...In the state of the roads at that time, swiftness was impossible." Then, on the following page, he refers to the deplorable state of the highways.

[1] *J., H. of C.,* Dec. 9, 1740, xxiii, p. 554. In *J., H. of C.,* xxiii, pp. 639–40, we are told that the road from Selby to Leeds was partly at least a good wheel carriage road, despite the petition for its repair; but the form of expression here used does not convey to us the impression that the witness who said these words was very fully convinced of their truth. It appears to be a weak statement, without very much conviction behind it. The ruinous condition of the roads in general in this locality is evident from the statements of very many witnesses. See *J., H. of C.,* xxiii, p. 620; also Report on River Dunn, in *J., H. of C.,* Jan. 31, 1739, etc.

[2] This is thoroughly substantiated by the Report of a Committee of the House of Commons, Jan. 26, 1740, on a Bill to amend the highways from Selby to Leeds, to Wakefield, to Halifax, to Bradford, and other roads in adjoining places. The evidence went to show that the roads were so bad as to be ruinous and impassable in winter; that the heavy loads of lime, woollens, wool, corn, coal, etc., cut the roads so as to make them impassable; that farmers could not get their corn to these markets in winter when the roads were so bad; that the lock dues on the rivers Aire and Calder being very high, the manufactures of the western parts, and also wool, corn, etc., from Lincolnshire, and from other places, could "be conveyed by land carriage on the said roads, *when they are passable,* at an easier expense" than they were at this time carried by water. In all the important testimony, the winter time was singled out as the time of year when the roads were impassable.

horseback in four days, it took the York coach six days[1]. It took less time, however, to return from London[2]. On May 25th, 1692, John Hobson set out from London in the Nottingham coach, and got to his home at Calverley, near Leeds, on the 28th of May[3], which was not more than four days on the road; and on July 3rd of the following year (1693) he started from London for Yorkshire in the coach, and reached home on the 6th, which again was not more than four days[4]. At another time he went, in the York coach, from Ferrybridge to London in three days[5]. But in winter it required more time, for in Nov. 1695 it took at least seven days to go from his home to London[6], and in January of the same winter it required eight days to come home from London[7]. It will be seen, therefore, that winter travelling required practically twice as long as summer travelling, although this statement is not universally true[8]. The time necessary for the performance of such journeys gives a good indication as to the quality of the roads[9].

---

[1] Atkinson, *Ralph Thoresby, the Topographer*, I, p. 184.

[2] Ibid., I, p. 191.

[3] *Yorkshire Diaries* (John Hobson's *Diary*), II, p. 49.

[4] Ibid., II, p. 55.

[5] Ibid., II, p. 68; also II, p. 64.
The four-days journey between London and York is well portrayed by an old coaching bill of the year 1706, which is as follows:
"York Four Days Stage Coach.
Begins on Friday, the 12th of April, 1706.
All that are desirous to pass from London to York, or from York to London, or any other place on that road, Let them Repair to the Black Swan in Houlbourn in London, and to the Black Swan in Coney Street, in York.
At both which Places they may be received in a Stage Coach Every Monday, Wednesday, and Friday, which performs the whole journey in four days (if God permits), And Sets forth at Five in the morning.
And returns from York to Stamford in two days and from Stamford by Hunting-don to London in two days more. And the like stages on their return.
Allowing each passenger 14 lb. weight and all over 3*d.* a pound.

Benjamin Kingman,

Performed by { Henry Harrison,

Walter Baynes.

    Also this gives notice that Newcastle Stage Coach sets out from York every Monday and Friday and from Newcastle every Monday and Friday." Harris, *Old Coaching Days*, p. 93. It took four days for the same journey in 1658—v. Harris, p. 106.

[6] *Yorkshire Diaries*, II, p. 68, Nov. 20, 1695. "Went from home for London, and, morning after, went into Wakefield coach, and got thither 27 Nov."

[7] Ibid., II, p. 68, 22 Jan., 1696. "Set forwards down from London in the Wakefield coach, and got home 30 Jan."

[8] Ibid., II, p. 49. This man took coach on Mar. 21, 1692, at Ferrybridge, and reached London on the 26th of the same month.

[9] A writer who had travelled this road from London to the north of England

But before the middle of the eighteenth century some improvement had taken place in this north road, for turnpikes had been to some extent replacing the natural roads in certain localities; and with the firmer bed and better drainage, these new roads were usually a decided gain over the older[1]. But the turnpikes were so detached that there was no long piece of road that was of this construction; hence it frequently occurred that a short stretch of good turnpike road would have a long reach of bad road to connect with it at each end. On the road from Glasgow to London, as late as 1739, there was no turnpike on the southward journey till Grantham was reached, within 110 miles of London[2]. Notwithstanding the difficulties, the cost of carriage on

in 1704, made the following significant remarks upon it: Mar. 31, 1704. "I sett out from Royston, and with a great deal of toyle, travelling about two miles an hour at most, thro' the worst and deepest ways I ever rode, and (I believe) is in England. I gott 9 miles to Caxton—but passed on about 4 miles further, in a road but little better, to Godmanchester." Following this statement of the bad roads, he adds: "This is sayd to be a place of the best husbandry in England" (Brit. Mus. 10,848. ccc. 56, *North of England and Scotland*, p. 2). Continuing along this road, "From Huntingdon I travelled nine miles, through a bad road, to Stilton" (Ibid., p. 5). It will be remembered that this was the very road, to improve which the first turnpike Act was passed, in 1663. Then "From Stilton I came 2 miles, through a very bad road, to Yaxley" (ibid., p. 6). "From Yaxley to Peterborough is still a very bad road of 3 miles" (ibid., p. 6). Such a succession of remarks, and others of like import, must have been elicited by travelling along a road that had few merits; but, of course, we must remember that this journey was performed during the winter.

In the summer of the following year, another traveller followed this route from London to Edinburgh, and describes the arrival at Northampton "after an intollerable journey through Hickley Lane." Other parts of the road are described as "very indifferent" and "dismal," while some were "pleasant." Taylor, *A Journey to Edenborough in Scotland* (in 1705), pp. 13, 18–19, 44, 69. He also speaks of the "excellent causeways" (p. 66) around the city of York, which were kept in good repair for some miles round.

[1] Turnpikes were not always good roads, however, for a gentleman in describing his journey from Kimbolton to Ormesby in 1743, said that the very worst three miles he ever went in his life was a turnpike road in the midst of summer. It was between Lincoln and Ormesby. Brit. Mus., Egerton MSS. 2235, p. 86.

[2] The following extract is from Dr Bannatyne's scrap-book, as given in Cleland's *Statistics of Glasgow*, p. 156: "The public have now been so long familiarized with stage-coach accommodation, that they are led to think of it as having always existed. It is, however, even in England, of comparatively recent date. The late Mr Andrew Thomson, Sen., told me that he and the late Mr John Glasford went to London (i.e., from Glasgow) in the year 1739, and made the journey on horseback. That there was no turnpike-road till they came to Grantham, within one hundred and ten miles of London. That up to that point they travelled upon a narrow causeway, with an unmade soft road upon each side of it. That they met, from time to time, strings of pack horses, from 30 to 40 in a gang, the mode by which goods seemed to be transported from one part of the country to another...."

this road was being reduced[1], but whether this was due chiefly to there being more good pieces of road, or to the carriers' ability to take larger loads on their broad-wheeled waggons, it is impossible to decide. It is certain that there was considerable agitation in 1750 and 1751 for reduction of tolls on this north post-road[2], which is evidence that many people believed some parts of the road had been sufficiently improved.

The roads in the southern counties do not seem to have received the same attention in the seventeenth century as those north of London, and accordingly were not so much repaired as the latter; but when turnpike legislation became more prominent, in the eighteenth century, these counties secured some share in its benefits. In the early years of the reign of Charles II, a French traveller[3], on his journey to the English capital, "went from Dover to London in a waggon," which "was drawn by six horses one before another, and drove by a waggoner, who walked by the side of it." In the reign of William and Mary the roads in the county of Hants were "very stony, narrow, and steep hills; or else very dirty as in most of Sussex[4]." In 1740, a report on the Kent and Sussex roads declared that they were not kept in repair, even by the tolls which they were authorized to take; and that occasionally a coach had to go round about, through fields, to avoid the danger of some roads[5] A report of the year before showed that because of the badness of the roads, it cost 1*s.* per load per mile to bring to market the timber with which these two counties abounded; and that in some years there was not more than one or two months when this

---

[1] *J., H. of C.,* Mar. 14, 1758, xxviii, pp. 133–45. The price of carriage from London to Wakefield had been lowered within three years from 14*s.* to 7*s.* 6*d.* a pack (of 240 lbs.); and the cost from Wakefield to London had been reduced from 1*s.* to 10*d.* a stone.

[2] See the petitions of 1751, in *J., H. of C.,* for reductions of tolls on the North Post Road.

[3] Sorbière, *Voyage to England,* p. 7.

[4] Fiennes, *Through England on a Side Saddle,* p. 20. Yet she describes the ten miles from Dorking to Kingston, in Surrey, as "a chalky hard road."

[5] *J., H. of C.,* Dec. 19, 1740, xxiii, p. 567. This report of the roads of Kent and Sussex is corroborated by the observations of Milles in 1743 (Brit. Mus., MSS. 15,776, Milles' *Tours in England and Wales,* pp. 178, 219). The road from Sevenoaks to Tunbridge through an enclosed country was of clay, cut very deep even at the dry time of the year, and must have been exceedingly bad in winter. This was a direct road from London. The roads through the Weald of Sussex he describes as "bad even in summer time; and in winter they must be intolerable." Macky, in his *Journey through England,* 1714, says that the country around Petworth, in Sussex, "being fat and fertile, makes the roads bad in winter." It is evident, therefore, that when such main roads were bad in winter, and sometimes in summer, the amount of traffic upon them must have been rather small.

timber might be taken away[1]. On account of the "monstrous expense" of carriage, timber contracted for had sometimes to lie two or three years before it could be removed[2]. From all we have been able to gather, it seems to have been impossible to find, at this time, many portions of good road in the counties which bordered the Channel.

Like the great north post-road, the western road from London to Bath, Bristol and Exeter was of great importance, for by it the manufacturing area in Wilts, Somerset and Gloucester was brought into direct communication with ʹLondon, and this was also the great thoroughfare from most places in Wales to the city of London. That the travel on this road was considerable is evidenced by the fact that from about 1670 on, there were five daily coaches from London to Bath, for the conveyance of passengers to the mineral baths of that great resort[3]. We have ample proof also that the carriers along this route had a trade of some magnitude, for as early as 1687 they came to London from Bath, Cheltenham, Tewkesbury and Devizes each once a week; from Bristol, Exeter, Gloucester and Stroudwater each twice a week; and from Worcester three times a week. This makes no mention of the carriers from the smaller places, many of whom came once a week, nor of the clothiers from the parts of Gloucester, Wilts, and neighbouring sections, who came with their waggons several times a week to London[4]. Doubtless this early traffic had increased immensely during the century which followed the reign of Charles I.

But what can we say regarding the condition of these western roads during all this period? Their general nature in Worcestershire, during the first half of the seventeenth century, may be gathered from the long list of presentments for non-repair and neglect before the Court of Quarter Sessions in that county[5]. We would not be justified in saying that all the roads were in ill repair, for some parishes took more pride than others in their roads and would certainly maintain them in a state befitting the material prosperity of their residents; but the testimony of those who had seen these roads and travelled

---

[1] *J., H. of C.*, Feb. 20, 1739, xxiii, pp. 469–70. In some cases, timber contracted for could not be got to the market—the Navy-yard was the chief market—and so had to be obtained elsewhere.

[2] *J., H. of C.*, Feb. 1, 1739, xxiii, pp. 443–4. The roads of Sussex seem to have been a synonym for all that was bad. Note the experience of the Emperor Charles VI and that of Horace Walpole and of Dr John Burton, as given in Blew, *Brighton and Its Coaches*, pp. 19–20. The clay roads of this county must have been abominable down to 1751, whenʹDr Burton made his journey on horseback through the county.

[3] Davis, *The Mineral Baths of Bath*, pp. 47–48.

[4] Taylor, *The Carriers' Cosmography*.

[5] Bund, *Records of Quarter Sessions*, Pt. ii, in the publications of the *Worcestershire Historical Society*.

over them is generally of the nature of adverse criticism[1]. On August 15, 1694, the Duchess of Marlborough, accompanied by Princess Anne and her husband, Prince George, set out for Bath. Their carriage was drawn by four horses, at the slow pace of about five miles an hour; but the roads leading to that city were found to be so bad that on approaching the town the horses were unable to draw the carriage over the hill until some of the occupants got out and walked[2]. The state of the roads around Bath, at a later date, may be gathered from a letter of Lady Irwin to Lord Carlisle, dated at Bath, Oct. 27, 1729[3], in which she says: "I design leaving Bath to-morrow, and propose getting to Altrop on Friday the last of October. The road between Altrop and this place is so extremely bad that the coachman wont undertake it under four days, though it is but 64 miles. Everybody here tells me I shall run great hazards in going that road, but the coachman that drives me has provided me a very good set of horses, and will engage to carry me safe, allowing four days to do it in...." Such was the road before winter came on. But in winter, a writer of the year 1742[4]

---

[1] The parish of Alvechurch was just outside the limits of the forest of Feckenham, which had until recently been enclosed, but was disafforested in 1629. In 1633, a petition of the Vicar of Alvechurch to the Quarter Sessions shows what the roads were like at that time. This petition is found in the *Records of Quarter Sessions of Worcestershire*, Pt. II, 1633 (255), LVIII, 79, p. 528, and is as follows: "The parish of Alvechurch has many roadways and thoroughfares for travellers both on horse-back and for carriages by wains and carts, and other common highways to divers market towns through sundry parts of the said parish but *all generally so ill and negligently repaired* that divers enormities redound therefrom not only to many of the parishioners themselves but also to many others travelling those ways in particular myself in this harvest time riding about my lawful and necessary oc-casions of tithes have been twice set fast in the mire in common roads and market ways not without danger. By occasion of these ill-repaired highways I am forced to sell much of my tithes far under value. Much of this ill repair is caused by some who 'staunch' up water in ditches and turning them out of their course to water and overflow the adjoining grounds and in some of the roads formerly used for passage on horseback and loaded waggons and cattle cannot be used for passage on horseback without danger of getting fast and myring." The roads here must surely have been very bad, when the rector had to sell his tithe corn and pigs at a reduced price, because of the difficulty of taking them farther to a better market.

[2] Colville, *Duchess Sarah*, pp. 99–100. "On approaching the town the horses had not strength to drag the heavy conveyance over Lansdowne Hill, so it ran back, much to the alarm of the occupants. Lady Marlborough put her head out of the window and ordered the servants who accompanied them—some on horseback and some on foot—to put their shoulders to the wheels, which had the effect of stopping further disaster. The coach being lightened, the horses managed to reach the summit in safety. But more difficulties were met with in the steep descent, so the occupants preferred to walk, while the horses were carefully led down into the town."

[3] *Historical Manuscripts Commission*, Report 15, Appendix 6, p. 61.
    A Compleat History of Somersetshire, p. 8.

tells us that, "it wants not its winter-like qualities, being moist, wet, marshy, and in the roads extremely dirty; from whence it is that they have this proverb among them, 'bad for the rider, but good for the abider';" and again he says that the roads in winter were exceedingly dirty and miry[1]. Despite such evidence, however, proving the decay of the highways, there are some gleams of better things which point to a slow advance that was going on. Even although the roads were in general bad, we have facts which enable us to affirm that some parts of them were satisfactorily maintained[2]; and an increase in the speed of some public coaches running between London and these western cities would indicate that the means of communication were being somewhat improved. By 1724, the old system which was in operation early in the reign of Charles II, of a three-days coach from London to Bristol[3], was replaced by two time-schedules for coaches, namely, the three-days coaches left London twice a week, and the two-days coaches left three times a week, during the summer[4]. But the three-days coaches continued for many years after this time, for in 1788 it still took three days to go from Gloucester to London[5], which is not any further than from Bristol to London.

We have now considered the two great roads from London along which traffic was probably the heaviest, that is, the northern and the western routes. In the next place, we must note the means of communication with Birmingham, Chester, and other towns along what is now called the Holyhead road. Our information regarding early travel on this road is very scanty. Lord Clarendon, on his journey to Ireland

[1] *A Compleat History of Somersetshire*, p. 125.

[2] Mrs Fiennes, when riding *Through England on a Side Saddle*, in the latter part of the seventeenth century, came to a part of Gloucestershire regarding which she says, "It gives you a good sight of the country about, which is pretty much inclosed and woods a rich deep Country and so the roads bad." But from there she returned, and ascended a "high hill and travelled all on ye top of ye hills a pleasant and a good roade." (Fiennes, *Through England on a Side Saddle*, p. 23.) And so her description goes on, mentioning some good and many bad roads. From Sutton to Oxford, a distance of fourteen miles, was "all in a very good Road;" from Abingdon to Ilsley (8 miles) and thence to Newbury (7 miles) was mostly on downs and very fair roads. (Ibid., pp. 24, 30.)

[3] Account book of Gore family of Flax Bourton, 1663. See Latimer, *Annals of Bristol in the Eighteenth Century*, pp. 22–23. Also, according to a coaching bill of 1658 (Apr. 26), it took four days to go from London to Exeter (Harris, *Old Coaching Days*, p. 106).

[4] *London Evening Post*, of May 23, 1724. The two-days coaches were in operation only during the summer season. The slower coaches seem to have occupied four days on their journeys in the winter months.

[5] Counsel, *History of Gloucester*, p. 209, quotes from an advertisement to this effect in the *Gloucester Journal*, of Nov. 23, 1788.

as Lord Lieutenant, halted at Newport, in Shropshire, Dec. 24, 1687, and says: "We came hither quickly after three in the afternoon, though we set not out from Lichfield till after nine, and it is near 20 miles[1]." This was a speed of between three and four miles per hour. Next night he was to lodge at Whitchurch, fifteen miles farther, from which we judge that the roads were very bad; but as this was in the winter, and before turnpikes were established there, we should not expect that the natural roads would be found in good repair. Some progress had been made, however, in the next fifty years, for in the winter of 1739, the Chester stage, with six, and sometimes eight, horses, by being out two hours before day and as late at night, was able to reach London in six days[2], which was the equivalent of thirty miles a day. This time would unquestionably be reduced during the summer half of the year. By April 1753, the "Birmingham and Shrewsbury Long Coach," with six able horses, went from Shrewsbury to London in four days, the equivalent of forty miles a day, charging eighteen shillings fare[3]; and in June of the same year a rival coach performed the journey in three and one-half days, the equivalent of forty-five miles a day, at a fare of £1. 1s. for inside passengers and half fare for outside passengers[4]. These, it will be noticed, were summer rates, and

---

[1] *Salopian Shreds and Patches*, iii, p. 79.

[2] Pennant, in his *Journey from Chester to London*, p. 137, says: "In March 1739–40, I changed my Welsh school for one nearer to the capital, and travelled in the Chester stage; then no despicable vehicle for country gentlemen. The first day, with much labour, we got from Chester to Whichurch, twenty miles; the second day, to the Welsh Harp; the third, to Coventry; the fourth, to Northampton; the fifth, to Dunstable; and, as a wondrous effort, on the last, to London before the commencement of night. The strain and labour of six good horses, sometimes eight, drew us through the sloughs of Mireden, and many other places. We were constantly out two hours before day, and as late at night; and in the depth of winter proportionably later."

[3] Owen and Blakeway, *History of Shrewsbury*, i, p. 515.

[4] *Salopian Shreds and Patches*, i, p. 7. The advertisement of this coach is as follows: "SHREWSBURY STAGE COACH,
in three days and a half.

Sets out from the George and White Hart Inn in Aldersgate-street, London, every Wednesday morning at 5 o'clock, and from the Raven Inn in Shrewsbury, every Monday noon at One o'clock; each passenger to pay 1 guinea, one-half at taking their places, the other at entering the coach; children on lap, and outside passengers, to pay half a guinea each; each passenger allowed 14 lbs. weight of luggage, all above to pay 2½d. a pound.   Performed (if God permit) by
John Fowler,
Turvil Drayson,
John Benson."

It is important to notice that up to this time no coach had gone from Shrewsbury to London more than once a week.

did not differ very much from those on the London and York road. It is evident, therefore, that the rate of travelling had increased, and that, too, with the substitution of the coach for horseback riding; which would indicate that there must surely have been some improvement in the roads.

As Birmingham was the most important place on this north-western route, its traffic with London warranted the establishment of a special coach between these cities in 1731, to run during the summer season. This coach made the journey one way in two and one-half days[1]. But toward the middle of the century, the desire on the part of the trading community for increased speed, led in 1742 to setting up a "flying coach," which reduced the time from Birmingham to London in summer to two days[2]. As before, it is necessary here also to remember that this coach went only once a week, and then only in the most favourable time of the year[3].

The roads in the eastern counties (Essex, Suffolk, and Norfolk) were described by Ogilby, in the time of Charles II[4], as "for the most part hard and gravelly, the lanes being here and there a little washy, but not incommoding the traveller." With this would seem to agree the record of a Yarmouth clergyman, who took the stage coach from London on July 1, 1689, about 8.30 a.m., stayed at Bury St Edmunds that night, and reached Yarmouth at 7.30 the next night[5]. It is difficult to believe that this rate of travelling, between sixty and seventy miles a day, was any other than unusually rapid, even in the summer time; and in the winter half of the year, it regularly took three days to perform the same journey[6]. In order to its accomplishment at all in the summer, the road must have been fairly good. That we are right

---

[1] For full advertisement of Rothwell's coach, see Harper, *The Holyhead Road*, II, p. 18.

[2] Aris's *Gazette, of Birmingham*, in May 1742, advertised Coles's coach as follows: "The Birmingham and Warwick Stage Coach Begins Flying for the Summer Season in Two Days, on Tuesday, the 5th of May, and sets out every Tuesday Morning at Three o'clock, from the Swan Inn in Birmingham, and from the George Inn in Aldersgate Street, London, every Friday Morning, and returns to Birmingham on Saturday. Performed by Robert Coles." (Dent, *Making of Birmingham*, p. 97.)

[3] But throughout all this period the carriers' waggons were on the roads, as well as pack-horses in great numbers. It was through these two agencies that the conveyance of goods took place; and, after all, the goods traffic was much more than the passenger traffic. Some idea of the magnitude of the goods traffic in 1637 may be gathered from Taylor, *The Carriers' Cosmography* and about 1710 from Brit. Mus. 796. c. 36, entitled *A Brief Director*; but by 1750 the amount of this trade must have been tremendously increased.

[4] Ogilby, *Itinerarium Angliae*, published 1674, quoted in Mason, *History of Norfolk*, I, p. 431. See also Ogilby, *Britannia*, p. 149.

[5] Davies, *Journal*, p. 29.   [6] Ibid., pp. 56–57, 64, 81.

in calling this speed extra fast, is obvious from the fact that, in the summer of 1712, along this same road, it took a traveller a day and a half to go by coach from Yarmouth to Bury, which was a speed of only forty miles a day[1]. It is incredible that the foregoing was the usual speed of a stage coach in 1689, when that of a hired coach, under the same circumstances, in 1712, was so much less. Furthermore, Ogilby's statement that the roads did not cause any incommoding of travellers, may be as meaningless or vague as many other of his statements[2], and was probably coloured by his desire to present as good a report as possible to the King, at whose expense he had made a survey of the chief roads of England. The roads leading from Norwich to London are described, in a petition presented to Parliament by the former city in 1725, as being "in a very ruinous and dangerous condition," the repair of which would require a large sum of money[3]; and this is in accord with the observations of Arthur Young, about fifty years later, who said he knew not one mile of excellent road in the whole county of Suffolk[4], and many of the turnpikes even were infamous. When we have made due allowance for all exaggeration in these last statements, they form a helpful corrective to any idea as to the roads in this section being particularly good. The fact would seem to be that along with many poor and bad roads, there were interspersed a few stretches that were good, and some that were indifferent[5].

That part of the kingdom where we should expect the best roads is the vicinity of London, for here the traffic was the densest, and the need of good roads the greatest. What, then, were the facts regarding this locality? Notwithstanding the immense number of coaches which plied on the London streets about the time of Charles II[6], it is evident that the streets were, at times, very unfavourable either for foot passengers or for coaches. People washed barrels and other vessels, rinsed their clothing, sifted ashes, fed their pigs, chickens, etc., in the streets[7].

---

[1] Macky, *Journey through England*, i, p. 8.

[2] See his *Britannia*, p. 10 et seq., for some of these vague or meaningless statements.   [3] Blomefield, *History of Norfolk*, iii, p. 441.

[4] Arthur Young, *Six Weeks' Tour Through the Southern Counties of England*, p. 319.

[5] Brit. Mus. MSS. 15,776, Milles' *Tours*, pp. 25, 54, 59, 74, 83–84, 101, etc.; Sir Thomas Browne's *Works*, i, pp. 23, 41, 53, 289, etc. From the latter reference and page, we learn that the journey from Norwich to London, in Oct. 1680, required three days, thus confirming our opinion respecting Davies' journey from London to Yarmouth. See also Brit. Mus., Add. MSS. 34,754, p. 8.

[6] Firth and Tait, *Acts and Ordinances of the Interregnum*, ii, pp. 922–4, Ordinance of June 23, 1654. See also Brit. Mus. 816. 1. 4 (21), beginning 'Robinson, Mayor, Commune Concilium.'

[7] Brit. Mus., E. 856 (4), 'An Act of Common Councell...for the better avoiding and prevention of Annoyances within the City of London, and Liberties of the

The latter were especially bad in the rainy seasons and in winter; for with defective drainage, they, during and after a heavy rainfall, were almost impassable[1]. When we say this of London we are merely repeating what may be said of the highways leading out of London: in dry weather the roads were frequently good[2], but when they became soaked with a heavy rain, or when they were softened by the melting of the snow or frost of winter, they were not only bad, but sometimes dangerous[3]. A copious fall of rain often prevented further travel on some roads for hours, for to pass through the increased current of

same. 1655.' The same conditions are given in the Report of the Commissioners of Sewers and Pavements, 1748 (v. Chamberlain, *History and Survey of London and Westminster*, pp. 403–5), showing the terrible condition of the streets.

[1] Samuel Pepys writes in his *Diary* for Mar. 20, 1660, that he went from London, "then to Westminster, where by reason of rain and an easterly wind, the water was so high that there were boats rowed in King Street and all our yard was drowned, that one could not go to my house, so as no man has seen the like almost, most houses full of water." *Pepys's Diary*, I, pp. 75–76.

In 1736, when Kensington had been the seat of the Court for nearly fifty years, Lord Hervey wrote to his mother that "the road between this place (i.e., Kensington) and London is grown so infamously bad, that we live here in the same solitude as we should do if cast on a rock in the middle of the ocean, and all the Londoners tell us there is between them and us a great impassable gulf of mud. There are two roads through the park, but the new one is so convex, and the old one so concave, that by this extreme of faults they agree in the common one of being like the high road, impassable." Hervey, *Memoirs of the Reign of George the Second*, II, p. 362, a letter of John, Lord Hervey, to his mother, dated Nov. 27, 1736.

[2] On Oct. 19, 1680, Ralph Thoresby, going from London northward, passed through Hoddesdon and Ware, "twenty miles from London, a most pleasant road in summer, and as bad in winter, because of the depth of the cart ruts" etc. (Thoresby's *Diary*, I, pp. 67–68); see also his journey by coach in 1714, from London to Cambridge (*Diary*, II, pp. 229–30).

Pepys, on Feb. 28, 1660, went through the Epping Forest to London, where, he says, "we found the way good, but only in one path, which we kept as if we had rode through a kennel all the way." (*Pepys's Diary*, I, p. 59.) The narrowness of many roads is attested by the observations of Arthur Young at a later time.

[3] On the morning of May 18, 1695, Thoresby has the following entry in his *Diary*: "rode to Edmunton (where we had our horses led about a mile over the deepest of the Wash) to Highgate, and thence to London. I have the greatest cause of thankfulness for the goodness of my heavenly protector, that being exposed to greater dangers by my horses boggling at every coach and waggon we met, I received no damage, though the ways were very bad, the ruts deep, and the roads extremely full of water, which rendered my circumstances (often meeting the loaded waggons in very inconvenient places) not only melancholy, but really very dangerous." (Thoresby's *Diary*, I, p. 295.)

On May 17, 1695, Thoresby writes: "Morning, rode by Puckeridge to Ware, where we baited, and had some showers, which raised the washes upon the road to that height that passengers from London that were upon the road swam, and a poor higgler was drowned, which prevented our travelling for many hours, yet towards evening adventured with some country people, who conducted us...over the

greatly swollen streams was a risk which many preferred not to under-take[1]. Occasionally the traveller came to a piece of road which was overflowed, and had to be helped over, because he could not tell where the road was[2]. The way from London to Oxford was probably one of the best in the neighbourhood of the metropolis; for during the summer half of the year 1669 a flying coach, for the first time, accomplished this distance in one day[3], going at the rate of between four and five miles an hour. But such announcements show us the increased demand for conveyances of this sort, rather than an actual general increase in the speed of travelling; for even at this time and on till nearly the middle of the next century the old two-days coaches were still continued[4].

From all the sources which I have been able to examine, with reference to the condition of the country roads in general during the period we have been considering, it seems that they were by no means favourable for a large traffic[5]. Without doubt, there were some

meadows, whereby we missed the deepest of the Wash at Cheshunt, though we rode to the saddleskirts for a considerable way, but got safe to Waltham Cross, where we lodged." (Thoresby's *Diary*, I, p. 295.) See also *Pepys's Diary*, I, p. 355. Compare similar facts as given in Thoresby's *Diary*, II, p. 5 (May 17, 1708); II, p. 12 (Dec. 28, 1708); II, p. 43 (Feb. 16 and 17, 1709); etc.

[1] See Thoresby's *Diary*, May 17, 1695, as given above. On another occasion, Thoresby was detained four days at Stamford by the state of the roads, and was then encouraged to proceed because of the coming of fourteen Scotch members of Parliament, who had to be in London at the time appointed, and who took Thoresby into their convoy. Thoresby's *Diary*, II, p. 16.

[2] Pepys, on Sept. 20, 1668, has this entry: "My wife and I mounted, and... we rode to Bigglesworth by the helpe of a couple of countrymen, that led us through the very long and dangerous waters, because of the ditches on each side...." *Pepys's Diary*, II, p. 320. Cf. also ibid., I, pp. 75–76, Mar. 20, 1660, and Thoresby's *Diary*, I, pp. 50, 61; II, p. 16, etc.

[3] In Anthony Wood's *Diary* we have the following memorandum:

"Monday, Apr. 26, 1669, was the first day that the flying coach went from Oxon to London in one day. A. W. went in the same coach, having then a boot on each side....They then (according to the Vice-Chancellor's order, stuck up in all public places) entered into the coach at the tavern door against All Soul's College, precisely at six of the clock in the morning, and at seven at night they were all set downe at their inn, at London." Robertson and Green, *Oxford during the Last Century*, p. 6. See also Clark, *Wood's Life and Times*, II, pp. 153, 155.

[4] Clark, *Wood's Life and Times*, II, p. 109; also Mason, *History of Norfolk*, I, p. 432; showing that in 1742 the (winter) coach from London to Oxford occupied two days on the journey.

[5] See also Littleton, *Proposal for the Highways* (1692), who says: "It is most certain that the High Wayes of England are extremely bad at present." See also Brit. Mus. 1879. c. 4 (28), 'Case of the Waggoners of England.' Brit. Mus. MSS. 12,496, pp. 263–91, 'Orders and Directions, together with a Commission for the better Administration of Justice,' speaks of the highways in all counties of England as being in "great decay" (p. 290).

portions of the great highways which had been considerably improved by 1750; but these were mere links in the longer thoroughfares, other portions of which were very bad. For the carrying and marketing of commodities, such as manufactured goods, which were produced all the year round, it was necessary, if the industry were to be progressive, that there should be a ready market available at all times of the year; but we have seen that, for half of the year at least, the majority of the roads were in such a condition as to prevent easy access to good markets. Taking into account the amount of the carrying trade of the country, which by the middle of the eighteenth century must have been of considerable importance, the extent of the good roads throughout England seems to have been wholly inadequate to the economic needs of the kingdom[1]. According to Hemmeon, the great post-roads of Tudor and Stuart times were important from a political, rather than an economic standpoint, in order to watch the Scotch and wild Irish and to keep informed in regard to French and Spanish politics[2].

We have now discussed in some detail the nature of the country roads. It is not necessary here to consider the condition of the roads in the urban centres at any great length, for we have already shown that the streets of the towns partook largely of the nature of the rural highways in their vicinity. In former pages, we have investigated this subject throughout the Tudor period[3]; and here let it be noted that

---

[1] De Saussure, a Frenchman, who, after personal observation, describes *England under George I and George II*, says (pp. 146–7): "The journey on the high roads of England, and more especially near London, is most enjoyable and interesting. These roads are magnificent, being wide, smooth, and well kept. Contractors have the care of them, and cover them when necessary with that fine gravel so common in this country. The roads are rounded in the shape of an ass's back, so that the centre is higher than the sides, and the rain flows off into the ditches with which the roads are bordered on either side." Then he goes on to speak of the way by which those who use the roads are compelled to contribute to the expense of their maintenance, through the establishment of toll-gates; and this he contrasts with the system in operation in France, by which the poor peasants are forced to make and keep up the high roads at their own expense and care. At first thought, we would be inclined to doubt our foregoing conclusion as to the English roads, for De Saussure's testimony seems to be unreservedly in favour of them; but a few pages later (pp. 165–6), he contradicts part of the above statement by saying that in London the pavement is very uneven, so that a passenger inside a hackney coach gets "most cruelly shaken." Apparently his object was to make a comparison of the roads of the two countries, to the detriment of those of France, and to make out a case against the corvée by showing that it did not produce as good roads as did the turnpike system of England. From the information we have already brought together, it is very evident that to describe the highways of England as "magnificent" was a gross exaggeration.

[2] Hemmeon, *History of the British Post Office*, p. 98.

[3] See page 86 et seq.

throughout the seventeenth century we observe little change from the conditions of the preceding century. Obstructions of all kinds were allowed to remain on the streets, until the court ordered their removal[1]. The pavements, which should have been kept in repair by those who occupied the adjoining houses[2], were in some cases so bad that the towns had to take action and engage a professional paviour to put the streets in repair[3]. Occupiers were required to regularly sweep up and carry away the dirt in front of their own houses[4]; but sometimes this work devolved upon public agents appointed for that purpose[5]. Pigs continued to wander around the streets of many towns, and at times they caused so much trouble that the town had to engage one or more swineherds to take them out to the town pasture and keep them there during the day[6]. All these and a multitude of other nuisances continued, despite the ordinances to the contrary. But we need not go further here, for what we have said regarding the towns in the Tudor

[1] Picton, *Liverpool Municipal Records*, i, p. 191; Markham and Cox, *Northampton Borough Records*, ii, pp. 267–8; Earwaker, *Manchester Court Leet Records*; Challenor, *Records of Abingdon Borough*, p. 153; Tate, *History of Alnwick*, i, p. 344; etc.

[2] Act 2 W. & M., c. 8; Earwaker, *Manchester Court Leet Records*; Challenor, *Records of Abingdon Borough*, p. 153; Chanter and Wainwright, *Barnstaple Records*, i, pp. 50, 70–71, etc.

[3] Markham and Cox, *Northampton Borough Records*, ii, p. 267; Picton, *Liverpool Municipal Records*, ii, p. 152; Hadley, *History of Hull*, p. 122; Gent, *History o. Hull*, p. 131. See also Brit. Mus., Harl. MSS. 2057, p. 116.

[4] Challenor, *Records of Abingdon Borough*, p. 153; *History of Guildford*, p. 21; Markham and Cox, *Northampton Borough Records*, ii, pp. 268, 269, 270; Brit. Mus. 816. m. 9 (13).

[5] Challenor, *Records of Abingdon Borough*, p. 153; Markham and Cox, *Northampton Borough Records*, ii, p. 265.

It was evidently a step in advance when a town appointed one or more persons to look after the cleaning of the streets, or the repair of the pavement, and paid them for their work. In some cases, the appointment of a scavenger or a paviour did not bring much, if any, immediate change in the condition of the streets (see Scott, *Berwick-on-Tweed*, p. 303); but since this was a work which, unremunerated, would be done by only comparatively few, we can see that the decision to pay for such disagreeable work was in the line of progress, even if immediate results were unnoticed. (See Brit. Mus., E. 856 (4), 'An Act of Common Councell for the City of London.')

With reference to paying for the work of a paviour, an interesting case is presented by Liverpool. In the records of that town we find it stated that: "In 1750 Edmund Parker was appointed paviour (who was the lowest proposer), to keep the pavement of all the streets of the town in repair at £90 a year, for seven years, allowing him the boon and statute work of the town (according to printed proposals); and 18*d.* a yard for all new work." (Picton, *Liverpool Municipal Records*, ii, p. 152.)

[6] Chanter and Wainwright, *Barnstaple Records*, pp. 70, 71. In Liverpool in 1654, an order was issued requiring swine to be kept off the streets on Saturdays and Sundays. On other days they could run as they pleased. In the following year a stricter order was made, requiring them to be kept off the streets altogether.

period will apply also during the succeeding century[1]. In the town records, however, we do not find in the seventeenth century so many presentments of individuals because of such offences, as we find in the sixteenth century; thus we conclude, either that there was more care taken of the streets, or else that the records are less complete in the later than during the earlier century. The latter alternative would seem to be wholly unfounded; so that, in all probability, a change was taking place in public sentiment against allowing these glaring nuisances to continue. We have instances of towns that were anxious, not only to see that their streets were paved, but also to preserve them when they had been paved; the latter was to be effected by requiring that all carts used on paved streets should have broad wheels not shod with iron, and should be drawn with only few horses[2]. Doubtless, the increasing use of coaches, in some towns, in the century after 1650, had some influence in the direction of improving the streets. But when we

In 1646, a swineherd had been appointed "to looke to the swyne all the day long from trespassing about the towne." Picton, *Liverpool Municipal Records*, i, pp. 191, 234.

In the borough of Reading, the amounts received as fines from people who allowed pigs to run on the streets were considerable (v. Guilding, *Records of the Borough of Reading*, iii, p. 243). Apparently these animals were a great nuisance on the streets, for in 1633, the fine imposed upon any man who let his pigs go in the streets or market-places, was increased from 12*d*. to 2*s*. (ibid., iii, p. 180). Here, too, they engaged "hogherdes."

We do not wonder that some localities became plague infected, because of their unsanitary conditions.

[1] With reference to London, an interesting side light is thrown upon the condition of the streets by an Act passed in 1690, Act 2 W. & M., c. 8, which shows that ashes and other annoyances had been cast into the open streets until they were both filthy and dangerous. This Act was passed, in part, to put a stop to such practices. An Act of Common Council had been passed Sept. 11, 1655, to put an end to these and other nuisances, but evidently, like much of the legislation of the time, it had proven ineffective. v. Brit. Mus., E. 856 (4). For 'Abstract of the Forfeitures and Penalties set and imposed on Offences done contrary to the Act of Parliament for Paving and Cleansing the Streets' of London, see Brit. Mus. 816. m. 9 (13).

[2] Act 2 W. & M., c. 8, passed in 1690, for the improvement of the London streets, required that where streets were paved, the wheels of carts, drays, etc., were to be not less than six inches wide, and must not be shod with iron, nor drawn with more than two horses, under penalty of 40*s*. for each offence.

In Hull, in 1718, for the better repairing of the streets, "all carts or carriages shod with iron, belonging to the inhabitants thereof, or any brick carts so shod, were forbid to be used within the same, from and after the 24th of the then current month, under the penalty of 5*s*., to be paid by every person so offending." (Hadley, *History of Hull*, p. 303.) A quarter of a century before, in 1692, the people of Hull had asked Lord Dunbar, at Burton, for a supply of 40 tons of cobbles from the mouth of the Humber for paving purposes (ibid., p. 286).

have said all we can in their favour, we are still compelled to decide that, particularly at certain seasons of the year, many of the highways through the towns lacked the characteristics of a good road[1].

After considering the character of the highways both along rural and urban lines, we next inquire into the methods suggested for improving them. It was fully recognized that the legislation was to blame for some of the evils connected with the performance of statute labour. With the roads as they were at that time, without a solid foundation or a substantial surface, it was preposterous to allow heavy loads of forty, fifty, sixty, and sometimes seventy hundredweight to be drawn along them, even with the payment of extra toll, for it was inevitable that these great weights should cause the formation of ruts in the soft material of the roads. The Act of 1662 provided against this[2], but there seems to have been much connivance against the law and evasion of it, permitting the heavier loads still to be carried[3].

Then, too, the law operated inequitably upon different classes of the people, and thus caused friction among those who had the greater burdens of statute labour. According to the law, every person living in each parish, for every plough land in tillage or pasture occupied by

---

[1] The means used for the maintenance and repair of the streets were practically the same as those adopted under the Tudors, namely, statute labour, tolls, assessments, money obtained by gifts from individuals, gilds, etc. In nearly every case, the community required the individual to attend to the paving, and other work, opposite his own house. But, as was shown above, in some towns corporate responsibility came to take the place of individual responsibility. Challenor, *Records of Abingdon Borough*, p. 153; Tate, *History of Alnwick*, I, p. 344; Baker, *Nottingham Borough Records*, v, p. 130; Markham and Cox, *Northampton Borough Records*, II, p. 267 et seq.; Chanter and Wainwright, *Barnstaple Records*, I, pp. 70–71, etc. Regarding gifts for repairing roads, v. Scott, *Berwick-upon-Tweed*, pp. 229–30, 301; Thoresby's *Diary*, IV, pp. 100–1; and a peculiar case recorded in Hudson, *Memorials of a Warwickshire Parish*, pp. 109–10, 125.

Not only were the streets neglected, but they were, in many cases, so narrow as to be inconvenient. This was the case even with some of the principal towns, like Cambridge, Bristol, Lynn, Norwich, and Ipswich. Brit. Mus. MSS. 15,776, Milles' *Tours in England and Wales*, 1735–43, pp. 27, 69, 92; Brit. Mus. MSS 22,926, *Tour through England*, 1742, pp. 9, 22, 160. Certain places, including some that were rather unimportant, had streets of a good breadth (Brit. Mus. MSS. 22,926, pp. 25, 27, 102, 140). It is probable that there were but few towns that were so well provided as Bury St Edmunds, which had sidewalks paved with broad, flat stones, or else with bricks and tiles turned upon their edges, and with posts to hinder carriages from coming upon them (Brit. Mus. MSS. 15,776, p. 47).

[2] Act 14 Car. II, c. 6. It directed that the load should not, at any one time, be more than twenty hundredweight between Oct. 1 and May 1, nor more than thirty hundredweight between May 1 and Oct. 1. See Proclamations of 1629 and 1635, for restricting the weight to be carried. Rymer's *Foedera*, XIX, pp. 130–1, 697.

[3] See page 55 and footnotes of that page.

him, and lying in that parish, had to send a cart or waggon, with horses
or oxen, and also two able men, during the same six days of the year,
for amending the highways.   Differences of opinion arose as to what
should be the content of a plough land[1], and an order of explanation
was made in 1619 that it should be regarded as one hundred acres;
but afterwards, in the time of Charles I, eighty acres were to be accounted
a plough land[2].   A man who usually went with two teams or draughts
and kept perhaps one hundred acres or more of tillage land, generally
sent but one waggon or cart to work on the roads; and a man who kept
but one draught and had only a small piece of tillage, had to send his
waggon or cart furnished to the same work.   This was thought to be
unreasonable, since the man who employed two teams on the road, and
thereby caused the greater damage to the roads, it was said, should
contribute the more to their repair.   Again, if a man occupied a plough
land in pasture for feeding cattle, but kept neither cart nor plough, he
was required by the statute to send a cart and two able men to the
road; likewise the man that kept a plough or cart only to do work for
other men, though he occupied little or no land or pasture in his own
name, was also required, in fulfilment of his statute labour, to send to
the roads a cart and two able men.   The burdensome inequalities which
were possible under the law became more evident with the passing of time;
and the necessity of some change in the legislation was recognized by
those who were acquainted with the operation of the existing statutes[3].

To remedy the foregoing conditions, it was suggested that, instead
of retaining the statute labour on the basis of the number of teams and
ploughs that a man kept, those who were liable for such labour should

---

[1] Meriton, *Guide to Surveyors*, pp. 35–36, shows the different views that had been
held as to the amount of a plough land.

[2] Ibid., p. 36.   Mather on *Highways*, p. 21, speaks of a " plow-land " and a hide
of land as the same thing, and says that (in 1696) the plough land was commonly
allowed to be as great a portion of land as might be yearly tilled with one
plough.

[3] Mather on *Highways*, p. 27, says: "To speak (then) plain, the way and manner
of charging persons by the old statutes still in force, seems to me as unequal and
oppressive as it is uncertain and obscure." See also Meriton, *Guide to Surveyors*,
pp. 38–39, showing some hardships imposed by the highway laws as they existed
in the last years of the seventeenth century.   It was absurd that every poor cottager
and labourer, who had scarcely bread enough for himself and his family, should
be charged with six days' labour for repairing the highways, when he did not use
them at all, and when his labour from day to day barely sufficed for his family's
maintenance.   On account of this, some surveyors, especially those who lived in
towns where the poor were abundant, admitted the children of these poor people to
work on the roads instead of their parents; and where they had no children to
send, their labour was frequently overlooked or only a small money payment asked
in place of it. (Mather on *Highways*, p. 19.)

be charged an assessment, levied according to the rate of the poor tax, and the amount of this assessment should then be expended in hiring the labour of men and teams to work on the roads. In the time of Cromwell, this suggestion had been incorporated into a royal Ordinance for the highways, but this had been repealed upon the restoration of the Stuarts. Now, at the close of that century, the idea was revived, with very strong reasons why the system should be adopted[1]. Apparently, however, the country was not ready for such a sweeping change, and it was not until after the first third of the next century that the old system was replaced by the other.

But beside proposed changes in the statutory provision for highway repair, there were suggestions for improving the mechanical construction of the roads. As early as 1610, it was recognized that much of the labour bestowed upon them had been expended "to little or no purpose," in fact, had been utterly lost, for want of proper material to be used, or because of improper use of the material that was available[2]; and although the plan put forward at that time was much in advance of the prevailing practice of the day, it was, nevertheless, very crude and of little value for making permanent road-bed[3]. The chief feature of the plan was the formation of a good earthen foundation, and then the making of a firm and highly convex surface, which would prevent

[1] Mather on *Highways*, p. 27 et seq. The chief reasons he urged for the adoption of the assessment, and the abolition of statute labour, were:

First, under the assessment plan, the labourers hired would carry full loads and work hard, whereas under the old system the parishioners pleased themselves as to how little they carried for a load and how hard they worked; so that under the old system it took several men to do as much as would be done by one man who was paid for his work.

Second, the parishioners would be assessed proportionately to what they occupied in lands, or possessed in goods, and thus the burden would be more equitably imposed.

[2] Procter, *A Worke on Mending the Highways*, p. A. As far as our present information goes, this man was the first to make any suggestions in regard to the construction of roads, from the mechanical point of view.

[3] The following summary of Procter's scheme will show, in outline, how he would handle the roads:

Since the one principal cause of the bad ways was that the water remained on the highways, making them easily cut through by wheels, the fundamentally important thing was to get the road so convex that the water would turn quickly to the ditches along the sides, and leave the road dry. His first thought, therefore, was in regard to the foundation. This was to be made at least a rod wide, that is, wide enough for two carts to meet and pass; and care must be taken that it should be firm and well bound together before allowing carts to go upon it. The foundation should then be covered with stones, gravel, sand, or any other hard or dry matter, and upon all there should be a good covering of small gravel, rubbish or sand, or any other stones broken small. The surface was to be formed so that the middle should be two feet higher than the sides, thus allowing water to run off quickly into the ditches.

any water from lying on the road. In 1675, another suggestion was added by Thomas Mace to the general features of the foregoing plan, for he would have men stationed along the roads to keep them smooth and well-rounded, and prevent them from being worn into deep ruts. This would be done day after day and year after year[1]. But we have no record that either of these plans was successfully applied.

An entirely opposite view was maintained by Phillips in 1737; he thought that water was a good thing for roads, because it washed out the loam or clay, and left only the sharp gravel or stones. These when consolidated would form a hard, good road, that would be free from dirt in winter and dust in summer. When such a road had been formed, water would keep it in good condition by washing off what would only produce mud if left on[2]. It is easy to go thus far with him, in his consideration of what was necessary to make a good road; but when he advocates that instead of raising the road above the level of the adjoining lands to keep off the water, they should be made lower so as to bring on the water to wash them, we are unable to follow him. His theory as to what was proper material for road building was correct, and was put in practice by the road engineers half a century after this time; but to have made the roads lower than the adjacent land, so as to admit water from the land to wash the roads, would have meant that while the surface would be clean, the foundation of the road would be softened by the water lying there, and so in a very short time the intended good would result in positive injury. It does not appear that this plan was ever put into effect, probably because its ultimate results were too clearly foreseen to cause much deception.

The proposal made by Littleton in 1692 introduces an entirely new principle from those already mentioned[3]. He was well aware that "the scrambling way of sending in carts and labourers" would never mend the roads, nor would it be accomplished by single parishes even though they were fined over and over again for not doing more than they were able. He favoured the policy of placing the burden of repair more upon the hundred and the county, and less upon the separate parishes; for the hundred could do what the parish could not, and

---

[1] Mace, *Highways of England*. Like Procter, Mace thought that water was the great corrupter of all highways, and that by each of these road labourers keeping the road so repaired on his own section that water would not lie there, the maintenance of the roads would be assured. He would have one surveyor-general to look after the men and work on each hundred or two hundred miles, and each labourer would look after ten miles of road in most soils, after it had been once well mended.

[2] His complete scheme is elaborated, with drawings, in Phillips, *High Roads of England*.

[3] Littleton, *Proposal for the Highways* (1692).

the county could do what the hundred could not. He would abolish statute labour, and substitute an annual money payment of 4*d.* in the £1 upon the land. The collectors of this tax in each parish should pay one-half of their receipts to the parish surveyor, and the other half to the surveyor of the hundred. The surveyor of the hundred was to employ one-half of his receipts upon the roads under his control, and pay over the other half to the surveyor of the county. The county surveyor, in his turn, was to pay one-fourth of his money to the surveyor-general of England, and to use the rest on the bridges and great roads of the county. The surveyor-general was to employ all his money on the London roads, since these were the nucleus of the highway system of the kingdom, and with them the whole kingdom was concerned. In this way, he thought that by the outlay of a large amount of money on the roads in and near the metropolis, and the expenditure of a lessening amount on the roads receding from the metropolis, the great roads would be put in good condition and on the more remote roads there would be expenditure according to their importance. In every case, surveyors should do their work substantially as far as they went. In addition to the ordinary work done on the roads, authority should be given for turning a road, for opening and widening of roads in enclosed counties[1], and for cutting down hills and raising hollows, so that by these three means physical difficulties might be overcome. Then, when a road had once been made good, tolls might be taken and used for the maintenance of the highway in good condition. But we find no case where anything approaching this system was adopted, possibly because the parochial sentiment was strong and the inhabitants of a parish wanted to see all money raised in that parish devoted to the local uses only, and not used in part to help wider areas[2].

When we compare the extent of the roads in the time of Henry VIII with that in the period of the later Stuart or early Brunswick sovereigns, we observe that the same general lines were followed by the main roads in both cases; but in the later period the chief roads leading from London to different parts of the kingdom had become extended where that was possible, and were more numerous than in the early period[3].

---

[1] See Brit. Mus., Harl. MSS. 2264, p. 272, for the text of a license granted to Thomas Vernon to enclose a certain highway within his manor, so long as another equally convenient highway made by him should remain a public road. This was signed by the Attorney-General by royal warrant, June 24, 1710. Another, similar to it, is found in Brit. Mus., Harl. MSS. 2263, p. 328 (1708).

[2] For what seems, at the present day, like a visionary project, see the pamphlet in the British Museum by J. P., entitled 'For Mending the Roads of England: it's proposed,' apparently written about 1715.

[3] The roads in the time of Henry VIII, as given by Harrison, we have previously

In regard to statistics as to the actual length of road that was open for traffic at these different times, we are at a loss to present any, since none have come to our notice. It almost seems as if we should be well within the truth if we should say that such statistics, for that period, would be impossible of attainment.

Travelling.

We have hitherto seen that down to the end of the reign of Elizabeth all long journeys, even those performed by the nobility, were accomplished on horseback[1]. As early as the reign of Richard II, wheel carriages, under the name of "whirlicotes," were in use in England[2]; but we cannot think that the people of that day who rode in these vehicles were any better off than those who walked on foot, unless it might be in name. Even Queen Elizabeth's journeys were usually made on horseback[3], and all the better classes of the people who could afford it took this means of travelling. It would appear that this

outlined. Those of 1675 are given in the introduction to Ogilby's *Britannia*; and those of 1719 are found in Ogilby, *Survey of the Principal Roads of England and Wales*, I and II. In Harrison no mention is made of the north road (from London to Berwick) having any connection with Edinburgh, nor of the north-western road (from London through Chester, to Carnarvon) having any connexion with Holyhead, although, in all probability, these extensions were as good as the rest of these roads in England. At the later dates we have spoken of, distinct reference is made to the more distant terminus in each case. On the other hand, Harrison extends the south-western road from London to St Buryan, in Cornwall, and the northern division of the north-western road from London, through Warrington and Preston, to Keswick and Cockermouth; whereas *The Traveller's Companion* (1702), Brit. Mus. 712. a. 4, pp. 54–58, places the terminus of these roads as Plymouth and Lancaster respectively. So we cannot put strict dependence upon the accuracy and completeness of these outlines of the roads; they give the general directions, but are not intended to be comprehensive, except in the case of Ogilby's survey. The number of main roads leading out from London at the latter time is considerably increased over that given by Harrison, but this is largely explained by partial duplication of routes near London. There must have been an augmentation of the network of subsidiary and cross roads, which would be still further increased by the middle of the eighteenth century; but the great roads were still those from London to Edinburgh, London to Holyhead, London to Bristol and South Wales, London to Cornwall, London to Dover, and London to Yarmouth.

[1] See page 43 et seq.; also Hentzner's *Travels in England*, 1598, p. 1.

[2] In Stow's *Survey of London* (written in 1598), p. 32, we read: "Of old time coaches were not known in this island, but chariots or whirlicotes, then so-called, and they only used of princes or great estates, such as had their footmen about them; and for example to note, I read that Richard II, being threatened by the rebels of Kent, rode from the Tower of London to the Myles end, and with him his mother, because she was sick and weak, in a whirlicote...."

[3] Nicholls, *Queen Elizabeth's Progresses*. This is not paged, and therefore I cannot cite the references more exactly.

system prevailed for at least a century and a half after this; and even down through the first quarter of the nineteenth century, riding post was, doubtless, as common as, if not more common than, coaching. The system under which horses were provided for travellers riding post continued to be about the same as that which we have noted as prevailing during the sixteenth century. In the Tudor times postmasters were not the only persons to furnish horses for this purpose, but they were subjected to competition from others. In 1603, however, the Privy Council granted postmasters a prior right in this matter, and it was only when they did not have a sufficient supply of horses that the traveller might apply elsewhere for them. At the same time the rate was fixed at $2\frac{1}{2}d.$ a mile, not including the guide's fee, for those who were on a governmental commission, and others were to settle their rate by agreement[1]. In 1609, the rate was raised to 3d. per mile, and the postmasters tried to enforce their monopoly more strictly[2]. But others were apparently getting some share of this business; and, to prevent this, an Ordinance was issued in 1654 providing that none but postmasters were to provide post-horses, and that the rate for each post-horse was not to exceed 3d. per mile. A prohibition was also made against riding post-horses more than one stage, without the consent of the postmaster from whom they were hired or of the owners of the horses[3]. With occasional alternations of rates, but always keeping them either $2\frac{1}{2}d.$ or 3d. per mile for each horse[4], the postmasters enjoyed the monopoly of letting horses to travellers until the middle of the eighteenth century[5]. For the longer journeys, the horses could be hired by the day, week, or month, at suitable rates[4]; but on the chief roads the common practice was to use the horses for a stage of

---

[1] Hemmeon, *History of the British Post Office,* p. 92.   [2] Ibid.

[3] Firth and Tait, *Acts and Ordinances of the Interregnum,* II, p. 1011, Section X of the Ordinance of Sept. 2, 1654.

[4] Fynes Moryson, *Itinerary,* Pt. III, c. I, p. 61, says that he who travelled upon the public business paid two and a half pence each mile for his horse, and as much for his guide's horse; but those who had no such commission paid three pence for each mile. By the day, the traveller's horse cost 2s. the first day and twelve to eighteen pence for each succeeding day that the horse was kept, and the traveller had to find him food both going and coming. These statements as to cost are different from what we find in the Acts of the Privy Council at an earlier date; for in the latter it is frequently stated that those who were doing the public business were to get their horses for one penny per mile for each horse (*Acts of Privy Council,* 1542–6, I, pp. 164, 333, 355, 465, 469, 527, etc.); and in some other cases, orders were given that horses should be provided at a "reasonable" price (ibid., I, pp. 384, 401, etc). This would seem to indicate that there must have been an increase in the amount of travelling in the seventeenth and early eighteenth centuries, which could bear the increased charge.

[5] Hemmeon, op. cit., 92–94; *Quarterly Review,* XCVII, 189–90.

ten to fifteen miles before they were changed for a fresh relay[1]. Travelling post was the fastest means of covering any distance; and none who could bear the expense of this ever engaged any other means[2].

But a new method of travel came in with the introduction of coaches. It is generally conceded that these vehicles came into England in the reign of Queen Elizabeth; but with regard to who brought them, or the exact time when they were brought, there is some disagreement. Anderson ascribes their introduction to (Henry) Fitz-Alan, the last Earl of Arundel, in 1580[3]; but this can hardly be true for that nobleman died in 1579. Stow says that coaches were not used in England until 1555, when Walter Rippon made a coach for the Earl of Rutland, which was the first ever made in England[4]. Taylor, the water-poet, informs us that coaches were not in use in England till 1564, when one was brought from the Netherlands and given to Queen Elizabeth[5]; but the source of his information was probably not very reliable, and we may be sure that coaches were in use before that date[6]. We have

---

[1] See introduction to Ogilby's *Britannia*, showing the number of stages on the great roads: On the Yarmouth road, the average length of each stage was about 9 miles; on the Dover road, 12 miles; on the Rye road, 20 miles; on the great western road, 15 miles; on the Bristol road, 15 miles; on the Chester road, 12 miles. Fynes Moryson's *Itinerary*, Pt. III, c. I, pp. 61–62, informs us that in southern and western England, and from London to Berwick, post-horses were established at every ten miles or thereabouts, and that they went about ten miles an hour. This, however, is probably overstated, for the average speed of the mail on the great roads was but three to four miles per hour (Hemmeon, op. cit., p. 100).

[2] In some cases, under pressure of the emergency of a great occasion, ground could be covered at rates varying from one to two hundred miles per day. Cary, *Memoirs*, pp. 149–50; Stow, *Abridgement of the English Chronicle*, I, p. 455.

[3] Anderson, *History of Commerce*, I, p. 421 and IV, p. 180, and Chalmers, *Caledonia*, IV, pp. 729–30, give the same date. The writer of 'Coach and Sedan Pleasantly Disputing for Place and Precedence, the Brewers-Cart being Moderator' (Brit. Mus. 012,814. e. 88) says that the first one was presented to Queen Elizabeth by the Earl of Arundel.

[4] Stow, *Summary of the English Chronicles*, p. 588—quoted in Adams, *English Pleasure Carriages*, p. 42.

[5] John Taylor, *The Old, Old, Very Old Man*, postscript thereto. Here, Taylor tells us that Thomas Parr, who was born in 1483, was eighty-one years old before there was any coach in England, for the first ever seen in England was brought from the Netherlands by William Boonen, a Dutchman, who gave a coach to Queen Elizabeth, for she had been seven years a queen before she had any coach. If this was the same coach that she had in 1572, it was quite an elaborate structure, for when she visited Warwick in that year, she "caused every parte and side of the coache to be openyed that all her subjects present might behold her. . . ." Nicholls, *Royal Progresses*, IV, Pt. I, pp. 57, 60.

[6] On Parr's authority, Taylor mentions this as the first coach ever seen in England. But either Parr had never heard of the earlier coach, or else his advanced age had caused him to forget. In the *Annals of the First Four Years of the Reign of Queen*

another definite statement as to the use of coaches in 1556; for in that year Sir T. Hoby offered the use of his coach to Lady Cecil[1]. In 1564 Sir William Cecil, in preparation for the coming of the Queen, went to Cambridge University, reaching there on the fourth of August, and because he was unable to walk, he rode thither in a coach[2]. Of course, it would be natural that coaches should have been used for some time in the city before attempting long journeys into the country, and we are not left without evidence that this assumption was borne out by fact, for at least as early as 1586 we find a reference to the coach and coach horses of the Lord Justiciary[3]. We would not presume to think that this was the first coach by any means; and it would probably be safe to say that the introduction of coaches dated from about the early years of the sixteenth century[4].

From that time onward, their use became much greater among the upper classes, especially in the metropolis, and not infrequently private coaches were found travelling on the country roads[5]. By 1584, the demand for them had become so great that a considerable trade in coach-making had sprung up[6]; and so great was the clamour against

Elizabeth, by Sir John Hayward, edited for the Camden Society, from Harl. MSS. 6021, by John Bruce, there is a reference to the fact that on a journey in 1559 the Queen was riding in a coach.

[1] *Burghley Paper,* III, No. 53, quoted by Markland, ' Remarks on the Early Use of Carriages in England,' in *Archaeologia,* xx, p. 462.

[2] Nicholls, *Queen Elizabeth's Progresses,* I. This is not paged consecutively, and therefore we cannot be any more specific in the reference.

[3] Brewer and Bullen, *Calendar of the Carew MSS., preserved in the Archiepiscopal Library at Lambeth,* 1515–74, p. 86.

[4] We must not suppose, however, that carriages of other names and description were not in use long before this. About the beginning of the sixteenth century, the domestic arrangements pertaining to one of the greater families of the nobility included many such carriages. For example, when the Percy household removed from one place to another they required seventeen carriages for that purpose, all of which they owned. (See *Northumberland Household Book,* pp. 386–91, where the arrangement of these carriages is given, on the occasion of a removal from place to place.) In fact, Lord Percy kept a servant whose duty was to attend to his carriages, chairs, chariots, close cars, and carts, and who was paid 10s. per year therefor (ibid., p. 351). Doubtless these facts concerning this one family could be repeated for many other such families at that time.

[5] We are told that "when Queen Elizabeth came to Norwich, 1578, she came on horseback from Ipswich...but she had a coach or two in her train" (Letter of Sir Thomas Browne to his son, dated Oct. 15, 1680, as found in Sir Thomas Browne's *Works*). See also Stow, *Annales,* p. 867; and also Dr Dee's *Diary,* p. 8, under date Sept. 17, 1580. Hentzner, *Travels in England,* 1598, p. 88, informs us that he and his party left London in a coach, in order to see the remarkable places in its neighbourhood, and went as far as sixty miles west (to Oxford) and north (to Cambridge).

[6] Stow, *Annales,* p. 867, says, with reference to coaches, that "after a while

this luxury that Parliament, in the later years of the reign of Elizabeth, had to take up the matter. A Bill was brought into the house in 1601, "to restrain the excessive use of coaches" in England, its alleged necessity being that, because of the greater use of horses among the common people, the Government would find it difficult to get enough horses for the army; but on the second reading this Bill was rejected[1], for Parliament was opposed to curtailing the liberty of those who could afford such a luxury. The number of coaches continued to increase, until early in the seventeenth century we are informed that the streets of London were almost stopped up with them[2]; and their rivals, the Thames watermen, who conveyed passengers to and fro by river, were loud in their denunciation of this innovation that was taking their living from them[3]. So many coaches were in use on the London streets,

divers great ladies, with as great jealousy of the Queen's displeasure, made them coaches, and rid in them up and down the countries, to the great admiration of all the beholders, but then by little and little, they grew usual among the nobility, and other of sort, and within twenty years (i.e., after 1564) became a great trade of coach-making." Sir Martin Frobisher, the discoverer, in his will, proved Apr. 25, 1595, bequeathed to his wife, among other things, his two coaches, and furniture and horses (v. Cartwright, *Chapters of Yorkshire History*, p. 131). William Lilly, in a play called 'Alexander and Campaspes,' printed in 1584, makes one of his characters complain of those who used to go to the battlefield "on hard-trotting horses, now riding in easie coaches up and down to court ladies."

That coaches were common in London about 1595, is evident from a stanza in Stephen Gosson's *Pleasant Quippes for Upstart Newfangled Gentlewomen*:

"To carrie all this pelfe and trash,
because their bodies are unfit,
Our wantons now in coaches dash
from house to house, from street to street.
Were they of state, or were they lame,
To ride in coach they need not shame.
But being base, and sound in health,
they teach for what are coaches make:
Some think, perhaps, to show their wealth,
Nay, nay, in them they penaunce take.
As poorer truls must ride in cartes,
So coaches are for prouder hearts."

[1] D'Ewes's *Journal of all the Parliaments during the Reign of Queen Elizabeth*, edit. 1683, p. 602.

[2] Fynes Moryson, *An Itinerary, etc.*, Pt. III, c. I, p. 62. He says that at that time pride was so greatly increased that there were few gentlemen of any account who had not their coaches, "so as the streets of London are almost stopped up with them." And even those who regarded "comeliness and profit," and were considered to be free from pride, found it more desirable to keep a coach than a number of horses. Taylor, in *The Old, Old, Very Old Man*, speaks of the way in which the coaches had multiplied and swarmed until they pestered the streets in 1605.

[3] In Dekker, *A Knight's Conjuring. Done in earnest: Discovered in jest*, c. VIII,

that on several occasions royal proclamations were issued for the restraint of their excessive use, not only because they blocked up the streets, but also because they were destructive to the highways[1]. These prohibitions, however, were often of little effect.

we find: "The sculler told him he was now out of cash, it was a hard time, he doubts there is some secrete bridge made over to Hell, and that they steale thither in coaches." This shows us the watermen's point of view as to coaches. It is still further exemplified by John Taylor, the water-poet, who also felt this jealousy at their introduction, and often railed against them:

> "Carroaches, coaches, jades, and Flanders mares,
>   Doe rob us of our shares, our wares, our fares:
>   Against the ground we stand and knocke our heeles,
>   Whilest all our profit runs away on wheeles:
>   And whosoever but observes and notes,
>   The great increase of coaches and of boates,
>   Shall finde their number more than e'r they were
>   By halfe and more within these thirty yeeres.
>   Then watermen at sea had service still,
>   And those that staid at home had worke at will:
>   Then upstart hellcart-coaches were to seeke,
>   A man could scarce see twenty in a weeke,
>   But now I think a man may daily see,
>   More than the whirries on the Thames can be."
>
>   Taylor's *Works*—'A Thiefe,' p. 111.

Again, in 'The World runnes on Wheeles' (*Works*, pp. 237–44), Taylor displays the watermen (in their opposition to coaches) in much the same tenor.

At a later time (1636) another writer expresses his invective in the following words: "Coaches and sedans (quoth the waterman) they deserve both to be throwne into the Theames, and but for stopping the channell I would they were, for I am sure where I was woont to have eight or tenne fares in a morning I now scarce get two in a whole day; our wives and children at home are readie to pine, and some of us are faine for means to take other professions upon us." ('Coach and Sedan Pleasantly Disputing for Place and Precedence, the Brewers-Cart being Moderator.') About this time also John Taylor in 'The Coaches Overthrow. Or A Joviall Exaltation of Divers Tradesmen and Others,' gives rhyming expression of the hope that soon the coaches will be no more in London.

---

[1] Rushworth's *Collections* (ed. of 1721), ii, pp. 46, 301, 318.

In a letter of Mr Garrard to the Lord Deputy, dated Jan. 9, 1633, he says: "Here hath been an order of the Lords of the Council hung up in a table near Paul's and the Black-Fryars, to command all that resort to the play-house there to send away their coaches...and not to return to fetch their company, but they must trot afoot to find their coaches, 'twas kept very strictly for two or three weeks, but now I think it is disorder'd again." (Strafford's *Letters and Dispatches*, p. 175.)

In another letter, dated June 20, 1634, Mr Garrard writes to the Lord Deputy as follows: "Here is a proclamation coming forth about the reformation of hackney coaches, and ordering of other coaches about London; 1900 was the number of hackney coaches in London, base lean jades, unworthy to be seen in so brave a city, or to stand about a King's court...." (Ibid., p. 266.) See also ibid., p. 507, for a proclamation to prohibit all hackney coaches from passing up and down in London streets.

As to the time when hackney coaches began to be used for the conveyance of the public we have no exact and well substantiated information. It was probably difficult for many years to know what were private and what were hackney coaches; but it would seem that the latter must have been introduced during the reign of James I, although the precise year is as yet unknown. By some we are told that they first appeared in the streets of London in 1625[1], by others in 1626[2]; but both of these appear to be erroneous, for in 1623 the Thames watermen complained vociferously that they, who had long enjoyed the monopoly of carrying the public to places along and near to the river, were now being ruined. From the number that must then have been in existence, we may reasonably infer that their introduction probably occurred in the early years of the seventeenth century. We are informed that in the first year of the reign of Charles I there were not above twenty hackney coaches to be had for hire in and about London, and these were kept in their stables, not at stands in the streets waiting to receive a call from those who wished to use them[3]. But if this were true, it seems incredible that there should have been about London nineteen hundred such coaches by 1634[4], or that, within London and its suburbs and within four miles compass without, there should have been anything like six thousand of them[5]. We have no figures from other sources by which to test those here given; but it would seem from the number of these coaches allowed in later years that the latter numbers must have been greatly exaggerated. In 1634, Captain Bailey sent four hackney coaches to stand for hire at the Maypole in the Strand, and this was the first time that a coach stand was found in the streets. The drivers were dressed in livery, and were to convey passengers to different parts of the metropolis at certain fixed fares[6]. This enterprise was found to be sufficiently remunerative to attract many others to it; and the increase in the number of hackney and private coaches was so great that they became a civic nuisance. In consequence, in 1634, it was seriously proposed that very

---

[1] *Letters and Papers of the Verney Family*, edited by John Bruce for the Camden Society, p. 185; Brit. Mus. 290. c. 30, 'A New and Compleat Survey of London,' I, p. 447.

[2] *Diary of Thomas Buxton*, I, p. 297 n.

[3] Rushworth's *Collections* (ed. of 1721), II, p. 817.

[4] Letter of June 20, 1634, written by Mr Garrard to the Lord Deputy, in Strafford's *Letters and Dispatches*, p. 266. See footnote 1, page 114.

[5] Brit. Mus. 012,814. e. 88, 'Coach and Sedan Pleasantly Disputing....'

[6] See letter of Lord Strafford to Mr Garrard, Apr. 1, 1634, as found in the Rushworth *Collections*.

strict limitations should be put upon their use[1]. Later in the same year, the London watermen strongly objected to the coaches taking away so much of their business; they were willing that any number of coaches should drive northward from London, towards Islington and Hoxton, but they regarded it as "intolerable presumption" on the part of the hackney coachmen to compete with the wherries on the river in carrying passengers from the city to Westminster[2]. For a time there was much outcry against the coaches, by reason of the fact that the great increase in their numbers tended to obstruct the streets[3]; but we do not note that this was directed against the hackney coaches any more than the private coaches. Nor could it reasonably be so, for the private coaches greatly outnumbered the others. So much objection was urged to the coaches, that in 1634, in answer to the petition of Sir Sanders Duncombe, which showed that in foreign countries the use of Sedan chairs had avoided the use of many coaches, the King, in order to recompense him for having these chairs introduced into England, and "for divers other good causes and considerations," granted him the

[1] In the *State Papers, Domestic*, CCLXVII, 36 (May 5, 1634), there is a paper in the handwriting of Lord Cottington, containing suggestions probably made to a Committee of the Council, to whom had been referred the question of regulations for hackney coaches. He suggested:

(a) No coach to be hired in or about London, but to go over three, four or five miles journey, and none at all to be used in the streets.

(b) None shall go in coaches in the streets but with four horses at least of his own.

(c) No sons of noblemen, nor gentlemen unmarried, shall go in the streets in coaches, except in company of their parents, after the age of ten, eleven, or thirteen.

(d) No coach to be lent to go in the streets but to such only as keep four coach horses of their own.

(e) No saddle horse to be used in the town and for the street with a snaffle and a Scot's saddle, but with a bit, etc. See also last reference in footnote 1, page 115.

[2] The Watermen's Petition, of June 8, 1634, is found in *State Papers, Domestic*, CCLXIX, 52. It shows:

(a) That hackney coaches were so many in number that they pestered the streets and made leather exceedingly dear; and, moreover, they carried every common person, to the great prejudice of the petitioners.

(b) That they plied in Term time at the Temple Gate, and sometimes carried three men for four pence each, or four men for twelve pence, to Westminster, or back again, which outrivalled the watermen. If the coaches would carry people north or south, there would be no grievance; but their carrying them east and west would ruin the petitioners, etc.

[3] Rymer's *Foedera*, XIX, pp. 572–4, speaks of the streets of London and Westminster and their suburbs as being "of late time so much incumbred and pestred with the unnecessary multitude of coaches therein used," that many people were thereby exposed to great danger. Brit. Mus. 012,314. e. 88, 'Coach and Sedan,' refers to the streets as having been stopped up with them. But in neither case is it said that there was any special hostility to the hackney coaches.

sole right, for fourteen years, to reap the benefit of his initiative.[1] By bringing in the use of these covered chairs, which were carried by men, it was expected that the congestion of the streets would be mitigated through doing away with the need for some of the coaches. In 1685, it was intended to add a further measure of relief, for the royal proclamation of that year, after declaring that the too extensive use of coaches would block up the streets, break up the pavements, and raise the prices of hay and provender, forbade the use of any hackney or hired coach in London or Westminster or the suburbs thereof, except such coach were to travel three miles or more out of these cities; and also commanded that no person should go in a coach on the streets unless the owner of the coach should constantly keep within these cities four able horses for His Majesty's service when required[2].

It would almost seem as if this proclamation would be prohibitive to the maintenance of many coaches, either hackney or private, for the requisition that every proprietor of a coach should be bound to keep four able horses always ready for the King's service imposed a considerable burden. Whether these restrictions failed to accomplish the desired end of putting down the "excessive" coaches, or whether they did it so effectually as to eliminate almost all of them, we have been unable to decide[3]; but, two years afterward, in 1687, the order was revoked, and authority was granted to James, Duke of Hamilton, Master of the King's Horse, for his life, to license fifty hackney coachmen for London and Westminster and the suburbs, each of whom was to keep twelve good and serviceable horses (not more) to be used on the coaches. For "other convenient places" in England and Wales, he was given full authority to license as many coaches as were necessary[4].

---

[1] Rymer's *Foedera*, xix, pp. 572–4, gives this patent in full. It is also given in Moore, *Omnibuses and Cabs*, p. 185.

[2] The heading of this proclamation was, 'A Proclamation for the Restraint of the Multitude and Promiscuous Uses of Coaches about London and Westminster.' It is given in full in Rushworth's *Collections*, ii, p. 316, in Rymer's *Foedera*, xix, p. 721, and in Moore, *Omnibuses and Cabs*, p. 186.

[3] That portion of the Commission to the Duke of Hamilton which required him to prohibit from being coachmen all others than those which he licensed, would seem to suggest that, despite the proclamation of 1685, coachmen were still carrying on their work. On the contrary, the initial part of the Commission, after referring to the fact that Charles I had restricted the hackney coachmen by the order of 1685, says that the King has now found it necessary to have a sufficient number for the use of the nobility, gentry, foreign ambassadors, strangers, and others; and this might imply that there were not enough of these coachmen left to meet the requirements.

[4] Rymer's *Foedera*, xx, pp. 159–60, 'A Special Commission touching Hackney Coaches.' Each of these licenses, of course, had to be paid for.

In addition to those which were licensed, there were probably some acting as hackneys which were not licensed, and also a vast number of private coaches[1]. Then the coaches had to compete with the sedan chairs, the use of which had become very common, and between these two kinds of carriage a vigorous rivalry sprang up[2]. Gradually, the old aversion to the hackney coaches and contempt for those who used them were subsiding; and the public were coming to recognize in them a very useful and convenient means of getting from place to place. In 1652, the number of hackney coaches for the metropolis and suburbs was increased from fifty to two hundred[3]; and in 1654 to three hundred[3]. By a proclamation of Oct. 18, 1660, Charles II forbade these coaches to ply for hire in the streets; but this was evaded, for Samuel Pepys, writing under date of Nov. 7, 1660, says that notwithstanding this proclamation he got a coach to carry him home[4]. In 1662 the number of hackney coaches licensed was increased to four hundred[5]; and when that Act expired, application was made to Parliament to have the number increased to five hundred[5].

[1] The writer of 'Coach and Sedan Pleasantly Disputing for Place and Precedence,' tells us that in that year (1636) the vehicles "in London, the suburbes, and within foure miles compasse without, are reckoned to the number of six thousand and odd." But this was probably only a guess, so that we cannot rely with certainty on the figures given. They tell us, however, that the streets were full of coaches, the great majority of which were doubtless private.

[2] From the first, the sedan chair was regarded with jealousy by the coachmen, and the watermen looked with suspicion upon both. (See statement of the waterman, in 'Coach and Sedan,' that both these vehicles deserved to be thrown into the Thames, since they were taking away the patronage and fares from the wherrymen on the river.) The rivalry between the coach and sedan gave rise to the publication of this pamphlet. See *Archaeologia*, xx, p. 468, as to the reason for the introduction of such chairs.

[3] Brit. Mus., E. 1064 (18), · Regulation of Hackney Coachmen in London' (1654). Also Cleland, *Statistics of Glasgow*, p. 154. See also Ordinance of 1654 as given in Firth and Tait, *Acts and Ordinances of the Interregnum*, 1642–60, ii, pp. 922–4. By the Ordinance of 1654, the number of hackney coaches was not to exceed 300, but the number of hackney coachmen was not to exceed 200. Of this two hundred, the Ordinance names thirteen who were to be master hackney coachmen. They were to meet and to present to the Court of Aldermen of London the names of two hundred persons, and out of these and such others as the Court of Aldermen might think fit, 187 more were to be elected.

[4] *Pepys's Diary*, Nov. 7, 1660.

[5] Act 14 Car. II, c. 2; also Brit. Mus. 816. m. 12 (154). By this Act, no license was to be granted to any person following another occupation, and nobody might take out more than two licenses. Licensed coachmen were to pay a yearly rental of £5. Preference was to be given to "ancient coachmen," and to such as had suffered for their service to Charles I or Charles II. Fares were specified according to time and distance.

It would almost seem that the terms of the statute must have been overstepped,

On the country roads, the use of coaches for the public conveyance of passengers from town to town seems to have been delayed for some time after their use in London had become general. This was, in all probability, due to the condition of the roads at that time. The statement is made that the first mention of a coach as a public accommodation is given by Dugdale in his *Diary*, from which it appears that a Coventry coach was on the road in 1659[1]. But a more careful reading here would show us that Dugdale refers to one journey out of London by coach in 1657[2], and by 1661 and 1662 stage coaches were frequent on the roads leading from that centre[3]. That stage coaches were running to many parts of the kingdom, some at great distances from London, by the year 1658, is evident from advertisements of that year, which show that there was regular stage coach communication from London to Devon and Cornwall upon the south-western road, to places as far north as Newcastle and Edinburgh, and to places in the north-west, like Chester, Preston, and Wigan[4]. Along these highways of travel,

and that many even as early as 1659, were exercising that calling without conformity to the law. A traveller who, in that year, gives us 'A Character of England' (Brit. Mus. 292. a. 43, pp. 8, 27), says that the London streets were pestered with hackney coaches, insolent carmen, and others. The statutory three hundred such coaches would not be anything like sufficient to produce the aforementioned results. Compare also Brit. Mus., T. 1860 (3), 'A Journey to England (1700),' p. 5, in which the writer, a Frenchman, speaks of the troublesome conditions in the London streets.

[1] Smiles, *Lives of the Engineers*, I, c. II.

[2] Dugdale, *Diary*, p. 102, Dec. 10, 1657, says, "I came out of London, with Mr Prescott, by coach, by Aylesbury."

[3] Dugdale, *Diary*, pp. 104, 105, 108, 109, 112, 117, etc. The *Diary* of Anthony Wood first mentions a stage coach under the year 1661.

[4] One of the advertisements here spoken of is found in the 'Mercurius Politicus,' of Apr. 1, 1658. It has also been transcribed in the pages of the *Quarterly Review*, XCVII, p. 189 et seq. It is as follows:

"From the 26 day of April 1658 there will continue to go Stage Coaches from the George Inn, without Aldersgate, London, unto the several cities and Towns, for the rates and at the times, hereafter mentioned and declared;

Every Monday, Wednesday, and Friday. To Salisbury in two days for xxs. To Blandford and Dorchester in two days and half for xxxs. To Burput in three days for xxxs. To Exmaster, Hunnington, and Exeter in four days for xls.

To Stamford in two days for xxs. To Newark in two days and a half for xxvs. To Bawtrey in three days for xxxs. To Doncaster and Ferribridge for xxxvs. To York in four days for xls.

Mondays and Wednesdays to Ockinton and Plimouth for ls.

Every Monday to Helperby and Northallerton for xlvs. To Darneton and Ferryhil for ls. To Durham for lvs. To Newcastle for £3.

Once every fortnight to Edinburgh for £4 a peece—Mondays.

Every Friday to Wakefield in four days, xls.

All persons who desire to travel unto the Cities, Town, and Roads herein hereafter

the coaches wended their way, and the passengers were compelled to submit to many discomforts and annoyances. That stage coaches connected London with "remote places" of the kingdom as early as 1654, we gather from the express statement of the Ordinance of that year[1]. But we are compelled to take issue also with those who state that "there was no means of forwarding passengers until the time of Cromwell[2]," for it is inconceivable that such a system as that which we have observed in 1658 should have been of such rapid growth as would be necessary if we did not go back before the time of the Protector. We are not left, however, to mere conjecture in this matter, for Chamberlayne, writing in 1649 with reference to the state of the country, informs us that there was then, by stage coaches, an "admirable commodiousness" for travel from London to the principal towns of the country, and that these arrangements were far in advance of the foreign post[3]. If that were the case, it must have required some years

mentioned and expressed, namely—to Coventry, Litchfield, Stone, Namptwich, Chester, Warrington, Wiggan, Chorley, Preston, Gastang, Lancaster, and Kendall; and also to Stamford, Grantham, Newark, Tuxford, Bawtrey, Doncaster, Ferriebridge, York, Helperby, Northallerton, Darneton, Ferryhill, Durham, and Newcastle, Wakefield, Leeds, and Halifax; and also to Salisbury, Blandford, Dorchester, Burput, Exmaster, Hunnington, and Exeter, Ockinton, Plimouth, and Cornwal; let them repair to the George Inn at Holborn Bridge, London, and thence they shall be in good Coaches with good Horses, upon every Monday, Wednesday and Fridays, at and for reasonable Rates."

With the great western road and the Dover road equally well provided with these facilities, as they certainly must have been, we can readily see that there was a framework of lines of travel, by means of which almost any of the important places could be reached, although the rate of travelling was slow and the discomforts many.

[1] Firth and Tait, *Acts and Ordinances of the Interregnum*, 1642–60, II, pp. 922–4.

[2] *Quart. Rev.*, XCVII, p. 189.

[3] Chamberlayne, *The Present State of Great Britain*. He says: "There is of late such an admirable commodiousness, both for men and women, to travel from London to the principal towns of the country that the like hath not been known in the world, and that is by stage coaches, wherein anyone may be transported to any place sheltered from foul weather and foul ways, free from endamaging one's health and one's body by hard jogging or over violent motion on horseback, and this not only at the low price of about a shilling for every five miles, but with such velocity and speed in one hour as the foreign post can make but in one day."

The records of the Corporation of Gravesend give us information that by the year 1647, stage coaches were travelling between that town and Rochester, for the regular conveyance of passengers. These coaches carried to Gravesend those who wished to take the "tilt-boat," or tide boat, from there to London; and they took back from Gravesend those who had just come in on the tide boat from London. Because they carried passengers to and from the tide, they were given the name of "tide coaches," but they were, in reality, the ordinary stage coaches. While, from the nature of their business, they did not travel far, but rather kept up

for the attainment of this degree of progress in the means of communication; and we should be safe in saying that stage coaches were performing their journeys at least as early as 1640. Even here, however, we cannot stop, for we can trace them back by positive evidence to 1637, when a coach went twice a week from St Albans to London. Taylor, in detailing the carriers that came to the capital in that year, says: "The carriers of St Albans do come every Friday to the sign of the 'Peacock' in Aldersgate Street; on which days also cometh a coach from St Albans, to the 'Bell' in the same street. The like coach is also there for the carriage of passengers every Tuesday[1]." This is a clear case of a regular coach service twice a week between these places. In this connexion, it may be urged that, in another place in the same work, Taylor uses the words "waggon" and "coach" synonymously, and that therefore they may be so used here[2]. But there is no possible way of reading such a meaning into the quotation above given, for it distinctly mentions two separate kinds of vehicle coming from St Albans, one for carrying goods, and the other for carrying passengers. Then, again, it is impossible to make any sharp distinction between hackney coaches, so-called, and stage coaches; for those hackney coaches which were used for the conveyance of persons between the city and their residences in the country, three to six miles out[3], were virtually stage coaches which travelled only short distances. It is, therefore, beyond dispute that the stage coaches were in use before 1635, for in that year there was a considerable number of them; but how much earlier than this we are unable to say[4].

With increase in the amount of this kind of travel[5], there was also

communication between adjoining towns, yet they were engaged in the stage coach business just as much as those that travelled longer distances (Cruden, *History of Gravesend*, p. 321, gives the quotation from the Corporation Records).

[1] John Taylor, *The Carrier's Cosmography*, p. 229.

[2] Ibid., p. 295. Taylor says in this place: "Also the Waggon or Coach of Hatfield doth come every Friday to the Bell in Aldersgate Street."

[3] See Royal Proclamation of 1635 (Rymer's *Foedera*, xix, p. 721); Brit. Mus. 012,314. e. 88, 'Coach and Sedan;' etc.

[4] One of the chief difficulties in deciding this matter is, that in some of the statements of the time no hint is given as to whether it was a stage coach or a private coach that was used in the performance of a journey. For instance, in Zinzerling's *Description of England* (written about 1610), p. 133, we read that travellers generally went on horseback, but sometimes in coaches, which were too dear. We incline very strongly, from the context, to the belief that stage coach is here meant, but, of course, it is impossible to substantiate this with absolute certainty.

[5] Tombs, *The King's Post*, p. 23, tells us that there were only six stage-coaches known in 1662. It is certainly difficult to make any positive statements of this kind; but I would judge it to be more correct to say that there were six great roads along which stage coaches travelled, with London as a centre.

a demand for increased speed, and the "flying coaches" were introduced. The first of those that we have seen mentioned was the Oxford Fly, which began to run from Oxford to London during the summer half of the year 1669[1]. During favourable weather and with a good road, these would sometimes cover twenty leagues or more in a day[2]. The advantages from this swifter conveyance were apparently not immediately appreciated; and either from lack of support, or else because of the badness of the roads in the winter season, the flying coaches often reduced their rate of travelling during that half of the year[3].

In addition to the stage coaches for the conveyance of passengers along the country roads, stage waggons also pursued their regular routes, carrying chiefly goods, but sometimes also passengers. These probably came into existence about the beginning of the reign of Queen Elizabeth[4]; and because of their advantages over the old pack-horse method of carrying, they were soon found in frequent use[5]. Their work was carried on as regularly as was possible, considering the bad condition of the roads and the lack of uniformity as to the amount of traffic offered from time to time. They were expected to depart from and arrive at each terminus of their route on certain days[6], but it was

---

[1] Quoted from Wood's *Diary*, in Roberson and Green, *Oxford during the Last Century*, p. 6; see also Clark, *Wood's Life and Times*, II, pp. 153, 155.

[2] See a few pages farther on, where we take up the subject of the rate of speed.

[3] See the statements made a few pages farther on, when we are considering the rate of travelling.

[4] Stow, *Annales*, p. 867, says that "About that time (1564), began long waggons to come in use, such as now come to London, from Canterbury, Norwich, Ipswich, Gloucester, &c., with passengers, and commodities."

[5] The pages of Taylor, *The Carriers' Cosmography*, will show us how many of these carriers' waggons were coming to London from different parts of England in 1637. Brit. Mus., C. 32. d. 8, 'A Direction for the English Traviller,' published in 1643, only a few years after Taylor's work, contains 'A Brief Director for those that would send their Letters to any parts of England, Scotland, or Ireland. Or, a List of all the Carriers, Waggoners, Coaches, Posts, Ships,' etc., that connected with London by land and sea. Like Taylor's work, it shows when the carriers came to London and when they returned, also at what inns in London they might be found. In 1690, the publication of Delaune's *Angliae Metropolis* (pp. 401–41), gave an alphabetical account of all the carriers, waggoners, and stage coaches that connected all parts of England and Wales with the metropolis, and their days of arrival and departure. It shows that there must have been a heavy carrying traffic for that time. It is not always possible in this work to distinguish what were "coaches" and what were "waggons," for while the two words were sometimes used interchangeably, at other times there is mention of what is called a "coach and waggon." The comparison of Taylor's list of 1637 and Delaune's list of 1690 shows considerable increase of traffic during this half century.

[6] See works just cited of Taylor and Delaune; also London *City Mercury*, Nov. 4, 1675, p. 1, giving the advertisement of the Coventry Carrier, which made

inevitable that they were not always able to adhere to their schedule, so they made it conditional upon "if God permit." At first, they were the only wheeled vehicle for effecting interior communication, for stage coaches had not yet come in[1]; and many who could not afford the expense of posting were glad to avail themselves of these long, covered waggons, even with their tedious journeying. The carriers themselves took charge of the delivery of letters, parcels, and heavier goods, as well as the carrying of passengers; so that they were an important class of people, so far as the social and business economy was concerned[2].

Like many other useful improvements, the stage coaches encountered much opposition from those whose interests were likely to be adversely affected. One of the most violent pamphlets against the stage coaches was written above the initials J. C. (John Cressett, or Cressel), and the

only one round trip from London to Coventry and return in a week, and that only "if God permit."

[1] How the stage waggons were affected by the proclamation of 20 James I and by those of Charles I in 1629 and 1635 which forbade any common carrier using a four-wheeled vehicle or carrying more than twenty hundredweight at a time (v. Rymer's *Foedera*, xix, pp. 130–1, 697), we cannot say. In all probability these edicts were not much enforced.

[2] In the *Life and Correspondence of Sir George Radcliffe*, we see very clearly how important was the office of the public carriers in carrying letters and parcels from friend to friend, when these were widely separated from each other. Writing from Oldham, in Lancashire, to his mother at Thornhill, in Yorkshire, under date of Aug. 1, 1607, he says: "I received yesterday two shirt-bands by George Armitage the carrier" (p. 13). Writing to her, under date of Aug. 7, 1607, he says: "I received of George Armitage the carrier my hat and...the cloth..." (p. 15). Writing to her again from Oldham, Feb. 29, 1607, he says: "These rude lines are to signify to you that I received by George Armitage a great pie and four little ones, four oate cakes, and a book, from you and my sisters" (p. 19). He was at that time going to school at a place far from home, and the pies and cakes from home were no doubt devoured with great avidity. See also ibid., pp. 20, 28, 29, 32, 33, 34, 36–37, 38, 39, 43, etc.

These "caravans," as they were afterwards called, were built for carrying heavy loads, and this, together with their clumsiness, and the bad state of the roads, will easily account for their slow motion. Persons going by these waggons had to start early in the morning, and reach their inns late at night; and none but those who were unable to pay for better accommodation ever travelled in this way. In the latter half of the seventeenth century an ordinary stage coach, with four horses, carried six passengers; whereas a caravan, with four or five horses, carried twenty to twenty-five passengers. (v. 'The Grand Concern of England Explained,' in the Harleian Miscellany, viii, pp. 562–3; also Sir Robert Howard's comedy 'The Committee,' which is printed in his *Four New Plays* (London, 1665), pp. 71, 72, 75.) See also the advertisement of the 'Coventry Carrier' in *The City Mercury*, Nov. 4, 1675, p. 1; and of the 'Canterbury Flying Stage Coach' in ibid., July 4, 1692, p. 2.

several papers written by this man, who was a member of the Inner Temple, present the case as viewed by the most radical opponents of the coaches[1]. He said that the extensive use of this means of travel would be detrimental to the watermen by taking away their living; that it would make men effeminate; and that it would destroy the breed of horses[2]. The introduction of the stage coaches furnished a means by which gentry and ladies could get from their estates in the country into the city, where they could spend extravagantly, and since he regarded this as contrary to their good he thought the coaches should be suppressed[3]. Then, too, because travellers by stage coaches would spend more time on the road and less in the inns, there would be less consumption of beer, ale, and other provisions, and this would be detrimental to the interests of the innkeepers, and to the national treasury[4].

[1] ' The Grand Concern of England,' printed in Harleian Miscellany, VIII, pp. 562–8; also Brit. Mus. 816. m. 12 (162). One sentence from the former may be given: "These coaches hinder the breeding of watermen, and much discourage those that are now bred; for there being stage-coaches set up, unto every little town upon the river Thames, on both sides of the water, from London, as high as Windsor and Maidenhead and so from London Bridge to and below Gravesend; and also to every little town within a mile or two of the water side; these are they who carry all the letters, little bundles, and passengers, which (before they set up) were carried by water, and kept watermen in a full employment, and occasioned their increase (whereof there never was more need than now); and yet, by these coaches, they of all others are most discouraged and dejected,......they having little or nothing to do; sometimes not a fare in a week......" In Brit. Mus. 712. g. 16 (17), entitled ' Treatise of Wool and Cattel (1677),' p. 33, after speaking of the great decrease in rents, on account of which noblemen and gentlemen were no longer able to keep their stables furnished with good, serviceable horses, the writer traces this to the prevalence of stage coaches, and says: "Had it not been for this happy conveniency (i.e. stage coaches),......a project, though it hath found some confidence to defend it, yet is so injurious and destructive both to our breed of horses, and to all inns upon the roads, and at London too, that it may well be reckoned among the public grievances of the nation......." See also Brit. Mus. 712. g. 16 (20), ' The Trade of England Revived (1681),' p. 27; and Brit. Mus. 08,226. aaa. 29, ' Reasons for Suppressing Stage Coaches,' p. 1, both of which are very strong in their denunciation of the stage coaches.   [2] See last footnote.

[3] In Brit. Mus. 816. m. 12 (162), we find his words: "These stage coaches make gentlemen come to London upon every small occasion, which otherwise they would not do but upon urgent necessity; nay, the conveniency of the passage makes their wives often come up, who rather than come such long journeys on horseback would stay at home. Here when they come to town, they must presently be in the mode, get fine clothes, go to plays and treats, and by those means get such a habit of idleness, and love to pleasure, that they are uneasy ever after at being at home, and unfit to look after their country-affairs." See also Brit. Mus. 1029. h. 4 (1), ' The Interest of England Considered,' p. 62, and Brit Mus. 08,226. aaa. 29, ' Reasons for Suppressing Stage Coaches,' pp. 4–5.

[4] Ibid. In another pamphlet, written by a country tradesman, entitled ' The

It would seem that Cressett and the innkeepers had combined their forces in this movement to suppress the stage coaches[1], and that they endeavoured to enlist in the same cause the companies of cutlers, cordwainers, and watermen, and the postmasters throughout the country. A letter was sent to each rural postmaster, giving directions as to how this design should be managed in order to make it successful[2]; but it was apparently allowed to rest. A strong refutation was made of the seeming arguments used by Cressett[3]; his deductions were shown to be futile; and despite the arraignment of the stage coaches their number increased and their utility became more widely recognized. As their enterprise increased, they began to make their schedule harmonize with the arrival and departure of the packet boats from the various ports of the kingdom, especially those like Harwich, Deal and Dover, that were in communication with the Continent; and in this way facilities for travel were made more convenient and acceptable[4]. In point of numbers, however, the stage coaches linking up the various parts of the kingdom with London were comparatively unimportant at the close of the seventeenth century[5], and even down to the middle of the eighteenth century, for most of the travelling was done on horseback, rather than in wheeled carriages.

In the latter half of the seventeenth century, there was much difficulty in London regarding the licensing of hackney coaches for the convenience of the people of the city. In 1654, when Cromwell, by an Ordinance, licensed two hundred hackney coachmen to keep not more than three hundred coaches, to ply in London and Westminster and within a radius of six miles about these cities, he made provision that, out of these two hundred, thirteen should be Master Hackney Coachmen within these limits, and no others than those licensed should be allowed

Ancient Trades Decayed, Repaired Again ' (Brit. Mus. 1138. b. 11), pp. 26–27, this plea is put forth.

[1] Brit. Mus. 816. m. 12 (163), p. 4, gives a reply from one of the postmasters, which was not at all encouraging, and accused Cressett and the innkeepers of connivance to suppress the coaches.

[2] Brit. Mus. 816. m. 12 (163), pp. 1–3, gives the printed letter sent by Cressett to the postmasters, dated Oct. 19, 1672.

[3] Brit. Mus. 816. m. 12 (162), ' Stage Coaches Vindicated.'

[4] In 1675 the coach left London on Tuesday and Friday to be at Harwich before the packet boat pushed off. Similarly for connexion with Dover, etc., *City Mercury or Advertisements concerning Trade*, Nov. 4, 1675, p. 2; *Protestant Mercury*, Jan. 12–14, 1697, p. 2; *City Mercury*, July 4, 1692. See also *City Mercury : or Advertisements concerning Trade*, Feb. 20—Mar. 7, 1677, pp. 1, 2.

[5] Brit. Mus. 1029. h. 4 (1), ' The Interest of England Considered (1694),' p. 62. The writer speaks of the rareness of a coach in the country, although there were plenty in London. See also Brit. Mus. 796. c. 36, ' A Brief Director (1710?),' showing that the number of "coaches" so-called that came to London was very small.

to keep coaches for hire.   The Court of Aldermen was to have the power of making by-laws for regulating these coaches[1].   All these were to pay on admittance 40s. a piece towards the common charges of the company of hackney coachmen.   This was the beginning of what was regarded by these licensees as royal privilege, and when these "ancient hackney coachmen" had once secured their licenses, they considered that they should ever after receive the first consideration, in any change that was to be made.   But how did this work out?   In the early part of 1668, an Act was passed, authorizing four hundred hackney coaches to be licensed, on payment of the yearly rent of £5 each, and the Act ordained that the governing power should be taken from the Court of Aldermen and given to a Board of Commissioners, who were to take care, first of all, to license the ancient hackney coachmen, and those who had suffered for their services in behalf of Charles I or Charles II[2]. On account of irregularities in the administration of the Commissioners, and upon complaint to Parliament, new Commissioners were appointed, who gave liberty to many unlicensed coachmen to drive coaches, and took away licenses from sixteen ancient hackney coachmen[3].   To secure the restoration of their privilege, these four hundred coachmen petitioned Parliament to incorporate them, under the control of the city of London, instead of under the Board of Commissioners, so that there might be "a just regulation of that calling;" and they also requested "that all stage-coaches within thirty miles of London may be suppressed," because this service could all be done at the same rates by the hackney coachmen, and thus free the streets from "a continual multitude of coaches[4]."   Evidently, they were not going to sit idly by and allow their coveted privilege to lapse.

After the expiration of the Act of 1668, the situation in London was in confusion.   When food for horses was dear, some coachmen found it impossible to continue their work, and had to lose their licenses. Also, innkeepers and others set up coaches, and by the great competition thus aroused drove out of business those who were ill prepared to stand the competition[5].   In order to enable the coachmen to carry

[1] Brit. Mus., E. 1064 (18) gives the Protector's Ordinance.

[2] Act 14 Car. II, c. 2; see also Brit. Mus. 816. m. 12 (152), 'The Case of the Antient Hackney-Coachmen, etc.'

[3] Brit. Mus. 816. m. 12 (151); Brit. Mus. 816. m. 12 (152); and Brit. Mus. 1865. c. 17 (28), ' The Case of Many Coachmen in London and Westminster.'

[4] Brit. Mus. 816. m. 12 (151), and Brit. Mus. 816. m. 12 (152).   It would seem as if these "four hundred" regarded themselves as public benefactors, and they regarded their licenses, when once granted, as to continue for their life.   Brit. Mus. 1865. c. 17 (28).

[5] Brit. Mus. 816. m. 12 (153), 'The Case of the Hackney Coachmen,' and Brit.

on their business until the time when this difficulty should be removed, tradesmen and others had given them credit for some years, in the hope that Parliament would put an end to the disability under which they were working, and thus allow them to support their families and discharge their debts[1]. By 1683 the number of these hackney coaches in London had so greatly increased, that the streets were rendered unsafe and inconvenient, both for foot passengers and for carts carrying goods; and in order to prevent this and to establish a convenient number of coaches, the Common Council of the city enacted that there should not be more than four hundred such coaches licensed, and if any tried to go unlicensed they should be given stipulated punishment. Each licensed coachman was, for good reasons, to wear his badge. The fares to be taken were expressly set forth[2]. So long as this measure was duly executed and observed, all went well; but, afterward, the old abuses and annoyances, due to the excessive number of coaches, reappeared, until in 1691 the Act was once more ordered to be immediately put into effect[3]. After another spurt of law enforcement, there followed a period of laxity and confusion for two or three years; and, to bring harmony out of these conditions, Parliament, in 1694, passed an Act authorizing the appointment of not more than five Commissioners for the metropolis, who should have power to license seven hundred hackney coaches, each of which should pay £50 for a license that was to continue in effect for twenty-one years, upon the payment of £4 a year[4]. For any distance within ten miles of London

Mus. 816. m. 12 (154), 'The Case of John Nicholson, Walter Storey, William Hudson, Richard Hatt and Samuel Walters, in behalf of themselves, and the First 400 Ancient Hackney-Coachmen, and the Widows of them.'

That the number of coaches in London at this time was large is shown by the fact that in 1667, the writer of 'England's Wants,' Brit. Mus. 517. k. 16 (3), p. 4, proposed an impost to be levied on a great many articles of luxury and pride, including among them coaches, chariots, litters and sedans; and the amount of these imposts was to be used by a Royal Commission for making or repairing highways or bridges, for making rivers navigable, or for other public works. Unless the coaches were quite abundant they would not have been thought of as a subject of taxation. Compare similar evidence in Brit. Mus. 1029. h. 4 (1), 'The Interest of England Considered' (1694), p. 62.

[1] Brit. Mus. 816. m. 12 (159), 'The Case of Divers Tradesmen, Creditors of the Hackney Coachmen in London, and Westminster, and Stagemen to several Places of England.'

[2] Brit. Mus. 102. k. 52, 'An Act of Common Council, for the better Regulation of Hackney-Coaches.'

[3] Brit. Mus. 1851. b. 2 (15), Beginning, 'Pilkington, Mayor' (a Proclamation of the London Common Council).

[4] Act 5 and 6 W. and M., c. 22. The wording of such a license may be found in Brit. Mus., Harl. MSS. 4115, p. 231.

they were not to charge above 10*s.* for a day of twelve hours; and by the hour, not above 18*d.* for the first hour and 12*d.* for every hour afterward.

Following the passage of this Act, however, and its execution by the Commissioners, some of the foregoing conditions began to reappear: the increase in the number of coaches from four hundred to seven hundred gave rise to so many licensed coaches on the streets, that many others were led to take up this same occupation without a license, in the hope of securing some of the profits[1]. Petitions were presented, urging that some better way be adopted of regulating the licensed coaches, and putting an end to the violations of the law[2]; and in 1710, permission was given to the Commissioners to license eight hundred such coaches from June 24, 1715, for a period of thirty-two years[3]. Each coach was to pay a license fee of five shillings per week, payable monthly. Authority was also given to license two hundred hackney (sedan) chairs for the same term, from June 24, 1711; such chairs to pay an annual license fee of 10*s.* each, payable quarterly[4]. Those who were guilty of driving a coach without a license within the limits of London and Westminster were to pay a penalty of £5 for every offence. The similar penalty for carrying a chair without a license was 40*s.* The rates to be taken for the use of the coaches were practically the same as those of 1694; and the rates for chairs were not to be more than those for coaches, within the same limits. The number of coaches authorized by the Act of 1710 remained unchanged, notwithstanding much agitation to have it increased, until 1768, when one thousand coaches were licensed to stand for hire in the streets, and no further change was made until 1832, when the number was limited to twelve hundred.

We must not imagine that even the Act of 1710 put an end to the discontent regarding the coaching business in London, or the irregularities in its administration. To the salaries of the five Commissioners were added those of a series of assistants[5]; and from the data at hand,

---

[1] Brit. Mus. 816. m. 12 (155), 'The Hackney Coachmens Case, Humbly presented to the Honourable House of Commons; etc.' See also Brit. Mus. 816. m. 12 (158).

[2] Ibid.

[3] Act 9 Anne, c. 16. Probably the number of coaches was increased with the thought that by so doing it might prevent in a large measure the use of unlicensed coaches.

[4] See complaint of some two hundred formerly licensed hackney chairmen, now left without a license, in Brit. Mus. 8223. e. 9 (12).

[5] Brit Mus., Add. MSS. 18,047, p. 20. In 1715 the salary of each of the five Commissioners was £150 per annum. Besides these five in the Hackney Coach Office, there were a receiver, at a salary of £62 per year; a register and clerk at

it seems as if the management of this service necessitated a constantly increasing expense[1]. But another thing that seems inexplicable on the

£80; a solicitor at £50; a housekeeper and two messengers at £40 each; and two street keepers at £35 each. In that year, therefore, the management cost £1032.

[1] See the revenue receipts from the licensing of hackney coaches, 1693–1701, in Brit. Mus., Add. MSS. 18,054, p. 9 et seq.

In Brit. Mus. 357. b. 9 (2), we have (pp. 11–47) given us in MS. 'A Particular State of the Receipts and Issues of the Public Revenue,' etc., for these same years, 1693–1701. From this we quote the following receipts from the licensing of hackney coaches:—

From Michaelmas 1693 to Michaelmas 1694, £34,500
  „   „   1694 „   „   1695,   1,400
    No returns given for three intervening years.
From Michaelmas 1697 to Michaelmas 1698,   1,550
  „   „   1698 „   „   1699,   900
  „   „   1699 „   „   1700,   1,500
  „   „   1700 „   „   1701,   1,300

It will be noted here that the larger return in the first year was due to the licensing of seven hundred coaches, each of which had to pay £50 for a license that was to continue in effect for twenty-one years; and, of course, there was a rush to take out licenses because they were profitable. It is evident, however, that not all the 700 coaches authorized took out licenses at that time, for in that case the total returns would have been £35,000. But even the £34,500 of receipts in the year 1693–4 was considerably greater than the gross receipts per year, one hundred years after this, as given below. For the intervening century up to 1809, I have been unable to secure the data; but from 1809–1822, the Treasury Papers give us the information we want, as to the revenue from these coaches, and the expenses of management of the office, from which it would appear that the revenues decreased and the cost of management increased during this period.

Year	Gross Receipts			Charge of Management		
	£	s.	d.	£	s.	d.
1809	28,753	12	6	2,776	18	8
1810	28,571	5	0	2,964	13	5
1811	28,739	16	0	3,067	13	6
1812	30,909	0	0	3,182	1	3
1813	27,860	17	6	3,419	17	4½
1814	25,181	10	0	3,306	13	1½
1815	27,401	2	6	3,677	15	9
1816	28,932	2	0	4,562	5	2
1817	30,802	5	0	4,515	7	10
1818	28,970	17	6	3,832	11	1½
1819	26,347	12	6	3,779	5	0
1820	26,534	17	6	4,150	14	9
1821	26,374	7	6	4,122	5	9
1822	26,248	2	6	4,099	10	11

(*Treasury Papers.* Miscellanea. Expired Commissions, etc. Hackney Coaches, etc. Number 2, Public Record Office.) Thus, while there was a decline of 9% in the gross receipts, there was an increase of 48% in the cost of management of the office, in the above fourteen years. Note also that the cost of management in 1715 was only £1032 (see note 5, p. 128), while in 1809 it was almost three times as much.

basis of present information, is that by 1742 there were four hundred chairs in London[1], whereas by the Act of 1710 the number was limited to two hundred, and that Act would not expire till 1743. Before 1822, it had been proposed to abolish the Hackney Coach Office entirely, and give the licensing and regulating of coaches to some other body, probably the police commissioners; but the manager of the office issued a lengthy memorandum showing the nature of their work, their charges of management, and why it was necessary to have such an office that was competent to deal with these matters. He said that the reason for the establishment of the office was regulation and protection, not revenue, and his statement seemed to prevail for the continuance of the functions of this Board[2].

The number of hackney coaches and chairs in London, however, was numerically insignificant in comparison with the number of private vehicles that were owned in the city. In the register containing the names of those who paid duty for coaches and other carriages, I have counted for the year 1754, and found that there were then in London 4255 four-wheeled vehicles and 2909 two-wheeled vehicles (including chairs). Each four-wheeled vehicle paid £4 a year, and each two-wheeled vehicle £2 a year, so that the total amount of duty imposed upon them was £22,838[3]. The comparison of this large number of private carriages with the eight hundred public hackney coaches is easily made. With correspondingly large numbers of such vehicles in other cities and

---

[1] In the Public Record Office *Treasury Papers*. Miscellanea. Various. Bundle 305 (anno 1742), we find a certificate from the Coach Office, dated May 14, 1742, showing the receipts and payments for the month of April of that year; and in it is the entry, "Received Rent for 400 chairs, Lady Day Quarter, £50."

[2] *Treasury Papers*. Miscellanea. Expired Commissions, etc. Hackney Coaches, etc. Nos. 3 and 4, Public Record Office.

[3] Public Record Office. *Treasury Papers*. Registers: Plate, etc., i. Of these 4255 four-wheeled vehicles in London in 1754,

there	were	36	persons who had each	4	four-wheeled carriages,	
,,	,,	24	,,	,,	5	,,	,,
,,	,,	6	,,	,,	6	,,	,,
,,	,,	9	,,	,,	7	,,	.,,
,,	,,	2	,,	,,	8	,,	,,
,,	was	1	person who had	9	,,	,,
,,	,,	1	,,	,,	16	,,	,,
,,	,,	1	,,	,,	31	,,	,,

Of the 2909 two-wheeled vehicles registered in the same year,

there	were	14	persons who had each	4	two-wheeled carriages,	
,,	,,	2	,,	,,	5	,,	,,
,,	,,	5	,,	,,	6	,,	,,
,,	,,	2	,,	,,	7	,,	,,

and no person had more than 7 of the two-wheeled carriages (including chairs).

towns[1], we can readily see what a great factor these were in the social life of the time.

That a hackney coach license was profitable to its holder, is apparent from a petition which the licensed eight hundred presented to King George I[2]. They asked to have their occupation rendered less precarious and uncertain[3]; and in return therefor, in addition to their rent of 5*s.* per week, they would immediately raise, for the King's use, the sum of £16,000, as a fine of £20 on each license. The lucrative worth of such a license was shown also by another petition presented to Parliament, apparently before the Act of 1710 was passed, by two men of high standing and integrity, who promised that if eight hundred hackney coaches were licensed at six pounds each, payable quarterly, and they were allowed to be farmers of these licenses for a term of twenty-one years, they would pay yearly to the Crown £2000, upon which the King could raise the sum of £20,000; they would pay to the orphans of London £500 a year; they would raise and maintain a regiment of thirteen companies of foot soldiers at a charge of £3000 a year; and they would remove two great impediments to the national welfare, at a cost of £1,000,000, without any prejudice to the subjects[4].

Before leaving the conditions in London, it may be noted that not only were the hackney coaches subject to regulation, but also the vehicles that were employed by the public for carting or draying goods were subject to similar control. In 1606, because of complaints made by merchants and others as to the excessive rates demanded by carmen, the Lord Mayor fixed the charges that could be taken by such carters, according to distance, weight and material[5]. By 1663, the streets were, apparently, much obstructed and dangerous because of the great number of drays and brewers' carts that were always passing to and fro; and, to restore order, the Common Council decreed that, from Michaelmas to Lady Day, draymen could not be on the streets with their carts at work after 1 p.m., and during the other part of the year

---

[1] See last footnote. For example, in Barnstaple there were 12 four-wheeled vehicles and 27 two-wheeled vehicles; while in York the corresponding numbers were 116 and 214.

[2] Brit. Mus. 816. m. 12 (157).

[3] Up to this time, when a hackney coachman died his license ceased and his family were likely to be deprived of all means of support. They now petitioned that this be changed, so as to make the license an asset in the hands of the family, so that either it might be sold, or might be used to produce the usual income (Brit. Mus. 816. m. 12 (156)).

[4] Brit. Mus. 816. m. 12 (161*), 'The Humble Proposals of James Lord Mordington, and Martin Laycock, Esq., for the Farming of the Hackney-Coaches.'

[5] Brit. Mus. 21. h. 5 (2), 'By the Maior. Orders set down by the right Honorable, Sir John Watts, Knight,' etc.

after 11 a.m.[1] Some three or four years after this, in order to put an end to the rudeness and great disorders on the part of those who drove these drays and carts, it was ordered that they should not drive their carts and waggons when empty any faster than when they were loaded[2]. By 1672, by reason of the carmen charging higher than the legal rates in London, the Justices of the Peace drew up a complete scheme of the rates that the carmen might take for different commodities, weights and distances; and these were to be printed and put up in all public places, so that all might know the facts[3]. In 1677, in order to prevent the streets from being pestered with carts and waggons, so that coaches might freely pass along them, the Common Council enacted that no more than four hundred and twenty carts should be allowed to work for hire within their jurisdiction. These were to pay specified amounts for their privilege, were to be under the control of the authorities of Christ's Hospital[4], and the funds thus contributed were to be applied toward the maintenance of the hospital[5]. These licenses, like those for hackney coaches, were regarded as valuable assets; they cost £100 a-piece when taken out and a yearly rental had also to be paid to the hospital[6]. From the foregoing, it is evident that there was as much need of regulating the means for the carriage of goods as for the conveyance of passengers[7].

In connexion with the subject of communication, we would observe

---

[1] Brit. Mus. 816. 1. 4 (21), 'Robinson, Mayor. Commune Concilium tentum in Camera Guihaldae Civitatis London. decimo die Octobris, Anno Domini 1663,' etc.

[2] Brit. Mus. 21. h. 5 (36), 'Ad Session Oier' & Terminer & Gaolae Domini Regis de Newgate,' etc.

[3] Brit. Mus. 21. h. 5 (52), 'Ad General' Quarterial' Session',' etc. These rates were higher in some cases than former rates; and further increases of rates were made in 1673, v. Brit. Mus. 21. h. 5 (57); and again in 1749, v. Brit. Mus. 816. 1. 5 (10), and doubtless in many other years.

[4] That there was need for control of carmen and carters, is evidenced by a statement made by a Frenchman in *A Journey to England* in 1700. Brit. Mus. T. 1860 (3). On pp. 4–5 he says that, in London, dirt and roots were frequently thrown at him and his companions by the children and apprentices, without reproof: "civilities, that in Paris, a Gentleman as seldom meets withal, as with the contests of carmen, who in this town domineer in the streets, o'erthrow the Hell-Carts (for so they name the coaches), cursing and reviling at the nobles." Then, on p. 14, he says: "......I return to the town, where they are pestered with hackney coaches, and insolent carmen......"

[5] Brit. Mus. 816. m. 12 (79), 'An Act of Common Council for the Government of Cars, Carts, Carrooms, Carters and Carmen, etc.'

[6] Brit. Mus. 816. m. 7 (131), 'A Proposal' (for regulating cars, càrts, etc., in London).

[7] Brit. Mus. 816. m. 7 (131) shows also some other annoyances to the successful operation of the licensed carmen.

also that when any great person, such as the King, a foreign prince, or a nobleman, was known to be contemplating a journey through part of England, great preparations were made by the people along the proposed route to have the roads and bridges made as secure as possible. Notice of such a journey, when known beforehand, was sent by the Privy Council to the Justices of the several counties through which the journey would be made, requiring them to examine the condition of the highways and bridges along the line, have them put in good repair, and report on the same to the Council[1]. In the towns through which the line of travel was to pass, such an event was made the occasion for a general cleaning up of the streets, making them presentable and amply satisfactory, so that the royal visit might not be marred by anything unsightly or by any accident[2]. The mayors and aldermen were anxious to have their towns present a good appearance, and furnish good entertainment for the distinguished guest; for it was known that, at such times, the honour of knighthood was conferred

---

[1] Cal., S. P. D., 1631–3; quoted in *Archaeologia Aeliana*, N. S., xxi, p. 83. When notice was sent to the Justices of the counties through which Charles I's journey to Scotland would carry him, the Justices of Northumberland reported as follows:

"Right honorable and other very good Lords: Uppon the receipt of yo: Lorships l'res dated at Whitehall the 16th day of January, 1632, wee appointed a meeting that we mighte consulte together how to devide ourselves within our hon'ble devisions, according to yor hon^r comande; for the speedy and present repaire of the bridges and highways. Att which meetinge we gave order for an exact survey and view of the bridges and wayes deficyent, and have nowe according to our best judgmente, taken speedy course for their present repaire. And wee doubt not but before his Matie shall come down, they wilbe sufficyently repayred, according to yo Lo^{ps} comande, which with all due obedience wee shall ever be readie to execute.

And so we humbly take our leves, and shall always reste.

Your Lordships, ready to be commanded.

	John Fenwick.	Cuthbert Heron.
Morpeth in Northumberland	John Deland.	Jo: Barring.
the 13th of March, 1632.	William Carnaby.	William Widdrington."

The Justices of Durham made a similar report (ibid.).

[2] From the records of the town of Lincoln, we have the following quotation, under date of Oct. 28, 1695:

"This morning, as soon as the post came in from Grantham, Mr Mayor and the aldermen received an account that his Majesty King William was on his journey from London and intended to be in this city to-morrow night......Presently after, Mr Mayor sent to the aldermen to meet him at the Guildhall to consult what was the best to be done, and accordingly they met together, and went clear down the street as far as the Little Goat Bridges, and as they went along they ordered all the parishes to get carts and laborers to cleanse the streets......which was done..... All the cross rails down the street were ordered to be taken up, and all stones, wood and other obstructions lying and being in the highway were removed......" Sympson's *Lincoln*, p. 142. For similar instances, note the items from Gateshead parish records, in *Archaeologia Aeliana*, N. S., xxi, p. 84.

upon the mayors, and sometimes upon other worthy citizens, of towns that were especially favourable to the King.

In any comparison of the means and rate of travelling during the period now under review, with that during the later Tudor reigns, it is essential to remember that most journeys in the time of Elizabeth were undertaken on horseback. This, of course, would not require such good roads as would be necessary when the use of coaches had come in; with roads of the same quality, the coach would be compelled to go more slowly. If, therefore, the rate of travelling by coach, in the later period, should be found to be faster than the rate by horse-back in the earlier period, it would strongly argue that an improvement had taken place in the condition of the roads. But our available information as to the rate of travelling in the time of Elizabeth we have found to be meagre indeed; and yet it is with such scanty records that our comparison must be made. Under these circumstances, the few statements we shall make at the present time must be regarded as only approximately conclusive, and subject to change should more detailed information be brought to light.

What, then, do we find when we institute such a comparison as that suggested above? We have seen that, in the later years of the Tudors, the usual rate of travelling by riding post was from twenty-four to fifty miles per day, when there was no cause for delay, and when relays of horses could be had every ten or twelve miles; but only about half this rate could be attained when frequent changes of horses were impossible. From what we have been able to discover, the speed of fifty miles per day was seldom attained. Immediately after the middle of the seventeenth century, when stage coaches had come into some little prominence, their regular rate of travel was not much above thirty miles per day, taking it the year round; but during the summer months a speed of sixty to seventy miles per day was occasionally reached on the best roads[1]. More often, however, the speed of the fastest coaches in summer did not exceed fifty miles per day[1]. Even at this rate, when

---

[1] In 1669, the one-day coach began running for the summer between Oxford and London, thus making about sixty miles a day. But, during the same time, and for a long time to come, the old two-days coaches continued to run, making only thirty to forty miles a day. Clark, *Wood's Life and Times*, ii, p. 153; Tombs, *The King's Post*, pp. 23–24. Even in 1742 the stage coach took two days for this journey (Tombs, op. cit., p. 25).

Davies, in the summer of 1689, made the journey from London to Yarmouth, by way of Bishop Stortford, Newmarket, and Bury, in two days, which would be between sixty and seventy miles a day. But that this was exceptional speed is shown by the rest of his *Journal*; it usually required three days, and consequently he must then have been going between forty and fifty miles a day. See page 89 ff., and footnotes.

Queen Anne, in 1702, went from Oxford to Bath, about sixty miles, in one summer day (*The Queen's Famous Progress*, p. 4). But in 1750 the regular stage coaches took two days, even in summer, to cover that distance (*Bath and Bristol Guide*, 1750, p. 7). About 1750, it took the stage coach three days in summer to make the journey between Bath and Exeter, which would be about 26–27 miles per day; and in the same year one day was occupied in going between Bath and Salisbury, which would be a summer rate of about 40 miles per day (ibid., 1750, p. 7).

In 1660, and for a long time after that, the stage coach from London to Newcastle required six days for the journey (*Archaeologia Aeliana*, N. S., III, p. 244). This would be at a rate somewhat above forty miles per day, along the great north road, which was one of the best at that time. This speed was continued as late as the year 1712 (Richardson, *Borderer's Table Book*, I, pp. 343–4; Welford, *History of Gosforth Parish*, p. 57). Even as late as 1732, there was but one coach a week from London to Newcastle, and a coach once a fortnight for Edinburgh (*London Evening Post*, Jan. 20–22, 1732, p. 2).

Macky, in 1712, went from Yarmouth to Bury, at the time of the Bury fair, in one and one-half day, which would be a speed of forty miles per day (Macky, *Journey through England*, I, p. 3).

The journey between London and Yorkshire towns in the latter part of the seventeenth century was not usually made in less than five to six days, which would be at the rate of thirty or forty miles per day (*Memoirs of Sir John Reresby*, p. 174, June 24, 1679; p. 124, Oct. 26, 1678; p. 159, Feb. 5 and 11, 1679; p. 185, June 3 and 8, 1680; p. 203, Feb. 7, 1681; p. 268, Feb. 9, 1683, etc. Also Bishop Nicholson's *Diaries*, under dates of Nov. 14, 1702, and Feb. 22 and 27, 1703. In the latter, we find that his usual rate of travel was thirty to forty miles a day—see under dates of Mar. 4, 5, 6, 7, 11, 12, etc., of the year 1702, in the Bishop's *Diary* as given in *Transactions of the Cumberland and Westmorland Antiquarian and Archaeological Society*, I, N.S., anno 1901. See also Lord Hervey's *Diary*, p. 36, Mar. 28, 1702; p. 40, Oct. 2, 1703; p. 54, Aug. 27, 1711, etc.).

But in the summer of 1706, a four-days stage coach was going between London and York, thus making an average speed of about fifty miles a day. See advertisement of this coach in Harris, *Old Coaching Days*, p. 93, and given in the footnotes to page 90.

From what we have here presented, it would seem that most of the coach travelling before 1750 was at a speed of less than fifty miles a day; and probably on the longer journeys the speed was seldom above forty-five miles a day.

By way of comparison we may note the rates at which the mail was carried in this early part of the seventeenth century; and here we must remember that this was all done on horseback, and by regular relays of horses supplied by the postmasters. In Brit. Mus., Add. MSS. 34,727, pp. 14–16, is found a letter from Sir William Monson to the Earl of Salisbury, written on June 6, 1611, and sent by mail from Sandwich. It has the endorsements of the times at the various stages from Sandwich to London, showing that it was carried this seventy miles in eleven hours (from 8 p.m. to 7 a.m.). This would make its average speed about six and one-half miles per hour. The order of 1603, that the postmen should travel at the rate of seven miles an hour in summer and five miles in winter, was recognized by Cromwell as impracticable; and he issued orders that in future only letters to and from high officials and public despatches should be carried at the fast speed of seven miles an hour from April to September inclusive, and five miles an hour for the other half of the year. (Hemmeon, op. cit., pp. 99–100.) But it is clear that up to the time of Charles II the mail on the routes between London and the important centres of the kingdom was carried at three and one-half to five and one-half miles an hour, varying, of course, according to the time of year and the condition of the road. (Hemmeon, op. cit., 101 n.) If this was the speed of the post, the rate of ordinary travelling would not be in excess of that; and at this rate, a day's journey (12 hours) would cover forty to sixty miles.

To further prove our statement in regard to the rate of travelling, we give more details, as follows:

In 1692, it took from Mar. 21 to Mar. 26 to go from Ferrybridge to London (*Yorkshire Diaries*, II, p. 49). This was at the rate of thirty to forty miles a day. But ibid., II, p. 55, and II, p. 63, show us that under favourable conditions the same distance could be covered in three days, which was at the rate of about sixty miles a day.

In 1695 and 1696, it took six to seven days to go, during the winter season, from Wakefield to London, or from London to Wakefield (*Yorkshire Diaries*, II, p. 68, Nov. 20 to 27, and Jan. 22 to 30). This winter rate would be about twenty-five to thirty miles a day.

In 1702, it required three days to go from Wolverhampton or Birmingham to London, which was equivalent to forty miles per day (Harper, *Holyhead Road*, I, p. 6, quoting from the announcement of the *Wolverhampton and Birmingham Flying Stage Coach*).

In 1714, from London to Cambridge by coach (about fifty miles) took only one day; but it required great exertion both of passengers and horses. Thoresby, *Diary*, II, pp. 229–30.

In 1724, the *London Evening Post* of May 23 announced that besides the usual three-days coach from London to Bristol, a flying coach would be operated for the summer on that road, to go the whole distance in two days. Thus the regular rate would be forty miles a day, and the summer rate with the faster coach would be sixty miles per day. Tombs, *The King's Post*, p. 24, gives the advertisement of the coach that went, in 1667, between London and Bath in three days. These rates do not seem to have been increased before 1750, for in that year the *Bath and Bristol Guide*, p. 7, shows us that the regular stage coaches between Bath and London took three days, thereby making 36 or 37 miles per day; but from April to Michaelmas there were coaches that went the distance in two days, thereby accomplishing 55 miles per day.

In 1731, between Birmingham and London, there was increasing demand for a faster service than that of 1702 mentioned above; and Nicholas Rothwell put his coaches on this road, which were advertised to reduce the time from three days to two and one-half days, during the summer (see this advertisement as reproduced in Harper, *Holyhead Road*, II, p. 13). This made the rate forty-four miles a day. This increased speed must have led to the desire for still greater speed, for in Aris's *Birmingham Gazette*, of May 3, 1742, Robert Coles advertised that "the Birmingham and Warwick stage coach begins flying for the summer season in Two Days." It was to begin on May 5. This would make a rate of fifty-five miles a day. But the advertisements of Rothwell's and Coles's coaches show that these rates were not undertaken in winter months; they began each year in May.

In 1726, on the great north road, a stage coach performed the distance from London to Derby in two days, which was about sixty-three miles per day (Leader, *Sheffield Burgery Accounts*, p. lvii). But it is almost certain that this was an exceptional, and not the usual rate.

Mason, in his *History of Norfolk*, I, p. 432, tells us that the Exeter Flying Stage, which was very fast for that time, went from London to Exeter in three days, thus making sixty-four miles a day. In 1742, the only coach from London to Oxford

the stage coach could accomplish as high a speed as was attained by posting on horseback half a century before, it is evident that there must have been some improvement in the roads during that time. This is further attested by the fact that the system of statute duty authorized in 1555 was made perpetual in 1586, and continued in effect throughout the next two centuries, which would have been improbable had it been of no use in improving the roads.

When we come to the first half of the eighteenth century, we do not have any evidence to show that there was much improvement of the highways in general, over those of the period of the Restoration. The ordinary rates at which the coaches travelled throughout the country were seldom much above forty to fifty miles per day, and were often below that figure; but on the better roads and during the more favourable seasons of the year, when there was less wet weather and the days

in winter took two days for the journey, which was at the rate of between thirty and forty miles a day. (See also Tombs, *The King's Post*, p. 25.) But from London to Norwich, which was twice the distance from London to Oxford, could be performed in the same time; that is, the rate here was sixty miles a day. This coincides with Defoe's statement (*Tour Through the Whole Island*, i, p. 30) that (in 1748) the coach went from Ipswich to London in one day, which was about 70–75 miles per day.

The *Bristol Journal* of Apr. 26, 1746, announced that a summer flying coach would perform the distance between Bristol and Gloucester (forty miles) in one day; but the summer coach from Bath to Oxford, exactly ten years later, in covering this distance (less than 60 miles) spent two days.

In June, 1753, a coach was set up to go from London to Shrewsbury in three and a half days, thus accomplishing forty-eight miles a day. Until then it took four days for a coach and six horses to go this distance, at the speed of forty-two miles a day. (*Salopian Shreds and Patches*, i, p. 7.)

Miege, in describing the *Present State of Great Britain* (p. 150), during the time of George II, says: "These (coaches) set out from London at certain times for all noted places in England, and return with so much speed, that some will measure sixty mile in a summer day......" Of course, he does not wish to have us think that this was the average rate of speed, by any means, but that it was the rate on the best roads during the long days of summer.

The *Gloucester Journal*, Nov. 23, 1738, advertised a coach to go from Gloucester to London in three days, which would be forty miles a day.

We might multiply such examples as the above, but think it needless to do so, to establish our contention.

N.B. In computing the distances mentioned in all this work, I have usually taken the measure of the straight distances between the important places on the road, as given in Philips's *Atlas of the Counties of England*, and increased these by an amount from one-fifth to one-tenth, so as to make up for the winding of the roads at that time. If it be said that this is not sufficiently accurate, I can only say that no strictly accurate figures are obtainable, but that I have endeavoured to form as close an estimate as possible from my knowledge of the course of each road in the earlier days.

were longer, it was still true that the stage coach rarely, if ever, travelled more than sixty to seventy miles a day[1]. The average rate was considerably less than this, even on the roads which were most frequented. Those who wanted to travel faster than the regular coach rate, especially when they had a long journey and were not encumbered with much luggage, usually rode post; fresh horses and guides were obtainable at convenient places along the great roads, at a charge of about three pence a mile for each horse, and four to six pence a stage (about twelve miles) for the guide.

The cost of travelling varied fully as much as the rate. It was not determined, as on the railways of a later time, according to any fixed rate on a mileage basis, for the fares paid, when computed by that standard, show wide variations. Nor were the rates fixed, as a rule, by means of competition, for at this early day the number of coaches on each road was so small as to exclude competition as a regulative agency. But the rates charged, although decided with some reference to the cost of travelling by other means, such as riding post, going in the stage waggons, etc., appear to have been settled largely by custom. Furthermore, the rates at the middle of the eighteenth century were little, if any, different from those of a hundred years before, which gives us some insight into how firmly entrenched custom was in the life of the people. On some roads the coach rate varied from one to one and a half pence per mile[2]; but, generally speaking, the passenger travelling by stage coach paid about two and one-half pence per mile in summer and about three pence per mile in winter[3]. This, of course,

[1] See last footnote.

[2] In 1743, the rate from London to Northampton was 6s., that is, 1d. a mile, and from Northampton to London it was 7s., that is, 1¼d. a mile (see advertisement of this coach as given in Bull, *History of Newport Pagnell*, pp. 14–15).

In 1725 the rate from Yarmouth to Norwich was 3s., which was 1¼d. a mile (Palmer, *Perlustration of Great Yarmouth*, I, p. 182).

In June, 1753, the rate of 21s. between Shrewsbury and London was the equivalent of 1¼d. a mile. Before this, the fare had been 18s., or 1¼d. a mile (*Salopian Shreds and Patches*, I, p. 7).

In 1741, the rate by stage coach from London to Norwich was 15s., or 1¼d. a mile, and from Norwich to Bury 5s., or 1¼d. a mile (Mason, *History of Norfolk*, I, p. 432).

[3] Thoresby's *Diary*, II, p. 148 (July 29, 1712), tells us that the fare by coach from London to Yorkshire was 40s., which would be about 2¼d. a mile.

In 1667, the fare between London and Bath was £1. 5s., which would be equivalent to 2¼d. per mile. (Advertisement of this coach is given in Tombs, *The King's Post*, p. 24.)

In 1712, the rate of £4. 10s. from London to Edinburgh was equal to 2¼d. or 3d. a mile (see advertisement of this stage coach in Welford, *History of Gosforth Parish*, p. 57).

was for inside passengers, for up to the middle of the eighteenth century the roads were not sufficiently smooth to warrant taking passengers outside as well as inside the coach[1].

When we come to consider the cost of conveyance of goods by road, we meet an entirely new element from that which determined the prices charged for the conveyance of passengers. Even had there been no coaches, the conveyance of passengers from place to place, according to their necessities, would still have gone on increasing, for post-horses were always ready at suitable stages and could be hired for this purpose. But unless there were some arrangements for the carriage of goods to and fro, at a reasonable rate, the country could not advance industrially. In the latter part of the seventeenth century we get some evidence that, even at that time, there were attempts by the carriers on certain routes to obtain a monopoly of the carrying trade, and to charge excessive rates for their services[2]; and such monopolies continued at a much later date, despite legislative efforts to prevent them[3]. In order to prevent the carriers from abusing their privilege by making unreasonable rates, an Act was passed in 1691[4], entitled 'An Act for

In 1673, the fare by coach from London to Exeter, Chester, or York, was 40s. in summer and 45s. in winter, which was 2½d. a mile in summer, and 3d. a mile in winter (Goodman, *Social History of Great Britain*, p. 83).

In 1731, Rothwell's coaches began their work between Birmingham and London. The fare between these two places was 21s., that is, 2d. a mile; and the fare from Warwick to London was 18s., that is, 2½d. a mile (see the advertisement of these coaches reproduced in Harper, *Holyhead Road*, II, p. 13).

Jeboult, in his *Researches in the History of West Somerset* (p. 34), says that the ordinary fare in that locality was about 2½d. a mile in summer, and somewhat more in winter. Similar results are obtained from calculations made from many other sources; and they show us that the rate more commonly charged was about 2½d. a mile in summer and about 3d. a mile in winter.

[1] As a general statement this is true, although there were occasional cases of stage coaches taking outside as well as inside passengers; for example the coach set up in 1753 from London to Shrewsbury took both classes (*Salopian Shreds and Patches*, I, p. 7). Sometimes the outside passengers had to travel in an open "boot" at the sides or back of the coach; but at a later time they were put on the top of the vehicle.

[2] We find papers in 1670 connected with an attempt, on the part of a private carrier, to break down the monopoly enjoyed by the carriers from Oxford University to London, under the sanction of the University (see Clark, *Wood's Life and Times*, II, p. 196).

[3] For instance, at the April Sessions in 1743, the Northamptonshire magistrates declared that divers waggoners and other carriers, by combinations among themselves, had raised the prices of carriage of goods in many places to excessive rates, to the great injury of trade (*Victoria History of Northamptonshire*, II, p. 291, quoted from Morton, *Natural History of Northants*).

[4] 3 W. and M., c. 12. This Act distinctly says that there were combinations among the carriers, and that excessive prices were charged.

the better amending the Highways and for settling the Rates of Carriage of Goods,' and this was amended in 1748. Under this authority, the Justices of the Peace were required at the Easter Quarter Sessions to assess the prices for all land carriage of goods, whether brought in or carried out of their respective jurisdictions, by any common carrier. The rates made by them were to be certified to all mayors and other chief officers of all market towns, and any carrier who charged more than this rate was liable to a penalty of five pounds, payable to the person who had suffered by such exorbitant charge. Owing, no doubt, to the fact that these rates were published, and were fixed on the basis of precedent, they were more uniform and stable than if left to be decided by each carrier according to local custom, as was the case with the rates for passenger travel; and from the small amount of information at hand, our conclusion is, that the cost of carriage of heavy articles was high, ranging from one-half penny per hundredweight per mile in summer, to three-fourths penny per hundredweight per mile in winter[1].

---

[1] Jeboult, in his *West Somerset*, p. 34, says that his investigations show that the carriage of goods by stage waggon cost about 15*d.* per ton per mile, which is ¾*d.* per hundredweight per mile.

At Sleaford, in 1696, the Justices fixed the following rates for the carriage of goods:

From London to Stamford and Deeping,	5*s.* 6*d.* per cwt.		Each of these was	
,,	,,	Bourn	5*s.* 10*d.* ,,	equivalent to
,,	,,	Grantham	6*s.* 0*d.* ,,	about ¼*d.* per
,,	,,	Sleaford and Spalding	6*s.* 8*d.* ,,	cwt. per mile.
,,	,,	Donnington	6*s.* 10*d.* ,,	
,,	,,	Boston	7*s.* 0*d.* ,,	

For every parcel of 7 lbs. or less, the cost of carriage was 6*d.* (*Victoria History of Lincolnshire*, II, p. 340.)

In 1717, the Justices of Derbyshire fixed the rates for the common carriers, as follows:

	From Lady-Day to Michaelmas	From Michaelmas to Lady-Day
Between London and Derby or Ashborne,	6*s.* 0*d.* per cwt.	7*s.* 6*d.* per cwt.
,, ,, ,, Bakewell	6*s.* 2*d.* ,,	7*s.* 8*d.* ,,
,, ,, ,, Chesterfield	6*s.* 3*d.* ,,	7*s.* 9*d.* ,,

The rates for distances between other places on the same or other routes were to be calculated proportionally. From London to Derby is 126 miles; so that the charge was about ½*d.* per cwt. per mile in the summer half of the year, and ¾*d.* per cwt. per mile in the winter half of the year. Cox, *Three Centuries of Derbyshire Annals*, II, pp. 236–7. He gives also assessments for later times, especially that of 1754. He says that in 1773 the county had a large number of printed sheets of the rates of carriage struck off, with blank spaces at the top for inserting the day, month and regnal year of George III, when the rate was annually voted. The last of these printed sheets was filled up for the year 1812.

What were the means by which this carriage of goods was effected? A certain amount of it was done by the employment of stage waggons or "caravans;" for we have already noted the extent of this business from the pages of John Taylor, the water-poet, and others after his time. But contemporary evidence points very strongly to the conclusion that by far the larger portion of the carrying was done by packhorses. Long trains of these faithful animals, furnished with a great variety of equipment, known by such names as saddles, panniers, crooks, dung-pots, bales, etc., wended their way along the narrow roads of the time, and provided the chief means by which the exchange of commodities could be carried on. Each of the large merchants of the principal mercantile centres had his horses for carrying his goods to the more remote, as well as the nearer, markets; and generally on the main roads, and almost universally on the by-roads, the conveyance of all kinds of products was done by the agency of these carriers. In order to enable them to do their work throughout the year, along roads that were frequently almost impassable otherwise, a narrow track or causeway from two to four feet wide, sometimes paved with flagstones, in other cases with round pebbles, was formed usually at the side of the road; and as these causeways were wide enough for but one horse it was customary for the front horse of each gang to carry a bell which could be heard at some distance by any approaching gang, and this warning enabled the approaching merchant to choose the best

---

In 1748, the Justices of Northamptonshire saw the necessity of a new assessment for that county, because the carriers had combined, and raised the rates. Accordingly, at their April Sessions, they decided upon the following rates:

Between London	and	Northampton, Brackley, Towcester, Daventry, Higham Ferrers, Thrapston, Wellingborough,	8s. 6d. per cwt.
Between London		and Kettering, Rothwell, Oundle,	4s. 0d. „
Between London		and Weldon, Rockingham,	4s. 6d. „

(*Victoria History of Northamptonshire*, II, p. 291.) These rates were about ⅔d. per cwt. per mile.

In the county of Chester, when the river Dee had become so silted up that vessels could not come up the river further than Parkgate, goods were carried by waggons from Parkgate to Chester, 8 miles, for 6s. per ton. This was equivalent to ½d. per cwt. per mile. Brit. Mus. 357. c. 1 (37), 'Case of the Inhabitants of Chester,' p. 1.

Houghton, *A Collection for Improvement of Husbandry and Trade*, May 12, 1693, informs us that before that time the carriage of malt from Derby to London was 10 groats to 5s. 4d. per cwt., or, in other words, ½d. per cwt. per mile.

The *General Evening Post*, of Mar. 7, 1744, quoted by Newball, *A Concern for Trade*, p. 23, is authority for the statement that turnpikes had reduced the price of carriage from Cambridge to London by fully one-half; but we have little evidence to support any such statement for the turnpikes generally.

place for turning out in order to meet and pass. Such stone causeways have been found, which, by long-continued travel in the one track, have become worn down in the centre in the form of a ditch. Without such aid, many roads that were deep with mire during the winter could not have been traversed at all. All kinds of products and materials were carried across the country in this way, in outfits that were suitable for carrying such varied articles as raw wool, fine woollens, coal, ore, salt, fish, pottery, etc.[1]

[1] Cleland, *Statistics of Glasgow*, p. 156, quoting from Dr Bannatyne's scrap book, tells us that in 1739 two gentlemen travelled from Glasgow to London, and found no turnpike until they came to Grantham, 110 miles from London. Up to that point they travelled on a narrow causeway alongside of a soft road, and "they met from time to time strings of pack-horses from thirty to forty in a gang, the mode by which goods seemed to be transported from one part of the country to another. The leading horse of the gang carried a bell to give warning to travellers coming in the opposite direction, and......when they met these trains of horses, with their packs across their backs, the causeway not affording room to pass, they were obliged to make way for them, and plunge into the side road, out of which they sometimes found it difficult to get back again upon the causeway."

Up to 1760, there was no road for wheel carriages into Liverpool. Not a single coach left the town, and there was not even waggon trade with Manchester. Long lines of pack-horses, laden with bales of wool and cotton, crossed the hills between Lancashire and Yorkshire. Picton, *Memorials of Liverpool*, II, p. 106; Baines, *History of Liverpool*, p. 418.

Jeboult, *West Somerset*, p. 45, says that during the reigns of the first two Georges, the coal used in the southern counties was carried inland, from the seashore or navigable rivers, by means of panniers on horses' backs. Compare Smiles, *Lives of the Engineers*, I, Ch. I, and *Diary of Celia Fiennes*, pp. 160, 199, 205, 207, etc.

A traveller in 1704 found the country people bringing coals to Darlington, three times a week, with two small sacks on each horse's back. Brit. Mus. 10,348. ccc. 56, 'North of England and Scotland,' p. 26. And Richardson, *Borderer's Table Book*, II, p. 20, says that the roads through the parish of Whitfield, in the county of Northumberland, in 1749, "were mere trackways, and the principal employment of the people was the conveyance of lead-ore to the neighbouring smelt-mills, in sacks, on the backs of ponies. There was not a cart in the country."

Up to the middle of the eighteenth century, before the introduction of canals, the material for and the products from the potteries of Staffordshire were carried by pack-horses and asses. Heavy loads of coal, tubs of ground flint from the mills, panniers of clay, crates of pottery ware, etc., were transported in this way by animals "floundering knee-deep through the muddy holes and ruts" in the almost impassable roads. Meteyard, *Life of Josiah Wedgwood*, pp. 267, 275.

Defoe, *Tour Through the Whole Island*, I, p. 94, and III, pp. 49, 121, gives us some idea of the vast amount of carrying that was done in this way. See also Whitaker, *Loidis and Elmete*, pp. 77, 80–81; Cox, *Derbyshire Annals*, II, p. 223; Worth, *History of Plymouth*, p. 335; Burden, *Memoirs of the Life of Elias Ashmole* (1717), p. 69.

On the fish trade, see Defoe, *Tour Through the Whole Island*, I, p. 8, and III, p. 268; Roberts, *Social History of the Southern Counties*, p. 489; Roberts, *Diary of Walter*

Among the dangers incident to the carrying of passengers and goods, there were others than those caused by bad roads and floods; and of these, perhaps the most frequent was that due to the presence of highwaymen, footpads and robbers, with which the highways were infested. Travellers usually went in company for safety, and even then they required to go armed. A blunderbuss was as necessary as a whip for the coachman and the carrier. A gibbet erected by the roadside, with the skeleton of a malefactor hanging upon it, was no uncommon sight[1]. Tradesmen who had failed, and even young men of position, who had ruined themselves by dissipation, took to the road in many instances; and if they could manage to regain their lost fortunes, and at the same time escape detection, they might subsequently return to respectable life[2]. Although all the roads were rendered dangerous by the presence of these men, the roads near the towns and cities were most frequented by them, for it was here that they could most successfully accomplish their desired end. It is known that these highwaymen were often in criminal connivance with the innkeepers, who aided them, and in return received aid from them[3].

Yonge, edited for the Camden Society, p. xxvii; Brome, *Three Years' Travels in England, Scotland and Wales* (1700), p. 274.

Brit. Mus., Add. MSS. 19,942, shows the construction of a great variety of these panniers, crooks, etc., that were used upon pack-horses.

[1] See the files of the *Annual Register*, under the headings "highwaymen," "highway robbery," etc., for many examples of this; also Andrews, *Eighteenth Century*, pp. 228–46, and Brit. Mus. 10,349. g. 11, *A Journey Through England* (1752), p. 81, where the writer speaks of having seen a great number of gibbets upon Finchley Common, near Barnet.

[2] For the life of a typical highwayman, see, in addition to the above references, Roberson and Green, *Oxford during the Last Century*, pp. 72–74, 112; *Monthly Chronicle of North Country Lore and Legend*, II, pp. 18–22, 114–15; Lecky, *History of England in the Eighteenth Century*, VI, p. 265, etc. Hawkins, *A Full, True and Impartial Account, etc.*, gives a good idea of the vast amount of highway robbery going on in the early part of the eighteenth century, and of the bold and daring way in which it was done.

[3] Jeboult, *West Somerset*, p. 34; also *The Devil's Cabinet Broke Open*, p. 12. The writer of the latter pamphlet was formerly a highwayman, and he relates his experience. He shows how the inns were in league with the fraternity of highwaymen. But the latter often had to spend at the inns the money obtained from travellers, in order to be allowed by the innkeepers to carry on their wickedness, without being made public. This man shows all the tricks of the "knights of the road," and how to avoid them. Hawkins, op. cit., also shows how innkeepers were in league with these highwaymen. The same fact is told in *The Discoverie of the Knights of the Poste* by E. S. (1597), and in Harrison's *Description of England in Shakspere's Youth*, II, p. 108.

Bridges.

As in the early period, so now under the Tudors and later monarchs, custom largely determined who should repair and maintain the bridges. In this respect, bridges differed materially from highways, since the repair of the latter almost uniformly fell upon the parishes; but if a bridge, according to immemorial custom, had been kept up by a certain landowner, or by a gild, or by a township, the law recognized the obligation of such an individual or body corporate to carry the burden thus imposed. And what we have here said as to bridges applies with equal force to ferries that were maintained instead of bridges to convey teams and passengers across the rivers[1]. But a few of the customs formerly in use regarding the repairing of bridges, such as the granting of indulgences, and the contributions from passengers and worshippers at the bridge chapels, seemed to fall into disuse after the accession of the Tudors, probably because of the suppression of the religious houses and the confiscation of their estates and revenues by Henry VIII.

Of the public bridges, some were maintained at the expense of one or more landowners who held land adjoining them[2]; while a considerable number were erected by private munificence, and a great many were supported by gifts and endowments of various kinds[3],

---

[1] These latter were by no means uncommon down to the later years of the eighteenth century, especially on roads which did not form the great thoroughfares, but were mainly cross-roads; but in the period we have now under consideration, the use of ferries was very common. Even on the great north road, in 1705, two such important rivers as the Trent and the Tees had to be crossed by ferry; in crossing the former, the ferrymen, instead of pushing it over with a pole, drew it over with a rope which ran across the river and was fixed on both sides; and for crossing at such an important place as Stockton-on-the-Tees, we greatly wonder that there was not a bridge erected or maintained under instructions of the Bishop, since bridges ranked high in works of piety and devotion.

[2] From the *Records of the Worcestershire Quarter Sessions*, we learn that in 1598 the Lords of the manors of Powick and Wick were found liable to repair the bridge over the Teme in the road from Powick to Worcester and within the parishes of Powick and Wick. *Worcestershire Quarter Sessions*, 1598 (27), XL. 9, p. 13; 1599 (73), VI. 12, p. 21, etc.

[3] Denne, in his *History of Rochester*, p. 45, says that in all probability the money required for building Rochester Bridge was raised in the same way as that used for its repair, viz., by taxation on the adjacent manors, places and bounds, according to their respective values. After the bridge had several times fallen into ruin, authority was given by the statute of 18 Eliz., c. 17 (1575–6) to appropriate for its repairs certain rents and revenues; but after nine years it was found that the new fund was inadequate, and in 1584 the wardens were given full authority to assess the lands for the repairs of the bridge, and to distrain in case of refusal.

The bridge over the Trent at the town of Burton-on-Trent was formerly looked

usually land or money. Occasionally the gild of a town undertook to aid in the maintenance of bridges at the expense of their own members[1]; and at times the gild, through its Master, had a share in the administration of trust funds or charities intended for this public purpose[2]. In

after by the monastery of that place; but when the monastery was dissolved and the abbey lands were granted to the Paget family, the latter were required to repair the bridge for the future. Shaw, *History of Staffordshire*, p. 15. See also Meriton, *Guide to Surveyors*, p. 86. By Act of 1746 (19 Geo. II, c. 24) authority was given to a certain Lora Pitt to erect a bridge or bridges over the river Frome. This she accomplished in a short time (1747). Hutchins, *History of Dorset*, I, p. 574.

By Act of 1747 (20 Geo. II, c. 22), Samuel Dicker was authorized to build a bridge across the river Thames, from Walton in the county of Surrey to Shepperton in the county of Middlesex. This bridge was to be regarded as extra parochial, and the counties of Surrey and Middlesex were exempted from repairing it; but it was to be kept in repair by the man who erected it, and he was authorized to take pontage according to rates specified in the Act.

A similar instance is given in Act 23 Geo. II, c. 37; but in this case a proviso was added that if, after the expiration of the term during which the bridge was to be in private hands, the King should pay the expenses of building the bridge, the rights of private parties and the tolls should cease, and the bridge was to vest in His Majesty.

On Jan. 5, 1735, the Corporation of Weymouth thanked Mr E. Tucker, mayor, who at his own cost had repaired the bridge, that had been damaged by unmoored vessels (Moule, *Records of Weymouth*, p. 188).

In 1568, Walter Tyrryl, mercer, of Tiverton, deeded to the Corporation of Tiverton property to the value of £1800, the net annual produce of which was in part to be devoted to the building and maintenance of Exebridge (Dunsford, *Historical Memoirs of Tiverton*, p. 110).

In Exeter, there was a bridge over the river Exe, built by the city. Its great benefactor gave lands and rents for its continual maintenance. Brit. Mus., Add. MSS. 28,649, p. 62.

In Stocks and Bragg, *Market Harborough Records*, p. 222, we learn that in the year 1523 Sir William Sotherey bequeathed by his will, dated Oct. 1 of that year, 6s. 8d. for "mending off briggs and causes (i.e., causeways) off the same towne off Bowdon."

Chanter and Wainwright, *Barnstaple Records*, pp. 226–8, give a complete list of bequests that were left for the maintenance and repair of the Long Bridge at the Town of Barnstaple.

In Ipswich, in 1564, Edward Gardner was made a free burgess, because at his own cost he had built Handford Bridge (Bacon, *Annals of Ipswich*, p. 267).

For other examples, see Briscoe, *History of the Trent Bridges at Nottingham*, p. 5 et seq.; Blomfield, *History of Bicester*, VI, p. 5; Cox, *Derbyshire Annals*, II, pp. 220, 222; Molyneux, *Burton-upon-Trent*, p. 86; Clark, *Wood's City of Oxford*, I, p. 411; Astle, *The Will of Henry VII*, pp. 20–21. Stow, *Summary of the Chronicles*, pp. 431, 434, 440, 449, 462, 471, shows many gifts made for the repair of bridges in the early Tudor days. See also Bearcroft, *History of the Charterhouse*, pp. 86, 119–22.

[1] See Blacklock, *The Suppressed Benedictine Minster*, pp. 383–4, where the gild was presented in 1635 and 1650 for not having repaired the bridges at Leominster, with which they were charged.

[2] The Gild of the Holy Cross at Birmingham, and its offshoot, William Lench's Trust, well exemplify this point. See Bunce, *History of Birmingham*, I, p. 30.

some instances, bridges were repaired wholly or in part by church contributions[1], and by funds, often small in amount, from various other special sources[2].

The history of Westminster bridge, which extends across the Thames from the city of Westminster to the opposite shore of Surrey county, gives us an example of a bridge constructed with funds obtained from a very special source. When the necessity for such connexion was fully demonstrated[3], an Act of Parliament was finally passed therefor in 1736[4], appointing two hundred Lords and Commoners as

[1] Chanter and Wainwright, *Barnstaple Records*, pp. 234–40, give extracts from the bridgewardens' accounts for different years. The church collections for Long Bridge at Barnstaple, especially in some years, formed a very important part of the receipts of the wardens.

See also the important contributions by the church in the parish of Ashburton, during the years 1546–80 (*Churchwardens' Accounts—Parish of Ashburton*, pp. 29, 43, 48, etc.).

The Church Register of Burton-upon-Trent shows that in 1664 the custom began of making collections in the church, to aid in repairing churches in different parts of England, to effect the reparation of bridges, etc. (See examples on page 71 of Molyneux, *Burton-upon-Trent*.)

From the money received from the sale of the plate, goods, vestments, etc., of St Botolph's church, Boston, £58. 16s. was spent on the repairs of the bridge in 1546–50 (Thompson, *History of Boston*, pp. 163–4).

[2] Such sources as the fines collected from those who failed to live up to the laws for highways and bridges; also fines and assessments from a variety of other causes, for an example of which see Blacklock, *The Suppressed Benedictine Minster*, p. 401.

[3] The petition of the inhabitants of Westminster and Lambeth parishes was presented to the House and referred to a committee which took evidence. It was shown that almost all sorts of provisions were dearer in Westminster than in Surrey, for want of a bridge; that the Lambeth ferry was wholly insufficient as a communication between the neighbouring counties, for it was both inconvenient and dangerous. Brit. Mus. 8776. a. 17, p. 6. On the other hand, many reasons were put forth against the building of a bridge there. It was said that it would injure the navigation of the river by retarding the flux of the tide, by increasing the number of shallows and sandbanks, and by creating delay and danger to the conveyance of goods and passengers. It was complained that such a bridge would cause danger to wherries and small boats, as well as to the larger barges, especially to those that were unwieldy and not easily governed by sail or rudder. From these evils there would follow an increase of wages to labour, a rise in prices of commodities, danger of losing valuable cargoes, decrease in number of watermen who were so useful for service on sea, danger to adjoining houses from overflowing of the river. The city of London would also be injured because of the reduction of its tolls on goods passing over London bridge and the lessening of its profits from markets; v. Brit. Mus. 357. c. 3 (69), 'Reasons against building a Bridge over the Thames at Westminster.'

[4] Act 9 Geo. II, c. 29. In addition to the provisions here given, the Act stipulated that no houses or sheds should ever be built on the bridge when finished; and also provided that the Company of Watermen should be compensated for the loss of their Sunday ferries near the bridge. The Commissioners were empowered to lay

Commissioners to direct the execution of the work. To defray the charges, and enable them to carry out this purpose, Parliament granted them a lottery, consisting of 125,000 tickets at £5 each (= £625,000), out of which a deduction was to be made of £100,000, or sixteen per cent. of the whole amount of money ventured, towards building the bridge and keeping it in repair. The Commissioners agreed to allow the Bank of England £2000 for their trouble in receiving and disbursing the money ventured in the lottery. The surplus funds of the lottery, after the payment of the prizes and necessary expenses, were to be applied to the cost of building the bridge. But at the end of the time, when the lottery was to cease, only £43,116 had been contributed; and therefore in the last session of the Parliament of 1736–7, a second Act was passed confirming the former Act, and granting a new lottery of 70,000 tickets at £10 each (= £700,000)[1]. If sufficient money could not be obtained by this means during the specified period, the King had authority to incorporate the Commissioners and grant them a seal, by which they might borrow any sums of money, on the credit of their toll, at a rate of interest not exceeding five per cent., and might assign over the said toll or any part of it. It is not our purpose, however, to follow out in detail the history of this bridge, which is a long record, but we merely wish to show the way in which the great amount of money necessary for its construction was obtained[2].

Having now spoken of the accessory methods employed in the repair and maintenance of bridges, we turn to consider the customary liabilities with reference to such works. As a general thing, bridges that were not due to private benefaction were erected and maintained by the towns, by the parishes, by the hundreds or townships, or else

open and widen the streets and ways leading to it, and in order to do this they could compel the owners of houses and lands to sell.

[1] Act 10 Geo. II, c. 16. Of the 70,000 tickets, 7000 were prizes and the rest were blanks, valued in the Act at £7. 10s. each; both blanks and prizes were subject to a deduction of fourteen per cent., amounting to £98,000, which was to be applied to building the bridge, etc. In case the £98,000 was not enough to build the bridge and keep it in repair, the Commissioners had power, by this new Act, to lay a toll on the bridge at the rates specified in the Act.

[2] This second Act was passed in 1737. The lottery was soon filled and the Commissioners agreed to allow the Bank £3000 instead of £2000. The Commissioners advertised for tenders for a wooden bridge, but this met with popular dislike, as it would, in the long run, be uneconomical: a waste of money, wholly out of keeping with the needs of traffic and with the intention of the legislature. Hence the Commissioners were forced to decide for a stone bridge. The further history of this bridge may be traced through a long series of Acts following those above mentioned, viz., 11 Geo. II, c. 25; 12 Geo. II, c. 33; 14 Geo. II, c. 49; etc.

by the counties; and not infrequently the responsibility for them was divided between two or more parties.

In the case of bridges whose maintenance devolved upon the towns, the funds required were often obtained by bridge tolls, collected from those who passed over or under the bridge. These pontage grants were allowed by the King either by virtue of a stipulation to that effect in the town charter, in which case the tolls were usually perpetual, or else were granted by royal letters patent, in which case toll could be collected for only a few years at a time, although the privilege might be renewed at the expiration of that time[1]. Sometimes towns obtained bridge money as the income from property that had been left for that useful purpose[2]; other towns received money from the rents of houses that stood on the bridges[3], and from tolls imposed on goods that were brought into the town to be sold[4]; but perhaps a greater number of towns supported their bridges in the same way as they paid their other expenses, that is, by a general contribution, in the form of taxation or assessment[5]. Money collected for the repairing of bridges was put into

[1] We have already referred to this in an earlier chapter. In further proof we cite Challenor, *Records of Abingdon Borough*, pp. 72–73; Atkyns, *Glocestershire*, p. 58; Briscoe, *History of the Trent Bridges at Nottingham*; Thompson, *History of Boston*, p. 251; Clark, *Wood's City of Oxford*, p. 485; Act 1 Henry VIII, c. 9.

[2] About the year 1553–4, Queen Mary endowed the Corporation of Boston with lands, etc., that they might be the better able to support the bridge and port of Boston, both of which appear, from the words of her grant, to have then been in a deplorable state, and needing almost daily repairs (Thompson, *History of Boston*, pp. 66, 251). See Act 18 Eliz., c. 17, for a similar fact with reference to Rochester bridge.

Various parcels of land were left for the repair of the Trent bridges at Nottingham, for which see Briscoe, *History of the Trent Bridges at Nottingham*; Orange, *History of Nottingham*, p. 644; Bailey, *Annals of Nottingham*, I, pp. 412, 432–3; Deering, *Nottinghamia Vetus et Nova*, p. 315; etc.

[3] See, for example, *Historical Account of Bristol Bridge*, pp. 9–10. The same was true of London bridge and many others.

[4] The charter granted to Abingdon by James I in 1620, as well as previous charters (granted by Mary, Elizabeth, and James I), and the charter granted by James II in 1686, gave the Corporation the right to levy tolls on all the things and wares brought into the borough to be sold. These tolls were to aid in paving and in repairing the bridges and ways of the borough (Challenor, *Records of Abingdon Borough*, pp. 72–73). The charters of many other towns gave like privileges; see, for example, the charter of the city of Gloucester, as given in full in Atkyns, *Glocestershire*, p. 56 et seq.

[5] In the case of Leominster, we have already seen that the craft gilds were responsible for the repair of some bridges; but it is evident that in some cases the borough was responsible. See the items culled from the accounts of the Borough Chamberlain of Leominster, as given in Blacklock, *The Suppressed Benedictine Minster*, pp. 384–5.

In 1608 an assessment of £40 was voted by the Assembly for the repair of

the hands of bridgewardens or surveyors, who were required annually to give to the Justices of the Peace a full account of all money received and disbursed[1].

We have noted many bridges the repair of which was an obligation upon towns, and there were many others which were repaired by the parishes or hundreds, or some similar subdivisions[2]. Cox tells us that "bridges differed materially from highways[3], inasmuch as the repair of them, save quite exceptionally, never fell upon the parish." In this statement he seems to be in error, if he means to apply it to all public bridges. He goes on to say that streams and rivers often form the boundary lines of parishes, and that bridges spanning such streams

the west bridge and other bridges within the town of Northampton. In 1615 a further sum of £20 was raised by assessment for the repair of these same bridges. From this time on, there was a constant series of assessments, amounting to considerable sums (Markham and Cox, *Northampton Borough Records*, II, pp. 432, 433, 434, 538, etc.). During the years immediately following the Restoration, the town of Northampton was several times indicted by the county authorities for the condition of the highways and bridges that formed part of the great roads that traversed the borough. On May 11, 1663, the Assembly, to prevent charges and troubles that had come upon the town through these indictments, ordered £100 to be raised for repairing the highways and bridges. Two years later, another £100 was raised, chiefly for rebuilding the south bridge. This bridge had to be rebuilt almost to the foundation, and while the work was in progress an extraordinary flood destroyed almost the whole bridge. On Jan. 17, 1666-7, the Assembly ordered an assessment on the town for the new bridge, amounting to £300. Other assessments were made later.

Sometimes bridges were built by towns without recourse to Parliament to secure an Act for that purpose; and when once a bridge had been erected and its utility had been shown, its maintenance was a matter of public concern, required by the law. For example, the Assembly of Liverpool, on Nov. 23, 1635, ordered that a bridge should be made at the Poole, on the south side of the town, at a convenient place appointed by the Assembly. (Picton, *Liverpool Municipal Records*, I, pp. 187-8, 315.)

For similar instances, see Guilding, *Records of the Borough of Reading*, II, pp. 15, 21, 31, etc.; Leader, *Sheffield Burgery Accounts*, pp. 111, 112, 114, 115, 435, 447, etc.; Atkyns, *Glocestershire*, p. 56; Thompson, *History of Boston*, p. 251; Briscoe, *History of the Trent Bridges at Nottingham*; Bacon, *Annals of Ipswich*, p. 265; Blomefield, *History of Norfolk*, III, p. 441.

[1] To obtain a more complete view of the accounts of bridgewardens, showing the sources of their receipts, and the purposes for which disbursements were made, see *Nottingham Borough Records*, III, pp. 2, 4 (particulars of expenditure given on pp. 241-4); pp. 6, 8, 10 (particulars of expenditure given on pp. 246-52); pp. 12, 14; pp. 106, 108, 110; also ibid., IV, pp. 38-40. Chanter and Wainwright, *Barnstaple Records*, pp. 234-40, give extracts from the bridgewardens' accounts.

[2] Brit. Mus. MSS. 11,052, pp. 2, 21, 22, 25, 26, 37, 39-49, etc. On page 2 of same is given a statement of the amounts apportioned to each hundred, Apr. 1, 1600, for the erection of a bridge at Wilton on the Wye.

[3] Cox, *Derbyshire Annals*, II, p. 214.

could not be said to be in either parish; from which we might almost conjecture that he was thinking only of such bridges, had he not added immediately, "The old common law is quite clear, that of common right the whole county must repair bridges." Such a broad generalization, even with the above restrictive modification, seems to be wide of the truth. The fact is that the Statute of Bridges[1], passed in 1531, made provision for such cases by requiring that where a bridge was partly in one parish and partly in another, each parish should be responsible for its share in the maintenance of such a bridge. Because the law upheld and the Justices of the Peace in Quarter Sessions enforced the liability of the parishes for the repair of many bridges, it is evident that the county was by no means universally liable for such work. On the contrary, those which were recognized by custom and right as parish bridges had to be repaired by the parish[2].

---

[1] Act 22 Henry VIII, c. 5.

[2] In the *West Riding Sessions Roll*, p. 129, the Justices recorded that "A pain is laid by us that those townships which of right ought to repair Humberhead-bridge do it before the first of August next upon pain of £10." Other like orders are given on the same page. See also the Memorial presented in 1601–2 by the West Riding Justices, showing that certain bridges had always been repaired by their respective wapentakes and parishes, despite the fact that certain persons had laboured to put an end to this custom (*West Riding Order Book F.*, quoted from an order made at Pontefract, regarding Pathorne Bridge, Apr. 4, 1654).

In the *West Riding Sessions Roll*, p. 38, we find this entry: "Forasmuch as Robert Littlewood, gentleman, and his fellow jurors have presented that there are four bridges of stone within the town of Bradford, so ruinous and in so great decay by reason of certain floods which hath happened of late years past, that without speedy amendment and reparation they will utterly fall down and be carried away with the water, which will be to the great hindrance and loss of all the whole country, and they have further presented that it is very requisite and necessary that a contribution of an assessment should be made through the whole stewardship of Bradford for the repairing thereof: It is therefore by this court (ordered?) that two of the next Justices shall take a view thereof and certify at the next Sessions what sum of money will repair the decays and ruins of the said bridges, that order may be then taken for the levying and collecting of such a sum of money within the said stewardship of Bradford as shall be thought meet for the speedy repair thereof." For similar cases, see pp. 57, 74, 97.

See also the case of Longroide bridge which was repaired by the townships of Huddersfield and Quarmebie, between which it stood. *West Riding Sessions Roll*, p. 38.

In the *Documents and Records of the Worcestershire Quarter Sessions*, there is a memorandum in 1637 as to the parishes in the hundred of Doddingtree which had paid and those which had not paid for the repair of Stanford bridge—see 1637 (237), LXIV. 89, p. 650. From these records we can find only one case in which the modern idea of "through traffic" is put forward as a ground for relieving the parish from its liability for repairing a bridge. The people of Tenbury in 1615

But although the county was not universally liable, the tendency was to have it more and more assume this liability. In many instances it could not be known upon whose shoulders the liability to repair rested, and in others it was only with great difficulty that this responsibility could be located[1]. In all such cases, the statute of 1531 said that these bridges, if within any city or corporate town, were to be repaired by the inhabitants of that city or town; but if they were outside a city or corporate town, they were to be maintained by the shire or riding within which they happened to be[2]. It will be noted that this relieved the parish of the liability to repair, in all cases where the liability could not be definitely proven; but at the same time it made the county the scapegoat upon which were loaded the defaults and negligence of such jurisdictions as parishes and hundreds[3].

petitioned that, as the great stone bridge and the wooden bridge over the river Teme, which had been damaged by a sudden flood of water, would demand the heavy cost of £30 for its repair, the adjoining parishes might be ordered to contribute, as it was the great thoroughfare between Wales and the city of London—see 1615 (133), xxii. 83, p. 212.

On parochial responsibility for bridges, see also Blomfield, *History of Bicester*, vi, p. 34; Baigent and Millard, *History of Basingstoke*, pp. 312, 344, 345; Cox, *Derbyshire Annals*, ii, pp. 216–17 (the presentment of Duffield parish in 1658); Act 22 Car. II, c. 12, sec: xiv.

[1] See for example Brit. Mus., Harl. MSS. 4115, p. 43, 'A Commission for the Repair of the Great Bridge in Cambridge' (1654). See also footnote 3 below.

[2] Act 22 Henry VIII, c. 5, sec. ii. For example, in 1586, the bridge at Wansford across the Nen river, which was part of a great road, had become impassable, and was ordered to be repaired by Northamptonshire. See copy of Lord Burghley's order in Lansd. MSS. 49, pp. 74–75.

[3] Of course, where a subdivision of a county, such as a hundred, or a parish, or a township, was liable by immemorial custom, it might still be indicted for non-repair of a particular bridge, or part of a bridge within its boundaries, on the ground of immemorial usage alone; and this still holds true down to the present time. The same thing is true as to private liability to repair.

In the West Riding of Yorkshire, it was so hard for the Justices to find who should repair the bridges, that in 1601–2 they framed a Memorial in which they distinguished forty-eight of the most important of the bridges "to be repaired of right and custom at the charge of the whole West Riding, and the rest by the respective wapentakes and parishes, which order hath been constantly affirmed and practised, albeit sundry persons for the ease of their own particular parishes and places of habitation, have laboured to infringe the same custom." (*West Riding Sessions Roll*, p. xxxvii.)

At the Epiphany Sessions of the Grand Jury of Derbyshire, Jan. 10, 1748, the presentment was made that Aston bridge, over the river Dove, was in immediate danger of falling; and although it was an important bridge, the presentment closes with the words, "that it is altogether unknown to us nor can we find any Persons lands Tenements or Body Politic ought of right or by ancient custom repair the same or any part thereof." Cox, *Derbyshire Annals*, ii, p. 223.

Among the *Acts of the Privy Council* (xvii, 1588–9, p. 301) we find a letter sent

J. T.   13

This tendency to saddle the county with obligations which could not be proved to rest upon the smaller municipal bodies, is noticeable in the history of many bridges, right down into the nineteenth century[1]; and it is the most marked feature we have to mention in our consideration of this subject. But although the Act of 1531 transferred to the county the burden of repairing bridges, the liability for which could not be determined, there were many bridges which had always been kept up by assessments upon the county[2]. In this way, the number for which the county bore a direct liability was constantly increased by the number for which it was compelled to assume liability; and thus the responsibilities of the county progressively increased, while those of the parish decreased.

In some instances, the sovereign gave assistance to the municipal authorities in supporting or rebuilding a bridge. This is well illustrated by the Berwick bridge across the Tweed river at that town. As the result of a flood in 1607 ten pillars of the wooden bridge that spanned

---

to the Justices of Assizes in Middlesex county, saying that complaint had been made to them about the bridge over the river Lea, in the parish of Hackney, which had been broken down and not repaired, and that there was doubt as to who should repair it. The letter urges the Justices to find the party liable and to see that the bridge is no longer neglected.

As to the failure to know who ought to repair certain roads and bridges, see further, Blomefield, *History of Norfolk*, iii, p. 441; Harl. MSS. 6166, 'The Defaults and Common Nuisance of Bridges and of Causeys,' etc., p. 242.

[1] In 1725, the city of Norwich presented a petition to Parliament, showing their inability, through exhaustion of revenues, to keep up their bridges and the great roads leading out from their city to London. Parliament passed an Act which commenced May 1, 1726, by which tolls were laid on all goods brought up the river higher than Thorp. This revenue was to be applied by the city for repairing its bridges, etc. But since the people of the county of Norfolk had to pay part of these tolls, it was agreed, in order to put an end to all disputes between the county and the city as to the maintenance of Trowse bridge, Harford bridge, Cringleford and Earlham bridges, that the city should pay the county £30 a year toward the repairs, and that the bridges should thenceforth belong to the county (Act 12 Geo I, stat. v, c. 15; Blomefield, *History of Norfolk*, iii, p. 442).

At Sturminster Newton in Dorset, the great bridge had been usually repaired by the town; but in 14 Car. I, it was ordered to be repaired by the county, as were also the little bridges in the year 25 Car. II (Hutchins, *History of Dorset*, ii, p. 410; for similar examples of bridges transferred to the county, see ibid., i, p. 383, and ii, p. 400).

[2] For explicit references as to the up-keep of bridges by counties, note Axon, *Manchester Quarter Sessions*, i, pp. 39–40, 48, 109, 115, 142; *West Riding Sessions Roll*, p. 105; *Acts of Privy Council*, ix (1575–7), pp. 89, 117, 120, 131, 135; and Cox, *Derbyshire Annals*, ii, pp. 215–25, which contains copious extracts from the records of the Quarter Sessions of that county, which will amply repay reading. See also Act 39 Eliz., c. 24; 12 Geo. I, c. 15; etc.

that river were washed away, and the rest was so badly shaken that it was finally decided to erect in place of it a new bridge. Since this structure was on the great north road leading from London to the Scottish capital, it was necessary to make it so that it should be capable of sustaining an increasing amount of traffic, and a stone bridge was ultimately decided upon as the most economical in the matter of repairs. Many of the estates that had been chargeable with the repairing of the bridge had been dispersed among various persons so that there was little hope of receiving much from that source. But, remembering that the Crown had paid in the preceding forty years £5372 toward the repair of this bridge, the town again appealed to James I, in their present emergency and necessity, for financial assistance in securing the requisite £14,000 to carry out their purpose. The king made a very generous contribution toward the rebuilding of this important structure joining the two portions of his kingdom[1]. How often similar aid was rendered by the monarch has not yet been ascertained.

In cases of divided responsibility for bridges, if this were not otherwise settled, the Statute of Bridges made provision for the adjustment of liabilities. If part of any bridge were within one, and the other part within another shire, riding, city, or town corporate; or if part were within the limits of any city or town corporate, and part without; then, in all such cases, the inhabitants of each shire, riding, city, or town corporate were to be chargeable to repair that part of the bridge that lay within their own limits, and the responsibility for the bridge included the responsibility for the highway for three hundred feet at each end of the bridge[2]. Such a provision was very necessary, because there were a great many bridges over streams that separated adjoining parishes, or other territorial jurisdictions; but even with statutory regulation, differences occasionally arose. About 1716, the county bridges over the Dove occasioned various disputes between the counties of Derby and Stafford. Each county accused the other "of carrying their water works too farr into the River to cast the current and weight of the same upon the other." This finally resulted in the formation of a joint committee of the Justices of the two counties, to control the repairs of the bridges of Mappleton, Coldwall, and Hanging Bridge, as well as of three bridges at Tutbury[3]. Sometimes several authorities were concerned with one bridge (or a group of bridges) and in these

---

[1] Brit. Mus., Lansd. MSS. 166, pp. 84–93.
[2] Act. 22 Henry VIII, c. 5, secs. 2, 7.
[3] Cox, *Derbyshire Annals*, II, p. 222.

cases very elaborate specifications were required as to the share of each, so that harmony might be preserved[1].

Another aspect of the repairing of bridges is that many of those which were at first narrow and made merely for the convenience of

---

[1] Regarding the divided authority in the case of the Tyne bridge at Newcastle, see *Acts of the Privy Council*, vii, p. 290, and also the history of the bridge as given in Mackenzie, *History of Newcastle*, pp. 204–14.

The history of the Trent bridges at Nottingham is fully given in Briscoe's account of them, already referred to.

The nine piers of Rochester bridge were to be repaired by nine different authorities. See Rye, *Collections for the History of Rochester* (in MS.), Brit. Mus. c. 55, g. 2; Brit. Mus., Add. MSS. 24,933, pp. 2–7; Brit. Mus. 1855. a. 17, 'Collection of Statutes concerning Rochester Bridge;' and also Denne, *History of Rochester*, pp. 42–44.

Another bridge which was maintained under divided authority, was the Leen bridge, over the Leen river at Nottingham. It was a long stone bridge of twenty arches, and was to be repaired at the charge of the town and the whole county. Among the town records of 36 Henry VIII, it is expressly stated that the Leen bridge had from time immemorial been repaired and upheld by the town of Nottingham and the several wapentakes or hundreds of the county, in the following proportions:

Town of Nottingham, was to repair the north end of the bridge, and the two arches next adjoining to the same, containing in length 46½ feet.

Broxtall Hundred, the three adjoining to the above-mentioned two arches, containing 81½ feet; and the middle column between the two arches was to be upheld and repaired at the joint expense of Nottingham and Broxtall.

Thurgarton a Lyghe, was to repair the five next adjoining to the three arches, containing 135½ feet; the middle pillar between them and the three foregoing was to be repaired at the common charge of Broxtall and this hundred.

Bassetlowe Hundred, was to repair the five arches next beyond the five before-mentioned, containing in length 169½ feet, which was as much as anciently six arches contained; and the middle column between these ten arches was to be repaired in common by this hundred and the preceding.

Newark Hundred, was to repair the three arches next adjoining the last five; and the middle column between these three and the preceding five was to be repaired in common by this hundred and Bassetlowe.

Byngham Hundred, repaired a certain parcel of this bridge, containing 105 feet, and the middle pillar in common with Newark.

Ryscliff Hundred, was to repair two other arches next to the aforesaid parcel and the south end of the said bridge, containing in length 57 feet; and the middle pillar between these two arches and the said parcel was to be repaired in common by the two last-mentioned wapentakes (Deering, *Nottinghamia Vetus et Nova*, p. 167).

Act 18 Eliz., c. 18 shows us how Chepstow bridge, between the counties of Gloucester and Monmouth, was to be repaired by the two counties. Brit. Mus., Harl. MSS. 6166, giving an account of the bridges in the county of Surrey, shows most of the bridges in bad condition (temp. 25 Henry VIII), and in a large number of cases there was divided responsibility. See also [Owen], *Some Account of Shrewsbury*, p. 84, showing that, after litigation, it was agreed that the English bridge over the Severn at that town should be repaired partly by the monks of the monastery and partly by the burgesses of Shrewsbury.

horse traffic had later to be widened when waggons, carts, caravans and other wheeled vehicles became more common. The carriage of one of these heavy lumbering waggons, drawn by five or six horses, with its great load of twenty to thirty hundredweight allowed by law, and frequently much more than that carried contrary to law, involved a strain upon bridges which was greatly in excess of that caused by pack-horses, each of which was not loaded beyond 240 to 500 pounds.

With the conveyance of heavier weights, it was more difficult, and sometimes impossible, to ford the streams, and necessity required the rebuilding of many bridges with more substantial material and wider dimensions[1]. How much of this was done before the middle of the eighteenth century must be purely a matter of conjecture; but we are impelled to say that since most of the carrying was still done by means of pack-horses, there was comparatively little widening of bridges up to that time.

It is impossible to get any comprehensive idea as to the condition of the bridges in general, for, unlike the roads, each was a separate entity, and cannot be considered in association with others in the manner that various portions of road form a continuous line. Nor does it fall within our province to describe their features from an architectural or engineering standpoint. Of one thing, however, we may be certain, that there was great waste in the maintenance of some of them. By reason of lack of attention on the part of the Justices, some bridges were not repaired at the proper time, and the increased decay into which they fell could only be made good by increased expenditure[2]. The want of knowledge and skill in construction, both as to form and material, exposed many to the destructive action of floods, and the loss of those that had houses upon them was accompanied by much destruction of life. Those who had the raising of funds for bridge repair were known to assess in some places greater amounts than were needed for the work[3]; and frequently the money was not applied wisely by the surveyors[4]. These things, taken in connexion

---

[1] Brit. Mus. MSS. 6707, Reynolds, *Derbyshire Collection*, p. 11; [Owen], *Some Account of Shrewsbury*, p. 84.

[2] *Acts of the Privy Council*, VIII, p. 290; XIII, pp. 77–78; XVII, p. 301; XXV, pp. 216–17, 429–30. The latter shows how dilatory the Justices had been in this case, for the bridge at Upton, in the county of Worcester, had been out of repair for three years or more. See case of the Swarkeston bridge, given in Cox, *Derbyshire Annals*, p. 215 et seq.

[3] Preamble to Act 1 Anne, c. 12.

[4] Act 1 Henry VIII, c. 9; preamble to Act 1 Anne, c. 12; *Acts of the Privy Council*, IX, p. 89; Axon, *Manchester Quarter Sessions*, I, p. 142.

with the wilful destruction of bridges, to prevent which legislation of a drastic nature had been passed[1], give us some slight idea of the uneconomical administration of the time.

[1] See especially Acts 8 Geo. II, c. 20; 9 Geo. II, c. 29; 15 Geo. II, c. 33. Bridges were destroyed by armies sometimes, as a part of the tactics against the enemy; as for example, the bridges at Nottingham during the Civil War of Charles the First's reign. See also preamble to Act 22 Car. II, c. 12.

CHAPTER III

RIVER NAVIGATION, 1500–1750

THE Magna Charta, chapter 23, builds upon the supposition that what was the common law before is still in effect: that all public rivers were the king's highways, and as such free for all his subjects; and therefore it forbids putting into these rivers kiddles, weirs, and other things for fishing purposes, for these were recognized at common law as nuisances. This statute, in this particular, as well as in many others, did not introduce a new law, but merely declared the old law. Since the Magna Charta, there were other Acts passed to enforce it[1], and to extend the law further than to apply to fishing weirs, by making it necessary to pull down mills, millstanks, or mill-dams, and such like, if erected in the time of Edward I or after that.

Every river was either private or public. Private rivers belonged entirely to their respective owners, who might use them as they pleased, and no stranger could come on them without their consent. But if a river were a public river, that is, a common highway, the King himself could not restrain the common use of it; as, on the other hand, if it were a private river, belonging to private persons, he could not by any grant give a right of passage, for he could not invade any private man's property. Thus, every river, notwithstanding the King's grant, was either private and free for nobody but the owner, or public and free for everybody. This, of course, is to be understood where the King himself was not owner, for then he might grant a passage in that portion of the river that he owned, but in no other part. So might any private man, in the part he owned. Similarly, if the King were owner of one part of a river, and at the same time other persons were owners of other parts and of the adjoining lands, and the King granted away his part, he could not give a right of passage on the other parts of the river, or of going on the banks to haul[2].

---

[1] Act 25 Edw. III, stat 3, c. 4; Act 1 Henry IV, c. 12; Act 12 Edw. IV, c. 7.

[2] Brit. Mus., Stowe MSS. 818, p. 86, Lord Chancellor Macclesfield's 'Notes concerning Rivers, Navigations, etc.'

Similarly, if the King were owner of one part of a river, and granted the fishery but not the water, the patentee by that means had only a right to fish and to go

It was difficult at a later day, except in the cases of the very great and remarkable rivers, to prove directly what were public rivers in the time of Edward I; but since statutes had been made for pulling down mills and other nuisances erected in such rivers since Edward I's time, it was strong evidence of a river being a public river then, if at the later time it was capable of being navigated, and had been actually navigated, and had been kept free from those nuisances which the statutes prohibited in such rivers[1]. But it is of the public rivers that we would treat here, for they alone were common to all.

In our survey of the period before 1500, we have seen that the free use of public rivers had been obstructed in several ways, and that it was hard to enforce the law against these obstructions; so that the people, though nominally enjoying freedom of navigation, were really prevented from exercising their privilege fully[2]. In 1472, after much struggle, the shipping and mercantile interests prevailed to such an extent that offenders were ordered to destroy the obstructions they had erected in rivers, especially the Severn[3]. But although legislation was passed against it, rivers continued to be impeded in their navigation through the hindrances placed in them by private persons in carrying out their own individual interests. Sometimes fishgarths, for catching fish, were set in the direct passage of the stream[4]; and sometimes rubbish was thrown into the rivers by those who occupied adjacent houses[5]. In other cases, the path on each side of the river, along which men walked when drawing their boats on the river, was barred by "covetous persons," who would not allow a boat to pass without first taking toll from those who had charge of it[6]. These and various other forms of impediments to river traffic were intended to be removed by the legislation proposed in each case, but in reality they

upon the river to catch the fish, but could not navigate it for other purposes any more than if he had no such grant. Therefore, if this patentee had been accustomed to navigating it for other purposes, it was an evidence that it was a public river, and that everybody else might also navigate it, for he had no right to anything there (except the fishery) but what everybody else had.

[1] There were many other forms of presumptive evidence that could be invoked to prove a certain river to have been a public river, but into these we do not propose to enter. See Brit. Mus., Stowe MSS. 818, p. 87; also Coulson and Forbes, *The Law of Waters*, p. 65 et seq.; and Wellbeloved, *Laws of Highways*, p. 20 et seq., discusses the question as to what is a public navigable river, giving detailed references to the cases that had been decided.

[2] See page 24 et seq.

[3] Act 12 Edw. IV, c. 7.

[4] Act 23 Henry VIII, c. 18.

[5] Act 27 Henry VIII, c. 18.

[6] Act 23 Henry VIII, c. 12.

were frequently not obviated by the measures passed therefor. In some instances, the nuisances complained of were for a time stopped, only to reappear at a convenient occasion.

That there were many difficulties in knowing the extent of the rights and privileges enjoyed in connexion with river navigation, is apparent from the history of the Severn. This subject we have already touched upon, to some extent, in our introductory chapter. We have seen that the people of Tewkesbury had suffered much by reason of the fact that their boats and trows, when fully laden, had been cut in pieces, and the goods taken, by the rovers of the Forest of Dean. Notwithstanding the good legislation of 1429 and 1430[1], these hardships still continued; for as late as 1534 a penalty of fine and imprisonment was imposed on keepers of ferries on that river, if they carried offenders into or from Wales or the Forest of Dean, between sunset and sunrise[2]. How futile this law proved, will be apparent to those who remember that such organized bands of robbers could seldom be withstood by a single keeper of a ferry.

But it was not alone from such lawless gatherers of spoil that troubles arose; for even in the interpretation of the law, regarding freedom of passage along the Severn for all the King's liege subjects, difficulties were found. In 1384–5, the city of Gloucester was authorized to levy 4*d.* on each ship coming to the town, by the Severn river, loaded with goods for sale[3]. This was to help in the repair of the paving of the town. But it was not to apply to goods that were not to be sold in that city, and certainly the tolls were not to be imposed upon those ships which were merely passing along the river, and using the latter, as it had always been intended, as a free highway of trade. It would seem, however, that the authorities of the city of Gloucester exceeded their privilege, for they charged toll to all who were taking passage along the river, whether their goods were for sale in that city or not, and in some cases took the merchandise from such users of the navigation, and arrested and imprisoned them. The Act of 1504[4] said that if any party could prove that he or they had a right to take toll from vessels and goods passing along the river, such toll would be allowed. Amongst the select cases in the Star Chamber, bearing upon this point, is that of Whyte *v.* the Mayor and Burgesses of Gloucester (1505). Thomas Whyte, who was a merchant, said he could disprove the supposed right of the town of Gloucester to take toll. What was

---

[1] Acts 8 Henry VI, c. 27 and 9 Henry VI, c. 5.
[2] Act 26 Henry VIII, c. 5.
[3] Stevenson, *Records of Gloucester*, p. 51.
[4] Act 19 Henry VII, c. 18.

done about this case we do not know, for the decree of the Star Chamber has been lost; but we are led to infer that the town was unable to prove its claim, for, from the Act of 1532[1], it appears that the dispute had been revived by the town's demanding a toll, not for passing upon the water, but for the use of the towing path. The Act of 1532 declared this demand illegal, and reserved a free towing path, which it affirms to have been the immemorial privilege of all[2]. For hindering passengers on the banks of the Severn, or for demanding tolls from them, the penalty was 40s.[1]. Another instance of a similar barrier to free navigation farther up the same river, was the toll levied by the city of Worcester upon all boats passing under their bridge across the Severn; for even as late as 1564 the bailiffs of that city were insisting on taking their customary dues from Tewkesbury men, despite the claim of the latter for exemption, and were seizing their cattle in default[3].

Another reason why the rivers were not so fully used as they might otherwise have been for the conveyance of goods and passengers, was the high, and even extortionate, amounts which watermen charged for their services, on their barges, boats and ferries. This was notably the case on the Thames and Medway rivers. But in 1514, a law was passed imposing on watermen or bargemen making excessive charges, or on the owners or occupiers of such barges, boats, wherries, etc., a penalty of three times the value of the lawful charge[4]. Doubtless, this had some immediate effect in curbing the extortion, but the same complaint comes up again and again in later times.

The use of rivers for navigation purposes was often prevented by the destructive action of the forces of nature, aided at times by human agency. The washings of the soil were constantly being carried into the rivers which served the purposes of the drainage basins; and where

---

[1] Act 23 Henry VIII, c. 12.

[2] This whole case is fully set forth in the *Selden Society Publications*, XVI, pp. 209–26. It was not until the city of Gloucester obtained its charter from Charles I, in 1626–7, that it obtained the right to take toll from all vessels laden with goods passing on the Severn. The tolls authorized were: for each dole of wine and each ton of other merchandise passing on the river, 3d. (except for the burgesses of Tewkesbury or others who had made agreements with the Mayor and burgesses of Gloucester); 4d. for every ship or boat laden with timber, board or lath; and 2d. for every vessel laden with firewood. These tolls were allowed because the city had spent much in building two quays, two bridges, a causeway, etc., at that part of the Severn. (Stevenson, *Records of Gloucester*, pp. 43–44.) This privilege was confirmed in the charter given to the city by Charles II. The text of this charter is given in full in Atkyns, *Glocestershire*, p. 56 et seq.

[3] *Victoria History of Worcestershire*, II, p. 251. See also Act 19 Henry VII, c. 18.

[4] Act 6 Henry VIII, c. 7.

the fall in the river-bed was but slight, it required a flood of tremendous volume to clear out the navigable channel[1]. Then, the banks of the rivers were continually being washed away, and soil thus removed was deposited elsewhere in the rivers, causing irregularities in the depth of water, while at the same time the river was widened and the water covered a greater surface[2]. Such natural phenomena always tended to obliterate any one channel along which barges could pass with their loads. This filling up of the channel was frequently aided by the conduct of those who used these waterways; for when a barge-master, in coming up a river, found the depth of water insufficient for his purposes, he had only to throw out his ballast into the stream to allow his barge to float higher in the water, and thus carry him over the shallows[3]. The effect of such things was cumulative, for when ballast was thrown out at places which were already shallow, the result was that where the banks of sand and gravel accumulated, shelves were produced which soon became large enough to completely block the passage. As the result of such filling up of rivers, boats and barges which formerly had access to towns many miles from the mouths of rivers were later prevented from continuing in their course, and had to anchor considerable distances down the river, from which places the goods had to be conveyed by land at much greater expense[4].

[1] Beverley Beck, flowing into the Humber river, had become choked up by weeds and mud; and in order to make and keep this river clean, it was proposed to use an "engine boat" like those used in Holland, to pull up the weeds and loosen the mud, after which the river was to be flushed out at low tide by water collected in one or more reservoirs in the upper part of the river. Brit. Mus., Lansd. MSS. 896, pp. 162, 164, 166. The objections to this method of cleansing the river are given in ibid., p. 163. See also Harrison's *Description of England in Shakespere's Youth*, I, p. xxxv; III, p. 427.

[2] Brit. Mus., Lansd. MSS. 41, No. 45, pp. 169–76, gives the order of the Commissioners of Sewers of Lincolnshire, anno 6 Eliz., requiring the cleaning of the river Glen from bank to bank, and the embanking of the river to a sufficient height and breadth.

[3] At Ipswich the town officials seem to have been very decided in their efforts to stop the throwing of ballast into the river, and the leaving of other nuisances there in the way of the navigation. Apparently, Ipswich people took great pride in their city, for we have already noted their efforts to put their streets in good order and keep a degree of respectability. In regard to their efforts to keep the river channel clear for navigation, see Bacon, *Annals of Ipswich*, pp. 164, 195, 207, 209, 261. Compare Hanshall, *History of Cheshire*, p. 77; Ormerod, *History of Cheshire*, I, p. 134; and note also Act 34 and 35 Henry VIII, c. 9. See also Brit. Mus., Add. MSS. 36,767, pp. 1–4.

[4] Note the case of the Exe river and the city of Exeter, as given in Act 31 Henry VIII, c. 4, and in Oliver, *History of Exeter*, p. 249. Also the case of the river Dee and the city of Chester, as given in Ormerod, *History of Cheshire*, I, p. 134.

Other obstacles to navigation, of which we have a great many examples, were the mills, with their mill-dams, weirs, and other accompaniments, that had been erected along the courses of the rivers and streams. These mills were privately owned, and many of them had been in existence in the same places for long periods of time, without any mention of the fact that they were a hindrance to the passage of barges up and down the rivers. Mill-owners had chosen their sites, built their mills, mill-dams, and other necessary equipment, without any opposition from anyone who represented the public interests; the title to these things had been passed on from generation to generation, and the succeeding owners had enjoyed all the benefits of their situations, while the public had been well served[1]. With a series of mills along a river, and the water damned up at each place, we can readily see that, especially in dry seasons, the river below the dam would be very shallow; and to allow boats to pass over these shallows the miller frequently had to open the gate in his mill-dam and allow the water he had penned back to rush down and float the boat over the beds of sand and gravel, for which courtesy he demanded payment[2]. This was called flushing or flashing. But frequently, when water was scarce, and the miller had barely enough for his own purposes, boats were prevented, for days together, from continuing their passage, because the miller refused to give them a flash of water; and even when he did give it, he charged them extravagantly for the favour. It is evident that merchants and mill-owners who had once established themselves in this way in business, opposed any attempts to make the river navigable lest it might hinder their business by turning the water from private to public uses[3]. These obstructions in the rivers were a fruitful source of trouble, and led to the decay of the navigation, not only of the smaller rivers, but also of the large ones, like the Thames and Severn[4]. In

---

[1] Note, for instance, the Dee Mills, part of the history of which is given in Brit. Mus., Harl. MSS. 2081, p. 168; and the condition of the Thames river as given in Harrison's *Description of England in Shakespere's Youth*, III, pp. 411–26.

[2] It was accomplished by means of stanches erected across a narrow place in the river; and a man standing on a foot bridge above was able to open or close them as required. This method was employed even on the Thames and Severn rivers; the stanches on the Severn were removed at the time of the improvements of 1842, but on the Thames above Oxford some are still in existence, and serve to keep up the water level in summer (Harcourt, *Rivers and Canals*, I, pp. 64–65; *Proceedings of the Institution of Civil Engineers*, IV, pp. 111–12).

[3] See the case of the mill-owners at Newark when there was the proposal to make the Trent navigable (*J., H. of C.*, XXIV, p. 108 et seq.).

[4] On Sept. 4, 1580, the Privy Council sent a letter to the Lord Mayor requiring him to send to the Lords of the Council, the water-bailiff, or some other officer of the Thames river, with a declaration as to how many weirs there were between

some cases, however, mill-owners were compelled to regularly open their flood gates, so as to scour out the channel of the river[1].

In the case of rivers that were connected with the ocean, there was another element which entered largely into their efficiency or lack of efficiency as the agents for the transportation of goods. Such rivers were, of course, subject to tidal action, and many rivers which contained only one or two feet of water when the tide was out, were so increased in depth by the incoming tide that they could carry barges of considerable size[2]. But if the barges were not ready to take advantage of the tide they were often left stranded in the river or at their moorings until the next tide. This was a cause of much delay and inconvenience to those who were depending on this means of shipment; and the exigencies of shippers were frequently made the occasion for heavy charges by the barge-owners. On account of tidal action, therefore, there were some disadvantages, as well as some advantages; but it would seem that the latter far transcended the former, especially when we remember

London bridge and Staines, how many there had been anciently, and how many had been erected in the last seven years, for that Her Majesty had been informed that by the many weirs, the river in many places was being choked up, and made unnavigable, and she was disposed to have present redress taken (*Acts of Privy Council*, XII, p. 185). The truth of this is confirmed by Harrison's *Description of England in Shakespere's Youth*, III, pp. 411–26.

Ten years later, the Privy Council sent another letter to the Lord Mayor, saying that Her Majesty was informed that in the Thames between Kingston and London, because of weirs, stakes, etc., it was "so shallowe in divers places as boates and barges doe sticke by the way, not only to the hindraunce of the provicion broght by water and the ordinary passage, but by gathering of gravell and sand together shelves are made to the decay of the river." He was ordered to have these removed, and to attend to his duty in that respect more perfectly, or else he would have his office taken away from him. *Acts of Privy Council*, XIX, pp. 406–7. See also Brent, *Canterbury in the Olden Time,* p. 71. See also how the river Wye was obstructed, as given in Brit. Mus. MSS. 11,052, pp. 80–82.

[1] Bacon, *Annals of Ipswich*, p. 286 (1570), says: "......and for the better maintaining of the haven, that the flood gates may be set open once a month, or otherwise, as need shall require." Again, on pp. 294–5, we find: "Stoke Mills, with the Marsh Mill at the Friar's bridge, except the drowned marsh, demised to John Faircliff at £42 rent, for 20 years, and a fine of 100 mks.; he shall repair the premises, with the flood gates and banks, twice a year,......and shall......four times in a year set open his flood gates for the scouring of the channel......."

[2] We would naturally conclude that with the increase of river traffic, the barges engaged in that traffic would be increased in size, and this would be one reason for some of the difficulties in the navigation of rivers. I am unable, however, to confirm this conjecture by actual statistics, in regard to the increase in the size and tonnage of the boats and barges used in the river trade; but that there was such an increase is clear from Act 34 and 35 Henry VIII, c. 9, preamble.

that the outflow of the tide tended to scour the channel of the river twice every day[1].

During the sixteenth century, we have only eight Acts dealing with the navigation of rivers[2]. These all, with two exceptions, had to do with the improving or maintaining of navigations that had already been in effective use[3]. It is quite evident, therefore, that under the Tudors there was little attention given to inland navigation for the purpose of extending it more widely through the country[4]. As to how much of the carrying was done on the rivers, and how much by land carriage, I am unable to make any satisfactory estimate. Most of the information upon this point is to be gathered from the statutes, and these are as a rule very indefinite regarding this subject[5]; but we may reasonably conclude that for the conveyance of heavy goods the rivers offered the better facilities and the lower expense, and that, therefore, they would take the larger proportion of this traffic.

In nearly all cases, the Acts passed to further the navigation of rivers authorized the pulling down of mills, weirs, fishgarths, etc., and the clearing of the channels from shoals, shelves, and other obstructions. As a special case, which exemplifies an entirely different operation for recovering a navigation, it may be of interest to consider the case of the river Exe. Previous to the reign of Henry III, the tide came up this river as far as Exeter, carrying with it barges and small craft laden with commodities for that city's use. About this time the Countess Weir was built across the Exe, but in it was left an opening of thirty feet for the passage of vessels. Other weirs were added by the Earls

---

[1] See, for example, Brit. Mus. 816. m. 8 (4), 'The Case of the Town and Port of King's-Lynn in Norfolk as to their Navigation.'

[2] These were in order as follows: 19 Henry VII, c. 28; 6 Henry VIII, c. 17; 23 Henry VIII, c. 18; 31 Henry VIII, c. 4; 34 and 35 Henry VIII, c. 9; 3 and 4 Edw. VI, PR.; 13 Eliz., c. 13; 13 Eliz., c. 18.

[3] The first exception here mentioned is Act 13 Eliz., c. 18, which was an Act for making the river Welland navigable; but I have been unable to obtain this Act, and hence do not know its provisions. The other exception is 13 Eliz., c. 18.

[4] Several Acts were passed in the reign of Elizabeth for improvement of harbours, which were constantly tending to be silted up, and filled up by the ballast that vessels threw overboard as they came further up the harbour. Note, for instance, Acts 27 Eliz., cc. 20, 21, 22; 31 Eliz., c. 13, etc. A noted example of such improvement was that of the harbour of Chichester in Sussex. For bringing that harbour closer to the city of Chichester, authority was given to cut a canal for about a mile inland from the then-existing harbour; and the Mayor and citizens of the city were given jurisdiction over this canal, the same as they had over the city and harbour.

[5] From the legislation of the reigns of Henry VI, Henry VII, and Henry VIII, it seems to be clear that the traffic on the rivers was considerable at that time. In all probability, the improvement of the navigation of rivers went on but slowly, because of the lack of capital to embark in these enterprises.

of Devon, and although legal proceedings were taken against this, and verdicts were gained, the power of the Earls was greater than the law. In 1539 an Act of Parliament was obtained for the restoration of this navigation, for by that time the barges and vessels had to anchor several miles down the river below Exeter[1]. This Act gave the Mayor and Corporation of Exeter authority to purchase ground for the purposes of the Act, in other words, to enable them to make a cut for some distance parallel with the river, to connect Exeter with the sea[2]. Many attempts were made, in this and the two succeeding reigns, to effect this desired end, but these were largely unavailing, for the new channel, being left open to the ebb and flow of the tide, was soon damaged. In 1563, the Corporation engaged a Welsh engineer, who, instead of clearing the river, rendered the city accessible by a canal. This work was a true pound-lock canal, similar in all essential particulars to canals of more recent days[3]. It was finished in 1567, but gave little satisfaction, for it could not be entered at all tides, and, besides, no barge canal could successfully compete with an ordinary road in a distance of only three or four miles, for the double transfer and the injury of the goods, would outweigh the advantage of the cheaper conveyance. About a hundred years later, new works were begun, to extend, widen, and deepen this navigation, but we cannot follow these here[4]. It is interesting to note that this was the first pound-lock canal made in England, at least so far as I have been able to discover.

To show the attitude of many in Elizabethan England toward the improvement of rivers, it may be instructive to briefly outline the history of the river Lee navigation at that time[5]. In 1571, an Act was obtained for bringing the river Lee (or Ware) to the north side of the city of London[6], by making a cut, at a suitable place, out of the river

---

[1] Act 31 Henry VIII, c. 4.

[2] The term "canal" was not, however, applied to this cut until long afterward.

[3] The evidence for this is given in full from the records of the city of Exeter, by Philip Chilwell De la Garde, in his 'Memoir of the Canal of Exeter, from 1563 to 1724,' as published in the Minutes of *Proceedings of the Institution of Civil Engineers*, IV, pp. 90–102; also in Oliver, *History of the City of Exeter*, pp. 249–68.

[4] See De la Garde's work, p. 98 et seq., for further improvements that were effected in the navigation of the Exe, from 1675 on. Also James Green, *Continuation of the Memoir of the Canal of Exeter*, from 1819–30, as given in the Minutes of the *Proceedings of the Institution of Civil Engineers*, IV, pp. 102–13.

[5] Of course, this river was navigable at a much earlier date, for Act 9 Henry VI, c. 9 (1430–1), speaks of obstructions in it, and gave power to appoint Commissioners who should have power to borrow money and take toll for cleansing the river, for three years.

[6] Act 13 Eliz., c. 18. For the previous history of the river Lee, see our introductory chapter, under the heading "Rivers and Navigation."

as it then was. It was felt that cleansing the course of this river and freeing it from obstructions, so that barges, tilt-boats and wherries could carry goods and passengers between London and Ware, would be a great advantage both for the city and the country[1]. Accordingly, the city was authorized to lay out ground, not exceeding one hundred and sixty feet in width, for this purpose, the ground to be paid for by a reasonable compensation, and to be vested in the city's representatives. The latter were to have the conservancy of the new river thus cut; were to have authority to punish transgressors; and were required to repair any breaches in the work. The work was to be finished in ten years, at the expense of the three counties of Middlesex, Essex and Hertford, and there was then to be a free passage through the new cut and the old river.

It seems certain that the navigation was completed within the appointed time, for, as early as 1581, and in later years, there was much complaint against the navigation, and many misdemeanours were committed to render it of no effect[2]. The cry of vested interests, which, as we shall see, played an important part in the eighteenth and nineteenth centuries, was very strong even at this time. For instance, in 1581, a petition was sent to Lord Burghley by the inhabitants of Enfield, Cheshunt, Stevenage, and other places along the course of this river, complaining that the living which they had obtained, through carrying grain, malt and provisions to London, was completely taken away from them, since these things were now carried by water[3]. Along with the plea that their living was gone, they said that the trade by water was in the hands of a few wealthy men; that they were no longer able to provide as many men for the service of their country, nor to pay the subsidies levied on them; and that the maintenance of the navigation had been a constant expense to them. On the other hand, before the navigation was made there were a few rich badgers, who controlled all the carrying trade; they purchased the grain in

---

[1] Preamble to this Act.

[2] Norden confirms our statement as to the shipping on the Lea river from the Thames to Hertford, for he says: "Barges have of late passed that way, to Ware, which was granted by Acte of Parliament about the thirteene yeer of the raigne of Queene Elizabeth, but for some causes of late discontinued" (Norden, *Speculi Britanniae,* p. 11).

[3] The full text of their petition is given in Brit. Mus., Lansd. MSS. 32, p. 110, and their nine points of complaint are there given. The first of these reads as follows: " Many thousands of her Majesty's subjects within the counties of Hertford, Middlesex, Cambridge, Bedford, and Essex, which lived by the carrying of corn and other grain to the city of London by land, are now utterly decayed by the transporting of corn and other grain to the said city by the water of Lee."

the country and sold it at their own price in London; but when the navigation was completed the monopoly which these men had enjoyed was gone, and their complaints were probably added to those of the men whom they had engaged to do the actual work of carrying for them, thus giving double emphasis to the evils which they said had resulted[1]   These complaints were answered very fully, so as to leave no doubt of the real benefits conferred by the navigation[2]; and when the complainants found that their statements of the case were not sufficiently strong to secure the desired object by peaceful and legal means, they, or others deputed by them, set to work to riotously demolish the navigation.   Locks in the river were burned[3]; in other cases, the men who had charge of the locks were intimidated, and forbidden to open them for the passage of boats[4]; and in many instances openings were made in the banks so as to draw off the water from the river and thus prevent its use for navigation purposes[5].   We cannot but see, even at this time, the jealousy and hatred manifested by vested interests against such improvements, no matter how much the latter might advance the general good.

Despite the efforts of the Commissioners to keep a clear passage for the navigation of this river, by 1588 there was a long list of defects: shelves, claybeds, bars, beds of gravel, and other similar obstructions were found in mid-stream; weirs had been unlawfully erected, or extended too far into the river; a ford had become too shallow, and required to be deepened; the streams of by-water leading out of the river wasted the water that was so much needed for the navigation[6]. Still the carrying by water continued, for the breaches in the banks were repaired and the channel was cleansed and put in navigable condition.   In 1589, the carriers of Enfield again presented their complaints, this time to Queen Elizabeth, showing that by providing grain for the city of London and carrying it there, and by furnishing the Queen with her subsidies and with teams for her royal service, they had fulfilled the part of good loyal subjects; but with her assent, the river Lee had been made navigable and in consequence their living had been taken away from them.   By reason of the navigation also, their lands, they said, were no longer kept fertile by the overflow of the land water, and the mills were unable to grind their corn, because they had not water enough to run them.   This necessitated carrying their corn sometimes ten miles to have it ground, which was a great hardship.

---

[1] Brit. Mus., Lansd. MSS. 82, p. 105.   [2] Ibid.
[3] Ibid., pp. 95–103.   [4] Ibid., pp. 93, 111–13.
[5] Ibid.
[6] Brit. Mus., Lansd. MSS. 88, p. 91 et seq.

They asked that some action should be taken to restore to them the full amount of their carrying trade, and to grant them other relief according to their need[1]. These complaints were answered, by setting forth the other side of the case[2]; and we do not find that anything was done at that time to relieve these inhabitants of the "decayed town of Endfield." Instead, the difficulties of navigation continued, because of the misdemeanours committed in all probability by those who were opposed to it; and in 1594 several bargemen instituted proceedings before the Star Chamber against several other men for the riotous stopping of the passage of vessels and barges on the river, by placing stones, timber, earth, and other impediments in mid-stream, and by destroying the navigation works. The matter was referred by this court to the two Lord Chief Justices for investigation, and they found that the principal cause of dispute in the stopping of the navigation was, whether the river anciently went by one branch or by the other through the town of Waltham Holy Cross. This question was settled, and the court decreed that the river should be kept open for the passage of barges and that those who were hauling the barges should have a suitable path along the banks[3]. But although this issue had been disposed of, we are safe in saying that the trouble did not cease, by any means, at this time; for about a century later, in 1688, there was still some difficulty between the bargemen and the lock and weir owners[4].

From the illustration of the river Lee, we see the great hostility that was aroused against river improvement, even in the case of a river that had formerly been navigable. But notwithstanding the opposition to such public benefits, there was considerable interest taken, during the later years of the reign of Queen Elizabeth and in the time of James I and Charles I, in securing increased river facilities for the carriage of heavy and bulky articles[5], although we do not find that much was actually effected at this period. Such projects frequently slumber long in the public mind, ready to take shape at a favourable occasion; and some of those brought forward at this time did not materialize until half a century had elapsed.

To further show the attitude of the public at that early time toward

---

[1] Brit. Mus., Lansd. MSS. 60, p. 96.

[2] Brit. Mus., Lansd. MSS. 38, pp. 84, 85.

[3] Brit. Mus., Lansd. MSS. 76, Document No. 55, 'Order of the Star Chamber concerning the right of navigation on River Lea,' June 20, 1594, pp. 125–8.

[4] Brit. Mus., Add. MSS. 33,576, p. 63, 'Suit concerning the Lock near Waltham Abbey (River Lea),' 1688.

[5] See, for example, Acts 3 Jas. I, c. 20; 7 and 8 Jas. I, c. 19; 21 Jas. I, c. 82; Brit. Mus., Add. MSS. 34,218; Brit Mus. MSS. 11,052; etc.

navigation improvement, we shall take two other instances, namely, those of the rivers Medway and Wye.

The Medway, in the county of Kent, passed through a section of country where agriculture was carried on, and lumbering was the most prominent industry. Here was to be found the best oak timber, which, on account of both quantity and quality, was highly serviceable for the Royal Navy-yard at Chatham. But on account of the difficulty of getting this timber transported over the bad roads, usually two years elapsed from the time the trees were felled till, as timber, they reached their destination at Chatham; and this necessitated the keeping of two years' supply on hand at that place[1]. But by a navigation, the timber might have reached its destination for use one year sooner, and thus avoided this and other inconveniences[2]. Besides, the navigation would aid in several ways in the carrying on of agriculture; for the use of boats on the river would save the use of cattle and horses for carrying purposes, so that they might be used in tillage. To the "grievous neglect" of husbandry, these animals had often been employed in carrying timber, iron and other articles, when they should have been employed on the land; and the land had been left to be sown out of season or not at all, which would not be the case if boats were used on the river. Further, the navigation would provide cheaper carriage of agricultural and other products to and from the markets[3]. These reasons for making the river navigable up to Yalding or

[1] Timber cut in the winter time could seldom get farther than Yalding that year. Then, if the boats did not go as high up the river as Yalding, the timber had to lie there all that next winter unless they would pay 6s. a ton for having it carried by land to Milhale, whereas the boats never charged above 3s. per ton for carrying that distance. If, however, it stayed till the next summer, it would cost 4s. a ton to carry; but then it could not be used that year; so that they must have at Chatham two years' supply for their building. Thus timber cut down in 1598 could not be at Chatham or Woolwich for use before 1600. Brit. Mus., Add. MSS. 34,218, p. 40; also *J., H. of C.*, XXIII, pp. 443–4.

[2] Sometimes when there was great haste to get trenails for Her Majesty's ships, and when the roads were so foul that no carriages could pass along them, the trenails had to be carried on horseback to Yalding, which cost nearly 20s. a load to bring to Yalding (Brit. Mus., Add. MSS. 34,218, p. 40). The advantage of the navigation at that time could not but be apparent, when the boats of Yalding in the winter time would carry to the ships in one day more timber than two hundred oxen and horses could carry in two days (ibid.).

[3] The boats usually carried stuffs to Maidstone 2s. per ton cheaper than they could be carried by land, and the farmers' corn could be taken to this market cheaper, while, at the same time, their teams could be profitably employed at home, to their own and the public advantage. Similarly, by boats the houses would be provided with their necessaries and the clothiers with facilities of trade—and all this cheaper than by land carriage. A further advantage of having the river

beyond, were reinforced by another, very commonly given at that time and later, that such public works would be of great benefit to the poor; for the ordinary labourer who had nothing but his handiwork to support his family and whose usual day's labour was worth only 8*d.* in winter, would work very badly on the boats if he did not receive twice as much.

In 22 and 24 Elizabeth, the Commissioners of Sewers had decreed that all weirs and other impediments to the free passage from Maidstone to Twyford bridge should be removed; but this was not done at that time; and when the agitation for the improvement of the river was begun in 1600, the chief source of trouble that was complained of was the number of weirs in it. These were owned by the landlords, and notwithstanding that they caused inundations of adjoining property, the landlords did not want them removed, for they preserved to them fishing privileges which were highly valued. The report of nineteen of the Commissioners of Sewers for the river showed that there were many more annoyances in the Medway than had been presented, and suggested that the weirs on the river should be pulled down and other obstructions removed[1]. The objections raised by the landlords against making the river navigable seem to us to carry no weight at all[2]; but the landed interests had sufficient authority to win the day. Yet, although the improvement of the navigation was at this time prevented, the agitation did not cease. Two juries were summoned to inquire into the conditions, and in 1627 their complete presentment showed a vast array of obstructions that were found in the river, which should have been removed. Mills, weirs, dams, and flood gates were very common between Maidstone and Penshurst; trees were allowed to grow in the channel of the river, and fallen tree trunks were permitted to block up the stream; bridges and banks of earth impeded and deflected the course of the water; and a great many other annoyances prevented the current having free

navigable was that boats passing up and down the river would keep it clear of shelves, banks, trees, logs, etc., for when a tree should fall into the water the boatmen would cut it up and take it away, so that it would not form an obstruction (Brit. Mus., Add. 34,218, p. 40).

[1] Ibid., pp. 45, 46.

[2] These objections may be briefly given as follows: To make the river navigable would put an end to the fishing of the weirs; and, besides, the river was navigable by prescription to above Penshurst, and as it belonged to the Crown it was as much open for people to travel on as were the highways by land. These and other objections of like trivial character were put alongside of the fact that, since the boats had stopped, "the carriage by land has spoiled the common market ways, so that many poor travellers have been constrained to call help to draw them out of the mire." To make this latter statement, and in the same breath say that the river should not be made navigable, was the height of folly (Brit. Mus., Add. MSS. 34,218, p. 39).

passage[1]. Several of the landowners about Yalding had complained that their lands were much injured because the river lower down, between Yalding and Maidstone, was not improved so as to allow the water to drain off easily and quickly. In accordance with the King's decree to the Commissioners of Sewers that the abuses along that river should be rectified, it was ordered that the owners or occupiers of adjacent lands should remove these annoyances, or that they should so correct them as to prevent the overflowing of the lands and facilitate the passage of boats, or that they should consent to have this done for them by a certain person who had agreed to do it if he were authorized. The Commissioners urged, for the general good of the country, that the river between Maidstone and Penshurst should be made navigable for boats of at least four tons' burden[2]. What the outcome of this was, we have not been able to ascertain; but it was not until 1664 and 1665 that an Act was passed to give more fruitful sanction to the navigation desired at this time[3]. The later history of this undertaking we give elsewhere[3]. From the above, we see that the opposition to this navigation was wholly different from that we have seen in the case of the river Lee; both of these rivers had been said to be "navigable," but the actual navigation was opposed, in the one case by the carriers by land, and in the other by the landowners. In the former they wanted the weirs and locks in the river pulled down; in the latter they wanted the weirs retained.

In the case of the Wye river navigation we have many more complications than in that of the Medway. This river was considered a public navigable river in the time of Edward I, and its rightful navigability was upheld by many general statutes after that[4], all of which

---

[1] Brit. Mus., Add. MSS. 34,105, pp. 189–92.   [2] Ibid., pp. 193–5.

[3] 16 and 17 Car. II, PR. Act No. 12. By this Act it was made known that the river Medway and its tributaries were capable of being made navigable, and that this would be of great use for the better, easier and faster transport of iron ordnance, balls, timber, and other materials for His Majesty's service, and it was enacted that the river and its tributaries should be made navigable. But for want of money, the Act was not carried into execution. In 1739 the project was revived, and the same arguments put forth as we have noted above. It was shown that in some years there was not more than one or two months when timber might be taken away by the high roads. In some instances, timber contracted for could not be got to market, and so had to be obtained in other places. If there were a navigation, transport could go on at all times of the year and at half the expense of land carriage. Consequently, leave was given to bring in a Bill for making this river navigable for 32 miles from Maidstone to Forrest Row. *J., H. of C.*, xxiii, pp. 443–4, 469–70; Act 13 Geo. II, c. 26.

[4] Brit. Mus., Add. MSS. 6693, p. 305, 'The State of the Rivers Wye and Lugg in Herefordshire.' v. Acts Magna Charta; 9 Henry III, c. 23; 25 Ed. III, c. 4; 45 Ed. III, c. 2; 21 Ric. II, c. 19; 1 Henry IV, c. 12; 9 Henry VI, c. 9.

were maintained by an Act passed in the twelfth year of the reign of Edward IV requiring navigations to be kept free of weirs and other hindrances under penalty of one hundred marks[1].   By an Act of 1531–2, the liberty of free passage along all the great navigable rivers was confirmed to all the King's subjects[2]; and in 1527, all the weirs upon the river Wye were thrown down[3], and so continued till 1555, when an Act was passed for the rebuilding of four mills and their weir near the city of Hereford[4], but no authority was given for erecting or rebuilding any other weirs.   After 1555 the rest of the weirs, pulled down in 1527, were by degrees rebuilt; but they did not remain long, for in 1588–9 they were all, by virtue of a Commission of Sewers, broken down to Hereford Mills, but the weir at these mills, because built by Act of Parliament, was allowed to stand.   Probably the weirs were once more rebuilt, for in 1621–2 the city of Hereford and other inhabitants of the counties of Gloucester, Hereford and Monmouth presented a petition to James I, urging him to have the Commission of Sewers go ahead with their work and pull down the weirs on the river[5].   In 1624 a Bill was brought into Parliament for the purpose of having the river made navigable, but the measure failed to receive the sanction of Parliament[6].

The agitation for a navigable river continued, however, and at the Epiphany Sessions at Hereford in 1640 the Grand Jury presented that the weirs in the river Wye, which according to Magna Charta and other statutes should have been put down, still remained as a great nuisance to the county, in that they hindered navigation, that they were destructive to the fry of fish, and that they were injurious by causing the overflowing of adjacent land[7].   A new Bill was brought before Parliament to secure their desires, and the Mayor and Corporation of the city of Hereford sent a petition to Lord Scudamore, asking him to lend his

---

[1] Act 12 Edw. IV, c. 7.

[2] Act 23 Henry VIII, c. 12.   This Act was made for the river Severn; but whatever privileges were given for that river held good also for all other great rivers of the kingdom.

[3] Preamble to Act 2 and 3 P. & M., c. 14.   The Act is given in Lloyd, *Papers relating to the History and Navigation of the Rivers Wye and Lug*, p. 1.

[4] Act 2 and 3 P. & M., c. 14 (1555).

[5] Brit. Mus. MSS. 11,052, pp. 80–81; Brit. Mus., Add. MSS. 6693, p. 305. They said that the weirs were detrimental to trade between Bristol and Monmouth, Hereford and Ross; that they permitted the catching of great quantities of fish, which were given to swine; etc.

[6] In Brit. Mus. MSS. 11,052, p. 82, we find 'An answer to the Reasons alleged in opposition to the Bill now preferred to the Ho^ble Court of Parliament, for Opening the River Wye' (1624), which gives us, by inference, the case against the navigation, and by direct statement the case for the navigation.

[7] Copy of this presentment is given in ibid., p. 86.

assistance to the project for the subversion of the weirs[1]. Notwithstanding that the great Sir Matthew Hales gave his opinion in
favour of making the river navigable, the Bill failed to pass into a
law[2].

Again, in 1649, this question came up, and once more, at their
Sessions at Hereford on April 3 of that year, the jury presented that
the weirs in the river were contrary to the Magna Charta and other
good statutes; and that they were opposed to the public good, in
preventing trade with Bristol and other parts, and in destroying the
fry of fish while contributing to the private gain of some few men[3].
It was at this time, when the movement was gaining force, that the
real questions at issue came to the front, and showed the nature of
the opposition to a measure which had been twice brought before the
public by the Justices of the Peace.

The Justices of the county, and the city of Hereford, strongly
advocated making the river navigable, for several reasons. They
showed that at least for four or five months of the year there was
sufficient depth of water to carry a barge of ten tons, if only a passage
could be obtained through the weirs and a towing path on the banks.
But many weirs, formerly erected and lately raised higher, were kept
so by private persons for their gain, and thus prevented fishing and
navigation from Hereford to Bristol. If the navigation were opened,
it would enable the county to send out its surplus of corn and fruit
to other places in times of plenty, and to help in providing the county
with necessaries in times of scarcity; it would permit the bringing of
coal and wood from the Forest of Dean and other sections, at cheap
rates, now that the supply of wood near to the city of Hereford was
becoming exhausted; it would allow the bringing in of lime, at low
rates of carriage, to the great benefit of agriculture; and it would add
to the wealth of the city of Hereford and near-by towns, by having
easy trade relations with Bristol, so as to be able to supply themselves
with foreign commodities, and by having the price of coal greatly
reduced. Besides, it would be a means of furnishing employment to
many poor people who were out of work. For these reasons they asked
for the passage of the Bill then before Parliament, the object of which
was to make a free passage on the river between Chepstow and the
city of Hereford[4].

---

[1] This petition is dated Feb. 1, 1640, and is given in full in Brit. Mus. MSS. 11,052, p. 87.
[2] The opinion of Sir Matthew Hales is given in ibid., p. 89.
[3] Brit. Mus. MSS. 11,052, p. 93.
[4] Ibid., pp. 95–96, 99.

On the other hand, the landholders along the river, and the owners of corn and grist mills and fulling mills, who were dependent upon the weirs for water power to run their mills, vigorously opposed the Bill that had been brought into the House, ordering that, by a certain stipulated date, every farmer or possessor of a weir should make in every such weir a suitable passage for barges and vessels, and that those who were going up and down the river should have the right of a towing path upon the lands of any person adjoining the river. They said that in the time of Henry VIII the weirs in the river had been pulled down, and continued down for over twenty years, yet there had been no use made of the river for navigation purposes, because of the swift current, the many turnings, and the rocks in the bed of the river; these impediments were still there, and the river for eight or nine months of the year was so low and shallow that it would not carry vessels with any burden[1]. If by default or necessity, the owners or possessors of some weirs should fail to make or maintain a navigable passage through those weirs, the latter would be pulled down; and then the mill-owners would be left without the necessary means of grinding corn, which would be an "insupportable loss" to them and a great damage to the farmers. Of course, the two corn mills and the two fulling mills, which were erected at the instance and for the benefit of the city of Hereford[2], were under the protection of an Act of Parliament, and could not be pulled down. The landholders further complained that if the weirs should be destroyed, their pastures, meadows and orchards along the river would decay in productiveness for lack of winter floods caused by the stops in the river at the weirs; that if the river should be made navigable they would no longer have a passage over its fords and shallows, for carrying loads to and from their land on the other side of the river; and that the fruit from their orchards would be taken by watermen and others passing along the towing path of the river. Some even feared that, by improved navigation, the scarcity and dearness of leather, corn, butter, cheese, etc., would be greatly increased, because of the greater opportunity for exporting these commodities, and thus the country would be likely to suffer[3].

Even with the great amount of agitation that we have noted in 1649, the Bill for making the Wye navigable failed to pass the Houses

[1] The truth of this was recognized by those who wanted the navigation, and it was answered by saying that during the remaining four months of the year all the heavy traffic could be accomplished, and that was the chief reason why they wanted the river made navigable.

[2] See preamble to Act 2 and 3 P. & M., c. 14.

[3] Brit. Mus. MSS. 11,052, pp. 100–1, 102–3, 104–6, 107–8, 112–16.

of Parliament in that year. Two years later, the Justices of the Peace for the county of Hereford, in the General Sessions of April 9, 1651, ordered a survey to be made of the river Wye, especially with regard to its weirs and other hindrances to a navigation, and the report that was brought in was favourable to making the river navigable[1]. It would seem, however, that such local benefits could not receive attention during the stormy time of the Commonwealth, and it was not until 1662 that an Act was passed authorizing this navigation[2].

Under this, the right was given to Sir William Sandys, who had made much improvement in some other river navigations, to undertake the works along the Wye. It would appear that he was empowered to borrow a certain amount of money that had previously been raised in the county of Hereford for this purpose, but the rest of the necessary funds had to be provided by him and his associates[3]. He was not altogether successful in this venture, however, for the same methods that he had employed on the sluggish Avon (Worcestershire) were not suited to the torrential Wye. The partial failure of this enterprise was deeply regretted by the people of Herefordshire, who loudly complained of the enormous price of coal, and of the great cost of conveying their corn and timber to Chepstow and their merchandise from Bristol. More than twenty years passed without any further effort to improve the river navigation. In 1688-9, the project was again taken up, this time by the leading noblemen and gentlemen of the county, who proposed to buy up all the weirs upon the river and then pull them down, so that no impediment might be left in the river below Hay; and since the work would be for the benefit of the whole county, the charge should be borne by the county[4]. Determined opposition to this plan was aroused on the part of the weir owners and those who were at some distance from the river, for the latter thought that since they would derive no benefit from the navigation they should not have to pay toward its expense[5]. So bitter was the struggle of opposing interests that the Bill was defeated in 1692, but after three years more of agitation

---

[1] Brit. Mus. MSS. 11,052, pp. 120-7.

[2] 14 Car. II, PR. Act No. 15, entitled an 'Act for making navigable the rivers Wye and Lugg, and the rivers and brooks running into same, in the counties of Hereford, Gloucester, and Monmouth.' The Act is given in full in Brit. Mus. MSS. 11,052, pp. 127-9.

[3] Lloyd, *Papers relating to the History and Navigation of the Rivers Wye and Lug*, p. 3.

[4] Ibid., pp. 12-18.

[5] Ibid., pp. 19-25, gives two different papers rehearsing the objections to the Bill that was before Parliament. They are largely the substance of the earlier objections.

the measure finally passed in 1695[1]. No time was lost in enforcing the provisions of this Act[2], and all the weirs were pulled down except one owned by the Earl of Kent[3]. Two later Acts were passed relative to this navigation in 1727 and 1809. The navigation of the Wye, through the uncertainty of its stream, was never made into a good waterway for regular conveyance; but yet it proved of much service to the county, in the century following its opening, by permitting the carriage of heavy materials and articles at an easier charge than by land.

We have endeavoured to give as accurate a picture as possible of the way in which different classes of people regarded proposed improvements in this river navigation, because the advantages and disadvantages set forth in connexion with the Wye recur constantly in regard to other rivers. It is only by knowing the attitude of all classes toward such changes that we are able to estimate the extent of the desire for them, and what difficulties had to be overcome before they could be secured[4].

We have already spoken of the efforts to make and keep the Severn open for navigation, so that the people of western England might have free access to the sea; and we have seen that certain towns, notably Worcester and Gloucester, had taken tolls, sometimes illegally, from those who were using the river. When we turn to the Thames we find that here also there was the taking of illegal tolls; not, however, by towns, but by individuals. This river had always been navigable, and in order that the navigation should be preserved, in 1489 the complete jurisdiction over it was given to the Mayor of London[5]. As in the other rivers, like the Humber and Ouse[6], the tendency was for private individuals to place impediments in the river for the purpose of furthering their own interests; and, to protect the Thames from such obstructions, a penalty of £5 was imposed, in 1535, upon persons committing such

---

[1] Lloyd, op. cit., p. 14, gives a list of the mills and weirs at the time on the Wye and Lug, which shows how these rivers must have been impeded by so many barriers. See Brit. Mus. MSS. 21,567, 'A Survey of the Rivers of Wye and Lugg in reference to Portation and Shipping,' p. 2, where a minute description of these obstructions is given. In Lloyd, op. cit., pp. 15–19, there is a document of 1690, showing how difficult it would be to make the Wye navigable from Hereford upward.

[2] Act 7 and 8 W. III, c. 14. It is given in full in Lloyd, op. cit., pp. 32–43.

[3] The names of some of the weir owners and the amounts paid them are given in Lloyd, op. cit., p. 44. The case of the Earl of Kent is presented in ibid., pp. 26–31.

[4] A summary of arguments for and against the Bill for making the Wye navigable is given in Brit. Mus. MSS. 11,052, pp. 116–18.

[5] Act 4 Henry VII, c. 15.   [6] Act 23 Henry VIII, c. 18.

nuisances[1]. But legislation was not enough to prevent these annoyances and they still continued encroaching on the navigable channel of the river. In 1584, orders were given for the conservation of the river[2], and it is probable that by the end of Elizabeth's reign a considerable extent of the river had been improved[3]. In 1605 an Act was passed to finish clearing the Thames, so as to make it completely navigable to the city of Oxford[4]; but this was not carried into effect, and was repealed in 1623 by another Act which gave authority to complete the navigation from Bercott to Oxford, so that the University and city of Oxford might have a convenient means by which to bring in coal, stone, fuel, and other necessaries[5]. For some reason, which I have been unable to ascertain, there was much opposition to the artificial aids in this navigation, and certain locks were pulled down; this prevented barges with provisions coming to London, and a committee of the three associated counties, Oxford, Bucks, and Berks, was appointed to examine into this misdemeanour, to consider a remedy, and to rebuild these locks[6]. To further aid in improving the navigation, which naturally tended to be impeded by silting up and by vessels throwing out ballast as they came up the river, a proclamation was issued by Charles I, in 1636, granting to certain individuals, for a term of years and under a yearly rent to the Crown, the right to take up the gravel in the river channel and to dispose of it at moderate rates for the ballasting of vessels on the river. Also, barge-masters were to purchase

---

[1] Act 27 Henry VIII, c. 18 (1535–6). See Brit. Mus., Add. MSS. 36,767, pp. 1–4, an Order of Privy Council in regard to the Conservation of the River Thames.

[2] In Brit. Mus., Lansd. MSS. 41, we find the following orders for the conservation of the Thames, 1584:

"(a) That there be no purprestures, wharfs, banks, walls, or building of houses, in or upon the Thames to the stopping of the passage.

(b) That no dung or other filth be cast into the Thames.

(c) That no posts or stakes be fixed in the Thames.

(d) That the faire way be kept as deep and large as heretofore it hath been,

(e) That no person shall sell......or take any fish contrary to the ancient assize set down by decree, viz." (Then follow specific directions as to the time for fishing for the different kinds of fish, what kind of net to use, etc.)

[3] The reason for making this statement is that the Act of 1605 says that the "river of Thames is from the city of London till within a few miles of the city of Oxford very navigable and passable with and for boats and barges of great content and carriage," and that by removing a few obstructions it would be passable to Oxford.   [4] Act 3 Jas. I, c. 20 (1605–6).

[5] Act 21 Jas. I, c. 32. Eight Commissioners from the University, and four from the city, of Oxford were to have charge of this work, and the people of Oxford were to be taxed for making and maintaining this passage in the river. River navigation was desired because of the badness and danger of the adjacent roads. The Commissioners were allowed to make weirs locks, etc., in the river.

[6] *J., H. of C.*, Jan. 2, 1644.

this river ballast, as long as it was available, from no others than the grantees of these letters patent[1].

The greatest hardships connected with the use of the Thames navigation, however, were the exactions demanded by owners and renters of the vast number of private weirs, locks, gates, etc., from those who wanted passage through these, and from others who wanted the use of the boats that were kept by these weir and lock owners[2]. The tenants or occupiers of such works had raised their tolls so high that the prices for river carriage had been much increased[3]. Abuses had also been committed by the bargemen in going up and down the river[3]. These annoyances were regulated by an Act of 1694[3], which empowered the Justices of the several counties through which the Thames runs to fix the rates for passing the locks, to settle the rates of water carriage on the river[4], and to make orders for regulating the conduct of the bargemen and the best interests of the navigation[5]. During the continuance of this Act for nine years the navigation of the Thames was carried on with success, and the charges were known with certainty; but after its expiration in 1703, the proprietors of the locks, weirs, and other works again charged large amounts for the passage of boats and barges, to the injury of barge-masters and of the markets in London[6].

[1] See Rushworth's *Collections*, II, p. 377, for this Proclamation of Nov 28, 1636, "for cleansing the River of Thames of Shells and Annoyances and for Ballasting of Ships with the Sand and Gravel thereof." It is also given in Rymer's *Foedera*, XX, p. 93.

[2] Harrison's *Description of England in Shakespere's Youth*, III, pp. 411-26, transcribes several extracts from the Lansdown MSS. which show the immense number of these locks and weirs in the river, and the abuses connected with them which deteriorated the navigation, 1580-5. John Taylor (the "Water Poet") published in 1632, *A Description of the two famous rivers of Thames and Isis*, in which he shows the river to be greatly obstructed by shoals, shelves, weirs, stops, etc.

[3] Act 6 and 7 W. and M., c. 16 (1694); also Brit. Mus. 816. m. 8 (49).

[4] No one might take more than these rates, under penalty of £5 for every offence. The orders of the Justices of the Peace were to be registered among the Quarter Sessions' records, and to continue in force for seven years, and even longer than that unless changed by new orders.

[5] To prevent mischief being done by disorderly persons managing the barges, barge-masters and barge-owners were to be responsible for any injury done to the locks, weirs, etc., by their bargemen.

[6] In Brit. Mus. 816. m. 8 (49), we get the following comparison of rates:

Rates as settled by the Justices, 1694			*s.*	*d.*	Duties imposed about 1720		£	*s.*	*d.*
Wittenham Lock	..	..	1	6	Present Duty			5	0
Benson	,,	.. ..	1	6	,,	,,	1	5	0
Cleeve	,,	.. ..	6	0	,,	,,		15	0
Goring	,,	.. ..	6	0	,,	,,		15	0
Whitchurch	,,	.. ..	6	0	,,	,,		15	0

However, the good effects of the Act of 1694 were revived in 1729[1] by legislation which was not repealed till 1751[2]. The changes introduced in 1729 were merely matters of detail, for adaptation to the changing needs of the navigation; but the general provisions of the law of 1694 remained in force for nine years after they were re-enacted in 1729. Following the year 1738, the old abuses again crept in and the price of water carriage was once more raised; until, in 1751, the Commissioners, who were collectively the chief officers of each town along the river, had their jurisdiction increased and were required to settle the prices payable by barges for the use of towpaths and other privileges; to make orders for the size and draughts of barges; to make regulations for the conduct of bargemen; to inquire into the state of the locks, and into the rates then paid as compared with former rates; to assess the prices of carriage on the river between London and Cricklade, and to make these public[3]. It will be noted throughout all these changes, that there was a constant tendency on the part of the owners and renters of weirs, locks, and other aids, to demand unjustly high charges from those using the locks and other appurtenances of the navigation.

But there were other considerations than those we have just mentioned that were detrimental to the navigation of the Thames, and, despite considerable sums expended upon it under the authority of the Lord Mayor of London and Courts of Conservancy, prevented its being kept open and free and of the greatest usefulness. In some instances, persons navigating the river would turn their vessels and run aground across the channel of the river, so as to obstruct all other barges passing up or down at that time; and, consequently, not only the owners of

Other similar examples are given in the above reference; and the writer said that unless these abuses were rectified, the owners of locks, etc., could charge what they pleased, and this would cause the navigation on the Thames to decrease. Other statistical data as to the increased charges for passing the locks, etc., are given in Brit. Mus. 357. b. 9 (77), pp. 2–3. These impositions tended to discourage the navigation, but this condition was wholly changed by the Act of 1729.

[1] Act 3 Geo. II, c. 11.

[2] Act 24 Geo. II, c. 8, sec. 26.

[3] Act 24 Geo. II, c. 8; also *J., H. of C.*, xxvi, pp. 18, 30–31.

Another undertaking, not strictly included in the subject of navigation, but related to it. was the bringing in of a stream of fresh, pure water to the city of London from the springs of Amwell and Chadwell in Hertfordshire. Authority for this was given by Parliament in 1605 (Act 3 Jas. I, c. 18) and 1606 (Act 4 Jas. I, c. 12). With this work the name of Sir Hugh Middleton is connected, and because of his great success in this important enterprise he has received an honoured place (*Remembrancia, City of London*, pp. 555–6). This New River continued the chief source of their supply during the seventeenth century.

the barge's cargo suffered loss by this delay, but the navigation itself was injured, for this inevitable turning of the current of the river naturally would remove large quantities of sand and gravel from the shallows into the river channel[1]. Sometimes vessels, on account of being overloaded with cargo, were stranded in the channel, especially where there were no locks to increase the depth of the water, and thus the detriment already mentioned would be produced as a result of the avarice of the shipper or the barge-owner[2]. Some persons caused additional injury by floating several pieces of large timber, tied in rafts, and fixed alongside of their barges; these would loosen up the sand and gravel over which they were dragged and then the water would draw the loose stuff into mid-stream with the invariable tendency of silting up the river bed[2]. These obstructions and impediments might have been easily rectified by proper regulations that were duly enforced.

We have hitherto noticed some instances of the improvement of rivers so as to preserve the navigation, the difficulties encountered in the execution of the work, and the results which followed therefrom. Neither in authorizing nor in carrying out such works was there any well-defined economic policy followed, but each case was regarded as a separate and detached enterprise. The sole object in each case was to maintain, in good navigable condition, rivers which were already recognized as being legally navigable. Occasionally, it was hard to ascertain who were responsible for the maintenance of some of the smaller rivers, for at times the work of cleaning a river, or repairing its banks or sluices, was performed by the owners or occupiers of adjacent lands, or by the churchwardens of the parish[3]. There was probably more of such repairing than that of which we have knowledge, for the maintenance of the rivers in proper condition was essential to the welfare of the agricultural classes. But, gradually, in the early Stuart times, when the utility of river navigations for reducing the cost of carriage of such heavy articles as coal came to be more clearly recognized, a few individuals began to see that if rivers naturally navigable were so useful in this respect, it would be desirable, if possible, to render navigable some rivers that were not so naturally. For example, in 1606, the river Nene, which was already navigable up to Allerton, was surveyed by Sir William Fleetwood, with the object of finishing the

---

[1] Brit. Mus. 982. b. 22, 'Description of the River Thames,' p. 15.

[2] Ibid., pp. 16, 17.

[3] Brit. Mus. MSS. 12,497, p. 360, 'The Presentment of the Jury touching Mordon (Wandle) River;' Hobhouse, *Churchwardens' Accounts of Croscombe, Pilton, Yatton, Tintinhull, Morebath and St Michael's Bath*, ranging from A.D. 1349 to 1560, pp. xv, 160, 170; Gasquet, *Parish Life in Mediaeval England*, pp. 106–7, 112.

navigation higher along the river[1]. In 1634 the town and county of Leicester were feeling the necessity of some means of bringing in coal and other supplies, and a royal grant was made to enable Thomas Skipwith to make the river Soar navigable from the river Trent up to the town of Leicester[2]. As soon as this was done satisfactorily, he could take from every user of the navigation a "reasonable recompense" for its use; and he was to pay to the king yearly one-tenth of all his net gain, over and above the cost of maintenance. The undertaker did the work for five or six miles from the river Trent, and then, for want of money, he was forced to desist[3]. In 1636, William Sandys undertook to make the river Avon navigable from Tewkesbury to Coventry, and also the river Teme on the west side of the Severn, so as to facilitate the bringing of coal and other commodities up the Avon to the places along its course. The scheme was approved by the principal nobility and landowners in those counties, and the Privy Council took steps for the execution of the work. But after the lower part of the Avon had been made navigable, the work was interrupted, and it does not seem to have been revived[4]. Another enterprise of this kind to claim attention was that of the river Wey in the county of Surrey, the initiator of which was Sir Richard Weston, who brought from Flanders the plan of establishing water communication between London and Guildford. The scheme was formulated in 1635, but on account of the excitement in the country at that time, and the cost of the work, nothing came of it[5]. Later, he seems to have sought the assistance of some of the important men of Guildford, in whose name an Act was obtained in 1651, authorizing them, at their own expense, to make the Wey navigable and, where necessary, to cut a new trench for this purpose[6]. If they were not able, then James Pitson, Richard Scotcher, and two other men, were to do it. The Corporation of Guildford declared their

[1] Morton, *Natural History of Northamptonshire*, p. 5.

[2] Rymer's *Foedera*, XIX, pp. 597–600.

[3] Houghton, *A Collection for Improvement of Husbandry and Trade*, 1692–1703, June 16, 1693.

[4] Birmingham Free Reference Library, No. 90,318, *River Avon. Orders in Council, His Majesty's Commission, and Certificate of the Commissioners in the Year 1636, relative to the Navigation of the River Avon*, pp. 1–27; Capper, *Proposed Birmingham Ship Canal*, p. 4; Lloyd, *Papers relating to the History and Navigation of the Rivers Wye and Lug*, p. 2.

[5] Brit. Mus. 816. m. 8 (57), Reply to a Paper entitled 'An Answer to the Pretended Case, etc.;' Scotcher, *The Origin of the River Wey Navigation*, p. 8.

[6] Firth and Tait, *Acts and Ordinances of the Interregnum*, 1642–60, II, pp. 514–17. They were to remove all impediments to the navigation and to erect all works necessary. The rates to be charged for the carriage of commodities and of passengers were stipulated.

inability and Pitson and Scotcher undertook it. Sir Richard Weston
was induced to put a large amount of money into it, and the work
was immediately begun and rapidly pushed forward. Scotcher was
induced to give up his own business and devote himself to this work,
until, involved in liabilities, he was imprisoned for a protracted period;
and Pitson, who knew how to work the others for his own personal
ends, put little money into the navigation but obtained almost absolute
control of it. The navigation was opened to Guildford in 1653; but
on account of the many differences that arose, giving occasion for great
legal suits, the public good intended by the navigation was almost
nullified, and those who had put most money and time into the work
received no profits from it, although eventually it proved very successful[1].
It is not until nearly one hundred years later that the rest of the navigable
portion of the river was opened from Guildford to Godalming.

There is no doubt, whatever, but that the Protector was alive to
the benefits of improved conveyance by river, for in 1650 he appointed
fifteen Commissioners as a standing council for the regulation of trade,
and among the duties of their office, they were to consider how the
rivers might be made more navigable and the ports more useful for
shipping[2]. But the times were too stormy to devote much thought
and effort to these activities of peace. After the Restoration of the
Stuarts, however, there was a considerable outburst of industrial
activity and with it the increased desire for the concomitant development
of the highways and waterways. Along with the introduction of the
turnpike system of maintaining roads, many Bills were introduced into
Parliament for making rivers navigable; especially was this the case
in the years 1661-4[3]. Then, in a few years, the interest in the improvement of the rivers almost vanished, as was the case also in regard
to turnpikes, to be revived in the reign of William III.

Up to the middle of the seventeenth century, the chief concern in
water conveyance was to preserve, restore or improve the navigation
of rivers which were formerly navigable, only a few detached but
mostly unavailing efforts being made to make rivers navigable. After
that, we pass to the time when rivers that had not previously been

[1] Brit. Mus. 816. m. 8 (56), 'Case of the Navigation of the River Wye;' Scotcher,
The Origin of the River Wey Navigation, pp. 8-9. Scotcher's pamphlet, which was
written when he was in prison because of his too great generosity, shows the
methods by which Major Pitson had won the game through his avarice and deceit.

[2] Firth and Tait, *Acts and Ordinances of the Interregnum*, 1642-60, II, pp. 403-6.

[3] See, for instance, the Private Acts in the years 14 Car. II (Nos. 14, 15), 16
and 17 Car. II (Nos. 6, 11, 12, 13). Some other Bills were introduced into Parliament,
but failed to pass, for example the Bill for improving the Navigation of the Mersey
and the Weaver, 1663 (v. Picton, *Liverpool Municipal Records*, I, p. 241).

navigable were made navigable under parliamentary sanction[1]. After the works on the river Wey, of which we have already spoken, the first instances of this were the efforts directed toward making navigable the rivers Avon (from Christchurch to New Sarum[2]), Stour, Salwerp, and Wye and Lug. There were also proposals for making the river Chelmer navigable, and in 1663 a Bill was introduced to improve the navigation of the Mersey and the Weaver[3]. It had been an important step to restore the navigation to rivers in which it had partly or wholly decayed; but the possibility of making a navigation where none had existed before, seemed to open up benefits and advantages which would greatly increase the wealth of that section which was served by it. It was foreseen that, by the formation of a navigation like this, trade would be greatly extended, a means would be provided for sending out and bringing in goods at cheaper rates and with less injury to roads than by land carriage, the value of lands along the navigation would be increased, work would be provided for the poor, and, lastly, new centres of distribution would arise, which would provide good markets for the sale of goods produced in the neighbourhood, and from which the surrounding country would be supplied with merchandise brought in from abroad[4]. This last advantage was perhaps the only one that was new, for the others had been obtained, in some measure, through the improvement of rivers that were formerly navigable; and it was

[1] Over the class of rivers which were navigable "from time immemorial" the public had rights which could not be claimed against the conservators of a river made navigable by Act of Parliament.

[2] Act 16 and 17 Car. II, c. 12. The people of Salisbury obtained this Act for making the river Avon navigable, but spent all their first vigour in talk. But by 1675 they seem to have got to the point when they meant business (S., *Avona*, p. 2).

[3] Picton, *Liverpool Municipal Records*, I, p. 241.

[4] Hely, *Benefits of making the River Avon Navigable*, pp. 6–16. The writer of *Avona* speaks of the advantages from making rivers navigable. He says that on account of the sea extending inland along the river there would be more shore than nature had afforded the country; and by reason of this interior situation of a city like Salisbury on a navigable river, there would be a great increase of shipping for its people, while it would be less exposed to danger from enemies than the cities along the coast (p. 19). Further, more employment would be available for the poor and for those of restless disposition. This employment would produce wealth, and that wealth would create more employment. This movement would go on gathering new force and vigour from its progress, until it would "raise up a power and felicity, indeterminate, and boundless as the causes which built it up" (ibid., pp. 20–21). Having then gained *safety* and *plenty*, and a *power* sufficient to assure them to her, England would have leisure to attend to the promotion of knowledge, arts, inventions, etc., by which human life would become more agreeable (ibid., pp. 22–24). See also Brit. Mus. 816. m. 8 (58), 'Proposals for making River Chelmer navigable;' and Brit. Mus. 816. m. 8 (50), 'Reasons for making Navigable Rivers Stower and Salwerp.'

largely the opening up of these new inland distributing points and the facilitation and cheapening of water communication to and from them, thus bringing them into closer relation with other and larger markets along the coast, that led to the increased output and exchange of the products of industry.

But while the possibility of opening up in this way new markets and new inland centres of trade promised good results, it encountered strong opposition from those whose interests were already allied with the older markets. As was natural, the trading classes of a town which had an established market opposed the rise of a new market near the old, for this, they thought, would not only draw away trade from them, but also cause a reduction in their wealth, resulting from a decreased valuation of property of all kinds. The idea that the establishment of a new market would be accompanied by such an increase of trade as would not cause the decadence of the old market, had not yet obtained much, if any, hold upon the mind of the man engaged in business. A notable instance of the hostility of one town to the rise of a rival market in its vicinity, is that of Bristol, which opposed the making of the Avon navigable to Bath, lest the latter city should draw away their trade; and from the records it would appear as if the merchants of Bristol had taken the initiative, and had ranged with them all the other interests that were hostile to the rise of Bath as a commercial town[1]. The carriers, too, whose routes and volume of traffic had become fixed, and who had adjusted thereto their manner and standard of living, opposed any change that would tend to take away their livelihood. Even some who were assured that the making of such a navigation would ultimately have good effects, were loth to undergo temporarily the disturbance of their settled conditions, and joined with others in the opposition[2]. In some cases, the owners or occupiers of mills along the river which was proposed to be made navigable refused their assent to the navigation; and based their objection upon the ground that their mills would be seriously prejudiced if the water of the river were to be used for the purpose of carrying barges, instead of being devoted to its primary use of furnishing

[1] Brit. Mus. 816. m. 8 (52), 'The Case of Making the River Avon Navigable.' Mathew, in his *The Opening of Rivers for Navigation*, p. 11, says that before 1656 some endeavours had been made by Sir John Harrington to make the river Avon navigable between Bristol and Bath, but the work was attempted on too small a scale, for each sass (lock) would contain only one boat of eight tons, where each should have been made to contain six boats of thirty tons each, at once.

[2] Brit. Mus. 816. m. 8 (52), 'The Case of Making the River Avon Navigable.' See also Brit. Mus. 816. m. 8. (51), 'An Answer' [to the Reasons for Making the Rivers Stower and Salwerp Navigable].

mechanical power[1]. But in the case of some of these projects for making rivers navigable, the natural barriers were as great as, if not greater than, the opposition from private interests. The current of a river, especially when the volume of water was swollen by freshets, would at times break down dykes and other artificial constructions, and overflow the land until the location of the river channel had become completely changed; in slowly-moving streams, alluvial banks were formed in the quieter parts; in summer the channels were frequently provided with too little water to admit of navigation, while at other periods the current was so strong as to render it impossible to ascend the river, which would have been at all times an expensive and laborious undertaking. Whether, therefore, it was due to these natural impediments more than to the refusal of vested interests to support such enterprises, we are unable to decide; but the fact remains that attempts at this time to make rivers navigable were usually not very successful[2]. The navigation of the Avon, which was made in 17 Chas. II, was used for a short time, and then a heavy flood washed it away. An attempt was made to reinstate it later, but it was never afterwards used and finally disappeared[3].

As yet, no attempt had been made to connect one river with another by making a cut through the land. The first hint of the possibility of such a thing, so far as I have been able to ascertain, was in 1641, when John Taylor, the water-poet, went up the Thames, then crossed

[1] Brit. Mus. 816. m. 8 (58), 'Proposals for making River Chelmer navigable.' See also Brit. Mus. 213. i. 1 (91), 'Case of the Landowners on each Side of River Owze,' showing the opposition of the landowners to having barges hauled by horses along the banks of the navigation. Heretofore, barges had been hauled by men, rather than by horses; but about that time the size of vessels and lighters had been increased, and frequently nine or ten large lighters had been fastened in a train: hence the necessity of horses to draw them. The bargemen and lightermen recognized that this was a more economical way of conducting their business; but the Norwich Assizes decided that they had no right to haul with horses along the banks. Application was made to Parliament to secure this right; but the landowners opposed the Bill, chiefly because of the danger that might ensue from the possible breaking down of the banks of the navigation, thus allowing the water of the navigation to inundate the adjacent land. It would seem that they failed to fully appreciate the fact that the bargemen were just as much interested in keeping up the banks as were the landowners; for, of course, if the banks were broken down the navigation would be damaged, as well as the land overflowed.

[2] These difficulties in the way of river navigation seem to have suggested, at a later time, the expediency of abandoning river channels, and of digging artificial channels parallel to them, in which the depth and the flow of water might be regulated by locks. In all probability, the Sankey Brook Navigation was the first effort of this sort that resulted in permanent benefit to a waterway.

[3] *Royal Commission on Canals and Waterways*, 1906–9, I, p. 358, Q. 10,877, Evidence of Mr E. A. Rawlence.

overland to a tributary of the Severn, and then made a voyage upon several of the western rivers.   In describing this journey, he says that by a cut of four miles the Severn and the Thames might be so joined that passengers and goods could be conveyed by water between London and the west at cheap rates and without danger[1]. Taylor's suggestion, however, did not materialize, and it was not until 1655 that the subject of cutting navigations to connect rivers was again taken up.   In a work by Francis Mathew published in that year, a full account is given of the benefits to be derived from such inland improvements; and he seems to have obtained his facts and opinions from the experience of Holland, which had risen to the commercial supremacy of Europe.   If England were to be a great nation, she must imitate Holland, and the latter country had made great use of her waterways. "Rivers," he says, "may be compared to statesmen sent abroad; they are never out of their way: so they pass by great cities, marts, courts of princes, armies, leaguers, diets, and the like theaters of action, which still contribute to the increase of their observation; So navigable rivers, the more places of note they pass by, the more they take up, or bring, still gleaning one commodity or other from the soil they pass through, and are supplied by every town they touch at with employments[2]."   He would have England take up this work, where practicable, because it would furnish a training school for watermen, who would then be ready to serve the state at any time; it would release a great many thousands of horses from the carrying of packs and other burdens, and make them available for war; it would facilitate and cheapen the transportation of commodities, without tearing up the highways; and if the work were undertaken by the state, the tolls imposed would soon bring in

---

[1] In *John Taylor's Last Voyage*, 1641, he shows the great difficulty of taking a sculler's boat up the Thames, because of the many weirs, mills, etc., upon it, and because of the great lack of water.

He hints at a connexion of the Thames and Severn by canal, for he says: "Stroud (brook) and Churne might be cut into one, and so Severne and Thames might be made almost joyned friends......So that 4 miles cutting in the land betwixt Churne and Stroud, would be a meanes to make passages from Thames to Severne, to Wye, to both the Rivers of Avon in England, and to one river of Avon in Monmouthshire, which falles into the River of Uske.........By which meanes goods might be conveyed by water too & from London, in Rivers at cheape rates without danger, almost to half the countyes in England and Wales."

Then he shows the difficulty in navigating Stroud brook, "with passing and wading, with haling over high bankes at fulling milles (where there are many) with plucking over suncke trees, over and under strange bridges of wood and stone, and in some places the brook was scarce as broad as my Boate."   The mills on the brook were a great hindrance.

[2] Francis Mathew, *The Opening of Rivers for Navigation*, p. 3.

a fair revenue, without any grievance at all to the people[1]. To the objection that such great works would bring great difficulties, he answered that the state should rejoice in difficult things, for their accomplishment would bring great honour to the state; and when it was said that most of the rivers in summer had not water enough to carry a boat, his reply was that the tributaries of such rivers as should be judged fit for navigation should be turned into the main rivers, so as to increase their volume, rather than turned out of their natural course to drive mills[2].

Having considered the subject from a general standpoint, Mathew

---

[1] On pp. 3–4 of Mathew's *The Opening of Rivers for Navigation*, he describes some of the benefits which would arise by opening such rivers as should be found beneficially capable of navigation:

1st. By this industry, as it increases, will be raised many a thousand of watermen, fit at any time to be taken into the service of the state.

2nd. Many thousands of horses, now only used for packs and burthens, would be spared, and so multiplied for warlike service, and their provender devoted to improvable stocks.

3rd. The facility of commerce from one place to another, and the cheapness of transportation of commodities, without so much grinding and ploughing up our highways, "which maketh them now in so many places so impassable." "You shall see western waggons, which they call plows, carry forty hundred weight; insomuch as between Bristol and Marlborough, they have been enforced at a hill they call Bragdown-hill, to put twenty beasts, horse and oxen, to draw it up. This great abuse by this means would be taken away, by keeping our highways pleasant and withal, by this transportation of commodities by river, the price of commodities would fail" (i.e., be reduced).

4th. The imposition, though easy, laid upon every such navigable passage, would, as rivers by degrees are opened, amount to a fair revenue, without any grievance at all to the people, but rather with much comfort, as it is embraced in other countries, where they cannot live without the help of these bilanders, passing and repassing daily from town to town, from market to market and from coast to coast.

[2] On page 5 he considers two objections to his plans:

1st. That these great works bring great difficulties.

Answer. "Difficulties are the boundaries of narrow hearts; such ought not to be the heart of the state, which should most rejoice in difficult things, in the overcoming of which so much honour is achieved."

2nd. That most of the rivers in summer will lack water to carry a boat, even the Thames sometimes is so shallow, that our barges get stranded.

Answer. "Such rivers as shall be judged fit for navigation must not be debarred the contribution which other springs, brooks, and rivers would give them. Hence all streams that would naturally fall into the said rivers designed for navigation, ought to be free, and not be bound up with weirs, sluices, mills, etc., which turn the waters from their natural course. These mills, either the state should buy of their owners, and erect for every water mill three horse mills; or else agree with the owners of them to pull up these water mills, and locate horse mills in places more convenient."

then went on to show how a cutting could be made between the Isis and the Avon, so as to provide a navigable communication between London and Bristol, which would greatly reduce the cost of bringing commodities, especially coal, to London[1]. This he urged upon Cromwell as a work to be undertaken by the state, because the expense of construction and the resultant profits would be too great for any private individual or corporation[2]; but Cromwell was too busy with other things to take any notice of this project, and it was allowed to rest till 1670, when Mathew again brought it to the attention of the country[3]. He was so sure of the benefits of this work that in his plans, which were fully elaborated, he would not suffer any obstruction, not even a corn mill, to impede the enterprise[4]. In addition to the above plans

---

[1] He seemed to regard coal as the most important trade along this route (Mathew, *The Opening of Rivers for Navigation,* p. 8). By having this navigation from London to Bristol, "all commodities may be brought from Bristol to London (even at one farthing per pound) we now paying all the winter-time for carriage by land between London and Bristol 4s. per cent. and so preserve our horses for the States service" (ibid., p. 10).   [2] Ibid., p. 10.

[3] Mathew, *A Mediterranean Passage by Water from London to Bristol......And from Lynne to Yarmouth, and so consequently to the City of York : for the great Advancement of Trade and Traffique.* On pp. 1–8 of this work he again dwells upon the benefits to be obtained by connecting such rivers, so as to have a continuous navigation; but these are the same as those mentioned above, and given in his earlier work.

[4] On page 5 (ibid.), he speaks of cutting a graft betwixt two rivers, "at such a level as the water of the one shall run into the other, so as a vessel may sail or be towed from one of the rivers unto the other at arbitrement."..He saw no way of doing this but by sasses which would pen up the water, and thereby raise it, "for use as by so many stairs or steps as is or may be wanting to become of convenient depth for the transport of vessels of so good burthen as may be to good purpose....."

"Nor can the pulling up of corn mills (the onely obstruction, as being of a more publick service than any other intervening good) be a sufficient pretence to impede this enterprize of making such rivers navigable,......seeing that the necessary office of such mills may be performed at an easier charge by horse-mills, by wind-mills, or by mills termed river-mills, which in the Low-Countryes, and some parts of Italy are familiarly made use of in flat bottomed boats, and thereby conveyed to service......from one place unto another......"

His plan for this navigation I give in his own words: " The River Avon of Bristol may be......made navigable from Bristol to Calne, or to Mamsbury in Wiltshire, and by cutting a graft of five miles, or thereabout, in length onely, through a ground which I found favourable by nature for such purpose: the same river may take its journey for the same use (planting sasses also aptly upon the same) from Mamesbury to Leshlade in Oxfordshire, and there salute the river Isis already navigable, which so delivers itself into the Thames, and bring the trade of Ireland, the rich fruits of Cornwall, Devon and Somerset, Mendip Hills, and Wales......as well as the intervening countryes, to the Cityes of Bristol and London, mentioned; and back again at will; by so much a shorter and safer cut......and so much lesser charge of portage than else can be (one boat upon the same carrying as much as an hundred horse) as must exceedingly abate the price......" (ibid., p. 7).

for connecting the south and west, he had a scheme for connecting the north and east by navigable inland waters, and had gone over this route, as well as the other[1], to assure himself of its practicability.

Mathew failed, however, in securing the execution of his plan and it was not till more than a century later that the Thames and Severn were connected by a navigable waterway. But although there was no response to Mathew's scheme, it was not allowed to drop; for in 1677 Andrew Yarranton endeavoured to win the nation's attention to the great importance of improving the internal navigation of the island[2]. The chief feature of his scheme was to improve the rivers, so as to make them navigable, and thus render the interior more accessible to commerce; but, unlike his predecessor, he did not propose to establish water communication between rivers that were not then connected by water. To make England self-sufficient and strong to oppose the Dutch, he would have granaries established at the head of the navigable rivers, so as to have a supply on hand for years of want or scarcity[3]. He would have London connected with the interior and west by the extension of river navigation; for, as he says, "London is as the Heart is in the body, and the great Rivers are as its veins[4]." He also proposed to connect the rivers Severn and Thames as closely as possible; but his plan was to have goods brought up the rivers Severn and Avon, then across by land to Banbury, then down the river Charwell made navigable to Oxford, and thence by the Thames to London. By this means goods could be brought to London at two-thirds the cost of land carriage. His scheme also failed. But both Yarranton's and Mathew's plans are interesting, as showing us that they had in mind the connecting of the rivers Thames, Severn, Trent, etc., such as was

The trade in Welsh coal would also be great, and would not clash with the Newcastle trade at all. By this way, the inland counties could get coal far cheaper than they were then getting it from Newcastle. "And all other commodities may be commerced in upon this river, at half the charge of what is paid for land carriage,...... much to the subjects ease and happiness, which also may afford a very considerable return to His Majesty for the same" (Mathew, *A Mediterranean Passage by Water from London to Bristol*, p. 8).

[1] This project was that of making the river Waveney navigable, "by cutting of a convenient graft near Loppam Bridge, sufficient for navigation, less than three miles (through grounds in their nature favourable for such an enterprize) into Little Ouse which carryes itself unto Linn in Norfolk." Then he would improve the other streams connected with these, so as to extend communication in the interior of these and adjoining counties. He would open out the old Fosse Dyke, built by Henry I, from Lincoln to the river Trent (seven miles) and so have interior connexion between the ports of Yarmouth and Lynn and the interior, even to York, Cambridge, Boston, etc., thus linking up the east and north (ibid., p. 8).

[2] Andrew Yarranton, *England's Improvement by Sea and Land*.

[3] Ibid., I, p. 116. [4] Ibid., I, p. 179.

accomplished a hundred years later on a much grander scale by the construction of the canals.

A work of vast importance, undertaken about the middle of the seventeenth century, was the draining of the fens and low lands of the counties of Lincoln, Norfolk, and adjacent districts; and although this was of most importance to the development of the agriculture of that area, it has some interest also for those who would study the inland navigation of the kingdom. I do not propose to go fully into the history of that great undertaking, which has been so carefully told by Sir William Dugdale[1] and others[2]; but my object is to show how it affected the rivers as channels of trade.

The amount of water that had to find its outlet through the Great Ouse, Nene, and Welland rivers, was often so great that their regular channels were not large enough to carry it off, and as a consequence there were frequent inundations of the land adjoining these rivers and their tributaries. The area that was drained off through these means of egress was so extensive in comparison with the facilities for drainage, that the waters which gathered in great volume in the upper part of these tributaries would sometimes sweep along the river courses, gathering force as they went, and destroying the river banks, until the rivers became unduly widened, their beds became silted up by deposits from the banks, and sections of considerable extent in the almost level reaches of the rivers were, at times, completely covered. Then, too, some of this territory along the shore was practically on a level with the sea, and hence there was great difficulty in draining at all. The Duke of Bedford finally came forward with a method for draining the fens and making them into good agricultural land, in return for which he was to get 95,000 acres of the land. His scheme was accepted; the company of adventurers of which he was the most important member, fulfilled their work, and the stipulated amount of land was given over into the possession of the Duke, in compensation for his expenditure on that work[3].

[1] Dugdale, *History of the Embanking of the Fens.*

[2] Wheeler, *A History of the Fens of South Lincolnshire*; Allen, *A History of the County of Lincoln*; Thompson, *History of Boston*; Badeslade, *History of the Navigation of King's Lynn.*

[3] In a tract in the British Museum, 816. m. 8 (10), we find that there was great opposition to the draining of the fens, and especially, it would seem, to giving the Duke of Bedford 95,000 acres in return for the expense he had incurred in the work. In this tract we find the following decree of the House of Lords, 1661, which is sufficiently self-explanatory to require no comment:

"18th May, 1661.

Upon reading the petition of William Earl of Bedford, Participants and Adventurers for draining of the Great Level of the Fens, showing that the said Level

While this was successful in draining the great reach of low-lying country and turning it into fruitful fields, it was not so successful in improving the navigations of these rivers. The results in the case of the Great Ouse river will suffice to exemplify what we have just said. This river had an extensive inland navigation and a large drainage basin; and by the port of King's Lynn at its mouth a large outward as well as inland trade was carried on. In the accomplishment of their work, the Undertakers for draining the fens of Bedford Level had, in 1652, constructed a sluice and dam across the river Ouse, at Denver Dam, fifteen miles above the port of King's Lynn, and had embanked the lower part and outfall of the river to keep the water from spreading out so wide[1]. Besides, they had cut a new river, one hundred feet wide and twenty-one miles long, from a point in the Ouse adjoining Denver Dam, straight across the country to connect with the river again at Earith, with the expectation that when the tide had a chance of following this straighter course, it would extend farther up into the country[1].

As we have already said above, there was some opposition to the

having been drained at the petitioner's charge, for the recompense of 95,000 acres, And that by an Act in the last Parliament, possession of the said 95,000 acres is settled and quieted as the same then was, until the twenty-ninth day of May instant, within which time it was conceived another Act of Parliament would have passed for a perpetual settlement of that business, which being not effected, and the time near expiring, some attempts have been lately made by cutting of banks, to endanger the drowning of great part of the said level, and it is conceived that farther attempts will be made after the said twenty-ninth of May instant, if it be not timely prevented, which is endeavoured to be done by presenting a Bill to the Commons House of Parliament, which is now in agitation there; It is ordered by the Lords in Parliament assembled, That all unlawful forces, riots, and assemblies, within the said level and parts adjacent, and the cutting of any banks, sasses, sluices, or other works made for draining of the said Level, and preservation of the same, are hereby straightly forbidden, and that the quiet possession of the said great Level of the Fens, and the work for draining of the same, be continued in the possessions of the said Earl of Bedford, participants and adventurers, and their assigns, as now they are, until the Parliament shall take farther order therein, or an eviction be had by due course of law. Provided nevertheless that nothing in this order shall be any-wise prejudicial to the King's title, nor to any claiming from His Majesty, but that the King may enjoy his rights. And hereof all Mayors, sheriffs, Bailiffs, Constables, and other His Majesty's officers are to take notice for the keeping of the peace, and quieting of the possession of the premises as aforesaid. And lastly, that this order being published in the several parish churches and chapels in and about the said Level, all persons are to take notice and yield their obedience hereunto, as they will answer the contrary at their perils.

Jo. Brown, Cleric. Parliamentorum."

[1] Brit. Mus. 816. m. 8 (5), 'The Case of the Corporation of the Great Level of the Fenns, relating to the Bill for better Preservation of the Port of King's Lynn;' also Brit. Mus. 816. m. 8 (4).

draining almost from the start[1]; but the results upon the navigation were not such as to arouse hostility at the first: it required some time for the full effects to be shown. After a time it was observed that the tidal action in the tributaries above the Denver Dam was not so strong as formerly, and that, therefore, there was a tendency for these streams to become silted up, as a consequence of which there was danger of the country becoming again overflowed and navigation hindered or prevented[2]. The town and port of King's Lynn complained to Parliament that the channel from Denver Dam down to the sea had become silted up, because the embanking of the river had left it so narrow, that the small amount of water that could get in at the flow of the tide was not sufficient to scour the river bed at the ebb of the tide. This left navigation unsafe and dangerous, and likely to be completely lost, unless soon prevented[3]. These representations were answered by the Company which had drained the fens; they endeavoured to show that their works had not prejudiced, but rather aided the navigation, and that if this sluice should be pulled up a great part of the land drained would again be inundated, to the ruin of those families which had taken the land[4].

Toward the middle of the eighteenth century, complaint was made to Robert Walpole, who was the member of Parliament for that section, and he was asked to see if some improvement could not be effected. As a result, he had surveys made, and ordered a description of the true condition of affairs to be drawn up, so as to give the public full information[5]; but nothing seems to have been done at that time to make any change.

Whether the draining of the fens benefited or injured the navigation most, it is impossible to say with certainty[6]. It seems to have been

---

[1] See footnote 3, p. 190.

[2] Brit. Mus. 816. m. 8 (4).

[3] Brit. Mus. 816. m. 8 (4). See also the petition of the University and town of Cambridge to the House of Commons, that the river below Cambridge toward King's Lynn might be cleansed and the navigation recovered. Brit. Mus. MSS. 5865, p. 183. Also *J., H. of C.*, XIII, p. 758.

[4] Brit. Mus. 816. m. 8 (5).

[5] See Badeslade, *History of the Navigation of King's Lynn.*

[6] From evidence at hand, it is clear that even the Duke of Bedford himself, the inheritor of this land, was not satisfied with the results of the drainage works; for about 1745 he asked Labelye, the engineer in charge of the construction of Westminster Bridge, to make a survey of the Fens with the object of determining what was best to be done for their improvement, and the Commissioners of Westminster Bridge gave him leave of absence for that purpose. (Labelye, *The Great Level of the Fens*, pp. 4–5.) The report of his view and survey shows that vast sums had already been spent in draining the Fens, but that the results were far

the opinion of the men who had charge of the draining of the level, that if they could construct a dam and preserve the height of the water in the river above that point, they would by that means maintain there regular and constant navigation, where before barges had to be wholly dependent upon the tide; but apparently they did not consider the danger of sediment filling up the river bed, where there was no current to keep it scoured out. Their action in narrowing the river below the dam by means of embankment, was probably based upon the supposition that by confining the action of the tide within closer limits of width its scouring power would be greater within those limits, and would result in keeping the channel flushed out, clear of all obstructions; but the scouring power is dependent on the volume of water that can be brought into and carried out of the river by the tide, and the narrowing of the river by embankment would lessen the amount of water that the river would contain for flushing purposes. On the whole, it is safe to say that relatively the navigation has decayed, in comparison with what it was in the mediaeval days; but whether this was due in part to the drainage works, or altogether to the working of natural agencies, it is difficult to judge.

In the four years following the Restoration, we have noted that there was some activity in regard to the improvement of former navigations and, more especially, in the making of new ones; but after that time these projects were neglected, until peace had been established by William III and the country had again settled down to a régime of commercial and industrial advancement. As there came a slight increase in the amount of turnpike building in the last decade of the seventeenth century, so also there was an increase in the amount of navigation opened up. Both of these movements were, doubtless, due to the same causes, namely, the bad state of the roads as a whole and the industrial growth of the country, which necessitated greater facilities for trade. The cost of transporting goods by land was a serious burden on those who were endeavouring to promote the progress of industry, for it not only added to the cost of the raw material necessary for manufacture, but also to the cost of the manufactured product before it could be got to the consumer. Some improved means for the conveyance of goods was essential, if economic activity were to proceed unhampered.

Since each of these navigations had to be authorized by a special

from satisfactory; and (ibid., pp. 1–10) he shows what course ought to be pursued to accomplish the two-fold end of draining the Great Level and restoring the navigation. The sanity of his statements is now evident to us in the light which has come from accumulated experience. He tells us that the Fens had grown worse, both as to the drainage and the navigation (ibid., p. 7).

Act of Parliament, we get a good index of the interest taken in such improvements from the Acts that were passed for that purpose; for although not all these Acts were carried into effect, yet no Act would be passed unless there was the need and desire for the improved facilities which its execution would give.   From this legislation which was passed in the later years of the seventeenth and the first half of the eighteenth century, we see that there were three sections of England which were particularly interested in increased opportunities for navigation, namely, first, that part which is between the counties of Lancaster and Chester in the west, and York and Lincoln in the east; second, the south-eastern counties of Essex, Kent, Surrey, Sussex and Middlesex; third, the part which centres in Gloucestershire, with the adjoining sections of Wiltshire, Somersetshire, and Berkshire.   It may be noted in passing that these were the very sections in which manufactures were most important.   In the first area we have the rivers Humber, Ouse, Don, Trent, Aire, Calder, Mersey, Irwell, Dee, Weaver, and others; in the second, we have the Thames, Medway, Lee, and some smaller ones; and in the third, we have the Severn, the three Avons, the Frome, the Isis. All of these, and many of less importance which we need not here mention, were improved during this period.

When a river was to be made navigable, the first thing was to obtain an Act of Parliament for that purpose.   The object of the promoters was brought to the attention of the legislature, and their petition, with the petitions of all others in favour of, or opposed to, the measure, was referred to a committee, whose duties were to carefully inquire into the need for the proposed navigation, and to hear complete evidence on both sides of the question.   If the proposals for the navigation were favourably received, permission was given to bring in a Bill for obtaining the object sought by the promoters; but before such a Bill received the royal assent, it usually required the expenditure of large amounts of money in its passage through the two Houses.   Such Acts were usually private, and, as a consequence, their provisions do not appear in the statutes of the realm.   It becomes necessary, therefore, to know how such works were to be carried on under the statutory provisions.

When such an Act was obtained, it usually contained the names of those who were to be the Undertakers for carrying out the work under the terms laid down by Parliament.   Sometimes the undertakers would be the corporation of a town, or the corporations of several towns, situated on or near to the river which was to be made navigable.   Sometimes they were a loosely united group of landowners, millowners, or other influential men, who were desirous, from their own standpoint, to have the navigation improved, in order to obtain easier and cheaper access to markets. Generally, in such a body of undertakers, the nobility and landed classes of the region to be served by the navigation were the prominent members; and the others were chosen from the list of promoters. In no case did the promoters become undertakers, simply because they were the men who were agitating for the navigation; but the Undertakers were appointed to their office by Parliament, chiefly because of their influence in the community. Occasionally the undertaker was only one individual, as for instance, a manorial lord, who was given full power to make the river navigable[1]. In a few instances, after the undertakers had been engaged in their work, it was found that more capital was required than they had at their disposal; and to obtain this, the undertakers were allowed, by special Act of Parliament, to incorporate[2].

Over the undertakers were the Commissioners of the navigation, who were also appointed by Parliament and named in the Act. They were to have complete jurisdiction over the river; if the undertakers could not agree with the landowners, millowners and others, as to the compensation which the latter should receive for any injuries to their property caused by the making of the navigation, the Commissioners were to mediate in such cases of dispute, and the valuation they placed on these was to be final, and accepted by both parties. They were to see that the work was effectively carried out by the undertakers; that the navigation was made serviceable, according to the tenor of the Act; and that sums sufficient for the maintenance of the navigation in good condition were expended upon it.

Before the undertakers began their work, they were usually required to raise a definite sum of money, in token of their good faith in taking up the enterprise. On rare occasions, the Commissioners were allowed to borrow money on the credit of the tolls[3]. This authority, which was customarily used in the construction of turnpike roads, was very seldom given in the case of navigation works; but why this was so, I have been unable to ascertain with certainty.

When arrangements had been made for financing the enterprise, the undertakers endeavoured, if possible, without the aid of the Commissioners, to suitably compensate the owners of property along the course of the river, for the purchase of the land and for any injury to such property that was traceable to the navigation. They purchased lands on which to erect wharfs, warehouses and all other necessaries

---

[1] See Acts 10 and 11 W. III, c. 20; 6 Geo. II, c. 30; 7 Geo. II, c. 28; etc.
[2] Act 14 Geo. II, c. 8.
[3] Act 24 Geo. II, c. 19.

and conveniences of a navigation; and proceeded to construct locks, weirs, pens for water, cranes, and other works for making the navigation as useful as possible. When the work was completed, barge- or boat-masters were held responsible for damages to the locks and other equipment, caused by themselves or their men; and they were also liable to the owners of adjoining lands for any injury done to their property. To recoup themselves for their expenditures, the undertakers were allowed to take toll, and sometimes also tonnage dues[1], from all vessels using their navigation; and, to protect the public from excessive charges, the tolls were fully specified in the Act[2].

From the foregoing general considerations as to the making of navigations, we pass on to give one or two instances which will show the magnitude of some of these enterprises, the difficulties in the way of securing and maintaining them, and the great complexity of the rights and privileges that were sometimes involved.

The river Dee was in early days a river of great importance; but it was ruined as a haven early in the fifteenth century. In 1449, a Commission of Inquiry was instituted to look into the state of the navigation, and a quay was formed near Shotwick Castle, about six miles below Chester, from which place troops were usually embarked for Ireland. After the middle of the sixteenth century, a new quay was built at Wirrall, about eight miles down the river from Chester, and for a long time goods being sent to or from Chester were there discharged or loaded[3]. In the early years of the reign of James I, there was much difficulty in regard to the river Dee, its navigation and its mills; but it cannot be said that much good came from the settlement of these disputes[4].

---

[1] Note, for example, Acts 10 and 11 W. III, c. 20; 24 Geo. II, c. 39.

[2] The statutory provisions here outlined will be clearer, perhaps, from referring to the full text of the Aire and Calder Navigation Act, and also from Act 7 Geo. I, stat. 1, c. 15, 'An Act for making the Rivers Mercy and Irwell Navigable,' as given in Brit. Mus. 1246. 1. 16 (1). See Act 12 Geo. II, c. 32, under which no toll was to be charged on the river Lee; for by Act 13 Eliz., c. 18 that river was made free of tolls. I have made no mention of any special provisions, although each Act had some provisions which applied specially to that particular navigation.

[3] In 1560 a collection for this new haven at Wirrall was made in the churches throughout the kingdom; and in 1567 the city of Chester was assessed for the same purpose (Hanshall, *History of Cheshire*, p. 77). In 1551, the citizens of Chester sent a letter to the Lord Treasurer of England, asking for money to aid them in building this new quay (Brit. Mus., Harl. MSS. 2082, p. 14).

[4] The nature of these disputes can be studied in the following references in the British Museum, viz., Harl. MSS. 2003, 2022, 2081, 2082, and 2084, which are full of papers on the Dee Navigation case. Other papers are found in Brit. Mus. 816. m. 8 (38), and Brit. Mus. MSS. 11,394, pp. 29–31. This case will repay more minute study.

In 1674, Andrew Yarranton's assistance was secured in a project for improving the Dee navigation. He tells us that at that time the river was navigable for vessels of only twenty tons burden, and these could not come up higher than Neston. In order to restore the ancient navigation, and allow vessels to come right up to Chester, he suggested cutting a new channel, a scheme that would also recover from the sea a large tract of the "white sands," which had been rendered useless by the inundation of the sea. It was a part of his plan also to cut a canal from the collieries at Aston, near Hawarden, to connect with this new channel, and thus facilitate the carriage of coal up to the city[1]. Neither of these plans was carried out at that time, although both of them were accomplished afterward.

In 1693, another proposal was made by Evan Jones to make the Dee navigable, so as to permit vessels of one hundred tons to be brought up. He offered to do the work at his own expense, provided that all the lands that he could recover from the sea should be vested in him, on paying the usual rent to the Crown, and one-fourth of the clear profits to the City Companies, and that he and his heirs should be entitled to certain duties on coal, lime and limestone. "His plan was rejected, largely, it would seem, on account of the duties on coal which he stipulated for[2]."

In 1698 the question of the Dee navigation was again prominent, for it was evident that the usefulness of the river was declining. The looseness of the sands in and near the river often changed the channel of the river and hindered navigation; the emptying of ballast of ships into the river added another impediment; but probably the chief reason for decay was the fact that the river was not confined so as to compel it to flow in one definite channel. To accomplish this latter purpose, a plan was submitted to the city of Chester by a Mr Gell, of London, whose offer was much like that of Evan Jones, but with some difference as to the duties on coal, lime, etc. At this time, vessels could not come farther up the river than Neston, whence goods had to be transferred to carts or else to open boats for the eight miles to Chester. His plan was to deepen the bed of the river and confine the course of the water by banks that would be high enough to keep the highest spring tide from overflowing the land on each side in what was called

---

[1] Yarranton, *England's Improvement by Sea and Land*, p. 192. Yarranton's new channel was to end opposite Flint; the cut that was made at a later time opened opposite to Wepra, somewhat farther in along the estuary. Yarranton's coal canal was to fall into the Dee near Flint; but the canal which was actually constructed afterwards approached the Dee about two miles below the city of Chester.

[2] Hanshall, *History of Cheshire*, p. 78.

the level of Saltney Marsh. He would make the river navigable *at all tides* for vessels drawing twelve feet of water, at an expense to himself of forty to fifty thousand pounds; and, in return, he was to be allowed a duty on coal and lime brought to be sold or used in the city of Chester, and was also to have the sands of the river (so far as it should be made navigable) which yielded no grass[1]. A Bill was brought into Parliament to this effect; and while there was no objection to allowing him the duty on coal and lime, there was strong opposition to granting him the land mentioned in the measure. Because of this opposition and his failure to make provision for any payment to the City Companies, his plan was at first rejected; but he made a second offer, expressing his readiness to deposit £2000 with the trustees for the completion of the work. This attracted the attention of the city, and in 1700 the people of Chester, upon a petition to Parliament, obtained an Act for recovering and preserving the navigation of the river[2]. Under this Act, the Mayor and Citizens of Chester were empowered to make the river navigable from the sea to Chester, for ships of 100 tons or more, and in order to accomplish this end the attempt was made to turn the river from its old course near the Welsh side, and to confine it to the Cheshire side; but the experiment convinced the citizens that it was impracticable, because of the shifting sands there.

In 1731 another petition was presented to the House by the citizens of Chester. Investigation by a Parliamentary committee revealed the fact that for about ten miles from Chester outward, the river-bed was being still further filled up with sand, so that, even with the aid of the tide, it was extremely difficult to reach the city from the sea[3]. Those who were most interested in the river considered that the only way

---

[1] Brit. Mus., Sloane MSS. 3323, pp. 267–9.

[2] The petition was sent to Parliament in 1699, and showed that the river Dee which had formerly been navigable for vessels of considerable burden from Chester to the sea, was becoming silted up, because of the tide wearing away the sand banks and leaving a very uncertain channel (*J., H. of C.*, xiii, p. 36). The Act passed in 1700 was Act 11 and 12 W. III, c. 24. It imposed taxes on coal, lime and limestone brought to the city by sea or land; and these taxes, together with the receipts from the public sale of this coal, etc., were to be applied to the improvement of the river, which was to be made navigable to the sea. The aforementioned duties were to be granted for twenty-one years to the corporation of Chester, for recovering and preserving this navigation.

In addition, as soon as the navigation had been completed according to the terms of the statute, the title to the "white sands" was to vest in the corporation, and they were at liberty to enclose and improve this land and to receive the rent and profit therefrom, upon the understanding that such sums received were to be applied to maintaining or extending the navigation works as necessity should require.

[3] *J., H. of C.*, xxi, pp. 812–13; also preamble to Act 6 George II, c. 30.

to improve its navigation was to cut a new channel for that distance, so that by confining the river within narrower bounds, the tide and the fresh water might keep it scoured out deeper[1].

This proposal for improving the Dee, so as to make Chester a useful port, led to opposition from her old rival Liverpool and also from Parkgate, whose inhabitants were anxious to assume the place of importance as the port of the river Dee, while throwing Chester into the background. Some landowners opposed the project, on the ground that it would be prejudicial to many thousand acres of land along the river[2]; and an engineer of great name, reasoning from the ill effects to navigation that resulted from the draining of the fens, and other like works, was very outspoken in the conviction that the project in view would hinder, rather than promote, the navigation of the river[3]. It was thought by others that it would be absolutely impossible for the undertakers to make the river sixteen feet deep at a moderate spring tide, as they had promised. Apparently, opposition was encountered also from the cheesemongers of London, whose ships were about the only ones that came into the river. These men seem to have so monopolized and engrossed the Cheshire cheese trade, upon their own terms, that they enjoined the masters of their ships not to take on board any cheese but what was bought for their use. If the Dee navigation were improved, this monopoly would be broken up by the other vessels that would come up the river to have a share in the trade; and, therefore, in order to prevent the farmers from having more choice as to the disposition of their cheese, the London dealers opposed the Bill for the improvement of the navigation on the ground that it would not answer the ends intended[4].

But the citizens of Chester answered the objections that had been made against the proposed improvement, and after a further investigation, Parliament passed an Act giving effect to their desires[5]. The work on the new cut was commenced in 1735, and was completed in

---

[1] *J., H. of C.*, xxi, pp. 812–13, Feb. 24, 1731. On condition that the undertakers would make the river navigable from Chester to the sea, the proposal was that they were to have the right to enclose a large tract of land, called the "white sands," which was at that time barren, on account of being overflowed by the salt water every time the tide came in.

[2] *Ibid.*

[3] See Badeslade on the *River Dee Navigation*.

[4] Brit. Mus. 357. c. 1 (37), 'Case of the Inhabitants of ......Chester......in Answer to the Cheesemongers of London.'

[5] The points brought out in the investigation, as to the condition of the river, and the expense of making it navigable, are given in *J., H. of C.*, xxii, p. 44. The Act passed was 6 Geo II, c. 30. 'The Case of the Citizens of Chester,' answering the objections to their plan, is given in Brit. Mus. 816. m. 8 (38).

1787, when the water was turned from the old channel into the new. After that, the undertakers were allowed to take the tonnage duties authorized by their Act. The new channel was about ten miles in length and about one hundred yards in breadth, and cost over £40,000 for its construction[1]. It was to be maintained so that there should be sixteen feet of water in every part of the river at a moderate spring tide[2]. The later history we shall not follow here.

An intricate case that came up for consideration and legislation was

[1] *J., H. of C.*, xxiv, pp. 527, 600.

By 1740 more money was needed to carry on the undertaking most effectively, and the Undertakers(Nathaniel Kinderley and his heirs and assigns)were incorporated by Act 14 Geo. II, c. 8, so as to allow them to borrow (see also *J., H. of C.*, Jan. 19, 1740). In 1743, the tolls and tonnage authorized by Act 6 Geo. II, c. 30, were found to be too high, and, on account of agitation by the citizens, merchants and traders of Chester, the rates were reduced (*J., H. of C.*, 1743, Jan. 31, Feb. 29, Mar. 5, May 12) by Act 17 Geo. II, c. 28.

According to their Act of Parliament, the Undertakers were allowed not only to take tonnage rates on the river, but also to enclose for themselves the lands they could recover from the sea on both sides of the river, the returns from which were to be devoted to the maintenance of the navigation. According to Hanshall (*History of Cheshire*, p. 78), as early as 1754 more than 1400 acres of land were recovered from the inundation of the tide; 664 acres were also recovered in 1763; and in 1769, 348 acres more. A further enclosure of about 900 acres of land was made in 1790, and other enclosures were made after that time. Ormerod, however, (*History of Cheshire*, i, p. lxxiii), says that at the time he was writing (1882) the embankment of the sands had been completed nearly as low down as Shotwick, and more than 2400 acres had been rescued from the sea. This latter figure is rather vague, and it is probable that Hanshall's figures, which are definite, are more nearly correct.

The sums spent in cutting this channel, and in embanking the "white sands" along the river, were very large; in consequence of which no profits were received till 1775, and then only two per cent. was divided on the principal (Hanshall, p. 78). Pennant, in his *Tour in Wales*, i, p. 199 (1784), says in regard to this work: "The expenses proved enormous, multitudes were obliged to sell out at above 90 % loss, and their shares being bought by persons of more wealth and foresight, at length the plan was brought to a considerable degree of utility, and a fine canal formed, guarded by vast banks, in which the river is confined for the space of ten miles, along which ships of 350 tons burden may safely be brought up to the quays. Much land has been gained from the sea, and good farms now appear in places not long since possessed by the unruly element."

[2] In addition to the references above-mentioned, Act 17 Geo. II, c. 28, preamble, gives much detailed information, especially as to the legislation, concerning the Dee. The provision as to maintaining sixteen feet of water was changed by this latter Act; and because of the liability of the river to become silted up with sand, it was enacted that there should be fifteen feet in every part of the channel at a moderate spring tide. If the company did not maintain that depth, the payments of tonnage, etc., were to cease; and in that case commissioners were to be appointed to take these duties and the revenues from the "white sands" and apply them to the purposes of the navigation.

that of the river Trent. In 1741 a petition from Newark-upon-Trent was sent to Parliament, stating that the navigation through their branch of the river Trent, because of the mills erected upon and dams constructed across that branch, was obstructed, and that lately sands and shoals had been formed there. Because of this lack of water in their part of the river, all goods to be sent from Newark up the river had to be transported two miles by land to the place where the two branches met, and left there, on the bank of the river, exposed to the weather, until a vessel was ready to carry them up-stream. Similarly, whatever goods they had to send down the river had to be taken by land carriage to a place below the Crankleys, and there left exposed for

a day or two, to await the arrival of a barge. The inhabitants of Newark, therefore, asked authority to make their branch of the river navigable, so that their carrying trade might be greatly facilitated[1].

The nature of the question at issue here is evident. The farmers and the inhabitants of small towns near to Newark were interested in having Newark's trade freely opened up and down the river, in which case this town would be a better market for the sale of their corn, wool and other products, and for the purchase of coal[2]. Opposition to the proposed scheme came from three sources: the mercantile towns along the river, such as Gainsborough, were afraid that part of their trade would go to Newark; the landowners along the Newark branch of the river feared that if the river were made navigable it would be prejudicial to them, by causing their lands to be overflowed; and the barge-masters and navigators of the river who were interested in the navigation of the other branch, would naturally oppose any deflection of the water from their branch[3]. In reality, it would seem that the avowed object of the people of Newark was to wholly turn the navigation from the other branch into their branch[4]. For this, there would appear to have been some justification, for the trade of Newark had greatly increased

---

[1] *J., H. of C.*, xxiv, p. 108.

[2] *J., H. of C.*, 1741, petitions of Mar. 16, 23, 24, 30.

[3] *J., H. of C.*, 1741, petitions of Mar. 25, Mar. 31, Apr. 1, Apr. 2, etc.

[4] See petitions of the towns of Gainsborough and Hull referred to in previous note.

and was in need of improved facilities for carrying[1], while on the other branch of the river there was no place of great importance and, therefore, no great need for a navigable river. This instance is given, not to be followed minutely to the close of its history, but to show how, at times, place discrimination entered very largely into the consideration as to the advisability of making a river navigable, and to exemplify the difficulties which were thus introduced in the determination of the best course of procedure.

Perhaps as interesting and complicated a case as we could mention is that of the river Don or Dun. The early history of this river we may pass over, to notice that in 1691 the borough of Doncaster contributed £5 toward making the river more navigable[2]. By 1697, the case was being brought before Parliament[3]; but it was not till 1708–4 that the corporation of Doncaster became active in promoting a Bill in Parliament for making the Don navigable, and voted for that purpose a sum not exceeding £100[4]. It does not appear that anything was done for the improvement of the river until 1721 and 1722, when the people of Sheffield took it up[5], in collaboration with the Company of Cutlers of Hallamshire and the corporation of Doncaster. Their scheme proposed that the river should be made navigable as high as Doncaster for vessels of thirty tons burden, and from Doncaster to Sheffield for vessels of twenty tons; the former part of the work was to be performed by the corporation of Doncaster, and the latter by the Cutlers' Company of Hallamshire, whose manufactures would thereby be carried more cheaply to Hull and thence to London[6]. Notwithstanding the general advantages emphasized by the promoters, the design was strongly opposed by the country gentlemen whose estates lay contiguous to the river, and who saw nothing in it but an unwelcome intrusion. Even the fact put forth by the promoters, that these owners of estates, instead of suffering any detriment from the measure, would receive a material advantage in being able to bring in the lime and coal they required

---

[1] See Defoe, *Tour Through the Whole Island*, III, p. 58.

[2] Hardy, *Records of Doncaster*, IV, p. 155.

[3] A letter showing this fact is given in Brit. Mus., Stowe MSS. 747. This Bill was lost (Leader, *Cutlers' Company*, I, p. 167).

[4] Hardy, *Records of Doncaster*, IV, pp. 161–3. This project also failed in 1704 (Leader, *Cutlers' Company*, I, p. 167).

[5] Among the Hunter MSS. in the British Museum, there is a petition, dated Dec. 11, 1721, "from the Corporation of Sheffield to the Rt. Hon. the Earl of Strafford in London" asking him to use his influence in favour of a Bill immediately to be brought into the House of Commons for making this river navigable (Leader, op. cit., I, p. 167). Another letter was sent to him in 1723 soliciting his assistance (Leader, op. cit., I, pp. 167–8).

[6] Brit. Mus. 816. m. 8 (39).

at a cheap rate, failed to make any favourable impression on the minds of the opponents. There was determined opposition also from the towns of Gainsborough and Bawtry, which felt that the new navigation would divert traffic from the time-honoured trade route through these places.

Gradually, concessions were made by the projectors to pacify some whose opposition was most dreaded, and finally, in 1726, a Bill was introduced in Parliament. The hostility to this Bill, although very decided, could not prevent its passage, for the public advantages were too evident and too forcibly represented to be brushed aside[1]. The Bill passed in that year[2], but a part of the original design had to be given up. The Act now passed enabled the Company of Cutlers to make the river navigable only from a place called the Holmstile, in Doncaster, to the most westerly limit of Tinsley, that is, to within three miles of the town of Sheffield; and empowered them to improve and keep in repair the highway from Sheffield to Tinsley[3]. The Act of 1727 authorized the corporation of Doncaster to improve the river below that town, from the Holmstile to Wilsick-house in the parish of Barnby-upon-Don[4]. The navigation shares began to pay dividends in 1731[5]; but still the calls on the shareholders continued. Other unexpected difficulties arose and in 1732 the two corporations were glad to relieve themselves

----

[1] The cause of the navigation was ably championed before committees of the House, and in conversation with Members of Parliament, by Mr Samuel Shore and Mr John Smith, two deputies sent by Sheffield to London. An account of the difficulty of securing the passage of the Bill is given in the letters of Mr Smith to his wife in 1726, extracts from which are given in Leader, *Sheffield Burgery Accounts*, pp. lvi–lvii.

[2] Act 12 Geo. I, c. 38.

[3] This part of the original project was probably the more willingly abandoned because of the difficulties in the way along the river between Sheffield and Tinsley. A canal was proposed to overcome these difficulties, but it was not constructed at that time. The Act of 1726 appointed the Cutlers' Company as undertakers of the navigation. They were authorized to make the river navigable from Doncaster to Tinsley for vessels of twenty tons burden, and to take rates and duties for tonnage.

[4] Act 13 George I, c. 20.

[5] Leader, *Sheffield Burgery Accounts*, p. 361 (Feb. 22 and Sept. 19, 1731).

from powers which they had taken so much pains to acquire. To accomplish this, it was proposed that they should consolidate their interests, and transfer them to private individuals, which was done under sanction of an Act of Parliament incorporating "The Company of Proprietors of the Navigation of the River Don[1]."

In 1739, a difficult situation arose, with reference to this river. The Company of Proprietors of the Navigation, in that year, presented a petition to Parliament, showing that in carrying out previous statutes for making this river navigable, they had spent £24,000, and had made a good navigation from Rotherham to Barnby; but between Barnby and Fishlock Ferry there were several shallows which made the navigation so imperfect that at neap tides, or in dry seasons, boats or vessels could not pass. This discouraged trade, on account of the loss of time in carrying goods up and down the river, and they asked for authority to impose additional tolls to be used in improving the navigation there[2]. The barge-masters and traders on the river and the Company of Cutlers of Sheffield reinforced the above petition, by showing how much the navigation had already been improved and what great advantages had accrued to trade, through cheaper carriage of all goods on the river[3]. But opposition was soon encountered from several sources: the city and also the traders of Doncaster objected to making the rates of carriage on the river more burdensome, because the existing rates were sufficient to defray the expenses of perfecting the navigation, and they were also against the proprietors of the navigation for imposing their rates in an unequal and oppressive manner, exempting some and charging others[4]. The cities of Hull and Manchester also objected to

---

[1] Leader, *Cutlers' Company*, I, pp. 167, 168–9, 171.

[2] *J., H. of C.*, XXIII, p. 427. In the investigation, it was shown that there were shallows in some parts of the river, so that vessels drawing 2 ft 4 in. to 3 ft 2 in., which were the usual boats on the river, constantly got stuck on the shallows in dry seasons, when there was only eight or nine inches of water. With spring tides, however, vessels of twenty to forty tons usually went on this same part of the river. The badness of the roads made it necessary to have the navigation of the river attended to; and for £3750 the navigation could be completed (including one lock two miles long and more than eight feet deep), and three bridges could be built over the said cut and river. *J., H. of C.*, Jan. 31, 1739.

[3] *J., H. of C.*, XXIII, p. 462, Feb. 15, 1739. The Company of Cutlers showed that their manufacture of cutlery had been widely extended in and around Sheffield; and that since the river Don was partly made navigable they could get their raw materials brought to Sheffield more cheaply, thereby reducing the cost of production; and that further improvement of the navigation would be of much more general and public service. See also *J., H. of C.*, XXIII, p. 459.

[4] *J., H. of C.*, XXIII, p. 462. Petition from the Gentlemen, Freeholders, Merchants, Tradesmen and others, inhabitants of the town of Doncaster; and also another

the proposed increase of rates[1]. But the most decisive opposition came from the city of York. In order to further the interests of that city and enable its industries to prosper, the river Ouse had been improved so as to admit an increased quantity of tidal water, and by thus raising the water there two feet, larger vessels could be used on the navigation for the conveyance of goods. But in the time of Charles II, the mouth of the Don river had been widened and deepened into a straight channel, called the "Dutch cut," nearly five miles long and one hundred yards broad at high water, so that more of the tide had been admitted into this river. The Dutch cut had been further widened by tidal action, and so the tide was prevented from flowing strong up to York. If permission were now to be given to deepen the river Don for some distance above the Dutch cut, still greater quantities of tide would be drawn into that river, and consequently the flow of the tide up to York would be rendered insignificant[2]. Here, then, was the problem to be decided: to deepen the river Don would cause more tide to come into that river and thus be detrimental to the industries of York and places adjacent; but to refuse authority to deepen the Don would be to impose a heavy burden on the carriage of materials to and from a rapidly rising place like Sheffield, which, with its environs, had a trade of considerable magnitude[3].

In the investigation of this case, it was found that the Company had let the navigation of the Don to three (probably Sheffield) men for fourteen years; and the people of Doncaster had not received as favourable rates as the cutlers of Sheffield. It was an instance of place discrimination by those who had the control of the transportation by river. Many efforts had been made by Sheffield and Doncaster to come to terms regarding the navigation, so that they might act in harmony, but all such efforts were fruitless, except in expenses[4]. This was the complicated situation which Parliament was called upon to face. After

petition from the Mayor, Aldermen and Burgesses of the borough of Doncaster. They complained of unequal rates, which were, presumably, lower rates to Sheffield manufacturers, and higher rates to the merchants of Doncaster.

[1] *J., H. of C.*, xxiii, pp. 467–8. The Hull petition said that they received great amounts of corn, coal, and other commodities from Doncaster and the neighbouring country; but if additional tolls were permitted on the Don, the price of sending these things to Hull would be greatly increased, and export trade would be consequently prevented. They said, also, that after all that could be done, the Don could not be made navigable at neap tides and in dry seasons for vessels of greater burden than those then used.

[2] *J., H. of C.*, Feb. 12, 1739, xxiii, pp. 455–6, 459.

[3] Brit. Mus. 816. m. 8 (39), 'Reasons for making the River Dunn in the West Riding of the County of York navigable........'

[4] Leader, *Sheffield Burgery Accounts*, pp. 350–407.

a full hearing, the Bill was amended in several particulars[1]; and since there was a good navigation from Rotherham to Barnby and from Fishlock Ferry to Hull, in spring tides, an Act was passed to enable the Company to complete the intervening part of the navigation by making the necessary cut in the river, and otherwise improving this two miles, after which they were allowed to take certain specified rates of tonnage and toll[2]. In order that the city of York might not be prejudiced, however, Acts were later obtained for improving the river Ouse and cutting off its bends, so as to secure greater tidal flow up that river.

Having thus outlined a few cases to show the difficulties in the way of obtaining increased facilities for inland navigation, we now turn to consider the cost of carriage of goods by rivers and to compare this with the cost of carriage by land. The latter we have more fully dealt with in a former part of this work, so that the results there reached need be only referred to at this stage of our discussion, in order to have our best information on that part of the subject. But when we try to obtain definite information as to the cost of river carriage, we are met at the very outset by widely varying statements, from which it is almost impossible to deduce any general conclusions. The fact noticed in connexion with the roads, that there was comparatively little long-distance traffic, but that most of it was local, applies with even greater force to the rivers; and because each navigation served its own locality and was not connected with other navigations so as to have uniform rates with them, we find considerable divergence as to the charges that were imposed for the conveying of goods by river. Another thing which tends to prevent our obtaining any accurate comparisons of the cost by land and by water, is that the river traffic, where possible, took charge of the heavy and bulky commodities, such as coal, salt, wool, etc., while the lighter goods and those of greater value in less bulk went by land. Now it has always been the case that goods of the former class have been charged at a different rate from those of the latter class, for any service of carriage; and this fact also helps to vitiate any results we might get as to comparative cost. But we give our results as we have been able to obtain them.

In 1655 the cost of carriage by land between London and Bristol was 4s. per cental (or hundredweight) in the winter time, which was

---

[1] *J., H. of C.*, Mar. 5, 1739.

[2] Act 13 Geo. II, c. 11. The later history of the Don river comes to view in connexion with the history of the Stainforth and Keadby canal and the Sheffield and Tinsley canal.

equivalent to two-fifths of a penny per hundredweight per mile[1]. This statement is corroborated to a degree by a writer in the last quarter of that century, who says that, along this same road, the cost of carriage from Oxford to London was £3 per ton[2]. But at this latter time the cost of transportation by river from Oxford to London was 20s. per ton, or one-fifth of a penny per hundredweight per mile[3]. It will thus be seen that here land carriage cost just three times as much as river carriage.

In some instances we get only bare statements as to the relative advantage of river carriage over land carriage, without the complete statistics which would enable us to verify the statements made. For example, when there was the agitation for making the river Wye navigable, in 1649, in discussing the benefits of the navigation it was said: "For it is well known that the carriage of a ton weight twenty-four miles by water does not exceed the charge of three miles carriage by land[4]." This looks like a somewhat exaggerated opinion, even though it was said to be "well known" by many people; and yet the statement that sometimes land carriage cost eight times as much as water carriage is borne out by other facts.

In 1699, the carriage of coal by land from Worcester to Droitwich (seven miles) cost 3s. 6d. and 4s. per ton, together with 5d. per ton custom: which would make the cost 6d. to 7d. per ton per mile[5]. At the same time, the carriage of salt over the same road from Droitwich to Worcester cost 5s. a ton, which would be eight and one-half pence per ton per mile[6]. But the carriage of the salt down the Severn river from Worcester to Bristol cost only 5s. a ton[7], and the distance here, on account of the sinuosity of the river, is more than seventy-seven miles, which would make the cost of carriage four-fifths of a penny per ton per mile. If, therefore, the salt could be brought down the river seventy-seven miles for the same amount as it cost to bring it seven miles by land, it is evident that land carriage cost eleven times as much as river carriage[8].

---

[1] Mathew, *The Opening of Rivers for Navigation*, p. 10.

[2] Brit. Mus., Stowe MSS. 877, p. 21. £3 per ton for the distance between Oxford and London would be almost exactly three-fifths of a penny per cwt. per mile.

[3] Ibid.

[4] Brit. Mus. MSS. 11,052, p. 96.

[5] Brit. Mus., Add. MSS. 36,914, p. 9. This reference calls the distance from Worcester to Droitwich five miles, when in reality it is seven miles.

[6] Ibid., p. 10. It is also said that some bargained for the summer half of the year, that is, from May-day to Michaelmas, and gave 10s. for the carriage of a ton of salt to Worcester and a ton of coal back to Droitwich.

[7] Ibid., p. 10.

[8] We here give some other material as to the relative cost of carriage by land

From the foregoing, it would appear that the cost of transporting by land was from three to eleven times that by water. But we cannot rely upon these statements as a basis of a generalization for the whole of England. For example, we have formerly seen that the cost of land carriage was from one-half penny per cwt. per mile in summer to three fourths of a penny per cwt. per mile in winter. Compare this now with the results for the Mersey and Irwell. In 1753 this Navigation Company advertised that, for persons who used their flats in summer as well as in winter, they would carry at rates which were the equivalent of from one-third to one-half penny per hundredweight per mile from Warrington to Manchester, and one-fourth penny per hundredweight per mile from Manchester to Warrington[1]. It will be observed that these rates were but very little lower than the rates for land carriage. A similar instance

and by water. We are told that, about the beginning of the eighteenth century, the land carriage from Derby to Wilden Ferry (about nine miles) cost twice as much as the carriage by water from Wilden Ferry to Gainsborough, which was about sixty miles by the Trent river. (See Brit. Mus., T. 100* (14), 'Case of the River Derwent.') If this were true, the cost of land carriage there was more than twelve times the cost of carriage by river. Compare Brit. Mus., Stowe MSS. 818, pp. 83–84. Of similar import are the words of Houghton, *A Collection for Improvement of Husbandry and Trade*, May 12, 1693, showing that, before that time, the cost of carrying malt from Derby to Wilden Ferry by land, which was five miles, was as much as from Wilden Ferry to Hull by water, which was sixty miles. This would make land carriage cost twelve times as much as water carriage. But, correcting these distances to nine miles and one hundred miles respectively, land carriage must have cost eleven times as much as water carriage.

Bradley, *Husbandry and Trade*, II, p. 287, says that in 1675 a discourse about water was read before the Royal Society in which it was stated that a chalder of sea coal, weighing about thirty-three hundredweight, was carried nearly 300 miles for 4s., but that the land carriage of this by waggon would be about £15, that is, 75 times as much as by sea, and on horseback about 100 times as much. According to this statement, the cost of carriage by waggon was about two-fifths of a penny per cwt. per mile (just what we have noted in Mathew, *The Opening of Rivers for Navigation*, p. 10). How much credence we are to place in his figures, it is difficult to decide.

In Brit. Mus. 357. c. 1 (37), 'Case of the Inhabitants of the County and City of Chester......," p. 1, we learn that the cost of carrying goods by land from Parkgate to Chester, a distance of eight miles, was 6s. per ton; or by small boats on the river, 2s. per ton. This would make the cost of land carriage one-half penny per cwt. per mile, and the cost of water carriage only one-third of that by land.

[1] Axon, *Annals of Manchester*, p. 90. The exact rates advertised by the Mersey and Irwell Navigation Company were:
　From Bank Key, at Warrington, to the Key at Manchester,
　6d. per cwt., from May 1 to Nov. 11 (= ⅓d. per cwt. per mile),
　7d. ,, ,, ,, Nov. 11 ,, May 1 (= ½d. ,, ,, ,, ,,).
　From the Key at Manchester to Bank Key, at Warrington,
　4d. per cwt., at all times (= ¼d. per cwt. per mile).

is that of the Medway, on which the cost of carriage by boats in 1600 was but 2*s.* per ton cheaper than that by land[1].

From what we have already shown, it is plainly impossible for us to make any comprehensive statement, as to the relative cost of carriage by land and by water, for the whole of England. It is certainly true that the cost of water carriage was lower than that by land, but as to how much lower it was, that must be determined by the information at hand for each locality; as yet, I have not been able to secure sufficiently complete data to warrant any definite conclusion on this point that would be consonant with the facts for all the different parts of the kingdom.

Another subject upon which I have been unable to form any accurate conclusion, because of lack of information, is as to whether the cost of land or river carriage in general increased before the period of canal building. If the facilities of transportation lagged behind the demand for them, as seems entirely probable at this period of expanding commerce, we would naturally expect that the cost of carriage would increase, for the carriers both by land and water could put up their prices and still be assured that the service they offered would be amply required. But in the case of river transport there was no limitation of the price of carriage by the Justices of the Peace, as there was in the case of land carriage; and therefore there would be all the more inducement for the river carriers to increase their charges. In the investigation of the Thames and Isis navigation, in 1750, it was said that the price formerly paid by full-sized vessels for passing the first lock on the Thames, at Sunning, had been not more than 7*s.* 6*d.*, but at the latter date (1750) it was 12*s.* 6*d.* and upward. At the same time, the price of water carriage had advanced about ten per cent[2]. Again, in the case of the river Avon, it is clear that the rates of carriage had been increased for a considerable time before 1760; and in such cases as the Avon, where the navigation was let out or farmed, the prices charged were variable[3]. So also when a committee of Parliament, in 1758, were looking into the desirability of permitting the extension of the Calder navigation from Wakefield to Halifax, it was found that the charges on the navigation were so high that the proprietors made thirty to forty per cent. on their investment[4]. From these detached pieces of evidence

[1] Brit. Mus., Add. MSS. 34,218, p. 40. Even this slightly lower cost of river transportation was rendered still more important from the circumstance that "the boats of Yalding in the winter time would carry to the ships in one day more timber than 200 oxen and horses can do in two days." (Ibid., p. 40.)

[2] *J., H. of C.*, xxvi, pp. 30–31.

[3] *J., H. of C.*, xxvi, p. 182 et seq.

[4] *J., H. of C.*, xxviii, pp. 133–44.

it would seem as if the cost of river carriage, at least in some places, had increased before the advent of the canals; and it is probable that sometimes the cost of land carriage had also increased[1].

---

[1] In the investigation of the Parliamentary Committee of 1758 into the Calder navigation, it was pointed out that the need for the extension of this navigation was apparent, from the fact that land carriage between Sowerby bridge and Wakefield, which was carried on almost wholly by pack-horses, had increased of late years by one-third to one-half; and even then, wool had been known to lie at Wakefield and Leeds three or four months waiting for carriage (*J.*, *H. of C.*, xxviii, pp. 133–44). The evidence as to this increased cost of carriage was not disproven.

On the other hand, it was shown that, in some instances, the cost of land carriage had been reduced. Land carriage from London to Halifax was performed by using broad-wheeled waggons drawn by eight horses, each waggon capable of carrying 80 packs of 240 pounds each. In 1758, we are told that this cost of carriage had been lowered in the last three years, from 14*s.* a pack to 7*s.* 6*d.* (ibid.).

CHAPTER IV

ROADS AND THEIR IMPROVEMENT, 1750-1830

In an earlier chapter we have considered the condition of the roads and the legislation regarding them, up to the middle of the eighteenth century. From this time on, the roads received more attention than had been given them before this; but we must not suppose that there was any radical change for the better in these highways of trade and travel. The increased effort devoted to their improvement was rather an index of the greater need for such facilities, than of any great change in the quality of the roads themselves. In what ways was this greater need shown?

About this time, England was securing a strong foothold in foreign trade, through the establishment of her colonies and the expansion of her empire in different portions of the world. The introduction thence of raw materials, and the supplying of these markets and those of Continental Europe with English manufactured goods, reacted favourably on her domestic industry; and the manufacturers were extending their businesses to take full advantage of the opportunities afforded them for increasing their wealth. The wide markets thus acquired were a preliminary to a further advance in manufacturing; for in order to supply them, goods had to be manufactured on a larger scale than by the old processes of the domestic system. Hence we find that even before 1750 some factories had been established, driven by animal or water power, so as to make the supply of these products more nearly commensurate with the demand[1]. England was awakening to a new industrial era; and to meet the requirements of this new era, it was

Mantoux, *The Industrial Revolution in the 18th Century*, for the development of the factory system. See also Hutton, *History of Derby*, pp. 161-72; Yates, *History of Congleton*, pp. 93-94; Brit. Mus., Add. MSS. 5842, 'Tour Through England in 1735,' p. 267; *Voyage of Don Manoel Gonzales (late Merchant) of the City of Lisbon in Portugal, to Great Britain* (1731), pp. 36-37; Dunsford, *Historical Memoirs of Tiverton*, pp. 216, 298; *Repertory of Arts, Manufactures and Agriculture*, 2nd series, XXXII (1818), pp. 79-83; *Memoirs of the Literary and Philosophical Society of Manchester*, 2nd series, III (1819), pp. 135-7. These references show that silk factories probably began in England about 1715-19, and cotton factories had their inception about 1733.

necessary to extend the facilities for the conveyance of the increasing quantity of raw materials and finished products. If this industrial advance were to continue, the cost of transportation must be reduced and the means of conveyance increased; and partly on account of this pressure, and partly as an accompaniment of it, there came the development of the roads and the construction of canals.

The change which we have noted in the commercial and industrial life of the nation was supplemented by a corresponding change in agriculture. From the middle of the century onward, the enclosure movement was proceeding rapidly, especially after 1760[1]. The enclosures at this time were for the development of arable farming, as distinguished from pasture farming[2]; the new methods of farming were increasing the productivity of the agricultural lands[3]; this increasing productivity required better facilities for communication with markets[4]; and as a result attention was turned to the improvement of the highways of trade, both by land and water. Agriculture, under the old, slow, dangerous and expensive means of conveyance, was rather a means of subsistence to particular families, than a source of great wealth to the

[1] In Anne's reign there were three private Acts for enclosure; in the reign of George I, there were sixteen; under George II, two hundred and twenty-six; and in the reign of George III, from 1760–75, there were seven hundred and thirty-four; from 1776–97, there were eight hundred and five; from 1797–1810, there were nine hundred and fifty-six; and from 1810–20, there were seven hundred and seventy-one. Besides this, there was a general enclosure Act in 1801 (Tooke, i, p. 72; Prothero, *Pioneers and Progress of English Farming*, p. 257).

[2] The enclosure movement which began about the middle of the fifteenth century was carried out to provide large sheep farms, so that in rearing sheep England might have an ample supply of wool for her great woollen industry. But after 1760 the great need was for food products, to furnish with sustenance the increasingly large manufacturing population. Hence the change in the character of the agriculture, from pasture to arable farming.

[3] The system of rotation of crops was coming into predominance over the old, wasteful methods of cultivation. The increasing use of artificial fertilizers led to the improvement of the soil and to the extension of the margin of cultivation by making it possible to bring into cultivation land that would not otherwise have been fertile enough to pay for use. Then, too, the quality and breed of domestic animals were being constantly improved, to provide larger returns in meat products.

[4] We must not suppose, however, that the trade had ceased to be intramunicipal, or intermunicipal, and had become national; for, in reality, a great deal of it was still of the local character that we have already noted. For example, in 1769, Bedfordshire had 11 market-towns; Berkshire, 12; Buckinghamshire, 17; Cheshire, 12; Dorsetshire, 22; Herefordshire, 7; Huntingdonshire, 6; Leicestershire, 11; Lincolnshire, 31; Norfolk, 32; Middlesex, 7, etc. See Brit. Mus. 577. b. 6–10, 'Description of England and Wales.' Of course, we cannot make the number of market towns in any county an exact measure of the amount of local trade that was carried on at these centres; but the fact that so many market towns could flourish shows us that there must have been a large amount of trade at such places.

kingdom; but under the later régime the lands advanced in value, the products of the soil could be shared more equally by all in the kingdom because of the increased facility of reaching markets, and every product of husbandry thus supplied became an article of national value in the support of the large population that was pushing England to the front in industry and trade. In the words of Adam Smith, "Good roads, canals, and navigable rivers, by diminishing the expense of carriage, put the remote parts of the country nearly on a level with those in the neighbourhood of a town: they are, upon that account, the greatest of all improvements."

Notwithstanding all the local prejudices and interests which tended to keep people from taking the wider view, there was gradually growing up a sentiment in favour of freer and easier communication between the different parts of the country. The more enlightened were beginning to see that it was an essential part of the domestic economy of any people, to furnish the means for transporting both passengers and goods at low rates. Easy communication would lessen the time, as well as the cost, of transport; a saving of time would mean a saving of money, and this in its turn would permit the employment of a greater amount of capital. They were coming to realize that many places in the kingdom might be rendered much more valuable if access to them were made easier, for the lowering of the expense of carriage would permit the profitable application of additional labour and capital to all soil under cultivation; and soil that had not as yet been cultivated might be made to pay under tillage if there were easy access to markets. Such conceptions as these were slowly gaining recognition, and showing that a large amount of capital was annually sunk in the transport of marketable commodities, which was not only an unproductive outlay, and consequently loss, to the seller, but also a burden upon the buyer in increasing the cost of every article of daily consumption.

From about the middle of the eighteenth century, therefore, we notice an increased interest taken in the roads, and, consequently, we have many more suggestions than before for their improvement[1]. To mention them all would be to show the inventiveness of minds that were thinking along this line, but it would also show the impracticability of some schemes put forward as a solution of the difficulty. The

---

[1] See, for example, Shapleigh on *Highways*, and Brit. Mus. 8776. c. 21, 'Proposals for the Amendment of Roads,' the former of which is summarized in Appendix 2. Also *Gentlemen's Magazine*, xxxiii, pp. 288–90; xxvii, pp. 404–5, 585–6; lv, pp. 168, 194, 254, 255, etc.; Marshall, *Rural Economy of the Midlands*, i, pp. 44–47; Marshall, *Rural Economy of Yorkshire*, i. pp. 180–9; Gentleman, *Proposals for Amendment of Roads*, p. 8 et seq.

promoter of each plan, of course, presented the matter from his standpoint alone, and we do not get a complete view of the situation regarding
the roads without taking into account the proposals of many writers.
One, in 1749, referring to the highways, said "that it has always been
found by experience, that the many laws, which have been hitherto
made concerning their repairs, have never met with the desired success;
hence there must be some fundamental error in these laws......"
He thought the fundamental error was in permitting parishes, towns,
or other small subdivisions, to be presented or indicted for not repairing
their roads; and besides doing away with such prosecutions, through
a change in the law, he would modify the latter so as to compel the
surveyors and the parishioners to do their statute duty[1]. He did not
think it just that the parish should be forced to bear the burden of
such proceedings, when the default was due to the surveyor; and the
whole system of indictments was erroneous, as shown by experience,
for such legal action seldom, if ever, answered the end intended[2]. The
negligence shown in the performance of the statute labour was apparently
one great reason why the roads were so unsatisfactory; for he says
that "The six days work have hitherto in most parishes been so much
neglected, and so slightly performed, that I believe, very few parishes
can truly say, from their own experience, that the six days work, duly
and properly attended to, and performed by all the parishioners, liable
by law, to work in the amendment of the highways, with due care, and
diligence, are not sufficient[3]."

A writer in 1753 urged that the chief means that should be employed
to improve the roads and then maintain them in good repair, was the
use of broad wheels on the vehicles used for conveying heavy loads[4].
He did not impute any blame to the Commissioners of the turnpike
roads, but it was his opinion that much larger sums than they had to
spend on the roads would not be sufficient, without some further
regulations. The narrow wheels, and the uncertain breadth or distance
between the wheels, cut up the roads and rendered them impassable.
He recommended that the wheels of the above-mentioned vehicles
should be at least nine inches wide, a provision which was incorporated
into the Act of that year[5]. He would have the ruts and holes in the
roads filled up and levelled before putting this proposed law into effect,
and when the roads had been put into good condition he was certain

---

[1] Shapleigh on *Highways*, pp. 4–6. See Appendix 2.
[2] Ibid., pp. 9–17. See Appendix 2.
[3] Ibid,, p. 56.
[4] Brit. Mus. 8776. c. 21, 'Proposals for the Amendment of Roads.'
[5] Act 26 Geo. II, c. 30.

that the broad wheels would·keep them so. His proposal that the roads should be widened to at least eighteen feet, so as to be wide enough for two carriages, while leaving room for foot-passengers, shows us that former Acts for widening the highways must have proved largely ineffectual[1]. The Act of 1662 had provided that the roads should be widened to eight yards from ditch to ditch, but this reformer is more modest in his stipulations, and would prefer a somewhat narrower road, kept in good repair. Then, too, in his suggestion that a surveyor-general should be appointed in every county, to direct the repairs of all highways in the county, and that he should receive a yearly salary and his travelling expenses, we have a decided step in advance in the way of suggestions for improved administration of roads. Under the system in force up to that time, the surveyors were appointed for each parish, and according to the enterprise and executive ability of the surveyor, the roads of that parish were well or ill repaired. But there had been no one to take thought for the roads over a wider area, and consequently there had been great diversity in the quality of these roads. Had there been a surveyor-general for each county, who would receive a regular salary, together with travelling expenses, he could have spent all his time in the performance of this public duty, could have looked after this work for the public good without having to stand all the expenses himself (as was required of the parish surveyors), and thus the services of a competent man could have been commanded and the work satisfactorily done. This suggestion, however, was not put into effect, and as late as 1817 Edgeworth complains that "it is in vain that one parish repairs its roads, if its neighbours will not do the same[2]."

From about the middle of the eighteenth century, when attention began seriously to be turned toward improvement of the highways, the utility of statute labour for the purpose for which it was intended had been greatly questioned. It was a fact too well known to be disputed that those who were liable for statute duty were negligent in the performance of their obligation, and, as we have just seen, the surveyors were not always able or willing to enforce the law's demands. The days for doing the work on the public highways had long been looked upon as holidays, as a kind of respite from accustomed labour, to be devoted to idleness[3], and consequently the public was defrauded by

---

[1] Acts 14 Car. II, c. 6 (1662), and 8 and 9 W. III, c. 16 (1696–7).

[2] Edgeworth on *Roads and Carriages*, p. 6. See also Brit. Mus., T. 1157 (4), 'Highways improved,' p. 12; *Parl. Papers*, 1772, xxxi, Rept. No. 12.

[3] Hawkins, *Laws of Highways*, p. 27. Burn, *History of the Poor Laws*, p. 239, refers to the impossibility of having the roads improved "under the care of those

such evasion of the law. Further, in the fulfilment of what work was done, the men of the parish were working at four or five different places, and therefore were not under the oversight of the surveyor, nor were they executing a well thought out plan. On account of these and other objectionable features of the system, many urged the abolition of statute duty entirely[1]; and a committee of the House of Commons, appointed in 1763 to report on the general laws for amending the roads, reported that, in their judgment, it would be better to repair the roads by means of assessments than by the six days' labour[2]. But notwithstanding the strong opposition to this method of repairing the roads, which had been the prevalent means for attaining that object during two centuries, it was destined to last for over three-quarters of a century longer before its abolition was brought about[3].

Since it is not our purpose to follow out all the proposals for improving the roads, in the early part of the period we are now considering[4], we turn to discuss the nature of the legislation that was actually in force to secure better roads. We have already traced this up to the year 1755, and have shown the encouragement that was given to extend the use of nine-inch wheels by those who were carrying heavy loads

spiritless, ignorant, lazy, sauntering people, called surveyors of the highways." See also Murray, *Agriculture of Warwick* (1813), p. 172; Crutchley, *Agriculture of Rutland*, p. 21; James and Malcolm, *Agriculture of Surrey*, pp. 62–64; Holland, *Agriculture of Cheshire*, p. 304; etc.

[1] *J., H. of C.*, xxx, p. 608. On this opposition to statute labour, see also *Gentleman's Magazine*, xxxiii, pp. 288–90 (1763), where it is stated that by abolishing the statute labour and substituting a pound rate, "more work could be done with three hired teams than with five statute teams, and more with five hired labourers than twenty others." In the headings of a Bill as drawn up in 1757 for the more equal and effective repair of the highways, it was proposed that all statute labour in the kingdom should cease, but that the surveyors should assess the parishioners according to a specified rate and use the money thus obtained for the work of improvement (*Gentleman's Magazine*, xxvii, pp. 404–5). In the same year, among the chief remedies which were proposed by another, was the abolition of statute labour, and the proper spending of the money obtained from the turnpikes (*Gentleman's Magazine*, xxvii, pp. 585–6). See also the references under footnote 3, p. 215, urging the abolition of the old system of statute labour.

[2] See first references under footnote 1 above.

[3] This was effected in 1835.

[4] A writer in the *Gentleman's Magazine*, xxvii, pp. 585–6, in speaking of the bad state of the Sussex roads, suggested, among other things, draining the soil before throwing up the bank for the road, and widening the roads, by taking in more ground, so that there would be more open space in the roadway, thus allowing the sun and wind full scope for drying the roads. This certainly would have been an important step in the case of most roads. A somewhat visionary 'Method to keep the roads good all the year round' is given in the *Gentleman's Magazine*, xix, pp. 218–19.

upon the roads. Some parts of this Act were altered two years later, but always in favour of the wide wheels and against the narrower wheels[1]. In spite of the benefits expected from these broad wheels, new difficulties arose from their introduction. On account of their additional strength, they were calculated to bear considerably heavier weights; and it was found necessary to limit their size and the width of their wheels apart. Accordingly, all waggons were prohibited having the wheels wider apart than five feet six inches from the middle of the fellies of the wheels on each side; and they were also directed to be drawn by the horses in pairs, although carriages with narrower wheels were to have the team at length.

In 1765 a new statute repealed the exemption from half toll within one hundred miles of London in Act 30 Geo. II, c. 28, and from over-weight in Act 26 Geo II, c. 23[2], and these exemptions were only to be allowed to waggons and carts with wheels constructed according to the following plan, namely: All waggons and carts employed in carrying were to have the axle-trees of such different lengths that the distance from wheel to wheel of one pair of the wheels should not be more than four feet two inches, and that the distance of the other pair should be such that the fore and hind wheels on each side of the waggon should roll a surface of at least sixteen inches in width, and the wheels should be nine inches wide. On waggons thus constructed a weight of six tons might be drawn, and on carts of like construction, three tons. For all overweight, the extra toll was the same as in former Acts, 20s. per hundredweight. But waggons built in this way, and especially when they were allowed to carry such heavy loads, would work havoc to the roads. When the latter were at all soft, as they too frequently were, the heavy weight on the front wheels would cause the waggons to sink into the soft soil of the roadway and displace it; and then when the hind wheels passed along they simply displaced this material still more, thus making the ruts in the roads wider and deeper. Consequently, in the following year, further regulations were made as to these sixteen-inch rollers, by which they were to be so constructed as to roll only one single surface, or path, of sixteen inches wide on each side of the waggon; and the exemptions from toll and overweight were not to be allowed where two different surfaces, or paths, were rolled in order to make up that width[3]. To discourage narrow wheels, the trustees of turnpike roads were directed to issue orders to their toll collectors not to allow any waggon or other

---

[1] Act 30 Geo. II, c. 28 reduced by one-half the toll on carriages with nine-inch wheels, within a radius of one hundred miles from London.

[2] Act 5 Geo. III, c. 38.

[3] Act 6 Geo. III, c. 43.

four-wheeled carriage, with wheels less than nine inches wide, to pass through any toll-gate, drawn by more than four horses, without seizing one of the horses.

Throughout the eighteenth century, up to this time, we have noted the great amount of legislation in regard to the roads, all aiming to limit excessive weights and to regulate the construction of waggons, but without securing the objects desired. These laws had been made at different times, as need required, and when taken in connexion with the great Highway Statute of 1555 and those which had been passed since that time, the result was a multiplicity of laws abounding in clauses which legal skill could not reconcile. Clauses in the older statutes had been left unrepealed, although these clauses had been altered and amended by subsequent Acts; different penalties had been inflicted, by different statutes, for the same transgression; and in these and other ways the highway laws had so accumulated as to be a subject of universal complaint[1]. Since the passage of the Act of 1555[2], the use of coaches, chaises, post-chaises, and similar vehicles, had become general; and the gentlemen who owned these got off with the same amount of road work as the poor cottager who had none of them, although the former used the roads much more than the latter. On account of the changing valuations since that law was made, it was often advantageous for the farmer to refuse to do the statute labour; for if he were to let out the cart, team, and men to a neighbour, instead of sending them to the roads, he would save the labour of one man each day by incurring the forfeiture imposed on account of his refusal[3]. With a statute of this kind, which furnished a strong motive for its disobedience, it is evident that non-compliance with its statute labour requirement would be very general. The above inconsistencies in the highway laws, and the difficulties in the matter of interpreting the laws on a uniform and reasonable basis[4], led to a vigorous agitation to have all these laws that were effective comprehended in one Act which would be clear, simple and definite.

Since there were now two distinct systems, the highways and the turnpikes, in 1767 two Acts were passed, one relating to each system, and by these all foregoing Acts were repealed[5]. This General Highway

---

[1] Hawkins, *Laws of Highways*, p. 43. Hawkins was a Justice of the Peace of the county of Middlesex, and was well posted in the matter of the actual way in which the laws worked out. I have, therefore, given in Appendix 3 a summary of his views.

[2] Act 2 and 3 P. & M., c. 8.

[3] Hawkins, *Laws of Highways*, pp. 24–25.

[4] See Appendix 3.

[5] Act 7 Geo. III, c. 40, entitled 'An Act to explain, amend, and reduce into one

Act, while it was an effort in the right direction, was not found to be exactly what was wanted, and in 1773 we have the passage of another to amend and explain the former laws[1]. As we should naturally expect, those that were regarded as the good features of the previous laws were retained in this, with the necessary modifications; and it was sought to entirely remove the discrepancies which had existed because of the many statutes. As in former Acts, provision was made that no trees should be allowed to grow in the highway[2]; that landowners should cut and prune their hedges so as not to shade the road; that they should keep the ditches open so as to easily carry off the water from the road; and that no nuisance should be left or placed in the highway[3]. In accordance with the desire of many for wider roads, it was provided that cartways to market towns should be made at least twenty feet wide, and every horseway at least eight feet wide, if the ground between the fences would admit of this; and Justices might order roads that were too narrow to be widened, but not to exceed thirty feet when widened[4].

Regarding the use of wide wheels, this Act followed the consensus of opinion at that time, and gave a preference to these over the narrow wheels. Hawkins, one of the Justices of the Peace for the county of Middlesex, who had had much experience in road matters, in 1763 said that the advantages from the use of broad wheels were so apparent that it was needless to insist on them; and that by this means the price

Act of Parliament, the general laws now in being for regulating the turnpike roads of this kingdom,' repealed the former turnpike Acts, except as to some provisions; but Act 7 Geo. III, c. 42, entitled 'An Act to explain, amend, and reduce into one Act of Parliament, the several statutes now in being for the amendment and preservation of the public highways of this kingdom, and for other purposes therein mentioned,' had to do with the highways which were not under turnpike laws.

[1] Act 13 Geo. III, c. 78.

[2] Ibid., section 6, decreed that no tree, bush or shrub was to be allowed to grow in the highway within fifteen feet of the centre of the highway (except in special cases). The owner of the land was to have such things removed.

[3] No person was to lay in the highway any straw, timber, dung, or other matter; and anyone leaving his waggon, plough, etc., in the highway beyond a reasonable time was to pay 10s. for stopping up the free passage (ibid., secs. 9, 11).

Surveyors, at proper times, were to view the highways and if they saw any nuisance or obstruction, they were to give notice to those doing or allowing the same. If the offender did not remove the nuisance in twenty days after the notice, the surveyor was to do it, and the offender to pay the charges of it (ibid., sec. 12).

[4] The desire that the roads be widened, and that all trees and shrubs along them be cut down, was, in part, of course, to allow the natural agencies of sun and wind to exert their influence in drying the road-bed. This was specially recommended as a remedial measure for the Sussex roads (*Gentleman's Magazine*, XXVII, pp. 585–6).

of carriage from York to London had been reduced forty per cent.[1]. With such important testimony from men of high position, it is not surprising that the broad wheels were favoured by legislative enactment. The Act of 1773 made another provision for broad wheels, by declaring that although a waggon with wheels six inches broad could not be drawn with more than six horses, yet if such a waggon with wheels six inches broad were made to roll on each side a surface of nine inches, it could be drawn by seven horses[2].

Concerning statute duty, this Act made one important change in that a more equitable amount was apportioned to each class than had been by former Acts. Landowners were required to perform their statute labour according to the yearly value of the lands they occupied. Those who did not work the land, but who were engaged as carriers with their horses and carts on the roads, were required to perform slightly less road work than the former class. Those who had not lands of a minimum value of £50 a year and did not have any heavy draught work, but kept a light carriage for driving on the road, were allowed to pay still less, in lieu of statute duty. Those who were mere cottagers, and occupied lands of a less valuation than £4 annually, were required to give nothing but their own labour[3]. In accordance with a practice

---

[1] Hawkins on *Highways*, p. 52. If this was so evident to a Justice of the Peace, it would be evident to all who had anything to do with the roads; and if the broad wheels were regarded as the cause of the decrease of the price of carriage, it is no wonder that the law of 1773 provided for their use.

[2] The other provisions regarding the use of broad wheels were practically the same as those of the Act of 1755 (Act 28 Geo. II, c. 17), namely: No waggon with wheels nine inches wide should be drawn with more than eight horses; no waggon with wheels six inches wide and giving a rolling surface of six inches should be drawn with more than six horses. No cart with six-inch wheels should be drawn with more than four horses. An exception was, of course, made for farmers' waggons, which, even with narrow wheels, might be drawn by five horses. With the wide wheels, the more horses that could be used the easier would be the draught or the greater the load that could be drawn.

[3] The provisions with reference to statute duty are found in section 34 of the Act. It was to be required according to the following scheme: Every person keeping a waggon, cart, wain, plough, or tumbrel, and three or more horses to draw the same, was regarded as keeping a team, draught, or plough, and was required for six days each year to send one wain, cart or carriage, with horses or other beasts of burden, and also two able men with such cart or wain. This was for persons with lands not exceeding the yearly value of £50. Every person keeping such a team, draught or plough, and occupying lands etc., of the yearly value of £50 beyond the yearly value of £50 in respect of which team duty should be performed; and every such person occupying lands, etc., of the yearly value of £50 in any other parish or place, besides that where he resided; and every other person not keeping a team, draught or plough, but occupying lands, etc., of the yearly value of £50 in any parish or place; was required in like manner, and for the same number of

that already prevailed, contrary to the law, those who were liable for statute duty might compound with the surveyor, if they so desired, and the payment of the money equivalent would discharge them from the performance of the work[1]. This new adjustment of the statute duty was probably effected as the result of a continued agitation for some such change, for it had long been felt that many who used the roads most, by driving on them with their coaches, escaped with very little obligation to repair them; while the poor who seldom used them were compelled to contribute an unduly large amount for their maintenance[2]. After the money had been properly spent and statute duty performed, if the highways were still out of repair, the Justices might call for an assessment upon the parish[3].

Another new feature of the Act of 1773 was that Justices of the Peace, with the consent of the landowners, were authorized to turn the highways, so as to make them more commodious, and to stop up, enclose and sell the soil of any old highways that were unnecessary. In this, we see the desire to have the important roads improved as

days, to send one wain, cart, or waggon, furnished with not less than three horses (or their equivalent in horses and oxen) and two able men to each wain, cart, or waggon. And in like manner, for every £50 per year respectively, which every such person should further occupy in any such parish or place, such wains, carts, or waggons to be employed in repairing the highways in the parish or place where such lands were situated. (Then it gives the regulations for those with other possessions.)

Section 35 (as altered by Act 34 Geo. III, c. 74) is as follows: Every person who shall not keep a team, draught or plough, but shall keep one or more cart or carts, and one or two horses or beasts of draught only, used to draw in such carts upon the highways, shall be obliged to perform his statute duty for the like number of days (6) with such carts, and horses or beasts of draught, and one labourer to attend each cart; or to pay money according to a specified scheme. Every person who keeps a coach, post-chaise, chair or wheel carriage, and does not keep a team, draught or plough, nor occupy lands of the annual value of £50, shall pay the surveyor 1s. in respect of every such day's statute duty, for every horse which he shall draw in such carriage, or else pay money according to the value of his lands, etc., which he occupies. Personal labour to be required from those who occupy lands of less than £4 annual value.. For the other regulations regarding statute duty, see the Act.

[1] See section 41 of the Act. This practice of permitting commutation seems to have been already employed by surveyors as an easy way out of a difficulty, for when there was an aversion to the performance of statute duty, the surveyors would accept a money payment instead of the work. A writer in the *Gentleman's Magazine* (May 1763, pp. 236–7), says: "Custom has......in many places converted the labour into a rate as the law now stands" although it was done illegally. See also Hawkins on *Highways*, p. 38.

[2] Hawkins on *Highways*, p. 2 et seq. This adjustment of the statute duty was one of the great things for which Hawkins pleaded (ibid., p. 47). See Appendix 8.

[3] See section 45 of the Act.

much as possible, so as to be sufficient for the largely increasing trade that was to be carried on them, and at the same time to cut off the expenditure of money and labour for roads that did not seem to be justifiable. It was for the purpose of utilizing as effectively as possible the means that were available, so as to secure the greatest public good. Some of the cross-roads had, doubtless, been laid out before there was much traffic, merely as a temporary accommodation; but to keep these open now, when there was little need for them on account of the changes of trade routes, and when the highways of greatest trade were still in need of improvement, would have been a wasteful policy[1]. Besides having authority to close up roads that were no longer required, Justices at their regular Sessions might order to be first repaired those roads that needed it most[2].

We shall not enter further into the details of this statute, for we are already familiar with the other effective provisions of former Acts which we would expect to find in this[3]. The new law was required to make definite what had been subject to confusion through the interpretation and administration of a multitude of former laws, and also to make the law reflect the best practice and the utmost justice consonant with the existing stage of progress; in other words, to make the law suit the conditions of the time. By 1794, another Act was found to be necessary for regulating the amount of statute labour to be performed by each person; and also showing who were to be exempt from, and who were to pay for, statute labour[4]. In passing, we may note that, in order to facilitate the postal service, statutory authority, in 1785,

---

[1] Hills, *History of East Grinstead*, p. 152.

[2] Section 25 of the Act. By the common law, an ancient highway could not be changed without the King's license, which had first to be obtained through a jury of the district under a writ of *ad quod damnum*.

[3] The Act of 1773 changed the method of appointing the parish surveyors. On Sept. 22 in each year, a list was to be made out of at least ten persons, at the usual place of public meetings, to be surveyors of the highways. The surveyor was required to have an estate in lands or tenements, in such parish or place, in his own right or that of his wife, of the yearly value of £10, or a personal estate of the value of £100; or else he had to be occupier or tenant of a house, land, or tenement of the yearly value of £30. The list of these ten persons was sent to the Justices of the county, and at their next sessions, if these persons were qualified, the Justices appointed them as surveyors. Each surveyor was to hold office for one year.

As we have seen, Shapleigh and others had shown the need for a change in the choice of surveyors, as well as in the execution of their work. The other and greatest objection which Shapleigh had to the existing highway law administration, was that the system of presentments or indictments was expensive and unproductive of good. Still the system continued to prevail, as we see by section 24 of this Act.

[4] Act 34 Geo. III, c. 74.

exempted vehicles carrying the mail from paying tolls at any turnpike gates in Great Britain, and the latter were to be open for the passage of the mail coaches, so that no delays might be encountered[1].

But while we are speaking of the highways, we must not omit to speak also of the turnpike roads. The much larger number of these which was authorized after the middle of the eighteenth century necessitated that all the laws which were in force for their regulation should be reduced into one Act, so as to be more clearly understood and more easily administered[2]. To this end a General Turnpike Act was passed in 1773, incorporating into one law the provisions that were applicable to all turnpike roads, and leaving to be inserted in each separate turnpike Act the special provisions that applied to that road only[3]. By this statute, for the first time, the turnpike roads were also put under Sessional control[4]. Probably the chief reason for this change was to have other than merely local interests directing the improvement of these roads, so that, if possible, improvement of longer lines of road might be obtained by giving the one body of men (the Justices) jurisdiction over connected groups of turnpike roads.

Some of the provisions of this Act may profitably be considered. In the first place, the stipulation of the Act of 1751[5], requiring that weighing engines should be erected and an additional toll of 20*s.* per hundredweight taken on everything over sixty hundredweight, was now made optional[6]. These weighing engines appear to have been the cause of much trouble, for some of the toll collectors were allowing

---

[1] Act 25 Geo. III, c. 57. It will be noted that this measure was passed in the year following the establishment of Palmer's system of mail coaches, which introduced a new era in the speed with which the mails were transmitted throughout the kingdom.

[2] It will be evident, therefore, that this movement for consolidating the laws governing turnpikes, was in harmony with the movement for consolidating the laws of highways.

[3] Act 13 Geo. III, c. 84.

[4] The highways had always been under control of the Justices of the Peace, during the period we have under consideration, that is, since the accession of the Tudors. But the turnpike roads had each been under the control of the trustees appointed by Parliament for that road; and it was only when legal action became necessary that any case was brought before the Justices. Now, however, while trustees were still appointed to have charge of the collection and application of the funds for each particular road, they were to act under the direction of the Justices for that county.

[5] Act 24 Geo. II, c. 43.

[6] As it had been under the Act of 1741 (14 Geo. II, c. 42). Section 1 of the Act of 1773 gives a schedule of the weights allowed to be drawn at different seasons and by different numbers of horses. The following is a copy of the notice as to

certain of those who were using the roads to carry excessive weights, without paying the additional toll. To detect this connivance of the collectors, Justices might cause waggons to be weighed; and, upon complaint, they might order these weighing engines to be erected at any places that they might think proper. The penalty of £5 imposed by former Acts upon anyone who unloaded before coming to a gate or weighing engine was still retained; and owners or drivers of waggons who turned out of the road to avoid being weighed were to forfeit not less than forty shillings nor more than five pounds. If such arrangements for weighing and taking extraordinary tolls were found to be productive of good, it is difficult to understand why they should not be compulsory on all the turnpike roads. Possibly the explanation is that some roads were good and sufficiently firm to carry heavy loads, while others were bad, and needed the increased amount of tolls in order to effect the desired improvement.

We have noted in several Acts before this time that much emphasis was laid upon securing the use of broad wheels, and this Act was no exception; for while it granted special privileges as to the number of horses that could be used and the amount of toll paid when waggons and carts had wheels six inches or nine inches wide[1], it also charged those using narrow wheels one-half more than the regular tolls, and after

weights posted on the East Grinstead toll-gate, about 1776 (Hills, *History of East Grinstead*, p. 156):

"Table of Weights Allowed in Winter and Summer (including the Carriage and Loading).

	Summer tons cwt.		Winter tons cwt.	
To every Waggon upon Rollers, of the breadth of sixteen inches	8	0	7	0
To every Waggon with nine-inch wheels, rolling a surface of sixteen inches on each side	6	10	6	0
To every Waggon with nine-inch Wheels	6	0	5	10
,, ,, Cart ,, ,, ,, ,, ...	3	0	2	15
,, ,, Waggon ,, six-inch ,, ...	4	5	3	15
To every Waggon with six-inch Wheels rolling a surface of eleven inches	5	10	5	0
To every Cart with six-inch Wheels	2	12	2	7
To every Waggon with Wheels of less breadth than six inches	3	10	3	0
To every Cart with Wheels of less breadth than six inches	1	10	1	7"

[1] Section 13 of the Act said that no nine-inch four-wheeled waggon was to be drawn by more than eight horses, nor two-wheeled carriage of the same width of wheel by more than five horses, and then in pairs. A six-inch four-wheeled carriage was not to be drawn by more than six horses, nor two-wheeled carriage of the same width of wheel by more than four horses, etc. But section 16 of the Act declared that waggons and carts might be drawn by any number of horses, when an engine was erected and the carriage weighed.

three years they were to pay double the ordinary tolls[1]. But there were wheels broader even than nine inches, for some waggons had been made to go on low rollers, thirteen to sixteen inches wide. Sufficient attention had been devoted to the latter, that, in order to encourage their greater employment, in this Act of 1773 it was enacted that carriages moving on rollers sixteen inches wide on each side, with flat surfaces, might be drawn with any number of horses, and might pass toll-free for one year, after which they were to pay only one-half the regular toll[2]. But despite the legislative encouragement that was

---

[1] Section 23 of the Act. Of course, these provisions were not to extend to waggons used in husbandry. Where extraordinary tolls were granted, the trustees might reduce them for those using four-wheeled waggons with wheels at least six inches wide. No composition for tolls was allowed unless the waggon wheels were six or more inches in width.

In 1776, a petition from several farmers, landowners and others, was sent to the House of Commons showing that by the Act of 1773 it was enacted that from and after Sept. 29, 1776, all waggons and carriages having fellies of wheels less than six inches wide were to pay double tolls for passing through any turnpike gate; and that the tires of all wheels of such waggons used on turnpike roads were to be countersunk so that the nails should not rise above the surface. The petition showed also that many farmers and others lived some distance from any turnpike road, and being obliged to drive through narrow lanes and over moors for several miles before reaching a turnpike road, which lanes and moors were of such soil that six-inch wheels could not be used, it would be impossible for many to get to market except with narrow-wheeled waggons and carts. Then, too, the heavy toll laid on narrow-wheeled carriages was so excessive that many would be driven to the greatest distress; and the additional expense to get their tires countersunk would be too heavy a burden unless further time were allowed to let the present tires wear out. Then having presented their case, they asked for relief (*J., H. of C.,* xxxv, p. 725). The House took the matter into consideration and made provision for suspending for a limited time this part of the Act which was complained of (*J., H. of C.,* xxxv, May 10, 1776). The double tolls to be imposed by Act 13 Geo. III, c. 84, on narrow-wheeled carriages, were suspended by Act 16 Geo. III, c. 39, and finally repealed by Act 18 Geo. III, c. 28.

[2] On the subject of waggons moving on rollers, see the report from the Committee of the House of Commons appointed in 1772 to consider the breadth of wheels used on turnpike roads, in which there was a resolution that waggons and carts having rollers sixteen inches wide should be encouraged, by exemption from payment of toll for three years (*J., H. of C.,* Feb. 26, 1772). See also Bourn, *Treatise on Wheeled Carriages,* showing the advantages of such broad wheels. This was answered by Jacob, in his *Observations on the Structure and Draught of Wheel Carriages,* and the latter brought out a response from Bourn, in his *Remarks on Mr Jacob's Treatise on Wheel Carriages.* Gentleman, *Proposals for Amendment of Roads,* p. 52 et seq., answers some objections to the use of broad wheels, and shows the ways in which narrow wheels are objectionable. In Brit. Mus. 213. i. 3 (100), 'Considerations about the Method of Preserving the Public Roads,' the writer goes into an elaborate argument to show how much better the wide wheels would be. In Brit. Mus. 213. i. 3 (101), we have 'Reasons against a Bill for permitting only Carriages with Broad Wheels, and those drawn by two Horses, to pass on Turnpike Roads, with regard

given to such waggons, their mechanical construction was detrimental to their general use.

We have previously noted that in the execution of the public trust imposed by turnpike Acts, there had been frequent misapplication of funds and sometimes great fraud by the officials. To obviate such things, it was thought that if men of wealth were put into office, there would be less temptation to dishonesty and a stronger desire to serve the public welfare. Consequently, in the Act of 1778 the stipulation was continued that, when the special Act did not fix the qualification of trustee, no person should be qualified to act in that capacity unless he were, in his own right or his wife's, in actual possession or receipt of the rents and profits of lands and tenements of the clear yearly value of £40; or unless he were possessed of, or entitled to, a personal estate valued at £800; or unless he were heir apparent of a person whose estate yielded a clear annual value of £80. To further safeguard this office, the law forbade any person keeping a public house from being a trustee, or from holding a place of profit[1]. This was in harmony with what had been enacted by the law of 1753; and probably this prohibition had been found advantageous after twenty years' trial, or it would not have been once more affirmed in this general turnpike law[2].

With the other provisions of the Act we need not be long detained. On some waggons and carts with wide wheels, the tires had been so constructed that there was a central band around the outside of the wheel, which formed a distinct projection. In this way, the wheel, instead of rolling on a flat surface nine inches wide, was really rolling on a surface only two or three inches wide (the width of this band), while at the same time the waggon was drawn by the maximum number of horses and carried the heaviest loads. To avoid the tearing up of the roads by this means, the Act of 1778 required all tires to be flat, and the nails countersunk, so as not to rise above the surface[3]. Trustees

to the Countries within Twenty-five or Thirty Miles of London,' which is an argument against putting into immediate effect the proposed legislation as to broad wheels. Wheelwrights objected to the broad wheels because, they said, they could not find timber enough for them, and because their hearths would not enable them to make the wide iron bands to go around such wheels. Wyatt MSS. (1761), ii, pp. 81–87, in Birmingham Free Central Library, Nos. 93,189 and 93,190.

[1] The reason for this provision will be apparent to those who remember what was stated in an earlier paragraph concerning innkeepers.

[2] By the law of 1753 (26 Geo. II, c. 30) the innkeepers were forbidden even to farm the tolls; but the Act of 1778 allowed them this privilege.

[3] Act 16 Geo. III, c. 39 repealed the clause compelling the countersinking of the nails of the tire, and explained the meaning of the sole of the wheel being flat, viz., that all wheels of 6 inches or more which did not deviate more than 1 inch from a flat surface would be regarded as flat.

were not to compound for tolls unless the waggons and carts had wheels at least six inches wide. All statute duty required by the turnpike Acts, and the compositions arising from the same, were to be performed and expended where they originated. Justices might farm out the tolls under particular directions and restrictions to the best bidder; and the last bidder was to be the farmer of the tolls[1]. In order the more readily to identify possible or actual transgressors against the law, the full name of the owner of every waggon, cart, etc., and also of every coach used for hire, was to be painted upon the sides of the vehicles. It must be borne in mind that the foregoing terms were applicable to all the turnpikes; the special provisions for each separate road were inserted in the special Act for that road.

These two Acts of 1773, with a few changes made in them by subsequent Acts[2], before the beginning of the nineteenth century, remained the basis of the highway law until well on into that century.

In the early years of the nineteenth century, there was apparently a much deeper interest taken in the roads, and year after year committees were at work investigating the whole system of administration and operation. The problems pertaining to the use of broad and narrow wheels, weighing engines, cylindrical and conical wheels, the relative advantages of statute labour and assessments, proposed changes in the appointment of surveyors, the consolidation of turnpike trusts, and, in fact, everything connected with the improvement of roads, came in for extended consideration. Evidence was taken from all sources which promised to yield valuable information, and many separate papers were handed in by the experts of the time, dealing with various phases of this national issue. Gradually, public opinion crystallized along some of these lines, so that there came to be a well-defined body of sentiment in favour of certain proposed remedies; but in the period prior to 1830 not many of these secured recognition upon the statute book.

In addition to the investigation of the topics noted above, the only important work done by Parliament before 1830 was a new codification

---

[1] This Act gave the Justices authority to farm the tolls, where previously this had been in the hands of the trustees. In reality, the trustees still performed this service in most cases (see, for instance, *Hereford Journal*, Apr. 20, 1803, p. 1; Dec. 4, 1805, p. 2, etc.).

[2] Note, for example, the changes made by Acts 25 Geo. III, c. 57, and 34 Geo. III, c. 84. The most important of these were changes in the statute labour requirements, in the rates that were allowed to be charged for hauling on roads, in the weights that might be drawn on the roads in summer and winter, and in the number of horses that might be used with each kind of vehicle.

of the laws pertaining to the roads. As we noted previously that there was a strong agitation before 1778 for reducing the various laws for highways and turnpike roads into one general law for each, so we may say here that, from 1810 on, there was recognized the necessity of again combining the old and new regulations into one general code, so that a clear-cut presentation of the laws for turnpike roads might be open to all who had to do with these[1]. Accordingly, after many attempts to frame a suitable statement of the law, a Bill was prepared in conformity with the plan recommended by the Committee of 1821, and this was finally enacted by both Houses in 1822[2]. The objects of the Act may be briefly summarized as: first, to embody the former Acts in one; second, to try to establish one uniform system of law applicable to turnpike roads; third, the encouragement of the use of carriages of a construction less injurious to the roads; fourth, the regulation of the officers of the trusts, and of the lessees and collectors of tolls; fifth, the checking of extravagant expenditure of the funds by providing for the proper keeping and publishing of the accounts; sixth, the reduction of the expense of passing the local Acts, and the curtailing of their length, by rendering unnecessary the insertion in them of those clauses that were applicable to every such road. This Act, as amended in the following year[3], presents to us the substance of the final turnpike legislation during the period we have under consideration.

A short analysis of the chief provisions of this Act may here be given, that we may see what changes, if any, were made under the additional light secured by the prolonged study of the needs of the roads in the first twenty years of the nineteenth century. It must not be thought, however, that all those improvements which by many were deemed desirable were enacted into law, for this supposition would be entirely erroneous. On the contrary, many which seemed to be of prime importance secured no recognition in this statute; and, on the whole,

---

[1] *Parl. Papers,* 1810–11, III, 855; 1819, V, 339, etc. The Committee of 1819 took up the recommendation of the Report of 1811, in favour of a new Act to combine the laws into one code for highways and another for turnpikes, and said "that unless this task, however arduous, be accomplished, the laws relating to roads must remain in an incomplete, uncertain and inconvenient state." This recommendation, so far as turnpike roads were concerned, was carried into effect by Act 3 Geo. IV, c. 126; but the consolidation of the Highway Laws was delayed till 1835, when Act 5 and 6 W. IV, c. 50 was passed.

[2] Act 3 Geo. IV, c. 126. It repealed all former Acts, and was to take effect at the beginning of the year 1823.

[3] Act 4 Geo. IV, c. 16 (1823). A few minor changes were also made by Act 7 and 8 Geo. IV, c. 24 (1827), but these were technical and legal trifles, rather than features of importance, and did not materially affect the Acts of 3 Geo. IV, and 4 Geo. IV, that were in force.

it would seem as if this law of 1823 were largely a compilation of previous laws, rather than the reflection of advancing thought concerning the basal principles of road maintenance and administration.

The Act of 1822 discouraged the use of narrow wheels, by enacting that, after Jan. 1, 1826, no waggon or cart with wheels of less breadth than three inches should be used on any turnpike road. This was one of the important provisions of the Act; but because of a great cry of the farmers, the clause was repealed in the Act of 1823. The adherence of this class to the old abuses was, apparently, a considerable obstacle to the improvement of roads; for if a provision considered beneficial to the roads was "an injury to farmers and others," it was very likely to be repealed. The two chief objections urged against this clause in the Act of 1822 were, first, the expense to which it would put the farmer in getting new wheels during this time of agricultural distress, and, second, the impossibility of broad wheels being used in many of the narrow parish highways. As to the first, this was not an expense imposed upon him suddenly, for three years were to be given before this clause would come into effect, and in that time provision might be made for it; besides, it seems to have been totally overlooked by the farmers that the sooner they put away their narrow wheels the sooner would their burden of road repairs be lightened. As to the second objection, it is almost inconceivable that putting on wheels of three or four inches in breadth would prevent a waggon from being drawn on any road on which it could then pass with its narrow wheels. But even should this have been the case, there were statutory provisions for widening roads that were too narrow, and the wider roads would certainly have been 'a great benefit.

An important provision of the Act of 1822 was that which directed that, after the beginning of the year 1826, all wheels of waggons and other carriages to be used on turnpike roads should be constructed so as not to deviate more than one-half inch from a flat or level surface on wheels exceeding six inches wide, or more than one-fourth inch on wheels less than six inches wide; and that the nails of the wheels were to be so countersunk as not to project more than a quarter of an inch beyond the surface of the wheels[1]. This was retained in the Act as amended in 1823. So much injury had been done to the roads by the use of wheels which, although nominally wide, were yet of such

[1] After Jan. 1, 1823, waggons having wheels of 3 inches and less than $4\frac{1}{2}$ inches in breadth were to pay one-half more toll than that paid by 6-inch wheels; and waggons having wheels of $4\frac{1}{2}$ inches and less than 6 inches in breadth were to pay one-quarter more toll than that paid by 6-inch wheels.

construction as to roll upon a surface of only two or three inches, and which often had the nails projecting too far beyond the tires, that this precautionary measure was a wise safeguard.   Where waggons or carts had the bottom of the wheels rolling on a flat surface, and were cylindrical, and had the nails of the tires countersunk, so that the whole breadth of the wheel bore equally on its flat or level surface, trustees might order that the toll to be taken should not be less than two-thirds of the full toll[1].   The weights to be carried, both in summer and in winter, by waggons and carts with various widths of wheels, were minutely specified, and also the additional tolls for overweights[2].

By the Act as amended in 1828, trustees might, if convenient, compound with parishioners and others for the payment of the ordinary tolls for a term not exceeding one year.   As to composition for the extraordinary tolls, this was forbidden by the Act of 1822, notwithstanding the agitation of many trades to obtain it; but in 1828 there was a reversal of this policy, and composition for the additional tolls for overweights was allowed, on the principle that the trustees would take care in entering into it to make the parties pay in proportion to the injury done to the road by the excess of weight.   This provision was left in the Act, in the face of evidence proving the fraud that was perpetrated in the administration of the system.

A cognate provision to this, in the Act of 1822, was that empowering trustees or commissioners to order the erection of a weighing engine at any of their toll-gates, or at any expedient distance from them, and to require every waggon or other carriage to be weighed that came within one hundred yards of it.   Penalties were imposed upon anyone who tried to evade the payment of the tolls[3].   This, too, was simply re-enacting the old law, after conclusive proof had been given that

---

[1] See Reports of Committees of 1807–8 and 1810 as to the relative merits of cylindrical and conical wheels.

[2] The following weights (including carriage and loading) were allowed:

							Summer tons cwt.		Winter tons cwt.	
For every	Waggon	with	9-inch	wheels	..	..	..	6	10	6	0
,,	Cart	,,	9-inch	,,	..	..	..	8	10	8	0
,,	Waggon	,,	6-inch	and less than 9-inch wheels				4	15	4	5
,,	Cart	,,	6-inch	,,	,,	,, 9-inch	,,	8	0	2	15
,,	Waggon	,,	4½-inch	,,	,,	,, 6-inch	,,	4	5	8	15
,,	Cart	,,	4½-inch	,,	,,	,, 6-inch	,,	2	12	2	7
,,	Waggon	,,	wheels less than 4½ inches wide					8	15	8	5
,,	Cart	,,	,,	,,	,,	,, 4½ inches	,,	1	15	1	10

(Act of 1822, sec. 12; Act of 1823, Schedule No. 1.)
For the additional tolls for overweights, see Act of 1822, secs. 15, 35.

[3] Act of 1822, sec. 41.

weighing engines were inequitable, oppressive, and of little or no utility[1].

Officers of turnpike roads were to send to the trustees, when required, a financial statement of money received and disbursed during a specified time[2]. Mortgagees who had seized the toll-gates in payment of their claims were also required to give to the trustees or commissioners an account of all money received, and a heavy penalty was attached if they kept the gates after their indebtedness was paid in full[3]. As another check to the finances, the surveyor and the clerk of the turnpike road, who were not to be the same person, were each to keep books, showing money received and how it was spent. The clerk's books were always to be open to inspection by trustees or commissioners or creditors; and trustees, at their annual meetings, were to audit their accounts and report the state of their roads[4]. It was thought that these regulations would protect the funds of the trusts from any designs of unscrupulous officers; and had they been carried out strictly they might have exercised much control in this way; but, as we shall see later, there were good reasons why these accounts were not kept strictly accurate; and when it was left to the trustees to audit and check their own accounts, we can readily imagine the indifference they would display to such (in their view) unnecessary work.

Still holding to a precedent of three-quarters of a century's standing, the importance of which had been shown very clearly long before it was embodied in the law, the Acts of 1822 and 1823 prohibited any person from holding any place of profit under the trustees, if he were engaged in selling wine, ale, etc., or provisions by retail[5]; and no trustee or commissioner was to have any place of profit under an Act of Parliament that he was appointed to carry into execution[6]. This, it will be noticed, was another method of safeguarding the finances of the trusts; and at this time when many of the latter were deeply in debt, and becoming more and more involved, every channel of outgo required to be carefully watched.

In addition to the labour that was employed upon the roads and paid for out of the revenues of the tolls, statute labour was still enforced upon those who had formerly been legally liable; and idleness, negligence, or refusal by parishioner or surveyor was punishable by penalty[7].

---

[1] *Parl. Papers*, 1806, II, 249, Appendix No. 9 (B); 1809, III, 431, Appendix A, Section VI; 1810, I, 233; 1820, I, 333; etc.

[2] Act of 1822, sec. 77; Act of 1823, sec. 47.

[3] Act of 1822, secs. 47, 48.

[4] Act of 1822, secs. 69, 73; Act of 1823, secs. 44, 45.

[5] Act of 1822, secs. 64, 75.

[6] Ibid., sec. 65.   [7] Act of 1823, sec. 80.

The turnpike roads were still left under the jurisdiction of the Justices of the Peace, and they might determine what part of the statute duty should be done, and what proportion of the composition money should be laid out, upon the road. As was the case in regard to several other features of the system of road administration, statute labour had been shown to be obsolete; and its retention in the face of the argument of facts against it is a seeming paradox.

With this brief outline of the turnpike legislation in its more essential particulars, we leave this aspect of our subject. It is evident that in the statutes of the third decade of the nineteenth century there were survivals of earlier centuries which should have been eliminated long before, for some of them had been known to be of doubtful utility for almost two hundred years, and others had been demonstrated to be useless or worse than useless more than half a century previous to this time. The most important issues were left unchanged; the repetition of old principles encumbered the statute book, and their application continued to produce some untoward results.

While the consolidation of the highway laws into one code does not strictly come within the period up to 1830, yet it may profitably be included in this survey of legislation. Since the passage of the General Highway Act of 1773[1], certain amendments had been made or alterations introduced by a succession of Acts[2]; and the opinion had long been entertained that these general laws might be improved and combined into one law[3]. We have seen that this was effected for turnpike roads in 1822–3, but the corresponding change for the highway laws was delayed until 1835[4].

One of the great changes effected by the Act of 1835 was the abolition of all regulations which referred to the limitation of weights of loads to be carried, and the construction of the wheels of waggons and carts. In the earlier days, it was considered impossible to keep the roads in good repair without restrictions of this nature; in fact, the idea prevailed that carts and waggons could be so constructed as to repair, rather than to wear out, the roads, and accordingly there was the agitation for converting wheels into rollers. Later investigations, however, and the results from improved road construction, had led to the view that all regulations as to weights and wheels of carriages were entirely subordinate, so far as the preservation of the road was concerned, and hence they were omitted in this Act.

---

[1] Act 13 Geo. III, c. 78.

[2] Acts 34 Geo. III, cc. 64, 74; 44 Geo. III, c. 52; 54 Geo. III, c. 109; and 55 Geo. III, c. 68.   [3] See Reports of Committees of 1810 and 1819.

[4] Act 5 and 6 W. IV, c. 50.

Another feature worth noting was the abolition of the proceeding by presentment for the non-repair of highways, and the substitution of a summary mode of proceeding before the magistrates.

But probably the most important changes made by this Act in the general highway laws were the abolition of the old system of statute duty (almost three hundred years old) and composition for statute duty, with the substitution of a rate in place of it, and the power to form parishes into districts with a view to the appointment of district surveyors[1]. These had long before been recommended, and the results of their adoption fully testified to the soundness of the principles.

Having now seen the nature of the legislation that was enacted after 1750, we next consider the significance of the amount of this legislation. Since each new turnpike authorized, and every change in an existing one, required a special Act of Parliament, the number of Acts passed for these purposes will give us a good indication of the interest taken in road improvement. Of course, not all the Acts giving authority for the repair of the roads were carried into effect; but those which were not were comparatively few, so that our statistics will give us a fairly reliable, though rough, estimate of the progress that was taking place[2]. Beginning, then, with the period from 1751–70, we note a vast increase in the number of road Acts, over the preceding

---

[1] The immediate reasons for these changes, as stated by the contemporary Report of the Committee of the House of Commons on the County Rates, were as follows:

That while the farmers were suffering under an expenditure amounting to about one million pounds, the general management of highways was exceedingly defective, partly owing to the incompetence of the surveyors, and partly because the system of statute duty interposed practical difficulties which the most experienced surveyors could not overcome.

That the system of statute labour produced a great waste of labour without corresponding public benefit; and rendered wholly impossible the adoption of the improvements in the management of highways which had been so successfully introduced in various parts of the kingdom, without resorting to a compulsory composition for statute labour, which was at best but a bad substitute for a highway rate.

That under the improved system of road making, there was necessarily a great increase of manual labour, and a proportionate diminution of team labour; and it was urged as an object well worthy of the consideration of Parliament, especially when so great a change was about to be effected in the administration of the laws for the relief of the poor, whether such an alteration in the highway laws as would have a tendency to create a great demand for labour ought not to meet with every possible encouragement.

[2] See Appendix 13, where this subject is dealt with at greater length, showing the full significance of the change.

years; for in these two decades alone, the number of these Acts was more than double the number passed in the preceding fifty years. This change is especially marked in some sections of England, as for example in the north midland counties, where the number of Acts increased from 55 in 1701–50 to 189 in 1751–70. This increase continued right down to 1830. The great interest that began to be taken in the means of internal communication shows itself very markedly in the decade from 1751–60, which would seem to point very decidedly toward the fact that the Industrial Revolution was at this time going on, in its incipient stage. The extent of the change may, perhaps, be more clearly seen by the fact that the number of road Acts passed in the forty years from 1751–90 was an increase of three hundred and eighty-eight per cent. over the number passed in the preceding fifty years, from 1701–50. From the legislation, we get no adequate conception as to the relative amount of attention devoted to the improvement of highways and of turnpike roads, for in many cases no hint is given in the statutes as to which class of road was to be benefited. That the turnpikes were the subject of more public interest than the highways, is indisputable, and, with their greater promise, it was but natural that this should have been so. But we must be careful not to mistake this greater interest in turnpikes for a greater amount of construction. In the constant investigations of the subject of road conveyance in the first three decades of the last century, the amount of consideration given to the turnpike roads far overshadowed that given to the ordinary parish highways; and on perusal of the reports of these investigating committees one would get the impression that the roads of England must have been largely turnpike. But a corrective of this possible error is found in the statistics; for in 1820, out of a total length of about 125,000 miles of road, only a little over 20,000 miles, or, roughly, one-sixth of the whole, was turnpike[1]; and even by 1838 there was only about 22,000 miles of turnpike, while the amount of ordinary highway was computed as not less than 104,770 miles[2]. The great industrial and commercial centres at this time were linked up by practically continuous turnpike roads, but it is certain that large and important sections of the country had still to depend for their conveniences upon the parish highways, which, according to a judicial pamphleteer of 1825, were generally in a bad state[3]. Even some of the counties near the metropolis had but a small part of their mileage

---

[1] 'Report from the Committee of the House of Commons to consider Acts regarding Turnpike Roads and Highways, 1821.'

[2] 'Report of the Royal Commission on the State of the Roads, 1840.'

[3] Brit. Mus., T. 1157 (4), 'Highways Improved,' p. 2.

turnpiked, for example, in 1815, Middlesex had 81 per cent., Suffolk and Essex only 10 per cent., Lincoln, 11 per cent.[1].

The interest in good roads was, of course, due primarily to the necessity of providing the means for the conveyance of the raw material and manufactured products of a rapidly developing industrial people, including the means for their subsistence. But good roads were not only a result, but also a cause, of prosperity, although their effect, as a separate element of prosperity, is usually very difficult to trace. The town of Walsall, for example, was formerly out of the regular line of trade and travel; but just as soon as the turnpike system became improved, the inns of Walsall (and consequently other kinds of business there) became active[2]. A similar result was noticeable at Carlisle, after the Military Road was built between Newcastle and that town. At first this road raised the price of provisions locally, by opening up new markets for their sale; but soon it diverted the line of cross-country traffic, which had previously gone from Newcastle to Dumfries, and now caused it to follow the Military Road to Carlisle, and thence to Whitehaven. Both these places thereby received an important accession to prosperity: the shipping trade at Whitehaven was much increased, and Carlisle became a commercial centre[3]. Another effect, of a wholly different nature, was that the easier, cheaper and faster communication by good roads led those who were wealthy to reduce the amount of their trading with local shopkeepers, and to increase the amount of their business with the larger but more distant markets[4].

In the history of the roads down to about 1750, we have seen that beside the regular statutory means, other aids were available for repairing them, such as gifts of money, income from grants of lands, etc., and that these were common sources of revenue. In the period

---

[1] 'Report of the Royal Commission on the State of the Roads, 1840.'

[2] An advertisement in Aris's *Birmingham Gazette*, Oct. 17, 1774, illustrates this in the following words:

"To be let and entered upon at Christmas next, all that new erected and compleat Inn, in Walsall, called 'The New Inn,' standing near the Bridge and New Road, and conveniently situated for the reception of noblemen and gentlemen travelling through Walsall. *The business of the said Inn is daily increasing on account of the turnpike roads being made exceeding good* from Walsall to Birmingham, Wolverhampton, Lichfield......"

[3] Ferguson, *History of Cumberland*, pp. 277–9.

[4] Phillips, in his *Tour through the United Kingdom* (1828), p. 30, says that at Newport Pagnell, and at other places within sixty or eighty miles of London, he heard a general complaint among shopkeepers, that their opulent neighbours gave no encouragement to the place from which they derived their fortunes, or their enjoyments; but by system sent to London for all their wants.

following, these additional funds became proportionally much less important. Rarely do we find that gifts of money were made by individuals; but occasionally the corporation of a town voted a certain amount to aid in securing a turnpike road, which would be of direct value in opening up trade with the town[1]. As a general thing, instead of depending upon such extraordinary aids, trustees relied upon the ordinary means provided by law for the amending of the roads, that is, turnpike tolls[2], statute duty or composition therefor, and assessments.

But while there were few adventitious aids to the upkeep of the roads, the burden imposed upon turnpike trustees for the maintenance of their section of road was greatly augmented by the heavy charges which had to be paid, at first, in order to obtain the turnpike Act, and later, to secure its renewal at the expiration of its term of duration. When such an Act was sought, a skilful negotiator had to be employed to carry the case through the Committee and before Parliament; opposition had to be placated or pacified, sometimes by sums of money; and the expenses incident to the maintenance of such a solicitor in London, and the payment of his services, usually made a heavy bill. This burden continued to increase, and since each turnpike Act was to continue for not more than twenty-one years, the cost of this frequent renewal was a great drain on the revenues of the trust. By 1821 it was felt that the existing system could not go on unchanged, for the fees for the renewal of turnpike Acts had so enormously increased, that

---

[1] In the *Sheffield Burgery Accounts*, from 1740 on, there are constant references to financial aid rendered to the improvement of roads.

In 1767 the corporation of Abingdon agreed to subscribe £40 towards raising and repairing the turnpike road between the town of Buscot, in Berkshire, and the town of Leachlade, in Gloucestershire, or such part thereof as the corporation might direct, in case a sufficient subscription could be raised for repairing the road so as to keep it from being overflowed by the usual floods there (Challenor, *Records of Abingdon Borough*, p. 206).

Many of the roads leading into Liverpool had merely a narrow strip of paving in the middle (a causeway) for pack-horses, but were impassable for wheel carriages. The corporation gave money for their improvement (Picton, *Liverpool Municipal Records*, II, p. 155). In 1770, they voted £100 towards obtaining an Act for making a turnpike road from Liverpool, through Ormskirk, to Preston (ibid., II, p. 257).

Doncaster seems to have been very liberal in such matters. In 1763 the corporation authorized the Mayor to subscribe £300 towards repairing the road from Doncaster to Tinsley, and £200 to repair the road from Chesterfield to Rotherham. In 1766 the corporation subscribed and gave a bond for £200, towards erecting a turnpike from Balby Pinfold to Worksop (Hardy, *Records of Doncaster*, IV, pp. 233, 237).

[2] On the farming of the tolls, see Appendix 4.

in England, Wales and Scotland, the whole receipts of the trusts for two years were not enough to meet the expense of their renewal[1].

In 1826 and 1827 there was a strong agitation, first, for the abolition of fees in the cases of bills that sought the renewal, or repeal and renewal, of turnpike trusts; second, for transferring to the public the expense of printing these bills; and, third, for having the term extended from twenty-one to thirty-one years, the latter of which alone would save half the expense of these bills. After subsequent consideration of these proposals, it was decided in 1831 to grant the first of these requests, by relieving trustees from the heavy charges hitherto imposed for renewal of such Acts individually, but the other conditions were left without any change.

Before proceeding to later phases of the road question, it is necessary for us to inquire into the way in which the intentionally beneficial laws worked out in actual practice; and here we find some statutory and administrative reasons why the roads too often merited the opprobrious epithets that were applied to them. The first of these which we would mention is that the appointment of surveyors annually or by rotation was decidedly detrimental to the roads, for in this way most of the roads were kept under the supervision of men who had no knowledge as to what course to take to secure their improvement. Nor had such men any inducement to study the situation so as to plan

---

[1] *Parl. Debates*, N.S., XVIII, pp. 1445–50. An ordinary amount of fees on a turnpike bill was £148, of which £83 was exacted in the House of Commons and £65 in the House of Lords. The average amount of costs on fifteen bills solicited by Mr Dorington in 1826 was £158 (£94 in the House of Commons and £64 in the House of Lords), which with the average cost of printing, £26. 5s., made the total charge, £184. 5s. These were some of the most favourable bills, with regard to cost. A gentleman of Devonshire had solicited three bills, on which he paid respectively £278. 10s., £280. 10s., and £273. 8s. In *Parl. Papers*, 1809, III, 431, 'The Third Report of the Committee on Broad Wheels,' Appendix B, it is said that to renew these Acts cost on an average £300; but in all probability this was simply putting the figures in round numbers, rather than giving them with exactness.

When turnpike Acts were first applied for, the idea seemed to prevail that it was only necessary to widen, straighten and substantially repair the roads, and continue the tolls until the principal sum borrowed and interest thereon were paid off; that then the toll-gates might be pulled down and these roads would thenceforward need no other than the ordinary attention of the parish surveyors and the statute duty. The first Acts were, therefore, to be merely temporary Acts, for twenty-one years at most. But when the twenty-one years expired, in nearly every case a petition came to Parliament stating that the debts still remained, and would be lost without a renewal. So it went on from one term to another, taking tolls from the public, which, instead of being used to pay off the debts, often went to pay salaries and fees to officers, attorneys, surveyors, witnesses, etc. The renewal fees were very heavy, as we have shown before. See also Farey, *Agriculture of Derbyshire* (1811), III, pp. 232, 234.

for the best adaptation of means to ends, since they knew that at the end of the year their office would be given over to another equally inefficient. Then, too, it was very difficult for a surveyor to rigidly enforce the law, and compel those who were his neighbours to do their most faithful work on the roads, for any undue pressure of this nature would render him unpopular in his own community, and therefore his life there would be far from agreeable. Add to this the fact that he, during the past, and in all probability in the future, would be doing his share, like the rest, to avoid his full quota of road work, and we can easily see that everything but moral considerations would point him to the policy of allowing a good deal of negligence or laxity in the enforcement of the law's demands. While there was a great amount of complaint as to the way in which the surveyor neglected to execute the work imposed upon him[1], the foregoing considerations show that there was some justification for his course of procedure, especially as the great amount of time he should spend in this work was lost to his farm and unremunerated by the public. For three reasons, therefore, the surveyors failed to perform the functions of their office: first, because of economic considerations; second, because it was morally impossible for them to wholly subject private interests to public duties; and, third, considerations of social expediency led them to be as easy as possible with their neighbours, so as to avoid being exposed to censure. Then, when the surveyor failed in the performance of his duty, there was usually no one who wanted to take action against him for his neglect because most of his neighbours would be likely to want similar clemency extended to them when their turn came to fill his office.

---

[1] As to these complaints, see Burn, *History of the Poor Laws*, p. 239, where he calls the surveyors "spiritless, ignorant, lazy, sauntering people;" Murray, *Agriculture of Warwick*, p. 172; Brit. Mus. 213. i. 3 (100), 'Considerations about the Method of preserving the Public Roads,' p. 3; Crutchley, *Agriculture of Rutland*, p. 21; James and Malcolm, *Agriculture of Surrey*, p. 62; Holland, *Agriculture of Cheshire*, p. 304; Strickland, *Agriculture of the East Riding of York*, p. 274; Plymley, *Agriculture of Shropshire*, pp. 273, 280; *Communications to the Board of Agriculture*, I, pp. 120, 121, 122; Mavor, *Agriculture of Berkshire*, p. 423; Brit. Mus., T. 1157 (4), 'Highways Improved,' pp. 2–10, etc.

At the meetings of the commissioners of a turnpike road, most of whom were country gentlemen, it too often happened that party influence ruled, and thus the stronger party entrusted the management of the road to some ignorant or pretended surveyor who was favourable to them. Thus the public business was neglected, or else got into the hands of the stronger party without regard to the community's interests. See *Communications to the Board of Agriculture*, I, pp. 121–2.

Because of the negligence of the surveyors, the statute labour was frequently misapplied or neglected, and the composition money was spent ill-advisedly. See Malcolm, *Agriculture of Bucks*, p. 44; Vancouver, *Agriculture of Cambridge*, p. 218, etc.

Moreover, if anyone complained of the neglect or inattention of the surveyor, the latter might break up the fields of the complainant and dig for gravel in accordance with permission granted by the Act of Parliament, and in this way render unsightly and inconvenient the land of any man who ventured to make trouble. In such cases, "it looks as if the surveyor were the master, and the commissioner his humble servant[1]." But, it was sometimes profitable for a commissioner to stand in with the surveyor as a matter of expediency; and the latter, instead of cutting down a difficult hill on the highway, thereby improving the road and getting good materials for repairing it, would leave this hill and bring materials for miles because the commissioner would thereby get the benefit of selling so much road material[2]. The advantages of carrying out any such policy of collusion would put an end to an effective execution of the law.

Another statutory reason why the roads were not better managed was the unlimited power given to the commissioners, who were usually country gentlemen, over the roads in the county or district in which they lived. Many of these men deserved credit for their benevolence and liberality; but they could not view impartially their own grand and beautiful estates. Hence they would not want a turnpike road to go through the estate they had adorned, even if that were the better direction for the road in the public interest. They would naturally use their influence to keep the road from going through and dividing their property, and would induce their friends to help them; consequently there was the formation of a party opposed to the public good. This policy necessitated making roads zigzag through a level country where they ought to have gone straight; and in other cases the roads were put over hills which could easily have been avoided if estate owners had been willing to forego personal aggrandizement to communal advantage[3]. But trustees and commissioners were too often careful

---

[1] Deacon on *Stage Waggons*, p. 51.   [2] Ibid., p. 52.

[3] In support of this, it is only necessary to refer to many parts of the chief thoroughfares, the directions and inclinations of which were determined by the personal bias of landowners. See *Communications to the Board of Agriculture*, i, pp. 121–2. The writer of these "Communications" would have the control and management of the roads entrusted to a special Board under which able surveyors and inspectors would be appointed. This, he thought (i, p. 124), would do away with the numerous abuses in the management of the turnpike trusts, and of the vast amounts of money levied at the toll-bars.

Clark, *Agriculture of Hereford*, pp. 53–54, shows the evil of having country gentlemen as judges of where roads should be put through and how they should be repaired. They invariably looked to their own interests. He shows also the evils resulting from discord at their meetings. See also Grahame, *Treatise on Internal Intercourse and Communication*, pp. 18–21.

of their own local private interests alone, without regard to the larger interests and the general good.

An administrative defect of great significance was the misapplication of funds, and the absence of proper inspection and accounting on the part of the officials of the roads. If this had been attended to, it would have done away with numerous abuses in the management of turnpike trusts and in the expenditure of the money received at the toll-gates; but large amounts of money were collected from the public, and expended without adequate responsibility and control, and therefore the resources of the country, instead of being devoted to useful purposes, were too often improvidently wasted[1]. Adam Smith regarded this as very important, for he tells us that the money levied at the turnpike gates throughout Great Britain was more than double the amount that was necessary for executing in the best manner the work which was often performed in a very slovenly way, and sometimes not at all[2]. Of course, there were various ways in which money was not employed for the good of the roads, such as the payment of salaries to those who did little or no work, payment of interest on large amounts of debt, hiring horses and men to water the roads to keep down the dust, putting too much material on the roads, and many others[3]. Even after the surveyor had used the money that was put into his hands in a way that did not best promote the interests of the trust, he could then take his accounts to his favourite magistrate and have them passed, without much possibility of appeal[4]. The abuses committed and allowed by the trustees in the management of the tolls were justly

[1] *Parl. Papers*, 1808 (225), II, 333, 'First Report from the Committee on the Highways,' p. 6. *The Grand Magazine of Universal Intelligence* (1758), I, p. 329, refers to the prodigious sums that were spent on the turnpikes and yet the latter were in an incommodious and disagreeable condition.

[2] *Wealth of Nations*, II, Bk. v, Chap. I, Pt. III, Article I, pp. 173–6, 'Of the Public Works and Institutions for facilitating the General Commerce of the Society.' See also *Communications to the Board of Agriculture*, I, pp. 124, 163–5; James and Malcolm, *Agriculture of Surrey*, p. 62; Malcolm, *Agriculture of Bucks*, p. 44; Dickson, *Agriculture of Lancashire*, pp. 607, 608; Farey, *Agriculture of Derbyshire*, III, pp. 235–6.

[3] In addition to references in footnote 2 above, see *J., H. of C.*, XXIX (1763), pp. 646–64, 'Report of a Committee appointed to inquire into the Management and Application of all Money collected in the last eleven years so far as concerns Kensington, Marybone, Islington, Hackney, Stamford Hill, New Cross and Surrey Turnpikes.' This report shows the way in which funds were misapplied, so that most of these trusts were increasing their debt. See also Middleton, *Agriculture of Middlesex*, pp. 395, 397; *Parl. Papers*, 1819 (509), v, 339, 'Rept. from Select Committee on the Highways,' evidence of J. L. Macadam, p. 21; *The Times*, Apr. 17, 1816, p. 3, letter from "X" on Turnpike Tolls and Retrenchment.

[4] Hansard's *Parliamentary Debates*, 1831, v, p. 1035.

complained of, and, despite the vast increase of the tolls in the last quarter of the eighteenth and the first quarter of the nineteenth century, the debt in which many of the trusts were involved was unaccountable to those who did not know the character of the management. Nor was there any proper remedy for such abuse of public trust, for a turnpike trust could not even be indicted for its default or negligence; and the only responsible authority that could be indicted or presented was the parish or township within which the bad road was found. By the passage of the General Turnpike Act of 1778 provision was made for apportioning the fine and costs between the parish and the turnpike trust, but this could only be done when the trust had money enough to pay without endangering the security of the creditors who had advanced money upon the credit of the tolls[1].

The system of statute duty, which, since its inception in 1555, had required the forced labour of men and teams upon the roads for a certain number of days each year, had remained practically unchanged in all that time; and it was too well known to require proof that days intended to be devoted to statute labour were regarded, not as days to be seriously employed, but as an occasion of fun and frivolity. Parishioners in many cases allowed the burden of public responsibility in regard to the roads to rest very lightly upon them; and the days spent in fulfilment of this obligation were regarded as that much time lost[2]. Evasion of the law was common, because it could easily be effected; and in the fulfilment of the statute labour the poorest horses and the boys or men who were not able to do a good day's work on the farm were sent to work upon the highways. Sometimes they were sent an hour or two too late in the morning, or recalled earlier than the proper time in the evening. To bring such cases, and many others of like nature, before the magistrates, would involve the loss of more time and arouse ill-will among neighbours, and, consequently, they were allowed to pass unnoticed[3]. Moreover, even if the people had been disposed to do their duty in this respect, it was more profitable for a man to refuse to do the work and thereby incur the penalty; for on account of the change that had taken place in the value of money and the prices paid for service, the man and his team could make more

[1] Act 18 Geo. III, c. 84, sec. 33. See also *Parl. Papers*, 1808 (275), II, 459, 'Second Report of the Committee on Highways,' Appendix A, p. 136, and Whitaker, *Loidis and Elmete*, p. 82.

[2] Strickland, *Agriculture of the East Riding of York*, p. 274; Billingsley, *Agriculture of Somerset*, p. 308; Clark, *Agriculture of Hereford*, p. 55; Mavor, *Agriculture of Berkshire*, p. 426; Brit. Mus., T. 1157 (4), 'Highways Improved,' pp. 3-5, etc.

[3] Plymley, *Agriculture of Shropshire*, p. 280; *Communications to the Board of Agriculture*, I, p. 147.

by hiring out to a neighbour than they would receive credit for if they voluntarily went upon the roads[1]. It is evident, therefore, that the law encouraged its infraction and placed a premium upon wilful disobedience.

The multiplication of trusts, under statutory authority, created an immense number of separate jurisdictions in control of the roads, a system which was inimical to their development from the national standpoint. The quality of each distinct piece of road depended largely upon the public spirit of those who had charge of it, and there was no relation between one piece and another. Of two trusts which might happen to join end to end, one might be firm and smooth, and the other ill-constructed and dangerous. Such a method of decentralization of control could not lead within reasonable time to the completion of long continuous lines of turnpike, and this lack of systematic improvement of the roads was a great barrier to any degree of facility of trade over widely extended areas. Then, too, these small trusts, each having control over a few miles of road, were almost universally poor; and the tolls contributed by the public were diverted from their intended purpose and used to pay salaries of trust officers and for other unproductive ends. This was regarded as one of the greatest evils in the existing management of the roads, for the trusts of narrowly limited range were uneconomical, politically objectionable, and an impediment to the extension of traffic[2].

After the opening of the nineteenth century, another administrative feature which militated against good roads was the employment of pauper labour upon them. Under the law, the parish had to support

[1] *Parl. Papers,* 1808 (275), II, 459, 'Second Report of the Committee on Highways,' p. 142, Appendix No. 4 (A).

[2] *Parl. Papers,* 1833 (703), xv, 409, 'Second Report of the Lords Committee on Turnpike Returns,' pp. iii, 489, 554, 567–8 (evidence of James Macadam, William Macadam, G. Hollis, etc.). The small trusts were uneconomical because they had to buy their materials in small quantities, because they paid salaries that were as large as those paid by other trusts with much greater revenues, and for other reasons, all of which were running them further into debt. They were politically objectionable, because local interests were sometimes strong enough to influence parliamentary votes to turn a road out of its best course from a national standpoint in order to promote the welfare of the few; and because when trusts were coming into close juxtaposition they frequently showed their activity by endeavouring to prevent Parliament from granting authority for new turnpike roads that threatened to compete with them (v. Clark, *General View of the Agriculture of Hereford,* p. 53; *The Times,* May 31 and June 16, 1828; Scott, *Digests of the General Highway and Turnpike Laws,* 1778, p. 317; Homer, *Inquiry into the means of Preserving and Improving the Publick Roads,* pp. 21–22; *Parl. Papers,* 1808 (275), II, 459, 'Second Report of the Committee on Highways,' p. 182, Appendix A; *J., H. of C.,* Jan. 25 and 28, 1780).

its able-bodied poor, and it was but natural that those who had the oversight of the roads and those who had to look after the poor should think that by setting the latter at work on the roads, instead of keeping them in idleness, the highways would be mended and some return would be secured from these "supernumerary poor," for whose maintenance the parish was responsible. These persons who would otherwise be chargeable to the parish, on account of lack of work, were handed over to the surveyor of the highways, with a polite note asking the latter to secure work for the men, for a certain number of days, upon the roads, under the usual conditions of labour in husbandry[1]. They often employed not only the man, but also his wife and all his family, girls as well as boys; and great numbers of women, and children from nine to ten years old and upward, were employed on the roads in breaking stones[2]. But men of this class were the poorest kind of labourers; for since they knew that the parish must maintain them, whether they worked or not, their inclinations usually led them to pass the time as easily as they could. Even when the surveyor's eye was upon them they worked very leisurely, but when his back was turned the amount of work they did was insignificant; and since the surveyor was usually a farmer who had other work to do, about the best oversight he could give them was to see them once a day and start them at work, leaving them to their own freedom for the rest of the day. It will be apparent, from what we have just said, that most of the money spent on pauper labour was that much wasted from the public treasury[3].

Turning now from the statutory and administrative defects which tended to retard or prevent the securing of the best roads, we consider, in the next place, those factors of a mechanical or engineering character which delayed or impeded the attainment of that result. Under this heading we would say that a very fruitful source of trouble was the unduly heavy loading that was allowed upon roads which were not fitted to bear such great pressure. Sometimes these loads were carried by carriages with narrow wheels, and in other cases by those with wide wheels; but in each case the effect was of much the same nature. Such ponderous weights as were allowed by law to be carried on narrow-wheeled vehicles caused the wheels to sink down and produce deep

---

[1] 'First Report of the Poor Law Enquiry Commissioners, 1884,' Appendix A.

[2] *Parl. Papers*, 1824 (892), VI, 401, 'Report of Select Committee on Labourers' Wages,' pp. 411-18, evidence of James Macadam; *Parl. Papers*, 1833 (708), XV, 409, 'Second Report of the Lords Committee on Turnpike Returns,' p. 570.

[3] 'First Report of the Poor Law Enquiry Commissioners, 1884,' gives great detail as to the working of this system; see, for example, Appendixes A and C.

ruts, except on the few roads which were able to withstand the burden; and even the wider wheels with a correspondingly heavy load would often but make the ruts wider[1]. In some kinds of soil, the ruts made by the narrow wheels became set by the roads becoming dry and hard, and could not be filled up or smoothed over by the passing of broad wheels along the road[2]. The effects of these heavy loads on narrow wheels were, of course, cumulatively injurious when, as frequently, they were drawn upon roads that had just been softened by rain, or that allowed the water to stand upon them. We are not here arguing for or against the wide or the narrow wheels, for both were in use, and there were great differences in the width of the wide wheels, as there were great differences of opinion in regard to them[3]. The fact is, that the law recognized that heavy loads were destructive to the roads and as a consequence weighing engines were erected to compel carriers to pay extra tolls for the overweight. We must not think that the heavy loads that had to pay extra tolls were always carried on broad wheels, for this was by no means true. Even in the case of the common carriers, the amount of load which they were allowed to carry per inch of bearing surface of the wheel was considerably greater on the narrow wheels than the weight allowed per inch of surface on broad wheels[4]. These narrow-wheeled waggons with their great burdens went along some roads and cut them from side to side into a series of ruts which made travel upon them difficult and sometimes dangerous[5]. Everyone who was interested in good roads seemed to be uniformly of the same opinion, that narrow wheels, especially when heavily laden, were a serious impediment to keeping the roads in fair condition[6]; and yet it was narrow wheels almost exclusively that were used in husbandry

---

[1] Malcolm, *Agriculture of Buckingham*, p. 44, refers to the opposition to the wide wheels, as well as to the narrow wheels. The wide-wheeled waggons with wheels in some cases as broad as eighteen inches, and loaded so heavily, with eight to ten tons' weight, while drawn by ten horses, simply acted like millstones grinding the flints, stones or gravel to powder, and sinking the road deeper than it was before.

[2] Deacon on *Stage Waggons*, p. 63.

[3] Contrast, for instance, *Communications to the Board of Agriculture*, I, p. 171, and I, p. 180.

[4] Deacon on *Stage Waggons*, pp. 23–24.

[5] Ibid., pp. 24–25. He says that for miles together the narrow wheels had cut the road into ruts or "quarters," ten or twelve of which composed the whole width of the road, and that the ruts were so deep as to render it extremely unpleasant, and sometimes dangerous, to ride on horseback on the top of the quarters. He thought it was truly wonderful how men travelling post, or in stage coaches, were able to proceed on such roads at the existing rate of speed.

[6] Ibid., pp. 25–26, 63–64.

and these were allowed to carry their heavy loads upon the roads without paying toll[1]. This overloading occurred also in the case of the stage coaches, with wheels generally about two inches wide, which were not under any restriction as to width of wheel or weight to be carried. According to some who were most familiar with them, their destructive effects on the roads were the greatest evil that the Legislature would ever have to combat in their efforts to improve and preserve the roads[2]. These coaches were not subject to weighing at the weighing engines and consequently the drivers had no hesitation in taking as many passengers as could find room. When this had gone on for some time, an Act was passed in 1788 limiting the number of outside passengers to six, in addition to two who could find room alongside the driver[3]; but it would appear that this was not effective, for in 1790 another Act was passed forbidding stage coach drivers to take more than four passengers on the outside, together with one on the box beside himself[4], under penalty of five shillings for every passenger above that number[5]. But this law also was set at open defiance, for sometimes as many as twenty persons were found on the roof of a stage coach, when the law allowed but six[6]; and in 1806 another Act was passed to facilitate the enforcement of the foregoing[7]. Violation of the law went on for many years after this and frequent complaints through the public press attest the continuance of this dangerous practice. With such heavy loads of passengers or of goods carried on narrow-wheeled vehicles, it would have been impossible to keep the existing roads with anything like a firm foundation or a smooth surface.

But the narrow-wheeled carriages were not the only sinners in this respect. If the introduction of wide wheels had not been accompanied by permission to carry greater weights, they might have proved in most instances very beneficial in precluding this "quartering" of the roads and in solidifying the road materials. But with the increasing width of wheels an increasing weight was allowed to be carried, and

---

[1] *Parl. Papers*, 1808 (275), ii, 459, 'Second Report of the Committee on Highways,' p. 140, Appendix No. 3 (A), gives a letter from the commissioner of a turnpike road near the city of Norwich, showing that the narrow-wheeled waggons carrying manure from Norwich to the farms, although exempted from the payment of toll, caused much greater damage to the roads than all other traffic together.

[2] Deacon, op. cit., pp. 31–33, 46, 73.

[3] Act 28 Geo. III, c. 57.

[4] If the driver had less than three horses, he had to take fewer passengers than the four and one just mentioned.

[5] Act 30 Geo. III, c. 36.

[6] *The Times*, Apr. 19, 1794, p. 3; 'First Report of Commons Committee on Broad Wheels and Turnpike Roads, 1806.'

[7] Act 46 Geo. III, c. 136.

at reduced tolls. This, along with the greater number of horses that were required and allowed to draw the heavier load, placed great pressure upon the roads, which they were usually unable to bear. For some time after the use of broad wheels came into effect, many regarded them as highly advantageous; and even Arthur Young wondered that they did not become more generally employed, "as the great convenience of them is evident and indubitable[1]." As time passed, however, experience demonstrated what theory had not foreseen, and by the end of the eighteenth and during the first third of the nineteenth century, it was clearly observed that the broad wheels with their heavy loads were highly objectionable, for their effects were either to crush to powder the small stones placed on the road for its improvement, or else to press this road material down into the soft substratum, while, at the same time, forcing the soft, earthy foundation up to the surface. One who had seen the results from the use of broad wheels, when carrying these extraordinary weights, said that it was impossible for even a road of adamant to withstand the crush of such destructive engines[2]; and with him agreed the great majority of those whose opinion was sought because of their experience[3]. The results were notably

---

[1] Young, *The Farmer's Letters to the People of England* (1767), pp. 271–8. He enters into a description of them, and proves by a mathematical calculation how much cheaper they would be for farmers than the narrow-wheeled waggons. He takes up the objections made to their use and in answering them shows the broad wheels to be much superior (ibid., pp. 280–2). See also *Oxford Gazette and Reading Mercury*, Jan. 11, 1768, p. 8, showing that according to existing tolls near Reading, a farmer with his great waggon, whose wheels rolled a surface of sixteen inches, could bring to market with six horses in one day with a toll payment of 8d., as much corn through one gate as a little farmer with narrow wheels and three horses could bring through the other gate in four days for 5s. He goes on to exemplify how the little farmer is discouraged. See also *Public Advertiser*, July 10, 1786, p. 2.

[2] *Communications to the Board of Agriculture*, I, p. 152.

[3] Ibid., I, p. 154; ibid., I, p. 171; ibid., VI, p. 182; Fry, *Essay on Wheel Carriages*, pp. 45–53, who mentions (p. 53) the use of eight-wheeled carriages with their enormous loads, between Bristol and Bath, in the latter half of the eighteenth century. See also Murray, *Agriculture of Warwick* (1813), p. 172; Brit. Mus. 213. i. 3 (100), 'Considerations about the Method of Preserving the Public Roads,' pp. 1–3; Stevenson, *Agriculture of Surrey* (1813), pp. 546–7; Monk, *Agriculture of Leicester*, p. 53; Lowe, *Agriculture of Nottingham*, p. 53; Bishton, *Agriculture of Salop*, p. 20; Wedge, *Agriculture of Warwick*, p. 28; Rennie, Broun and Shirreff, *Agriculture of the West Riding of York*, p. 86; Pitt, *Agriculture of Leicester* (1809), p. 310; Priest, *Agriculture of Buckingham* (1813), p. 340; Farey, *Agriculture of Derbyshire*, III, p. 244; Dickson, *Agriculture of Lancashire*, p. 612; Middleton, *Agriculture of Middlesex* (1798), p. 396; *The Times*, June 3, 1816, p. 2; Deacon on *Stage Waggons* (1807), pp. 13–15; *Parl. Papers*, 1819 (509), V, 889, 'Report from Select Committee on the Highways,' p. 7; *Parl. Papers*, 1821 (747), IV, 348, 'Report of Select Committee on Turnpike Roads and Highways,' pp. 4, 6. Contrast the views of "A Road Trustee" and

bad where the roads were so moist that the wide wheels only made the ruts of the narrow wheels the wider, and also where the wheels had to be locked, for in such places the road was torn up to a depth and width that rendered it disgraceful[1].

In addition to the great weights they carried, the broad-wheeled waggons were baneful in their effects also because of the way in which the wheels were usually constructed. Instead of being, as at present, cylindrical, they were made conical, with the smaller circumference of the cone most distant from the body of the waggon. It was thought that the sixteen-inch wheels would act on the roads like a garden roller upon a plot of ground[2]; and for this cause they were granted special exemptions from toll in the Acts of 1766 and 1773. These Acts do not state whether such wheels were to be cylindrical or conical; but in reality the conical wheel on a dished axle was the kind of construction that was brought into effect, and some of these conical wheels were made as wide as eighteen inches[3]. One of the most important Scotch road surveyors informs us that the dished axle (i.e., bent down at the end) was probably adopted in order to allow people to increase the breadth of their carriages and yet have the wheels run in the same track, for the wheels ran wider above than below[4]. Whatever was the reason for the adoption of this principle, it was generally applied to both broad and narrow-wheeled waggons that were intended for common use; in fact, we learn that it was rare to meet a carriage of any kind, whether used for pleasure or for carriage of goods, that did not have something of the cone in the shape of the wheels[5]. One of the great carriers out of London measured one of these conical wheels, and found the small (outer) circumference thirty-six inches less than the large (inner) circumference[6]. Such a conical wheel, if moved forward by the axle-tree, would have to partly roll and partly slide on the ground, for the smaller circumference could not advance as far as the larger

"Anti-Caput Mortuum" in *Public Advertiser*, May 29, 1788, pp. 1–2, and June 12, 1788, p. 1, who thought the broad wheels beneficial. Favourable to broad wheels are also letters in *Public Advertiser*, Aug. 17, 1789, p. 1; Jan. 15, 1790, p. 2; Jan. 30, 1790, p. 1; Apr. 27, 1790, p. 2.

[1] Stevenson, *Agriculture of Surrey* (1813), p. 547.

[2] Deacon on *Stage Waggons*, p. 5.

[3] *Parl. Papers*, 1821 (747), IV, 343, 'Report of Select Committee on Turnpike Roads and Highways,' p. 4. Only a few of these cumbrous wheels were still found carrying out of London in 1821; the nine-inch wheels generally prevailed on the carriers' waggons at that time.

[4] Paterson, *Practical Treatise on the Making and Upholding of Public Roads*, pp. 79–80.

[5] Ibid., pp. 6, 79–80.

[6] Ibid., p. 6.

in one revolution of the wheel. The team would, therefore, have to drag the small part of the wheel along a certain distance (in the above case, thirty-six inches) for each complete turn of the larger inner side of the wheel. The effect of this upon the roads can be well imagined; for instead of consolidating them, it had the very opposite result, namely, that roads with a fairly hard and smooth surface became so loose, and their hard materials so ground down, that water could readily percolate through and soften the foundation. Then, when the narrow wheels were driven along this road, they very easily and very quickly sank into its substance and made the ruts of which we have already spoken[1]. Since these waggons with conical wheels necessarily required more power to draw them along the road, on account of this partially dragging movement, it was a fortunate thing for their owners that the Legislature permitted the widest 16-inch wheels to be drawn by any number of horses, otherwise they might frequently have got stuck on the roads. For five years after the passage of the Act of 1773 sanctioning the sixteen-inch wheels, they were allowed to pass on the roads toll-free, and afterwards they could use the roads by paying only half the usual toll paid by other waggons. The wheels that were nine inches and six inches in width did a proportionate amount of injury to the roads; they had to carry correspondingly less weights than the sixteen-inch wheels, and be drawn by fewer horses, although they had to pay full toll[2]. Not only were these heavy weights allowed on broad wheels, but frequently the construction of these wheels, by having a projecting rim around the centre of the tire, made them bear all the weight upon this narrow middle rim. In this way, the practical effect was the same as if such weights as six to eight tons were drawn upon a waggon the tire of which was only two or three inches wide[3]; and these heavy loads, drawn by any number of horses from eight to

---

[1] Paterson, *Practical Treatise on the Making and Upholding of Public Roads*, p. 80; Deacon on *Stage Waggons*, pp. 11–12; 'Second Report of the Committee on the Use of Broad Wheels, 1806;' *Parl. Papers*, 1819 (509), v, 339, 'Report from Select Committee on the Highways,' p. 7; *Parl. Papers*, 1833 (703), xv, 409, 'Second Report of the Lords Committee on Turnpike Returns,' p. 540, evidence of John Macneill; *Parl. Papers*, 1821 (747), iv, 343, 'Report of Select Committee on Turnpike Roads and Highways,' p. 4; *Parl. Papers*, 1795–6, xlviii, Report No. 132, on the Turnpike Acts 13 Geo. III and 14 Geo. III, evidence of Mr Cumming, p. 5, and of John Lewes, p. 2; *Parl. Papers*, 1797–8, lii, Report No. 147; *Parl. Papers*, 1808 (225), ii, 333, 'First Report from the Committee on the Highways,' p. 3.

[2] On the various widths of wheels and their effects, see Deacon on *Stage Waggons*, pp. 14–19.

[3] Ibid., pp. 15–17; *Parl. Papers*, 1795–6, xlviii, Report No. 132, on the Turnpike Acts 13 Geo. III and 14 Geo. III, p. 5; *Parl. Papers*, 1797–8, lii, Report No. 147, p. 2.

twelve, went ploughing their way along the roads, loosening, rather than consolidating, them[1].

In connexion with this subject, we must also consider the administrative machinery designed to prevent the carriage of excessive weights. We have already noted that although narrow-wheeled vehicles had to pay extra tolls yet their use continued; and although the wide-wheeled vehicles were authorized to carry heavy loads, they were continually seeking to increase those amounts. Some means had, therefore, to be found by which those who carried more than the legal maximum would have to pay for the amount of excess; and in 1741 parliamentary consent was given to erect weighing engines at suitable places along the roads, for the purpose of weighing the waggons and loads[2]. A complicated scale of tolls to be taken for overweights was put into effect; and those carriages carrying more than the legally specified amount were required to pay for the extra weight. The amount of these extra tolls varied according to the kind of vehicle, the width of wheels, and the season of the year; and because the weighing machines were regarded as oppressive by those who wished to use the roads, many tried to evade them by various plans. These cranes became especially numerous in and near the metropolis[3], and those gardeners and others who were engaged in trading with London found the operation of the weighing engines very uncertain and unsatisfactory. For example, the weights, as given by different engines, were not uniform and a load that would pass at one engine would not pass at another. Some loads that were not greater, but sometimes less, than the maximum allowed when the carrier started from home, became much heavier before the destination was reached, on account of rain or wet weather[4]. The vexatious, troublesome, uncertain and expensive operation of these machines caused many to turn out into

---

[1] Of course, we must not suppose that, even in the first quarter of the nineteenth century, these broad wheels did not have their advocates and supporters; for there were some who strongly favoured these broad, conical wheels, while others favoured wide wheels that were slightly rounded on the surface, and still others preferred those that would roll a perfectly flat surface. See 'First and Second Reports of the Committee on the Use of Broad Wheels, 1806;' also the reports of 1809 on the same subject. As to the kind of carriage to be used, some favoured large waggons carrying as much as eight tons, and drawn by eight, twelve and sometimes more, horses (see evidence of Mr Russell, of Exeter, before the Committee of 1806); others recommended lighter waggons and loads (v. *Parl. Papers*, 1806, II, 249, under heading A. 1, 'Carriages;' also *Parl. Papers*, 1809, III, 411, 425, etc.).

[2] Acts 14 Geo. II, c. 42; 7 Geo. III, c. 40; 13 Geo. III, c. 84.

[3] *Parl. Papers*, 1795–6, XLVIII, Report No. 132, p. 2; Deacon, op. cit., p. 47.

[4] *Parl. Papers*, 1795–6, XLVIII, Report No. 132, pp. 9, 11; Deacon, op. cit., pp. 47–48.

by-roads and lanes in order to avoid them; and even the waggons belonging to the nobility were found to defeat the purpose of the law by this means[1]. On some of the roads about London there were no weighing engines at all; and, farther from that centre, some roads had them every few miles, while on others, equally public, sixty miles might be found without any[2]. In this way the burden of extra tolls was not levied with an impartial hand upon all. In order to escape the payment of these impositions, it was not uncommon for a waggon to drop part of its load before coming to the engine, and then go back for it after passing that hated obstacle; or else employ a cart to bring, say, four or five hundredweight behind the waggon until this had passed the weighing machine, and then load it all upon the waggon for the rest of the journey[3]. But among the evils connected with the weighing engines, perhaps one of the greatest was the practice of compounding for overweights. The keeper of the weighing engine was often in league with some or all of the carriers and for a certain weekly sum would allow their waggons with extra weights to pass through without weighing[4]; and sometimes the waggon masters themselves had rented the weighing engines[5]. Since the latter were usually farmed out by the year at high rents, it was the lessee's interest to secure as great returns as possible. In order to do this his only plan was to compound for the overweights; for if he had been too rigid in exacting every penny, he would have put an end to all overweighted waggons, and, in doing so, would also have put an end to his own profits[6]. Thus ponderous weights continued to be carried upon the roads, despite the restriction of the law, which provided a strong incentive for its own infraction.

In not a few cases, roads were bad because of defects in their location. They frequently followed the same courses that had been the bridle-paths for many generations; this made them longer and more circuitous than they should have been, and with the constant traffic upon this natural road-bed, the surface of the road became lower than that of the adjoining land[7]. The result was that the

---

[1] *Cambridge Chronicle and Journal*, Oct. 29, 1813, p. 2.

[2] *Parl. Papers*, 1795–6, xlviii, Report No. 132, p. 2; Deacon, op. cit., p. 48.

[3] *Parl. Papers*, 1808 (275), ii, 459, 'Second Report of the Committee on Highways,' pp. 167–8, Appendix No. 6 (N), 'On the Means practised to evade the Weighing Machines.'

[4] *Parl. Papers*, 1795–6, xlviii, Report No. 132, p. 4; *Parl. Papers*, 1809, iii, 431, Appendix A, Section VI.   [5] *Parl. Papers*, 1810, i, 233.

[6] *Parl. Papers*, 1809, iii, 431, 'Report of Commons Committee on Broad Wheels and Turnpike Roads,' Appendix A; Deacon, op. cit., p. 44.

[7] Edgeworth on *Roads and Carriages*, p. 4. He said: "The system of following

water from the adjacent land flowed into the roads, and because many
of these were not drained, and were so narrow and shaded that the
sun and wind could not get a chance to dry them, their foundations
softened under the standing moisture[1]. Sometimes, because of the
opposition of the landlords, roads could not be put through the most
suitable places, lest they might divide or disfigure the estates; and,
therefore, they either had to take a roundabout course, or else traverse
lines over hills or marshy ground along which it was much more
difficult to make and maintain a road[2].

Not only were the location and direction of some roads unfortunate,
but in many cases the methods of construction were very imperfect.
Along with a poor foundation to begin with, the roads were frequently
lacking in drainage, and the water, instead of being drawn off to drains
of suitable depth at the sides of the road, was deprived of that outlet
and allowed to lie, softening the subsoil of the road[3]. In order to keep

the ancient line of road has been so pertinaciously adhered to, that roads have been
sunk many feet, and in some parts many yards, below the surface of the adjacent
ground; so that the stag, the hounds, and horsemen, have been known to leap
over a loaded waggon, in a hollow way, without any obstruction from the vehicle."
See also Brit. Mus., Add. MSS. 17,398, p. 53; Clark, *Agriculture of Hereford*, p. 51;
Malcolm, *Agriculture of Buckingham*, p. 44; Leonard W. Clarke, *The History of
Birmingham* (in MS.), VII, pp. 211–13; Hutton, *History of Birmingham* (1st ed.), p. 21.
Sometimes this lowering of the roads was due to floods washing away the loose
soil of the road; v. Clark, *Agriculture of Hereford*, p. 51; Marshall, *Rural Economy
of Yorkshire* (1788), I, p. 180; Farey, *Agriculture of Derbyshire*, III, p. 266; Coxe,
Historical Tour through Monmouthshire, p. 35.

[1] All those who showed some sanity of judgment were strong in their convictions
that the roads should be widened, and the hedges and overhanging trees cut low,
so as to let the natural agencies have a chance to dry the roads. Rudge, *Agriculture
of Gloucester* (1807), p. 335; Brit. Mus., T. 1157 (4), 'Highways Improved,' p. 8;
Vancouver, *Agriculture of Hampshire* (1813), p. 391; Parkinson, *Agriculture of
Huntingdon* (1813), p. 276; Marshall, *Rural Economy of the West of England* (1796),
I, p. 285; Marshall, *Rural Economy of Yorkshire*, I, p. 188; Marshall, *Agriculture
in the Southern Counties*, I, p. 10; Dickson, *Agriculture of Lancashire*, p. 608. On
the contrary, the roads of Hertfordshire seemed to get plenty of sun and wind to
dry them; v. Walker, *Agriculture in Hertford*, p. 86.

[2] Hassall, *Agriculture of Monmouth* (1812), p. 101; *Communications to the Board
of Agriculture*, I, pp. 121, 123, 128, 147; Brown, *Agriculture of Derby*, p. 43; Priest,
Agriculture of Buckingham (1813), p. 339; Vancouver, *Agriculture of Devonshire*
(1808), p. 368; Marshall, *Rural Economy of the West of England* (1796), I, p. 30;
Warner, *Tour Through Cornwall* (1808), pp. 86–87; Granger, *Agriculture of Durham*,
p. 26; Bailey and Culley, *Agriculture in Northumberland*, p. 56; Tuke, *Agriculture
of the North Riding of York*, p. 84.

[3] *Communications to the Board of Agriculture*, I, p. 129; Brit. Mus., T. 1157
(4), 'Highways Improved,' p. 4; Foot, *Agriculture of Middlesex*, pp. 68–69; Pitt,
Agriculture of Worcester (1813), p. 260; Parkinson, *Agriculture of Huntingdon* (1813),
p. 275; Marshall, *Rural Economy of Gloucester* (1789), p. 14; Marshall, *Rural Economy
of Yorkshire* (1788), I, pp. 182, 188.

water from lying on the surface of the roads, both those which were turnpiked and those which were not, the practice arose of making the top of the roads too convex. The driver naturally took the highest part of the road, and others followed him in the same place; and this caused ruts to appear and to increase in depth. This would continue until the road from side to side had become "quartered;" then when rain fell into these ruts five or six inches deep they would be made still deeper. The surveyor, in the fulfilment of his duty, would then throw loads of gravel or other material into the centre of the road to make a decent cone on top of the other, and thus the roads became more convex than before. Such roads were too round and too narrow to be safe; and in turning out when meeting teams there was great danger of overturning[1]. Sometimes the earthen road was left too flat, in which case the water would lie long upon it[2]; and in other places concave roads were adopted, chiefly to utilize the effect of water in their improvement. The latter plan for the formation of roads seems to have had some ardent advocates among those who had most intimate knowledge of its results. It would seem that, where the road-bed was sandy or gravelly, the concave form allowed the water to stand

---

[1] Crutchley, *Agriculture of Rutland*, p. 21; Murray, *Agriculture of Warwick* (1813), p. 178; Strickland, *Agriculture of the East Riding of York* (1812), p. 267; *Communications to the Board of Agriculture*, I, p. 181; Brit. Mus. 578. k. 30, 'Observations on a Tour in England, Scotland and Wales,' (1780), p. 101; Malcolm, *Agriculture of Buckingham*, p. 43; Monk, *Agriculture of Leicester*, p. 53; Wedge, *Agriculture of Warwick*, p. 30; Pitt, *Agriculture of Northampton* (1809), p. 230; Pitt, *Agriculture of Leicester* (1809), p. 310; Vancouver, *Agriculture of Devonshire* (1808), p. 369; Parkinson, *Agriculture of Huntingdon* (1813), p. 276; Marshall, *Rural Economy of Gloucester* (1789), p. 14; Warner, *Tour through Cornwall* (1808), pp. 85–86; Dickson, *Agriculture of Lancashire*, p. 608. Marshall, *Rural Economy of Yorkshire* (1788), I, p. 182, tells us that the general method of forming roads in many districts was to raise the road too high and leave it too narrow. This confined carriages to one track on the ridge of the road, and formed two deep ruts. The method of repairing was equally absurd; for instead of the ruts being closed, by pecking in the ridges on either side of them, or by filling them with a few additional stones, the entire roadway was covered with a thick coat; and as often as fresh ruts were formed, so often was this expensive and therefore doubly absurd method of repairing repeated, until, having laid coat over coat and piled ton upon ton unnecessarily, a mound of earth and stones, resembling the roof of a house rather than a road, was formed. See also Deacon on *Stage Waggons*, pp. 35–87; *Oxford Gazette and Reading Mercury*, May 18, 1767, p. 2, Letter from J. Smith regarding the road from London to Marlborough.

[2] Stevenson, *Agriculture of Surrey* (1813), p. 546; Malcolm, *Agriculture of Buckingham*, p. 43; Foot, *Agriculture of Middlesex*, pp. 68–69; Tuke, *Agriculture of the North Riding of York*, p. 84; Dickson, *Agriculture of Lancashire*, p. 608; Marshall, *Rural Economy of Yorkshire* (1788), I, p. 188; Middleton, *Agriculture of Middlesex* (1798), p. 395.

longer and thus rendered these materials more adhesive[1]; but in other cases this method would have been very unsatisfactory. The proper method to be applied in road construction was a subject upon which little thought had been bestowed, as a general thing, by the surveyors; and as a consequence of their lack of knowledge, they consulted their own present convenience rather than the public good.

In many instances, roads were bad because they were not mended at the proper season or with proper materials. It very often occurred that the mending of the highways was done in the winter, when the farmers had most leisure; but this did not produce the desired results, for when the stones and gravel were put on the roads that had been softened by the snow and rain, these materials were crushed down into the soft earthen foundation by the heavy loads that were drawn upon the roads, and so the surface of the road was left as soft mud until it dried up in the spring[2]. If, on the contrary, the stone and gravel had been applied when the roads were dry, so that they would have had time to consolidate before the wet season came on, there would then have been some chance for the roads to be improved by the application of these materials. But, further, in order to fill up the ruts and holes in the roads, stones of various sizes were brought and dumped into them without any preliminary draining of the road or levelling of its inequalities; they seemed to think that it was the quantity of materials that made good roads, and that it was not necessary to put the roads into a state fit for profiting by a proper distribution of the materials used[3]. Stones were sometimes gathered from a neighbouring

---

[1] *Communications to the Board of Agriculture*, I, pp. 188–4; Marshall, *Rural Economy of the Midlands* (1796), I, pp. 37, 41, 47. This plan of road making was probably seldom applied. In other places, the road was made of one inclined plane, with just enough inclination to throw off the water. See Dickson, *Agriculture of Lancashire* (1815), p. 609.

[2] The efforts at road repair in the winter season were almost universally spoken against. Stone, *Agriculture in Lincoln*, pp. 45, 89, 96; Crutchley, *Agriculture of Rutland*, p. 21; *Communications to the Board of Agriculture*, I, p. 147; Malcolm, *Agriculture of Bucks*, p. 43; Middleton, *Agriculture of Middlesex* (1798), p. 395.

In *Communications to the Board of Agriculture*, I, p. 167, we are told that the surveyors in the vicinity of the metropolis covered their roads with ballast from four to fourteen inches thick; they filled up no holes or ruts before this stuff was laid on; and, of course, it was usually reduced to sand or jelly by the constant action of all the wheeled vehicles, even before the winter set in. The roads were then one continued slough, or had the appearance of a canal of loose dirt. This prevented material binding together into a hard road. Compare ibid., I, pp. 168–70, and Middleton, *Agriculture of Middlesex* (1798), p. 395.

[3] We have abundant evidence that this was a common method: Horace Walpole, in a letter to Richard Bentley, dated at Wentworth Castle, Aug. 1756, says, "During my residence here I have made two little excursions, and I assure you it requires

common or from adjacent fields; but often they were broken from the near-by quarry and put down on the roads in these large masses, with the expectation that the heavy broad-wheeled waggons would grind

resolution; the roads are insufferable: they mend them—I should call it spoil them—with large pieces of stone." (Horace Walpole's *Letters*, III, p. 445.)

In Brit. Mus., Add. MSS. 28,802, p. 22, the writer says that "After travelling (from Coventry) over four miles of loose stones, we entered Coleshill " (anno 1821). This was one of the best sections of England.

Arthur Young, in his *View of the Agriculture of Oxfordshire*, p. 324, says that about 1760, the roads of Oxfordshire "were in a condition formidable to the bones of all who travelled on wheels. The two great turnpikes which crossed the county by Witney and Chipping Norton, by Henley and Wycombe, were repaired in some places with stones as large as they could be brought from the quarry, and when broken, left so rough as to be calculated for dislocation rather than exercise." See also Young, *Six Weeks' Tour*, p. 319.

The editor of the *Monthly Magazine* (XVIII, Pt. 2, p. 1) tells us that in 1804 "The usual method of making or mending roads in stoney countries is a great nuisance to the traveller. It consists in breaking stones taken out of the neighbouring quarries into masses not much less than a common brick, and spreading them over the line of road. It may be conceived with what pain and difficulty a poor horse drags a carriage over such a track."

In *Parl. Papers*, 1810–11 (240), III, 855, Appendix C, there are given some 'Observations on Highways,' by J. L. Macadam. He says the roads were too often repaired by using gravel. In Staffordshire and Shropshire, the roads were chiefly made of rounded pebbles, averaging ten to twelve pounds weight, mixed with sand; such roads were nearly impassable and very expensive. Many other roads were made of rounded stones from the size of a hen's egg to that of a man's head.

In Brit. Mus., Add. MSS. 28,793, 'An Account of Rev. John Skinner's Tour of Cornwall,' etc., in 1797–8, p. 57, we read: "The roads from Crockern Wells to this place (i.e., Oakhampton) are very rough, owing to their bad method of making them, since they lay no foundation, but throw down large stones, which, as the surface wears, project in the middle of the way. This is the more inexcusable as they have the finest materials at hand, and with a little trouble might have as good turnpikes as any in the kingdom."

Rev. S. Shaw, in his *Tour to the West of England* (1788), p. 219, says that the road near Landenabo was "intolerably rough, but might easily be mended by breaking their hard materials smaller."

In Brit. Mus., Add. MSS. 17,398, p. 51, the writer says: "The roads in general (i.e., between North Leech and Gloucester) were pretty good, but in many places had been lately mended with large stones digged from the way side." In speaking of the roads from Ross to Hereford, he says they are "rough, narrow & hilly. In many places scarcely passable by a carriage. This is in a great measure owing to the method of Amending the Roads, which is by throwing down large Stones which the horses & carriages are to grind to a proper size " (ibid., p. 55). Again, on p. 107, he says: "From Bolverhide over a tollerable gravelly road to Hastings Mill. Here missing our road & turning up to the left by the mill we passed a dreadful piece of road like a step ladder consisting of huge loose stones buried in mud." All this was in the month of June, 1775.

See also Crutchley, *Agriculture of Rutland*, p. 21; Hassall, *Agriculture of Monmouth*,

them down into a suitable road-bed. This method was applied, not only on the ordinary highways, but even on the turnpikes of some of the best counties[1]. At some places, these rough stones, when laid down in the road, were covered over by gravel or sand; but the inevitable tendency was for this finer material to work down and leave the coarser material at the surface of the road[1]. In other instances gravel mixed with earth was thrown into the holes, and as soon as the road became soft with rain, the teams passing over this pressed the gravel down through the soil of the road and left the latter as bad as before[2]. Sometimes, the scraping of the road to remove the mud left piles of this ooze along the sides of the road, while the roadway became flat and lower than the fields, so that the water was allowed to lie on it and keep it soft[3]. In getting an accurate picture of the methods of repairing the roads, we must, however, in all fairness, remember that materials for putting on the roads were not always available in the immediate neighbourhood, and this, in some cases, will account for their condition[4]. In some localities of this kind, the roads were repaired by stones, flints, or other materials, brought considerable distances and wisely applied; in others, they were left in their natural condition, with utter disregard for surface or gradients[5]. Where good roads were difficult to construct or to repair, we sometimes find only narrow causeways, just wide enough to be traversed on horseback, and

p. 104; Donaldson, *Agriculture of Northampton*, pp. 48–49; Billingsley, *Agriculture of Somerset*, p. 260; Bailey and Culley, *Agriculture in Northumberland*, p. 56; Pitt, *Agriculture of Northampton* (1809), p. 231; Duncumb, *Agriculture of Hereford* (1805), p. 142; Stevenson, *Agriculture of Dorset* (1812), p. 439; Vancouver, *Agriculture of Devonshire* (1808), p. 369; Warner, *Tour through the Northern Counties* (1802), ii, p. 139; Warner, *Walk through the Western Counties* (1800), p. 123; Marshall, *Rural Economy of Yorkshire*, i, pp. 47, 182; Farey, *Agriculture of Derbyshire*, iii, p. 266.

[1] See last footnote.

[2] *Parl. Papers*, 1810–11, iii, 855, Appendix C, 'Observations on Highways,' by J. L. Macadam; *Communications to the Board of Agriculture*, i, pp. 167–70; Holland, *Agriculture of Cheshire*, p. 302; Strickland, *Agriculture of the East Riding of York*, p. 267; Gooch, *Agriculture of Cambridge* (1813), p. 291; Rennie, Broun and Shirreff, *Agriculture of the West Riding of York*, p. 86.

[3] *Parl. Papers*, 1819(509), v, 339, 'Report from Select Committee on the Highways,' p. 14, evidence of Wm. Horne.

[4] Stone, *Agriculture in Lincoln*, pp. 45, 89; Rudge, *Agriculture of Gloucester* (1807), p. 333; Gooch, *Agriculture of Cambridge*, p. 291; Vancouver, *Agriculture of Cambridge*, p. 218; Wedge, *Agriculture of Chester*, p. 26; Wedge, *Agriculture of Warwick*, p. 28. Davis, *Agriculture of Wiltshire*, p. 156, says that the private roads were good or bad according to the scarcity of materials. See also Pitt, *Agriculture of Northampton* (1809), p. 231; Duncumb, *Agriculture of Hereford*, p. 142.

[5] Marshall, *Rural Economy of the Southern Counties* (1798), ii, p. 98; Brit. Mus., Add. MSS. 32,442, 'Tour from London to the Lakes' (1799), p. 5.

these were the only means by which the soft, springy soil might be crossed[1].

Having now indicated the chief factors tending to prevent the greatest improvement of the roads, we next turn our attention to the remedies proposed in the way of administration and construction; and here we shall follow the same order as that in which we have discussed the defects.

The complaints against the system of local surveyors and the way in which their negligence proved detrimental to the highways, were met by the proposal that the surveyors employed, instead of being, as too often they were, ignorant and incompetent, should be men of superior ability and experience[2]. The necessity for skilful surveyors was recognized much earlier[3]; but when the law was such that each man held the office but one year and might be required to fulfil these duties without compensation, there was no inducement to the development of skill. In order that the work of the surveyors should be directed and rendered more effective, it was recommended that parishes should be grouped, and that a paid surveyor should be appointed over each district or over each county. By his being able to study these problems and to give all his time to this work, he would be qualified to guide the parish surveyors; and, not being a local official, he would be able to put pressure upon those who, with impunity, refused or neglected

---

[1] Holland, *Agriculture of Cheshire* (1808), pp. 302–8; Wedge, *Agriculture of Chester*, p. 26. Boys, *Agriculture in Kent*, p. 98, says that the highways of the Weald were perhaps the worst turnpike roads in the kingdom. In winter, it was frequently impracticable to ride on horseback along the main roads; consequently, narrow paved tracks, called horse-paths, were paved with stone or formed by sea-beach, at one side of the roads.

In Dr Pococke's *Travels*, we have frequent references to this means of making roads passable. In his 'Journey into England from Dublin' (original MS.), p. 14, he says, "They have in all this country from Skipton & on to the South east a sort of causeway made of hewn free stone about 18 inches broad and a yard long, which are laid across ways, so the road is but three feet wide, & not very secure for horses, not used to it—though not apt to slip by reason of the softness of the stone; on these they ride when the roads are bad, as they are in most parts after a rain."

In his *Travels in England* (1764), I, p. 88, he tells us that near Lavenham, in Suffolk, "They make the roads by causeways, I found them tolerable good." And on I, p. 20, he says: "There is a road to Smarden......impracticable in winter except by a narrow stone causeway which they make where they have stones, but where they have not they form tracks ˄bout two feet wide and half a foot deep. They put into them either gravel from the sea beaches or other gravel or pebbles or bavins laid along and covered with earth where they have no other materials."

[2] *Parl. Papers*, 1772, xxxi, Report No. 12, from the Committee appointed to consider the Highway and Turnpike Acts of 7 Geo. III, p. 4; *Parl. Papers*, 1819 (509), v, 339, 'Report from Select Committee on the Highways,' p. 342.

[3] Shapleigh on *Highways* (1749).

to obey the behests of the parish surveyor[1]. The work of the surveyor was so important that it was even recommended that surveyors-general should be appointed by a parliamentary commission, and each be placed over twenty to fifty miles of road, with full power to purchase all necessary materials, hire labourers, and oversee all the repairs[2]. In any case, it was recognized that the only effectual remedy for improving the standing of the surveyor was to appoint those who would be active, and free from local prejudices and attachments, to superintend the repairs over a larger area, and in this way firmly and impartially administer the laws for the public good[3].

For the improvement of the finances of the roads, there were several suggestions that received a good deal of attention. In the first place, it was a potent means of squandering the public funds to have a vast number of small trusts, each with its own paid officers; and to eliminate this element of waste many were in favour of some means by which these small principalities might be consolidated and their interests merged, so as to do away with a host of parasitic officials who were draining the trusts of their funds. By lessening the number of clerks and other officers, and the expense of renewing so many local Acts, there would be a reduction of useless expenditures; and by enabling the consolidated trusts to employ more competent officers, the country would secure more uniform and efficient administration of the turnpikes. Through this means, also, the increase of the great debts under which many trusts were struggling might be prevented and some of these heavy obligations might even be liquidated[4]. In connexion with this,

[1] *Parl. Papers*, 1808 (275), ii, 459, 'Second Report of the Committee on Highways,' p. 129; Brit. Mus., T. 1157 (4), 'Highways Improved,' p. 12; Brit. Mus. 8245. bb. 14, 'Letter to the Inhabitants of Hertford,' which shows the way in which petty local jealousies prevented adjoining turnpike trusts from working together for the repairing of a line of road, and the necessity of larger jurisdictions; *Parl. Papers*, 1772, xxxi, Report No. 12, from the Committee appointed to consider the Highway and Turnpike Acts of 7 Geo. III, p. 4; *Parl. Papers*, 1809, iii, 431, Appendix A, Section VII; *Parl. Papers*, 1819 (509), v, 339, 'Report from Select Committee on the Highways,' p. 342 et seq.

[2] *Parl. Papers*, 1810–11 (240), iii, 855, p. 48, Appendix (C) No. 4, 'On Surveyors.'

[3] See references in footnote 1 to page 238; also Wedge, *Agriculture of Chester*, p. 63; Bailey and Culley, *Agriculture in Northumberland*, p. 56; Bishton, *Agriculture of Salop*, p. 19; Gentleman (pseud.), *Proposals for Amendment of Roads* (1753), p. 8 et seq.; Whitaker, *Loidis and Elmete*, p. 81.

[4] *Parl. Papers*, 1833 (703), xv, 409, 'Second Report by Lords Committee on Turnpike Returns,' pp. iii–iv, and the evidence of James Macadam, pp. 488–9. Macadam would consolidate the trusts, not in continuous lines, but in groups in the form of a star and to the extent of 100 to 150 miles in each. See also *Parl. Papers*, 1808 (225), ii, 333, 'First Report of the Committee on the Highways,' p. 7. *Parl. Papers*, 1833 (703), xv, 409, 'Second Report by Lords

it is to be noted that many of those best qualified to judge advocated the abolition of the local administration of roads and the substitution of a more general system of control[1]. This proposal had been advanced as early as 1763[2], with the object of securing uniformity in the quality of the roads of all the parishes, but nothing had been done toward its realization. In 1808, and again in 1809, a committee of the House of Commons recommended that the roads of the kingdom be placed under a parliamentary commission, whose duty it would be to superintend their maintenance, to audit the accounts of the trusts, to make suggestions to the local officials for their improvement and to appoint county surveyors over the actual performance of the work[3], but the work itself was to be left in the hands of the local bodies. In 1817, Edgeworth, after showing the inadequate repairs done by the parishes, declared that nothing but a general system for all the roads of the kingdom could be effectual[4]. His remedy was in substance much the same as that of the committee of 1809: he would have the appointment of a body of commissioners, resident in London, who should annually audit the accounts of all the roads, and whose sanction should be required before any proposal for improvement of an existing line or construction of a new line could be carried into effect. On the basis of a survey, a map should be made of each road, giving all details; and guided by these maps, with accompanying instructions, the commissioners should exercise complete control[5]. Evidently parochial repair of highways, causing great differences in different parishes, was not a system that would produce lines of continuous good roads throughout the kingdom, and many were casting about for some means

Committee on Turnpike Returns,' p. 586, gives a statement of the debts, incomes and expenditures of the Turnpike trusts for the years 1821 and 1829, showing the increase in net debt from £5,049,433 in 1821 to £7,304,803 in 1829.

[1] See Hawkins, *Laws of Highways*, p. 42.   [2] Ibid.

[3] *Parl. Papers*, 1808 (225), ii, 333, 'First Report on the Highways,' p. 7; *Parl. Papers*, 1808 (275), ii, 459, 'Second Report on the Highways,' p. 131; *Parl. Papers*, 1809, iii, 431, Appendix A, Section VII. Ibid., p. 459, gives a summary of the improvements suggested.

[4] Edgeworth on *Roads and Carriages*, p. 6. He says: "The system of parish repairs of highways is in itself sufficient to prevent improvement" of the poorer roads. "In fact, nothing but a general system for all the roads of a kingdom can be effectual; it is in vain that one parish repairs its roads if its neighbours will not do the same; for the neglect of any one part renders the whole impassable, without a team of sufficient strength to get over what has been neglected."

[5] Before the commissioners authorized any plan for a new or an old road, in every case of importance they were to send an eminent engineer to report on the subject. Details were to be given regarding each road; and a book was to accompany each map, explaining the advantages and defects of the existing road and pointing out such improvements as should be attended to in future. Ibid., pp. 7–11.

to supersede this. The Report of the committee of 1821 also recommended having the accounts of the trusts audited by a parliamentary committee, in order to secure the judicious application of the trust funds. Again, in 1833, another committee declared their opinion that trustees of roads should not be allowed to borrow money on the security of the tolls exceeding in amount three years' revenue of such tolls; and recommended, upon the universal testimony of witnesses, a system of general control over the roads for the purpose of effecting an economical and skilful management to reduce the existing debt[1]. It is clear, therefore, that the people were gradually waking up to the waste that was going on and that earnest minds were at work endeavouring to secure a less prodigal and more fruitful application of the public money contributed for the roads.

The strong desire was also expressed that the system of statute duty, which had been in effect since 1555, should be abolished. Many were the evils connected with it, and its lack of effectiveness for the purpose intended was well known. We have seen that there were some persistent advocates of the overthrow of the system by the middle of the eighteenth century; and the baneful effects accompanying it had by no means abated since that time. In 1763, a Bill had been introduced to substitute an assessment for the statute labour[2], and in a Bill of 1810 to amend the highway laws it was proposed that those who should refuse to do statute duty on being regularly summoned should pay double its ordinary estimated value. This shows us the attitude of some toward the performance of this work; and there must have been many who took that attitude, else there would have been no need for making this a part of the proposed law for the whole country[3]. As a compromise, in a Bill of 1813–14, it was proposed that if the roads could be better repaired by a composition in money than by performance of statute duty, the Justices should be at liberty to require the money payment[4]; and the same provision was incorporated in another Bill of 1821[5]. As already noted, composition for statute duty was in effect in some places and its good results were known: then why should it not be adopted universally? Moreover, it was apparent to thoughtful people that statute labour was a remnant of personal servitude; it

---

[1] *Parl. Papers*, 1833 (703), xv, 409, 'Second Report of Lords Committee on Turnpike Returns,' pp. iv, 458, 463; also Hansard's *Parliamentary Debates*, 1835, xxix, pp. 1183–92; Horsfield, *History of Sussex*, i, pp. 96–97.

[2] *House of Commons Papers*, Bills Public, 1760–5, iv, No. 133.

[3] *Parl. Papers*, 1810, i, 233. The same provision was inserted in a Bill three years later, *Parl. Papers*, 1813–14, ii, 611, 615.

[4] *Parl. Papers*, 1813–14, ii, 611, 615.

[5] *Parl. Papers*, 1821, ii, 913. See table No. 2, p. 982, for composition rates.

might as well be argued that rents paid in kind were easier and more equitable than money rents as to defend the custom of mending the highways by compulsory labour. Other services formerly required had been commuted for money payments and it was foreseen that commutation for this service also must eventually take place; it was, therefore, insisted that it was better to reap the advantage of such a change as soon as possible[1]. Although both argument and the logic of facts pointed to the desirability of such a change, its actual accomplishment was not realized until 1835.

Another recommendation in furtherance of good roads was that the custom of employing pauper labour on the highways should be discontinued. This evil did not make its appearance till about the beginning of the last century, so that it was not so firmly entrenched as some others. Investigation and observation had disclosed the fact that money was squandered in this way, with little or nothing to show for it, and Macadam was very outspoken in his opposition to its continuance[2]. The study of the working of the Poor Laws revealed their iniquity; and the payment of pauper labourers for work which they did not perform was strongly declaimed[3].

The operation of the toll-bars or turnpike gates was by no means satisfactory, either to the trustees or the public, for in the collection of the tolls there was much evasion, partiality and downright fraud. If they were not farmed out, the tolls would be reduced in amount by the embezzlement of the collectors and so it was the almost universal custom, which was later made obligatory[4], for the trustees to let them for a definite amount. The statutory list of exemptions from the payment of toll was made the basis upon which others, through wilful misrepresentation, also claimed exemption. Injustice was felt to result from the great number and variety of ways in which the tolls were cancelled or relaxed in favour of certain occupations, notably agriculture. A farmer's teams carrying manure, or the implements of husbandry, or the product of the harvest field, no matter how heavily laden, or how much injury they did the road, might pass toll free. Sometimes, in certain localities, their particular industries would also secure release from the payment of the usual tolls[5]. Probably a more

---

[1] Edgeworth on *Roads and Carriages,* p. 33; also Macadam's statement in *Parl. Papers,* 1819, v, 339, p. 368.

[2] *Parl. Papers,* 1819, v, 339, p. 368; *Parl. Papers,* 1824 (392), vi, 401, 'Report of Select Committee on Labourers' Wages, pp. 411–13.

[3] On this whole subject, see especially the 'First Report of the Poor Law Enquiry Commissioners,' 1834, with evidence and appendixes.

[4] Act 13 Geo. III, c. 84, sec. 31.

[5] Acts 17 Geo. II, c. 18; 20 Geo. II, c. 6; 24 Geo. II, c. 11.

significant fact in the administration of the tolls was the great amount
of composition that was allowed, for this seriously reduced the revenues
of the trusts[1]. Some were in favour of retaining the toll-bars, and
even a parliamentary committee in 1808 declared that the excellence
of the roads was due to the application of the principle, which was
dictated by both justice and expediency, that those who used the roads
should sustain them[2]. But the fact is that it was not alone the carriers,
but the public as a whole, that reaped the benefits from good roads,
and therefore the upkeep of the roads should not be a charge upon
those who used the roads, but upon the public treasury, for all derived
the advantages from them[3]. It was, therefore, inevitable that in
time the turnpike gates should be taken down and a more equitable
method adopted to secure the end desired; but this did not come
until after the middle of the nineteenth century, and then not without
agitation accompanied by riot and destruction[4]. Gradually, by Acts
of Parliament, the trusts were reduced in number, until in 1895 the
only remaining one ceased to exist. It may be remarked in passing
that, almost simultaneously with the extinction of the last turnpike
trust, the authority of the parish over the highways ceased.

From those who were interested in road improvement there was
much outcry, especially after 1800, against the use of the existing
broad wheels, narrow wheels, and heavy weights, the evils of which
we have formerly noted. As has been said, about the middle of the
eighteenth century public sentiment and legislation were in favour of
wide wheels; and notwithstanding the fact that in 1767 very few of
these waggons were used by farmers[5], it seems that their use, at least
on some roads, was becoming well recognized[6], although probably not
very common. With the increasing use of broad wheels the opposition
to the narrow wheels was the greater; and, of course, the more favour

---

[1] 'Report of House of Commons Committee to inquire into the application of
Money,' etc., *J., H. of C.*, xxix (1763), pp. 646–64. The Committee of 1808 recommended that keepers of weighing engines and toll-bars should be prohibited from
compounding with proprietors of waggons, *Parl. Papers*, 1808 (275), ii, 459,
'Second Report of the Committee on Highways,' p. 142, Appendix No. 4 (A).

[2] *Parl. Papers*, 1808 (275), ii, 459, 'Second Report of the Committee on
Highways,' p. 131.

[3] Pagan, *Road Reform* (1845), pp. viii, 200, 223–6; Hansard's *Parliamentary
Debates*, 1835, xxix, pp. 1183–92.

[4] Pagan, *Road Reform* (1845), pp. 200, 204, 226–8.

[5] Young, *Farmer's Letters*, p. 271.

[6] *Oxford Gazette and Reading Mercury*, May 18, 1767, p. 2, letter from J. Smith
regarding the road from London to Marlborough. In all probability, both oxen
and horses drawing together on the same load was not at all uncommon, v. ibid.,
May 11, 1767, p. 3.

that the wide wheels secured the more hostility did the narrow wheels encounter[1]. But, in time, with more extended observation and knowledge of the effects of broad wheels, opinions regarding them began to change; and as early as 1809 a Bill was framed, giving the owners of waggons, the wheels of which were sixteen inches wide, two years to wear out their wheels and make them conformable to the nine-inch width[2]. While this Bill did not become law, it shows us the tendency of opinion at that time. Three years before, the committee appointed to report on the use of broad wheels had said that no roads could bear the pressure of such enormous weights as eight tons, drawn by eight to twelve horses, especially when the wheels were sixteen inches broad and of a conical shape[3]. In 1819, a parliamentary committee declared against exempting wide wheels from the payment of tolls, because, although intended to consolidate the roads, they really did more harm than good in that they made narrow ruts wider and deeper[4]. The consensus of public opinion, therefore, was evidently opposed to the wide wheels; but it would seem as if the opposition were not so much because they were wide as because they were conical in construction. From the opening of the last century especially the best thought turned away from conical and toward cylindrical wheels, for the reason that with the latter there could not be any dragging motion such as we noted in the use of the conical wheels. The adoption of the cylindrical wheels, therefore, was strongly advocated, because of their effects in the preservation of the roads and in rendering the draught easier[5]; and

---

[1] See a letter from "A Road Trustee," in the *Public Advertiser*, May 29, 1788, pp. 1–2, stating that if Parliament would compel the abolition of all narrow wheels it would be highly advantageous to the farmers and of great economy to the public. If his postscript were true, that it was common for a stage coach to carry five tons' weight on wheels an inch and a quarter wide, it would be no wonder that the roads were cut into deep ruts. With him seem to agree "Anti-Caput Mortuum," in his letter in ibid., June 12, 1788, p. 1, and "The Original Projector," in his letter in ibid., Aug. 17, 1789, p. 1; also ibid., Jan. 15, 1790, p. 2, letter from "A Road Trustee," and ibid., Apr. 27, 1790, p. 2, letter from "W. J."

[2] *Parl. Papers*, 1809, I, 679. At a much earlier time than this some realized the disastrous effects of the heavy loads that were drawn upon these wide wheels. See, for example, Brit. Mus. 213. i. 1 (44), p. 2, and also the preambles to many Road Acts contained in the collections, Brit. Mus. 213. i. 1, 213. i. 2, 213. i. 3, etc.

[3] *Parl. Papers*, 1806, II, 249, 'Second Report of the Committee on Acts regarding the Use of Broad Wheels,' etc., under the heading A. 1, "Carriages." This committee said it was better to restrict wheels to a width of nine to twelve inches. Again, in Appendix No. 9 (B), they said that the great injury done to the roads was by rolling waggons, carrying from eight to ten tons, which pressure the hardest road could not stand. See also *The Times*, June 3, 1816, p. 2.

[4] *Parl. Papers*, 1819, v, 339, 'Report on the Highways,' p. 342.

[5] *Parl. Papers*, 1795–6, XLVIII, Report No. 132 on Turnpike Acts 13 Geo. III

wheels six to nine inches wide progressively replaced those of greater width.

Closely related to this, there was the suggestion that the weighing engines should be abolished. The reasons for their failure to acceptably serve the purpose intended by Parliament we have discussed elsewhere, and need not now recapitulate. Before the end of the eighteenth century, they had been shown to be both inadequate and vexatious, and their abandonment and removal had been recommended[1]. They had already been taken down on the Surrey roads[2], and so inconvenient, troublesome and grievous were they on the highways of the metropolis that by 1828 they had all been abolished there[3]. In several cases it was proposed to do away with all of them; for example, in 1820 it was provided in a Bill to amend the turnpike laws that as soon as possible after Jan. 1, 1822, all weighing engines should be sold for the best prices that could be obtained and the money appropriated to repairing and maintaining the roads[4]. But the next year the same Bill was amended to allow the setting up of these machines and the weighing of all waggons[5]. As early as 1795, and again in 1806, a committee of Parliament had shown that they were of little or no use, when placed at such extreme distances; that they could easily be evaded, or contracted for, as suited the weigher and the carrier; and that if they were to be kept, there should be one every fifteen miles at least, with proper persons to attend them[6]. Such nuisances, however, die slowly and weighing engines were retained for many years after this strong declaration against them. In 1833, again, a committee of the House of Lords said that the purpose of these engines had been so much defeated by the practice of compounding for overweight

and 14 Geo. III, p. 5; *Parl. Papers*, 1797-8, LII, Report No. 147, p. 2; *Parl. Papers*, 1808 (225), II, 333, 'First Report of Committee on the Highways,' p. 3; *Parl. Papers*, 1808 (275), II, 459, 'Second Report of Committee on Highways,' p. 130; *Parl. Papers*, 1819 (509), V, 339, 'Report from Select Committee on the Highways,' p. 7; *Parl. Papers*, 1820 (301), II, 301, 'Report of Select Committee on Turnpike Roads and Highways,' p. 10; *Parl. Papers*, 1833 (703), XV, 409, 'Second Report of Lords Committee on Turnpike Returns,' p. 540; Deacon on *Stage Waggons*, pp. 20-23; Paterson, *Practical Treatise on the Making and Upholding of Public Roads*, pp. 79-80.

[1] *Parl. Papers*, 1795-6, XLVIII, Report No. 132, pp. 2-3, 9, 11; *Parl. Papers*, 1797-8, LII, Report No. 147, pp. 1-2.

[2] *Parl. Papers*, 1795-6, XLVIII, Report No. 132, p. 6, evidence of George Thakston, Clerk of the trustees of the Surrey roads.

[3] *Parl. Papers*, 1828 (311), IX, 23, 'Second Report of the Commissioners of the Metropolis Turnpike Roads,' p. 25.

[4] *Parl. Papers*, 1820, I, 333.   [5] *Parl. Papers*, 1821, II, 913.

[6] *Parl. Papers*, 1806, II, 249, Appendix No. 9 (B).

that they urged their abolition. But by that time their number had greatly diminished, for no overweight tolls were taken on the Shrewsbury and Holyhead road, which Telford had re-made, and within a radius of fifty miles around London there was only one weighing engine left[1].

Many suggestions were made for the improvement of road construction, most of which are to us self-evident because of the results which have been secured by their practical application. But we must keep in mind that it was not until the latter part of the eighteenth century, and, more particularly, the first quarter of the nineteenth century, that road building was taken up as a regular profession, and, until then, the proposed methods of improvement were tentative and experimental, rather than based upon scientific principles. The old plan of scraping the road and leaving the drift piled along the sides until it could be carted away, was still carried on in some places and, notwithstanding its evils, had still a few supporters[2]. It needed constant reiteration that if roads were to be maintained in good condition the hedges and trees at the sides must be kept low in order to allow the natural agencies of sun and wind to dry the road surface[3]. The extreme convexity, which, as we have seen, was the cause of great inconvenience and was largely contributory to the roads becoming "quartered," was the occasion of much complaint; and while it was recognized that roads should have their surfaces slightly convex in order to allow the water to run off, too much of the round form should be avoided in their construction[4]. With the desire for greater speed there came also the desire to see the roads shortened by cutting off corners, reducing hills, and lessening the number of turns and the serpentine course which many of the roads had to take, when refused the straight course across the estates of landlords. A notable case was that of the connection from London to Brighton; after the latter had become a fashionable watering-place and the number of coaches had greatly increased, three routes were chiefly used, namely, one through Lewes, 58 miles in length, one through Horsham, 57 miles, and one through Cuckfield, 54 miles, which was the shortest and most fashionable of the three. The latter was chosen by post-chaises,

---

[1] *Parl. Papers*, 1833 (703), xv, 409, 'Second Report of Lords Committee on Turnpike Returns,' pp. iv, 436, 496.

[2] *Grand Magazine of Universal Intelligence* (1758), I, p. 327; Middleton, *Agriculture of Middlesex* (1798), p. 399.

[3] Middleton, op. cit., p. 398.

[4] Deacon on *Stage Waggons*, pp. 39–42. In advocating the reduction of the convexity, he went too far the other way, and urged that the road surface should be flat and that nothing but flat (cylindrical) wheels should be used on them.

private carriages and most of the coaches; but in spite of the opposition of established interests along these roads the directions of the roads were changed in order to cause a still greater reduction of the distance[1], and we are told that in 1818 the shortest distance between these places was 50 miles[2]. This is only one illustration of what was going on all over England, especially on the great highways of trade and travel[3]. Along with this shortening of roads there was a corresponding desire to have the gradients made more favourable. Instead of putting a road over the top of a hill, or cutting through a hill, merely to get the road as straight as possible, it was realized, by those who had most experience as road surveyors, that it was of greater importance to have the road somewhat near the level, even although the distance should be a little longer, and, therefore, it was usually better, where possible, to go round the hills than to go over them[4]. This principle found its way by degrees into practice and roads were turned from their old courses over hills and made to pass, even at the expense of circuity, along lower levels; while in the construction of new roads the general level was carefully considered at the outset[5].

It would take too long to describe all the various plans, the application of which, it was thought, would result in better roads. Our discussion of the evils in connexion with the construction and administration of the highways and of the vehicles which were used upon them, will suggest corresponding remedies, which we have not specifically taken up[6]. But probably the greatest lack in securing

---

[1] Blew, *Brighton and its Coaches*, pp. 21–23.

[2] *The County Chronicle and Weekly Advertiser*, Nov. 10, 1818, p. 3.

[3] See also *Public Advertiser*, July 10, 1786, p. 2; *The Times*, June 23, 1802, p. 3; ibid., Sept. 25, 1811, p. 1; ibid., Nov. 11, 1818, p. 3; ibid., Sept. 10, 1821, p. 2; ibid., Feb. 2, 1824, p. 3; Brit. Mus., Add. MSS. 35,691, pp. 68; Paterson, *Practical Treatise on the Making and Upholding of Public Roads*, p. 13; Herepath's *Railway Magazine*, N.S., v, p. 105.

[4] Paterson, *Practical Treatise on the Making and Upholding of Public Roads*, pp. 13–15. He did not want perfectly level roads, but those with easy gradients, for where a horse dragged a load over a long stretch of road, having here a gentle acclivity, and there a declivity, it would not fatigue the animal so much. Ibid., pp. 17–18.

[5] Whitaker, *History of Craven*, 3rd edition, p. 188.

[6] It may not be amiss to mention one or two more suggestions that were made in the interest of the roads.

As early as 1806 it was recommended that railways should be used for carrying heavy goods, thus leaving only light traffic for the roads, since this would tend to prevent the cutting of the roads into ruts (*Parl. Papers*, 1806, II, 249, under heading "Preservation of the Roads"). It was early recognized in the history of the iron railways that they would be the best means for conveying goods by land, but that

better highway facilities was that before the period we are now studying
no engineer had turned his attention to road construction; consequently, the proper materials for road building, the best methods
for draining roads, their proper formation, and similar subjects, had not
been carefully investigated by anyone competent to take them up.
The earlier method had been to try to regulate the weight of the load,
the kind of vehicle and the number of horses according to the nature
of the road, in other words, to make the traffic suit the road.   But
a new era was at hand when, by the application of skill, the road would
be made in accordance with the traffic which it had to bear; and
instead of thinking of devices for keeping the surface of the road smooth,
a method of construction was adopted which began with the foundation,
and upon a firm substratum built an enduring superstructure that
would be equally satisfactory for pleasure and for permanence.

The distinction of being the first road engineer belongs to John
Metcalfe[1].   He was born at Knaresborough, in Yorkshire, in 1717,
the son of poor working people.   He became blind when only six
years of age; but he grew up active and strong and soon knew every
place in the vicinity of his home.   He early learned to make his way
around the town and then would take long walks in the lanes and
fields alone, to the distance of three miles or more, and return.   His
father kept horses, and the boy learned to ride and in time became
a skilful and fearless horseman.   Then he learned to hunt and to
follow the hounds.   He had great confidence in himself and was
always ready for any adventure, hunting, swimming and diving being
some of his pastimes.   He travelled widely, sometimes walking, sometimes on his horse, but he never missed his way even over the bad

even stone railways were deserving of a fair trial (see also *Parl. Papers,* 1809, III,
481, Appendix B on "Roads").

The expenses of securing a turnpike Act and of renewing it at the expiration
of the customary twenty-one years' duration were burdensome (see for example
Brit. Mus., Add. MSS. 85,691, pp. 67–68, 94, 116), and it was vainly sought for many
years to reduce these heavy charges, by comprising in one general Act such clauses
as applied to all trusts and especially by dispensing with the attendance of witnesses
in London to prove the contention of the petitioners for the Act.   The legal fees
connected with the passage or renewal of an Act were also very heavy.   An effort
was made to have all expenses on renewals of turnpike Acts abolished, and this
was attained in 1831 by having all expiring Acts grouped into one annual renewal
Bill.   See p. 237.

[1] Our knowledge of this man is contained in his autobiography, *The Life of
John Metcalfe, commonly called Blind Jack of Knaresborough.*   It gives an account
of his various contracts for making roads, erecting bridges, etc.   From this smal
work, Smiles obtained his material for Metcalfe's life and achievements, as given in
Lives of the Engineers, I, Chap. v.

roads. It would take too long to follow his early career, which he describes in his autobiography[1], but it is sufficient to say that by his adventures he learned the roads perfectly and could even act as guide[2].

About 1765, an Act of Parliament had been obtained for making a turnpike road from Harrogate to Boroughbridge. Metcalfe saw that many similar undertakings would be carried out in this and the adjoining counties in the following years and he decided to take up this new line of business. He got into the company of the surveyor and agreed to make about three miles of the road. He obtained the contract and went to work; and by his diligent attention to business he completed the work much sooner than was expected, to the entire satisfaction of the surveyor and trustees[3].

Soon there was a mile and a half of turnpike road to be constructed between Knaresborough bridge and Harrogate, which Metcalfe also agreed for and completed[4]. But we need not here describe the many

---

[1] *Life of John Metcalfe*, pp. 1–50.

[2] Ibid., pp. 20–25. A few other particulars about this man may not be un-interesting. In London, he met with a North country man who played on the small pipes, and who frequented the houses of many gentlemen there. By his intelligence, Metcalfe found out several who were in the habit of visiting Harrogate, and among others, Colonel Liddell, who was a Member of Parliament for Berwick-upon-Tweed. The Colonel lived near Newcastle-upon-Tyne and on his return home from London he had stopped for three weeks at Harrogate, for a number of years successively (p. 55). The May following his acquaintance with Metcalfe, Colonel Liddell notified his friend, who was then in London, that he was going to Harrogate and that he might go down with him, either behind his coach or upon the top. Metcalfe politely declined the offer. saying that he could walk as far in a day as the Colonel could travel. His statement was proved true, for on the journey down he stopped every night with the Colonel (pp. 56–58). But he arrived in Harrogate before the Colonel's coach, for the latter stayed over Sunday in Wetherby.

After his marriage, he began the business of fishmonger, carrying fish out through the country and selling them (p. 75). This was unremunerative and he gave it up. Then next he took part in the rebellion of 1745 (p. 75 et seq.).

In 1751, Metcalfe commenced a new employment: he set up a stage waggon between York and Knaresborough, which was the first one on that road, and conducted it constantly himself, twice a week in the summer season and once a week in the winter season (p. 123). This business, together with the occasional conveyance of army baggage, employed his time till he began making roads.

[3] *Life of John Metcalfe*, pp. 124–5. About the time of finishing this road, the contract for the building of a bridge at Boroughbridge was advertised. The same surveyor that Metcalfe had worked under before was appointed to survey the bridge; and Metcalfe told him that he wanted to undertake the work, though he had never done anything of the kind before. The plan of the bridge he had fully worked out in his mind and the job was allotted to him (p. 126). The men he employed gave him the utmost aid and he soon had the arched bridge completed.

[4] On this line of road, between Forest-Lane head and Knaresborough bridge, there was a bog, in a low piece of ground, over which to have passed was the nearest

roads which he afterwards constructed and the difficulties he met with and overcame[1]. Most of these roads were in Yorkshire and Lancashire, and some of them were very important in connecting these two counties at various points. He was also employed on roads in Derbyshire and Cheshire. At one time he had about four hundred men employed on a certain piece of road nine miles in length, working at six different parts of the road[2]. The difficulties of drainage and construction of this road were very great, for the ground was soft; but Metcalfe got his men to lay down bundles of heather or ling in the right way in the bottom of the road and then covered it with stone and gravel, until the whole road was made firm and durable[3]. The total length of turnpike roads built under his direction was about one hundred and eighty miles, for which he received in all about £65,000[4]. He completed his last road in 1792, when he was seventy-five years old.

Metcalfe was the first, and by no means the least important, of the road engineers. With him we have the first application of scientific principles in the construction of highways. Wherever possible, provision was first made for drainage of the roads, by having ditches along the sides and the roads elevated. A firm foundation was next to be looked to, and, if necessary, the soft surface soil of the road was dug out, until the stones could be placed on a firm bottom. These things being attended to, his final work was to secure a smooth convex surface, so that the water might readily flow off to both sides and not lie on the road to soften it. It will be evident that this was a great advance in road-building over that which prevailed before his time, and, although the amount of road which he constructed was relatively small, his name comes down to us as justly celebrated.

The work of the engineers Smeaton and Rennie, though to some extent directed toward the improvement of the highways, was comparatively unimportant in that line: their activities being chiefly concerned with hydraulic engineering and bridge work. The most distinguished names connected with the development of the modern system of road construction are those of Thomas Telford[5] and John

---

way; but the surveyor thought it impossible to make a road over it. Metcalfe, however, assured him that it could be easily done, and the surveyor told him that if he would put the road across he would be paid for the same length as if he had gone round. Metcalfe put whin and ling in the bottom of the road, and made it as good as any he had constructed (ibid., p. 127).

[1] These are given on pp. 128–52 of his *Life*.

[2] Ibid., p. 134.   [3] Ibid., p. 135.

[4] Making these roads involved also the building of many bridges, culverts and retaining walls for them.

[5] Our information regarding Telford is largely obtained from his papers, etc.,

Loudon Macadam, the latter of whom has had his name perpetuated down to the present day by the macadamized roads. To the labours of these two men, successively, we now turn our attention.

Toward the end of the eighteenth century, increased interest was aroused in the improvement of the longer routes within the kingdom and in perfecting the connexion of London with the chief towns of Scotland and with Ireland. Telford, whose name in bridge building had earned for him a worthy reputation, was also a great road-maker[1]; and he was early called upon to advise as to repairing the road between Carlisle and Glasgow, which had fallen into a wretched condition, and also as to the formation of a new road from Carlisle, across the counties of Dumfries, Kirkcudbright and Wigton, to Portpatrick, for the purpose of ensuring a more rapid communication with Belfast and the northern parts of Ireland[2]. The road between Carlisle and Glasgow was ruinous; the tolls had been raised as high as they could be with any prospect of increasing the funds for its improvement. The trustees seemed to be helpless and could do nothing; a local subscription was tried and failed, for the district passed through was very poor[3]. The road was also in debt; but as it was absolutely required for more than merely local purposes it was finally determined that its reconstruction should be undertaken as a work of national importance. Under the Act of 1816[4], the sum of £50,000 was granted by Parliament for this purpose and the works were placed under Telford's charge.

The road was intended to vastly facilitate the mail-coach traffic and also to make it easier for heavy coaches and waggons carrying from one to four tons. Consequently, the middle of the road had to

which have been edited by Rickman, under the title of *The Life of Telford*, and from his evidence before Parliamentary Committees.

[1] Telford was born in 1757. In 1780 he visited Edinburgh, where for two years he studied drawing and the architecture of abbeys, castles, etc., after which he came southward, where his work would be greater and more remunerative (*Life of Telford*, p. 14 et seq.). He began his early life as a mason in his district of Eskdale, where he was born. "Wherever regular roads were substituted for the old horse tracks, and wheel carriages introduced, bridges (numerous, but small) were to be built over the mountain streams" (ibid., p. 2). His practice in bridge building and in other like work gave him great experience in the details of road construction (ibid., p. 3).

[2] "It was only in the latter part of the last century (the eighteenth century) that the western border or march between North and South Britain was rendered productive or valuable by a regular system of improvements," when the Duke of Buccleuch caused it to be intersected with roads (ibid., p. 4).

[3] See Report of 1814–15 on the Carlisle and Glasgow road, in Vol. III of the British Documents of that year.

[4] Act 56 Geo. III, c. 83.

be made solid and the ascents regular and easy[1]. Of this new line, Telford had charge of sixty-nine miles; the rest was. already in charge of local trustees, under former Acts of Parliament[1]. Near Glasgow, no material for covering the road existed, nor had any such ever been used, except hammer-broken rock. But Telford was convinced of the necessity of having, first of all, a solid foundation, upon which this broken rock ("metalling") might be laid as a covering; and this firm foundation he obtained by laying solid masses of stone in the most compact manner[2]. He paid special attention to two points: first, to lay out the road as nearly as possible upon a perfect level, so as to reduce the draught for horses hauling heavy loads—one in thirty being about the heaviest grade at any part of the road; and, second, to make the working or middle portion of it as firm and substantial as possible, so as to bear, without shrinking, the heaviest weight likely to be carried along it[3]. On these principles the road was constructed; the old road, one hundred and two miles, was shortened by nine miles; and the sixty-nine miles made by Telford had probably no equal in any part of the kingdom[4]. Of this series of changes, Telford said: "To persons who were in the habit of travelling in Lanarkshire previous to these improvements, the change was surprising as well as gratifying; instead of roads cut into deep ruts through dangerous ravines, jolting the traveller, and injuring his carriage,—or leading him, if on horseback, plunging and staggering, circuitously over steep hills, the traveller

---

[1] *Life of Telford*, pp. 178–9.

[2] Ibid., p. xx. A book was published about that time recommending in preference a substratum of vegetable earth, or even elastic bog, for any line of road. By having elasticity, it was thought that a road would be easier for travel. This doctrine was asserted at some length by Macadam, in his evidence before a Select Committee of the House of Commons on Highways in 1819 (v. their Second Report, June 25, 1819, pp. 23–24). This plan had some supporters, but Telford saw that by such construction there would be a perpetual uphill draught and his decision in favour of a solid foundation has been proved judicious.

[3] Ibid., pp. 179–80, 212. Telford's method of forming a road was as follows:

It was composed of three layers. The lowest layer was of hard stones, seven inches in depth, carefully set by hand, with the broad ends downwards, all cross-bonded or jointed, and no stone more than three inches wide on the top. The space between these was then filled up with smaller stones, packed by hand, so as to make a firm and even surface. The layer above this consisted of properly broken hard stone laid to a depth of five or seven inches: none of the pieces to weigh more than six ounces, or to be more than two and one-half inches across. The upper layer was a binding of gravel placed over all, to the depth of one ihch. Under the bottom layer, at a distance of one hundred yards apart, drains crossed to the ditches. The result was an easy, firm and dry road, capable of use in all kinds of weather and requiring little repair; for even in wet weather the curvature of the surface allowed the water quickly to drain off. [4] Ibid., p. 179.

has now smooth surfaces, with easy ascents, rendered safe by protecting fences. Such advantages being equally beneficial to all ranks of society, are of the first importance in a civilized nation[1]."

The reconstruction of the western road from Carlisle to Glasgow, which Telford had thus satisfactorily carried out, soon led to similar demands from the people on the eastern side of the kingdom. They wanted roads which would enable both passenger and mail coaches to perform their journeys in much less time than hitherto. There was a desire for the wider and more rapid dissemination of the increasingly important political and commercial intelligence from both the English and the Scottish capital; and, urged by the public, the Post Office authorities were aroused to unusual efforts in this direction. Surveys were made and roads laid out, so as to improve the main line of communication between London and Edinburgh and the intermediate places. The first part of this road that was constructed was that from Edinburgh, through Coldstream and Wooler, to Morpeth, which saved more than fourteen miles between the two points and secured a line of road of much more favourable gradients[2]. In 1824, under the direction of the Post Office and with the authority of the Treasury, Telford proceeded to make detailed surveys of an entirely new post-road from Morpeth to London. These extended over several years and all the arrangements had been made for beginning the work, when the advantages and possibilities of the locomotive as a new means of travel were brought before the public and the road was given up.

The most important road improvements actually carried out under Telford's immediate superintendence were those on the western side of the island, the purpose of which was to shorten the distance and facilitate the communication between London and Dublin, by way of Holyhead, as well as between London and Liverpool[3]. The shortest route between London and Dublin was by land to Holyhead and thence by packet boat across the Irish Sea. The road from Holyhead, across the island of Anglesey, to the Menai Strait was rough and circuitous; the passage of the strait by ferry was difficult and treacherous; the road through North Wales from Bangor to Shrewsbury

[1] *Life of Telford*, p. 183.

[2] Ibid., p. 184 et seq.

[3] At the time of the union of Great Britain with Ireland, at the beginning of the nineteenth century, the chief lines of communication between the two islands were: the southern route, by Milford Haven and Waterford; the middle route, by Holyhead and Dublin; and the northern route, by Portpatrick and Donaghadee. The middle route has always been the most frequented, because it is the most direct route between the two capitals (ibid., p. 204).

was rough, narrow, steep and unprotected; and even after reaching smoother ground, from some distance beyond Shrewsbury to London (one hundred and eighty miles), the mail coach road was in a very imperfect state[1]. Thus public business was impeded and the Irish members of Parliament suffered much personal inconvenience in attending their duties in London and in resorting at Easter, as magistrates, to their respective counties in Ireland.

As early as 1808 the Post Office determined to put on a mail coach between Shrewsbury and Holyhead[2], but after several attempts, a regular mail service was found to be impracticable. Efforts were made to enforce the law that the parishes should repair their roads, but this section was too sparsely settled, and the inhabitants too poor, to fulfil this obligation by providing a sufficiently good road[3]. Many complaints continued to come in from the Irish members, regarding the delays and dangers to which they were exposed in their journeys to and from London, and Parliament was forced to take up the matter as a national concern. By instructions, under date of May 4, 1810, Telford was authorized to survey the road from Holyhead to Shrewsbury, a d that from Bangor to Chester, and to report as to the best line for a perfect mail coach road, with the expense necessary for its construction[4]. The surveys were completed and the report made in 1811. The district through which the surveys were carried was mountainous, and the existing roads very imperfect. The old road through this district was not only hilly but crooked and with steep ascents. It was frequently not more than twelve or thirteen feet wide and passed along the edges of precipices which were very dangerous for wheel carriages[5]. The committee to whom Telford reported urged on the House the necessity for all the improvements he had recommended, but nothing further was done for four years.

In 1815 Sir Henry Parnell, one of the members for Ireland, vigorously took the matter in hand; a Board of Parliamentary Commissioners was appointed; the necessary surveys and estimates were made under their direction; the aids granted by Parliament were administered through them; and the contracts and payments were made immediately under their control. The Commissioners were

---

[1] *Life of Telford*, p. 204.

[2] The small amount of traffic on the road from Bangor to Shrewsbury was carried by a cart which went but once a week in summer.

[3] Brit. Doc. 1817 (313), III, 179, Appendix No. 1, pp. 15, 16.

[4] *Life of Telford*, p. 207.

[5] Ibid., pp. 208–9. The ascents were of various degrees, from 1 in $6\frac{1}{2}$ to 1 in 15, that is, one foot ascent in a length of six and one-half feet, etc.

responsible to Parliament and made annual reports of their work[1]. They appointed Telford to direct and superintend all the practical operations. He had one principal assistant on the part between London and Shrewsbury and another between Shrewsbury and Holyhead; besides whom, in the latter district (which was wholly under the management of the commissioners) four inspectors were also employed. But as the part of the road in England remained—with the exception of the new improvements—in the hands of the turnpike trustees, only occasional inspectors were there necessary. All the work in both districts was performed by contract[2].

By 1819, many of the most dangerous portions of the road through Wales were rendered commodious and safe by means of many bridges, cuttings, embankments, fence walls, etc., and the rough country was being made wonderfully level and accessible to traffic[3]. As soon as there was a considerable distance of road ready to be travelled with facility and safety, it was thought that the road would be better cared for if the newly-made parts were left in the hands of the local turnpike trustees; but in the year 1819 the seven trusts between Shrewsbury and Holyhead were consolidated and their management vested in the Parliamentary Commissioners. Within fifteen years from the time this work was commenced, all this western part of the road was completed, including a bridge across the Menai Strait, and a new road across the island of Anglesey (about twenty-two miles) to Holyhead[4].

Of the route from Shrewsbury to London a careful survey was made and the short line by Coventry was the one selected to be improved to the utmost[5]. On the whole length of this road from Shrewsbury to the capital there were in existence seventeen separate turnpike trusts. With these the Commissioners did not materially interfere, either as to letting, collecting, or applying the authorized tolls; and each trust had its own surveyor and managed the usual repairs, so that the Commissioners could not get funds, nor enter into any contract, until they had obtained the approval of the local trustees for each proposed

---

[1] *Life of Telford*, p. 209.

[2] Ibid., pp. 209–10. The drawings, specifications, and estimates were made by Telford; then advertisements were put in the newspapers inviting persons experienced in road-making to transmit proposals to the Board of Commissioners, and stating where the plans might be seen. The Commissioners accepted the lowest tender, if supported by character and security.

[3] Brit. Doc. 1819 (78), v, 115, 'First Report of the Commissioners of the London and Holyhead Roads.'

[4] *Life of Telford*, pp. 212–13.

[5] Down to 1819 this road between London and Coventry was in a very bad state, judged from the standpoint of the engineer. In Brit. Doc. 1819 (549), v, 223, Appendix No. 2, a full account of it is given.

improvement. For removal of this obstruction, Parliament sanctioned the levy of an additional toll, not exceeding one-half the previously existing toll, for defraying the expense of intended improvements. The performance of the contracts was effected in the same manner as for the Welsh roads; and the portions first completed were given over to the care of the local trustees. But it was found that these improved parts of an old road did not get the attention which newly-made roads need until they are perfectly consolidated; hence another Act was passed authorizing the Commissioners to retain the management for two years after each improvement was completed[1]. Under Telford's oversight this road was kept thoroughly repaired, to the great convenience of travellers and facilitation of business.

The people of Liverpool, seeing the change in the road from London to Holyhead, and eager to render the communication between Liverpool and London better, applied to the Government for similar surveys. The result was that Telford got instructions to examine and report on this road, but no further steps were taken to accomplish the object sought[2].

The work of Telford in England was only a small part of his work; for the network of roads and bridges in the Highlands of Scotland was also made under his direction and from his specifications. But we must not limit the influence of Telford in this direction by the amount of road which he actually constructed; for by the changes which he effected, a stimulus was given to many parts of the kingdom to improve their highways, when they saw how much might be accomplished by applying the principles that he introduced. The spirit of reform went abroad when the excellence of his work became the theme of universal praise.

From what we have seen of the work of Metcalfe and Telford, some general features of scientific road building will be at once apparent. Not only was there the formation of a well-drained and solid road-bed, but care was taken to make the road as level and as smooth as possible[3].

---

[1] *Life of Telford*, p. 215.   [2] Ibid., p. 245.

[3] See, for instance, Telford's London and Holyhead line through Wales, which we have described.

Phillips in his *Tour through the United Kingdom* (1828), p. 14, says: "The road from Ampthill to Bedford is literally as even and smooth as the gravel walk in any royal gardens......I afterwards observed and was informed that every road in Bedfordshire is preserved in the same state of perfection......Really (these roads) cannot be too much praised." Again, on page 79, he says: "On the way northward (i.e., from Leicester) all was new, and roads were widened, turned, straightened, and levelled......Beyond Rothley, I expected some miles of Mountsorrel stone, and its horrid pavement; but I found the best macadamized road of the county. The stones, once a nuisance, were now the glory of the road."

Hills were cut down so as to provide an easier ascent and descent over them; in other cases, the course of the road was altered so as to avoid difficult hills, or places where the foundation of the road would be spongy or boggy[1]. Many Acts were passed for widening the highways and it was the constant care of the engineers that the roads should be increased in width to accord with the amount of traffic which would pass along them. By straightening some roads, the distances between their termini were reduced and consequently the expense of their maintenance was also greatly lessened[2]. But, in addition to this work which was accomplished by the engineers, another feature of the highway improvement was the closing of by-roads which were of little or no use and the consequently greater attention given to the main roads[3]. The purpose of all these changes was to have a series of good

[1] Gilbert, *Parochial History of Cornwall*, II, p. 104, says, in referring to the town of Gluvians, that "the main street descending with the ridge is scarcely safe for carriages; and the great road from London, through Truro, to Falmouth, passing directly across the ridge, has to go up and then down through streets so steep and narrow, and in parts so turned, as to make the safe passage of the mail coach a matter of wonder. These defects have been, however, completely remedied by a road carried round the point......thus reducing the road to a level....." See also Smith, *History of Dunstable*, p. 112; Leader, *Sheffield in the Eighteenth Century*, p. 96; Brit. Mus., Add. MSS. 27,828, IV, p. 16, etc.

Another effect of such changes in the roads was that sometimes when one route fell into disfavour the towns on that line relatively declined, while the opposite was true on the rival line which had come into favour. For example, soon after 1770, the present main road from London to Brighton obtained the preference, and gradually the through coaches, which had formerly passed through East Grinstead on a different road, ceased to visit that town and the latter was left to provide its own communication with the outside world (v. Hills, *History of East Grinstead*, p. 152). Again, Dr Pococke, soon after 1750, said that Salisbury had been made to flourish by having the western road turned from its course through Wilton to pass through Salisbury. Wilton had decayed and Salisbury flourished (Brit. Mus. MSS. 23,000, p. 45).

Northampton was another example. Dr Pococke, in 1751, said of it that it lived by its fairs, and by its being a great thoroughfare. "But the Yorkshire road being carried another way & as they have lately neglected repairing the roads the Chester post road is now through Daventry, so it is now only the great thoroughfare to Leicester and Nottingham & consequently begins to be on its decay" (v. Brit. Mus. MSS. 22,999, p. 4).

[2] Telford had reduced, in this way, the length of the line from Carlisle to Glasgow by about one-tenth. He also shortened the London to Shrewsbury road. See also Place MSS., IV, p. 16.

[3] Authority was given for this by Act 13 Geo. III, c. 78. That there was much stopping up of unnecessary highways, is shown by Cox, *Derbyshire Annals*, II, p. 232; *Gentleman's Magazine*, LVII, p. 879; Tate, *History of Alnwick*, I, p. 463; Fishwick, *History of Lancashire*, pp. 255-8. Great evils also seemed to have come from this source, one of which was the facility with which public and most useful highways might be stopped up by the order of two magistrates, and the difficulty of getting

roads, as great arteries of communication, and then to continue and perfect the lateral branches where they would be of most service.

While Telford was engaged on the roads in Scotland, and later in England, John Loudon Macadam had been studying road-making, as practised in Scotland by such men as Paterson and Lester, keeping in mind the essential conditions of a smooth surface and a durable foundation[1]. He had been a road commissioner in Scotland since his arrival from America in 1783 and had occasion in that capacity to see a great deal of road work. At that time, legislation was largely concerned with the breadth and the shape of wheels that should be used on roads and with the weight that should be carried by different kinds of vehicles; while Macadam was of the opinion that these were of secondary importance and that the main points to be attended to were the foundation, the materials and the construction of the roads which were to bear so much traffic. He first began to claim public attention

such an order quashed by appeal to Quarter Sessions. This evil called loudly for remedy. Hansard's *Parliamentary Debates*, 1831, v, pp. 1035–6.

[1] Evidence goes to show that Macadam was not the inventor of the system he employed for road construction, but that he perfected it. Before Macadam, Lester had been working on a similar plan; and long before Macadam was heard of, Paterson had constructed a number of roads on the principles which Macadam afterwards employed (v. Hansard's *Parliamentary Debates*, N.S., xiii (1825), p. 595). Paterson was a road surveyor who had secured good results on the road between Dundee and Montrose; and out of his experience he wrote *A Practical Treatise on the Making and Upholding of Public Roads*, which tells of his methods. These may be briefly summarized as follows: It is more important to secure roads somewhat near the level, although the distance should be a little longer. The road-bed should be thoroughly drained to ditches at each side. The surface should be slightly convex but where the road is level and the bottom wet, the convexity should be greater. This form, besides being better for turning the water, will require less material for repairs. The stones used for the metals of any road should be the hardest and most durable, but durability will depend largely on the dryness of the road. The bottom should be composed of stones broken large, and, of course, the softer the bottom the larger they should be. The top metals should be small broken stones, the size of which should be regulated by the situation of the road and the nature of the ground over which it is formed. For a road formed on a sloping bank or on a very dry bottom, the top metal should be six to eight inches deep, composed of broken stones two to three inches in size. The width of the road and the breadth of metals depend largely on the amount of traffic on the highway, for example, they would be the greater near a large city; but, as a general thing, a width of thirty-five to forty feet of road and fourteen to sixteen feet of metals would be sufficient. When a road is newly made, or when it is repaired with a coat of metals more than six inches thick, the materials on the top should be bound together by throwing on a little earth or small sharp gravel. Roads should be repaired by small broken stone, laid on in amounts just equal to the decay of the road; and it will be more economical to keep the roads from becoming rutted than to allow them to decay and then apply the material. In any case, broad cylindrical wheels should be used.

in 1810, when he sent a communication to the Board of Agriculture, of which Sir John Sinclair, a brother Scot, was chairman, outlining his plan for changing what he regarded as the radically unsound system of road-making, and requesting permission to demonstrate his own method. The chairman was interested in the proposals and had them brought to the attention of the parliamentary committee that was then considering the highways[1]. It was not long before his skill in road construction was recognized and his evidence before successive committees of Parliament was regarded as the last word concerning the building, repair and administration of roads.

His plan was as follows: the road should not be sunk below, but rather raised above, the ordinary level of the adjacent ground and care should be taken to have a sufficient fall to take off the water, so that it should always be some inches below the level of the ground upon which the road was to be placed. This must be done, either by making drains to the lower ground, or, if that were impracticable, because of the nature of the country, then the soil upon which the road was intended to be laid must be raised, so as to be some inches above the level of the water[2]. Having drained the soil from *under* water, the road-maker should next secure it from rain water, by a solid road, made of clean, dry stone, or flint, so selected, prepared and laid as to be perfectly impervious to water. This could not be effected unless the greatest care were taken that no earth, clay, chalk, or other matter that would hold water, should be mixed with the material used for repairing the roads[3]. When the road had been well drained, he would cover this to a depth of six to ten inches with stones broken into small angular fragments; and during the time these were consolidating, any inequalities due to traffic should be filled up[4]. Macadam's debt to

[1] *Memoirs of Sir John Sinclair, Bart.*, by Rev. J. S. Sinclair (1837), II, pp. 95–98.

[2] Brit. Doc. 1819 (509), V, 339, evidence of J. L. Macadam, pp. 17–33, gives a full account of his system. Brit. Doc. 1810–11, III, 'Report from the Committee on the Highways and Turnpike Roads in England and Wales,' pp. 885–6, also gives Macadam's directions for repairing a road.

[3] If the surface of the road had an admixture of earthy substance that would hold water, the water would percolate down through the whole bed of the road and thus cause it to be loose and easily heaved by frost, rather than firm. In the case of Telford's roads the top was covered with a layer of gravel an inch thick, which would allow the water to soak down through the substratum of the road; but Telford prevented his roads from becoming soft and yielding by having heavy stones packed together by hand in the bottom of the road, upon which was a thick bed of broken stones, which became firmly consolidated.

[4] J. L. Macadam, *Remarks on the Present System of Road-making*, 8th edition, pp. 50–51; Brit. Doc. 1810–11, III, 'Report from the Committee on the Highways and Turnpike Roads,' Appendix C.

Gravel, he said, was too poor for roads, for it contained rounded stones mixed

Paterson will be readily conceded upon comparison of the methods employed by the two men.

What difference was there, then, between the method used by Macadam and that used by Telford in the construction of roads? Both of them insisted on good drainage and carefully prepared materials and adopted a uniform cross-section of moderate curvature, instead of the very round surface given before. But while Telford was specially particular to obtain a solid foundation for the broken stone, Macadam passed that by as of little consequence, maintaining that the subsoil, however bad, would support any weight if made dry by drainage and kept dry by a covering of broken stones which water could not penetrate[1]. By comparison of the two methods, it is clear that the amount of material and labour required to make a macadamized road must have been much less than that required in making the same length of road by Telford's plan; and since they were equally substantial the macadamized road was more economical than the other.

In 1816, Macadam was made general surveyor of an extensive turnpike trust around the city of Bristol and by 1819 he had under his care one hundred and eighty miles of turnpike roads in that neighbourhood. "The admirable state of repair into which the roads under his direction were brought, attracted very general attention[2];" and it is noteworthy that his improvements were attended with an actual reduction in the expense of repairing. The maintenance of the first one hundred and forty-eight miles of road around Bristol and the making of many expensive permanent improvements and alterations were accompanied by a considerable reduction of the principal debt and the discharge of a floating debt; all in the three years from 1816 to 1819[3].

Macadam's method proved so successful that application was made

with earth, and this would not consolidate to form a hard, smooth surface. Such roads would be loose, hard to draw on and constantly needing repair. Small rounded stones, too, were useless, because they would be easily displaced by the carriages passing over them and the roads would constantly be put out of repair. This would not occur if the stones were broken into pieces about an inch long. The method, too commonly used, of throwing rounded stones, from the size of a hen's egg up to that of a man's head, into the ruts, was a very expensive and reprehensible practice (v. Appendix C mentioned above).

[1] Macadam would use the labour of women and boys to break up the small stones for use on the road. He was engaged more with the repairing of old roads than with the making of new ones.

[2] Brit. Doc. 1819 (509), v, 339, 'Report of the Committee of the House of Commons on the Highways.'

[3] Brit. Doc. 1819 (509), v, 339, 'Report of the Committee of the House of Commons on the Highways,' p. 4; also evidence of J. L. Macadam, pp. 19–20.

to him for assistance and advice; and in a few years the practical example set by him was followed in many other parts of the kingdom. Before the committee of 1819 he testified that already three hundred and fifty-two miles of road had been put into a very good condition, and three hundred and twenty-eight miles more were under repair, all of which had been, or were being, mended under directions given by him or his family[1]. It was his system of road construction which, consistently pursued, brought the chief English roads to a state of comparative perfection and elicited great praise for the decisive improvement wrought[2].

But it was not alone because of his having perfected a method of scientific road building that Macadam merits such ample recognition. His name is connected also with a system of administration which went hand in hand with good formation of roads. In order to secure capable men to attend to the roads, a living salary should be offered, and this would attract into the public service men of skilful training and executive ability who would accept these positions with the object of making them permanent and developing the best professional talent. He also strongly insisted upon economy and was unsparing in his hostility to statute labour and to the use of pauper labour on the roads. As the final arbiter of this administrative system, Parliament, he thought, should have control over all the trusts, so that the latter would have to hand in annual statements of their finances to this central authority, which in turn should supervise and direct the work.

When Macadam's success in improving the roads had brought him prominently before the public, he was invited to become surveyor-general of the metropolis turnpike roads; and by scientific formation and prudent administration of these roads, at first by himself and

---

[1] Brit. Doc. 1819 (509), v, 339, evidence of J. L. Macadam, pp. 18–19. Some years after this, he was placed in charge of the metropolis roads, and under his careful administration the work of reform was substantially begun, to be carried on by his son (afterwards Sir) James Macadam.

[2] In reference to Macadam's system of constructing and repairing the public roads, a letter from Mr Johnson, the Superintendent of Mail Coaches, sent to the House of Commons' Committee of 1823, shows how the postal service was affected by the change. He says: "As I travel rapidly over great distances......I feel myself well warranted in stating, that whenever I have found anything done under Mr Macadam's immediate direction......or even in imitation of his plan and principles, the improvement has been most decisive, and the superiority over the common method of repairing roads most ·evident;......I have abundant reason to wish that Mr Macadam's principles were acted upon very generally: if they were, a pace which in winter, or any bad weather, cannot be accomplished without difficulty, would become perfectly easy" (v. Brit. Doc. 1823 (476), v, 53, 'Report of the Committee,' p. 3).

afterward by his son, James Macadam, great and lasting improvements were made. After 1820, there was a movement for the consolidation of small turnpike trusts into larger ones, the chief object of which was to economize the revenues[1]; and this was especially the case in and around the capital, where the trusts and toll-gates were multiplied excessively[2]. The delays, as well as the charges, at the many toll-gates were obnoxious[3]; the funds were largely used in supporting trust officials; and much less than the requisite amount was available for improving the roads[4]. The debts of the trusts had been increasing to such an extent that some of them were breaking down under this increasing burden[5]; and in other cases money had been borrowed, even at annuity interest, to provide the means for discharging the debt[6].

[1] Macadam said that the principal evil in the road system was "the number of trusts, their small extent, and their limited means and powers." He was strongly in favour of consolidating them. *Parl. Papers*, 1833 (703), xv, 409, pp. iii, 489. Thomas Grahame, *Treatise on Inland Intercourse and Communication*, p. 18, deplored this condition and said that the system was much the same as if a large town were to put the management of each street under the sole control of a few persons located in such street, irresponsible to the general body, and looking after their own local interests without regard to the general good of the larger whole.

[2] From *Parliamentary Debates*, N.S., xii (1825), pp. 529–30, we learn that for three and one-half miles of road to the north of London there were three Acts of Parliament, three sets of commissioners, and ten turnpike gates. See also *Parl. Papers*, 1820, ii, 301. In the Bill for consolidating the trusts for ten miles around London, not less than 120 Acts of Parliament were recited in the preamble, being the number under which these turnpike roads were at that time maintained (Dehany, *General Turnpike Acts*, p. xxxv).

[3] The multiplication of toll-gates and charges made travelling and carrying expensive, and the delays incident to stopping at so many gates, especially on a rainy day, were a great tax on patience (*Parl. Debates*, N.S., xii, p. 530). In 1791 a German traveller in England said that on the roads near London there was something to pay nearly every moment and that in travelling three miles from London to Vauxhall there were twelve turnpike tolls to be paid. But he said that this feature applied not to London only. Brit. Mus. 567. e. 7, 'Beyträge zur Kenntniss vorzüglich des Innern von England und seiner Einwohner,' i, p. 41. See *The Times*, Feb. 8, 1816, p. 4, showing the inconveniences of turnpike gates around the metropolis, and ibid., June 3, 1816, p. 2, showing how funds might be procured for the abolition of these toll-gates, and what results would be secured therefrom.

[4] Within ten miles of the city of London, not less than £200,000 was annually collected in various directions and about one-half of this sum was consumed in salaries and perquisites (*Parl. Debates*, N.S., xii, p. 529). See also *Annual Register*, lxvii, p. 283, and *Parl. Papers*, 1825 (355), v, 167.

[5] *Parl. Debates*, N.S., xviii, pp. 1445–50; *Parl. Papers*, 1820, ii, 301, pp. 323–5; ibid., 1821, iv, 343, pp. 374–681, etc.

[6] *Parl. Papers*, 1820, ii, 301, 'Report of Committee on Turnpike Roads and Highways;' *Parl. Debates*, N.S., xii, pp. 529–30. In the case of the Stamford Hill trust, this money was borrowed by the trustees from some of themselves, with interest at 10 % (see latter reference).

Other irregularities also prevailed in the management of the trust funds[1]; and Macadam and others strongly recommended the consolidation of the trusts around London under a single Board of Commissioners as the most effective means of securing the desired results[2]. The advantages which they claimed would accrue from such consolidation were: first, that the expenses connected with the maintenance of the roads would be much lessened; second, that such necessary expenses could be thereby collected with less inconvenience to the public; and, third, that the roads would be materially improved because there would be more money to spend on them[3].

After several failures to merge under one body of overseers the

[1] See Report of the Committee of 1820, on Turnpike Roads and Highways; also the Reports of the Committees of 1821 and 1825.

[2] *Parl. Papers*, 1819 (509), v, 339, 'Report on the Highways,' pp. 365–8; 1820, II, 301, 'Report of Committee on Turnpike Roads and Highways,' and Appendix A of same; 1821, IV, 343, 'Report of Committee on Turnpike Roads and Highways;' *Annual Register*, LXVII (1825), p. 283.

Sir Henry Parnell, to whom England was mainly indebted for the improvements in the London and Holyhead roads, on being examined before the Committee of the House of Commons in 1820 observed: "From the experience I have had as Chairman of the Holyhead Committee during four sessions and a constant communication with all the trusts between London and Holyhead, I am convinced, that the leading defect of the existing turnpike system consists in the limited extent of the trusts, and the number of the commissioners, and that almost all the evils which prevail so generally in road management, may be traced to this source: I think I may refer to the evidence I have already given, of the complete success which has followed a consolidation of the trusts between Shrewsbury and Holyhead, as a convincing proof of the soundness of the general principle of consolidation; and therefore I have no hesitation in saying that a consolidation of all the London trusts would be a means of the greatest utility to the public."

[3] *Parl. Papers*, 1825 (355), v, 167, 'Report of the Committee on the Metropolis Turnpike Trusts, with the Minutes of Evidence.' See also Reports of the Committees of 1820 and 1821.

The committee of 1825 appointed to inquire into the receipts, expenditures and management of the turnpike trusts within ten miles of London made a careful inquiry and report. They observed that many of the accounts of the different trusts were found to be in a very confused state and the clerks of the trusts utterly incapable of affording the information which the committee required; and it appeared that in some instances no regular accounts had been kept till within recent years. The evidence showed that the amount of income raised was much larger than was necessary to keep the roads in the best state of repair; and that if those funds had been skilfully applied and proper materials obtained and used for the last seven years, according to the recommendation of the committee which first inquired into this subject, the roads would have been in a much more perfect state of repair and the debts of the trust much reduced and the tolls consequently lowered.

The committee had a survey made by Macadam who, with an associate surveyor,

approaches to London on both sides of the Thames, it was decided to limit the consolidation at first to the roads north of the Thames; and in that form the measure was passed in the year 1826[1]. Its execution brought the good results anticipated[2]: the toll-gates were greatly reduced in number, and placed in more suitable positions; weighing engines were abolished; the elimination of useless offices and of the salaries wasted on their incumbents left a larger net surplus to be devoted to the improvement of the highways and the reduction of the debt. By purchasing in quantity, the cost of material was reduced 25 per cent. to 30 per cent. The system of supervision by a central

pointed out that great improvement might be made if the trusts were under uniform and better management and if better materials were used in repairing the roads.

"The misapplication of money collected by the tolls is manifested in many cases; and in some instances large sums are paid out of them for the maintenance of pavement, which ought to be defrayed from the rates" of the different parishes. The committee said that a great part of the money borrowed was still at five per cent., which they disapproved of as exorbitant, when the security of the revenue from the tolls was considered.

By consolidation, the whole of the revenue of the trusts might be placed in the Bank; this would put an end to the disjointed interests of the several treasurers and would also abolish "floating debts," which were a further waste in the public funds, by making payments when they were required.

The committee showed the impolicy of granting separate Acts of Parliament for short lines of road and mentioned four separate trusts, City Road, Bethnal Green, Old Street and Shoreditch, and all the expenses attending four distinct establishments, within only four and one-half miles. Some legislative remedy was necessary for the more efficient and economical application of the immense revenue which was collected from the public on the roads immediately around London.

The committee's opinion was, "that a consolidation of all the trusts adjoining London is the only effectual method of introducing a proper and uniform system of management in the roads, economy in the funds, and of relieving the public from the present inconvenient situations and obnoxious multiplicity of turnpike gates, with which the inhabitants are now fenced in every direction. The important object of procuring a durable material for constructing the roads in the suburbs of London, and parts immediately adjoining, can only be obtained by dealing on an extensive scale." They finally recommended the consolidation of the sixteen trusts in the county of Middlesex.

[1] Act 7 Geo. IV, c. cxlii. Under this Act the former trustees of the several roads were to turn over their accounts to the Board of Commissioners and all amounts that were paid to or by these trustees were, under this Act, to be accepted by the said Board. The important rights and obligations under former Acts were to be re-enacted under this law, except that one Board of Commissioners took the place of sixteen sets of trustees.

[2] *Parl. Papers*, 1828 (311), ix, 23, 'Second Report of the Commissioners of the Metropolis Turnpike Roads,' especially pp. 25–28; *Parl. Papers*, 1833 (703), xv, 409, 'Second Report of the Lords Committee on Turnpike Returns,' pp. 457–8. See also the annual reports of the Commissioners of the Metropolis Turnpike Roads from 1827 to 1872; *The Times*, Nov. 10, 1826, and June 12, 1828; Hansard's *Parliamentary Debates*, Mar. 31, 1829, and Mar. 12, 1830.

Board of Commissioners, the appointment of an efficient engineer to take charge of the improvements, and the requirement of regular and exact statements of the finances from those who had the spending of them, put a stop to the needless draining of the funds and brought the roads of the metropolis and its immediate vicinity into excellent condition.

With the good results obtained from the consolidation of fourteen of the metropolitan trusts north of the Thames, renewed efforts were made to extend the application of this method to the country generally. The reports of the committee of 1838, after showing the great benefits that had arisen from the merging of the interests of the trusts around the metropolis, recommended every consolidation of trusts which their localities and other circumstances would permit, with the objects of securing more uniform and efficient administration, reduction of useless expenditure, liquidation of the debts which had been rapidly piling up, and increasing the confidence and security of the creditors of the roads. Viscount Lowther, the chairman of the Board of Commissioners of the Metropolis Roads, was very enthusiastic over the results which his Board had obtained and was urgent that the trusts throughout the country should link up in similar aggregations of one hundred to one hundred and fifty miles each. James Macadam, the surveyor-general of the metropolis roads, and many other witnesses were clamant for such a change. It was shown what consolidation had done in three cases in Scotland and Wales, in placing all the roads of a county under one set of trustees[1]. But it seems that the agitation ended in words, for we have no record of the extension of consolidations, probably because public attention was then diverted from the roads to the railways.

We have now considered the various methods which were employed for the benefit of the roads, and the difficulties which were encountered in aiding their systematic development. We have seen that the vast increase in the amount of legislation after 1750 furnishes an indication of the much greater interest which was taken in the roads; and consequently, at this stage, our concluding subject of inquiry is, What are the actual facts as to the condition of the roads of England in the latter half of the eighteenth century and the first third of the nineteenth century? Was there an improvement in them at all commensurate with what we would expect from the progress of the country industrially and from the greatly increased amount of legislation?

---

[1] *Parl. Papers*, 1838 (703), xv, 409, 'Second Report of the Lords Committee on Turnpike Returns,' pp. iii–iv, 457–64, 470–1, 488–9, 505, 554, 567; Hansard's *Parliamentary Debates*, 1835, xxix, pp. 1188–92; Humphreys, *History of Wellington*, p. 222.

A decisive answer to this question cannot be given until we come to consider whether there was any change in the rates of travel as compared with those of the period before 1750. We know there was a greater amount of traffic upon the roads during this period, for the changes in industry and agriculture would be accompanied by that result; but this, of itself, would not assure us that the roads were better, for the greater traffic might be handled by simply increasing the number of carriers, without improving the roads. Neither would the much larger number of Acts passed after 1750 be a sure proof that the roads were greatly benefited, for some of them were Acts merely to amend or continue former Acts, and some others were, doubtless, not carried into effect at all. Before making any general statement, therefore, let us look at the condition of the roads in different portions of the kingdom, as illustrated by the material at hand; and here we shall follow much the same plan as that which we adopted when examining the same subject for the period before 1750.

First of all, we shall look at the roads leading from London, through Derbyshire and Yorkshire, towards Edinburgh and the north generally, including not only the Great North Road, but those adjoining or tributary to it. On this group of roads connecting the northern capital and the northern manufacturing section of England with London there would be more travel than on most of the other roads and we would expect that it would be correspondingly improved. Mrs Calderwood, who, in June 1756, travelled over this route from Belford, in Northumberland, to London, completing the journey by chaise in five days, describes the roads as "good indeed;" but it is evident that in making this statement she was comparing the English roads with those of Scotland, for she says in the same sentence that "the levelness of the country makes travelling much quicker[1]." Just how much she intended to mean by the word "good" is not clear; but she evidently did not intend it to apply to all the route along which she travelled, for a later observation is to the effect that the general benefit of made roads had not yet reached them[2]. In accordance with this view, that only parts of the road were good, are the words of Horace Walpole, who said, in one of his letters of that year, that this road from London to Stamford was "superb," but north of that it was "more rumbling[3]." Yet even

---

[1] Mrs Calderwood's *Letters and Journals*, p. 18.   [2] Ibid., p. 15.

[3] Horace Walpole's *Letters*, III, p. 442, "The Great Road as far as Stamford is superb; in any other country it would furnish medals, and immortalize any drowsy monarch in whose reign it was executed. It is continued much farther, but is more rumbling." Perhaps it was the goodness of the roads that Mrs Calderwood is hinting at when she says that the miles about London are not above one thousand yards (v. her *Letters and Journals*, p. 18).

this part north of Stamford was, apparently, in a few years put into good condition, so that Arthur Young could describe the road from there to Tuxford as "excellent, and very well kept[1]." It would seem, therefore, that the poorer parts of the road were being improved as rapidly as possible, with the object of securing a continuous line of smooth and substantial road[2]. So great was the improvement that the time required in 1754 to go from York to London (four days) was reduced until in 1785 the whole journey from Newcastle to London occupied only three days. A corresponding change was made in the highway from London to Sheffield and Leeds, which was practically parallel with the Great North Road. Before the reign of George III, there was no communication by coach between Sheffield and London[3] and the

---

[1] Young, *Northern Tour*, I, p. 104.

[2] I give here the testimony of other observers, upon which I have come to this conclusion. The roads of Middlesex, both public and parochial, were in general good (Foot, *Agriculture of Middlesex*, pp. 68–69), and those who travelled this north road out of London were loud in their praises of it. One who made a tour from London to the Lakes in the Summer of 1799 spoke of it as being "in excellent order" (Brit. Mus., Add. MSS. 32,442, p. 2); and a noted French traveller, in the same year, described the stage from London to Barnet as "a superb road," and said that nothing could surpass the beauty and convenience of the sixty-three miles from London to Stilton (Faujas de Saint-Fond, *Travels in England, Scotland, and the Hebrides*, I, pp. 127, 128). The roads of Hertfordshire at the beginning of the nineteenth century were in general excellent (Walker, *Agriculture in Hertford*, p. 86; Young, *Agriculture of Hertford*, p. 221). The public roads of Bedfordshire were generally in good condition (Stone, *Agriculture of Bedford*, p. 46). The turnpikes of Cambridgeshire, except in the higher parts of the county, were only moderately good, due, of course, to the fen country through which they were carried (Vancouver, *Agriculture of Cambridge*, p. 218; Gooch, *Agriculture of Cambridge*, p. 291). The many turnpikes of Huntingdonshire were generally very good (Parkinson, *Agriculture of Huntingdon*, pp. 274, 276); but this sort of unqualified praise could not be given to those of Northamptonshire (Pitt, *Agriculture of Northampton*, p. 230). The roads of Yorkshire were, many of them, very good; and of many others it was said that they were indifferent, or that they were the subject of much complaint (Rennie, Broun and Shirreff, *Agriculture of the West Riding of York*, p. 36; Tuke, *Agriculture of the North Riding of York*, p. 84; Leatham, *Agriculture of the East Riding of York*, p. 15; Marshall, *Rural Economy of Yorkshire*, I, p. 180; but contrast ibid., I, pp. 181, 182). So good were many parts of the Great North Road between Gunnersbury Hill, in Lincoln, and Ferrybridge, in York, that Marshall (op. cit., I, p. 185) regarded them as good models for road surveyors to study. The turnpike roads of Durham and Northumberland about the close of the eighteenth century were said to be mostly in good order, but the writers who made these statements had so many reservations to make that we cannot but see that they were still lacking in smoothness and some other characteristics that were desirable; for example, they were encumbered with many hills (Granger, *Agriculture of Durham*, pp. 26, 49; Bailey and Culley, *Agriculture in Northumberland*, p. 56).

[3] Hunter's *Hallamshire*, p. 157; Leader, *Sheffield in the Eighteenth Century*, p. 99.

roads around Sheffield were in bad condition[1]. It took four days to go from London to Leeds and three days from London to Sheffield, by the best coaching arrangements that prevailed in 1760[2]; but by 1787 the distance between London and Sheffield was covered in twenty-six hours, by leaving Sheffield at five o'clock one morning and arriving the next morning at seven o'clock in London[3]. This route was so much improved that a gentleman who was making a tour of the United Kingdom in 1828, after having passed along this line, said of part of the road that it was "literally as even and smooth as the gravel walk in any royal gardens," and that all the roads of Bedfordshire really could not be too much praised[4]. The continuation of the road, between Harborough and Leicester, was "in the finest order[5];" but when he got a little to one side of this through route, he found part of the road was, by comparison, wretched[6]. Northward from Leicester, again, he found the roads widened, straightened, levelled and macadamized, and from Derby onward to the north the road was "excellent[7]." It will be readily seen, from what we have shown regarding the condition of these roads and the increased speed of travel, that considerable change had taken place in them during the period we have now under review.

When we get off the main road and look at the subsidiary and cross roads, we do not always find as favourable conditions. The tendency was for the roads radiating out from important towns and

[1] Horace Walpole, in his *Letters*, III, p. 445, says that during his residence in Wentworth Castle (1756) he made two little excursions and that it required resolution to do so, because the roads were "insufferable." In the same letter (p. 446) he speaks of the roads around Leeds as "very bad black roads." Leader, *Sheffield in the Eighteenth Century*, p. 96, informs us that for coaches entering the town from the south, extra horses had to be sent to Heeley Bridge to get them up to Highfield and across Sheffield Moor, where the road ran in a sort of broad ditch.

[2] See advertisement of Glanville's coaching arrangements, in Ward's *Sheffield Public Advertiser*, of Nov. 4, 1760.

[3] Leader, *Sheffield in the Eighteenth Century*, p. 100, quoting from the advertisements of the coaches.

[4] Phillips, *Tour through the United Kingdom*, p. 14.

[5] Ibid. See also Brit. Mus., Egerton MSS. 2235, p. 89, where we are told that in 1752 there was a very good turnpike road from Harborough to Leicester; Pitt, *Agriculture of Leicester* (1809), p. 308; Monk, *Agriculture of Leicester*, p. 53.

[6] Phillips, *Tour through the United Kingdom*, p. 82, where he says: "The state of the road (from Newport Pagnell to Northampton) as compared with those I had recently passed in Bedfordshire, was wretched. I was told at Northampton that they considered it a very good road." It could not have been very bad, in reality, for in 1799 this same piece was "a good limestone turnpike road" (v. Brit. Mus., Add. MSS. 32,442, 'Tour from London to the Lakes,' p. 5).

[7] Phillips, *Tour through the United Kingdom*, pp. 79, 115. This was different from what it was in 1766, when, we are told, it was very dreary (Brit. Mus. MSS. 6767, p. 57).

cities to be well repaired in the neighbourhood of these places and to be less carefully attended to the more remote they were[1]; but this was by no means universally true, for the roads in the immediate vicinity of some towns were bad[2], while those that passed through sparsely populated sections were sometimes excellent[3]. The by-roads had often much less skilful attention devoted to them, and many of them were still in a condition but little better than that in which nature had left them, so that in wet seasons it was almost impossible, in travelling on them by coach, to make even moderate speed[4]. We notice, however,

[1] Hutton, in his *History of Derby*, pp. 8–9, informs us that in 1791, when he was writing, eight roads, all turnpike, proceeded from Derby to the adjacent places, and that these were excellent and used with pleasure. Brown, *Agriculture of Derby* (1794), p. 43, says, that, in the limestone districts of that county, the materials were so durable that generally both the public and parochial roads were good, except on steep ascents where access was difficult. But this was entirely contrary to the statement of a traveller who, in the summer of 1799, found the roads in Derbyshire in very bad condition, although the turnpikes were numerous and expensive (Faujas de Saint-Fond, *Travels in England, Scotland, and the Hebrides*, II, pp. 266, 309, 388). Arthur Young, in his *Northern Tour*, II, p. 320, paid a compliment to the small town of Swinton by saying that the roads which branched every way from it were admirable, for there were very few towns of that size of which this could be said.

[2] See, for instance, Twining, *A Country Clergyman of the Eighteenth Century*, pp. 46–47. He describes the "execrable" roads leading out from Sheffield.

[3] In the thinly peopled section (16 miles) between Winster and Buxton in Derbyshire, they had the "benefit of an excellent turnpike" (v. Brit. Mus. MSS. 6668, p. 909). Arthur Young, *Northern Tour*, IV, pp. 573–86, gives a long account of the roads in the north of England, where population was by no means dense, some of which were "very good," "excellent," "much superior to many turnpikes," etc.

[4] Dr Pococke, in 1760, travelled on a road from Yarm, through Stokesley, to Guisborough, which is just off the course of the Great North Road in Yorkshire. He says of it: "All this road from Yarum is mostly a clay ground without stones; the roads in winter are excessive bad; and they have narrow paved causeways for one horse." Speaking of the very populous country around Patrington, near Hull, he says: "the roads are very bad in winter, and they have narrow pav'd causewaies which are very disagreeable riding;" but the road from Patrington, through Hedon, to Hull was all a fine turnpike road. Between Beverley and Hull the way was "kept in very good order;" but in Nottinghamshire, after they had left the good road through the marshes, they came to roads of which he says that "nothing can be imagined worse; they have neither stone nor gravel, and the soil is very deep" (v. Pococke, *Journey Round Scotland to the Orkneys*, IV, pp. 98, 120, 125, 136, 151). These roads were all close to the Great North Road, and in a part of England which was populous and neither rough nor broken. Pennant, in his *Tour from Downing to Alston Moor* (1778), has much the same things to say of the cross roads in York and Durham; a few were good, but many were not good and some were "dreadful" (v. pp. 106, 111, 117 of Pennant's description). In Pennant's *Tour from Alston Moor to Harrogate* (1778), p. 54, he says that after a tedious ride from West Tanfield, through fields and intricate by-lanes, he reached the little

much difference in the cross roads at the end of this period from what was customary before 1750, for in the earlier period there were very few of them that were frequented by travellers in their coaches, but most of the travelling upon them was performed on horseback. On the other hand, after 1750, the introduction of turnpikes extended also to the cross roads, and the improvement caused some of them to be, in parts, as good as the main roads[1].

The next group of main roads includes those leading north-west from London to Holyhead, Chester and Liverpool. These roads led

village of Gruelthorp. For similar statements, see Twining, *A Country Clergyman of the Eighteenth Century*, pp. 37, 46–47, 164, 166, 167; Brit. Mus., Add. MSS. 32,442, pp. 297, 303, 310, etc.

But it will be well for us to note the character of these cross roads and parish roads at a later time also, and we will adduce evidence to show what they were like about the beginning of the nineteenth century. Stone, *Agriculture of Bedford*, p. 46, tells us that the private or cross-country roads were generally much neglected and, of course, in a very bad state; and the same thing was true of the cross roads of Cambridge, which were described in 1818 as "miserably bad" (Gooch, *Agriculture of Cambridge*, p. 291; compare also Vancouver, *Agriculture of Cambridge*, pp. 179, 184, 218). The parish roads of Northampton (1809) were in many places ruinous and generally so narrow as to admit of only one track (Pitt, *Agriculture of Northampton*, p. 230). The private roads of the county of Huntingdon were very much neglected in many parts and frequently were in an incredibly bad state (Parkinson, *Agriculture of Huntingdon*, pp. 275–6, 278). In Leicester, they were, in some cases, "indifferent and miry," and in some others they were "infinitely bad" (Monk, *Agriculture of Leicester*, p. 53; Pitt, *Agriculture of Leicester*, pp. 308, 311). As to Yorkshire, it seems impossible to decide where the emphasis should be placed; but, taking all in all, we must conclude that, while some parish roads were good, there were many that were greatly neglected (Tuke, *Agriculture of the North Riding of York*, pp. 84–85; Leatham, *Agriculture of the East Riding of York*, p. 15; Rennie, Broun and Shirreff, *Agriculture of the West Riding of York*, p. 36; Marshall, *Rural Economy of Yorkshire*, I, p. 181). The same thing may be said of the township or parish roads of Durham and Northumberland: the greater part of them could by no means be called good (Granger, *Agriculture of Durham*, p. 49; Bailey and Culley, *Agriculture in Northumberland*, p. 56).

[1] See the instances that have been given above; also Arthur Young, *Northern Tour*, I, pp. 45, 46, 119; IV, pp. 573–86. Young, *Agriculture of Hertford* (1804), p. 221, tells us that there were many cross roads in that county that were nearly as good as turnpikes. Of course this was near the metropolis. The roads of Middlesex, both public and parochial, were generally good, considering the flatness of the surface of many parochial roads (Foot, *Agriculture of Middlesex*, pp. 68–69). In Walker, *Agriculture in Hertford*, p. 86, we learn that the roads of that county were generally excellent. Lowe, *Agriculture of Nottingham*, p. 53, shows that, by imitation of the turnpikes, the roads of that county were lately much improved, although bad in many places in the clay soils, and especially in the coal districts where heavy loads were carried. The parochial roads of Derbyshire were said to be generally good, except on the steep hills where access was difficult; and yet the traveller was easily convinced that on many of them all lawful statute labour was not done. Brown, *Agriculture of Derby* (1794), p. 43.

through the midlands where the enclosure movement had made most progress in the advance of agriculture and where such places as Birmingham and The Potteries were flourishing. The state of the road in 1740 may be inferred from the fact that it took six days in winter to go by stage coach from Chester to London, and then they were out two hours before day and as late at night. There seems to be no doubt, however, but that this road was much better in the summer; for the amount of travel and traffic on it, about the middle of the century, was the subject of comment by several writers[1]. In 1763, the carriers on this road, between Birmingham and London, gave notice that they intended to raise their rates, because they could not carry more than two-thirds as much in weight as they formerly had done, on account of the badness of the roads[2]. Of course, the carriers had a monopoly of the trade at that time and could venture on this advance; and there seems to be no reason to accept as true the statement they made in apparent justification of their decision. In the next twenty years, however, a great change must have been made, for Pennant in 1782, comparing the conditions of travel then with the difficulties experienced in 1740, says that, in the later year (1782) the "enervated posterity sleep away their rapid journies in easy chaises, fitted for the conveyance of the soft inhabitants of Sybaris[3]." By 1782 the journey from Birmingham to London, which was performed in the summer of 1742 in two days[4], was accomplished in nineteen hours[5]; this also seems to be confirmatory of the improvement that had taken place, even by that time[6]. But the work of Telford on this line from London, through

---

[1] Defoe, *Tour through the Whole Island of Great Britain*, I, pp. 94, 97, III, pp. 49, 50; Brit. Mus. 10,349. g. 11, 'A Journey through England' (1752), pp. 82, 103, 127.

[2] Bunce, *History of Birmingham*, I, p. 49.

[3] Pennant, *Journey from Chester to London*, p. 138.

[4] See advertisement of Coles's coach in Aris's *Birmingham Gazette*, May, 1742.

[5] Account of Hutton's Journey from Birmingham to London, 1785, as found in Brit. Mus., Add. MSS. (Place MSS.) 27,828, IV, p. 11.

[6] Even before Telford was engaged to put this Holyhead road into good condition, the great roads in this direction from London to the midlands must have been much improved, although it is by no means certain that there was a corresponding improvement from Birmingham, the centre of the midlands, to the terminus of this route in Wales. In 1794, Wedge, *Agriculture of Warwick*, p. 28, tells us that the turnpike roads of that county were "tolerably good;" and, in 1813, Murray, *Agriculture of Warwick*, p. 172, says that the turnpike roads were generally good. On the other hand, Marshall, *Rural Economy of the Midlands* (1796), I, p. 47, in speaking of the road between Nottingham and Loughborough, says that, considering the good materials and the publicness of the thoroughfare, this road might be deemed one of the worst-kept roads in the kingdom. When, however, we analyze this latter statement, taken in connexion with the accompanying circumstances he mentions, it may not mean what, at first reading, it would seem to mean; but it was probably

Shrewsbury, to Holyhead and Chester, brought the improvement to its culmination and the road to a high degree of perfection[1].

A few words may not be out of place here regarding the cross and by-roads in close proximity to this route. In the west, for example, the roads near Nantwich, which was on the main line to Chester, had not an enviable reputation, for, in 1754, the way between that place and Cholmondeley was a "very bad road and troublesome paved causeway;" while from there into Wales the roads were "excessively bad, with a pitched causeway on one side about a yard wide," on which the traveller was obliged to go[2]. Nor can we find evidence that around

intended to mean that because of such excellent materials by which to make a good road, and with such a great amount of traffic requiring a good road, this road should have been far better than it was. In accordance with our declaration at the beginning of this footnote, we note the increase in the number of coaches and waggons traversing the roads that led out from London and Birmingham, in the later years of the century. For instance, in 1740, there was only one coach going once a week from London to Birmingham; but in 1783 there were four coaches that left London for Birmingham every day and two that left three times a week (v. Guides to London for 1740 and 1783). In other words, leaving out Sunday, coaches left London for Birmingham thirty times a week in 1783, but only once in 1740. It may be said that this need not necessarily indicate that the roads were better at the later date; but the fact is that people would not have increased the amount of their coaching, some of which was for pleasure, if the roads had not been considerably improved from what they were in 1740. Besides this, the rate of travel had also increased, as we shall show later; and all these facts taken together point with emphasis to a decided improvement of the great highways between the midland metropolis and London.

We cannot be so certain about a co-ordinate change in the quality of the western division of this great road system. In 1752, the road from Liverpool to the midlands, even in the summer, was in many places very bad (Brit. Mus. 10,349. g. 11, 'A Journey through England,' pp. 24, 26, 27, 57, etc.). The roads in the county of Shropshire in 1794 were described as "generally bad" (v. Bishton, *Agriculture of Salop*, p. 18); but, in 1803, Plymley, *Agriculture of Shropshire*, p. 273, seems to give his assent to Bishton's earlier report, although in another place (p. 279) he says the turnpike roads of the county were generally tolerably good when they did not have too much heavy carriage. What this latter statement may mean, it is hard to see; it may convey the meaning that most of these roads were bad, because of the heavy loads drawn along them. The public roads of Cheshire in 1794 were generally not very good, since they were most commonly rough pavement, called causeways, or deep sand (Wedge, *Agriculture of Chester*, p. 26); but in 1808, while they were generally far from good, they were better than they had been twenty years before, so that they were regarded as being in a "state of progressive improvement" (Holland, *Agriculture of Cheshire*, p. 302). On the whole, therefore, the great roads of this western extension were probably not very good, but were being little by little improved (see also Dickson, *Agriculture of Lancashire*, p. 607).

[1] We have already shown what this road was like before and after Telford took it in hand, when we were considering the work of that engineer.

[2] Brit. Mus. MSS. 22,999, 'Pococke's Travels in England,' pp. 44, 47.

Nantwich there was much change for the better as late as 1823, for a gentleman's diary of that year informs us that the roads there were "very generally paved with rounded stones, and withal very uneven, so that I was in a constant state of most annoying concussion[1]," and that the roads in Cheshire were "jolting and wearisome[2]." Farther east, in Staffordshire, it appears that, until about 1760, the roads were in a wretched condition and most of the traffic was carried by pack-horses; after that, by the initiative of the pottery manufacturers, like Wedgwood and Bentley, and other enterprising citizens of the towns in the neighbourhood of the potteries, turnpikes began to be constructed, roads were widened and in every way possible the means of communication were brought close to the requirements of an industrial people[3]. In the centre of the midlands, some of the roads were of poor

[1] Brit. Mus., Add. MSS. 32,442, p. 297 (Aug. 25, 1823).

[2] Ibid., p. 303. In general agreement with this are the statements of those who reported to the Board of Agriculture in 1794 and 1808. Wedge, *Agriculture of Chester* (1794), p. 26, shows us that the parochial roads in the clay parts of Cheshire were generally bad for carriages, but that a narrow horse pavement on one side of the roads rendered them passable for horsemen. Holland, *Agriculture of Cheshire* (1808), pp. 302–4, tells us that, some time before his date of writing, pavements formed of such boulder stones as could be got from marl and gravel pits were the most frequent in the county. Because of the scarcity of these stones, the expense of forming roads with them and the rough and unpleasant roads thus formed, they had been superseded by gravel (sometimes mixed with broken stones) and this had proved to be good where there was a good foundation for the road; but on wet or clayey bottoms, and especially where the road was kept wet by the shade of the hedgerows, it had been necessary again to resort to pavements. These stones for paving the roads had lately been brought in from the coasts of Wales or from Derbyshire. Even with this paucity and poor quality of material, some of the private roads had been improved and made convenient for carriages. This paving of roads with large cobblestones was frequent also in Lancashire (Warner, *Tour through the Northern Counties* (1802), I, pp. 185–6, II, pp. 132, 139; Dickson, *Agriculture of Lancashire* (1815), p. 608); and while such roads were said to be "the most expensive and most disagreeable of any," there was no other kind of material that would stand the heavy cartage of these coal regions. Dickson tells us (p. 608) that the expense of making these paved roads was very great; and that some of them had not cost less than £1200 to £1500 and even £2000 per mile.

While the roads of Worcestershire were not directly connected with this north-western system, yet they were as closely related to it as to any other. Many of the cross roads of this county were described in 1813 as "scarcely passable from Christmas to Midsummer, either on horseback or with a loaded carriage" (Pitt, *Agriculture of Worcester*, p. 260). But while these cross roads could not be called good, they were being gradually improved (ibid., p. 261; Pomeroy, *Agriculture of Worcester*, p. 22).

As to the parish roads of Shropshire, Bishton, *Agriculture of Salop* (1794), p. 18, says that they, as well as the turnpike roads, were generally bad; and that the private ones especially were almost impassable. Plymley, *Agriculture of Shropshire* (1803), pp. 273, 280, 281, complains that they were very much neglected.

[3] Meteyard, *Life of Wedgwood*, I, pp. 266–8, 271–4. The writer of this biography

quality[1] and some were nothing but bridle-paths[2], but many of them, especially in the third decade of the nineteenth century, became fine specimens of the art of road-making[3]. The increasing postal facilities from 1760 on is another evidence that gradually even the cross roads were being improved[4]. Nearer London, of course, the roads had received greater attention, and their condition showed the care that had been bestowed upon them. Arthur Young, in 1804, found many of the cross roads in the counties between London and Oxford nearly as good as turnpikes, and said that it was almost impossible for the roads to be bad in a county so near to the metropolis as Hertfordshire[5].

had access to the private correspondence and family papers of Wedgwood, as well as to other private original sources.

[1] Bray, *Tour into Derbyshire* (1783), p. 346, says that he came "into the turnpike road from Oxford to Banbury, at Adderbury......in a bad country, and surrounded by execrable roads." On page 349, he says: "Leaving Edgehill, go through Pillerton and Edington, and turning on the right through Wellesburn and Barford, to Warwick. It is something round to go by Edge-hill from Banbury to Warwick, but the road by Keynton is so bad, that it would be worth the additional trouble..... From Edge-hill to Edington the road is tolerable; from thence to Wellesburn, very good, and from thence to Warwick excellent." On page 374, he tells us "There is another way by Duffield, which leads into the turnpike road from Derby to Matlock......but neither of these (roads) are good for a carriage." See also Brit. Mus., Add. MSS. 28,802, p. 22; Brit. Mus., Add. MSS. 32,442, p. 5; Brown, *Agriculture of Derby*, p. 43; Wedge, *Agriculture of Warwick*, pp. 28, 30; Murray, *Agriculture of Warwick*, p. 172.

[2] Bray, *Tour into Derbyshire*, p. 386; Brit. Mus. MSS. 22,999. p. 2, etc.

[3] See Bray, op. cit., p. 349; Brown, *Agriculture of Derby*, p. 43. Wedge, *Agriculture of Warwick* (1794), pp. 28, 30, and Murray, *Agriculture of Warwick*, p. 172, give us to understand that, chiefly because of mismanagement and mis-application of statute labour, the by-roads of that county were mostly bad. But by 1821 we are told that "The roads in Warwickshire have been compared to a bowling green, but since the improvement of Mr Macadam in other parts of the country, these are by no means entitled to any pre-eminence" (Brit. Mus., Add. MSS. 28,802, p. 16; see also ibid., pp. 7, 39, 86, 90). It is possible, however, that this latter statement was made more particularly with the turnpike roads in mind, rather than the private roads. The spirit of improvement came into that section to some extent about 1775 (Marshall, *Rural Economy of the Midlands*, I, p. 35).

[4] Bunce, *History of Birmingham*, I, pp. 49–50.

[5] Young, *Agriculture of Hertford* (1804), p. 221.
His testimony regarding the roads between the metropolis and Oxford is equally conclusive. About 1760 "the roads of Oxfordshire were in a condition formidable to the bones of all who travelled on wheels;" but by 1809 he says: "A noble change has taken place, but generally by turnpikes which cross the country in every direction, so that when you are at one town you have a turnpike road to every other town. This holds good with Oxford, Woodstock, Witney, Burford, Chipping Norton, Banbury, Bicester, Thame, Abingdon, Wallingford, Henley, Reading......and in every direction and these lines necessarily intersect the county in every direction. The parish roads are greatly improved, but are still capable of much more. The

The "noble change," however, which he witnessed before 1809 was, in many cases, still further changed when Macadam's principles began to be more generally applied.

The third group of roads to be considered are those which connected London with the western parts of England. The main lines of road along this route, leading to Bath, Bristol and Gloucester, were in such a condition that, in 1754, the coach from London to Bristol went only three times a week during the summer and took two days for the journey[1]. By 1784, some coaches were performing the same journey daily in sixteen hours[2]. We get a good idea as to the condition of this western trunk from a letter written by a gentleman who, in 1767, travelled from London, through Reading and Newbury, to Marlborough; and although he called the road very good from London to Reading and said that the commissioners could not be sufficiently praised for widening it very judiciously in many places, yet he had to pass through water which was so deep as to threaten to come into his post-chaise. It was strange that on this highway of western trade he should so often have to refer to the fact that the narrowness of the road and its great convexity made it dangerous to meet other carriages lest they should be overturned[3]. In the summer of 1775, it took only nine and one-half hours to go from London to Oxford, a speed of six miles per hour, which was fast for that time; and the road westward from Oxford, through Witney and Northleach, to Gloucester, also merited commendation, except in some places where it had been but recently mended[4]. In 1794, the part of this great western road between London and Oxford, we are told, was kept in good condition[5]; and by 1823, the remoter portion of this trunk route, from Dorchester, through Abingdon, Faringdon, Lechlade and Fairford, to Cirencester, had been made by Macadam into a most excellent road[6]. Under these conditions, we may reasonably conclude that the conveniences afforded by this route for trade and travel were among the best of that time. The great highway from Gloucester still farther west to Newnham, near the

turnpikes are very good, and, where gravel is to be had, excellent" (v. Young, *Agriculture of Oxford*, p. 324).

[1] It took two very long days to cover the distance from London to Bath in 1752, by being out about 2 a.m. each day (Brit. Mus. 10,349. g. 11, 'A Journey through England,' p. 139).

[2] Bonner and Middleton's *Bristol Journal*, of Sept. 4, 1784, p. 3. These were not the mail coaches. See also *The New Bath Guide*, 1784, p. 72.

[3] *Oxford Gazette and Reading Mercury*, May 18, 1767, p. 2.

[4] Brit. Mus., Add. MSS. 17,398, pp. 50–51.

[5] Malcolm, *Agriculture of Bucks*, p. 43; Young, *Agriculture of Oxford* (1809), p. 324.   [6] Brit. Mus., Add. MSS. 32,442, p. 282.

Forest of Dean, was level and excellent[1]; and, farther south, the roads around Bath, even in winter, were kept in good repair[2], whereas immediately after 1750 they were bad[3]. The cross roads in this part of the kingdom, during the first quarter of the nineteenth century, were largely of the same quality as those in the midlands; but when we get to the remote western counties bordering on Wales, namely, Hereford, Monmouth, and Gloucester, they were, as a rule, rough and bad[4].

[1] Shaw, *Tour to the West of England* (1788), p. 232. The roads in the vale of Evesham were good; and that from the west into Ledbury was a "smooth, winding road" (ibid., pp. 204, 210).

[2] Simond's *Travels in Great Britain* (1810–11), I, p. 16. He says, "this part of England (i.e., around Bath) is a great bed of chalk full of this singular production (flints). They are broken to pieces with hammers, and spread over the roads in deep beds, forming a hard and even surface, upon which the wheels of carriages make no impression. The roads are now wider; kept in good repair, and not deep, notwithstanding the season (January)......Our rate of travelling does not exceed six miles an hour, stoppages included, but we might go faster if we desired it." See also Brit. Mus., Add. MSS. 33,683, p. 45.

[3] Brit. Mus. MSS. 22,999, p. 75. This was the case even in August. Compare Defoe, *Tour through the Whole Island,* II, p. 297, who says that the approaches to Bath were growing better daily, and the use of coaches was coming in. See also Aris's *Birmingham Gazette,* Feb. 3, 1752, p. 1.

[4] Brit. Mus., Add. MSS. 17,398 (1775). The road from Crickley Hill to Gloucester was "extremely narrow" (p. 51). The road part of the way from Gloucester to Ross was "narrow and bad," in many places through hollow ways where even a horse could not pass by a carriage (p. 53). The roads from Ross to Hereford were "rough, narrow and hilly" (p. 55). The road from Hereford to Aconbury was "very bad" (p. 55). The road from Hereford to Monmouth was "hilly and very stony" (p. 58). And this was during the summer when the roads would be at their best. See also ibid., pp. 59, 60, 61, 62, 90, 92, 96, etc. The same thing may be noted in Brit. Mus., Add. MSS. 30,172, pp. 4, 5, 27, and in Shaw's *Tour to the West of England* (1788), pp. 200, 209, 219, 232, 293, from which we learn that the roads were mostly "intolerably rough." Bridle-paths were very common as late as 1760–80 (v. Brit. Mus. MSS. 22,999, pp. 2, 6, 8; Brit. Mus. MSS. 23,000, p. 83, etc.). To further exemplify the truth of what we have said in this connexion, see Turner, *Agriculture of Gloucester,* p. 40; Marshall, *Rural Economy of Gloucester* (1789), I, p. 14, who says that even the great public road between Gloucester and Cheltenham was scarcely fit for the meanest subject to travel on—AND PAY FOR ; Fox, *Agriculture of Monmouth,* p. 20; Hassall, *Agriculture of Monmouth* (1812), pp. 100, 104, 105; Clark, *Agriculture of Hereford* (1805), p. 142; Duncumb, *Agriculture of Hereford* (1805), p. 142; Brit. Mus. 578. k. 30, 'Observations on a Tour in England, Scotland and Wales,' p. 101.

But even in the district between London and Oxford some of the by-roads were like those in the west; for example, Priest, *Agriculture of Bucks* (1813), p. 339, says that the by-roads of that county were extremely bad, some of them dangerous, and ought to be used cautiously; and after giving many instances, he concludes: "Bad, very bad, are best." In his description he is in close accord with Malcolm, *Agriculture of Bucks* (1794), p. 43, who said that the cross roads and also the parochial roads of that county were equally bad; that in some parts where they travelled in the chaise it took more than four hours to go little more than ten miles.

The roads in the southern counties early acquired a notoriety for badness which seems to have transcended that of any other section of England, except the extreme north. In all probability, such a record came from considering the roads of Sussex as typical of all the roads in the counties bordering on the Channel. Horace Walpole, in 1749, wrote to a friend that if he desired good roads never to go into Sussex[1]. In 1752, another writer tells us that the roads of Sussex and the adjoining part of Surrey were "bad for travellers, so bad, indeed, as to have become proverbial," and that a Sussex road was "an almost insuperable evil[2]." Probably the chief reasons for these roads being so bad were that the soil was clayey, that it easily retained the moisture, and that when wet it was very sticky and easily cut into ruts which, when dry, made the roads very rough[3]. But *all* the roads, even in Sussex,

It is difficult to understand why there should be such a difference between the parish roads of Buckinghamshire, as above described, and those of Oxfordshire, which Arthur Young (*Agriculture of Oxford* (1809), p. 824) described as "greatly improved," although still capable of much more improvement.

[1] Horace Walpole's *Letters*, II, p. 406 (1749). Writing to George Montagu, he says: "If you love good roads, conveniences, good inns, plenty of postilions and horses, be so kind to yourself as never to go into Sussex. We thought ourselves in the northest part of England......Coaches grow there no more than balm and spices; we were forced to drop our post-chaise, in which we were thrice overturned, and hire a machine that resembled nothing so much as Harlequin's calash, which was occasionally a chaise or a baker's cart."

[2] Brit. Mus. MSS. 11,571, 'Journey through Surrey and Sussex,' 1752. On p. 116, he says: "Either of them leads to the vale of Surrey and thence into Sussex, a miry country, having a fertile soil, & feeding many oxen......but bad for travellers; so bad, indeed, as to have become proverbial; & it might justly be said that a Sussex road is an almost insuperable evil. And yet that of the neighbouring part of Surrey, lying also low, is no better; for the soil being free from stones, & very retentive of moisture, absorbs nothing......" On p. 118, he says that after walking on a causeway or Roman road, extending out eight miles from Arundel toward Horsham, which was one of the great roads leading to London, "we fared badly in that respect, coming into a country inhabited, whether by men or beasts we could not decide, and passing over the road (to give it you in one word) of Sussex. These did not appear to be public ways, but bye-roads, or rather the haunts of kine, for the oxen had left everywhere deep marks of their cloven feet, and we too on horseback treading in their crooked steps seemed to slip back again as much as we advanced; the roads likewise still retained their wintry appearance, and the puddles yet remaining on the clayey surface......so that from the slipperiness and roughness of the way our horses had not firm footing, but slipping & missing their steps, & almost off their legs, made but sorry speed." On the following page he tells us that they tilled their land and drew their waggons with oxen in preference to horses. See also pp. 119, 123, 125.

[3] Ibid., p. 119, gives a humorous turn to this circumstance: "Now, my Friend, let me pose you with a question in the way of Aristotle: Why have the women, & oxen, & swine, & other animals in Sussex such long legs? Is it because of the difficulty of pulling them out of the muddy soil, which will not let them go,

were not of this nature; for the above traveller, when he got down to the land near the coast, could not but remark, " what easy and delightful riding on so smooth a turf[1]." Despite the tendency to the contrary, we are not warranted in continuing to apply this early reputation for an indeterminate period. Dr Pococke, who travelled through these southern regions in 1764, frequently speaks of the roads and turnpikes, but it is very seldom that he says anything against them; from which we may infer that, during his tour, which was in summer, they were generally fairly good[2], although he speaks of many as being impassable in winter[3]. We may say with a considerable degree of assurance that, during the last quarter of this century, the great roads in the southern counties, except those in the extreme south-west[4], and in some particular sections like the Wealds of Kent and Sussex[5], were being greatly

by which means the muscles are extended too far, & the bones lengthened?" On p. 123 he speaks of a man being "settled and buried in that impracticable clay of Sussex."

[1] Brit. Mus. MSS. 11,571, ' Journey through Surrey and Sussex,' 1752, p. 126.

[2] Pococke's *Travels in England*, 1764, I, pp. 12, 13, 19, 73, 75, etc.

[3] Ibid., pp. 20, 35, 36, 38, 46. He says that in Sussex they were content to walk through the fields in winter (p. 20).

[4] The roads in the south-western peninsula down to the beginning of the nineteenth century were usually rough, hilly and altogether unsuited for fast travel. Dr Davy, in describing that section about 1780, says that Cornwall was then without great roads, and that the prevailing manner of travel was, not by carriage, but on horseback. In the account given by the Rev. John Skinner of his tour in Cornwall, in 1797–8, after he passed Somerset he nowhere mentions one good road, but frequently speaks of them as rough and bad (v. Brit. Mus., Add. MSS. 28,793, pp. 45, 46, 52, 57, 63); and yet he tells us (p. 124) that he travelled from Penzance to Helston in a little more than two hours, an equivalent of at least seven miles per hour, which, on the very doubtful supposition that he was travelling in a coach, was certainly a good speed at that time. In 1808, Warner travelled through Cornwall and when he got off the main roads he found himself, in most cases, on ways which were narrow, hilly and often precipitous and dangerous (Warner, *Tour through Cornwall*, pp. 85, 87, 110, 112, etc.). In 1800, Warner walked through these western counties, and in speaking of Devon, he says: "The North Devonians take the best possible means of preventing strangers from visiting this part of England, by the execrable state in which they keep their turnpike roads;" and again he speaks of the "abominable and intricate roads of North Devon." (See Warner, *Tour through the Western Counties*, pp. 123, 129. This was in the summer.) By 1808, however, some improvement had taken place in these roads by the rigid enforcement of the highway laws (Vancouver, *Agriculture of Devonshire*, p. 374). By 1817, most of the great thoroughfares of Cornwall were in fairly good condition, although many of the lesser cross roads, both of Cornwall and Devon, were narrow, badly made and hilly. Gilbert, *Historical Survey of Cornwall*, I, pp. 395, 396. See also Vancouver, op. cit., pp. 369–70; Worgan, *Agriculture of Cornwall* (1811), pp. 161–2; and Pococke's *Travels in England*, 1764, p. 77.

[5] Boys, *Agriculture in Kent* (1794). p. 98, says that the highways of the Weald were perhaps the worst turnpike roads in the kingdom. In winter, it was frequently

improved[1]; but in order to reach places off the main highways, the coaches had usually to be left at the nearest point of the turnpike, and the by-roads to be travelled on horseback[2]. This work of improvement continued through the first third of the nineteenth century, until at the end of that time the routes leading directly to and from the important centres were mostly firm, smooth and well suited to the carrying trade and the work of coaching, while the parish roads in many places were being repaired so as to make them useful adjuncts for bringing traffic to and from the main roads[3].

impracticable to ride on horseback along the main roads. Consequently pavement was formed at the sides of the roads to enable the rider to pass along (compare also Marshall, *Rural Economy of the Southern Counties* (1798), I, p. 343, who expresses this fact in other words). Marshall (ibid., II, p. 98, and I, p. 365) describes in similar language the facts as to the roads in the Weald of Sussex and in Romney Marsh.

[1] Moritz, *Travels through England* (1782), pp. 539, 678, etc. He frequently speaks of the good roads and in referring to his journey through Kent, he says: "These carriages are very neat and lightly built, so that you hardly perceive their motion, as they roll along these firm, smooth roads." This was in the month of June. Note also Brit. Mus., Add. MSS. 17,398, pp. 107, 108, 110, 112, 119, 120, etc.; and Brit. Mus., Bibl. Eger. MSS. 926, pp. 9, 70. Boys, *Agriculture in Kent*, pp. 83, 90, says that the roads that were most used were kept in "excellent" or "tolerably good" order. Marshall, *Rural Economy of the Southern Counties* (1798), I, p. 20, II, pp. 6, 229, 308, 393, comments very favourably on the great roads of these counties. See also Marshall, *Rural Economy of the West of England* (1796), I, pp. 31, 285, II, pp. 5, 13, 107.

[2] Brit. Mus., Add. MSS. 17,398, pp. 103, 107, 110, 111, 112, 119, 120, etc.

A significant comment on what was called a good road is found in Brit. Mus., Add. MSS. (Place MSS.) 35,143, p. 189, in a letter of Aug. 6, 1811. It reads as follows: "In Albin's map, the way from Arreton (near Southampton) is marked as a good road, but do not suppose it like the roads in the vicinity of Cockney Shire or you will be disappointed—it is no more like them as a road than those roads are like it in diversity and pleasant prospects—in fact like all the roads in the Island (i.e., Isle of Wight) except that from Ryde to Newport and from Newport to Cowes—it is in many places a very narrow lane only in which two carriages cannot pass."

[3] Arthur Young, *Agriculture in Sussex* (1818), pp. 416–19, describes the changes that had taken place in the roads of the southern counties, in the twenty or twenty-five years preceding the time when he was writing. Since the roads of Sussex had formerly been among the worst in England, we can see that changes for the better in that county would mean still greater changes in the adjoining counties. Young says: "The turnpike roads in Sussex are generally well enough executed;...... turnpikes are numerous and tolls high; in some places in the east, they are narrow and sandy. From Chichester, Arundel, Steyning, Brighton, Bourne, the roads to the metropolis, and the great cross road near the coast, which connects them together, are very good." In the Weald, he says, the cross roads were in all probability the very worst that were to be met with in any part of the island; and he attributed this to the fact that the trees prevented the wind from drying up the moisture just when it fell. The road at Horsham (Sussex), which was part of a through road to London that was made in 1756, had very few equals, in his estimation. Before it

In the opinion of some, the chief highways in the eastern counties in summer would seem to have been more uniformly good than in the other section we have just considered; and, from the last quarter of the eighteenth century[1], we find much said in their praise[2]. But many of the minor, and some of the great roads, were mere lanes through

was put through, travellers were forced to go round by Canterbury; but since its construction rents had risen from 7s. to 11s. per acre, and there was "a general spirit of mending the cross roads."

In further amplification of our statement as to the improvement in the great highways, see Marshall, *Agriculture in the Southern Counties*, I. p. 9, who says that the roads around London were "proverbially good" and ought to be followed as a pattern by the road-makers of the kingdom at large; also Vancouver, *Agriculture of Hampshire* (1813), p. 391, who said that the great roads were generally good, and some were the very best in the kingdom: that the turnpike roads were nowhere better than what might generally be met with in Hampshire. Billingsley, *Agriculture of Somerset*, pp. 91, 159, 260, speaks very highly of the roads of that county, some of which, he says, are "comparatively speaking, as smooth as a gravel walk" (p. 260). See also Stevenson, *Agriculture of Dorset* (1812), p. 439; Mavor, *Agriculture of Berkshire* (1808), p. 422; Davis, *Agriculture of Wilts*, p. 156. We must not here be regarded as giving indiscriminate praise, however, for while much had been done, there still remained much to be done in most of the counties. See, for example, Mavor, *Agriculture of Berkshire* (1808), pp. 422–3; Stevenson, *Agriculture of Surrey* (1813), p. 546.

The by-roads and parish roads were, in some parts, in fairly good condition, where materials were abundant and the necessary inducements to and supervision of the work were found (Davis, *Agriculture of Wilts*, p. 156; Stevenson, *Agriculture of Surrey*, p. 547; Vancouver, *Agriculture of Hampshire*, p. 392). But, in the majority of cases, the cross roads suffered much from neglect; the care taken and the time spent on the chief thoroughfares left very little to be devoted to the lateral roads. See Boys, *Agriculture in Kent*, p. 90; Stevenson, *Agriculture of Surrey*, p. 547; Vancouver, *Agriculture of Hampshire*, p. 391; Stevenson, *Agriculture of Dorset* (1812), p. 439; Mavor, *Agriculture of Berkshire* (1808), p. 426; Davis, *Agriculture of Wilts*, p. 156; Marshall, *Rural Economy of the West of England*, I, p. 31.

[1] If the description of Wheatfield, in the county of Suffolk, in 1761 were typical of many such villages, there would have been little need for good roads, but fortunately other places had a spirit of progress in their community life. Regarding Wheatfield, at the above date, it was said, "Neither post, coach, nor stage-waggon, set out from hence, nor are they in the least wanted; for the waggons, tumbrels, and horses of the place are always sufficient to carry out the inhabitants and their commodities, as far as they have ever occasion to go; and the single postage of a letter to London will amply pay a messenger to the utmost extent of their correspondence" (Clubbe, *History and Antiquities of the Ancient Villa of Wheatfield*, p. 90; Brit. Mus. MSS. 19,200, p. 16).

[2] In Grose's ' Tour in Suffolk, Norfolk, etc., 1777,' we find such expressions as the following: "Roads still level and good" from Ipswich to Blackendone; "the roads hither (to Bury St Edmunds) from Stow Market very fine;" the road near Bury, on the way to Thetford, was "fine;" farther on, "the road lies here through large cornfields, without hedges or any divisions. Come into a road less frequented, but good." This road to Thetford "is in general fine, though here and there interspersed with a sandy spot." Left Thetford......and set out for Cambridge, "the

fields, and both lesser and greater roads were often impassable in winter[1]. The sand used in some places for mending them, when moderately wet, made a very good road; but when it became dry, the road was heavy, and when it became very wet, or in time of thaw, the road was like mortar[2]. It was evidently at such a time as the latter that Arthur Young travelled through Suffolk and Norfolk, when he uttered his maledictions against the "execrable muddy roads" or the "infamous turnpikes[3]." In 1799, Young referred to the roads of Lincolnshire in general as "below par[4];" and he said that he knew "not one mile of excellent road in the whole county of Norfolk[5]." This latter statement might be perfectly true, for there might not be any "excellent" roads, although there might be some good ones; and according to his own declaration five years afterwards the roads of that county were in general "equal to those of the most improved counties[6]." It is probable that both his statements were overdrawn, for it is not likely that the roads in Lincolnshire would vary widely from those of the adjoining counties of Norfolk and Suffolk; and it is inconceivable that in 1804 the roads in every part of the county of Suffolk could be "uncommonly good[7]," and the roads of Norfolk be as a rule "equal

road fine and over common fields" (Brit. Mus. MSS. 21,550, pp. 2, 4, 10, 11, 20, 39). He mentions only one piece of bad road.

In Ord's 'Tours in Norfolk and Suffolk (1781–97),' we find similar testimony to the good roads (Brit. Mus. MSS. 14,823, pp. 21, 29, 96, 101, 118). Arthur Young, *Farmer's Letters* (1767), p. 282, said that in Suffolk "exceeding good roads" were everywhere met with.

[1] Brit. Mus. MSS. 21,550, pp. 20, 21, 36; Brit. Mus. MSS. 14,823, pp. 21, 29, 96, 120.

[2] Young, *Agriculture of Lincolnshire*, p. 405; Ord's 'Tours in Norfolk and Suffolk' (Brit. Mus. MSS. 14,823, pp. 29, 120). Arthur Young, in his *Six Weeks' Tour*, p. 319, speaks of "the execrable muddy road from Bury to Sudbury in Suffolk: in which I was forced to move as slow as in any unmended lane in Wales; for ponds of liquid dirt, and a scattering of loose flints, just sufficient to lame every horse that moves near them, with the addition of cutting vile grips across the road, under pretence of letting water off, but without the effect, all together render, at least twelve of these sixteen miles, as infamous a turnpike as ever was travelled."

[3] Ibid.   [4] Young, *Agriculture of Lincolnshire*, p. 405.

[5] Young, *Six Weeks' Tour*, p. 319.

[6] Young, *Agriculture of Norfolk*, p. 489.

[7] Young, *Agriculture of Suffolk* (1804), p. 227, says of the roads there: "These are uncommonly good in every part of the county, so that a traveller is nearly able to move in a post-chaise by a map, almost sure of finding excellent gravel roads; many cross ones in most directions equal to turnpikes. The improvements in this respect, in the last twenty years, are almost inconceivable." Young, *Agriculture of Essex* (1807), II, p. 384, says, "It is impossible to say too much in praise of the roads of most of the districts in Essex;" and he speaks of them as "excellent," "incomparable," etc.

to those of the most improved counties," if those of Lincolnshire, only five years before, were "below par."   In 1818, again, he speaks in the highest praise of some of these eastern roads, for of one he says that it was perhaps "the very best road in England," and of another that "no road in the world surpasses it[1]."   We are probably justified, therefore, in saying that, making all due allow-ance for exaggeration, the eastern counties were not better supplied with good roads the year round than were the midlands and the south.

Having briefly examined each of the great systems of roads, what shall we say in regard to the roads as a whole?   From what we have formerly noted, as to the vast increase in the number of road Acts passed after 1750, we would naturally expect that this increased interest in the highways would have brought an immediate and startling change for the better in the character of the English roads.   To some it seemed that even the changes effected before 1770 were like a revolution. A writer in 1753 spoke of the "almost impassable state of the roads," which was a grievance "very justly and universally complained of[2];" while another, fourteen years afterwards, said that "There never was a more astonishing revolution accomplished in the internal system of any country, than has been within the compass of a few years in that of England....Everything wears the face of dispatch[3]."   We do not mean to impugn the motives of the writer of this latter statement, but the view he gives us is entirely too roseate to accord with the facts as given to us from other sources.   In all probability, his experience was concerned with the district between London and the midlands, and certainly in that area parts of the great roads had been much im-proved; but Young's description of some of these roads, about the

---

[1] Young, *Agriculture of Sussex*, p. 417.

[2] Brit. Mus. 8776. c. 21, p. 1, 'Proposals for Amendment of Roads (1753).'   See also Shapleigh on *Highways* (1749), p. 4.

[3] Homer, *Enquiry into the Means of Preserving and Improving the Publick Roads* (1767), pp. 7–8.   He says: "Dispatch, which is the very life and soul of business, becomes daily more attainable by the free circulation opening in every channel, which is adapted to it.   Merchandize and manufactures find a ready conveyance to the markets"......"There never was a more astonishing revolution accomplished in the internal system of any country, than has been within the compass of a few years in that of England.   The carriage of grain, coals, merchandize, etc., is in general conducted with little more than half the number of horses with which it formerly was.   Journies of business are performed with more than double expedition. Improvements in agriculture keep pace with those of trade.   Everything wears the face of dispatch; every article of our produce becomes more valuable; and the hinge which has guided all these movements, and upon which they turn, is the reformation which has been made in our public roads."

same time, shows us that no general revolution had taken place[1]. The testimony of a clergyman in 1808 was that the roads throughout England in 1778 were very bad, but that three horses could do in 1808 as much as five horses could do thirty years before[2]. In reality, what we find was not a sudden change; for this, like most other great social and economic movements, went on gradually, though much more rapidly now than during the preceding hundred years. On the other hand, we differ from some who reason on the basis of Arthur Young's statements that almost all the roads were bad[3], for we have already observed very much to the contrary[4]. The truth of the matter seems to lie in a moderate view, so that before the end of the eighteenth century, it was possible, for those who took a wider retrospect of the preceding fifty years, to note great improvements within that time, such as made the English roads "the admiration of foreigners[5]." This great improvement went on during the first third of the next century, so that not only the main arteries of communication, but also many of the cross roads, assumed a quality befitting a nation which had attained industrial and commercial supremacy[6]. While this vast

---

[1] Young, *Agriculture of Hertford*, p. 221; Young, *Agriculture of Oxford*, p. 324. To the same effect is a statement in the *Gentleman's Magazine* (1785), LVII, p. 879, which says: "The turnpike roads in several parts of England are so narrow, contrary, I apprehend, to the Acts in that particular, that two carriages cannot possibly pass each other; they are also, in several parts, in such wretched condition (though the usual tolls are levied) as to be scarcely passable."

[2] *Parl. Papers*, 1808 (275), II, 459, 'Second Report of the Committee on Highways,' p. 142.

[3] Toynbee, *The Industrial Revolution*, p. 52; also Brit. Mus., T. 1157 (4), 'Highways Improved,' the writer of which speaks of the generally "bad state of a great proportion of the public highways in the kingdom" (p. 2).

[4] See under the descriptions previously given of the great road systems.

[5] A contributor to the *Gentleman's Magazine* (1792), LXII, p. 1161, says: "The great improvements which, within the memory of man, have been made in the turnpike roads throughout this kingdom, would be incredible did we not actually perceive them." In 1798, another writer says that the public roads in England, though they are much abused, "are the admiration of foreigners; and, it must be allowed, where materials are to be had, are, on the whole, well constituted and kept in good repair......To turn our eyes back to fifty years ago, it is with wonder and delight we view the improvements on every approach to the great city" (*Gentleman's Magazine*, LXVIII, Part II, p. 647). See also letter from "Themistocles" in the *Public Advertiser*, Jan. 30, 1790, p. 1.

[6] Brit. Mus., Add. MSS. 35,147, VI, p. 253, tells us that at that time (1829) the roads of England were the "marvel of the world;" and that the improvements that had been effected during a century would have been regarded as almost miraculous, were it not considered that they had been produced by the spirit and intelligence of the people. For the road network in 1764, see Owen, *Britannia Depicta: or Ogilby Improved*.

improvement is reflected in many ways, especially in the increased rate of travel, much had yet to be done before most of the roads could be called good by the road engineers[1].

A few words regarding the streets of the towns and cities must suffice to conclude our account of the condition of the highways. The industrial advance which was made during this period brought many people from the country to the towns, and because of the increased amount of business and travel on the streets improvement was imperative. In some cases the streets seem to have been but little improved during the preceding century[2]; but now it was recognized

---

[1] See, for example, the testimony of J. L. Macadam before the Parliamentary Committee of 1819 on the Highways.

Edgeworth on *Roads and Carriages*, p. 4, tells us that, in some places, up to that time, the cross roads of England continued in a wretched state of repair. Twenty years after that, McCulloch says that some of the turnpike roads were still in a very bad state, and that most of the existing roads were far from being in the state that might be expected and in which they ought to be (McCulloch, *Statistical Account*, II, p. 178). In 1825, the writer of 'Highways Improved' (Brit. Mus., T. 1157 (4), p. 2), after mentioning the improvements effected by Telford, Macadam, and other engineers, said that this improvement had extended to many parish roads also; but of these parish roads, a very large proportion was in such a state as to be inconvenient and even dangerous to all who travelled on them. See also *Communications to the Board of Agriculture*, I, p. 120. Similar statements were made by Sir Henry Parnell and Thomas Grahame (Grahame, *A Treatise on Internal Intercourse and Communication*, p. 21).

[2] Warner, *Tour through Cornwall* (1809), pp. 110, 112, 319, shows that the streets of the Cornwall towns were narrow and irregular and generally paved with pebbles from the shore, the points of which were turned upward, thus forming a footing neither safe nor pleasant. See also Gilbert, *Parochial History of Cornwall*, II, p. 104.

But we need not go so far from the industrial centres to find such conditions. Sheffield is described by Horace Walpole, in a letter of 1760, as "one of the foulest towns in England, in the most charming situation." The same place is spoken of by Twining, in his *A Country Clergyman of the Eighteenth Century* (1776), pp. 46–47, as having pavement which was "execrable" and reaching almost two miles from the town. Samuel Roberts, who was born in 1763, remembered the streets of Sheffield as in a very disagreeable condition. A few dirty, dull, oil lamps, just within sight of one another on dark nights, served to show how very dark it was. The bellman was the watchman; pigs were the principal scavengers, and they rooted in the garbage accumulated in the open channels, or gutters, that ran down the middle of the streets. "Her highways were then rather low ways," and the footpaths were flagged with "grindle kouks" and stones of all kinds and shapes, except square (Samuel Roberts's *Autobiography*, p. 24).

Liverpool, also, whose streets were narrow, ill-paved and tortuous, did not enter on the work of improvement till about 1785 (Picton, *Liverpool Municipal Records*, II, pp. 258–9, 364–6). Manchester in 1776 and 1777 began the widening and repairing of some of her streets, which had long been felt to be a disgrace to the town (Axon, *Annals of Manchester*, p. 104). When the streets of Rye were first paved, round boulder stones were laid down, so as to cause a slope from the houses to the centre

that, from an economic and sanitary point of view, great changes must
be made.   Many of the streets were narrow at best and the obstructions which we have before seen to have existed had to be removed
so as to permit the fullest use of the available space[1].   Frequently, the
streets which were narrow had to be widened to accommodate the increasing traffic[2], but in other cases the houses were built so close to the
streets that the latter could not be widened.   In such cases, the only
things to be done were to regulate the driving of waggons, carts, etc.,
to widen the foot pavements for the safety and convenience of pedes-
trians and then to pave the carriage ways in a better manner[3].   The
latter problem was solved at the time when Macadam's principles
became the rule of practice; but we may say that it was not until
about 1830 that the streets of the chief towns and cities, including
London, assumed that form and firmness of construction, accompanied
by comparative smoothness of surface, that is characteristic of the best
urban highways at the present time[4].

of the street, where ran an open gutter to carry off the rain and filth.   This continued
down to 1819 (Holloway, *History of Rye*, p. 465).   Note also the conditions in
Westminster, as given in Brit. Mus. 213. i. 2 (53), p. 1, 'Reasons for the Petition
for Better Paving, Cleansing and Lighting the Streets of Westminster.'

But, on the other hand, some towns were well paved and kept very neat (Brit.
Mus. MSS. 14,259, pp. 129, 136; Brit. Mus., Add. MSS. 33,684, p. 92).

[1] For instance, by the Act of 1772 for the city of Chester, the people were
forbidden to lay or leave any ashes, rubbish, soil, timber, boards, stones, filth, etc.,
in any open street or lane, so as to obstruct the way; and no cart or waggon was to
be allowed to remain in the public ways any longer than necessary.   Hutton, *The
Scarborough Tour in* 1803, pp. 90, 91, said that the people of the city of York used
a set of the worst streets he ever saw; they were so very narrow and had so little
room for use.

[2] Picton, *Liverpool Municipal Records*, and Axon, *Annals of Manchester*, as
above referred to.

[3] In the closing days of 1815 and the early part of 1816, a series of eight letters
was published in *The Times*, dealing with the pavement and roads of the metropolis,
showing their present condition, the reasons why they were so bad, the legislative
enactments that had been passed for their improvement and discussing the best
methods of improving them.   See issues of Dec. 26, 1815, p. 2; Dec. 27, 1815, p. 4;
Jan. 2, 1816, p. 2; Jan. 5, 1816, p. 4; Jan 26, 1816, p. 4; Feb. 8, 1816, p. 4; June 3,
1816, p. 2; and June 25, 1816, p. 3.   *The Times*, May 24, 1830, p. 2, and Oct. 19,
1831, p. 4, showed the wretched condition of the streets and pavements in some
parts of London like the Strand and Charing Cross.

[4] Even the streets of Morpeth, on the Great North Road, were not changed
from a "rough hog-backed pavement," very dangerous and inconvenient, to a good
macadam, until the winter of 1827–8 (Hodgson, *History of Northumberland*, Pt. II,
II, p. 529).   On Dec. 2, 1811, the Chairman of the Grand Jury for the county of
Middlesex said that the streets of London were, in many of the most populous
parts, so decayed, that they were not only highly inconvenient but absolutely
dangerous (v. *Gentleman's Magazine*, LXXXII, Pt. I, p. 85).   This is in accord with

Travelling and Conveyance.

In the earlier period, we have seen that a large part of the carriage of goods on the highways was effected by pack-horses[1] and this method was in use to a considerable extent even in the last half of the eighteenth century. When Josiah Wedgwood and his partner Bentley gave so much support to securing the construction of the Trent and Mersey canal through the district in which their potteries were located, it was for the purpose, among other things, of obtaining a better means of carriage than the existing system of "pack-horses and asses heavily laden with coal,—tubs full of ground flint from the mills, crates of ware or panniers of clay," and "floundering knee-deep" through the miry roads of Staffordshire and Warwickshire[2]. Most of the produce of the Sheffield manufacturers was carried weekly to the metropolis by some of Newsom's pack-horses[3]. This kind of conveyance was also the most common throughout the next county to the south, for a traveller, in speaking of Derbyshire, says: "The road (from Buxton to Matlock Bath) is a continuance of the same scene, naked hills and desart dales: nothing worth notice occurred, except the vast number of pack-horses travelling over the hills, of which we counted sixty in a drove; their chief lading is wool and malt, which they carry across the country from Nottingham and Derby to Manchester[4]." Even down to the beginning of the nineteenth century this same method prevailed in that district[5], as it did farther north in Cumberland and Westmorland[6], and in the western counties like Devon and Cornwall[7].

their condition at a later time (1820–2), when their improvement was actively taken up (v. Brit. Mus., Place MSS., vi, p. 264).

[1] Defoe, *Tour through the Whole Island* (1748), I, p. 94, III, p. 121. He shows us the great importance of pack-horse carriage to those who were carrying on a large amount of wholesale trading, especially at the times of the great fairs and markets. See also Hassall, *Agriculture of Monmouth*, p. 101.

[2] Meteyard, *Life of Josiah Wedgwood*, pp. 267, 275.

[3] *Sheffield Local Register*, p. 45; Whitaker, *Loidis and Elmete*, p. 81.

[4] *Four Topographical Letters* (1755), p. 42. Muir, *History of Liverpool*, p. 256, tells of the 70 pack-horses which daily left a single inn in Liverpool for Manchester.

[5] Strickland, *Agriculture of the East Riding of York*, p. 266.

[6] Nicholson and Burn, *History of Westmorland and Cumberland*, I, p. 66; Walker, *Remarks made in a Tour from London to the Lakes* (1792), p. 25. See also Jeans, *Jubilee Memorial of the Railway System*, p. 5.

[7] Dr Pococke, in his *Travels in England* (1764), I, pp. 77, 133, says: "Though they have turnpike roads in most parts yet they do not use carriages in Devonshire and Cornwall on account of the steep hills and of the by roads which are so narrow that they will not do for wheel carriages. They have large wooden forks on each side of their horses, which carry a large quantity of goods. They call them crooks.

Somersetshire and Gloucestershire had similar provision for the carriage of coal to Bristol, Bath and other centres[1]. But after 1750, on account of the great changes which were taking place in industry and agriculture, there were much larger quantities of material to be transported and a great change took place in the carrying trade, both by land and water. The former means of conveyance by land had to be either wholly

They are of different sizes according to what they are to carry, and they have a sort of baskets made in different ways but mostly of hoops in which they carry earth etc. and call them dung pots."

Shaw, in his *Tour to the West of England* (1788), p. 263, observed the same thing at a considerably later time. He says: "The common traffic and business of this county (i.e., Devon) is mostly done by horses with panniers and crooks; the former are well known everywhere, but the latter are peculiar to the West, and are simply constructed, with four bent heavy sticks in the shape of panniers, but the ends awkwardly projecting above the rider's head; with these they carry large loads of hay or garden vegetables. The country people ride in a prodigious large boot of wood and leather, hung instead of stirrup to the horse's side, and half open, which they call gambades." From what he says above, that the panniers "are well known everywhere," we would naturally conclude that in most parts of England this means of carrying survived, alongside the waggons, even to this late date; and the evidence in substantiation of this is fairly convincing. But while the panniers and crooks were not unfamiliar to that generation, we must not suppose that their use was at all common in the central, southern and eastern parts of the kingdom. They were familiar as survivals of the earlier period. Marshall, *Rural Economy of the West of England* (1796), II, p. 227, seems to be telling the truth more accurately in saying that "Carriage on horseback may now be said to belong to the extreme west of England." Here, however, we must keep in mind that he was not speaking of the rural economy of the whole of England, but only of the western part; and we must also remember, what we have already shown, that these means were employed in other parts of England.

Fraser, *County of Cornwall*, p. 46, says: "Carts are not used in this county. Everything is carried on the pack-saddle, for which both horses and mules are used." Lipscomb, *Journey into Cornwall* (1799), pp. 149–50, says that, in the part of Devonshire near Honiton, the produce of the land, as well as merchandise, was chiefly carried upon a "crook." These horses ran loose in troops, consisting of five or ten, having either one or two men mounted upon other horses to drive them. Loads of hay, straw, wood and furze were carried in the same way; and the horses that carried them were no small annoyance to any unfortunate passenger whose horse happened to be restive, or who might chance to get too near the crooks, of which in the narrow roads it was very difficult to keep clear. Compare also Vancouver, *Agriculture of Devonshire* (1808), pp. 371, 373, and Hassall, *Agriculture of Monmouth* (1812), pp. 100–1.

In Brit. Mus., Add. MSS. 19,942, Dr Jeremiah Milles has sketched for us a great variety of horse-panniers, pack-saddles, crooks, etc., which were widely different according to the material they were intended to carry and the sex of the person who was to use the saddle. Marshall, *Rural Economy of the West of England* (1796), I, pp. 121–8, describes the various kinds of equipment for a pack-horse.

[1] Mathews, *Remarks on the Cause and Progress of the Scarcity and Dearness* (1797), pp. 83–84.

abandoned for, or made supplementary to, a more economical arrangement, namely, the more extended use of stage waggons[1]. These were large carriages, of great width and length, usually with four wheels,—though occasionally with eight,—that, according to statutory requirements, frequently had their wheels from nine to sixteen inches broad. The waggon had a very spacious box which was arched over with cloth; and when filled it was drawn by six or eight horses, whose neck-bells warned of their gradual approach on the narrow roads[2]. In these capacious vehicles, the lighter articles of freight were transferred from place to place, while the heavier freight was taken on the canals and rivers. In some cases, the merchants did their own carrying; but for the great majority of small traders it was impossible to economically carry their own goods, and thus, in accordance with the greatly augmented amount of trade, there was a progressive increase in the number of that highly important class of carriers, who carried most of the traffic on the routes connecting the great towns. Of course, this was merely a development of the system that was already in operation, but on a greater scale. Some of these carriers were business entrepreneurs, each with many waggons and teams, who owned warehouses at the important places along their carrying routes, and not only collected the goods to these warehouses but also distributed from them[3]. Their waggons were found on the roads at all hours of the day and night and the wide area which they covered is proof of the enormous extent of their trade. For instance, the raw materials and the manufactured products of the potteries at Burslem and adjoining places had to be carried from and to places as far south as Bristol and

[1] How much more economical waggon carriage was than carriage by pack-horse is indicated in Aris's *Birmingham Gazette,* Dec. 13, 1824, p. 1, in a letter from F. Finch regarding the Birmingham and Liverpool Railway, in which it is said that when the slow and irregular conveyance by horses was relinquished for waggons, six horses were enabled to do the work of thirty.

[2] Espriella, *Letters from England* (1807?), p. 18. He says: "Carrying is here a very considerable trade; these waggons are day and night upon their way, and are oddly enough called flying waggons, though of all machines they travel the slowest, slower than even a travelling funeral." See also Marshall, *Rural Economy of the Southern Counties* (1798), ii, p. 393.

[3] Perhaps the most famous of all the carriers was Messrs Pickford & Co., whose waggons traversed all the great roads, but whose greatest business was done between London and Liverpool and Manchester. Some of their teams collected and distributed goods to and from their London warehouse for residents of London, others were engaged in the same business at Coventry, Birmingham, Manchester and Liverpool; and the rest of their teams were on the road carrying to and fro along this route.

From Manchester, another of the chief carriers was Mrs Ann Johnson, who carried on the business built up by her husband (v. Slugg, *Manchester Fifty Years Ago,* pp. 225–6).

Bewdley and as far north as Liverpool and Manchester[1]. The waggons from Manchester regularly visited places as remote as Bristol, London and Edinburgh[2]; and Birmingham had communication in this way with one hundred and sixty-eight other towns, so wide apart as to include York northward and Bristol southward, Welshpool westward and Lincoln eastward[3].

[1] Meteyard, *Life of Wedgwood*, I, p. 268; Picton, *Memorials of Liverpool*, II, p. 106; Shaw, *Rise and Progress of Staffordshire Potteries*, pp. 148–9.

[2] Slugg, *Manchester Fifty Years Ago*, p. 225. Over one hundred waggons and carts left Manchester, some of them daily, and others two or three times a week. In an advertisement of Ann Johnson's at that time, it was stated that the waggon for Liverpool left every evening at seven o'clock and arrived there at nine o'clock the following morning. Her waggon for Birmingham left Manchester every Wednesday and Saturday evening at eight o'clock, reaching there in two days, whence goods for Bristol were forwarded by Gabb and Shurmer, arriving there on the fourth day after leaving Manchester.

[3] Bunce, *History of Birmingham*, I, p. 49, quoting from the *Birmingham Directory* of 1763 and 1764. So perfect was the monopoly enjoyed by the carriers between Birmingham and London, that in 1763 they announced their intention to raise their prices, on account of the bad roads. How wide were the connexions that Birmingham had with other places in the kingdom, may be seen in Swinney's *New Birmingham Directory*, ca. 1774 (unpaged); *The Birmingham Directory of 1777*, pp. 57–64; Chapman's *Birmingham Directory* (1803), pp. 132–7. They carried to all parts of England and Scotland.

To further illustrate the great extent of the carrying trade, note the following examples:

In 1760, Richard Whitworth wrote: "There are three pot waggons go from Newcastle and Burslem weekly, through Eccleshall and Newport to Bridgnorth, and carry about eight tons of potware every week, at £3 per ton. The same waggons load back with ten tons of close goods, consisting of white clay, grocery, and iron, at the same price, delivered on their road to Newcastle. Large quantities of potware are conveyed on horses' backs from Burslem and Newcastle to Bridgnorth and Bewdley, for exportation, about one hundred tons yearly, at £2. 10s. per ton. Two broad-wheel waggons (exclusive of 150 pack-horses) go from Manchester through Stafford weekly, and may be computed to carry 312 tons of cloth and Manchester wares in the year, at £3. 10s. per ton. The great salt trade that is carried on at Northwich may be computed to send 600 tons yearly along this (proposed) canal, together with Nantwich 400, chiefly carried now on horses' backs at 10s. per ton on a medium" (Jewitt, *The Wedgwoods*, p. 171, quoting from Whitworth).

Phillips, *Tour through the United Kingdom* (1828), pp. 78–79, shows us the change in the amount of carriage at Leicester in the fifty years preceding that time. He says: "About half a century ago, the heavy goods passing through Leicester for London to the south, and on the great northern lines to Leeds and Manchester, did not require more than about one daily broad-wheeled waggon each way. These ......were also fully adequate for the supply and transit of goods for all the intermediate towns, of course including Leicester. One weekly waggon, to and fro, served Coventry, Warwick, Birmingham, and on to Bristol and the west of England; the return waggon being capable of bringing all from that quarter that was directed to Leicester, and all the northern and north-eastern districts beyond. At present, there are about two waggons, two caravans, and two fly-boats, daily passing or

We must remember that this great trade by land was only a part of the total traffic, for all the heavy materials were conveyed by river and canal navigations, where these were possible.

With the increasing extent of the carrying trade, there came also the desire for regularity of service; and the waggoners endeavoured to meet this demand by forming and adhering to a fixed schedule of arrival and departure. The difficulty of keeping to such fixity of movement will be evident to anyone who considers the fluctuations in the amount of goods offered for transport and the great variety of other conditions over which the carriers had no control but which would affect their movement. When, therefore, the carriers advertised to set out from and return to a certain place at definite times[1], we must

starting from Leicester for London and its intermediate towns: the same number ......extend the connection not only to Leeds and Manchester, but by means of canal conveyance to the ports of Liverpool and Hull. There are at least six weekly waggons to Birmingham, independent of those to Bristol three times a week, and the same to Stamford, Cambridge, Wisbeach, and the eastern counties; to Nottingham to the same extent, exclusive of carts; and at least two hundred and fifty country carriers to and from the villages, many twice a week, necessary to keep up the conveyance of material and manufactured goods between the workmen and the hosiers, and the wholesale and retail dealers in other articles of necessity......"

Battle, *Hull Directory* (1791), pp. 65–73, gives a list of the carriers leaving Hull for other parts of the kingdom. *Guide to London*, 1740, pp. 94–119; *Guide to London*, 1772, pp. 132–76; *Guide to London*, 1783, pp. 135–89—these three give 'An Account of all the Stage Coaches and Carriers in England and Wales' that came to or left London for each of these years. The coaches are not differentiated from the carriers' waggons very clearly and it is impossible to know exactly the number of each at these successive intervals. Furthermore, the same coach or carriage is often listed under two different destinations, thus leading to manifold complications if we wished to secure numerical results. But as the size of the pages in these three *Guides* is the same, we may roughly trace the increase in the amount of the coaching and carrying business by the increase in the number of pages devoted to giving these details at these three times. In such a view of the matter, we find that the number of pages in the guide of 1740 is 25; in that of 1772, 44; and in that of 1783, 54. So that these numbers, 25, 44, 54, may represent the relative increase in the coaching business and carrying trade in the corresponding years, 1740, 1772, and 1783. The range of the carrying business between London and the rest of the country, in 1802 and 1809, is given in Holden's *Annual List of Coaches, Waggons, Carts, Vessels, etc.*, from which it is almost impossible to get any adequate conception of the complex structure of the carrying trade. Mathew's *New Bristol Directory* (1793–4), pp. 93–98, gives facts for Bristol corresponding to those above for London. See also *Four Topographical Letters* (1755), p. 1. *The York Guide*, 2nd ed., 1796, pp. 45–46, gives a list of carriers for that city and shows how great was the traffic between York and other places at that time. *The Chester Guide*, 1795, pp. 62–64, 66, reveals a wide series of connexions between Chester and the rest of the kingdom.

[1] Brit. Mus. 579. c. 43 (3), 'The Ancient and Modern History of Portesmouth, Portsea, Gosport and their Environs,' pp. 117, 120–1; *Cambridge Chronicle and Journal*, Aug. 6, 1813, p. 1.

not think that necessity did not cause any variations from these arrangements. In the case of the coaching business, barring emergencies, the times of arrival and departure could be depended upon, for that was merely the taking up and setting down of passengers; but the nature of the carrying business must have precluded anything like unvarying regulation. It would seem that there was considerable complaint against some of the carriers because they did not forward within reasonable time goods committed to their care, but left them for weeks and sometimes months unnoticed in their warehouses[1]. If this prevailed at all widely, it would be a serious impediment to the conduct of business and must have lent emphasis to the agitation for canals. In some instances, stage waggons were displaced and vans substituted, in order that passengers' luggage, as well as meat, farm and dairy produce, might be more speedily conveyed along the roads[2]. In the improvement of the carrying service, there came to be also a closer co-ordination of the road and water facilities, for the carrying establishments were linking themselves on to the water routes by making their arrangements to harmonize with those of the waterways. Not only were the waggons timed to connect with the boats so that there might be as little delay as possible in the forwarding of goods; but there were also working arrangements by which the through rate was a little less than the sum of the two rates by road and water[3]. A complete

[1] *Gloucester Journal*, Jan. 18, 1790, p. 3, shows a letter from the traders of Worcester protesting against this practice, which, they said, was general.

[2] Blew, *Brighton and Its Coaches*, pp. 165–6. These vans carried some passengers, as well as the fast freight. On the Brighton road there was much competition among these vans, as there was among the coaches.

[3] *Leeds Intelligencer*, July 16, 1792, p. 4, shows the close working agreement between the packet boat on the Leeds and Liverpool Canal and the road facilities. Ibid., Apr. 21, 1794, p. 2, in the advertisement entitled, 'Expeditious Conveyance of Goods to and from Manchester and Hull, by way of Wakefield and Huddersfield,' gives the working arrangements which Richard Milnes & Co. had for the most rapid transport of goods by his waggons and the boats on the navigations. The rates of freight and carriage were as follows:

From Manchester to Huddersfield.. 1s. 3d. per cwt.
    „   „   „ Wakefield .... 1s. 6d. „
    „   „   „ Hull .... 2s. 3d. „
From Hull to Wakefield .... 0s. 9d. „
    „   „ „ Huddersfield .... 1s. 3d. „
    „   „ „ Manchester......... 2s. 3d. „
From Huddersfield to Wakefield .. 0s. 6d. „
    „   „   „ Hull........ 1s. 6d. „
    „   „   „ Manchester.. 1s. 3d. „
From Wakefield to Hull .... 1s. 0d. „
    „   „   „ Huddersfield .. 0s. 6d. „
    „   „   „ Manchester .... 1s. 6d. „
It will be noted from this table that the freight rate by waggon from Manchester

study of the whole question, however, leads us to the conclusion that this did not very often occur, for the canals were too jealous of their power and too eager for profits to enter into such negotiations very often.

Leaving now the subject of the carrying trade, which we have considered as to its extent and its organization, we next examine the amount of passenger travel and the means adopted to facilitate and expedite it.

After 1750, with the great development of England in her national industries, there was an increased demand for the facilities of travel; and it was not long before coaches became much more numerous on some roads[1]. Not until the last quarter of that century, however, did the coaching business assume any great magnitude over the country as a whole[2]; and the first third of the nineteenth century was the hey-day of coaching, when the chief roads were active with the great variety of vehicles upon them while the jack boots and variegated doublets and hats of the drivers made an animated scene. A few details will serve to show the vast change in the means of travelling. About 1760, one Manchester and one Leeds or Sheffield coach passed in each direction, north and south, through Leicester. In 1828, from Leicester there were at least twelve daily opportunities of going to London, five opportunities of going to Manchester, five to Birmingham, three to Sheffield and Leeds, six to Nottingham, two to Derby (independently of the Manchester coaches), and two to Stamford—in

---

to Huddersfield, 1*s.* 3*d.*, plus the water rate from Huddersfield to Hull, 1*s.* 6*d.*, would make a total of 2*s.* 9*d.*; but the through rate was only 2*s.* 8*d.* Other combinations give similar results.

See also the advertisement of the 'Important New Line of Conveyance for Goods and Packages,' given in Macturk, *History of Railways into Hull*, pp. 18–14.

[1] Brit. Mus. 10,849. g. 11, 'A Journey through England' (1752), pp. 103, 127 ff., speaks of "the incredible quantity of coaches" on the road north of London and shows the importance of having seats booked ahead for the coaches.

[2] Brit. Mus. 567. e. 7, 'Beyträge zur Kenntniss vorzüglich des Innern von England' (1791), I, pp. 39–40, speaks of the great number of coaches between Dover and London, and says, "one meets one of them at every glance of the eye." Referring to the amount of travelling on the English roads, he says that it surpasses that of any European country. *Sheffield Local Register*, p. 88, gives the coaching arrangements into and out of Sheffield in the year 1797. For the amount of the coaching and carrying trade of London, see the three London *Guides* of 1740, 1772 and 1788, before mentioned, which show the great increase of this business in the later years over the earlier. For instance, in 1740 there was only one coach once a week from London to Birmingham; but in 1788, destined for Birmingham, four coaches left London every day and two others left three times a week. Thus, in 1740, only one trip and in 1788 thirty trips were made weekly from London to Birmingham. See also Mathew's *New Bristol Directory* (1793–4), pp. 93–98; *Hull Directory* (1791), pp. 65–78.

short, there was a daily arrival and departure of between forty and fifty stage coaches for the conveyance of persons and parcels[1]. In the latter year there was not a turnpike road leading out of Leicester but had from one to twelve coaches daily entering and leaving[2]. In 1748, it was an easy day's journey from Colchester to London and one coach went each way daily, except Sunday[3]; but by 1825 there were eight regular coaches which ran daily between these two places, four each way, besides fourteen others which passed through, to and from London, daily[4]. In 1801, seven coaches left Chester daily, while in 1881 there were twenty-six[5]; and of the latter eight went to Liverpool, whereas in 1784 not a single coach went from Chester to Liverpool[6]. In 1818, thirty-five regular coaches left York daily, to say nothing of the great number which arrived at that city; and, besides, there were extra coaches to all parts of the kingdom at almost every hour of the day[7]. On the road from London to Brighton, in 1756, there was one stage coach that went in one day, and in the following year a "two days' stage coach" was added to this service[8]; in 1787, there were three light post-coaches, two heavy coaches, three machines, and three waggons running on this road[9]; in 1811, there were twenty-eight coaches daily between London and Brighton[10]; and during the summer and autumn of 1828 there were twenty-one coaches each way[11]. In 1822, there were sixty-two coaches which daily came into and went out of Brighton[12]; and the numbers which served in London were

---

[1] Phillips, *Tour through the United Kingdom* (1828), p. 77.
[2] Ibid., p. 78.
[3] Description of Colchester, in Morant, *History and Antiquities of Essex* (1748).
[4] Cromwell, *History of Colchester* (1825), p. 408.
[5] Hemingway, *History of Chester*, II, p. 334, gives the following comparative statement of the number of coaches leaving Chester daily:

				In 1801	In 1831
To London	..	..	..	2	5
„ Manchester	..	..	..	1	4
„ Liverpool	..	..	..	2	8
„ Shrewsbury	..	..	..	1	4
„ Welshpool	..	..	..	0	2
„ Holyhead	..	..	..	1	2
„ Wrexham	..	..	..	0	1
				7	26

[6] Ibid., p. 335.
[7] Hargrove, *History of York*, II, Pt. II, pp. 671–5, gives a full list of these coaches.
[8] Blew, *Brighton and Its Coaches*, p. 35.
[9] Ibid., p. 38.   [10] Ibid., p. 83.
[11] Ibid., p. 158. Of these twenty-one, sixteen went throughout the year.
[12] *The Times*, Sept. 30, 1822, p. 3.

almost beyond computation[1]. Between London and Birmingham, in 1829, there were thirty-four coaches daily, seventeen each way[2]. In 1830, between Leeds and Bradford there were not less than fifteen coaches each way daily[3], which was a very dense passenger traffic for cities of that size. About 1810, it was said, there were a dozen coaches arriving at and departing from Preston in Lancashire; by 1830, that number had been increased to sixty-seven, and the bathing season added still more[4]. Between Stockport and Manchester, in 1828, there were over one hundred coaches going and returning every day[5].

Then, too, the competition which arose in the local coaching trade at a few centres, soon after 1760[6], increased greatly with the development of the passenger traffic; the more intense rivalry among coach proprietors brought increased speed and some reduction of cost to the traveller[7]; and when, in the first third of the nineteenth century,

[1] Holden's *Annual List of Coaches, Waggons*, etc., for 1809 shows the vast system of coaches that were working from London at that time; and this must have been greatly augmented by 1830.

[2] *Birmingham Journal*, Nov. 28, 1829, p. 3.

[3] *Manchester Guardian*, May 29, 1830, p. 3.

[4] *Manchester Guardian*, June 12, 1830, p. 3, quoting from *Preston Chronicle*.

[5] *Sheffield Iris*, Oct. 14, 1828, p. 4, quoting from *Stockport Advertiser*. Of these, forty centred at the office of one proprietor, the same number with three other proprietors, and more than forty coaches ran through or near Stockport, destined for London, Birmingham, Nottingham, Sheffield, Liverpool and Buxton. Mackenzie, in his *History of Newcastle* (1827), p. 3, says that at that time two mail coaches ran daily from Newcastle to the south and one to Carlisle. There were also coaches which set out daily to London, York, Leeds, Lancaster, Carlisle, Edinburgh, Berwick, Alnwick, Morpeth, Hexham, Durham and Sunderland, and a gig three times a week to Blyth. Daily communication with all these places would mean a great amount of coaching. There were also ten coaches and twenty-eight gigs constantly employed in conveying passengers to and from Tynemouth and North Shields; while forty years before only one gig was employed on this road.

The number of coaches going from Shrewsbury to London increased from two, going twice a week, in 1776, to seven, going daily, in 1822. Besides, at the latter time, there were daily mail coaches from Shrewsbury to Chester, Hereford, etc., and thirteen other coaches to Chester, Manchester, Worcester, Birmingham and other places (Owen and Blakeway, *History of Shrewsbury*, I, p. 519). See also Worth, *History of Plymouth*, p. 340; Baines, *History of Liverpool*, p. 526; Picton, *Memorials of Liverpool*, II, pp. 116–17; Smith, *Dunstable*, p. 112. These instances might be multiplied almost indefinitely but it would not give us any clearer conception of the comparatively enormous extent of this business during the golden age of coaching.

[6] Hills, *History of East Grinstead*, p. 150; also Latimer, *Annals of Bristol in the Eighteenth Century*, p. 367.

[7] Picton, *Memorials of Liverpool*, I, pp. 106, 203–4; II, pp. 116–17; Owen and Blakeway, *History of Shrewsbury*, I, p. 515; *Salopian Shreds and Patches*, I, p. 55; Worth, *History of Plymouth*, p. 340; Baines, *History of Liverpool*, pp. 418, 444, 460.

competition reached its highest point[1], and the comfort[2] and speed of this mode of travelling attained their climax[3], stage coaches were run on scheduled time, from which they seldom varied[4]. Sometimes this competition was so fierce that fares became ruinously low for a time; and then, upon the arrangement of a truce, the fares would be once more increased. In 1831, for example, the spirit of opposition was so great among the coach proprietors of Manchester, that some of them, apparently, were carrying passengers to London for 16*s.* inside and 8*s.* outside[5]; and on the Brighton road in 1828, competition reduced the regular fares between London and Brighton from 21*s.* inside and 12*s.* outside, to 7*s.* from London to Brighton and 5*s.* from Brighton to London[6], after the settlement of which prices went back to their former figures. At times the rivalry was so intense that passengers were carried for nothing and on some occasions were even treated with a dinner and a bottle of wine[7]. In order to facilitate travel, arrangements were often made among different coaches to have their schedules of arrival and departure harmonize for the convenience

[1] For the great extent of the coaching business that centred at Birmingham, see West, *History of Warwickshire* (1830), pp. 468–71. A complete list of the coaches, both mail and ordinary, that had their headquarters at Birmingham in 1803 is given in Chapman's *Birmingham Directory* (1803), pp. 114–25, 126–31.

That competition was very keen in some cases, is shown by the fact that the practice of "speeding up" became prevalent, frequently to the danger point. Aris's *Birmingham Gazette*, Mar. 26, 1821, p. 3, advertisement of the Aurora Day Coach to London; ibid., Apr. 2, 1821, p. 3, advertisement of the Crown Prince Day Coach to London; ibid., June 11, 1821, p. 1, advertisement of the Fly Van.

[2] A good idea of the kind of coach that was then in use may be found in Harris, *Old Coaching Days*, pp. 43–45.

[3] The speed of the fastest coaches sometimes exceeded 10 miles per hour. Harris, who knew the coaching system at first hand, has given us the names of six of these fast coaches (*Old Coaching Days*, p. 149). Some coaches went as fast as 12 miles per hour in summer (*Liverpool Mercury*, July 9, 1819) and occasionally a speed of 14 miles was attained (Baines, *History of Liverpool*, p. 624). See Appendix 5 for additional facts.

[4] Smith, *Dunstable*, p. 112. Harris, *Old Coaching Days*, pp. 152, 153, gives the time-bills of several coaches from London to Manchester and Shrewsbury. Harris, *The Coaching Age*, p. 3, says that so punctual was the Shrewsbury "Wonder" coach that many people of St Albans regulated their watches by that coach as it entered the town.

[5] *Manchester Guardian*, Nov. 26, 1831, p. 2.

[6] Blew, *Brighton and its Coaches*, pp. 138, 159. Aris's *Birmingham Gazette*, Feb. 15, 1808, p. 1, shows that by the establishment of an additional coach from Birmingham to Sheffield the other coaches had to reduce their charges from £1. 10*s.* to 8*s.* for inside fare and from £1. to 5*s.* for outside fare. See also ibid., Feb. 4, 1811, p. 1, showing reduced coach fares from Birmingham to Leicester.

[7] *The Times*, Jan. 2, 1813, p. 3, under heading "Stage Coaches;" *London Morning Post*, Jan. 2, 1813, p. 3, under heading "Stage Coach Fracas."

of passengers, so that there might be as few delays as possible and the least necessity of stoppages on the road over night[1]. In case passengers should find themselves fatigued, certain coaches on the long roads made arrangements that their patrons might rest as long as they pleased and then be assured of their seats when they wanted to proceed, without the payment of any extra fare[2]. In certain cases, the proprietors of one coach made agreements with the proprietors of some other coaches to transfer their passengers in preference to all others[3]. These methods were comparable to "through booking" by railways over each other's lines, and must have been of decided advantage to the patrons of the coaches. How wide these arrangements sometimes extended is evident from an announcement made in 1833 by the proprietors of the Hull and London Mail Post Coaches that they in conjunction with the extensive coach proprietors on the western roads had opened a speedy and direct conveyance through Nottingham, Derby, Tamworth, Coventry, Leamington, Warwick, Birmingham, Worcester, Gloucester, Cheltenham, Bristol, Bath, Shrewsbury, Manchester, Liverpool, etc., combining as great a facility to the traveller as the most direct conveyance in the kingdom[4]. Not only were there these inter-coach provisions for increasing the convenience and expedition of travel, but, in some cases, coaches were changing their times of arrival and departure, to make them accord with the schedules of the packet boats and steamers[5]. To what extent this practice

[1] *General Advertiser*, Dec. 8, 1786, p. 1, and Dec. 22, 1786, p. 1; *Bath Chronicle*, Jan. 11, 1787, p. 1; Bonner and Middleton's *Bristol Journal*, Dec. 20, 1783, p. 1; *Morning Post and Daily Advertiser*, July 20, 1791, p. 1; Oxford Historical Society, *Collectanea*, IV, p. 278, advertisement of the "Guide" coach; Malet, *Annals of the Road*, p. 24; *Cambridge Directory* (1796), p. 161; also advertisement of the "Beehive" coach in Harris, *Old Coaching Days*, pp. 43–45.

[2] *Morning Post and Daily Advertiser*, July 20, 1791, p. 1, advertisements of the York, Newcastle, and Edinburgh Mercury Post Coach and the Carlisle and Penrith Rapid Post Coach.

[3] In the *Newcastle Courant*, of Apr. 16, 1774, p. 2, where is given the advertisement of the York and Newcastle Post Coach, which began to run that year between these two places, three times a week, it is said that the coach "Sets off from Mr Sanderson's the Coach and Horses, in the Oat-market, Newcastle, at twelve o'clock at night, breakfasts at Darlington, dines at Easingwood, and will be at the George, in Coney Street, York, at six, where six places will be kept in the London Fly, which sets out at eleven o'clock that night" for London. The advertisement shows that similar arrangements were made for returning passengers, for six places would be kept in the Newcastle coach for passengers coming back from London, through York, to Newcastle. For other examples, see J. Hodgson Hinde, 'The Great North Road,' found in *Archaeologia Aeliana*, N.S., III, p. 249.

[4] Macturk, *A History of the Hull Railways*, p. 12.

[5] *Leeds Intelligencer*, July 16, 1792, p. 4; Macturk, op. cit., pp. 13–14.

prevailed does not now appear, for it is seldom mentioned in the records of the time; but we incline to believe, from the way in which the coaches were eager to adapt their business to the existing conditions, that probably many of them connecting with port towns made their arrangements to dovetail with those of the coast vessels and packets.

When speaking of the carrying trade, we learned that, with its increase, the business often fell into the hands of great entrepreneurs who had much capital embarked in it[1]. The same tendency is apparent in the coaching trade. In some instances, those who were engaged in agriculture found it profitable to enter also into the coaching business, since they had all the necessaries for keeping a large number of horses at less cost than those who had to buy all the food required[2]. But, as a general thing, it was the innkeepers who put the coaches on the road. Most of the London coaching-inn proprietors confined their business individually to one district, or to one of the great systems of roads; and they would have their inns at all the important places along the route which their coaches travelled, so that they could change horses at places where it was most desirable. But a few of them had several establishments each, and sent their coaches in all directions from the metropolis[3]. A few of the proprietors, like Chaplin and Horne, had many hundreds of horses a-piece; and the yards of such

---

[1] See footnote 3 to page 309 for references; also Picton, *Memorials of Liverpool*, ii, p. 106.

[2] Hills, *History of East Grinstead*, p. 150.

[3] Harris, *Old Coaching Days*, p. 5. In Liverpool, in the early years of the nineteenth century, Bretherton and Company had their connexions with all parts of the kingdom and provided a superior style of conveyance at a high rate of speed. This firm worked the large proportion of the (nearly) fifty coaches that left Liverpool daily in 1805 (Picton, *Memorials of Liverpool*, ii, p. 117). In London, the firm of Chaplin & Co., about 1834, had from 1300 to 1500 horses and 64 coaches. They were the largest establishment engaged in this business; and from small beginnings, conducted with great success, they had worked up until they had many good hotels in various parts of the country and their annual returns from their business were £500,000 (Fay, *A Royal Road*, p. 28). It must have been of this company that Sir Charles W. Dance was speaking, when, in his letter to *The Times*, Sept. 26, 1833, p. 4, he says he has been told that one great coach proprietor alone employed about 2000 horses. Horne and Sherman, the two next largest coach proprietors in London, had about 700 horses each in 1841 (*Annual Scrap Book*, 1841, p. 75). A German traveller in England in 1791 speaks of the great establishments that he found for horsing the coaches, like as if the horses belonged to some lord (Brit. Mus. 567. e. 7, 'Beyträge zur Kenntniss vorzüglich des Innern von England und seiner Einwohner,' i, p. 39). Harris, *The Coaching Age*, Chaps. VII and VIII, gives much detail as to coach proprietors and their methods of horsing the coaches, and he knew the system from his own experience. He abstracts from original records some facts to show the financial returns from various coaches about the time of their greatest prosperity (ibid., pp. 191–212).

inns presented scenes of great activity at almost every hour of the day, when the night or day coaches were returning or leaving.

Some facts pertaining to the financial operation of stage coaches must be given if we would understand the conditions under which they carried on their business. The turnpike tolls were a heavy item in the expenses of a coach and, according to the experience of some of the largest coach proprietors, averaged not less than 11s. 6d. per mile a month[1]. This, to a coach that travelled one of the longer routes, like that from London to Edinburgh, was a burden which made a large deduction from the revenue obtained by the carriage of passengers. Another charge, almost equally heavy, was the stage-coach duty, which was really a duty on passengers; it depended on the number of individuals the coach was licensed to carry, and had to be paid whether the coach was loaded or empty. At one time it was fixed at 8d. per mile for a coach licensed to carry four inside and eleven outside passengers; but an additional passenger would have added another half-penny per mile to the duty[2]. At different times, however, these duties were changed, both as to the amount to be paid and the number of persons to be carried, so that no general statements can be made as to the exact amount of this expenditure. Coach proprietors felt the necessity of reducing this to the lowest possible amount, and, accordingly, it was customary, when winter was coming on, to lessen the number for which the license was taken out at the Stamp Office. In this way a coach might be licensed for the winter to carry only four inside and eight outside; but when summer returned and business increased the license might be altered at the Stamp Office, and by the payment of the additional duty the number of persons that the coach could carry might be increased from twelve to eighteen[2]. This charge, although burdensome, was regarded by coach proprietors as a protection, preventing others from recklessly starting coaches without the means of carrying them on, which would tend to depress the business of those already established. On the other hand, a coach which had secured a license for a smaller number of persons could not with impunity carry a larger number, because there were informers along the roads who made their living by laying complaints against any driver who carried more than his legal number of passengers. The number of individuals that the coach was licensed to carry had to be conspicuously painted on the coach; and if the informer, who was watching for infractions of the law, complained to the magistrate, he received one-half, or, perhaps, in some cases, more, of the fine imposed[3].

---

[1] Harris, *The Coaching Age*, p. 195.
[2] Ibid., pp. 195–6; Act 2 and 3 Will. IV, c. 120.   [3] Harris, op. cit., pp. 197–8.

While these two elements, the turnpike tolls and the stage-coach duty, were the largest items of expense, apart from the maintenance of men and horses, in the operation of the coaches, some others deserve to be mentioned and of these probably the next largest was the mileage duty. This was payable according to the number of miles the coach ran, but was not dependent upon the number of persons it was licensed to carry[1]. The amount of the mileage duty varied from two to three pence per mile, according to the agreement that the coach operator could make with the coach builder, the latter of whom had his coaches on the roads throughout the year, earning him money. In a few instances, coach proprietors bought their coaches outright, while sometimes one proprietor would have the coach and would arrange with his partners to hire it of him at a mileage rate, as was ordinarily done from a builder. In addition to the mileage duty there were assessed taxes of various kinds: a license tax of £5 had to be paid on each coach kept for the road and this had to be paid whether it ran only a few times or every day of the year. In the same way, the assessed tax on every coachman and guard had to be paid, irrespective of the continuity or constancy of their employment[2]. When we combine these with the great expenses connected with "horsing" the coaches, wages of coachmen and guards, advertising, booking offices and many incidentals, it is very evident that the operation of the coaches was expensive; and yet a net profit of from four to eight pounds per mile of line, after deducting all expenses of operation, gave a good return upon the investment[3].

While coaching upon fine smooth roads, with all the glitter and show of elegant equipage, had much to interest and to stimulate those who were thus rolled along, there were some features connected with it which lent a darker aspect to the otherwise bright picture. One of these was the recklessness of the drivers, which caused a great many accidents. In the intense competition which prevailed, speed was of prime consideration and the coach which could reach its destination in the shortest time usually attracted the greatest number of passengers. Certain coaches, such as the Shrewsbury "Wonder," acquired a reputation for celerity which was an asset of much advantage; in some cases this reputation was safeguarded by careful driving, but in other cases the speed mania overshadowed all considerations of prudence

---

[1] Harris, *The Coaching Age*, pp. 198–9.

[2] Brit. Doc. 1837 (456), xx, 291, 'Report of Committee on the Taxation of Internal Communication, Minutes of Evidence,' p. 3; Harris, op. cit., p. 195.

[3] Harris, op. cit., pp. 201–12, gives statements of coach accounts which show such returns.

and the violent manner in which this business was conducted caused considerable loss of life[1]. Two coaches which started from the same place and were going in the same direction were often found to race their horses at break-neck speed, without paying any heed to the protests of the passengers who were in momentary danger of instant death through the overturning of the coaches[2]. Even the imposition of the fine that the law authorized for this transgression was not sufficient to deter drivers from committing the same offence again, for when the proprietors had paid the fine they would tell the coachmen that as they had once beaten the opposition they could do so the next time. A kindred evil to this, and one which was frequently the cause of indiscretion on the part of the drivers, was the indulgence of the latter in liquor at the many inns along the roads. The practice of tippling was very prevalent; and as the driver drew up before an inn he would commonly leave his horses untied while he went in to quench his thirst for the intoxicating beverage. This gave occasion for the horses to run away while the reins were loose; and it also disturbed the mental balance of the driver until he was unfit to carry his load of passengers with security through the country. When time had been lost at the numerous alehouses, it had to be made up by galloping and the instability of the driver was responsible for many of the accidents that were constantly occurring[3]. The drivers were not only reckless but insolent and treated any individual who remonstrated against their unbecoming and unwise conduct with an air of arrogance and disdain that aroused opposition. Along with this lack of courtesy, the levies which the coachmen and guards, with the sanction of the coach proprietors, were allowed to make upon passengers, in order to supply the lack in their wages, formed a tax of considerable magnitude and

---

[1] *The York Herald, County and General Advertiser*, Dec. 10, 1814, p. 2; Whitaker, *Loidis and Elmete*, p. 81; Blew, *Brighton and Its Coaches*, p. 135; Deacon on *Stage Waggons*, pp. 73–75; *Parl. Papers*, 1808 (315), ii, 527, 'Third Report of the Committee on Highways,' p. 199.

[2] *The Times*, Sept. 10, 1803, p. 2, shows that long, narrow coaches were particularly liable to be overturned and gives some instances. See also ibid., June 11, 1816, p. 2; Oct. 29, 1822, p. 2; Oct. 8, 1825, p. 2; Nov. 18, 1829, p. 4, letter of S. M. G. See also index of *The Times* under the heading "Accidents" to get a correct idea as to the great dangers from stage coach travelling.

[3] Whitaker, *Loidis and Elmete*, p. 81, deplored the fact that under the better condition of the roads "the lives of thirty or forty distressed and helpless individuals are at the mercy of two intoxicated brutes;" and said that under such circumstances a journey from town to town resembled a voyage from Dublin to Holyhead, short indeed but extremely perilous. See also Blew, *Brighton and Its Coaches*, p. 135; *The Times*, Feb. 7, 1811, p. 4; Oct. 7, 1823, p. 2; Dec. 13, 1823, p. 3; Oct. 21, 1825, p. 4; Hutton, *Scarborough Tour in* 1803, pp. 13–14.

a grievance that evoked resentment[1]. Nor did these impositions
stop with those who had charge of the coaches on the road; for at the
inns where the coaches stopped for refreshments, notwithstanding the
unsatisfactory food and service, the charges were frequently extortionate[2]. Time after time, new concerns would place their coaches on
the road and advertise that means were being taken to protect the
public from the above-mentioned dangers and rapacity[3]; but history
repeated itself and the evils attending the system still prevailed.

Another of the evils connected with stage coach travelling was that,
in contravention of law, the drivers persistently carried more passengers
and luggage than their licenses allowed. On the road, they would
take up and set down persons who wished to go but short stages and
appropriate the money paid to their own private use. They would
also agree with the passengers thus taken that, before the coach reached
a turnpike gate, they should dismount so that no extra toll would be
charged and after it had passed this place of payment they could
remount[4]. Around the metropolis, it was understood, there was a
pecuniary arrangement between the coachmen and the toll-takers,
by which the purposes of legislation were easily defeated[4]. With
such heavy loads, especially when carried on the outside, there was
great risk of the coaches breaking down or being overturned, and
many were deterred from travelling in these vehicles because of their
attendant danger[5]. Notwithstanding that various Acts had been
passed for limiting the number of passengers or the weight of baggage to
be carried on the outside of stage coaches[6], the grievance continued,
apparently unabated, even when it was well known that informers
were on all the roads eager to lay complaints and secure convictions
for these offences[7]. Similar complaints were made against the return

---

[1] *The Times*, June 11, 1816, p. 2; Feb. 7, 1811, p. 4; Oct. 21, 1825, p. 4; Blew,
Brighton and Its Coaches, p. 135.

[2] Blew, *Brighton and Its Coaches*, p. 135.

[3] Ibid., pp. 135, 137.

[4] *Parl. Papers*, 1808 (315), II, 527, 'Third Report of the Committee on Highways,' pp. 197–8; Blew, *Brighton and Its Coaches*, p. 137.

[5] *Parl. Papers*, 1808 (315), II, 527, 'Third Report of the Committee on Highways,' p. 199; *The Times*, Feb. 7, 1811, p. 4; July 31, 1818, p. 3; Aug. 28, 1818,
p. 3; Harris, *The Coaching Age*, pp. 196–7.

[6] *Parl. Papers*, 1808 (315), II, 527, 'Third Report of the Committee on Highways,' p. 197; *The Times*, Aug. 31, 1818, p. 2; Act 50 Geo. III, c. 48; Blew,
Brighton and its Coaches, p. 142.

[7] *Leeds Intelligencer*, July 8, 1830, p. 3, shows the work of men who were professional informers and who had others working for them in the same employment.
So also Harris, *The Coaching Age*, pp. 197–8. The informer got one-half the
fine.

post-chaises, which likewise carried many inside and outside passengers, for the benefit of the boy's pocket, but to the detriment of the stage coaches. Since the latter paid considerable revenue to the government, the committee of 1808 thought that the post-chaises should not be allowed to take away their passenger business from them; and their report recommended that a heavy penalty should be inflicted upon the driver of any return post-chaise who carried any inside, or more than one outside, passenger, except on roads where there was no stage coach[1].

Several other objectionable features accompanied the system of stage coach travelling which we may but mention. We have already referred to the great multiplication of toll-gates around London and other cities, which caused much annoyance and delay to those who were using the coaches. Then, a traveller was sometimes compelled to take a vehicle which he did not want, and to which he had an aversion, because the one for which he had engaged a seat some days in advance had departed with a full load a few minutes before his arrival. Occasionally coaches got stuck fast in the snow, and all their occupants had to put up with whatever accommodation could be secured. From the standpoint of the coach proprietor, too, there was one very serious drawback, namely, the difficulty and cost of keeping a supply of horses. On the fast coaches operating the first fifty or sixty miles out of London, the life of a horse, according to the testimony of those who kept the coaching inns, was three, or not more than four, years[2]. On the roads that were more distant from London, the work was lighter and the food and lodging better, and there the horses would last probably twice as long[3]. But even six years as the average life of a horse made the cost of renewing the stock a very heavy burden, and this wholesale destruction of animals is a dark spot upon the picture of the glory of coaching.

Along with the great increase in the amount of coaching, there was an equivalent, if not a greater, increase in the amount of posting. We have seen that in the period up to 1750 those who travelled post

---

[1] *Parl. Papers*, 1808 (315), II, 527, 'Third Report from Select Committee on the Highways,' pp. 199–200; also ibid., p. 215, Appendix No. 8, entitled 'Remarks on the Mischiefs arising from Return Post-Chaises, etc., being permitted to carry a number of outside and inside passengers.'

[2] *Parl. Papers*, 1819 (509), v, 339, 'Report from Select Committee on the Highways,' evidence of Wm. Waterhouse, Wm. Horne, and John Eames, pp. 13–15. Also *Remarks upon Pamphlet by "Investigator" on the Proposed Birmingham and London Railway*, p. 25.

[3] *Parl. Papers*, 1819 (509), v, 339, 'Report from Select Committee on the Highways,' evidence of Wm. Waterhouse and Wm. Horne, pp. 13–15.

usually rode on horseback; but in the period succeeding that, with the improvement of the roads, the use of post-coaches gradually became much more prominent than before, although riding post still seemed to claim the pre-eminence among the means of travelling. While formerly the postmasters were the regular providers of horses for this purpose and the innkeepers were interlopers seeking to get a share of the business, at this time it seemed that a considerable amount of the trade had got into the hands of the innkeepers, who, it was said, made more by that means than by their regular inn trade[1]. But the post-masters in some places were also active in securing as much as possible of the money to be derived from this source, and even gave a large fee to the boys who would bring them post-chaise customers[2]. It would appear as if the postmasters were not infrequently also innkeepers, and most innkeepers were also farmers, so that by combining all these functions they were able to carry on their business most successfully[3]. It is certain that those who furnished the facilities for posting received ample remuneration, and the charges that were made for post-horses and post-chaises proved very lucrative[4]. These charges were supposed to fluctuate according to the prices for hay, oats and other food; but the records of the time show that they did not consistently follow any such course[5]. On the other hand, when the prices had once been raised, they were not afterwards reduced, as a general thing, unless some pressure were brought to have them lessened. Sometimes it was the pressure of public sentiment, at other times the influence of declining business or of competition, that caused the charge to be lowered; while occasionally it was due to the sense of what was just and reasonable. We are well within the truth in saying that, on the whole, the charges for posting were high, and in many cases extremely high; rarely did they go below 1s. per mile for a chaise and pair[6]; and often

---

[1] Brit. Mus. 567. e. 7, 'Beyträge zur Kenntniss vorzüglich des Innern von England und seiner Einwohner,' I, p. 40.

[2] *The Times*, Jan. 1, 1808, p. 3.   [3] Ibid., Aug. 20, 1801, p. 2.

[4] Ibid., Jan. 1, 1808, p. 3.   [5] Ibid., Jan. 24, 1823, p. 2.

[6] Ibid., May 29, 1822, p. 3, and July 1, 1822, p. 3, informed the public that the postmasters at Maidenhead Bridge and Exeter had reduced the price of posting to 1s. per mile; and in the issue of that paper Nov. 1, 1822, p. 3, it was said that the opposition among the coach proprietors in Cornwall had caused the price of posting in some parts of that county to be reduced to 9d. per mile. It seemed to be generally recognized that 1s. per mile was amply remunerative, and many of the papers were carrying on an agitation for that charge (Ibid., Jan. 24, 1823, p. 2). See letter from (Lord) "Deerhurst," written to the *Worcester Journal*, and copied in *The Times*, Mar. 18, 1823, p. 3, which said that the price of oats and other provender did not justify a charge of 9d. per mile and at 1s. innkeepers were making a good profit, but that as long as the rich would pay what was asked, the

1*s*. 6*d*. per mile and more was charged[1].   Perhaps the safest figure for the country *tout ensemble* is 1*s*. 3*d*. per mile; but in addition to this charge for the horses and chaise the traveller had to pay on the average 3*d*. per mile for the postilion to bring back the horses, and the tolls that were demanded along the way[2].

Reiterated and persistent complaints were made against the extortionate charges for posting demanded from travellers by postmasters and innkeepers; and it seems that there was much justification for this continuous and vociferous outcry.   Travellers were at the mercy of the innkeepers, who could charge what they pleased for their horses and for their house accommodation; and the voice of the public demanded some regulation and superintendence over these charges[3]. Where there was but one posting establishment at a place the rates charged were usually high, and often these became still higher by the application of fraudulent methods[4], such as charging for more than the actual mileage, changing the number of stages on a certain journey, charging more in one direction than another, and by various similar devices.   Even where there were several potential competitors for the business along the same line of road, the charges were frequently kept high by reason of a combination among the interests involved[5], and there was reason to believe that in a few instances opposition had been

imposition would go on.   In a footnote to that letter it is stated that the Grand Jury of Worcester, after considering the circumstances, decided to support those who let their horses at 1*s*. per mile, which they considered a full remunerative price. *The Times*, Apr. 26, 1802, p. 3, reports that the Grand Jury of Bristol resolved to discountenance such innkeepers as did not immediately reduce the price of posting to 1*s*. per mile.

[1] For example, *The Times*, Sept. 6, 1813, p. 3, said that the postmasters of Bury, Botesdale, Stowmarket, Newmarket, and some other places, had lowered the price of posting to 1*s*. 6*d*. per mile, except when carrying more than four persons.   How high the charges were before they were lowered does not appear.

[2] London and Birmingham Railway Bill.   Extracts from the Minutes of Evidence given before the Committee of the Lords on this Bill, evidence of Mr H. Cheetham, p. 23; Carter, *Letters from Europe, comprising the Journal of a Tour through Ireland, England, Scotland, France, Italy and Switzerland*, I, p. 184; *The Times*, Aug. 29, 1801, p. 3; Sept. 3, 1801, p. 2; Oct. 17, 1814, p. 2; Feb. 21, 1815, p. 3; Jan. 24, 1821, p. 4; Apr. 11, 1821, p. 3; May 16, 1823, p. 4; *Hampshire Telegraph*, Sept. 4, 1815, p. 3.

[3] *The Times*, Aug. 20, 1801, p. 2; Aug. 26, 1801, p. 2; May 10, 1802, p. 3; May 29, 1802, p. 3; May 16, 1823, p. 4; Aug. 27, 1825, p. 3; etc.

[4] Ibid., Aug. 26, 1801, p. 2; Nov. 17, 1801, p. 3; Nov. 20, 1801, p. 3; May 13, 1802, p. 3; May 29, 1802, p. 3; *Leeds Intelligencer*, Mar. 4, 1793, p. 3, letter from "A Yorkshire Man," and Mar. 25, 1793, p. 3, letter from Samuel Peech.

[5] *The Times*, Aug. 26, 1801, p. 2; May 29, 1802, p. 3; Jan. 1, 1808, p. 8; Jan. 24, 1823, p. 2; Mar. 18, 1823, p. 3; May 14, 1823, p. 1; May 16, 1823, p. 4.

bought off and induced to desist[1]. Occasionally there were public meetings at particular places to see what could be done to bring a reduction of posting charges[2]; but we do not find that anything substantial ever came of such meetings. The recommendation so often made, that a new rival should come into a place where monopoly prevailed, in order to effect a cut in the rates, was seldom carried out; and so long as the wealthy would pay the prices asked so long would they be demanded.

In connexion with this subject, it is important for us to notice the changes which took place in the postal facilities afforded to the country generally. Up to the last quarter of the eighteenth century the only significant improvement in the transmission of mail had been that introduced by Ralph Allen in 1720, by which the old system was abandoned, and mail was carried between the chief centres of population by means of post-boys who rode on horseback, supposedly at the average rate of five miles an hour, and carried the mail-bags with them. But occasionally the post-boys were old men and it was necessary to provide them with light carts for carrying the mail-bags. Allen's scheme was considered by the Postmaster-General and the Government so beneficial to the trade of the country that he was granted during the rest of his life (1720–62) the farm and exclusive management of the cross posts; and since he received nearly the whole of the profits during these forty-two years, he accumulated considerable wealth. In 1741 the increase of trade and population encouraged Bristol citizens to appeal to the Ministry for improved postal communication with London; and to give additional facilities, Allen changed from going three times a week to going six times a week between London and Bristol. Of course, all intervening towns enjoyed these benefits as well as the two termini. Other improvements were made in succeeding years; but notwithstanding these, the results secured left much to be desired. The service was in the hands of those who were frequently irresponsible—boys or old men—and incapable of defending themselves and their treasures against the attacks of highwaymen who thronged the roads. Robberies of the mail-bags by this predatory class of outlaws were very common; delay was occasioned by the drunkenness of the mail-man, by storms and other contingencies; and the conveyance of letters between the principal towns was more or less desultory. By this system the mail service between Bristol and London took thirty to forty hours, according to the state of the roads; but the stage coach would reach London earlier than the mail and, although the charge was much

---

[1] See, for example, *The Times*, Apr. 26, 1802, p. 3.
[2] Ibid., Feb. 9, 1815, p. 3.

greater, Bristol sent their most urgent and valuable letters by the stage coach[1].

The people of two large cities like London and Bristol could not submit to these conditions, which led to vexatious delay and often loss of mail[2]; and in 1788, John Palmer submitted to Pitt his proposed method for increasing the postal facilities and the revenue obtainable therefrom[3]. He showed that the post was about the slowest conveyance of the country and urged the Government to establish mail coaches, protected by well-armed guards, the working cost of which would be defrayed by travellers desirous of increased speed and security. By this plan the revenues of the Post Office would also be benefited, through the recovery of the business that had fallen into private hands. Palmer promised that if his plan were accepted the mail and passengers taken by his mail coaches would be carried between London and Bristol in sixteen hours[4], but Members of Parliament refused to admit

[1] The details of Allen's system, including the conditions he found at the inauguration of his plan and the improvements he effected, are given minutely in Ogilvie, *Ralph Allen's Bye, Way and Cross-Road Posts*. Note especially pp. 5–86, giving 'A Narrative of Mr Allen's Transactions with the Government for the Better Management of the Bye, Way and Cross-Road Posts,' from 1720–62. Allen died in 1764. Tombs, *The Bristol Royal Mail*, pp. 1–18, gets the material he used from the above original and authentic source.

[2] Bonner and Middleton's *Bristol Journal*, Mar. 5, 1785, p. 1.

[3] The story of John Palmer's mail coaches receives full treatment in Joyce, *History of the Post Office* (1893), Chap. XII; Palmer, *Papers relative to Government Agreement*; Bonner, *Mr Palmer's Case Explained*; and in the Bristol newspapers of that time.

[4] In Felix Farley's *Bristol Journal* of Oct. 2, 1784, p. 1, we find the advertisement of Palmer's mail coach:

"Mail Diligence commenced Monday August the 2nd.

The Proprietors of the above Carriage, having agreed to convey the Mail to and from London and Bristol, in sixteen Hours, with a Guard for its Protection, respectfully inform the Public, that it is constructed so as to accommodate four inside Passengers in the most convenient Manner,—that it will set off every Night at Eight o'Clock, from the Swan-with-two-Necks, Ladlane, London, and arrive at the Three Tuns Inn, Bath, before Ten the next Morning, and at the Rummer-Tavern, near the Exchange, Bristol, at Twelve. Will set off from the said Tavern at Bristol, at four o'Clock every Afternoon, and arrive at London at Eight o'Clock the next Morning.

The Price to and from Bristol, Bath, and London, £1. 8s. for each Passenger. No outsides allowed.

Both the Guards and Coachmen (who will be likewise armed) have given ample Security for their Conduct to the Proprietors, so that those Ladies and Gentlemen who may please to honour them with their Encouragement, may depend on every Respect and Attention.

Parcels will be forwarded agreeable to the Direction, immediately on their arrival at London, etc., and the Price of the Porterage as well as the Carriage,

the possibility of such a rapid rate of travelling. Finally, however, he triumphed; his scheme was adopted by the Government and Palmer himself was installed in the London Post Office to superintend the working of this new departure[1]. In spite of much opposition, the experiment was successful; the revenue of the Post Office immediately increased and the desired result was achieved in causing the delivery of letters at least twelve to eighteen hours earlier than before. This was promoted by having each coach arranged to carry only four inside passengers and none outside[2]; whereas the coaches that were already on the road usually carried six inside passengers and also some outsiders[3].

on the most reasonable Terms, will be charged on the Outside, to prevent imposition.

N.B. Any Person having reason to complain of the Porters' delay, will oblige the Proprietors by sending a Letter of the Time of delivery of their Parcels to any of the different Inns the Diligence puts up at.

Performed by { Wilson and Co. London, Williams and Co. Bath.

The London, Bath, and Bristol Coaches, from the above Inns as usual."

This advertisement is interesting not only because of the change in speed which it announced but also because of the glimpse it gives us, by implication, into existing conditions of coaching.

[1] The papers describing the condition of the postal facilities when Palmer brought forward his scheme, and his understanding of the arrangements with the Government and the remuneration he was to receive from the Government, are fully set forth in Palmer, *Papers relative to Government Agreement*. It was his understanding that, if his scheme succeeded, he should receive during his life two and one-half per cent. on the future net increased revenue of the Post Office from the inauguration of his plan; but if it did not succeed he was not to receive anything. Evidently he had much difficulty with the employees in the Post Office, during the decade of his superintendence 'there; but he had the support of the Lords of the Treasury in this trouble and was retained in his office. Finally, however, these relations with his employees became more strained and he was suspended from the office. Because of the cessation of payment to him of the two and one-half per cent. of the revenue, he appealed to the Lords of the Treasury to see that justice was done him in the matter; but they refused to grant him more than £3000 a year during his life in return for his services rendered to the country. Bonner, *Mr Palmer's Case Explained*, took the other side from Palmer and contradicted the basic statement on which Palmer rested his case, namely, that in regard to the $2\frac{1}{2}$ per cent. This controversy we leave others to settle; suffice it to say that the revenues of the Post Office increased almost immediately (Palmer, ôp. cit., p. iii).

[2] See above advertisement of Palmer's mail coach.

[3] Up to this year (1784) it was customary for the coaches to carry six inside passengers, as well as outside passengers (Felix Farley's *Bristol Journal* of Nov. 2, 1776; Aug. 14, 1779). But some of the more progressive coach proprietors, as early as 1775, had decided to limit the number of passengers inside to four (Bonner and Middleton's *Bristol Journal*, Jan. 14, 1775, p. 1), and by April 1784 one of the coaches on the road from Bristol to London had reduced the number of passengers it carried to four inside and one outside (Bonner and Middleton's *Bristol Journal*,

As soon as the practicability of Palmer's coaches had been demonstrated, on Aug. 2, 1784, other coaches began to realize that their passenger business was in danger unless they could offer as good service as the mail coaches and immediately they began to speed up. They inserted advertisements in the newspapers that their coaches also were provided with a guard for protection from highwaymen and that they went to London in sixteen hours[1]. "London Balloon Coaches" were claiming public attention[2]; and every means was used to retain their business. If we mistake not, this was a difficult situation for many coach proprietors to meet; many of the coach advertisements almost ceased in the Bristol papers about the close of the year 1784 and the early part of 1785, and the mail coaches were about the only ones that were kept in this way before the public[3]. The inability to meet the increased speed will be more readily seen if we remember that before Aug. 2, 1784, most of the coaches took two days to accomplish this journey[4], and the fastest of them required a full day (twenty-four hours)[5], whereas now the time occupied by the fastest coach had to be cut down by one-third.

No sooner had Palmer's scheme been realized between London and Bristol, and its financial results seemed to justify its extension, than he experienced much trouble from those officers in the London Post Office who were in authority there[6]. Many of the staff worked against him and so serious was the opposition that it appeared as if the fast mail service to Bristol would be discontinued. But the commercial interests of Bristol drew up a memorial showing the advantages of Palmer's system and presented it to Pitt, urging the continuance of these benefits and expressing the hope that he would put an end to the opposition that had been aroused in the General Post Office[6]. Knowing the temper

Apr. 17, 1784, p. 1). Probably the idea of increasing the speed by limiting the number of passengers carried was the result of Palmer's agitation, which had been going on for some years before it assumed its final form.

[1] Bonner and Middleton's *Bristol Journal*, Sept. 4, 1784, p. 3.

[2] See, for example, Bonner and Middleton's *Bristol Journal*, 1784, Oct. 30, Nov. 13, etc.

[3] Note the issues of Bonner and Middleton's *Bristol Journal* at that time.

[4] Felix Farley's *Bristol Journal*, Nov. 2, 1776, p. 2; ibid., Aug. 14, 1779, p. 1; etc.

[5] Ibid., Aug. 14, 1779, p. 1. The 'Bath and Bristol New London Post Coach' advertised in Sarah Farley's *Bristol Journal*, Aug. 3, 1782, p. 2, that it would reach London in eighteen hours, but we have no record that it kept its promise.

[6] The papers, letters and memorial in connexion with this case are given in full in Bonner and Middleton's *Bristol Journal*, Feb. 12, 1785, p. 3, and Feb. 19, 1785, p. 3. See also Palmer, *Papers relative to Government Agreement*.

of the people of Bristol and their determination not to submit to any detention of their letters, Pitt finally gave his voice for the perpetuation of this rapid conveyance of the mails. By April 1785 there was a further extension of Palmer's plan authorized for the mail between Bristol and Southampton and Portsmouth[1]. From that time on, this system of mail coaches was gradually amplified until its service included all parts of England, both on the main roads and cross roads, and with it there came a new era in both postal and travelling facilities[2], which lasted until the fourth decade of the nineteenth century, when the railways received the preference.

[1] Bonner and Middleton's *Bristol Journal*, Apr. 30, 1785, p. 2.

[2] Bonner and Middleton's *Bristol Journal*, June 4, 1785; Gore's *Liverpool Advertiser*, July 22, 1785; Baines, *History of Liverpool*, p. 468. With the increased speed at which coaches travelled, there came also a greater development in the systematic arrangement of schedules, so that the time tables on subsidiary coaching routes might be harmonized with those on the main lines, in order to facilitate the transfer of passengers (see, for example, Bonner and Middleton's *Bristol Journal*, July 30, 1785, p. 1). See also Sarah Farley's *Bristol Journal*, Nov. 11, 1786, p. 2. Bird, *Laws respecting Travellers and Travelling*, pp. 85–88, gives the mail coach routes in England in the year 1801.

Many inflammatory paragraphs and advertisements had been published in the newspapers, and hand-bills had been exhibited in inns and other public places along the roads, condemning the mail coaches as the cause of many accidents and their fast speed as the ruin of horses. Great pains were taken to expose these coaches as unsafe on account of their expedition. Mr Bonner, the London agent to Mr Palmer's post plan, in order to silence such misrepresentations, published a letter showing that the mail coaches did not travel any faster than the post-chaises and that their contracts required them to go only eight miles an hour, including stoppages. Felix Farley's *Bristol Journal*, Oct. 15, 1785, p. 4. See also *The General Advertiser*, Feb. 26, 1785, p. 3.

The Morning Chronicle and London Advertiser, Apr. 1 1786, p. 2, gives a complete list of all the mail coaches established up to that time and shows the service extended to Exeter, Bristol, Gloucester, Hereford, Shrewsbury, Holyhead, Chester, Liverpool, Manchester, Carlisle, Leeds, Derby, Ipswich, Norwich, Dover, Southampton, and many intervening places.

In order that the mail coaches should not be hindered in accomplishing their work, all the toll-gate keepers were required, under penalty, to have their gates open when the mail arrived, and they were also given the right of way on all roads, for as soon as the horn of the mail coach guard was heard other travellers were required by law to immediately turn out of the road (v. the mail coach advertisement in Gore's *Liverpool Advertiser*, July 22, 1785). This speed was made possible also by the fact that time was not lost in making the relays of horses at the several stages. On the Great North Road we are told that one minute's delay was all that was required to make the change of horses (J. Hodgson Hinde, 'The Great North Road,' in *Archaeologia Aeliana*, N.S., III, p. 254).

In contrast with the arrangements for the speed of the mail coaches, and later for some other coaches also, it may be stated that Espriella, in his *Letters from England* (1807?), p. 24, says of one of the stage coaches that he saw, "The passengers sit sideways; it carries sixteen persons withinside, and as many on the roof as

We cannot close this account of the means of travelling without describing the efforts that were made to establish the use of steam carriages on the common roads; and here we are taken back to the early years of the nineteenth century to find the initial stages of what has recently become a highly important phase of the transportation problem.

In the later part of the preceding century, William Murdock, the assistant of Boulton and Watt, made a small model locomotive which worked successfully in his house and on a straight and level walk outside[1]; but he received neither encouragement nor financial help from his principals, and was, therefore, unable to go far. Trevithick, in a few years, took up the matter and he did more than any of his predecessors to make steam locomotion on ordinary highways an accomplished fact[2]. In 1801 he built his first steam carriage at Camborne, in Cornwall, and on Christmas Eve of that year this carriage conveyed the first load of passengers ever moved by the force of steam. It went faster than a man could walk, and it also went half a mile up

can find room; yet this unmerciful weight with the proportionate luggage of each person is dragged by four horses, at the rate of a league and a half within the hour." It would seem, however, that he must have been thinking of the stage waggons rather than of the stage coaches or mail coaches.

[1] Messrs Richard and George Tangye, of Birmingham, bought this model locomotive in 1883, and it is now in the Birmingham Art Gallery. It can still travel at some speed when placed under steam.

Murdock has seldom been given the credit due to him for the great results he accomplished. Samuel Timmins tells of Murdock's experiments with his model engine in performing the great feat of road locomotion, and says that this "was practically the first example of steam locomotion on roads, and on rail-roads also" (*Birmingham Miscellaneous Pamphlets*, vol. H. 2, in Birmingham Free Central Library, No. 145,025, 'Samuel Timmins' Account of William Murdock,' p. 4). He tells us that Watt had no great faith in Murdock's results and wished that the latter could be brought to give up this "hunting shadows" (ibid., p. 5). Murdock was anxious to introduce steam on common roads and in his house at Redruth, in Cornwall, was to be seen in 1784 a small model high-pressure locomotive drawing a model waggon around the room; but he had not the facilities with which to prosecute his work and his employers, Boulton and Watt, were not interested in the development of the engine for locomotive purposes, but rather for industrial uses. Cowper, *The Steam-engine*, p. 94, says that Watt was so strongly opposed to steam locomotion on common roads that when letting his house, Heathfield Hall (now owned and occupied by Mr George Tangye), he actually put a covenant in the lease that no steam carriage should on any pretence be allowed to approach the house.

[2] I shall not follow out in detail the names of those who were engaged in experiments with steam locomotion. They are given in Fletcher, *Steam Locomotion on Common Roads*. Smiles, in his *Life of George Stephenson*, says that Trevithick was a pupil of Murdock's and learned from him the knowledge of the steam carriage. This view is wholly untenable (see Trevithick, *Life of Trevithick*, I, pp. 145–6).

a steep hill, which horses could not ascend at more than walking speed. This experiment was successful so long as steam could be kept up; but the attempt to maintain steam pressure for any considerable time proved a failure[1]. Several successful experiments with the steam carriage were made in the next two years, on the Camborne road; and with the increase of the power of the engine unprecedented results were secured[2]. In 1802, Trevithick and his partner, Vivian, took out a patent for their steam coach; and in 1808 their best locomotive was taken to London, where it travelled from four to nine miles per hour[3]. These demonstrations in London for the six or eight months beginning with January 1808, and the expense involved in connexion with them, drained the pockets of the inventors and for a time put an end to their locomotive experiments[4]. But by their London experience, Trevithick had learned that a smooth road was better than a rough pavement; and from that time on he turned his attention to the adaptation of locomotives for running on railways[5].

A letter of Trevithick's, written in 1808, shows that he lost no time in applying his locomotive engine to tramways, as well as to other purposes[6]; and, in February 1804, the first tramroad locomotive in Wales, drawing ten tons at a time, ran with facility up and down inclines of one in fifty[7]. In a few weeks more the load had been increased from ten to twenty-five tons; and with that weight to draw, the engine travelled at the rate of four miles an hour for a distance of almost ten miles, while without a load it went at the rate of sixteen miles an hour[8]. These experiments were continued and Trevithick even applied the exhaust steam, as in later engines, to increase the draught of the furnace[9]. With these improvements in construction,

---

[1] Trevithick, *Life of Trevithick*, I, pp. 106–10. The Falmouth paper in 1801 reported that the first common road locomotive carried persons amounting in weight to at least one and one-half ton, against a hill of considerable steepness at the rate of four miles per hour, and on the level at eight to nine miles per hour. This was called Trevithick's "puffing devil" (ibid., I, p. 119). This, probably, was exaggerated, for a later better engine went only four to nine miles per hour (ibid., I, p. 143).

[2] Ibid., I, pp. 111, 117, 121, 124.

[3] Ibid., I., pp. 127, 139, 141, 143–4.   [4] Ibid., I, pp. 144–5.

[5] Brit. Doc. 1831 (324), VIII, 203, Evidence of Richard Trevithick. The history of his connexion with the application of steam as a tractive power is set forth therein. See also Trevithick, *Life of Trevithick*, I, p. 145.

[6] Ibid., I, p. 149.

[7] Ibid., I, pp. 160–2. Boulton and Watt tried to get a Bill through Parliament to stop these engines, on the plea that they endangered the lives of the public; but they were balked in this by the fact that Government engineers reported favourably on Trevithick's engines.

[8] Ibid., I, p. 182.   [9] Ibid., I, p. 191.

Trevithick's engines drew loads on railways, showing that mere friction or gravity was enough for locomotion[1]. It seems very clear, therefore, that Trevithick, and not William Hedley[2] nor George Stephenson[3], was the "father of the locomotive engine[4]."

Others continued to experiment with varying success; and between 1820 and 1830 much was accomplished by the efforts of Hancock, Gurney, Summers, Ogle, Maceroni, and several whose names we need not include in this list of contributors to progress. So different are the opinions as to which of these was in the front rank, and of so little value is it for our purpose here, that we shall omit a discussion of this question[5]. It is generally conceded that Goldsworthy Gurney and Walter Hancock were the most noteworthy among those who brought the steam carriage into practical operation; but to decide between these in the matter of priority is not within the scope of our present inquiry[6].

---

[1] Trevithick, *Life of Trevithick*, i, p. 201. He demonstrated these facts in London for some weeks, on a circular railroad of his own construction, on a waste piece of ground, now Torrington Square (ibid., i, pp. 192–4).

[2] Hedley claimed to be the man who established the principle of locomotion by the friction or adhesion of the wheels upon the rails. See his letter in ibid., i, p. 203.

[3] About 1846 Peel gave George Stephenson this compliment (ibid., i, p. 203).

[4] Ibid., i, pp. 203–6.

[5] On this subject, see Fletcher, *Steam Locomotion on Common Roads*; also Brit. Doc. 1831 (324), viii, 203, especially the evidence beginning with page 17; and Brit. Doc. 1834 (483), xi, 223.

[6] The history of Gurney's experiments and results are given in Brit. Doc. 1831 (324), viii, 203, Evidence of Mr Gurney, p. 17 et seq., and in Brit. Doc. 1834 (483), xi, 223, Evidence again given by Mr Gurney. From 1824–31, Gurney devoted his time exclusively to the subject of steam locomotion. His investigations in heat enabled him to construct a generator with much smaller surface than those hitherto used for the purpose of raising a given quantity of steam and in that way to reduce the weight of the carriage. His first successful carriage weighed four tons; his second, three tons; his third, two tons; and the one he was using in 1831 weighed only thirty-five hundredweight. At the latter date he was having one built to weigh not more than five hundredweight, the object being to carry two or three people and to develop speed.

In substantiation of Mr Gurney's claim to precedence in the practical application of steam locomotion to common roads, we refer to the evidence of Mr Gordon, an engineer, before the Committee of 1834 (Brit. Doc. 1834 (483), xi, 223), whose statement was: "It is the opinion of all other engineers, as well as of myself, that the subject was altogether *in nubibus* till Mr Gurney took it up." None of the engineers of that time denied this claim to Mr Gurney. Further, the Report of the Select Committee of 1835 (Brit. Doc. 1835 (373), xiii, 489) says, that they were unanimously of the opinion that there was no one to dispute the claim of Mr Gurney to the merit of having been the first person to apply steam carriages successfully on common roads. Sir Charles Dance, another noted experimenter and successful inventor, gives full credit to Mr Gurney for the introduction of steam carriages. See

The earliest of these successful carriages was heavy, and required the expenditure of much fuel to produce the necessary tractive power. The drawing around of so much coal and water only added to the burden, and the efforts of the inventors were bent to the attainment of greater power with less weight. Trevithick recognized the difficulty thus stated, that as the power increased the weight increased in nearly the same ratio[1]; but he and the later experimenters were able to produce a high pressure engine with the use of much less fuel and water, and consequently produced an engine that could travel up-hill as well as on the level road.

Some of the results attained by the steam carriage, even at this stage of its development, are worthy of record. In one case, one of Gurney's engines weighing but little over two tons drew a weight of eleven tons, on a road with an inclination of one in twenty-five, at the rate of five or six miles an hour, and this was by no means its greatest power[2]. In 1829, Gurney undertook a journey in his steam carriage from London to Melksham (thirteen miles from Bath) and back, a distance altogether of, in round numbers, two hundred miles, which was accomplished with only one slight accident, and the rate of speed returning was twelve miles per hour[3]. Before the Committee of 1831 he testified that he had run his carriage safely at a rate of from eighteen to twenty miles per hour; but twelve miles per hour he considered perfectly safe and practicable[4]. Similar results were obtained by Hancock. With his steam carriage he ascended Pentonville Hill, which had an inclination of one in eighteen or twenty, after a frost had glazed the road so that horses could scarcely keep their footing. This trial had been made in the presence of witnesses and its success was demonstrated by the fact that the carriage gained the top of the hill, while his competitors with their horses were only a short distance from the bottom of the hill[5]. Another ascent of the same hill in 1833 was made at the rate of six to eight miles per hour, while on the level road the carriage travelled ten to twelve miles per hour[6].

his letter in *The Times*, Sept. 26, 1833, p. 4. But Fletcher, in his *Steam Locomotion on Common Roads*, p. 97, gives Mr Gurney an entirely subordinate place in regard to this issue.

[1] See Report of Select Committee of 1831 (Brit. Doc. 1831 (324), VIII, 203), Evidence of Richard Trevithick, p. 63; also Trevithick, *Life of Trevithick*, I, pp. 123–4.

[2] Brit. Doc. 1831 (324), VIII, 203, Evidence of Mr Stone, engineer for Gurney.

[3] Brit. Doc. 1835 (373), XIII, 489, Report of Select Committee on Mr Gurney's case. A full account of this trip is given in Brit. Doc. 1834 (483), XI, 223, pp. 79–81.

[4] Brit. Doc. 1831 (324), VIII, 203, Evidence of Mr Gurney.

[5] Hancock, *Narrative of Twelve Years' Experiments*, p. 21.

[6] Hancock, op. cit., pp. 40–45; *Morning Advertiser*, Apr. 26, 1833.

Sir Charles Dance secured corresponding success, for in 1833 his carriage rolled along the road from London to Brighton at a rate that varied between ten and twelve miles per hour[1]. Colonel Maceroni gave fifteen miles per hour as the fastest speed he attained and that was on the Edgware road[2]; while we are informed that on the Huddersfield and Manchester turnpike a steam carriage, with ten passengers, went on level ground at a velocity of eighteen miles per hour, and ascended a hill with a rise of three and one-half inches to the yard at the rate of six to seven miles per hour[3]. Summers' experiments demonstrated that he could carry ten persons and travel at the rate of nine miles per hour[4]. Before the Committee of 1831, Mr Ogle said that the greatest velocity he had attained with his steam carriage was from thirty-two to thirty-five miles an hour, but that it could have been increased on a good road to forty miles. He testified that he had ascended one of the highest hills near Southampton, when his carriage was loaded with people, at a rate of twenty-four and one-half miles an hour[5]. But his figures were so far beyond those of other steam carriage proprietors that it is difficult to look upon them as anything else than greatly exaggerated. From all the evidence at hand, this Committee seem to have been conservative in making their report that carriages could be propelled by steam on common roads at an average rate of ten miles an hour; that at that rate they could convey more than fourteen passengers; and that they could ascend and descend hills of considerable inclination with facility and safety[6]. Not only was the steam carriage used for demonstration purposes, but one had actually been in use as a public conveyance between Gloucester and Cheltenham, where it went four times a day for four months and carried three thousand passengers, at one-half the regular coach fares and in less time than the coaches[6].

[1] *Brighton Guardian,* quoted by *Birmingham Advertiser,* Oct. 10, 1833, p. 4; *The Times,* Aug. 26, 1833, p. 1.

[2] Maceroni's results are given in *Life of Col. Maceroni,* II, p. 474 et seq. The greatest obstacle to his success was poverty. See also *The Times,* Oct. 7, 1833, p. 3.

[3] *Manchester Guardian,* June 26, 1830, p. 3.

[4] Brit. Doc. 1831 (324), VIII, 203, Evidence of Mr Summers.

[5] Ibid., Evidence of Mr Ogle.

[6] Ibid., Summary of the Committee's findings. Some further results secured by Hancock, given in his *Narrative of Twelve Years' Experiments,* pp. 48–82, describe some of his journeys to Brighton, Marlborough, Birmingham, etc., and show the details of the steam carriages he built, the last of which would accommodate 22 persons inside.

Gurney, *Observations on Steam Carriages,* p. 8, shows his success in ascending hills, as witnessed by hundreds of persons. The details of his four months' service between Gloucester and Cheltenham are given on pp. 36–39. For his results, see

The substitution of inanimate for animal power in draught on ordinary highways was regarded as one of the most important improvements ever introduced in the means of internal communication; and the practicability of the new system was regarded as fully established by 1831[1]. It was fully agreed by those competent to judge, that this would be a much speedier and cheaper mode of conveyance of passengers than by the stage coach; and, in fact, this had been actually proved by experiment[2]. So deeply was this fact realized, that when, in the autumn of the year 1827, it was announced that a steam carriage would soon commence running between London and Southampton, the several daily coaches along this route considerably reduced their fares to meet, in some degree, the low prices at which it was understood the steam conveyance would carry passengers[3]. The use of steam power would release from work on the roads a great number of horses which could be productively employed in other ways to meet the necessity of human wants; and land that had been used for growing provender for horses could be used to supply human needs. It was recognized that in three respects, namely, safety, speed and economy, the steam carriage was pre-eminent over horses[4]; and so great were the benefits to be derived from the use of steam power on common roads that to some it seemed as if even railways were inferior[5]. But

also *The Times*, Aug. 5, 1829, p. 2; Aug. 13, 1829, p. 2; Sept. 8, 1829, pp. 3–4; Mar. 16, 1830, p. 4; June 3, 1830, p. 2. In Ibid., Mar. 16, 1830, p. 4, and Apr. 18, 1831, p. 3, are given two letters from John Herepath, showing the utility of the Gurney steam carriage; so successful was its operation that he saw no limit to what might be accomplished by such vehicles.

[1] Brit. Doc. 1831 (324), viii, 203, Summary of the Committee's findings. See also Brit. Doc. 1834 (483), xi, 223, Evidence of John McNeil, Civil Engineer; and the Report of the Committee of 1835. Confirmation given also in *The Times*, Mar. 16, 1830, p. 4; June 3, 1830, p. 2; Apr. 18, 1831, p. 3; *Birmingham Advertiser*, Oct. 10, 1833, p. 4.

[2] Brit. Doc. 1831 (324), viii, 203, Summary of the Committee's findings. The test here mentioned is given in Brit. Doc. 1834 (483), xi, 223, Evidence of Mr Gurney. Before the starting of the steam carriage between Gloucester and Cheltenham the fare was 4s. each person. This was reduced to 1s. per passenger by the competition of the Gurney carriage (*The Times*, May 12, 1831, p. 4). Moreover, the latter frequently accomplished this nine miles, with ten or twelve passengers, in forty minutes, which was considerably faster than the mail coach (*The Times*, Apr. 18, 1831, p. 3). See also Gurney, *Observations on Steam Carriages*, p. 11; *Life of Col. Maceroni*, ii, p. 480.

[3] *The Times*, Oct. 30, 1827, p. 2, and Dec. 3, 1827, p. 2.

[4] Ibid., Sept. 26, 1833, p. 4.

[5] *Birmingham Advertiser*, Oct. 10, 1833, p. 4, gives the following points in which the superiority of the steam carriage over the railway is shown: first, the great injury done to property through which railways may pass, by disfiguring it with

many objections and much opposition appeared to this new means of locomotion. It was urged that their great weight would cause these carriages to be injurious to the roads; but the falsity of this was clearly shown by Telford, McNeil and other engineers[1]. It was also said that their use on roads would prevent the use of horses, as no horse would bear the noise and smoke of the engine; but this, too, had been proved to be untrue and a useless subterfuge[2]. The fact seems to be that the new invention evoked the prejudice of the road commissioners, who were the landowners, and who thought that such machines would be destructive to the roads, objectionable to their use of horses and subversive of the peacefulness and composure of their estates[3]. Through the influence of these men in Parliament, in some cases tolls to a prohibitive amount were imposed on these vehicles; and in other cases the charges, if not prohibitory, were very unfair as compared with those imposed on ordinary coaches[4]. Because of these conditions, the use of steam carriages was discontinued to a large extent and the factories for their manufacture were compelled to close down[5]. The Committee of 1831 advocated an equitable adjustment of the tolls and many others urged the abolition or reduction of them[6]; but notwithstanding all the recommendations and protests the tolls were allowed to remain unaltered.

After 1828, when the practicability of the steam carriage was demonstrated, many contracts were made for supplying these for use embankments, hedges, and other erections; second, the great expense of constructing railways and keeping them in repair; third, the immense amount of valuable land that they throw out of cultivation and cause to be lost to the country; fourth, the inconvenience to the public of abandoning their present habits and adjusting themselves to the monotonous and limited accommodation offered by railways. Against steam carriages, properly constructed and managed, none of these objections arise.

[1] See evidence of Gurney, Farey, Trevithick and Ogle before the Committee of 1831; and evidence of John McNeil before the Committee of 1834.

[2] Brit. Doc. 1831 (324) VIII, 203, Report of the Committee. Hancock, op. cit., pp. 40–46, shows the malignant attempts of drivers of horse vehicles to impede and baffle the course of their new competitor. See also Gurney, *Observations on Steam Carriages*.

[3] Brit. Doc. 1831 (324), VIII, 203, Report of the Committee; Gurney, op. cit., pp. 12, 21.

[4] Brit. Doc. 1831 (324), VIII, 203, Report of the Committee. Also Appendixes C and D to this Report. Note the evidence of Mr Gurney, showing that in the session of 1831 fifty-four private bills were introduced in which steam carriages were specially taxed and some of them passed into law. See also *The Times*, Apr. 18, 1831, p. 3.

[5] Brit. Doc. 1831 (324), VIII, 203, 'Report of the Committee on Steam Carriages.'

[6] In addition to the Report of the Committee, see Cundy, *Inland Transit*, p. 80; Hansard's *Parliamentary Debates*, 1832, XIV, pp. 824–5, 1300–2; and 1834, XXIII, pp. 203–7.

on the roads, to take the place of the stage coaches; but as soon as Parliament showed a disposition to unduly tax them there was an immediate cessation in the placing of contracts and a discontinuance of the experiments[1]. Whether it was due to the fact that the tolls charged were excessive and unequal, or that the railways overshadowed them, steam carriages on the common roads, though showing great possibilities, were soon almost entirely abandoned[2].

Before touching upon the general rate of travelling, we shall take some specific cases to see how the improvement of the roads reacted upon the speed at which distances could be covered; and, since we have facts for the main roads chiefly, we shall confine our examples to these. In 1754 the journey between London and Edinburgh required ten days in summer and twelve days in winter[3]. In the summer of 1776 the flying coach performed the same distance in four days[4]. In 1818 the mail coach took only fifty-nine hours, and the stage coach sixty-one hours[5]; but even these were still further reduced, for in 1836 the mail coach was timed through in forty-five and one-half hours, at an average speed of nine and one-half miles an hour, exclusive of stoppages for meals and official work[6]. Thus it will be seen that the time required in 1836 was practically one-fifth of that required in 1754. This is confirmed by the facts regarding the journey between London and York, which followed the same line as that above. In

---

[1] Brit. Doc. 1835 (373), xiii, 489, 'Report of Select Committee on Mr Gurney's case.' The difficulties experienced by Mr Gurney, culminating in the closing of his factory, are also given in full here. See also Gurney, op. cit., pp. 22–29, which contains frequent quotations from the Report of the Committee.

[2] It was the universal opinion at that time that their decline was due to the heavy tolls to which they were subjected (v. Conclusions of the Committee of 1831). Country gentlemen, road commissioners, road trustees, farmers, coach proprietors, and others who were interested in the continuance of horse power, opposed the steam coach by every means, even to throwing heaps of stones in the road where it was to pass (Gurney, op. cit., pp. 40–41, testimony of engineer Stone, in his letter of June 23, 1831). Such formidable opposition soon secured the result they desired. Adams, *Practical Remarks on Railways*, p. 8, said they were put out of business because the road was not hard and smooth enough to carry the weight of the engine, but this scarcely seems tenable.

For further information on the development of this kind of carriage down to the present time, see J. E. Homans, *Self-Propelled Vehicles*, Chap. ii; Young, *The Economy of Steam Power on Common Roads*, pp. 157–219.

[3] See advertisement of the Edinburgh stage coach given in Malet, *Annals of the Road*, p. 13.

[4] Armstrong, *Post Roads between London and Edinburgh* (1776), p. 5. The distance between London and Newcastle took three days, and from Newcastle to Edinburgh one day.

[5] Hargrove, *History of York*, ii, Pt. ii, pp. 671–5.

[6] Harris, *Old Coaching Days*, p. 93.

1706 it took four days to travel this distance[1] and this time had not been cut down before 1754[2]. By 1761 the time had been reduced to three days in summer[3] and by 1774 and 1776 the regular coaches were going in two days, while the flying coach went in thirty-six hours[4]. In the years following these the rate was accelerated, until in 1818 the mail coach made the distance in twenty-nine and one-half hours and the stage coach in thirty and one-half to thirty-one hours[5]. But still the acceleration continued so that in 1825 the time required was but twenty-four to twenty-five hours[6] and in 1836 only twenty hours[7]. Here, also, it will be observed that the time required in 1836 was only about one-fifth of that required in 1754[8].

We shall next look at the road leading from London to the northwest. From Liverpool no coach left for London till the year 1760; but in 1757 one travelled between Warrington and London, performing the journey in three days[9]. In 1766 the stage coaches travelled between London and Liverpool in two days in the summer and three days in the winter[10]. In 1781 there were three coaches going from Liverpool to London: one went in forty hours, and the others in two days[11]; but in 1785, when Palmer's mail coach was set up, this time was reduced from forty to thirty hours[12]. About 1830 the time occupied could not have been more than twenty-two hours, for from Manchester to London required only twenty hours[13]. The same reduction of time we see on the road from Manchester to London; in 1760 this journey was performed in three days[14]; in 1772 it took

---

[1] See Harris, *Old Coaching Days*, for the advertisement of the coach.

[2] *Archaeologia Aeliana*, N.S., III, p. 247.   [3] Ibid., p. 248.

[4] *Newcastle Courant*, Apr. 16, 1774, p. 2, gives the advertisement of the coach in that year. For the facts of the year 1776, see Armstrong, *Post Roads between London and Edinburgh*, p. 5.

[5] Hargrove, *History of York*, II, Pt. II, pp. 671–5.

[6] *London Magazine*, N.S., I, 1825, p. 36.

[7] Harris, *Old Coaching Days*, p. 93.

[8] It must not be supposed, however, that this represents the relative amounts of time actually taken in travelling on the road; for in the earlier time many coaches stopped for the night while at the later time most of them travelled both day and night. The figures we have given show the time consumed by the journey and not always the time spent in actual travel.

[9] Williamson's *Liverpool Advertiser*, June 16, 1757, gives the advertisement.

[10] Picton, *Memorials of Liverpool*, II, p. 116, quoting from the first Liverpool Directory published.

[11] Picton, *Memorials of Liverpool*, II, p. 116.

[12] Gore's *Liverpool Advertiser*, July 22, 1785, gives the advertisement of this first mail coach between Liverpool and London.

[13] Advertisement of "Beehive" coach, in Harris, *Old Coaching Days*, pp. 43–45.

[14] Axon, *Annals of Manchester*, p. 98.

only two days in summer[1]; in 1788 it took only twenty-eight hours[2]; about 1880 it took but twenty hours[3]; and about 1886 it required only eighteen hours and fifteen minutes[4]. We may, therefore, conclude that the time occupied in reaching these cities from London was, in 1880, only one-third to one-fourth of the time required in 1760.

Along the same line of road, for part of the distance, the coach between London and Shrewsbury pursued its course. In the month of April, 1758, the time required for this journey was four days[5], but in June of that year another coach was put on which required only three and one-half days[6]. In the summer of 1764 a new carriage started, which performed the journey in two days, but this could not be kept up in winter[7]. In 1772 this time was reduced, for the summer season, to one day and a half[8]. In the summer of 1788 the time required was only twenty-two hours[9] and in 1822 only eighteen hours[10]. In the early 1880's the time-bill of the "Wonder" coach allowed only fifteen hours and forty-five minutes[11]; and in June 1835 even that speed was excelled, for the famous "Wonder" went the whole distance in twelve hours and forty minutes, which included delays of one hour and thirty-four minutes on the road[12]. From this it is apparent that the time required to cover this distance in the years immediately following 1880 was but one-fifth or one-sixth that of the year 1758.

In further exemplification of this increased speed in travelling, we note also the change along the great western route. The journey from London to Bristol in 1754 required two days[13]; in 1765 a new post coach could complete it in about thirty-five hours, but of that time one night was spent at Andover[14]; in 1776 both stage coach and

---

[1] *Manchester Collectanea*, in Chetham Society Publications, LXVIII, p. 127.

[2] Ibid., p. 153.

[3] Advertisement of "Beehive" coach mentioned above.

[4] Harris, *Old Coaching Days*, p. 152, gives this time-bill in full.

[5] Owen and Blakeway, *History of Shrewsbury*, I, p. 515.

[6] *Salopian Shreds and Patches*, I, p. 7, gives the advertisement of this coach.

[7] Owen and Blakeway, *History of Shrewsbury*, I, p. 515.

[8] Ibid. The same time was occupied in the summer of 1774, as seen in the advertisement in the *Shrewsbury Chronicle*, July 9, 1774, p. 1.

[9] Advertisement is given in *Salopian Shreds and Patches*, I, p. 55. See also Owen and Blakeway, *History of Shrewsbury*, I, p. 519.

[10] Owen and Blakeway, *History of Shrewsbury*, I, p. 519.

[11] Harris, *Old Coaching Days*, p. 153, gives this time-bill.

[12] *Salopian Shreds and Patches*, II, pp. 26–28.

[13] Latimer, *Annals of Bristol in the Eighteenth Century*, p. 309.

[14] In Brit. Mus., Add. MSS. 27,828, IV, we have the advertisement of a new post coach which left Bath at 7 a.m. and reached London on the following day at 4 p.m.

post coach required about the same time as in 1765[1]; in 1779 the distance was travelled by the fast coaches in one day, which in this case would probably signify twenty-four hours[2]; in 1784 Palmer's mail coach performed the journey in sixteen hours[3]; and by 1836 only eleven hours and forty-five minutes were required[4]. The time spent in going between these two places in 1836, therefore, was only one-fourth of that necessary in 1754.

A similar increase of speed was obtained along the Oxford road. In 1742 the stage coach travelled this distance in one day (thirteen hours) in summer and two days in winter[5]. By 1775 the same distance could be covered in nine or nine and one-half hours in summer[6]; and in 1828 the time required in winter was only six hours[7]. This change, from two days in the winter of 1742 to six hours in the winter of 1828, shows us that at the latter date the *actual speed* of the coaches on this road was three times what it was in the middle of the eighteenth century[8].

In order that we may arrive at some general conclusions it will not be necessary for us to follow out these particular cases any further.

This would be thirty-three hours. But from Bath to Bristol took two hours, so that the time from Bristol to London would be thirty-five hours. On the journey, one night was spent at Andover (say ten or eleven hours), and thus the actual time on the road would be twenty-four or twenty-five hours.

[1] The advertisement given above in Brit. Mus., Add. MSS. 27,828, IV, says that, setting out from Bath, the coach reached London at 4 p.m. If we add on the extra two hours for the time from Bath to Bristol, the time required for the whole distance from Bristol to London would be from 7 a.m. to 6 p.m. of the following day, i.e., two days. This corresponds with what is stated in the advertisement in Felix Farley's *Bristol Journal*, Nov. 2, 1776, p. 2, where it is said that the time from Bristol to London is two days.

[2] Felix Farley's *Bristol Journal*, Aug. 14, 1779, p. 1. This same advertisement tells us that the two-day coaches were also on the road. In 1782 a post coach was advertised to go from Bristol to London in eighteen hours (Sarah Farley's *Bristol Journal*, Aug. 3, 1782, p. 2), but seven years after that there were coaches going this distance in one day, one and one-half day, and two days (*Bath Chronicle*, Jan. 8, 1789, p. 2).

[3] Felix Farley's *Bristol Journal*, Oct. 2, 1784, p. 1.

[4] Harris, *Old Coaching Days*, p. 96, gives distances travelled and time allowed for the fastest day coaches out of London.

[5] Place MSS. (Brit. Mus., Add. MSS. 27,828), IV, p. 16; also see the advertisement in Clark, *Wood's Life and Times*, II, p. 153.

[6] Brit. Mus., Add. MSS. 17,398, pp. 50, 102.

[7] Advertisement is given in Oxford Historical Society, *Collectanea*, IV, p. 278.

[8] In 1742 the coaches travelled only during the day, and the two days, according to the advertisement, were of ten hours each, or twenty hours in all. So that if the time required in 1828 was only six hours the rate of travelling must have been three times as much as at the earlier period.

What we have shown, in the instances already adduced, leads us to make the conservative statement that on the great highways of trade the time consumed on a journey between the termini of the longer routes was, in 1880, only from one-third to one-fifth of what was required in 1750. Of course, on the shorter routes, which could be accomplished without the necessity of spending one or two nights on the road, the time occupied at the two periods was in inverse proportion to the speed of the coaches.

But when we consider in the next place the actual rate of speed, we are met by the fact that before John Palmer's coaches were set up and timed to hours and minutes there was seldom an exact account kept as to the time required for the longer journeys, except in days and half days. Along with this rough way of giving information in regard to time, the report of a journey or the advertisement of a coach rarely gave the time spent on the road at nights and for meals, so that, in such cases, the number of hours necessary for travelling a certain distance (exclusive of the time otherwise spent on the journey) is entirely unknown. The available data on this subject have been brought together elsewhere in tabular form[1]; and from the statistics there given it is well within the limits of accuracy to say that, in 1830, the average rate of speed of the fast mail and other coaches was nine to ten miles per hour[2]. By comparing this with the results of our examination of the rate of speed during the period ending with the middle of the eighteenth century, the conclusion we arrive at is that, considering each class of coach and road in 1830 in the light of analogy with its corresponding class in 1750, the rate of travelling while on the road in 1830 was fully twice as fast as that of eighty years before[3].

---

[1] See Appendix 5.

[2] Macadam, in his testimony before the Committee of 1833, said that on the north road the fast coaches usually travelled nine to ten miles an hour (*Parl. Papers*, 1833 (703), xv, 409, 'Second Rept. by the Lords Committee on Turnpike Returns,' p. 514). *The County Chronicle and Weekly Advertiser*, Nov. 8, 1825, p. 2, referred to the fact that on this road the coaches were timed at ten miles per hour. Shaen, *A Review of Railways and Railway Legislation*, p. 31, says that before the railway's advent the best average travelling was eight to ten miles an hour; and this was confirmed by Mr Chaplin in his evidence before the Committee of 1838.

[3] Thrupp, in his *History of Coaches*, p. 109, published in 1877, says: "...in 1784 coaches became universal at the speed of eight miles an hour." His statement is certainly far from true, according to the information which is presented in our tabular view in Appendix 5.

In making the inference which I have here given, I am also met by the statement of Homer (*Inquiry into the Means of Preserving the Publick Roads*, p. 4) in 1767, that journeys of business were at that time "performed with more than double [the] expedition" of a few years before. The casual reader would judge from this

A question closely related to the rate of travel is that of the cost of travel and to this subject we now turn our attention. In speaking of the period ending with the middle of the eighteenth century, we noted that, as a general rule, the roads were too rough for the coaches to carry outside passengers; but in the period following that time it was the usual custom for them to carry both inside and outside, the latter class usually paying only half the fare paid by inside passengers[1]. Of course, about 1784, when it was desired to greatly increase the speed of the mail coach, outsides were limited to not more than one and in some cases none of them were taken. Another difference from what we have seen in the first half of the eighteenth century is that during that period there was considerable difference between the cost of travelling in summer and in winter, but following that time the seasonal variations fade away and the price paid seems to have been chiefly determined by the character of the accommodation furnished.

We must not omit to mention another feature, namely, that all coaches carrying passengers were not on an equality, as seen by the fact that the stage coaches on some roads were compelled to make

that the actual rate of travelling while on the roads was at this time double what it had been shortly before; but no greater mistake could be made than to take this meaning from Homer's words. In the case of a long journey, much time had been spent at nights in the inns, so that the time required *for the journey* was greatly prolonged; but if we compare the time actually spent *on the road* in Homer's comparison we get a result entirely different from what is implied in his statement. The same misunderstanding seems to have gained credence from the pages of those who have used Homer's opinions as statements of fact; for example, Dr Cunningham, *Growth of English Industry and Commerce in Modern Times* (1903), Pt. I, p. 539, after quoting Homer, says: "There is ample evidence to confirm this account of the improvements." I have been unable to accept the implication of Homer's statement, from the evidence which Dr Cunningham adduces; on the contrary, it appears to me that part of the evidence proves the very opposite from that which he intends it to prove; for when he quotes from Defoe and Arthur Young that "corn was usually taken in bags on horses," it is to me almost conclusive evidence that if strong waggons were seldom used on the roads, the latter must have been in poor condition. Homer's cheerful statement (*Inquiry*, etc., p. 4), that "our very carriages travel with almost winged expedition between every town of consequence in the kingdom and the metropolis," appears, in the face of the detailed facts in Appendix 5, as a purely rhetorical flaunt.

    [1] *Shrewsbury Chronicle*, July 9, 1774, p. 1; *Newcastle Courant*, Apr. 16, 1774, p. 2; Felix Farley's *Bristol Journal*, Aug. 14, 1779, p. 1; *Bristol Gazette and Public Advertiser*, Aug. 7, 1777, p. 1—these and many more give advertisements to prove our statement. The fare for children was the same as that for outside passengers.

    [2] This, of itself, would indicate that the seasonal differences in the roads were by no means so great as in the earlier period and therefore we can be reasonably certain that the roads were becoming much improved in comparison with what they had been.

heavy disbursements on account of turnpike tolls[1]; and according as the gates were greater in number these payments became more burdensome. It was in and near the large cities, like London and Manchester, that these gates were most numerous; and we have already found that in the neighbourhood of London they had been multiplied almost beyond endurance, so that a consolidation of the trusts here was demanded. While the separate turnpike Acts stipulated the tolls that were to be taken at the toll-bars, there came to be considerable uniformity in this matter throughout the kingdom; and the toll usually paid for a horse ridden or led through the gate was one and one-half penny, for a horse drawing a vehicle four and one-half pence, for two horses drawing a carriage nine pence, and so on. In a few instances there was observable a tendency to increase the tolls[2]; but no case has been found where this led to charging higher fares.

[1] Harris, who knew intimately the coaching arrangements about the year 1830, gives (in his *Old Coaching Days*, p. 141) the following table of tolls paid by some of the coaches running out of London:

				£	s.	d.
London and Brighton	per day	..	..	1	4	6
London and Manchester	,,	..	..	5	18	5
London and Birmingham	,,	..	..	3	11	9
London and Liverpool	,,	..	..	5	4	7
London and Cambridge	,,	..	..		17	6
London and Portsmouth	,,	..	..		12	9

From East Grinstead to London, 30 miles, the fare in 1756 was 6s. and the turnpike dues along the line were 2s. (Hills, *History of East Grinstead*, p. 147.)

[2] Hemingway, in his *History of Chester* (1831), II, pp. 235–6, speaks of the excellent repair of the turnpike roads leading from Chester, and then says: "In several directions, however, the tolls are particularly heavy, which necessarily operates as a serious disadvantage on the conveyance of goods [and likewise in regard to passengers] by land carriage. I am at a loss to account for the great increase in these tolls, which in some instances have been advanced within the last three years not less than two-thirds, which will be shown by the following table:

A Table of Tolls taken on the different turnpike roads leading to and from the City of Chester, with the tolls formerly taken. The tolls are for four horses and waggon with six-inch wheels.

	Distance from Chester	Present tolls		Former tolls		Difference more than formerly	
	miles	s.	d.	s.	d.	s.	d.
Chester to Preston Brook and back ..	14	8	4	1	8	6	8
,, ,, Frodsham ,, ,, ..	10	5	0	1	8	3	4
,, ,, Wrexham ,, ,, ..	12	8	0	2	8	5	4
,, ,, Mold ,, ,, ..	12	7	4	4	0	3	4
,, ,, Eastham ,, ,, ..	10	1	4	1	4	0	0
,, ,, Whitchurch ,, ,, ..	20	2	8	2	8	0	0
,, ,, Northwich ,, ,, ..	18	10	0	4	4	5	8"

The toll from Wrexham to Shrewsbury and back, 28 miles, each way, was only 6s.

From the payment of tolls the mail coaches were free, since they were engaged in the public service and were required to perform their journeys with expedition. It has been said that, on account of the increased speed and the greater cost of maintenance, the mail coach fare was "considerably higher than in other stages[1]." This is what we should expect under ordinary conditions; but we have found very little to justify any such conclusion. If there was any time when an increased charge was likely to be made, it would be just when the faster (mail) coaches were set up and before others were accelerating their speed in competition. But what do we find? After much research, only one clear case of this kind has come to our attention, and that was on the London to Liverpool road, where in 1781 the fare between these two places was £2. 15s., but in 1785, on Palmer's coach, it was £3. 18s. 6d.[2]   In contrast to this, and as typical of most of the other roads, we may refer to that between London and Bristol. In 1776, the "London, Bath and Bristol Machines," which performed this journey in two days, charged for each inside passenger £1. 8s.; and at the same time the post coaches, which also occupied two days, charged £1. 8s.[3]   In 1779 the two-day post coach went as usual, but there was also a one-day machine, which charged £1. 5s., and a "light post coach," which charged £1. 12s. for each inside passenger[4]. Taking the post coach, the best vehicle of travel before the establishment of the mail coach, as a basis of comparison, we see that its charge was fully as high as, if not higher than, that of Palmer's mail coach, which charged £1. 8s.[5]   Of course, by comparing the mail coach with the poorer grades of accommodation of the other conveyances it is recognizable that there was some difference in cost, but even then the difference of from three to five shillings, for a distance of about one hundred and twenty miles, was comparatively small. Probably the chief reasons why the mail coaches did not put up their rates were that they could go toll-free, and that other coaches almost immediately developed as great speed as the mail and hence the latter had then no advantage in that respect.

---

[1] Espriella, *Letters from England*, p. 158.

[2] The advertisement of Palmer's coach is given in Gore's *Liverpool Advertiser*, July 22, 1785, showing that the mail coach system would begin on that road July 25 of that year. Possibly the same thing applied to the road from London to York, on which the fare in 1776 was not more than £2. 10s., while by 1818 it was £3. 13s. 6d., but I have not the facts regarding the intervening years that would enable me to make any positive statements (v. Hargrove, *History of York* (1818), II, Pt. II, pp. 671–5).

[3] Felix Farley's *Bristol Journal*, Nov. 2, 1776, p. 2.

[4] Ibid., Aug. 14, 1779, p. 1.

[5] Ibid., Oct. 2, 1784, p. 1.

What effect competition had in the settlement of the fares usually paid for coaching it is very hard to determine, for sometimes an increase in the number of coaches was soon accompanied by a reduction of the cost and at other times it was accompanied by a slight increase of cost. It seems almost certain, from the usually insignificant changes in the cost of coaching which followed any change in the number of coaches along a particular road, that competition, as a regulative agency of the price charged for the service over any lengthened period of time, was largely ineffective[1]. In all probability, custom, which looked to an adequate return for the service rendered, was the most important factor in deciding what the fares should be[2]. Occasionally there were cases where the proprietors of coaches along a certain route got together and, exercising their monopolistic privilege, raised the fares for all who wished to travel on that line. In other similar cases, these coach proprietors would charge passengers a higher rate for coming back from London than for going to London, because passengers in London had to get home and they had to take the only means that was available for that purpose. When such a situation was presented the competition of a new coach would invariably break up the monopoly and re-establish normal prices[3]; but it was only in such unusual

[1] As illustrations of this fact, I cite the following cases, the authorities for which will be found in Appendix 6, under the respective dates. On the London to Liverpool road, the fares for the whole journey were: in 1760, £2. 10s.; in 1781, £2. 15s.; and in 1785, £3. 13s. 6d. On the road from London to Newcastle, the fare in 1761 was £3. 5s.; in 1774, £3. 3s.; and in 1776, £3. 6s. Between London and Shrewsbury, the fares were: in 1753, £1. 1s.; in 1764, £1. 10s.; in 1772, £1. 14s.; in 1774, £1. 10s.; in 1776, £1. 16s., and in 1788, £2. With a constantly increasing amount of coaching along these routes (and others that we might mention), we can see that there was neither a constant decrease nor constant increase in cost, but a series of fluctuations, without any relation to the number of coaches in use.

[2] See, for instance, Sarah Farley's *Bristol Journal*, Apr. 27, 1782, p. 2, where the proprietors of the different stage coaches on the road leading north from Bristol announced that, because the existing fares were not high enough to pay the expense of their establishments, they would be compelled to charge more. This was not a case of extortion, for even the increased fare was not more than was charged on the great majority of the roads.

[3] As confirmatory of our position here, the following advertisement is quoted from the *Hereford Journal*, Aug. 14, 1805, p. 2:

"A New Telegraph Post Coach, called The Accommodation, from the Greyhound Inn, Hereford, through Ross and Glocester, to the Bolt-in-Tun, Fleet-street, London.

The Proprietors of the New London and Hereford Coach beg leave to return their sincere Thanks to the Inhabitants of Hereford, Ross, and their Vicinities, for the very liberal encouragement they have received, and, at the same time, assure them, every possible exertion shall be made to merit a continuance of their favours.

From the decided preference which has been shewn to the New Coach, the

circumstances that competition proved a boon to protect the public.

In seeking to ascertain the exact cost of travelling, some evidence of a very unsatisfactory character is obtained. In 1798 one of those who had been employed by the Board of Agriculture reported that the whole expense for each person travelling by coach was sixpence per mile and that of this amount two-thirds was the advertised fare and the remaining one-third was made up of gratuities and fees to coachmen and guards[1]. Perhaps that may have been accurate enough for the roads of the county of Middlesex near the metropolis where the amount of travelling was very heavy; but our results do not warrant any such statement for the country as a whole. Nor can we accept the above implication that the gratuities were usually one-half as much as the regular advertised fare; they may have been in some cases when extortion was practised[2], but such treatment was probably seldom accorded to travellers. Had there been much of it the volume of complaint over exorbitant charges would have found vehement voice[3]; but, as a matter of fact, among the multitude of other complaints, we scarcely ever note that of unduly high charges. We are, therefore, compelled to dismiss the above generalization as inapplicable to the whole of England. Another, writing in 1844, said that about 1830 the general average fares by mail coaches were 5d. per mile inside and 3d. per mile outside; and by the stage coaches 3d. inside and 2d. outside[4]. This, too, can scarcely be accepted, for after the mail coaches were established their regular inside charges for short distances were 4d. per mile[5], and the mileage charge for a longer distance was usually

Proprietors of other Coaches have determined to lower their Fares, hoping thereby to regain that encouragement they have so deservedly lost.

The Public need not be reminded of what Fares they *have* paid, as that must live in every person's memory that has ever travelled previous to the New Coach coming to Hereford; but it is necessary to remind them, *what* in all probability they *must* pay, provided the old system should ever be encouraged so as to prevent the New Coach from running. It ever has been the intention of the Proprietors of the New Coach to act upon *liberal* principles; to be content with a *fair profit*, and to gain the preference shewn them, by treating their Friends with Civility.

*   *   *   *   *   *   *

The Proprietors of *this* Coach charge the *same* FROM as to London."

[1] Middleton, *View of the Agriculture of Middlesex*, p. 394.

[2] *The Times*, Aug. 27, 1825, p. 3, gives such an instance.

[3] From the evidence given before the committee who had in charge the Great Western Railway Bill and the Oxford and Didcot Railway Bill, we would conclude that the gratuity amounted to one-eighth or one-ninth of the advertised coach fare.

[4] Galt, *Railway Reform*, 2nd ed., 1865, p. 71.

[5] Felix Farley's *Bristol Journal*, July 28, 1787, p. 1.

less than for only a few miles. About the same time the regular inside fare charged by the stage coach for short distances was 8*d.* per mile[1], and the mileage rate for longer journeys would tend to be less than this. Moreover, this writer makes too much of a difference in the charges of the mail and the stage coach; for if there were any difference, which we have shown was not commonly the case, it must have been in particular instances; and the natural tendency would have been for the stage coaches to charge higher fares than the mail coaches, since they sometimes went at the faster speed. A third writer, in 1847, said that before the introduction of railways, the fares per mile were three and one-half pence inside and two and one-half pence outside[2], and he confirmed his statement by the similar affirmation of Chaplin, the greatest coach proprietor of the time, before the Committee of 1838. But, as we have already shown, the mileage rate for short distances was three pence, and that for longer distances would tend to be somewhat below this figure. There was some show of truth also in the declaration of a pro-railway man in 1831 that the coach fares were then at least four pence inside and two and one-quarter pence outside[3]; for on the road from London to Newcastle the mileage rate inside was three and three-fourths pence[4], and on the Brighton road it was four and one-half pence[5], and on the great western road between London and Bristol it was five pence per mile[6]. But if he intended the above assertion to be taken for the whole kingdom, he, too, failed to give us accuracy of statement. In another part of this work statistics have been brought together from the most reliable sources, to see if some definite mileage cost could not be determined which would apply over the greater part of the kingdom[7]. By reference to that table it will be seen how wide were the differences in this respect and how impossible it would be to fix upon any one figure as comprehensive. Having regard to all the variations, the most accurate statement we can make is that, speaking in general terms, each inside passenger travelling by coach paid from two and one-half pence to four pence per mile[8]. In computing our figures we have usually taken

---

[1] Bonner and Middleton's *Bristol Journal*, Mar. 20, 1784, p. 1; Sarah Farley's *Bristol Journal*, Aug. 3, 1782, p. 2; *Bristol Gazette and Public Advertiser*, Aug. 7, 1777, p. 1.

[2] Shaen, *A Review of Railways and Railway Legislation*, pp. 31–32.

[3] *Birmingham Journal*, Jan. 22, 1831, p. 1.

[4] Harris, *The Coaching Age*, p. 194.

[5] Blew, *Brighton and its Coaches*, p. 138.

[6] Brit. Mus. 8235. ee. 4 (1), 'Oxford and Didcot Railway Bill,' pp. 7, 26.

[7] See Appendix 6.

[8] Out of 157 entries in Appendix 6, 22 % include fares of 4*d.* or more; 87 % include fares of 3*d.* to 3*s.* 9*d.*; 38 % include fares of 2*d.* to 2*s.* 9*d.*; and 3 % include

the fare as advertised; and the actual cost of travelling would be the advertised rate increased by, say, one-eighth of itself, to make up for gratuities. By comparing this with the cost of posting, which we have previously considered, it will be seen how much more expensive was the latter[1].

Having now considered the speed and the cost of travelling by coach, our next subject is the speed and the cost of conveyance of goods. First, then, at what rate of speed did the waggons travel in carrying goods from place to place? It must be acknowledged at the outset that wide diversity prevailed, according to the character of the country through which, and the roads along which, the products were carried, and also according to the nature of the load. Certain things had to be carried more quickly than others, else they could not be carried at all; and for this purpose "fly vans" were established on some roads. Between London and Brighton, for example, the coaches could not carry all the luggage of passengers, besides the meat and farm and dairy produce sent to the London market, and, consequently, vans were adopted for this purpose[2]. They did not travel so fast as the coaches nor so slowly as the stage waggons; and in addition to carrying goods they also took passengers who were not able or willing to pay the stage coach fares. As common carriers, they were responsible for what was entrusted to them and their business was sufficiently remunerative to attract much opposition. What speed they maintained on this road does not appear; but we learn that on the road from London to Birmingham their speed was five miles per hour[3]. On other roads, vehicles that were comparable to the fly vans in carrying both passengers and goods, but which were called "post waggons," travelled at the rate of forty-seven to fifty-three miles in a day[4]. But waggons that carried no passengers went at a more leisurely speed. On the road from Manchester to Leeds, over a somewhat mountainous country, it took twenty-four hours to go this forty-five miles[5]; and over a similar country between Sheffield and Manchester it took the waggons forty

---

fares below 2*d*. But the fares of 2½*d*. to 4*d*. inclusive take in 61 % of all the entries; and I have fixed upon this as the best general average for the whole country.

[1] Passengers might go cheaper still by travelling in the slowly moving stage waggons or vans used for the conveyance of goods (*Leeds Intelligencer*, Jan. 9, 1792, p. 4).

[2] Blew, *Brighton and its Coaches*, pp. 165–6.

[3] *Birmingham Journal*, Jan. 22, 1831, p. 1.

[4] *Leeds Intelligencer*, Jan. 9, 1792, p. 4, states that it took these waggons four days to go from Leeds to London, 190 miles, and two days to go from Leeds to Birmingham, 106 miles.

[5] *Manchester Guardian*, Jan. 29, 1831, p. 1.

hours to accomplish this distance of forty to forty-five miles[1]. The "fly waggons" going from Chester to London in 1795 travelled about thirty miles per day[2]; and this probably agrees fairly well with the statement of a writer in 1847, who said that before the coming of the railways the ordinary speed of land carriage of goods was about two and one-half miles per hour[3], for this rate per hour effected through twelve hours in the day would just make thirty miles per day. But, doubtless, it was frequently the case that the speed was below that, as on the road from Newcastle to Carlisle, where in 1829 the ordinary speed was nineteen to twenty miles per day[4]. From the foregoing, it is evident that no conclusion can be reached as to the speed of waggon carriage that would be applicable to any considerable part of the country. There were too many variable elements entering into it to enable us to make anything but specific statements.

The last subject to be considered in this connexion is the cost of carriage of goods by land, after 1750. We have seen that in the earlier period, from the year 1691 onward, the rates of carriage could be fixed by the Justices of the Peace at their Quarter Sessions; and this arrangement was found to be so productive of good results, at least in some cases, that the system was continued during the period we are now studying[5]. Where the Justices of a county fixed the rates to be charged for carrying goods into or out of that county, they did it without conferring with the Justices of any other county; and this of itself would naturally account for some differences in the rates that were charged. But we have seldom found that this assessment of the rates of carriage, which the Justices were empowered to make, was fulfilled. In the Act of 1691 the preamble gives just ground for supposing that the reason the Act was passed was to prevent combinations of carriers from raising rates to an unduly high figure, to the detriment of the public; and although the statute required the Justices to assess the charges, it appears to us, from the few instances we have found of such a practice, that the Justices must have considered it their duty to make such an assessment only when the carriers were endeavouring to charge more than public opinion thought right[6]. There is good reason for

---

[1] *Sheffield Iris*, May 10, 1836, p. 2.

[2] *Chester Guide*, 1795, pp. 62–64, 66. They performed the 180 miles in six days.

[3] Shaen, *A Review of Railways and Railway Legislation*, p. 32.

[4] 'Copy of the Evidence before a Committee of the House of Commons on the Newcastle and Carlisle Railway Bill. Taken from the Shorthand Notes of Mr Gurney,' p. 2.

[5] In addition to the Act made for this purpose, which we have formerly noted, see also *J., H. of C.*, xxx, p. 608.

[6] Occasionally we notice a tendency of carriers who had a monopoly of the carrying trade to raise their prices, despite the fixing of rates of the Justices. This

our assertion that the activity of the magistrates in fulfilling the letter of the law was a very minor factor in fixing the rates that the carriers charged; and, in all probability, custom—what the public thought to be right—and competition of one kind and another, including competition of waterways with highways and of one carrier with another, were the deciding factors in the determination of rates. As in the case of the railways to-day, the recognized freight rates are departed from according to special circumstances, so was it in regard to the roads of that time: it was not always possible to charge a fixed price for carrying a certain amount of traffic for a certain distance. The details as to the prices for carriage we have elaborated elsewhere[1], and even a casual perusal of these will show how great were the variations. But a more minute examination of the particulars there presented shows us that, in general, the cost of carriage was from one-half to three-fourths of a penny per hundredweight per mile[2]. Of the total number of entries which we have made, less than sixteen per cent. show a figure below one-half penny, and less than nine per cent. show a figure above three-fourths of a penny. We may, therefore, regard these limits as defining the price at which goods were carried. It will be recalled that the rates of carriage before 1750 were fixed within these same limits; and the question may be asked, Why, with the improvement of the roads, did the cost of carriage not decrease?

we have so seldom found, however, that it need not concern us much (v. Bunce, *History of Birmingham*, I, p. 49).

[1] See Appendix 7.

[2] In order that we may have some basis of comparison, I have frequently had to assume that the hundredweight rate would be the same in proportion as the tonnage rate; and have made the hundredweight-mile the standard by which to express all the rates. This assumption is sometimes contrary to fact; but the hundredweight-mile basis is the closest approach I have been able to obtain to a useful standard of comparison.

Shaen, *A Review of Railways*, p. 32, said that before the time of railways the carriage of goods often amounted to 1s. per ton per mile. This would be the equivalent of three-fifths of a penny per cwt. per mile, which would come within the limits we have fixed.

Jeans, *Jubilee Memorial of the Railway System*, p. 6, informs us that the heavy cost of transporting by pack-horse was greatly reduced by the introduction of macadamized roads to 8½d. per ton per mile, which would be approximately one-half penny per cwt. per mile.

On the contrary, a writer in *The Gazetteer and New Daily Advertiser*, Feb. 27, 1787, p. 4, said that the price of carrying was something over three-fourths of a penny per cwt. per mile. His letter asserted that the common load of a waggon was three tons and this moved at the expense of 5s. per mile. This would make the expense of carriage 1d. per cwt. per mile. By reference to our tabular view, it will be seen that this writer's figures were too high as an average for the whole country.

In reply, the only answer that seems to fit the facts is that the means of conveyance were still far behind the necessities of the time; and even with the aid of the canals the demand for carriage, arising by reason of the Industrial Revolution, outran the existing facilities for the transportation of commodities.

Bridges.

In a former section we have brought this subject down to the middle of the eighteenth century; and the same conditions which existed at that time continued for over eighty years longer. The use of ferries was being more and more discarded and often their places were being taken by bridges of strong construction and architectural beauty. In other cases, both ferry and bridge continued; but in such instances the value of the ferry was usually lessened and it became necessary, by arbitration, to make amends therefor to the party who had the title to the ferry[1].

With reference to the construction of the bridges, it was becoming more evident that nothing but the most approved design and substantial workmanship could suffice to render them durable. Before the work of enclosure had made much progress the water of a river could often spread out over the adjoining land in times of flood and in that way lessen the force of the flood; but when the lands along the rivers had become enclosed and the banks were raised to prevent inundation, the narrower channel for the waters increased the force with which they moved along and thus endangered many bridges which, up to that time, had been able to withstand the current[2]. It, therefore, became necessary that the subject of bridge construction should receive particular attention; and as it was during this period that the great road engineers came into prominence, so it was at this time that the noted bridge-builders, a few of whom were also road engineers, began their work[3]. These two aspects of highway making are inseparably connected; and as the work of the road engineers resulted in a more economical administration of both funds and material, so also the erection of enduring bridges under competent men put an end to fraud and misuse of the public funds that were raised for that purpose.

---

[1] See, for instance, Wearmouth Bridge Act, 32 Geo. III (1791), secs. 13, 14, 15, etc.

[2] As examples of the havoc caused by floods in destroying bridges, see Sykes, *Local Records*, I, pp. 283–9; Mainwaring, *Annals of Bath* (1809), pp. 83–87; Rickman, *Life of Telford*, pp. 28, 29, 30, etc.; *Archaeologia Aeliana*, N.S., XII, p. 142.

[3] Men like Thomas Telford, John Rennie, William Smeaton and others became proficient in this art.

It is unnecessary here to refer to the different ways in which money was received for these structures, for these have been already considered in a former part of this work.   Our chief object is to advert to two tendencies which became well marked at that time, one of which deals with the repairing of bridges and the other with their control.

In accordance with the great increase in the amount of trade, which was an accompaniment of the Industrial Revolution, it was desired to have the bridges widened to accommodate the pressure of traffic on the roads.   Until that time some bridges were so narrow that it was dangerous for foot-passengers to try to cross them on a market day and those who were walking across sometimes had to take refuge from vehicles by stepping into the angular recesses on both sides of the bridge[1].   Other bridges had houses built upon them, which lessened the space for the roadway.   As late as the beginning of the nineteenth century, some of the bridges on the great thoroughfares were so narrow that two carriages could scarcely pass on them[2].   But as occasion offered, many of these bridges were widened, their piers were made more substantial and the arches were made of a wider span so as to facilitate the ready passage of a greater volume of water.   This was the more easily accomplished after 1777–9, when cast iron had been found to be serviceable for this work[3].

The other tendency was the increasing desire to bring the important public bridges under the jurisdiction of the county.   By the Statute of Bridges[4], the liability for repairing these rested upon the county; but although this statute was in force the repair of many bridges of public utility had been left in the hands of the parish, or hundred, or other municipal corporation—often without good results[5]—or was

---

[1] Axon, *Annals of Manchester*, p. 104;   Cox, *Derbyshire Annals*, II, p. 224; Aston, *A Picture of Manchester*, p. 199.

[2] This was the case with the great Trent Bridges at Nottingham (v. Briscoe, *History of the Trent Bridges at Nottingham*), London Bridge, bridge at Bristol, and many others.

[3] Rickman, *Life of Telford*, p. 29.   On the historical development of cast iron bridge building, see Telford and Douglass, *Improvements of the Port of London*, pp. 6–9. See also [Owen], *Some Account of Shrewsbury*, p. 84; Aston, *A Picture of Manchester*, p. 199 ;   Brit. Mus. MSS. 6707, Reynolds' *Derbyshire Collection*, p. 11.

[4] Act 22 Henry VIII, c. 5.

[5] Dickson, *Agriculture of Lancashire* (1815), p. 111, shows us that in that county the bridges maintained by the hundreds were not generally in as good repair as the county bridges; and the parish or township bridges were worse again than the hundred bridges.   In all probability this was true also in other counties; and this would be a determining factor in having the county take over the care of all important bridges.

imposed by immemorial usage on private individuals or other parties[1]. With the opening of the eighteenth century and the gradually increasing amount of trade, there had arisen the necessity for rebuilding and widening some of the bridges; and as the local community had some-times declared their inability of themselves to make these changes the Quarter Sessions had come to their aid and given them money to assist in carrying out these improvements. In every case, however, it was understood that the immediate aid from the county was for only the one particular time and did not render the county liable for any other repairs. It would seem that the Justices had been too free in employing county bridge funds for such purposes; and in 1739 Parliament forbade them applying any money to the repair of bridges "until presentment be made by the Grand Jury, at the Assizes or Sessions, of their insufficiency, inconveniency, or want of reparation[2]." We are unable to say how far this statutory change went in limiting the contributions of the Quarter Sessions to these so-called "gratuity bridges;" but we are safe in saying that other devices would be employed to nullify the intention of the Act. In 1780 a far-reaching decision was handed down, which did much to transfer to the county the responsibility for bridges. The township of Glasburne in the West Riding of Yorkshire had recently substituted a carriage bridge for an old foot-bridge which it had always maintained. The carriage bridge became decayed and in this year the West Riding was indicted for its non-repair. The latter denied any liability, but the unanimous verdict of the Judges was that the Riding was liable. The Court's decision was that if a man built a bridge and it became useful to the county in general the county should repair it[3]. The fact that the township had constructed the bridge did not render it liable for repair, and the bridge had been too recently constructed to make any authority liable by prescription. Hence, by falling back on the Statute of Bridges, the liability was rolled upon the county. Under this interpretation of the law, many of the smaller jurisdictions eagerly went to work to transfer their supposed liability for maintenance and repair of certain bridges to the counties; and whenever any dispute arose as to who should maintain a particular bridge, unless a definite responsibility could be otherwise established,

---

[1] The Court of King's Bench, in 1833, held that a parish might be in-dicted for non-repair of a bridge, without stating any other ground of liability than immemorial usage (Rex *vs.* Inhabitants of Hendon, 4 Barnewall and Adolphus 628).

[2] 12 Geo. II, c. 29, sec. 13.

[3] This Glasburne bridge case was Rex *vs.* W. R. of Yorkshire, 1780, in Sir William Blackstone's *Reports*, II, p. 685.

that liability was imposed upon the county[1]. Even when a bridge had been erected for private benefit and it was constantly used by the public after its erection, the law required its maintenance by the county as a public bridge[2]; and where a parish bridge, on account of change of time and circumstances, was no longer sufficient for the public use without being enlarged or otherwise improved, such bridges might be taken over and all repairs done by the county[3]. In these ways, the county became saddled with the burden of maintenance of many bridges which had formerly been supported by local authorities.

In 1803, Lord Ellensborough's Act was passed, which enacted that no bridge erected after that time, at the expense of any private person or persons, should be regarded as a county bridge unless it should be erected in a substantial and commodious manner, under the direction or to the satisfaction of the county surveyor[4]. After the passage of this Act, the satisfaction of the county surveyor was a necessary preliminary to making the county liable to repair new bridges. But where a bridge was built before 1803 the county could escape this common law liability to repair the bridge only by proving that this liability was on some one else[5].

---

[1] A few cases may serve to illustrate this more fully:

Rex *vs.* Glamorgan (1788), 2 East 356 n. The County of Glamorgan was indicted for not repairing a certain public bridge, erected in the King's highway, across the river Tave. In the evidence it was shown that the bridge had been built by the owner of certain tin-works, for his own use and benefit, as a suitable way to his tin-works, and that he and the tenants of the tin-works enjoyed a way over the bridge for their private benefit. It further appeared that the business of the tin-works could not be carried on without the bridge. But it was also shown that the public had constantly used the bridge from the time when it was built. Held, that the county was liable for repairs.

Rex *vs.* Bucks (1810), 12 East 192. Queen Anne, in 1708, for her greater convenience in getting to and from Windsor Castle, built a bridge over the Thames at Datchet, on the highway from London to Windsor, to avoid the use of an ancient ferry, with its toll, belonging to the Crown. She and her successors maintained the bridge till 1796, when, being partly broken, the whole structure was removed and the King re-established the ferry for the use of the public, toll-free. Thirteen years afterward, the county was indicted, when it was held that the bridge had become a common public bridge and that, therefore, the county was liable to rebuild and repair it.

Rex *vs.* Kent (1814), 2 M. & S. 513. In this case a man erected a mill and dam for his own profit, thereby slightly deepening the water of a ford through which there was a public highway. The passage through the ford, before it was deepened, was very inconvenient and at times unsafe to the public. The miller afterward built a bridge over it, which the public used constantly. Held, that the county was liable to repair the bridge.   [2] Ibid.

[3] Act 3 Geo. IV, c. 126 (1822), sec. 107.   [4] Act 43 Geo. III, c. 59, sec. 5.

[5] Vide Attorney-General *vs.* W. R. of Yorkshire County Council (1903), 67 J.P. 173. The duty of repairing included the duty of rebuilding when necessary (v., for

In 1795 Chief Justice Lord Kenyon handed down a decision that if a bridge, used for carriages, though formerly adequate to the purpose intended, were not now sufficiently wide to meet the public demands, owing to the increased width of carriages, the burden of widening it must be borne by those who were liable to repair the bridge. But under Lord Ellensborough's Act the Justices at Quarter Sessions had power to order narrow bridges to be widened and improved, or, if necessary, rebuilt by the county[1].

From what we have just said, it is evident that it was coming to be recognized more and more that the county should assume the obligation of the upkeep of the bridges and that the common law liability in this respect should be more carefully complied with[2].

The administration of the affairs of a bridge was, like that of a turnpike road, placed in the hands of a body of commissioners who were required to put the Act into effect. As a corporate body, they could sue or be sued in the name of their clerk and had complete jurisdiction over the purchase of necessary adjoining lands, the erection of toll-houses, the appointment of officers, such as the surveyor, collector of tolls, etc. The money required for the construction of the bridge was obtained in various ways: sometimes it was received by direct subscription or by the levying of a rate: sometimes the commissioners were empowered to borrow money on the credit of the tolls, in which case the tolls were disposed of as security for the money borrowed; in other cases funds were obtained by the issuance of shares, as was done for some of the largest bridges[3]. As soon as the

---

instance, Rex *vs.* Bucks (1810), 12 East 192; Rex *vs.* West Riding of Yorkshire (1770), 5 Burrows 2594, etc.).

[1] There were certain exceptional cases in which bridges built by private persons for the use of the public were not repairable by the county, but by those who had built them for their private uses (v. 1 Roll. Abr. 368; Rex *vs.* Kent (1811), 13 East 220; Rex *vs.* Lindsey (1811), 14 East 317; Rex *vs.* Kerrison (1815), 3 M. & S. 526; etc.).

[2] We shall not go into detail as to the repair of bridges by towns, by private munificence, etc. Regarding the former, see Blacklock, *The Suppressed Benedictine Minster*, pp. 383–4, 387; Thompson, *History of Boston*, p. 251; etc. On the latter, see Green, *History of Worcester*, II, p. 15; *The Chronicle of Bristol*, Aug. 1, 1829, to Jan. 1, 1830, p. 53; etc. These phases of the discussion have been more fully treated in the earlier chapters of this work.

[3] In the erection of the cast-iron bridge over the Trent at Dunham in 1832 the cost of the work was raised by a proprietary of £50 shareholders (Bailey, *Annals of Nottinghamshire*, p. 386). See also Brit. Mus. 1890. e. 4 (57), 'Proposals for Raising by Subscription the Sum of £400,000, in Shares of £100, for the Purpose of Building and Maintaining a Bridge over the River Thames;' and Brit. Mus. 8223. e. 10 (151), 'Proposed Tontine Bridge across the Swale, at or near King's Ferry, Isle of Sheppey;' Aston, *A Picture of Manchester*, p. 200.

trustees had completed the bridge and the amount of the debt had
been discharged by the tolls received, the tolls were usually to cease.
In some cases no tolls were allowed to be taken and the expense of
construction had to be borne by some other form of contribution or
taxation[1]; while in other cases lands were set aside and income was
obtained from this source rather than from bridge tolls[2]. The desire
of the public was to have these tolls abolished as soon as possible, so
as to make the highways of communication as free of obstructions as
could be.

[1] Oliver, *History of Exeter*, p. 172. On the details with reference to bridges,
see Wearmouth Bridge Acts, 82 Geo. III and 54 Geo. III, c. 117; Horsfield, *History
of Sussex*, I, pp. 96, 98; [Owen], *Some Account of Shrewsbury*, p. 84.

[2] *The Times*, Apr. 16, 1810, p. 8.

CHAPTER V

WE have noted in a preceding chapter some of the efforts which were made in the seventeenth and in the first half of the eighteenth century to improve the inland navigation of England, by resorting to the canalization of rivers where that promised to be of utility in developing the trade of the kingdom. Some good results were secured through the efforts of that time, but the work then accomplished was insignificant in comparison with what was accomplished in the later period. We must not think, however, that after the middle of the eighteenth century there was any radical departure, so far as the method of the improvement of rivers was concerned, from the earlier practice, for the same system of locks and weirs that had already been in use was continued in all river improvement subsequent to that time.

In addition to the instances which we have formerly considered, there were some other rivers the improvement of which pertains to the period before 1750, but has bearings also upon the later period, and these we must note briefly. By an Act of 1720, certain undertakers were empowered to make the rivers Mersey and Irwell navigable from Liverpool to Manchester and to take the stipulated tonnage dues for all goods carried between Bank Quay, near Warrington, and the city of Manchester[1]. This was done by cutting off many turns of the river by collateral cuts or canals. This has been called "the first canal ever made in England[2]," but the statement is not true; for in our modern sense of a canal, as an artificial navigable waterway apart from a river, this waterway has no claim to such a designation. Even when considered in its connexion with the river it cannot claim precedence, for then the so-called Exeter Canal has the position of priority. In 1737, Parliament passed an Act for making Worsley Brook navigable to its junction with the river Irwell, near Manchester[3]; and in 1755 an Act was obtained for making Sankey Brook navigable

---

[1] Act 7 Geo. I, stat. 1, c. 15, 'An Act for making the Rivers Mercy and Irwell Navigable.'

[2] *Gentleman's Magazine*, XCI, Pt. II, p. 491.   [3] 10 Geo. II, c. 9.

from its junction with the Mersey[1]. The latter work was the more advanced, but even here the waterway was constructed alongside the brook, and the water of the latter was to feed the new waterway[2]. Although the Worsley Brook Navigation was probably the beginning of the Bridgewater Canal enterprise, we can readily see that it was in no sense a canal, in our present meaning of that word.

England was by no means among the first nations to introduce canals, for the Chinese, the Romans, the Italians, the French and the Dutch had early constructed works of this kind[3], and some of these early canals have persisted and are in active operation to the present day. It was from these countries, especially Holland and other countries of continental Europe, that Englishmen got their ideas regarding the benefits to be obtained from canals[4]; so that when England introduced these improvements she was simply following the lead of some other nations that had received great advantages from such means of conveyance.

In the latter half of the seventeenth century the suggestion had been made that the rivers Thames and Severn might be connected by the construction of a short portion of canal[5], but nothing came of it for more than a century. The first modern canal projected in England seems to have been one to connect the Trent river with the Severn in the county of Stafford. This route was surveyed and described by Thomas Congreve in 1717. The canal was to be made from Burton-on-Trent, via Penkridge and Prestwood, to the Severn river. Its engineering features were fully worked out, and its economic advantages elaborated; but why the work was not undertaken does not appear[6].

---

[1] 28 Geo. II, c. 8. The powers and privileges usual in such cases, of purchasing land and other things necessary for the navigation at a fair estimate to be made by commissioners named in the Act and the removal of all the impediments and obstructions, were granted to the undertakers in the fullest way. In view of the expenses connected with the work, the undertakers were allowed to take certain duties specified in the Act.

[2] When the undertakers of the Sankey Brook Navigation applied to Parliament for an Act, in 1754, the corporation of Liverpool agreed to advance £300 toward obtaining the Act and the survey of the navigation. This was to be paid back if the Bill passed into a law (Picton, *Liverpool Municipal Records*, II, pp. 144–5).

[3] On the early canals of other countries, see Phillips, *History of Inland Navigation*.

[4] Francis Mathew constantly refers to the experience of the Low Countries with canals.

[5] This has already been referred to in the 'earlier chapter on navigations, in connexion with the work of Francis Mathew, 1670. On this subject, see also Brit. Mus., Stowe MSS. 877, pp. 21–22.

[6] For full account of this, see Congreve's pamphlet addressed to "William Ward, Esq., Knight of the Shire for the County of Stafford," in Brit. Mus., Stowe MSS. 877, pp. 16–20.

Then, too, in the same year in which application was made to Parliament for power to put through the Sankey Navigation, the corporation of Liverpool had under consideration a much larger work, namely, the uniting of the Trent and the Mersey so as to open up water communication between the ports of Liverpool and Hull. Surveys of this route had been made but the enterprise had not been pushed.

These few details show us that the increased need for the means of conveyance was attracting attention and the reason for this is not far to seek. The northern counties of England, especially Lancashire and Yorkshire, were coming to be the centre of an Industrial Revolution. The work of manufacturing was actively progressing there, and there was the movement of population from the agricultural sections of the south and south-east to the industrial north. Manchester and Liverpool were gradually rising in importance and increasing in population; the growth of manufactures and the increased amount of the necessaries of life for the greater number of people, required greater facilities for the transportation of raw and manufactured products, as well as for the bringing in of larger quantities of the means of subsistence. Manchester had long been noted for its manufacture of coarse cottons, in addition to which it manufactured also fustians, mixed stuffs and small wares. Its growing prominence is evidenced also by the increase in the number of its inhabitants, from no fewer than twenty-four hundred families in 1724 to a population of 20,000 in 1757, for the united towns of Manchester and Salford[1].

The growth of Liverpool, too, had been rapid, considering that time. In the year 1760 the income of the corporation was over £4700 a year, a nearly fourfold increase from the year 1720, when it amounted to a little more than £1200. The town dues, raised by a small tax on imports and exports, were £1022 in 1760, having increased more than threefold over the year 1720, when they amounted to £305. The dock dues were £2388 in 1760, which was nearly three

---

[1] Aikin, *Description of the Country from Thirty to Forty Miles round Manchester.* It is evident here that this is simply stating the estimated population in round numbers, and not with exactitude. In 1773, the number of people in Manchester and Salford and their contiguous vicinity was 27,246, according to Corry, *History of Lancashire*, II, p. 459; and this would seem to be fairly in accord with the statement of Enfield, *Essay towards History of Leverpool*, p. 25, who says, on the basis of a comparative computation, that the population of Manchester alone in that year was 24,533. If the population of Manchester, therefore, in 1773, were about 24,000, the 20,000 figure given by Aikin for 1757 would probably be a little over-stated. But while the accuracy of these figures cannot be vouched for, the underlying fact is that there had been a large addition to the population.

times the amount of these in 1724, at which time they were £810[1]. The amount of shipping that entered and cleared from Liverpool in the seven years ending 1716 averaged over 18,000 tons, and in the seven years ending 1765 it averaged over 62,000 tons, a more than threefold increase[2]. The customs at the port increased rapidly during the eighteenth century, from £50,000 in 1700 to more than £248,000 in 1760[3]; and in harmony with the development of trade there was the corresponding increase of population[4].

The poor state of the roads leading to and from these populous centres about 1750 we have formerly noted. This made it difficult, especially at certain seasons, to provide food for so large a population, for it must be remembered that the working classes seldom have the means which will enable them to lay up in advance for their future needs, but must obtain their sustenance week by week. In the winter season, when the roads were bad, and sometimes closed, the price of food rose exorbitantly; and even in summer, on account of the

---

[1] Baines, *History of Liverpool*, pp. 406–7. The amount is stated only in pounds, and the shillings and pence are omitted. See also Enfield, *Essay towards History of Leverpool*, p. 69.

[2] Baines, op. cit., p. 491. Enfield, op. cit., p. 67, gives the figures for each year, showing the number and tonnage of ships that arrived at and sailed from Liverpool, 1709–72.

[3] Baines, op. cit., p. 492. Enfield, op. cit., p. 70, gives an account of the number of ships that yearly sailed to and from the ports of Liverpool and Bristol, 1759–63, showing how the trade of Liverpool had pushed away ahead of that of Bristol; and on pages 73–87 he gives a detailed account of the imports and exports for one year beginning Jan. 1, 1770, showing the nature and amount of each article exported and imported. While he does not pretend that his account is perfectly accurate, yet it shows the tremendously wide sweep over the world that was taken by the trade of this port.

[4] On the basis of "a calculation proportionate to the bills of mortality," Corry (*History of Liverpool*, p. 79) made it appear that the population of Liverpool in 1700 amounted to 4240; and according to his statement the population in 1730 was "upwards of 12,000" (ibid., p. 80). He gives nothing more definite than that, and gives no authority even for that; but in the same paragraph he added that this trebling of the population in thirty years was a striking proof of the regular and rapid progress of the improvements in manufacture and of the city's increase in foreign commerce. On the basis of the number of inhabited houses, Enfield (*Essay towards History of Leverpool*, p. 26) computed that the population in 1758 would be about 20,000 and in 1760 about 25,000; while, for the latter year, Corry (op. cit., p. 119) says that the population was estimated at 25,787—practically the same figure as that given by Enfield. In 1770, according to Corry (op. cit., p. 81) the population of Liverpool amounted to 35,000; but, in all probability, he simply made into round numbers the figure given by Enfield, who said (op. cit., pp. 24–25) that, by an *actual survey* made in the beginning of 1773, the number of inhabitants was found to be 34,407. As in the case of Manchester, the population statistics for Liverpool are merely approximate; but they show us roughly the great increase that had taken place in the number of people who had become residents there.

comparative infertility of the soil in that section, these large cities were poorly supplied with the ordinary necessities. These things were often brought from considerable distances by means of panniered horses and the tendency was for them to reach a very high price.

For the same reason, the supply of coal was scanty in winter; and although there was abundance of it within a few miles of Manchester, in nearly every direction, the cost of those few miles of transport, in the existing state of the roads, was a great barrier to the use of coal by the working classes. On account of the expense of carrying the coal by pack-horses, the price of it at the pit was usually doubled before it reached the consumer in Manchester[1]. Nor was the expense less by using the river for part of the distance; for in that case the coal had to be carried on horses' backs or in carts from the pits to the river Irwell, there to be loaded into boats and taken to Manchester, after which it had to be unloaded and carried to the houses of the consumers. So much loading and unloading, in the use of this route, could not effect any saving; and, besides, the high charges of the Mersey Navigation Company were an effectual barrier against using this river for the carriage of coal[2]. In addition to the cost of carriage, a further difficulty was encountered in securing a supply of fuel, in that there was a combination among the colliers; and unless their demand for gratuities were first met by the carter, the latter could not obtain any coal at the pits for carriage to Manchester[3].

The same difficulty, namely, the cost of carriage, stood in the way of the transit of goods between Manchester and Liverpool. By road the charge was 40s. a ton and by the Mersey and Irwell Navigation it was 12s. a ton[4]; and, besides, there was great risk of delay, loss and damage by the way. Delay was inevitable from the fact that

---

[1] Coal was sold at the mouth of the pit at so much per horse load, which was as much as an average horse could carry in two panniers on its back, or about 280 lbs. The carriage alone was, as we have seen, about 9s. per ton for 10 miles (see *The History of Inland Navigations*, i, p. 59).

[2] Meteyard, *Life of Wedgwood*, i, p. 275. The minimum charge insisted upon by the Mersey and Irwell Navigation Company was 8s. 4d. a ton for even the shortest distance.

[3] *Description of Manchester*, pp. 5–6. The custom of the colliers was to bring the coal and keep it in the pit while they went up and idled away their time, or were drinking at the alehouse, until they obtained their demand for gratuity, which they levied as a tax on the carters who were waiting for loads. When this exaction was complied with, they went down and brought up the coal. How this combination was to be broken up could not be imagined; but at last the Bridgewater Canal effected this desired end.

[4] Meteyard, op. cit., i, p. 275. Also Brit. Mus. B. 504. (4), 'Advantages of Inland Navigation,' p. 18; and *The History of Inland Navigations*, i, p. 28.

boats had to be drawn along portions of the river by the labour of men; and those who had this work in charge were never known to be scrupulously honest, so that goods sent in their care usually reached their destination very much reduced in weight[1]. It was obvious that unless some means could be devised for facilitating and cheapening the transport of goods between the seaport and the manufacturing towns inland, there was little prospect of any further considerable development being effected in the industry of the district.

In the face of this situation the Duke of Bridgewater turned his attention to the relief of the social and industrial life of the city of Manchester and its surrounding towns. He was the owner of a large property and coal mines at Worsley, about ten miles from Manchester, and he set about to devise some plan for bringing this coal into Manchester, so that it could be sold there at a price that would be within the reach of all. For securing this end, he contemplated constructing a canal from his mines at Worsley Mill to Salford and in 1759 introduced into Parliament a Bill to give effect to his plans. In this Bill the Duke promised that if he were allowed to make this canal, he would deliver coal at Salford at not more than four pence per hundredweight, where formerly it had been sold for five and one-half to seven pence per hundredweight[2]. The Bill was so strongly supported by the towns of Manchester and Salford that it passed both Houses without opposition[3]. In the next year another Act was passed which authorized the Duke to carry this canal over the river Irwell to the town of Manchester, but so as not to obstruct the navigation of that river[4].

---

[1] Meteyard, *Life of Wedgwood*, i, pp. 275–6.   [3] *J., H. of C.*, Nov. 25, 1758.
[2] Act 32 Geo. II, c. 2. This is given in full in Cary, *Inland Navigation*, pp. 1–3. It empowered the Duke to make and maintain the canal at his own cost, to enter lands, dig and remove obstacles, and then to make towing paths on the sides of the canal, having first given satisfaction to the owners of lands and grounds through which the canal was carried. Any disputes regarding the value of these lands were to be settled by commissioners; and after payment of the sum assessed, the lands were to vest in the Duke. Coal from the Duke's mines was not to be sold in Manchester or Salford for more than 4d. a hundredweight. The navigation was to be free on payment of the tolls. The Duke was empowered to fix the tonnage rates, which were not to exceed 2s. 6d. per ton. No water was to be taken from the river Irwell for this canal.

A canal differs from a river navigation chiefly in the fact that the company or proprietors working it do so for their own profit and usually have the soil of the canal vested in them by the terms of the Act; while the trustees of a river made navigable by Act of Parliament appear usually to have a mere possession of the soil for the purpose of improving the navigation and are bound to apply the profits of the navigation for the future benefit of the public using the river.

[4] Act 33 Geo. II, c. 2, PR. By this Act the canal was constructed to Longford Bridge.

In both these cases the Legislature carefully guarded against any infringement of public or private rights, or trespassing upon private property.

Having now secured Parliamentary sanction for his canal, the Duke employed as his engineer James Brindley, a man with no learning but with abundant resourcefulness in overcoming difficulties[1]. Brindley surveyed the route, and was assured that a perfectly level canal could be constructed for the more than ten miles between Worsley Mill and Manchester. In order to make this, valleys would have to be filled up, hillsides cut down, the earthworks made impervious to water by means of clay puddling, and the soil of the canal so drained that it would be able to withstand the effects of heavy land floods. Engineers attempted to turn Brindley from the formation of what they regarded as a "castle in the air;" but he was convinced of its practicability, and nothing could turn him from his course until the work was all completed.

The engineering difficulties that Brindley surmounted in this vast undertaking, it is not our purpose to describe[2]; but it may be well merely to refer to one part of that work, the Barton aqueduct, which carried the canal over the river Irwell at a height of thirty-nine feet above the water in the river. There was no other such wonder in England at that time; and it was a great triumph to make it possible for vessels to sail along the river, while, at the same time, in the canal, other vessels were crossing the river, so high above the latter. The first boat-load of coal was towed by horses over the Barton viaduct to Manchester on July 17, 1761; and from that time a regular supply was obtained at greatly reduced price[3].

Although an abundant supply of coal had now been opened up to Manchester, the carriage of raw materials for manufacture from Liverpool to Manchester was still as much impeded as before. There were two means of communication between these two cities: one by the road and the other by the rivers Irwell and Mersey. On account of the bad road, the cost of conveyance by that means was not less than £2 per ton; and this was practically prohibitive[4]. By the river also the conveyance of goods was tedious and difficult, for the upper part of the navigation was full of fords and shallows, and Liverpool

---

[1] The history of this man, which forms a romantic chapter in the development of inland navigation, is given to us in Smiles, *Lives of the Engineers*, i, Pt. v.

[2] These are well given in *The History of Inland Navigations*, i, pp. 36–55, where full details are included; Arthur Young, *Six Months' Tour in the North of England*, iii, pp. 251–91; *Gentleman's Magazine*, xxxvi, pp. 31–33; also Smiles, *Lives of the Engineers*, i, Pt. v.

[3] *Annual Register*, iv, p. 123.

[4] Brit. Mus. B. 504 (4), 'Advantages of Inland Navigations,' p. 18.

could be reached by Manchester vessels only by the assistance of the spring tides.

These circumstances were wholly inconsistent with the rapid commercial and industrial growth of this district and, in order to improve the transportation facilities, the Duke of Bridgewater, after completing the canal from Worsley to Manchester, decided to continue this canal from Manchester to Runcorn, at the mouth of the shallow part of the Mersey, at which place he could easily connect with Liverpool. This would furnish a navigable waterway that would be entirely independent of winds and tide and would permit of regularity of sailings. Brindley was immediately engaged on the survey and found that a level canal could be put through from Longford Bridge, near Manchester, to Hempstones, near Runcorn, where they would have to lock down into the tideway of the Mersey. The Duke then applied for statutory authority to extend his canal according to this scheme and in 1761 this permission was given by Parliament[1].

Many objections were urged from different quarters against the construction of this canal. It was said that the landowners would suffer by having their lands cut through and separated, and that a great number of acres would be covered with water and thereby lost to the public. Some thought that there was no necessity for this new waterway because the river navigation was sufficient to answer all the public needs, and that to make a new navigation parallel with and often close to the old would be no advantage to the public. Others regarded it unfavourably because the water needed for the canal would be taken out of the river, which would prejudice that navigation; and because the proprietors of the old navigation had spent large sums of money thereon, their property ought not to be taken from them without full compensation[2]. But probably the strongest opposition came from the proprietors of the Mersey and Irwell Navigation, who saw that if this canal should be constructed their monopoly would be at an end. At first they tried to buy off the Duke by offering him certain concessions[3]; then, as a conciliatory policy, they offered him some exclusive advantages of their navigation. But the Duke was determined to go on with his canal; it would be shorter, because

[1] Act 2 Geo. III, c. 11, PR.

[2] These objections, and the answers that were made to them by those who favoured the canal, show the early attitude towards canals. They are taken from *The History of Inland Navigations, particularly those of the Duke of Bridgewater in Lancashire and Cheshire*, pp. 24–31.

[3] Brit. Mus. 08,235. f. 77, 'Observations on the General Comparative Merits of Inland Communication by Navigations or Railroads,' pp. 6–7.

straighter, than the old navigation; it would not be subject to interruptions by floods or by droughts; and it offered the public a much lower freight rate. The proprietors of the Mersey and Irwell Navigation had for years carried on business with a very high hand, had extorted the highest rates, and in cases of loss or damage of goods in transit had refused all redress. Thus there was a strong feeling in favour of the canal and it was begun immediately after the Act was obtained.

We cannot here enter into details regarding the various aspects of the work, and of the financial expedients to which the Duke had to resort in order to complete it[1]. In 1767, after the lapse of about five years from the time the work was begun, the entire level length of the new canal, from Longford Bridge to the upper part of Runcorn, was finished and opened for traffic. The locks requisite for passing between the level part of the canal and the Mersey river were not completed until some years later[2]; and by that time the receipts from the sale of coal and from the canal traffic enabled the Duke to complete them with comparatively little difficulty.

The success of the Bridgewater Canal was almost immediate. Besides handling freight, the Duke put on two packet boats for carrying passengers[3], and in the year ending Oct. 4, 1776, he cleared £950 by passenger traffic between Runcorn and Manchester and £12,500 by carrying goods between Manchester and Liverpool[4]. By 1792 the passenger traffic amounted to £1500 a year and the whole revenue from the freight traffic on the canal was £80,000 a year[5]. With such annual revenues from the canal, the total cost of which was £220,000[6], we can easily see what a paying investment it was for its owner. But it is not from this individual standpoint that the canal is most worthy of note; the new line of navigation galvanized into activity hitherto unknown both Manchester and Liverpool; and formed the starting-point for the whole system of inland water communication, by means of which the Industrial Revolution might be made effective.

---

[1] These are given in Smiles, *Lives of the Engineers*, i, Pt. v, Chaps. vi, vii.

[2] The fall to be overcome by locks was given in Brindley's note-book as 79 feet (v. Smiles, i, p. 378), although the *Annual Register*, xvi, p. 65, gives the fall as nearly 90 feet. The *Annual Register*, xix, p. 127, says that the canal was completed, so that vessels could pass to Liverpool, on Mar. 21, 1776.

[3] *Annual Register*, xvii, p. 145 (1774).

[4] Ibid., xix, p. 184 (1776).

[5] Brit. Mus., Add. MSS. 30,173, 'Journey made by William Phillipps from Broadway to Manchester and Liverpool, p. 14. The facts here given were confirmed by the Captain and others in one of the Duke's boats.

[6] *Annual Register*, xix, p. 184.

The works which resulted from the genius of Brindley and the enterprise of the Duke of Bridgewater attracted public attention throughout the kingdom and were followed by a succession of great canal schemes. The project of connecting the east and west coasts by making a canal from the Mersey to the Trent was much discussed during these years; and among those whose interest in this work was greatest, Earl Gower, brother-in-law of the Duke of Bridgewater, stands out prominently. At his request, Brindley made a survey of the route in 1758, but as canals at that time were still untried in England such a vast enterprise was not to be entered upon hastily. In 1759 Brindley proceeded with his survey, but when he began the Bridgewater Canal this larger scheme was allowed to rest.

But the interest in this work did not flag, for the salt manufacturers of Cheshire and the pottery manufacturers of Staffordshire, all of whom were greatly hampered in the development of their business by the lack of adequate facilities for conveyance[1], actively set to work to secure this much-needed navigation. Generally speaking, the manufacturing classes and the commercial community supported this measure, while many of the landed interests and some others opposed it[2]. By the co-operation of Earl Gower, Josiah Wedgwood and his business partner Thomas Bentley with the leading landowners of the district, an account of the intended navigation from Liverpool to Hull was published, showing the advantages which would accrue from its completion[3]; and this served to evoke a storm centre around which

---

[1] Meteyard, *Life of Wedgwood*, I, p. 275. The pottery industry, which was so greatly advanced through the skill and enterprise of Josiah Wedgwood, was dependent on materials brought from a great distance: flint stones from the southern parts of England and clay from Devon and Cornwall. The flints came by sea to Hull and thence by boats. up the Trent to Willington. The clay was brought partly to Liverpool, whence it came up the Weaver river to Winsford, and partly to Bristol, Bewdley and Bridgnorth. Both clay and flints had then to be carried overland, mostly by pack-horses, and this was very expensive. Manufactured goods had to be shipped along the same routes, subject to heavy cost and at great risk of breaking and pilferage. So the expansion of the industry was prevented.

Salt, manufactured in Cheshire, was carried by pack-horse into adjoining counties, and the cost two or three counties away was almost prohibitive of its use. Its cost before shipment was also increased by the necessity of using coal from Staffordshire for boiling it down, and this was brought in, chiefly by pack-horse, at great expense.

Other industries suffered in like manner and were retarded by this same cause.

[2] *J., H. of C.*, xxx, pp. 520, 613, 627, 638, 643, 649, 683, 707, 708, 720.

[3] This is given in full in *The History of Inland Navigations*, I, pp. 55–77 (published in London, 1767).

the contest was waged[1]. A long controversy was carried on[2], until finally, prominent men, by concerted action, adjusted their interests

---

[1] *The History of Inland Navigations*, Pt. II, pp. 33–67, gives the arguments against such a canal, and pp. 67–72 give the 'Case in Behalf of the Bill for making a Navigable Cut or Canal from the Trent to the Mersey.' Part II of this pamphlet gives a good idea of the wordy contest going on at that time regarding this canal.

Meteyard, *Life of Wedgwood*, I, pp. 345–55, 385–6, 406–37, describes very fully the opposition that Wedgwood and his associates encountered in trying to obtain an Act of Parliament for the Grand Trunk Canal. Ibid., I, pp. 448–56, 497–502, and II, pp. 239–40, 246–50, give the later history of the navigation.

[2] One writer of the time was very anxious that such a great enterprise should be undertaken by the Government and not be entrusted to those whose interest in it was the pecuniary advantage that they could reap. Brit. Mus. 218. i. 5 (96), 'Navigation,' pp. 2, 3.

There were some who favoured this project of connecting the east and west seas, but who wanted the canal to begin at Burton-on-Trent, instead of Wilden Ferry, and to terminate at Northwich, on the river Weaver, whence there could be cheap river conveyance to Liverpool and Manchester. They declared that there was no just reason for extending the proposed canal to Wilden Ferry (16 miles farther down the Trent below Burton), nor to the river Mersey (14 miles below Northwich). It was believed that the Duke of Bridgewater was behind the project in getting it brought, independently of the river Weaver, to Preston Brook, so that, by a slight diversion of his canal, he could connect it with the proposed Trent and Mersey Canal at Preston Brook. Then when this junction had been effected, and the Duke had constructed that portion of the canal from Preston Brook to the Mersey, he would have control of a large traffic over both the canals. It was said that, at the other end of the route, the navigation of the Trent river from Wilden to Burton was good, and that the extension of the canal here was like that to the Mersey, a needless expense. But those who pushed these claims most vigorously were men who were financially interested in the old navigations of the Trent (at Burton) and the Weaver. (*Seasonable Considerations on a Navigable Canal from River Trent to River Mersey*, pp. 1–38.)

On the other hand, the promoters of the canal put forth 'Facts and Reasons tending to shew that the Proposed Canal from the Trent to the Mersey ought not to terminate at Northwich and Burton' (Brit. Mus. B. 263 (4)). They said that since this was to be a great trunk canal, with lateral branches, a great part of the kingdom would receive the advantage of the new conveyance; and therefore its termini ought to be such as would facilitate commerce in the best manner. For this reason they desired to end in a free part of the river Trent at one end, and in the Mersey at the other end, so that there would be no interruption at neap tides. They urged that the utility of this canal would be greatly limited by ending at Burton and Northwich. By a junction of the Grand Trunk with the Duke's canal at Preston Brook, a better, cheaper and more speedy communication could be made with Manchester; the trans-shipment at Northwich, with its delay and expense, would be avoided; and the accidents of wind and tide would be robbed of their influence. As the other end of the proposed canal, Wilden was chosen in preference to Burton, because the river navigation between Wilden and Burton was not sufficiently improved and was in the control of lessees who could interrupt the trade on the river at any time and could charge monopoly prices for the privilege of using the navigation (ibid., pp. 9–12).

and secured an Act of Parliament to effect the desired object[1]. The Duke of Bridgewater had petitioned for authority to join his canal at Preston Brook with the proposed Trent and Mersey Canal[2]; and according to agreement between these two canal interests, the Act provided that the Duke of Bridgewater might take part of their line from Preston Brook and thus carry his navigable canal into the river Mersey below Runcorn Gap. The six miles from Preston Brook to the Mersey, therefore, belonged to the Duke of Bridgewater; and this arrangement was intended to facilitate the interchange of commodities from one canal to the other.

This canal, which by Brindley was called the Grand Trunk[3], and which is also known as the Trent and Mersey, runs from the Bridgewater Canal at Preston Brook, through the salt-manufacturing districts of Cheshire, to its highest point at Harecastle hill, which it pierces by a tunnel; thence it passes southward through the pottery district of Staffordshire, and from there through Derbyshire to join the Trent river at Wilden Ferry. From this point, there was continuous navigation on the Trent to the Humber, so that by this interior waterway Hull was connected with Liverpool. The uniting of these two ports on the east and west seas attracted much attention, and to secure this advantage the town of Liverpool voted two hundred pounds toward the project[4].

It is hard for us to realize the great benefits which this canal conferred upon the districts through which it passed. So great was the change in the neighbourhood of the Potteries that an English clergyman, writing in 1781, was able to say: "How is the whole face of this country changed in about twenty years! Since which, inhabitants have continually flowed in from every side. Hence the wilderness is literally become a fruitful field. Houses, villages, towns have sprung up; and the country is not more improved than the people[5]." For the first time coal was abundantly supplied by water

---

[1] Act 6 Geo. III, c. 96. This was an Act for making a navigable cut from the river Trent, at or near Wilden Ferry, to the river Mersey, at or near Runcorn Gap. It authorized the proprietors to raise £180,000 (650 shares of £200 each) for carrying out this work and no person could take more than twenty shares. But subsequent Acts enabled them to raise a further sum of £194,250 on a mortgage of the tolls. These Acts are given in full in Cary, *Inland Navigation*, pp. 25–26; also Brit. Mus. 1246. l. 16 (8).

[2] *J., H. of C.*, xxx, p. 649.

[3] So called because he expected many lateral canals would run off from it.

[4] Picton, *Liverpool Municipal Records*, ii, p. 244.

[5] Ward, *History of Stoke-upon-Trent*, p. 33, quoting from John Wesley's Letters.

carriage and upon such reasonable terms as to be within the reach of all; a plentiful food supply could now be obtained for the increasing population without having to pay the extortionate prices hitherto demanded by the monopolizers of corn; and thus both social amelioration and industrial efficiency were the results[1].

When the midlands had thus obtained access by waterway to the ports of Liverpool and Hull, the next objective point with which connexion was sought was Bristol, the purpose being to unite the navigations of the three rivers on which these three ports are situated. Accordingly, an Act was obtained sanctioning the construction of the Staffordshire and Worcestershire Canal, from the Grand Trunk Canal at Haywood, in Staffordshire, to the river Severn, near Bewdley. This canal was completed in 1772, but instead of joining the Severn at Bewdley it was made to terminate at Stourport[2]. In reality, the main line of the Grand Trunk was not finished until five years after this canal was in operation; but navigation was open on part of the Grand Trunk for some years before the whole line was finally completed by the construction of the Harecastle tunnel[3]. From the history of the Wolverhampton (or Staffordshire and Worcestershire) Canal, it

---

[1] Pennant, in his *Journey from Chester to London*, in 1782, describes the change in the following words:

"Notwithstanding the clamors which have been raised against this undertaking (i.e., the Grand Trunk Canal) in the places through which it was intended to pass, when it was first projected, we have the pleasure now to see content reign universally on its banks, and plenty attend its progress. The cottage, instead of being half-covered with miserable thatch, is now secured with a substantial covering of tiles or slates, brought from the distant hills of Wales or Cumberland. The fields, which before were barren, are now drained, and, by the assistance of manure, conveyed on the canal toll-free, are cloathed with a beautiful verdure. Places which rarely knew the use of coal, are plentifully supplied with that essential article upon reasonable terms; and, what is of still greater public utility, the monopolizers of corn are prevented from exercising their infamous trade; for, the communication being opened between Liverpool, Bristol, and Hull, and the line of canal being through countries abundant in grain, it affords a conveyance of corn unknown to past ages. At present, nothing but a general dearth can create a scarcity in any part adjacent to this extensive work."

[2] The canal cost £200,000 (Brit. Mus. B. 504 (2), p. 24). When Brindley projected this canal, he proposed that it should enter the Severn river at Bewdley, because that town was flourishing and opulent. But the inhabitants of that place held a meeting and decidedly rejected the boon intended for them. The canal was therefore altered as to its direction so as to enter the river at Stourport, which was then composed of only a single cottage. The consequence was that Bewdley greatly declined and Stourport became a flourishing town. Many other instances could be mentioned to show similar results. (Aris's *Birmingham Gazette*, Mar. 28, 1825, p. 1.)

[3] *Annual Register*, xvi, p. 97. The Grand Trunk was not completed till 1777.

would appear that it early aroused the envy of other canal projectors in that section, probably because of its favourable connexion with the shipping of the Severn[1].

When Liverpool, Hull and Bristol had thus been linked up by inland waterways, the next place to be reached by the canal network was London. In 1767 a Bill was brought in for the construction of a canal from Coventry, to join a branch of the Grand Trunk Canal near Lichfield[2], and in 1768 the Act was passed empowering the company to make a canal from Coventry to Fradley Heath on the Trent and Mersey Canal then partly completed[3]. In the following year (1769) an Act was passed for a canal to connect with the Coventry Canal near that city and pass through Rugby and Banbury to the city of Oxford[4]. These two canals, the Coventry and the Oxford, would

---

[1] So much was this the case that another canal, the Birmingham and Worcester, was later constructed to take advantage of connexion with this river.

In 1786 another canal was intended to be made, evidently as a rival to the Staffordshire and Worcestershire. Its projectors tried to show the impropriety of improving the Severn from Worcester to Stourport, where connexion was made with the Staffordshire and Worcestershire Canal. But their real object seems to have been to open up a new canal, so as to be able more successfully to put on the markets the coal and other produce of their estates. Their petition for an Act was unsuccessful, however (v. Brit. Mus. B. 504 (2), 'Address to the Public on the New Intended Canal from Stourbridge to Worcester').

[2] *J., H. of C.*, Nov. 26 and 27, 1767.

[3] Brit. Mus., Maps 88. d. 13, No. 6, 'Case of the Coventry Canal Company.'

[4] Blomfield, *History of Bicester*, VI, p. 26; Dutens, *Mémoires*, p. 7. There was much opposition to the Oxford Canal, because it furnished a valuable connexion between London and the great midland section. As soon as the project was brought forward, the people of Yarmouth sent a letter to Parliament opposing the canal, for the reason that, if it were constructed, coal would be brought to London from the midlands to such an extent as to cause a reduction of the number of ships required along the east coast for carrying coal from Newcastle to London. They went into this very fully, showing the great evil of such a possibility. They evidently regarded their coasting trade as a strong claimant upon the public favour, for they said that the injury of the coasting trade "will, we apprehend, be deemed by Parliament a very considerable objection to the passing the Bill" (Brit. Mus. 214. i. 4 (119), 'Letter from Yarmouth regarding the Canal from Coventry to Oxford now depending in Parliament,' pp. 1–3). They tried to show also that the injury would extend further to the whole coasting trade between London and Liverpool. This diminution of the number of vessels or seamen was to them a calamity, which they sought to avert.

On the contrary, those who were desirous of having the canal constructed made light of any great competition of inland carriage of coal with that by sea; and endeavoured to assure the opponents that, instead of a decrease of vessels and seamen, there would be increased employment for them because of the greater amount of exportation from and importation into London, when the latter's trade should be greatly increased by better traffic connexion with the manufacturing and agricultural sections of the midlands. The value of adjacent lands would be

then provide a waterway from the Grand Trunk Canal to the Thames at Oxford and thence there would be river connexion with London. Owing to many financial difficulties, however, this midland system of canals was not opened to Banbury until 1778 and was not fully completed until 1790[1].

While these great arteries were in process of construction, the two great manufacturing counties in the north, Lancashire and Yorkshire,

enhanced, the number of horses employed in land carriage could then be employed productively on the land and so the country would derive untold benefits by the construction of this proposed canal (Brit. Mus. 214. i. 4 (103), 'Observations on the Effects of the intended Oxford Canal Navigation,' pp. 1–2).

The Report of the Parliamentary Committee said that "The only ground upon which this Court could...decently pretend to oppose the Bill in question is, the general objection to all inland navigations: viz. that they lessen the coasting trade, and consequently the numbers of seamen." But by showing that the cost of carrying by sea was so much less than by land, the Committee seem to have set at rest the minds of the people of Yarmouth, and the Bill passed (Brit. Mus. 214. i. 4 (120), 'Report of Committee on Oxford Canal,' pp. 1–4).

When this canal was proposed, but before it had been sanctioned by Parliament, a pamphlet was published suggesting the desirability of suspending the further construction of canals until it would be seen whether their advantages to the public were greater than the injuries due to the loss of so many hundred acres of land; and also whether, supposing the advantages greater than the injuries, it would not be better, on account of the high price of labour, to allow the completion of some of those under way before taking up a lot of new construction. These suggestions, however, did not prevail (Brit. Mus. 214. i. 4 (104), 'Queries on the Intended Canal from Coventry to Oxford').

[1] Dutens, *Mémoires*, p. 7. In Brit. Mus., Maps 88. d. 18, No. 6, 'Case of the Coventry Canal Company,' we have the inner history of this company, showing some of its difficulties. Brindley's estimate of the cost of the canal was £50,000. By 1772 this was all spent, but the canal was completed only from Coventry to Atherstone. During the next ten years, 1772–82, nothing was done, for the company could not get money. In the summer of 1782, with a view to the completion of the canal, there was a meeting of delegates of the Coventry, Oxford and Grand Trunk Canal Companies and of certain promoters of a canal from the coal mines near Wednesbury to Fazeley, intended to form a junction with the Coventry Canal. As a result of this meeting, the Coventry Canal Company gave up five and one-half miles of their line of canal, between Fazeley and Fradley Heath, to the Birmingham and Fazeley Canal Company. In 1786 the Coventry Canal Company obtained an Act to enable them to borrow £40,000 on the credit of their tolls, to complete the unfinished part of their canal; and with this the canal was finished in the midsummer of 1790. They were induced to engage in this additional expenditure not only to form the junction with the Trent and Mersey Canal at Fradley Heath, but also to connect with the Birmingham and Fazeley Canal at Fazeley, by which access would be secured to the Staffordshire collieries in order to supply with coal the different wharfs on the Coventry and Oxford Canals. These connexions considerably increased the amount of the trade and revenue of the canal.

As to the Oxford Canal and its difficulties, see Jackson's *Oxford Journal*, Nov. 8, 1781, p. 3, and Mar. 25, 1786, p. 3.

and the manufacturing district of the north midlands, Staffordshire and Warwickshire, were having a series of canals put through to facilitate their access to sources of raw materials and food products, as well as to provide more ready means of shipping their manufactured products to the ports and coast towns. In the north midlands Birmingham was becoming the Kremlin from which canals radiated in all directions, as, for instance, the Birmingham Canal, from Birmingham to Bilston and Autherley[1]; the Birmingham and Fazeley, to connect these two places[2]; the Birmingham and Warwick, joining these two towns[3]; the Warwick and Napton, from Warwick to connect with the Oxford Canal near Napton[4]; the Birmingham and Worcester, uniting Birmingham with the Severn river near Worcester[5]; and others which we need not mention. In Lancashire and Yorkshire we have the Leeds and Liverpool Canal, going by a circuitous route from Liverpool, by way of Wigan, Blackburn, and Skipton, to the city of Leeds[6]; the Liverpool and Wigan, which was opened in 1774; the

[1] Act was passed in 8 Geo. III—given in Cary, *Inland Navigation*, pp. 85–88.

[2] Acts passed in 11 Geo. III (1771) and 23 Geo. III (1783). See Cary, *Inland Navigation*, pp. 89–41, where these Acts are given.

[3] Act passed in 1798.   Authorized in 1794 (1793–4).

[5] Sanctioned by statute in 1791. See also *J., H. of C.*, Feb. 18, 1791. The Birmingham and Worcester Canal was projected to connect Birmingham with the Severn at deep water. It was undertaken almost in defiance of opposition, expense and difficulties of construction. They had to solicit their Act through two or three sessions of Parliament; they had much difficulty in securing funds for the work of construction; untoward circumstances and lack of cash caused several cessations in the work, and it was still uncompleted in 1818 (Pitt, *Agriculture of Worcester*, p. 271). After its opening, it was still unfortunate: it was too shallow for the most effective service; it suffered seriously from lack of sufficient water supply; and its record of operation is one of failure and loss. In this respect it differs entirely from the Staffordshire and Worcestershire Canal, which, on account of fortunate connexions at both ends and a favourable situation, has been a profitable undertaking (ibid., p. 270).

[6] The Act was obtained in 1770, but the canal was not opened until June 4, 1777, and then only in part. Its length was somewhat more than 108 miles (*Gent. Mag.*, XLII, p. 8). The first Act incorporated the proprietors, with power to raise £320,000 in shares. The work was carried on until all the money was spent. By the subsequent Acts of 1788 and 1790 the Company was enabled to vary the line and raise further amounts of money. The summit of the canal is near Colne, where it was made to pass through a tunnel 1630 yards long, 18 feet high, and 17 feet wide. The fall eastward from the summit to the river Aire at Leeds was 409½ feet, which was accomplished by 44 locks; the fall westward to Liverpool was 431 feet, which was effected by 47 locks, but the Liverpool basin was over 50 feet above the river Mersey at low water (Cary, *Inland Navigation*, pp. 18–20). Liverpool had formerly given £200 toward this intended canal; and in 1768 it voted £50 more toward a re-survey of the route, showing that this city expected to reap considerable advantage from the canal (Picton, *Liverpool Municipal Records*, II, p. 244).

Huddersfield Canal, constructed by Sir John Ramsden, who was the sole proprietor of the town of Huddersfield, from that town to Cooper's Bridge, where it joined with the Calder Navigation[1]; the Manchester and Bolton, authorized in 1791; the Kendal and Lancaster, and the Manchester and Ashton-under-Lyne, authorized in 1792; the Rochdale, connecting the Calder Navigation with Manchester[2]; the Barnsley and the Stainforth and Keadby Canals, authorized in 1793; and the important improvements that were made in the Aire and Calder and connecting navigations[3].

Another feature of the history of navigations in the midlands at this time was the beginning of the process of consolidation among the canals of that district. The Birmingham Canal and the Birmingham and Fazeley Canal communicated with each other at Birmingham; nearly all the shares of these canals were held by the same persons; and many inconveniences might be avoided if the two undertakings could be merged under the same control. Application was made to Parliament for an Act to unite the interests of these two concerns, and in 1784 statutory provision was made to effect this result[4]. Under this Act, these two companies were consolidated and the name was changed[5]; but none of the powers conferred by Parliament upon the proprietors of one canal were to be regarded as extending to the other canal. Another group of waterways the consolidation of which was sought was the chain extending between the Thames at the south and

[1] The Act was passed in 1774 (*Gent. Mag.*, XLIV, p. 200).

[2] Act was passed in 1794, but the canal was not opened till 1804. Its opening was of much importance, as a means whereby vessels could sail over the mountain ridge of England, and thus unite the North and Irish Seas (*Monthly Mag.*, XVIII, Pt. II, p. 556).

[3] *J., H. of C.*, XXVIII, pp. 188–44. "The general system of inland navigation" is described in Dupin, *Voyages dans la Grande-Bretagne*, I, Pt. III, pp. 159–67. Then, on pp. 167–84, he describes the canals that centre at Manchester; on pp. 184–206, those that centre around Liverpool; on pp. 206–24, those that centre around London; and on pp. 224–89, those that centre around Birmingham, Bristol and Hull. Dutens, *Mémoires*, pp. 1–20, traces the formation of the canal network, and pp. 28–79 are taken up with the engineering features of their construction and operation; while pp. 91–101 give tables of the canals that were carried out, showing the dates of their Acts of Parliament and the length of each canal, also the many canals which were projected but never executed. There is also given a table of the principal rivers of England that have been rendered navigable and the length of their navigations.

[4] Act is printed in full in Cary, *Inland Navigation*, pp. 42–48; v. *J., H. of C.*, Feb. 18, 1784.

[5] The new Company was called "The Company of Proprietors of the Birmingham and Birmingham and Fazeley Canal Navigations;" but by Act of 1794 the name was changed to "The Proprietors of the Birmingham Canal Navigations," which is still retained.

the Birmingham and Grand Trunk navigations in the midlands. We have already mentioned the financial difficulties under which the Coventry Canal was lying prostrate between 1772 and 1782; and it would appear that the Oxford Canal was burdened in like manner. For many years these two concerns were distracted by petty prejudices and animosities, until they had been brought nearly to the brink of ruin; and instead of working amicably together for the public good so as to furnish the utmost benefit in the facilities of conveyance, their relations were such as to impede the flow of traffic. At different times they had produced plans of undertakings that were calculated to improve the trade of one or both the canals, but these had always proved abortive. To the north and west there were the Birmingham canals, present and prospective, connecting with the Staffordshire collieries, upon which the traffic in coal would produce large returns, and the Grand Trunk Canal, which passed through the manufacturing districts of Staffordshire and Cheshire and connected with the rapidly expanding industrial section of Lancashire. If the Coventry and Oxford Canals could lay aside their discordant attitude toward each other, and could secure connexion with these more powerful companies so as to command a share of their trade, the reasonable expectations of profit for themselves and service for the public might be realized. In 1781 they were urged to desist from their previous hostility and adopt a more enlightened policy in the interests of their own and the public welfare[1]; but the suggestions made at that time were not immediately carried out. The possibility of some such union did, however, occupy the attention of those who were most directly interested; and in 1785, for administrative purposes, the Grand Trunk, the Coventry and the Oxford Canal were consolidated. The junction between the Birmingham and the Coventry Canal was completed in 1790, after which boats prepared to pass twice a week between London and Birmingham[2]. These mergers of canal companies were allowed only upon a very strict basis, much more strict than in the case of railways when they came to the front. In reality we find very few canal mergers, probably because the companies were jealous of each other's success and each was anxious to maintain its own advantage[3].

---

[1] Jackson's *Oxford Journal*, Nov. 3, 1781, p. 3, letter from "Publicola." This letter well shows the jealousy and dissension that prevailed among the proprietors of these canals.   [2] *Annual Register*, xxxii, p. 210.

[3] The only other cases of canal consolidation before 1830 that have come under notice are: In 1813 the Chester and the Ellesmere Canal were united under one control (*J., H. of C.*, Dec. 7, 1812, and Feb. 25, 1813); in 1819 the Grand Junction and the Regent's were united; and in 1821 the North Wilts Canal (from Swindon to Latton) and the Berks and Wilts Canal were consolidated. In 1793 a deputation

Even working agreements for passing vessels from one canal to another were very seldom entered into.

Before the framework of the inland navigation system could be completed the two great rivers at the south, the Severn and the Thames, must also be united. This was a project which had been considered as early as the reign of Charles II and in that reign a Bill was brought into the House of Commons to connect these rivers by a cut from Lechlade on the Thames to Bath on the Avon. Andrew Yarranton proposed a similar policy, namely, to unite the Thames, by its tributary the Cherwell, with the Avon, by its tributary the Stour[1]. But nothing came of either of these plans[2]. About 1775 the people along the Stroud valley, labouring under the disadvantage of a high price for coal, allied their interests in making the Stroudwater Canal from the Severn up to the town of Stroud, by means of which there was effected an annual saving of £5000 in coal. With such a benefit from a canal only eight miles long, it was easily conceivable that there would be a much greater benefit from uniting the Thames and Severn; in fact, this was deemed a probable consequence of the Stroudwater navigation when that was undertaken, for by this extension there would be formed the most favourable line of communication between these two rivers[3]. In 1781, after several meetings of Gloucestershire citizens had been held to promote this design, a subscription was made to carry forward the undertaking under the direction of a committee. The surveys made by authority of this committee were favourable to this course, in preference to another route that was being supported by way of Tewkesbury, Cheltenham and Lechlade, not only because there would be fewer difficulties of construction, but also because there would be a greater volume of traffic, and so the tide of public opinion set in favour of this southern route. The proprietors of the Stroud Navigation gave assurance that, in case this waterway were made to connect with theirs, their tonnage rates would be reduced and made satisfactory to

of the proprietors of the Basingstoke Canal attended the meeting of the proprietors of the Kennet and Avon Canal Company and proposed the junction of these two canals. It would appear that the representatives of the Kennet and Avon Canal consented to the proposed junction of their canal with the Basingstoke Canal; but they would not agree to unite in a petition to Parliament for legislative authority to effect this junction. Consequently, we have no record that the two canals were ever united (vide *Reading Mercury*, Jan. 20, 1794, p. 4, letter from "X.Y.", and Feb. 10, 1794, p. 4, letter from "A Friend to Trade.").

[1] Yarranton, *England's Improvement by Sea and Land*, p. 64.

[2] For a 'Historical Account of the Thames and Severn Canal,' see *Gentleman's Magazine*, LX, Pt. i, pp. 389–92.

[3] Brit. Mus. 8775. f. 20, 'Considerations on the Idea of Uniting the Rivers Thames and Severn,' pp. 2–8.

the undertakers of the proposed junction[1]. The coal miners of
Newcastle opposed this canal, on the plea that since coal would be
brought to London from the West, and the Newcastle coal trade would
consequently decline, therefore the nursery for seamen for the navy
would be largely at an end. This objection, which was also advanced,
as we have seen, against the Oxford Canal project, was shown to be
of little weight; for since the coasters could carry a much larger
tonnage at a lower rate than could be conveyed on inland navigations
the amount of this competition would be comparatively insignificant[2].
At the same time, the benefits to be derived by the far inland counties
along the proposed navigation would be important: fuel would be
more easily and cheaply obtained; there would be between counties
the exchange of surplus products which could not stand the expense
of land carriage; and, because of the lower cost of carriage each way,
the products of agriculture would find a market in London in exchange
for manufactures. In the year 1782, upon a survey and report by
engineer Whitworth, and at the risk of several persons, especially some
wealthy merchants of London, the project was undertaken; and the
Bill was introduced into Parliament, which passed into a law in 1783[3].
The undertaking received so much support that the Act was readily
obtained for making the canal from Lechlade, on the Thames, to connect
with the Stroudwater Canal at Wallbridge and thus to communicate
with the Severn at Framilode. This distance of over thirty miles was
completed in 1789 and barges laden with coal at once began to pass
from the Severn, by way of the Thames and Severn Canal, to London[4].
A rather unusual circumstance in connexion with this canal was that
the landowners in general through whose lands the canal was designed
to pass favoured the plan[5], whereas in many other instances the
landowners were vigorous opponents of such artificial waterways.
The anticipated advantages from the Thames and Severn Canal were,
however, not fully realized; on the cost of over £250,000 no dividends
were paid for some years[6], and at no time did the returns on the capital
expended warrant the outlay that had been made. The canal had been
constructed with the understanding that the Commissioners of the

[1] Brit. Mus. 8775. f. 20, p. 9.   [2] Ibid., pp. 13–15.
[3] Act 23 Geo. III, c. 38.
[4] *Gentleman's Magazine*, LIX, Pt. II, p. 1189. For description of this canal, and
of the engineering difficulties connected with its construction, see *Annual Register*,
XXXI, p. 228; also *Gentleman's Magazine*, LIX, Pt. II, p. 1189, and LX, Pt. I, pp. 109–
10.
[5] Jackson's *Oxford Journal*, Feb. 8, 1788, p. 3.
[6] *Parl. Papers*, 1793, XIII, 'Miscellaneous Reports No. 109 on Thames and Isis
Navigation,' p. 31.

Thames Navigation would have that navigation completely improved by the time the canal would be opened[1]; but the Thames Commissioners, although repeatedly implored to carry out their work, refused to do so until the canal should begin pouring its traffic upon the river[2]. This neglect of the Thames Navigation could have only one result, namely, to act as a barrier to the promotion of trade on the canal.

A rival route, from the Severn tideway to London, was soon started, by the construction of the Kennet and Avon Canal. When this had been put through, the line from the Severn lay along the Avon river, past Bristol and Bath, to near Winsley; then by the Kennet and Avon Canal and Kennet river navigation to Reading; thence down the Thames river to London[3].

But this route had scarcely been well begun when a petition was sent to Parliament asking for authority to construct a canal from Abingdon, to join with the Thames and Severn Canal, so as to avoid the many impediments in the upper part of the Thames and to shorten the course of the navigation[4]. It was not until 1795, however, that this was carried into effect by the passing of an Act sanctioning the construction of the Berks and Wilts Canal, from Abingdon, on the Thames, to join the Kennet and Avon Canal near Semington, and of a branch from Swindon, to connect with the Thames and Severn Canal near Latton[5]. This third through route, therefore, passed from the Severn, up the Avon river to Winsley, thence along the Kennet and Avon Canal to Semington, thence by the Wilts and Berks Canal to connect at Abingdon with the Thames, along which the rest of the passage to London was taken.

[1] *Parl. Papers*, 1793, XIII, p. 30.

[2] Ibid., p. 31. See also *St James Chronicle*, April 7–10, 1792, p. 2, letter from "A Preconsiderator," who declared that sensible men should never have embarked money in this undertaking until the Thames "had been reformed upon a regular system without locks." His principle was that the canal should have been sunk upon a horizontal level with the beds of the two rivers. But the impracticability of this is evident from the fact that the upper level of the Thames was not at all the same as the level of the Stroud or Severn river.

[3] Brit. Mus. 8235. cc. 41 (1), 'An Authentic Description of the Kennet and Avon Canal,' gives full details of this canal, including the dimensions, the size of locks, the rise and fall, the embankments and aqueducts and tunnel on the canal, the difficulties surmounted in the execution of the work, etc.

[4] *J., H. of C.*, XL, p. 592. Interesting arguments for and against such a canal are found in *J., H. of C.*, XL, pp. 751, 785, 825–8. The relative advantages of the Wilts and Berks Canal and the Kennet and Avon Canal are presented in the *Monthly Magazine*, XXVIII, Pt. II, pp. 554–7.

[5] *J., H. of C.*, L, April 30, 1795.

We have now described the framework of the canal system of England; we have seen how the interior counties obtained navigable connexion with the four great ports on the east and west coasts, through connexion with the Thames, Severn, Humber and Mersey rivers; how several navigable communications were made between Hull and Liverpool, through the counties of York and Lancaster which were then becoming great seats of industry; and how, parallel to the latter, other connexions were made in the southern counties, between the Thames and Severn, thus joining England's two largest rivers and also her two greatest ports, London and Bristol. After these interior waterways had been constructed, by 1792, it became necessary to considerably improve the route between the midlands and the metropolis in order to accommodate the increasing traffic. Early in 1792 the Marquis of Buckingham caused a survey to be made of the country between the Oxford Canal at Braunston and the river Thames near London, to find the shortest and best line. Since the trade of the kingdom required faster conveyance, he wished to avoid the circuitous route by Oxford and the uncertainty of the Thames Navigation. The report of the engineer whom he employed to make the survey was acceptable and a petition was sent to Parliament asking for authority to enable this scheme to be carried out. But there was also another party, composed of men equally respectable and equally zealous for the public good, who wanted the canal to join the Oxford Canal at Hampton Gay, near the city of Oxford, and proceed thence to London. Much activity was displayed by those interested in each canal and both enterprises were of acknowledged utility; but they could not both be constructed. A scheme was brought forward with the object of harmonizing these two interests, so as to prevent unnecessary waste of land and money and the useless multiplication of canals running side by side[1]; but in the end the Braunston terminal was fixed upon and in 1793 the first Act was passed giving authority for the building and equipment of the Grand Junction Canal between Braunston and Brentford, near London[2]. Although it was not finished till 1805, this canal has been one of the most important in England and has been among the most

---

[1] *St James Chronicle*, May 5–8, 1792, p. 2, letter from "Amicus" on "Canal Navigation."

[2] This Act is given in Cary, *Inland Navigation*, p. 85 et seq. Other Acts with reference to this canal were passed in 1795, 1796, 1798, 1801, 1803, 1805, 1810, 1811, 1812, 1818, 1819, 1826. The course of the canal may be seen on the map, running from Brentford, on the Thames, near London, past Uxbridge, Leighton Buzzard, Fenny Stratford, to join the Oxford Canal at Braunston.

successful in resisting the paralyzing influence of the railroads, as we shall show in a later chapter.

It would require too much space to enter into a full description of the whole network of canals, by means of which the industrial expansion was being greatly aided in all parts of the kingdom. On one of the appended maps the reader will be able to see the extent of the canals for which legislation was obtained; and, as a matter of fact, only very few of these failed to be constructed. By a comparison of this map with one showing the canals of the present day, we can see that the whole system was practically complete at the end of the period we are now considering.

But while we are here dealing chiefly with the canals, we must not omit to state that the improvement of some of the river navigations was also of importance during the canal-building era. In certain cases the canals could not be efficient agents of transportation without having the rivers with which they were connected put into good condition[1], and the building of the canals furnished the fitting occasion for the improvement of the river navigations. In other instances the depth of water and the capacity of vessel for which a navigable river would provide was not sufficient to meet the requirements of a constantly increasing trade, and these rivers would have to be improved to accommodate the larger vessels that were found necessary for the greater volume of traffic. Many of the river navigations, like most of the canal navigations, were not constantly improved to keep up with the demands of the increasing amount of business that was done upon them; but, on the contrary, their facilities were altogether inadequate for the economical conduct of traffic. Perhaps the only river navigations which have kept pace with the requirements are those of the rivers Weaver, and Aire and Calder[2]; but as no new features are manifested here, except those of engineering, it is unnecessary for us to follow thcm.

---

[1] This is well exemplified by the cases of the Thames and the Severn river. The Stafford and Worcester Canal, connecting with the Severn at Stourport, could not fulfil its purpose most successfully so long as the upper part of the river was full of shallows and had a swift current in a tortuous channel. Fortunately, the construction of the Gloucester and Berkeley Ship Canal, 1792–1827, straightened the navigation and shortened the distance. Its 16¼ miles were made all on one level and when full it was 70 to 90 feet wide and 18 feet deep. Its history is given in *Annual Register*, LXIX, p. 87 (1827). The Thames and Severn Canal and the Oxford Canal, as we have seen, could not render acceptable service until the upper and lower parts respectively of the Thames were improved.

[2] The improvements of the rivers Aire and Calder to accommodate an increased tonnage are given by the engineer of that navigation in Brit. Doc. 1883 (252), XIII, 1,

The history of the Thames Navigation, which we have in a former chapter traced down to 1750, is so unlike that of many other rivers that we shall endeavour to follow it further. In the latter half of the eighteenth century we find many complaints as to the unnavigable condition of this river. In 1770 there was a petition sent to Parliament by the Commissioners of the Navigation of the Thames and Isis and other prominent men in the counties bordering on these rivers, showing that from London to Cricklade the navigation was, at certain times of the year, impassable, because of the bad condition and construction of the locks and weirs on the rivers, and for want of a proper depth of water in many places. They were taxed sixpence per chaldron on all coal imported into the port of London, and yet got no benefit from this levy because of the impediments to the carriage of the coal inland; in consequence of which they urged that the channels of these rivers might be made navigable at all seasons, so that the cost of water carriage might be lessened, trade increased and many local benefits produced[1]. At the same time there was a petition before Parliament asking for permission to bring in a Bill to make a navigable canal from Reading to Monkey Island, which, if made, would permit of more regular voyages, in shorter time, and would give London cheaper food, while the country would get cheaper coal[2]. The Common

Q. 785 et seq. The improvements of the river Weaver are described in the same volume of these documents.

For a long report on the Calder Navigation, in the early part of the last half of the eighteenth century, see *J., H. of C.*, xxviii, pp. 183–44.

For previous petitions from the Aire and Calder for making improvements, see *J., H. of C.*, 1772, Dec. 9; 1774, Jan. 17, Feb. 22, Mar. 3, Mar. 21, Mar. 23. Originally the locks of this navigation were 60 feet × 15 feet × 3 feet 6 inches. Under the Act of 1776, the locks were made 66 feet × 15 feet × 5 feet. Under the Act of 1828, the locks were made 72 feet × 18 feet × 7 feet. Since 1860, the locks have been made 215 feet × 22 feet × 9 feet. These latter improvements were completed about the year 1886.

In 1828 the Undertakers of the Aire and Calder petitioned for authority to make a long series of cuts or canals, improvement of docks, harbours, locks, building of railways, etc.; and this permission was granted (*J., H. of C.*, Feb. 14, 1828).

[1] *J., H. of C.*, Dec. 10, 1770.

[2] *J., H. of C.*, Dec. 18, 1770; *Gentleman's Magazine*, xli, p. 56. In Brit. Mus. 215. i. 1 (105), 'Thames Navigation,' we have the objections made to this canal and the answers given to these objections, so that both sides of the subject are presented in full. The chief objections were: *first*, that this proposed canal, even if practicable, seemed to be calculated to benefit a few and prejudice many; *second*, that if the canal were built the Thames would be neglected, and in time its channel would be so choked up by diverting the water, that towns and mills that depended on it would be deprived of its facilities and conveniences; *third*, that some nearby villages would be in danger of being deluged and some of the rich land adjacent would be subjected to inundation in flood times, while above the intended canal the lands

Council of the city of London were so much interested in this work that, in 1770, they asked the celebrated Brindley to make a survey of this part of the Thames and report upon it. His report showed that long experience had proved the river to be impassable for barges during times of flood and of drought, which lasted several months of the year; and he urged that, since the expense of improving the river navigation by artificial works would be five or six times as great as that of making a navigable canal, and when done would be far from being as safe and rapid a conveyance, the construction of a canal would be more economical and more useful[1]. The parliamentary committee to whom the above petitions were referred ascertained that there was a real grievance; and that the delay and expense of navigating the river were a great detriment to its usefulness. But in opposition to the statement of Brindley and other eminent engineers, and probably because of the great influence wielded by the Thames Commissioners, in 1771 the improvement of the river was authorized[2]. Whether much was done to ameliorate the conditions along the river we can only conjecture; but by 1785 it was again represented to Parliament that, despite the Acts passed for that purpose, the improvement of the Thames between Abingdon and Lechlade had been very

would become rushy and swampy. The replies to these objections were: *first*, that the plan was not "calculated to benefit a few" because no private property was allowed, and it could not "prejudice many" because ample satisfaction would be made for every injury or damage sustained; *second*, that the more binding the obligation imposed on the projectors of this canal to keep the old channel in good repair, the more agreeable it would be to them, since the effectual repair of the river was the foundation of their proposal, while the mills and waterworks along the river would have a more regular and constant supply of water than before; *third*, that none of the adjacent land could be in greater danger of overflow because the canal would neither add to nor diminish the floods. The great reasons why such a canal was desired were, to make the navigation shorter, cheaper, regular and independent of the inconstancy of the amount of water in the river. But the plan was not carried out for it did not receive Parliamentary sanction.

[1] Brindley, ' To the Committee of the Common Council of the City of London.'

[2] The report of the committee is found in *J., H. of C.*, Feb. 11, 1771. It showed that in some places there was always a lack of water and in dry seasons these places were impassable. The largest barges had to wait from one to two days for water to carry them from one lock to another. Of the locks, some had been blown up, others were so constructed that vessels had to be drawn up with a cable, and through each of the eight locks between Reading and Monkey Island it took on an average three hours to drag a barge. Instead of using horses to tow the barges, the labour of men was employed, at more than double the expense; for the same work that required 70 men could be done by nine or ten horses. The report recommended the erection of twenty pound-locks upon the river, which would render it navigable for barges at all times, and the putting of the channel into proper condition so as to render navigation safer and more constant.

slight[1], and that there were many impediments in that part of the river.

These complaints continued during the following years with vociferous reiteration. Barges had to pay heavy tolls at the locks for the privilege of passing through, and their masters had to strongly supplicate and richly compensate for the flashes of water that enabled them to float their loads of freight over the shoals in the river. Even with this help, the barges sometimes lay stranded on the shallows of the river for a considerable time, thereby causing delay and loss to those who would naturally use this cheaper means of conveyance[2]. The tolls demanded by the owners and lessees of the artificial navigation works connected with the river were so excessive as to greatly hinder the development of trade[3]. Some were in favour of improving the navigation by deepening the bed of the river so that it would have a regular uniform section throughout its length[4]. By this means they would avoid the necessity for pound-locks and would thus keep the navigation open, free and unhampered by any transfluvial barriers. But the task of transforming a river of seventy-five or eighty feet fall into a navigation of the proposed uniform depth of ten to fifteen feet did not win the approbation of the most discerning engineers. About 1790 the Commissioners of the Navigation appointed engineers to survey the upper part of the river from Lechlade to Day's Lock and to report on its condition. Jessop, who had charge of this work, found the river in many parts very crooked and greatly obstructed with weeds. He complained that the depth of water was not sufficient to enable barges to carry their burden and showed that the only way in which a greater depth could be secured was by removing the shallows, by continuing the flashing, and by the erection of six pound-locks in this part of the river. These latter he would have placed where the greatest obstructions existed, so that their use would not be merely local, but would be beneficial throughout the whole length of the river. He showed that another great disadvantage of the navigation was the loss of time in waiting for the flashes; for although boats, by flashing alone, might pass these particular obstructions, yet they

---

[1] *J., H. of C.*, Mar. 7, 1785. Petition of the landowners, traders, manufacturers, etc., of Berkshire, in favour of the Wilts and Berks Canal.

[2] *Public Advertiser*, Oct. 11, 1786, p. 1, letter from "An Esteemer of Steam Engines;" ibid., Oct. 4, 1791, p. 1; Oct. 29, 1791, p. 3.

[3] Ibid., Nov. 21, 1789, p. 2, letter of "A Citizen." See also Brit. Mus. B. 508 (5), pp. 57–60, showing the amount of the tolls.

[4] *Public Advertiser*, May 29, 1786, p. 2, letter from "W. J.;" ibid., Nov. 21, 1789, p. 2, letter from "A Citizen;" ibid., Aug. 27, 1798, p. 1, letter from "A Thames Conservator."

could not pursue the water fast enough to get over other similar obstructions and hence must wait many days for repeated flashes. He also recommended necessary improvements in the towing paths, which were very narrow and inconvenient. By these changes he showed that a barge would often make five voyages with full lading, where under existing circumstances it made only one voyage with half a full load[1].

It would seem, however, that Jessop's suggestions received but scant consideration and it is almost certain that they were not put into effect, for in 1791 a petition was presented to the London Common Council by the owners and masters of barges that navigated the river from Lechlade to London, stating that, despite the large sums that had been spent to improve the river, the navigation was still so bad that trade thereby suffered great inconvenience, injury and loss. They urged that instead of laying out the large amount of money that would be necessary to permanently improve the navigation, the only easy and practicable means of remedying the inconveniences of the river was to construct a canal from Boulter's Lock, near Maidenhead, to Isleworth, so as to avoid the tedious navigation of the most difficult part of the river. Another petition, supplementing the foregoing, was sent by "the gentlemen, tradesmen and other inhabitants".of Lechlade, Abingdon, Reading and other towns along the river, praying that application might be made for authority to construct this canal[2]. It was shown, too, that the river below London needed to be improved; the channel had become too narrow because of embanking the river too far out into the stream, the bed of the river had become silted up, and London Bridge was too low to permit the free passage of boats. To remedy these defects, it was suggested that the shores of that part of the river should be sunk so as to cause the removal of the mud and allow the ships to float at low water; that the elbows or points of the river should be cut off; and that London Bridge should be demolished and rebuilt with larger arches[3]. None of these proposals, however, were carried out; and a writer in the early part

---

[1] See Report of the Engineers appointed by the Commissioners of the Navigation, as given in Brit. Mus. B. 503 (5), pp. 11–16. On pp. 57–60 of this pamphlet there is given an account of the tolls payable on the Thames Navigation between Cricklade and Staines.

[2] *Public Advertiser*, Oct. 29, 1791, p. 3. They said that by such a canal a barge drawn by eight or ten horses which at best took 48 hours by the river and frequently took four to six days or longer would be conveyed by two horses in six hours and at one-third of the expense paid at this time.

[3] Ibid., Dec. 23, 1791, p. 1, letter of "Mercator;" ibid., Jan. 20, 1792, p. 1, letter from "A Projector of Maritime Improvements."

of 1792 stated that the reform of the Thames from Lechlade to Gravesend by means of a canal made in a straight line between Maidenhead and Brentford also appeared to have been abandoned[1].

In the year 1793 much attention was focussed upon the improvement of this navigation. In the House of Commons the neglected condition of the navigation was brought out and a committee was appointed to inquire into the actual facts as to its financial standing and the progress made in its improvement[2]. The report of this committee showed that, according to Act 11 George III, the navigation had been divided into six districts and this division had been continued by subsequent statutes. At this time the first district, from the city of London to the city stone above Staines bridge, was under the jurisdiction of the city of London; while each of the other districts was under the control of a special body of commissioners composed chiefly of landowners of the counties through which the river runs[3]. From 1785 to 1792 the annual produce of the tolls had been on the average over £300 less than the disbursements and the total debts contracted during these eight years in carrying on and improving the navigation amounted to almost £15,000[4]. Many tolls were collected at locks and weirs by private individuals who did nothing to improve the navigation, and this imposition made an additional burden upon the trading interests. The depth of water in the river, especially in dry seasons, was insufficient for ordinary purposes, and the process of flashing was highly objectionable, because after a flash the river was left almost dry for twenty-four hours and barges and mills were brought to a standstill[5]. Most of those who were desirous of seeing the river kept open for navigation saw no other way of accomplishing this than by removing the shoals, widening the arches of bridges, fixing up old locks and weirs, and adding new pound-locks with their accompanying weirs where these would be most useful. But even if these changes had been made at the great cost that would be incurred therefor, there would still have remained good reasons for not using the river, because of the time and the expense required in navigating the barges upon it[6]. The committee of investigation were strongly

[1] *St James Chronicle*, Feb. 16–18, 1792, p. 2, letter from "The Thames Reformer."

[2] *J., H. of C.*, April 23, 1793; *Public Advertiser*, April 24, 1793, p. 2.

[3] *Parl. Papers*, 1793, XIII, Report No. 109, from the Committee appointed to inquire into the Thames and Isis Navigation, pp. 3, 4. The later Acts were 15 Geo. III and 28 Geo. III.

[4] Ibid., p. 4.   [5] Ibid., pp. 5, 6, 7, 20, 23, 24.

[6] In summer time, with favourable water, it took two days to go from Isleworth to Boulter's Lock, 37½ miles; but barges were frequently detained five or six days and sometimes longer on account of lack of water. From Lechlade to Oxford,

convinced of the futility of past methods to improve the navigation and they were surprised at the little progress that had been made in that direction. While they did not wish to see the river navigation completely abandoned, they made an insistent plea that the navigation in two places should be ameliorated and shortened, through the construction of a canal from Boulter's Lock to Isleworth, by which eighteen miles of the river would be saved, and another from Hart's Ferry to Abingdon, by which about nineteen miles would be saved. This proposal was directly contrary to the attitude of the commissioners of the five upper districts, who had opposed every attempt at improvement by canals. These canals, together with the removal of shoals and other impediments in the connecting parts of the river, would make the navigation more certain and expeditious. As necessary accompaniments of these changes, the committee urged that the tolls on the river should be so regulated as to place all parts of the river on equal terms and thus give encouragement to trade; that the old locks and weirs which had not yet been purchased from private individuals should be now purchased and placed under the control of the Commissioners of the Thames Navigation; and that the practice of flashing, to the continuance of which there were strong objections, should be abandoned[1].

In the years 1793–5 the improvement of the Thames was a subject which elicited much discussion. It is evident that there were two rival camps: one composed of those who favoured the retention of trade upon the river and were hostile to any deflection of the water out of its old accustomed channel; and the other composed of those who were eager to make the navigation better by constructing one or more canals, to obviate the great circuity and the shallowness of the river at certain places. One of these proposed canals, namely that from Boulter's Lock to Isleworth, had been under consideration for more than twenty years; and upon each occasion when it was brought up the land-owning classes had opposed it on the ground that any turning of the water out of its regular bed, or any interference with the customary agencies by which trade was carried on along

---

where there were frequent shallows in the river, barges ordinarily needed two or two and one-half days, and at flash times three days, to perform this journey of about 28 miles. With favourable water, it took one day to go from Wallingford to Oxford, and two and one-half days from Oxford to Wallingford, but at times they had been three weeks and from London to Oxford they had been eight weeks. *Parl. Papers*, 1793, XIII, Miscellaneous Reports No. 109, on 'Thames and Isis Navigation,' pp. 7, 23, 26. See also pp. 27, 28.

[1] *Parl. Papers*, 1793, XIII, Miscellaneous Reports No. 109, pp. 32–35, gives their conclusions.

the river, would be an invasion or despoiling of vested rights, for which
there could not be any justification. It would seem that this opposition
was headed by the Commissioners of the Thames Navigation, who were
mostly owners of adjacent or proximate lands, and who did not want
their estates injured in appearance or value by any possible abstraction
of water from the river or any diversion of traffic from its old-established
course. They opposed the canals ostensibly because the latter would
not be of so much public utility: to repay their cost of construction they
would have to charge higher rates than those on the river; like other
canals they would become "green mantled pools of stinking water,"
except in the track of the boat, and their filth would "contaminate
the passing breeze with noxious exhalations," thus rendering them
objectionable from the standpoint of public health; and they would
be liable, because of the action of drought and frost, to cause the
complete cessation of traffic, which would not occur on the river[1].
They tried to make the public believe that these causes of opposition
were sufficient to put all thoughts of a connecting canal into the
background; cheapness was the one great benefit to be secured by
inland navigation and such a result could be more effectually obtained
on the natural waterway than by an expensive artificial waterway.
On the contrary, the commercial community were very solicitous for
a navigation that would permit trade to be carried on with certainty
and speed, as well as with economy of expense; and, in conjunction
with the great engineers like Brindley, Jessop and Mylne, they favoured
the abandonment of the river where the obstacles to its usefulness
were too many and the construction of one or more canals to accommodate the necessities of traffic. The corporations of London and many
other large towns sent petitions to the House of Commons showing
that although the river had been considerably improved by their
expenditures, the natural defects could not be remedied either by
pound-locks or any other means, and that the only way to secure
a safe, regular and permanent navigation was by the construction of
artificial waterways at those places where the natural impediments
were to be overcome[2]. The navigators of the river were also eager
to have the barriers to the conveyance of their barges removed and

---

[1] *Reading Mercury*, Nov. 25, 1793, p. 4, letter on "Thames Navigation;" Dec. 2,
1793, p. 2, letter from "A Proprietor of Land near the Thames;" Dec. 9, 1793, p. 2,
letter from "A Commissioner" of the Thames Navigation; Dec. 23, 1793, p. 4;
Jan. 6, 1794, p. 4; etc.

[2] *Reading Mercury*, Mar. 23, 1795, p. 4, gives the text of the petition from the
corporation of London. This document contains the names of other places that
had petitioned in favour of the canal, so that trade between London and the West
might not be hampered.

the course straightened, through the construction of suitable canals;
they deprecated all attempts to improve the river by the mere making
of locks, for in that event the navigation would be embarrassed with
more difficulties, the charges for freight transport would be higher,
the intercourse between the country and the metropolis would be
carried on across a greater distance, and more dangers and inconveniences would be incurred both to goods and barges in the west country
trade[1]. From an unprejudiced point of view, it seems that the
supporters of the canal left no objection to this means of improvement
unanswered; and they had ranged with them the best expert skill
of the time[2]. But notwithstanding the strength of the interests
favourable to a canal, it would seem that landlord opposition was
too powerful in Parliament to permit the authorization of the canal;
so that throughout the remaining years of the eighteenth and a large
portion of the nineteenth century whatever improvement was effected
in the navigation was but the carrying out of the former policy of
piecemeal expenditure. The divided responsibility which we have
already noted was altogether unfavourable in the way of securing
harmonious action throughout the whole length of the river[3]; and
it was equally impossible for one part of the river to be improved
independently of the others[4]. Under these conditions of decentralization of control no well-concerted measures were adopted for the
amelioration of the navigation as a whole[5].

[1] *Reading Mercury*, Dec. 9, 1793, p. 4, and Dec. 30, 1793, p. 4, two letters from
"An Old Navigator;" ibid., Jan. 6, 1794, p. 4, letter concerning the Thames
Navigation.

[2] In addition to the last named references, see also ibid., Jan. 13, 1794, p. 4,
"Considerations on a proposed Line of Canal from Reading to London, through
Windsor."

[3] By Acts of 24 Geo. II, 11 Geo. III, 15 Geo. III, 28 Geo. III and 30 Geo. III,
the management of the upper part of the Thames, from Staines to Cricklade, was
placed in the hands of supposedly disinterested commissioners consisting of the
gentlemen residing in the counties bordering on the river, the number of whom in
1865 was between six and seven hundred. Brit. Doc. 1865 (399), xii, 611, 'Report
of Select Committee on the better Management of the Thames above Staines;' also
ibid., evidence of T. H. Graham, p. 2.

[4] For instance, in navigating the upper part of the river they often suffered from
a lack of water, due to the waste of water in the lower district on account of the
flashes required to float barges over shallows and other obstructions (Allnutt,
Navigation of the Thames, p. 11). The making of improvements in the upper part
was useless if the water there were drawn off to perform the services of the lower
part of the river; and improvements in the lower part of the river would be of little
avail if the current in the upper reaches were allowed to remove the gravel from the
bottom and sides of the channel and with it create shoals and obstructions elsewhere.

[5] The chief improvements were the construction of pound-locks and weirs, the
object of which was to dam up the water in long reaches and thus make it available

From the standpoint of inland navigation at this time the Thames was perhaps the most important river of the kingdom and we have shown in some detail the efforts that were made with a view to its improvement. Although these efforts were largely unavailing, yet it is evident that this was not because the interest in the river was allowed to flag. Of the next largest river, the Severn, we cannot say as much, for it did not occupy the field of public attention to anything like the same extent as the Thames. It came to be of greater importance after the construction of the Staffordshire and Worcestershire and the Worcester and Birmingham canals had opened up the trade with the industrial Midlands; and no sooner had these canals secured connexion with the Severn than there was an immediate desire to have this river improved, in order that the greatest benefit might be obtained from the canals. Accordingly, in 1787, notice was given that a meeting would be called to consider the expediency of applying to Parliament for an Act to improve this navigation. It was shown that, because of the delay in forwarding goods, much damage resulted to perishable commodities; and because of the uncertainty of the navigation many of the dealers in the interior who used to bring their goods from Bristol up the Severn had changed and were now getting them from Liverpool and other places by means of the recently constructed canals. A late survey had shown that between Coalbrookdale and Gloucester there was a fall of over one hundred and four feet in the bed of the river; and this, of itself, was sufficient to show that during the dry weather of summer and autumn there would be frequent interruptions and sometimes total cessation of the navigation for want of enough water over the shallows. Partial amendment, it was said, would be of little use; an extensive plan of improvement by which greater depth of water would be secured through the erection of locks was advocated, in order that regular and constant intercourse might be kept up between Bristol and the manufacturing counties of the Midlands[1]. We have no record, however, that anything further was done at this time to promote the interests of the navigation before Parliament; but in 1790 the Staffordshire and Worcestershire Canal Company took up the matter and obtained an Act to enable them to improve the navigation of the river from Stourport to Diglis. That

for flashing (Allnutt, *Navigation of the Thames*, pp. 13–21). How little the progress that was made and how much need there was, even at a much later date, for something to be done to benefit this navigation will be apparent from the 'Report of the Select Committee of 1865 on the Thames Navigation,' Brit. Doc. 1865 (399), XII, 611.

[1] Felix Farley's *Bristol Journal*, Sept. 15, 1787, p. 3, letter from "A Constant Reader."

company's works did not prove satisfactory, and upon action being brought against them for putting a nuisance in the river they were compelled to take away the jetties they had constructed. The unsatisfactory condition of the navigation continued till 1835, when there was a general movement for its improvement. A report was drawn up as to the best way of obtaining a depth of but four feet six inches of water and the fight for this improvement continued through the years 1837, 1838 and 1841. Some were desirous of handing over this work to a joint stock company for accomplishment, but Parliament refused to allow so important a navigation to be handed over to a private company. Finally, in 1842, the Severn Navigation Act was passed, by which very extensive authority was conferred upon a body of commissioners to enable them to dredge and deepen the river, to purchase land for locks and weirs, to remove shoals and other obstructions and to borrow money for the completion of the comprehensive improvements that were planned for this river[1].

Of the other great rivers we may refer to the Tyne, since one of the most enterprising cities and shipbuilding yards in the kingdom is situated along Tyne-side. From early times the conservatorship of the river, throughout the extent of its tidal flow, had been vested in the Corporation of Newcastle, but very little had been done to improve the river. About 1790 there was the removal of some projecting rock and stones adjoining a certain part of the river; but the Corporation did not seem to be very active in pushing the work of improvement. Probably this was due to the fact that by charters and grants the Bishops of Durham had been given authority over one-third of the river adjacent to the Durham shore; and because their authority was inconsistent with the extensive powers conferred upon the Corporation of Newcastle, the resultant disputes between these two jurisdictions imposed a barrier to any harmonious action tending toward river improvement. About 1800 an attempt was made to establish, in place of the Corporation of Newcastle, a commission of conservancy to assume the control over the river; but this plan of administration was never carried to a successful issue[2]. A forward step was taken in 1813 when the Corporation employed Rennie to report upon the possibility of improving the Tyne, which had become filled up with sand and ballast; but after his report was handed in, in 1816, nothing was done to carry out his recommendations[3]. In 1833, upon a complaint that Newcastle had a monopoly of the river,

---

[1] 'Report of Royal Commission on Canals and Waterways,' 1906, i, Pt. ii, evidence of Thomas Southall, Q. 2786 et seq.

[2] Guthrie, *The River Tyne*, p. 54.   [3] Guthrie, op. cit., pp. 59–61.

that it neglected the best interests of the port and that it used the river dues for corporate and town purposes, an investigation was made, under Royal authority, of the municipal corporation's affairs. It is probable that some of the funds received were used for the town, rather than for the river; but as the harbour was naturally a good one for the vessels then in use and there was no great demand for enlarged facilities to meet the requirements of modern commerce the Corporation ignored the river and its capabilities of improvement[1].

From what we have said concerning the Thames, the Severn and the Tyne, it will be apparent that the improvement of the rivers in the latter half of the eighteenth century received much less attention than the construction of canals. In fact, the importance of the rivers seemed to be overshadowed by the results that were accruing from the network of canals that were joining together all parts of the country. If this were true in the case of the larger rivers, it must have been still more true in regard to the smaller rivers. Many forces were at work tending to cause the deterioration of the latter to a greater extent than the former. All the rivers that were navigable were tidal; but in the case of the large rivers a much greater volume of tide could be admitted and consequently the scouring effect would tend to be the greater. On the other hand, the tide which came into the less capacious channels of the smaller rivers could not, in many instances, be confined within the banks, but spread out over the adjoining lands. The inevitable result was that the river channels, instead of being cleansed, were silted up, the banks were broken down or worn away, and where formerly barges were able to ride securely for considerable distances up the rivers, now the navigation was totally or partially lost. The changing of the position of the river bed from year to year, the increasing slowness of the current allowing sand and mud to settle, and in some cases the absence of sufficient declivity to cause the water to speedily recede aided the other natural agents that were at work in bringing about the decline of the smaller rivers as agents for the carriage of traffic. The case of the Stour in the county of Kent, in 1774–5, may be taken as typical of other small rivers; from Fordwich bridge to the mouth of the river at Sandwich harbour the Stour was one and one-half mile broad; for several months of the year some thousands of acres of land were under water because of the river overflowing its banks; and so great had been the growth of weeds in the harbour that the latter had become silted up and the river had frequently to change its mouth[2]. The foregoing were by

---

[1] Guthrie, *The River Tyne*, pp. 61–63.
[2] Brit. Mus. MSS. 5489, pp. 108–21, two reports on the river Stour.

no means the only elements entering into the decay of the river navigations; but we have given sufficient detail to this phase of our subject, so that we may leave it at this point[1].

There is another aspect of inland navigation to which we think it proper to refer here, which at one time seemed to have the possibility of effecting a great change in the conveyance of goods by water in England. It had the greater influence because the name of Robert Fulton was prominently associated with it[2]. He thought that the country would be much better served by small canals than by large ones; for by means of small boats on small canals and rivers goods could be brought right into the remoter parts of the interior of the country. He would prefer ninety miles of navigation for four-ton boats to thirty miles for forty-ton boats. Boats of twenty-five to forty tons could bring the goods up the river, and the freight should then be transferred at once to the small boats, which would be able to convey it inland to its destination without any further transfers. Again, "in every situation where a canal is to be formed for forty-ton boats, one-third of the sum necessary for that purpose would pay the expense of a canal for boats of four tons. Hence, if a company are about to expend £300,000 where £100,000 would answer the purpose, £10,000 per annum is sunk to save transfer" of the goods at the coast or other point where the larger vessel had to leave them[3]. His opinion was that no large canal could rival a small one, for evident reasons. He supposed, for instance, a large and a small canal running side by side, the large canal costing £300,000 (or, in proportion, three times the expense of the small one), and the small one £100,000. One penny per ton per mile to the small canal would be as good interest as three pence to the larger one[4]; consequently, the small canal could lower its tonnage rates so as to favour the shipper and render the expense

---

[1] As a special case, different from most other instances, we refer to the history of the navigation from Norwich to Yarmouth and to the agitation, about 1818–27, to make "Norwich a Port." A general survey of this agitation is given in Brit. Mus. 2064. a., *History of Norfolk*, i, pp. lxxxiii–lxxxvii; and Brit. Mus. 08.235. h. 12 contains 'Reports and Pamphlets on the subject of Norwich a Port, from the year 1818 to the passing of the Norwich and Lowestoft Navigation Act in 1827.' This navigation was partly completed, but why it was not finished I have been unable to ascertain.

[2] Fulton, *A Treatise on the Improvement of Canal Navigation.*

[3] Ibid., chap. iv.

[4] Fulton's reasoning here is, of course, fallacious. He seemed to think that a small canal could be run as economically as a large one, which may not, and under ordinary circumstances would not, be true. With small canals, more tolls would have to be paid and more horses or men would be needed to draw the same weight of goods along the canals than when the canals were of large size and could take large loads of goods. To us, such facts are too apparent to require any proof.

of transfer of no consequence; they would even grow rich by lowering the tonnage dues, for thereby they would draw the trade from the large canal and leave it as a stagnate and useless pool. Then, having proclaimed the above propositions, he did not hesitate to prognosticate the annihilation of lock canals, by improved science, in the same way as improvements in machinery render old apparatus useless.

Fulton, then, based his argument for the small canals on two facts: first, that the cost would be less, and second, that a much greater network could be obtained for less money. He thought river navigations ought to be extended as far as convenient for large boats and from that point onward the carriage should be effected by small boats on small canals. The boats should be of a particular construction and, by means of machinery, they would be drawn up the single or double inclined plane from one level of the canal to the next higher[1], so that thereby there would be a great saving of water. This was an important particular, for many of the canals of that day, which were operated from reservoirs of water that were likely to dry up in summer, were unable to carry on their work during those times of the year. The significance of this feature we can scarcely realize, for the only water supply that some of the canals had was what was obtained by surface drainage. In such cases, the preservation and most economical use of all the water of these basins was of paramount interest and if no water were lost in moving a boat from one reach of a canal to the next there would be no waste of the one essential for the conduct of the navigation. To Fulton, this system of small canals "meandering the hills," and capable of extension to nearly all parts of England, held out "assistance to the sun-burnt fields," and promised "some hope of progressive improvement."

That Fulton regarded this plan as a most plausible solution of the problem of inland navigation is clear from what he says in the latter part of his book: "As I venerate liberality and the light of reason, I despise the pusillanimity of the individual, who, like a dark lantern, conceals the light he receives. Therefore, whether this is a gleam radiating from a brilliant reflector, or the pale glimmering of inflammable vapour, I am determined it shall not be confined; and my reason is, that many useful improvements sleep for ages, for want of the fire of energy in the projector, while the only mode of proving their utility is to bring them to the test of discussion: I, therefore, feel myself quite ready to meet every objection to this system of small canals;

---

[1] In certain cases the boats could be taken by a perpendicular lift from the lower to the higher level of the small canal. Of course, this same principle could be applied to a large canal in transferring from one level to another.

and for this purpose, I here call on engineers, or others, who think proper to answer the arguments in their favour[1]."

The challenge thus given was taken up and at least two engineers expressed their views in regard to Fulton's plan. Chapman was in favour of small canals, but he opposed Fulton's idea that locks would, in the future, be found ineligible in all cases[2]. Tatham followed Fulton in wanting small canals and inclined planes, but the means by which he would carry out this principle were different from that of his predecessor[3]. Despite the fact that this plan had much that, in theory, recommended it, we have little evidence that the principle was practically applied by the construction of small canals.

It may not be amiss to note some other special plans for the improvement of inland navigation which were entirely different from those that were commonly used. In certain cases inclined planes were in use, instead of locks, to connect upper and lower reaches of a canal. These were probably introduced into England about 1789

---

[1] As a very interesting and ingenious plan for extending the inland navigation, this of Fulton's is well worthy of perusal.

[2] Chapman, *Systems of Canal Navigation*, p. 2. He says that the system of inclined planes was introduced into England by William Reynolds, of Ketley, Shropshire, on the Ketley Canal, about 1789; and that the boats were drawn up by a horse, which drew 15 boats, 20 feet × 6 feet × 4 feet, connected to each other by a few links of chain (ibid., p. 4). For description of these Shropshire canals, see Plymley, *Agriculture of Shropshire*, pp. 291–9. Chapman's examination of Fulton's plan, and his suggestions, are interesting, but they are not important for us here.

[3] Tatham, *The Political Economy of Inland Navigation*, p. 86 et seq. He says: "It follows here to contrast the advantages which are offered in the later improvements of the inclined plane; the use of machinery instead of locks; and a system of universal extension at cheap rates, that, I trust, offers the means of facilitating transfer and locomotion into the remotest corners of the earth, in a way which may enable the poorest man in the most sterile countries to partake of a common use and profit, from which the expence and size of the lock system has heretofore excluded them, in favour of peculiar affluence only...." He would use only the double inclined planes in passing from lower to higher levels, for this would enable a vessel to go the other way at the same time.

P. 121 et seq. give his criticisms of English canals, the chief of which were:

(*a*) The old-system engineers in many cases constructed larger canals than were needful for the trade of the country. Hence there had been unnecessary sinking of capital in construction.

(*b*) They had thus created an excessive demand for water, in the same proportion, and this often where the supply was inadequate to the demand.

(*c*) Unnecessary expenditure of money and waste of water in lockage. Tatham would use steam power for passing the boats up and down the inclined planes, and when not used for this purpose it could be applied to manufacturing purposes.

(*d*) Considerable delay in passing locks.

(*e*) Difficulty of making junctions between canals of different dimensions.

by Reynolds, one of the great iron-masters of Shropshire, upon the Ketley Canal in that county[1], and their use spread to other canals in the same county and to similar hilly sections in some other parts of the country[2]. In South Wales this method was used for lifting and lowering barges between two adjoining levels of canal; and in Devonshire the Bude Canal and its branches were joined together by a series of inclined planes[3]. On the Tamar Canal, in Devonshire, the engineer planned to use the inclined planes, instead of locks, and to have the lifts from one level to another from nine to twenty fathoms in perpendicular height. The boats were to be carried up the inclines by a hydraulic machine[4]. On the Monkland Canal, in Scotland, the inclined planes were adopted and were found to be very serviceable[5]. It is evident from a consideration as to the localities in which this system prevailed that they were adaptable to rough, uneven areas, where the regular lock canals would not have been suitable; but this method of overcoming the natural impediments of the surface of the country never attained any great significance in comparison with the usual method of canal construction. Another plan of operation was carried into effect on the Great Western Canal, where the boats were raised and lowered from one level to another by means of perpendicular lifts[6]. On the upper part of the Wye navigation, a unique scheme for overcoming the Monnington falls was by hoisting the loaded barges over the falls by means of pulley blocks[7]. We may also refer here to a rather chimerical project, proposed in 1796, in the nature of a "circular canal," which was to circuit Britain, passing by way of London, York, Edinburgh, Inverness, Glasgow, Carlisle, Chester, Worcester, Oxford, London. This was all to be on a level and was to be carried out by the State, to which the immense revenue therefrom would accrue[8].

---

[1] See footnote 2, p. 391.

[2] Plymley, *Agriculture of Shropshire*, pp. 291–9.

[3] Moore, *History of Devonshire*, pp. 48–49. In 1829, at the town of Bude, this canal would admit vessels of sixty to seventy tons. After running about two miles inland it reached the first inclined plane, 826 feet long and with an elevation of 122 feet. From there the canal accommodated boats of only five tons on a level of more than two miles to another inclined plane, 907 feet long and 225 feet elevation. About a mile beyond that it diverged into two branches: the Bude and Holsworthy Canal and the Bude and Launceston Canal, each of which had several inclined planes.

[4] Leach, *Treatise on Inland Navigation*. Leach was himself the engineer.

[5] For the history and description of the working of this device, see Leslie, *Inclined Plane on the Monkland Canal*.

[6] See Green's description of this in the *Transactions of the Institution of Civil Engineers*, 1838.

[7] Lloyd, *Papers relating to the Navigation of the Rivers Wye and Lug*, p. 47.

[8] Hibbard, *Utility of a Circular and Other Inland...Canal Navigation*.

Another proposed improvement in the construction of inland navigations was given some impetus in the later years of the eighteenth century, when it was advocated that, instead of proceeding on the haphazard plan hitherto followed, a systematic scheme should be first devised and then all improvements should be made in harmony therewith. The general outline of this method included the abandonment of locks and weirs, the deepening of the important rivers so as to render them navigable without locks, and then the inauguration of a well-planned and consistent programme of canal construction, to connect, usually on the same level, with the rivers. It would seem that it was the uselessness of such canals as the Thames and Severn and the Basingstoke, and the failure to make the greatest river of the kingdom, the Thames, navigable by the means already employed that gave point to the proposals just mentioned[1]. It was thought that if the rivers were first deepened the whole question as to drainage of the land and the prevention of inundation would be solved; and then if a comprehensive canal system, devised by men of skill, independence and public spirit, were to be joined with the improved rivers in such a way as to secure national advantages, rather than private emolument, the public would support the project in its entirety and all discouragement or hostility would be at an end[2]. In the case of a river, like the Thames, with great declivity, it would have been a stupendous task to reduce that declivity until there would be a uniform and perpetual depth throughout the whole length of the navigation; but those who favoured this plan thought that by the application of power it could be effected and the bed of the river sunk to a general level. In consequence of this there would be no floods in winter and no drought in summer; all shoals in the river would disappear; the commerce of the river would be uninterrupted and would constantly increase; the better navigation would render transport easier and cheaper; and trade would be emancipated from lock dues and other impositions and would thereby yield more liberal advantages to all[3]. While this reform of the natural rivers would be going on, so as to secure not less than ten to fifteen feet of water, parliamentary commissioners could be planning a toll-free and lock-free

---

[1] *Public Advertiser*, April 8, 1786, p. 2, letter from " W. J.," 'On Internal Navigation;' April 11, 1786, p. 2, letter from "Foresight," on the 'Danger of Canals upon wrong Principles;' *St James' Chronicle*, May 26–29, 1792, p. 2, letter from " W. J."; etc.

[2] *Public Advertiser*, June 26, 1786, p. 2, letter from " W. J." on 'Canals.'

[3] Ibid., May 29, 1786, p. 2; July 17, 1786, p. 2; Mar. 17, 1788, p. 2, letter from "Mercator;" Oct. 20, 1789, p. 2, letter from "Anti-Brindley;" Nov. 21, 1789, p. 2, letter from "A Citizen;" Aug. 13, 1791, p. 1, letter from "An Engineer."

system of canals to be constructed by the Government for the general benefit[1]. The engineering difficulties in the way of carrying out this proposed method of improvement are obvious, even at a casual glance; and, whether this were the reason or not, the plan, even in part, was never put into execution.

In the thirty years following the introduction of canals very important results were secured, financially, commercially and socially; and the contemplation of the success achieved by some canals[2], together with a large amount of capital available for investment, due to the rapid increase of the wealth of the country, served to unduly stimulate the projection of new schemes. This movement had been more rapidly gaining momentum in the years immediately preceding 1792[3]; but in the years 1792 and 1793, there was wild speculation and a perfect ferment about canal shares. Plans were brought forward for canals to parallel, or to invade the territory of, other canals[4]; speculative adventurers noted with satisfaction the attitude of the public toward these waterways and arranged to take advantage of the conditions of the time. Schemes of all kinds were advocated; and because certain canals that were favourably situated with regard to supplies of coal and manufactured products and in the midst of an area of large population had proved profitable, the public were duped into thinking that similar success might be achieved by an immense number of other canals. These promoters, in many instances, were anxious to cause the prices of the shares of their projected canals, even before the work of construction had been begun, to rise to an unduly high figure; and then they would unload their stocks upon unsuspecting purchasers so as to themselves net a great profit[5]. Canals were a

---

[1] *Public Advertiser*, Feb. 2, 1790, p. 1, letter from "The Inland Navigation Reformer;" April 20, 1792, p. 1, letter from W. J.; Oct. 19, 1792, pp. 1–2, letter from "A Projector of Reform in River Navigation;" Aug. 27, 1793, p. 1, letter from "A Thames Conservator."

[2] So great were the dividends earned by some canals that a writer in 1792 suggested limiting the rate of interest on money invested in canals to from ten per cent. to twelve per cent. (*Gentleman's Magazine*, LXII, Pt. II, p. 1162). For some years before 1790 the proprietors of the Staffordshire Canal paid more than twenty per cent. per annum on their investment (Publicola, *Utility of Inland Navigations*, p. 11). See also West, *History of Warwickshire*, p. 100, and Momsen, *Öffentlichen Arbeiten in England*, pp. 33–34.

[3] *Morning Chronicle*, April 1, 1791, p. 4, letter of " F. F." on 'Navigable Canals.'

[4] Examples, the Hampton Gay and Grand Junction canals.

[5] *Morning Chronicle*, April 1, 1791, p. 4; *Reading Mercury*, Dec. 2, 1793, p. 2; Felix Farley's *Bristol Journal*, Jan. 4, 1834, p. 2, letter of "Caveat Emptor;" *The Star*, Oct. 13, 1792, p. 4.

lottery and there was much gambling going on in the buying and selling of canal shares. No sooner had the plan been brought forward for another canal than the subscription list therefor would be immediately filled; and no matter how unlikely a canal was to pay good returns upon the capital invested there were many people who were intensely eager to put their money into it[1]. During this "canal mania," which reached its climax in the latter part of 1792 and the early months of 1793, the premiums on some canal shares rose to exorbitant heights[2]; and in the years 1791–4 no less than eighty-one canal and other navigation Acts were passed. It was seriously proposed in 1792 and again in 1793 to limit the returns on canal investments to ten or twelve per cent., in order that some check might be imposed upon riotous speculation in these undertakings[3]; and so great was the interest in canals that in 1793 a Bill was introduced into and discussed in the House of Commons to prevent the cutting of canals during corn harvest, lest this work of construction should absorb so many men that there would not be enough left to gather in the wheat crop[4]. As an outcome of this speculation in worthless schemes, much ruin was brought to many who could ill afford to lose[5]; but, on the other hand, some enterprises which were of great public utility were entered upon at this time[6]. In all probability, these would ultimately have been taken up without the interest enkindled by the canal mania, although the latter doubtless helped to bring them earlier to the attention of the people. Many individuals lost nearly all they had in the mirage of speculation, but the public gained

[1] *Reading Mercury*, Nov. 12, 1792, p. 3; ibid., Nov. 19, 1792, p. 1; etc.

[2] Baines, *History of Liverpool*, p. 488, gives the following values of canal shares in October, 1792: Shares of the Trent Navigation sold for 175 guineas (£183. 15s.) each; those of the River Soar Navigation, for 765 guineas (£803. 5s.) each; of the Erewash Canal, 642 guineas (£674. 2s.) each; one share of the Oxford Canal sold for 156 guineas; one share of the Cromford Canal sold for 130 guineas; one share of the Leicester Canal sold for 175 guineas; ten shares of the Grand Junction Canal (which was not yet dug) sold at 355 guineas premium; one single share of the Grand Junction Canal sold at 29 guineas premium; ten shares of the projected Mersey and Severn Canal sold for 29 guineas premium. The Leicester Canal shares, which had sold in October for 175 guineas, were enhanced in price to 324 guineas; and the Grand Junction shares had increased in price during the same time until they reached 420 guineas. *Reading Mercury*, Nov. 12, 1792, p. 3.

[3] *Gentleman's Magazine*, LXII (1792), Pt. II, p. 1162; *Public Advertiser*, Mar. 19, 1793, p. 2.

[4] *Public Advertiser*, Mar. 22, 1793, p. 1, and April 11, 1793, p. 1. Needless to say, this Bill was finally lost.

[5] Bull, *History of Devizes*, p. 468.

[6] Such as the Grand Junction, Gloucester and Berkeley, Grand Union, Kendall and Lancaster, and Birmingham and Warwick canals.

much by the opening up of communication in all directions. The evil effects of this crisis were not so great nor so long continued as were those of the "railway mania" fifty years afterward, nor did the country take so long to recover from them. The seventy years preceding 1830 may properly be called the canal era in English transportation; and the importance of the closely inter-related network of navigable inland waterways in furthering the Industrial Revolution, which was transforming the face of England, has never been exaggerated.

Having dwelt at sufficient length upon the construction of the system of interior navigation, we turn now to examine the nature of the objections which were urged against such works, for it is only as we see them in the light of that day that we can understand and appreciate their significance.

Opposition came, in the first place, from the landed classes, who claimed that the water would be drained off their land in order to serve the purposes of the canal and that, therefore, there would not be enough left to water their meadows and to provide for their animals pasturing upon the higher lands. Then, since the canals would be put through the low-lying ground where possible, part of the most fertile land would be devoted to the navigation and rendered useless for agriculture. By the digging of canals estates would be severed and their occupiers subjected to various inconveniences which could not be removed by building bridges across the waterways; and the operation of the canal would permit the passing to and fro of a rough class of men who might commit all kinds of depredations upon the property, thus rendering the adjoining land less valuable than it would otherwise be. Anything which would tend to destroy the quietness and seclusion of the landlord's domains was to be carefully rejected, irrespective of whether it promised to be of great public benefit or not[1].

---

[1] There, doubtless, was some truth in these objections; but the pecuniary amount of the injury from such inconveniences could be estimated, and proper compensation was required to be made to those who were thereby affected.

It would seem as if objection might have been made to canals on the ground that the large areas of land required for water reservoirs for the operation of the canals were withdrawn from cultivation; but thus far we have found no instance of such objection. For example, the Grand Junction Canal was fed by ten reservoirs, with a total capacity of seven and one-half millions of cubic yards. The Birmingham Canal Navigations had five reservoirs, of a total capacity of about 5,250,000 cubic yards. The Rochdale Canal's seven reservoirs had a capacity of nine and one-half million cubic yards (Harcourt, *Rivers and Canals*, II, p. 366). The reservoir of the Bude Canal in Cornwall covered sixty acres (Wallis, *The Cornwall Register*, p. 193). The Grantham Canal had two reservoirs covering seventy-nine acres (Allen, *History of Lincoln*, II, p. 307).

See also Brit. Mus. 215. i. 1 (105), 'Thames Navigation;' Brit. Mus. 213. i. 5 (94),

Mill-owners opposed any deflection of the water that would naturally flow into their stream, or any turning of the stream itself to feed the canal. Some mills were located on small streams where the water had to collect for several days before there would be sufficient pressure to drive the mills and in consequence these could do their work only intermittently. To take away any portion of their water supply would be to impede or cause the cessation of the mills; and, of course, the millers' opposition would be supplemented by that of the farmers whose interests would be adversely affected by the curtailment of the usefulness of these near-by conveniences. Even those who had mills along the great rivers, like the Thames, would object to the cutting of a canal at one side of the river lest the water required therefor would have to be shared by them out of their pens of water. Their opposition was still greater, however, in the case of the river navigation, when they would have to open their sluices and send down flashes of water to permit the navigation of barges over the shallows in the bed of the river, because then they would have to suspend the working of the mills entirely until a sufficient head of water once more collected[1].

Road trustees objected to the construction of a canal near the roads which they had under their care lest the canal should draw to itself most of the trade and thus cause a reduction of the road tolls to such an extent that the income of the trust would not be sufficient to pay the interest, much less the principal, of the amount borrowed on the security of the tolls[2]. Closely related to this was the opposition of the carriers by land, who were afraid that their living would be taken away by reason of the bulk of the goods being carried by the canals in their vicinity. But although this was their prevalent opinion they frequently veiled their opposition under other pretences and cloaked them under their apparently beneficent interest in others who would

---

'State of the Case regarding Several Navigations,' p. 3; 'Seasonable Considerations on Navigable Canals,' p. 5; *Parl. Papers*, 1826 (309), iv, 631, 'Minutes of Evidence taken before the Committee on the Birmingham and Liverpool Canal Bill,' p. 3; *J., H. of C.*, xxx, pp. 613, 627, 683, 707, 708. A significant statement is made by Jackson's *Oxford Journal*, Feb. 8, 1783, p. 3, for in speaking of the Thames and Severn Canal the editor says that it was "rather uncommon" for landowners in general to favour a canal designed to pass through their land.

The opposition of the mill-owners to the construction of the Rochdale Canal was very strong in their effort to prevent the diversion of water from the mills to the canal. *The Oracle*, Mar. 22, 1792, p. 2; Brit. Mus., Maps 88. d. 13, 'Documents and Plans relating to Canals of England,' No. 32 (1793).

[1] See references under last footnote.

[2] The only answer necessary to this complaint was that the roads were under the care of the Legislature in a more intimate way than the canals and that that body would not be likely to wantonly sacrifice the public good by allowing the roads to decay.

be injured by the building of the canal. For instance, it was not uncommon for them to assume the farmers' interests and say with them that if canals were constructed fewer horses would be necessary for the carrying trade; hence there would be the destruction of a large part of the demand for the farmers' hay and oats. In other cases they appealed to the popular prejudice by declaring that if canals were built the traffic would be nearly all carried on them and thereby there would be the development of a destructive monopoly[1].

The opposition of "vested interests" was always vigorous and in many instances prolonged. It was repeatedly said that surely Parliament would not sanction one means of conveyance that would injure or destroy another which, at an earlier time, had been favoured by parliamentary authority, assistance, or protection. Each navigation seemed to regard itself as the favourite child of Parliament, to be jealously guarded from any adversity due to possible or actual competition; and any upstart rival project ought to be put down, so as to avoid anything that might be detrimental to property or other interests that had formerly been created under legislative sanction. On account of this attitude earlier navigation companies interposed such difficulties in the Legislature as would impede or prevent the passage, or increase the expense, of an Act intended to authorize a rival waterway[2]. We may exemplify this by the case of the Worcester and Birmingham Canal. Early in the history of English canals, the Staffordshire and Worcestershire Canal was constructed to connect the river Severn at Stourport with the Grand Trunk Canal near Great Haywood, and thus bring the industrial section of Staffordshire into connexion with the great markets along the Mersey and the lower Severn. Notwithstanding the fact that the Severn below Stourport was in need of improvement, this canal was prosperous and the country it served was given remunerative outlet for its coal. But in 1785 and 1786 a company was formed to make another canal from Worcester, lower down on the Severn, to Birmingham, ostensibly with the object of avoiding the difficulties of the river navigation between Worcester and Stourport and of supplying the markets of the lower Severn with cheaper coal from the Midlands. Immediately this new project

---

[1] Mercator, *Tonnage Rates of the Grand Junction Canal*, pp. 8, 10; Brit. Mus. 215. i. 1 (105), 'Thames Navigation;' Brit. Mus. 213. i. 5 (94), 'State of the Case regarding Several Navigations,' p. 3.

[2] *Seasonable Considerations on a Navigable Canal from River Trent to River Mersey*, pp. 1–38; Brit. Mus. 215. i. 1 (105), 'Thames Navigation;' Brit. Mus. 213. i. 5 (94), 'State of the Case regarding Several Navigations,' pp. 2, 3; Farey, *Agriculture of Derbyshire*, III, p. 291.

encountered opposition from the earlier canal; and the latter used every available resource to prevent the new company from securing an Act of Parliament, lest the construction of this additional canal might jeopardise their existing interests[1]. Another instance of this opposition of one navigation to another is that of the river Cam, the conservators of which in 1811 wanted to be sure that the proposed canal from Cambridge to Bishop Stortford should not be allowed to deflect any of the water from the river into the canal, and for this purpose insisted that a special clause should be inserted in the Bill, giving effect to this desire[2]. A similar case is noted in connexion with the Coventry Canal, the history of which we have hitherto briefly outlined. Here the issue was the opposition of the Coventry Canal to the proposed Ashby-de-la-Zouch Canal. Near Ashby-de-la-Zouch, in Leicestershire, there were thousands of acres of excellent coal, but there was only a limited demand for it over a narrow area where it could be supplied by land carriage, and hence only a small part of the coal was worked. The people of that section were desirous of

---

[1] Felix Farley's *Bristol Journal*, Jan. 21, 1786, p. 1, letter entitled 'Worcester Intended Canal;' also ibid., Jan. 28, 1786, p. 1, 'An Answer to the Worcester Letter;' ibid., Feb. 4, 1786, p. 1, 'An Answer to the Worcester Letter;' ibid., Feb. 4, 1786, p. 1, 'Canals. To the Querist of the Worcester Intended Canal;' ibid., Feb. 4, 1786, p. 4, 'Canals. Further Queries to the Promoters of the Intended Worcester Canal.' The issues in this matter are not clear and it seems as if there were underlying motives on each side which were more potent than those which were apparent. From the discussion presented in the above references, it looks as if the Worcester and Birmingham Canal may have been actuated by purely selfish motives and thought that by this canal they would find a better market for their coal and would prevent their rival from enjoying all the benefit of this trade. Possibly, even, they may have been desirous of securing, through their connexions, a monopoly of the carriage of Dudley, Tipton and Stourbridge coal to the Severn markets. But, on the other hand, their opponents, who upheld the claims of the Staffordshire and Worcestershire Canal, were probably acting in like manner from interested motives; and under the guise of working for a great public benefit, namely, the improvement of the river navigation, they were hostile to the making of the proposed new canal. They endeavoured to sweep away all objection to the existing state of the Severn, which formed the chief argument put forth for the necessity of the new canal, by taking active steps to secure the co-operation of Shropshire in a movement for remedying the condition of the Severn from Coalbrookdale to Gloucester. In this way, they would benefit the coal mines and foundries of Shropshire and the coal mines of Staffordshire, while the amelioration of the river navigation would be of great public advantage and would do away with the incentive for the new canal. But under this seemingly altruistic exterior, the Staffordshire and Worcestershire Canal was, doubtless, trying to prevent the establishment of a competitor which might take away some of the benefits that had already accrued to the older canal.

[2] Brit. Mus., Add. MSS. 35,689, p. 21, letter from people who were interested in maintaining the river Cam navigation.

profiting by their resources, and realizing that if they could get their coal taken to the Coventry and Oxford canals and by them to the more distant markets there would be greater opportunities for increasing their wealth and opening up to the kingdom a larger supply of good coal, they sought the privilege, in 1793, of constructing a canal to join the Coventry Canal near Griff, and collateral cuts and railways were proposed to be built to other collieries to serve as feeders for this canal. But the Coventry Canal, after 1790, when it had obtained communication with the Staffordshire collieries, had enjoyed great benefit from the carrying of coal from Staffordshire to supply the places along its own line and that of the Oxford Canal; and this, together with the additional merchandise carried, had considerably increased its revenues. If the proposed canal should be allowed to pour its coal into the traffic of the Coventry Canal there would be less demand for Staffordshire coal, and therefore the revenues of the Coventry Canal would be decreased by carrying the Leicestershire coal the shorter distance rather than the Staffordshire coal the longer distance. For this and other reasons they argued that the plan of the Ashby-de-la-Zouch Canal should be rejected[1]. As a final illustration of this opposition of one navigation to the construction of a possible competitor, we have but to refer to the noted case of the Thames, which has been already considered; the commissioners of this navigation actively opposed all attempts to improve the facilities for the conveyance of goods by the making of lateral canals to avoid the obstructions in the river. They even decided, in 1798, to oppose the proposed Hampton Gay and Grand Junction canals, from the Oxford Canal to the Thames, near London, lest either of these might deprive the Thames of its trade and its water[2].

Referring to the antagonism of vested interests, it may be said that, in the early part of the canal era, there was, apparently, a general principle, acted upon to a very great extent, that when Parliament had sanctioned the execution of a navigable river or canal and limited, as it always did, the tolls or tonnage charges to be taken on such works, a precaution which was uniformly regarded as a sufficient guarantee against monopoly, any rival line which presented greater facilities to the public should guarantee to those which had been executed such compensations as would prevent the diminution of their revenues. So important was

---

[1] Brit. Mus., Maps 88. d. 13, 'Documents and Plans relating to Canals of England,' No. 6, on the Ashby-de-la-Zouch Canal.

[2] *Reading Mercury*, Jan. 7, 1793, p. 3; also ibid., Nov. 25, 1793, p. 4; Dec. 2, 1793, p. 2; Dec. 9, 1793, p. 2; Dec. 23, 1793, p. 4; Jan. 6, 1794, p. 4; Jan. 13, 1794, p. 4; Mar. 23, 1794, p. 4.

the application of this principle that a considerable part of the revenue of some canals was derived, not from the direct returns they received from the carriage of goods, but from the compensation paid by other canals that had been made for the purpose of extending or cheapening the means of conveyance[1]. In later years, however, this principle of compensation was abandoned[2], and if the plan brought forward showed decided advantages over those then existing the great parliamentary authorities favoured the adoption of the new and better without compensating the older and less efficient.

Some of the strongest protests against inland navigations came, as we have already seen, from another vested interest, namely, those who were carrying on the coasting trade. From Tudor times the rulers had made every effort to develop the merchant marine as a most important auxiliary to trade and a nursery for seamen. Indeed, so much thought and attention had been given to this aspect of the national welfare that those who were engaged in the coasting trade began to regard their well-being and perpetuation as the subject of supreme importance to the kingdom and apparently believed that other claimants to public consideration should be relegated to a distinctly secondary place. Many of this class of seamen did not seem to understand that the inland and the coasting trade were complementary to each other and that the more the former was developed the greater were the opportunities for the latter. Having their point of view, they opposed the construction of the Thames and Severn Canal and the other improvements in inland waterways between the West and London, lest coal should be brought from Wales and the western counties of England to London for such prices as would injure the Newcastle coal trade. The coasting trade likewise opposed the formation of the canal between Coventry and Oxford, on the ground that this would probably lead to London being supplied with coal from the interior, and so the coasting trade, which developed so many seamen, would decline. Even as late as 1803 prejudice still prevailed in some parts and the popular objection that inland navigation tended to diminish the number of seamen frequently influenced the minds of those who were not biased by any particular private interest[3].

[1] Brit. Mus. 08,235. f. 77, 'Observations on the General Comparative Merits of Inland Communication by Navigations or Railroads' (1825), p. 2. The writer of this letter said that the instances of the application of this principle were numerous; and he referred particularly to the Oxford, the Coventry and the Bridgewater canals.

[2] Ibid., pp. 2–3.

[3] Brit. Mus. 214. i. 4 (119), 'Letter from Yarmouth,' pp. 1–3; Brit. Mus. 214. i. 4 (120), 'Report of Committee on Oxford Canal,' p. 2; Brit. Mus. 214. i. 4 (103),

Some canals encountered hostility because of the fact that money invested in other enterprises of a like nature had not been sufficiently remunerative to fulfil the expectations of the subscribers. The opponents of the proposed Romford Canal, for example, while recognizing that much good had been done by canals in the Midlands, asserted that Romford was an agricultural district, and pointed to the unsatisfactory results of such canals as the Croydon, the Basingstoke, the Kennet and Avon, and the Salisbury, which were similarly situated to the contemplated canal[1]. Of course, wide latitude had to be allowed as to what constituted similarity of situation when a comparison was being made between different canals, for even apparently insignificant differences were sometimes sufficient to account for one canal being profitable and another less remunerative or unprofitable. But if some canals could be pointed out as failures, others could be shown to have succeeded in the highest degree[2]; so that this kind of opposition, while it may have determined the course of action of some individuals in particular cases, was, on the whole, inconclusive and unsatisfactory.

In certain cases there was an agitation against a proposed canal on the ground that it was unnecessary or undesirable. Sometimes a navigable river was found within a few miles of the place where the projected canal was to be built; and it was asserted that, since the river navigation, with little or no toll to pay, was not used to any extent, the construction of a canal would be a work of supererogation. It was said that if carriage by water in these districts had been profitable or desirable, the people of that vicinity would have availed themselves of the facilities of existing navigations in sending their products to market. Instead, farmers and others occasionally preferred to have

'Observations on the Effects of the Intended Oxford Canal,' pp. 1–2; *J., H. of C.*, xxxii, pp. 183, 274, 289, 315; Phillips, *History of Inland Navigation*, 4th ed. (1803), p. viii. To act on this principle alone would have been to put a stop to all progress; and Parliament, while safeguarding as far as possible the rights of individuals in this respect, endeavoured to secure by the navigations it authorized the greatest amount of public benefit.

[1] *The County Chronicle and Weekly Advertiser for Essex, Herts, Kent, Surrey, Middlesex*, etc., Nov. 10, 1818, p. 4, shows the argument of those who were hostile to this canal. According to the statements there made, the Croydon Canal, on its £100 shares, had paid a dividend of one per cent. for the first few years, but nothing had been received for the several subsequent years; and these shares were, in 1818, selling for about £4 each. Likewise, the Basingstoke Canal £100 shares were selling for less than £10. The Kennet and Avon shares, of the original value of £130, paid no dividend from 1800 to 1814; 11s. was paid for 1814; 15s. for each of the two succeeding years; and for the two following years the average profits on a £130 share amounted to 8s. 9d. per year.

[2] See, for example, *Hampshire Telegraph*, Oct. 14, 1816, p. 3, letter from "Vetus" regarding 'The Intended Canal.'

their products taken to market by waggon, if they were within reasonable distance of the market, for the expense of land carriage was thought to be generally compensated by greater security against the damage and pillage that constantly accompanied water conveyance. If, therefore, it were the general opinion that water carriage was more expensive or less advantageous than land carriage, whether this were true or not, there would be a strong protest against the cutting of a navigation in that vicinity[1].

When the formation of canal companies was first proposed this movement was resisted by the trustees and mortgagees of turnpike roads, on the assumption that these canals would become monopolies in the hands of the private corporations. By making their charges acceptable to the traders, a large part of the traffic would be diverted from the roads to the canals, and through the consequent decrease of the revenues of the turnpike roads the trustees of the latter would become financially embarrassed and the mortgagees of the tolls would either lose all they had loaned to the road or else the security upon which the loan was based would be seriously impaired. Then when the canals had taken to themselves the major part of the trade they would be able to increase their rates and reap monopoly profits at the expense of the public. The advocates of the roads urged, therefore, that the public interests would be best subserved by refusing to grant such possibilities of monopoly to private concerns, and by keeping the public roads as the highways of the nation's commerce[2].

In addition to the above reasons for opposing canal construction, many trivial excuses were given for antipathy to particular inland navigations. Sometimes certain gentlemen would oppose such a plan because, they said, they had not had reasonable notice in regard to it, nor had they enough time to examine the proposed plan in order to judge how far their estates might be affected by it[3]. This would cause the postponement of the designed undertaking. In another instance a canal was frustrated by a nobleman merely because he thought the promoters of it had not treated him with becoming deference and respect[4]. A small legal quibble was sufficient, in other

---

[1] *County Chronicle and Weekly Advertiser*, Nov. 10, 1818, p. 4; *Cambridge Chronicle and Journal*, April 5, 1811, p. 2, and many other letters published in the same paper at later dates against the proposed North London Canal to connect the Cam and the Stort.

[2] *The Times*, June 17, 1836, p. 3.

[3] *Seasonable Considerations on a Navigable Canal*, pp. 5–7; Brit. Mus. 215. i. 1 (105), 'Thames Navigation;' *Reasons for Extending Navigation of River Calder*, p. 1.

[4] Billingsley, *Agriculture of Somerset*, p. 159.

cases, to cause deferred action, lest a slight deviation from parliamentary rules should "produce a precedent fraught with alarming dangers[1]." Others opposed a measure because there had not been enough public discussion of it and its probable results. Another ground of opposition was that in order to dig a canal there would be a great influx of strangers from other parts of the country; and if these were to become burdensome to the parishes, through the death of their husbands, the poor rates would be greatly augmented, the workhouses would be filled and the burden of maintaining these paupers might be greater than any advantage from the canal[2]. The complaint was made against the construction of the Chichester and Arundel Canal that the inhumanity of making that canal might curtail labour by throwing eight coasting smacks out of their employment and prevent sixty or seventy horses from being harnessed to the waggons carrying between London and Portsmouth[3]. Many other objections were made in specific cases; but we have shown in sufficient fullness the character of the opposition to these artificial waterways and shall leave the subject at this point[4].

The benefits which were anticipated from the construction of canals were so great that to many they seemed to contain the possibility of transforming the world. Farmers would now be able to take their produce to market by water at all seasons, and consequently a fair price for this produce would prevail throughout the year, so that the earlier known bread riots, due to high prices, would be no longer possible, unless there should be a universal scarcity or failure of the crops[5]. The opening up of wider markets which could be conveniently reached would tend to maintain uniformly good prices for farm products; while, at the same time, the cheaper cost of carriage would help directly to reduce the prices of provisions and the cost of living[6]. The expense

---

[1] *Seasonable Considerations on a Navigable Canal*, p. 6.

[2] *County Chronicle and Weekly Advertiser*, Nov. 10, 1818, p. 4.

[3] *Hampshire Telegraph*, Oct. 14, 1816, p. 3, letter of "Vetus." This complaint was met by showing that the closing of one channel of industry meant the opening of another and that the proprietors of waggons could change their occupation and become carriers on the canal or forwarders of goods in other ways.

[4] Many of these complaints against canals are illustrated in Publicola, *Utility of Inland Navigations*, p. 5 et seq.; Brit. Mus. B. 504 (2), 'Address to the Public on the New Intended Canal from Stourbridge to Worcester;' Meteyard, *Life of Wedgwood*, I, pp. 345–55, 385–6, 406–37.

[5] Whitworth, *Advantages of Inland Navigation*, p. 31.

[6] Brit. Mus. 213. i. 5 (94), 'State of the Case regarding Several Navigations,' p. 3; Brit. Mus. 213. i. 5 (95), 'A State of Facts to shew the Utility of Navigation from Witton to Manchester,' p. 2; Provis, *Suggestions on Canal Communication*, p. 5; Phillips, *Plan for a Navigable Canal*, pp. 20, 21; Brit. Mus. 8235. h. 44,

of carriage by road varied according to the season, the character of the roads, or the number of carriers; but the cost of carriage by canal was expected to be the same at all seasons and under all conditions for any commodity[1]. The opening up of the country in these ways would increase the value of the lands and promote their improvement, since by more extended markets better prices would prevail for agricultural produce, and from the towns manure could be brought toll-free on the canals with which to enrich the adjacent lands[2].

Another advantage which was expected to accrue from the use of canals was the decreased employment of horses for hauling, because one horse could draw in a barge on a canal as much as many horses could draw in waggons on the roads. On account of this presupposed decrease in the number of horses required for conveyance of products, it was assumed that land which had formerly been used for growing hay to feed horses could now be used for the production of wheat and other food products for the use of mankind. From four to eight acres were required to produce the hay that was necessary for every horse; and when we consider the great number of horses engaged in the work of carriage on the highways, it is evident that an immense amount of land was thus devoted to so-called unproductive purposes. It did not seem to be considered that the expansion of trade and industry, which was going on simultaneously with the development of the canal network, would make so much more work for horses in other forms of employment as to prevent any decrease in their number; for even as late as 1812 we find it said that one of the less obvious, yet solid, advantages of canals was the saving in the maintenance of horses and the consequent economizing of the provisions of the country to an extent that would appear surprising if reduced to calculation[3].

'Remarks relating to a Canal intended to be made from the City of Chester...,' pp. 4, 5; Brit. Mus., Maps 88. d. 13, 'Documents and Plans relating to Canals,' Nos. 5, 8, 12, 25; *Public Advertiser*, Oct. 29, 1791, p. 3, on 'New Canal;' *Cambridge Chronicle and Journal*, Feb. 7, 1812, p. 2, and Feb. 21, 1812, p. 2; *J., H. of C.*, xxx, p. 520; *J., H. of C.*, xxxii, p. 725; etc. A writer in 1817, in urging the construction of a canal between Newcastle and Carlisle, gave as a sufficient reason for such a canal that it cost more to convey corn by road between these two towns than to bring it from the Cape of Good Hope to Newcastle. Brit. Mus. 1302. g. 8 (3), 'Canal between the Eastern and Western Seas,' p. 4.

[1] Whitworth, op. cit., p. 11.

[2] Brit. Mus. 213. i. 5 (94), 'State of the Case,' p. 3; Provis, *Suggestions*, p. 5; Plymley, *Agriculture of Shropshire*, pp. 302, 305–6; Brit. Mus. 8235. h. 44, 'Remarks relating to a Canal intended to be made from the City of Chester...,' p. 6; *Cambridge Chronicle and Journal*, Nov. 12, 1813, p. 2.

[3] *Cambridge Chronicle and Journal*, Dec. 25, 1812, p. 3. Phillips, *Plan for a Navigable Canal*, pp. 20–23, shows a calculation that the amount of corn required

Along with this reduction of the number of horses that were required for the carrying trade, it was thought that there would be a corresponding decrease in the number of men who, as drivers, were not in the ranks of producers but were consumers. It was asserted that by drafting into agriculture those who would not be required in their former employment after the construction of canals, the productive forces of the kingdom would be increased and, therefore, much assistance would be given in supplying the markets with the necessary food-stuffs to keep all classes from suffering for want of bread[1].

The elimination of the wastes and the reduction of the cost of land carriage, together with the possibility of increasing the productivity of the land, were very substantial reasons for the prosecution of canal construction; but the urban centres also and the industrial establishments expected to secure important advantages from these greater facilities of conveyance. One of the constant complaints made by the poor to the magistrates was the hardships which they had to endure because of the high price of coal[2]. The inhabitants of the towns had been largely dependent upon the immediate vicinity for all the means of subsistence, because many articles could not stand the cost of carriage

to feed the horses then (1785) employed in carrying along that road would necessitate the use of nearly 1760 acres of land for that purpose. See also *Hampshire Telegraph*, Oct. 14, 1816, p. 3, letter from "Vetus" regarding 'The Intended Canal.'

Ralph Dodd, a well-known engineer, writing in 1800 about the proposed Grand Surrey Canal Navigation (p. 13), made this computation: "As one horse on an average consumes the produce of four acres of land, and there are 1,350,000 in this island that pay the horse-tax, of course there must be 5,400,000 acres of land occupied in providing provender for them. How desirable any improvement that will lessen the keep of horses, and save thousands of pounds in importing grain into the kingdom for their subsistence." We would not to-day make this distinction between productive and unproductive labour, for we regard all labour as productive which issues in utility; and certainly the movement of goods from one place where they have but little value to another where they have a high value gives place utility to the goods, and thus this service was productive.

[1] *Cambridge Chronicle and Journal*, Nov. 23, 1810, p. 2, letter from William Leworthy on the North London Canal. He said it was computed that one horse and three men were able to transport at once by barge as much as sixty horses and ten men could carry at once by waggons, and so there would be a saving of fifty-nine out of sixty horses and seven out of ten men. In ibid., Feb. 21, 1812, p. 2, there is a letter from the Earl of Hardwicke in favour of the Cambridge and London Junction Canal, in which he said that he had been assured by well-informed persons that waggon-horses in constant employ cost £50 each and consumed the produce of eight acres of land annually. If the canal should be the means of releasing 1000 horses from this employment, £50,000 and 8000 acres of land, together with the abour of their drivers, might be applied to more useful purposes, which would help to keep the labouring poor from suffering for want of bread.

[2] *Cambridge Chronicle and Journal*, Feb. 21, 1812, p. 2, letter from the Earl of Hardwicke, himself a magistrate.

by land over more extended areas. But with the opening of canals and the lower cost of carriage upon them, it was easily foreseen that the necessaries of life could be brought from more distant sources at a lower charge, and thus the urban population along and adjacent to the canals would be provided with a cheaper and more regular and abundant supply for the satisfaction of their needs[1]. By this means also manufacturing sections might be able to draw their coal, iron and other products from greater distances; and this would relieve the difficulty that was experienced by some towns in procuring the requisite amount of raw materials for manufacture[2]. These encouragements, it was confidently expected, would induce existing manufacturers to enlarge their facilities, to work with greater energy and productive effect, and to cause their communities to throb with a new vitality; while they would be effectual also in influencing other manufacturers to locate their establishments alongside of, or near to, the canals[3]. The effect of such increased industrial activity would, of course, be very marked in enhancing the value of land adjacent to these works; and thus both industry and agriculture would receive a new impetus in their development.

Another very significant reason why canals were desired in some places was that they would break up an existing monopoly of carriage by land or water, or would avoid the growth of such a monopoly, because of the fact that the waterway would be open for everybody's use, upon payment of the tolls[4]. We may exemplify this by the

[1] Brit. Mus. 213. i. 5 (94), 'State of the Case,' etc., p. 3; *Cambridge Chronicle and Journal*, Nov. 2, 1810, p. 3; Nov. 9, 1810, p. 2; Nov. 23, 1810, p. 2; Aug. 30, 1811, p. 2; Feb. 21, 1812, p. 2. Brit. Mus., Maps 88. d. 13, 'Documents and Plans relating to Canals,' Nos. 5, 6, 25. See the figures given in the prospectus of the London and Cambridge Junction Canal, in *The Times*, Nov. 8, 1811, p. 1.

[2] Brit. Mus. 213. i. 5 (95), 'A State of Facts,' pp. 1-2; Provis, *Suggestions*, p. 5; Plymley, *Agriculture of Shropshire*, pp. 291-2, 305-6; *Cambridge Chronicle and Journal*, July 10, 1812, p. 2; Brit. Mus. 8235. h. 44, 'Remarks relating to a Canal intended to be made from the City of Chester, to join the Navigation from the Trent to the Mersey, at or near Middlewich,' pp. 5, 7.

[3] Whitworth, op. cit., p. 35; Brit. Mus. 8235. h. 44, 'Remarks relating to a Canal to be made from the City of Chester...to Middlewich,' p. 7.

[4] Brit. Mus. 213. i. 5 (95), 'Remarks on the Observations on the Intended Navigation from Witton Bridge,' p. 4. The writer of this pamphlet warned the public lest the Duke of Bridgewater's exclusive privilege of water carriage, under the control of unprincipled agents, might not in future be "the instrument of the most oppressive exactions." This, as a matter of fact, came true, and was one of the chief reasons for projecting the Liverpool and Manchester Railway. But at the time the Duke's canal was constructed it resulted in breaking up the earlier monopoly enjoyed by the Mersey and Irwell Navigation Company. See also Brit. Mus. 213. i. 5 (94), 'State of the Case,' p. 3.

conditions which prevailed when, in 1810–12, there was an agitation in favour of a canal to join the river Cam, near Cambridge, with the Stort Navigation leading toward London. The Cambridge merchants had long enjoyed a virtual monopoly of the carriage of the produce of the Fen country tributary to it to the London market by waggon. As an emporium for the grain of the adjacent Fens, this town had a position of great advantage; and the merchants, who were also warehousemen and carriers by land to London, seem to have utilized their advantageous location to pay the farmers but a low price for their grain and to charge the public as high a price as possible for this and other produce consumed. But not all the grain from the Fens went by land carriage from Cambridge to London; part of it went from the farms down to Lynn and thence to London by sea. In this case, Lynn also became a monopolistic emporium like Cambridge; and there seems to have been a partial partnership or mutual understanding between the merchants of these two places so that they could hold the trade of some of the richest counties of England. In the above-mentioned years, efforts were made to secure a canal to connect the Cam and the Stort, so that the producers of the grain, instead of being compelled to sell at low prices to these merchants, could float their produce in barges directly past Cambridge and either sell it at places en route or else take it to London, where it would command the highest prices. But although parliamentary sanction was given for the construction of this proposed canal it never materialized[1].

Some canals were put through in order to rectify the abuses or the inadequate facilities of river navigations. In this connexion we have already shown the earnest desire on the part of many to see the Thames Navigation improved and straightened by cutting one or two canals which would overcome the shallows or obviate the circuitous course of the river[2]. Similar considerations were effective in the case of other large rivers. The Bridgewater Canal from Longford bridge to

---

[1] *Cambridge Chronicle and Journal*, Nov. 23, 1810, p. 2; Sept. 13, 1811, p. 4; July 3, 1812, p. 2; July 10, 1812, p. 2; Nov. 13, 1812, p. 2; Sept. 10, 1813, p. 1; Oct. 8, 1813, p. 3; Nov. 5, 1813, p. 2; April 15, 1814, p. 3. A study of these sources reveals to us what was evidently a strongly entrenched monopoly on the part of the carrier-merchants, which for some time prevented the passage of hostile legislation and after its passage prevented its execution.

[2] See, for example, *Reading Mercury*, Nov. 25, 1793, p. 4, on 'Thames Navigation;' Dec. 9, 1793, p. 4, letter from "An Old Navigator;" Dec. 30, 1793, p. 4, letter from "An Old Navigator;" Jan. 13, 1794, p. 4, 'Considerations on a proposed Line of Canal from Reading to London, through Windsor;' Mar. 23, 1794, p. 4, Petition from the Corporation of London, to the House of Commons, in favour of the Canal from Datchet to Isleworth. *Public Advertiser*, Oct. 29, 1791, p. 3, on 'New Canal;' Dec. 23, 1791, p. 1, letter of "Mercator."

Runcorn was intended to make a safer, shorter and more regular communication between these termini, by avoiding the shallows, the shifting channel, the uncertainty and the meandering of the Mersey and Irwell. In the Ouse river up to York efforts were made to straighten the navigation by lateral cuts, so as to permit a greater amount of flood tide to come up the river, by which larger barges might be borne up and down on its surface. The corresponding improvement of the Severn, the Bristol Avon, the Trent, and the Great Ouse in the Fen district will be sufficient illustration of the desirability of remedying the natural defects of rivers by means of artificial lateral cuts, sometimes adjoining and at other places more remote from the river.

Other reasons that were given for the necessity of canals reflect the local conditions and the spirit of the time. It was thought that taking the heavy traffic from the roads and carrying it by canals would tend to the preservation and repair of the highways, many of which were in bad condition because of the ponderous loads that were drawn along them[1]. Nearly all of the canal Acts state in the preambles that this was one good result that was expected to be accomplished by the waterways[2]. In other sections the roads were bad, not only because of the excessive traffic but also on account of the great difficulty of getting good materials for repairing them; and the prospect of being able to bring in such materials by canal and to take off the roads the heaviest of the commodities to be carried furnished an inducement to the building of canals[3]. In addition to the foregoing, many other advantages were expected to accrue: the construction of canals would give work to the poor; their operation would furnish employment for and increase the number of watermen and seamen; and, in fact, it was usually stated that these inland waterways would provide a veritable nursery for sailors and seamen, so as to be effective contributors to one of the great aspects of England's power[4].

---

[1] Brit. Mus. 213. i. 5 (94), 'State of the Case,' p. 3; Phillips, *Plan for a Navigable Canal*, p. 23.

[2] See, for instance, Act 7 Geo. I, stat. 1, c. 15, 'An Act for making the Rivers Mersey and Irwell Navigable.'

[3] *Cambridge Chronicle and Journal*, July 24, 1812, p. 2, "A Cantab's" argument continued. He speaks of the unparalleled bad state of the roads in districts south of Cambridge, which included one of the great north roads. By the proposed Cambridge and London Junction Canal, it was presumed, the trade would cease to be on the roads; and none but the inoffensive wheels of the traveller would be left to pass that way.

[4] Act 7 Geo. I, stat. 1, c. 15, preamble; Phillips, op. cit., p. 23; Whitworth, op. cit., p. 35.

With the prospect of securing such results, we cannot wonder that large sums of money were spent in order to obtain the requisite authority for the construction of canals, and that the completion and opening of canals were occasions of the greatest rejoicing. At these times, there was usually an assemblage of distinguished public men gathered together for the celebration of the great event; and as the first boats and barges, bearing the honoured guests and the initial load of traffic, passed along the route, accompanied by bands of music and other elements of popular demonstration, the acclaim of the multitude rose to a high pitch. The ringing of bells, the concourse of people, the speeches of the day, and the grand fête which frequently formed the climax showed that, in the public estimation, a new era had dawned for that locality[1].

When we turn from the anticipated advantages to look at those which were actually obtained from the use of canals, we must keep in mind the great difficulty, before 1760, in the carriage of goods by land and the consequent expense connected with it. These conditions were inconsistent with any great development of trade and with the maintenance of a large industrial population; and so the first and most obvious effects of inland navigations were the great diminution in the cost of carriage and the opening of easy communication between the distant parts of the country, and from those parts to the sea. In order that England should take front rank in commerce she had to sell her products and manufactures at the lowest prices; to be able to sell her manufactures at the lowest prices her raw materials had to be obtained as cheap as possible and her finished products conveyed to market at the lowest expense; and this, in turn, depended on cheapness of carriage[2]. Along with the expansion of transport

---

[1] Note, for example, the celebration of the opening of the Rochdale Canal and of the Worcester and Birmingham Canal as given in *The Times*, Dec. 27, 1804, p. 3, and April 11, 1807, p. 3.

[2] The question as to how much cheaper water carriage was than land carriage will be more fully discussed later; but for the present we shall merely say that the former cost by land carriage was greatly reduced by the development of water routes. What may be implied by the statement that the cost of carriage by canal from Birmingham to the Severn, Trent and Mersey was "at an easy expense," for both raw materials and manufactured products, it is difficult to conjecture. (Birmingham Free Central Library, No. 73,742, Timmins' *Collections of Views, etc., of Birmingham and District*, p. 27.)

The *Morning Chronicle*, April 1, 1791, p. 4, printed a letter on 'Navigable Canals,' in which it was stated that, on the whole, they had been of great service to the country in reducing the price of carriage and supplying large districts which were almost destitute of fuel with large quantities of that necessary commodity at reasonable rates. See also Cowdroy's *Manchester Gazette*, Feb. 19, 1825, p. 4,

facilities there was the simultaneous industrial advance of the country, although the latter was probably not wholly due to, but correlative with, the former.   Certainly without the improvement in transportation the Industrial Revolution by which England took the lead of all the European nations in establishing industry on a modern basis could not have proceeded so rapidly.

Inland navigations not only aided existing manufactures in the ways above mentioned but also occasioned the establishment of many new manufactures in places where formerly the land was of little value and almost destitute of inhabitants.   When a new factory was looking for a suitable place at which to locate, the cheapness of a site was a desideratum; and this could be obtained at a more reasonable price in the country than in the populous city.   Moreover, the latter frequently had no more facilities for transportation than the former; for it was only in the few cases where a town had more than one navigable waterway that there was any advantage of location there rather than in the country[1].   In this way industrial establishments were sometimes induced to get out into the country along the bank of a canal, where the operating expenses would be cheapest and the shipping facilities the greatest; and when once settled there they did not want to move again when the railways came through the towns.

Both domestic and foreign trade felt the impetus from improved means of conveyance; and the merchants who resided at the ports where these canals or their connexions terminated were also benefited by having a wider range of country upon which to draw, not only for the products which were consumed in the port towns but also for those which were exported to distant places.   In like manner, these merchants had a wider territory to serve with foreign goods imported. The reduction in the expense of carriage for long distances put the remoter parts of the country more nearly upon an equality with those near large towns; and in this way, while the towns could serve and be served by wider areas, the volume of trade, due to the importing of a larger quantity of goods and the exporting of a greater surplus, would be considerably augmented.   The increasing amount of foreign

showing that by the Leeds and Liverpool Canal a plentiful supply of Wigan coal was brought into Liverpool to meet their heavy demand, increasing from 50,000 tons in 1780 to over 91,000 tons in 1786 (almost double) and to an average of 200,000 tons per year in the four years ending 1823.

[1] Phillips, *History of Inland Navigation* (1803), p. vi.   The Aire and Calder Navigations afforded good illustrations of the utility of inland navigation in increasing established manufactures, and in the encouragement of new ones.   Brit. Mus. B. 504 (4), p. 3.   See also Collins, *Treatise on Inland Navigation*, p. 10.

trade and the enlarging internal trade stimulated the growth of industry and the development of the country's resources, until in many instances they had entirely changed the appearance of the counties through which the canals were built[1]. Commercially and industrially, then, a new activity was imparted, and all felt the benefit in cheaper coal, iron, lime and other raw materials, and in the diffusion of produce and manufactures[2]. The wider distribution of products, due to the lower cost of carriage, enabled all classes to be better provided; and the increasing volume of trade permitted merchants to receive larger returns in the way of profits from their business[3].

Much was said at the time about the desirability of inland navigation in order to render unnecessary the maintenance of the vast numbers of horses which were, in the current opinion, employed unproductively on the coaches and stage waggons throughout the country. We have hitherto shown some of the results that were hoped for from the introduction of canals in the way of preventing the continuance of such great waste for unproductive purposes. There is no doubt but that the substitution of water conveyance for land conveyance in the case of heavy, bulky commodities would reduce the number of horses that were necessary for carrying on this phase of the transport service, for one horse drawing a canal barge would do as much effective work as many horses attached to carriers' waggons. Peel, in 1825, testified that in the neighbourhood of the place where he lived scarcely one waggon was at this time employed in the carriage of goods where formerly twenty to thirty waggons were constantly in use[4]. But while horses were being thrown out of employment along

---

[1] Phillips, *History of Inland Navigation* (1803), p. vii.

[2] An editorial in *The Times*, May 9, 1846, pp. 4–5, said that the inland navigation of England had been a main essential of its prosperity. Canals had made towns, opened secluded regions, peopled solitudes and communicated vast advantages to numerous inland provinces of the country. Canals had joined the factory and the port, the country and the town, and seemed to open a bright prospect to national fortunes. The enthusiasts of the eighteenth and nineteenth centuries saw in canals the pledge of millennial regeneration and happiness. To the canals the people referred with pride and their advantages seemed the *ne plus ultra* of improvement.

[3] Defoe, *Tour through the Whole Island,* ii, p. 331, iii, p. 58. In 1781 Hull began to feel the most important effects from the canals which communicated with that place. In that year the Hull custom-house paid into the Exchequer the net sum of £101,393; and because of the increase of the canals in that part of the country this sum was augmented in 1784 to £143,467. In 1791 it was £171,000 and in 1792 it was £200,000. This would seem to be a proof that canals did increase trade (Dodd, *Report on the Intended Grand Surrey Canal Navigation,* p. 11). The whole subject as to the influence of the canals, especially in the matter of rates, is discussed in the 'Second Report of the Select Committee on Railways and Canals Amalgamation, 1846.'   [4] *The Times,* Mar. 3, 1825, p. 2.

the great canal routes the volume of internal trade was continually increasing; and it must have required a larger number of horses to operate along the lateral lines of trade in bringing the goods to and from the canal depôts. Moreover, the amount of travel in the later eighteenth and early nineteenth centuries must have been considerably in excess of that which formerly prevailed, because the necessities of the great industrial, commercial and agricultural revival must have quickened the national pulse; and, in consequence, instead of a decrease there doubtless was a vast increase in the number of horses that were requisite for the adequate fulfilment of this public service. On the whole, therefore, it is but reasonable to conclude that canals were instrumental in producing a greater amount of employment for horses, although the particular direction in which their labour was exerted was inevitably materially altered[1].

In addition to the benefits to the industrial and trading classes, the agricultural interests were profited. The reduction of the cost of carriage, which put the distant parts of the country more nearly upon an equality of advantage with those near the large urban centres, would open wider markets for agricultural produce and encourage the cultivation of such ulterior sections. By bringing into cultivation these areas most remote from the markets, rents were thereby increased; so that while the farmers were benefited by high prices for their produce which they could sell in a more widely extended market, the landlords were also benefited by higher rents[2]. While these distant areas were thus aided, the people of the towns were also benefited by breaking up the monopoly of the country in their vicinity. When speaking of the turnpike roads, we noted that the extension of these roads back into the country was opposed by the landowners near the metropolis, because they were afraid that foodstuffs would come to the city from the more remote parts and thus overthrow their supposedly secure monopoly of the supply of that great market. The same considerations applied in the case of the canals[3]; undeveloped regions were given access to markets and the advantages of the more favourably situated

---

[1] The same thing was made the subject of complaint at a later time when railways were being introduced, but with this difference, that now it was thought expedient to release horses from the "unproductive" work of carrying, while at the later time it was one of the bitter complaints of those who horsed the coaches that the railway would remove the necessity for horses—which was not true.

[2] The writer of the *History of Inland Navigations, particularly those of the Duke of Bridgewater*, p. 10, tells us that "in many instances their lands have been improved to tenfold value;" but it would seem as if this must have been somewhat exaggerated.

[3] Phillips, *History of Inland Navigation* (1803), p. xi; Dodd, *Report on the Intended Grand Surrey Canal Navigation*, p. 12.

portions of the country were soon extended to those areas which had previously been more backward. As an illustration of what we have just said, we may refer to the fact that Bagshot Heath, formerly bleak, miserable, uninhabited for miles and scarcely capable of supporting a few sheep during a small part of the year, was transformed into a rich arable country studded with villages within a few years after the Basingstoke Canal was constructed[1]. Formerly an unfavourable situation with regard to carriage often prevented the surplus of heavy and bulky commodities from being of any value to their owners, since they were not valuable enough to stand the expense of transportation; but by the great reduction of this expense a deficient supply in one section might be amply made up by the superabundance in another. Then, too, manure, lime, marl, and other things for fertilizing the soil, could be conveyed at slight expense and used for bringing into cultivation the poorer and waste lands, all of which were necessary for furnishing food products to the continually increasing industrial population.

The network of inland navigations which overspread the country proved to be the best means for carrying the heavy commodities, such as coal, stone, corn, and the like; and by taking the conveyance of these off the roads the waterways must have contributed materially to the stability and perfecting of the public highways. As we have seen, there was much complaint that narrow wheels and heavy loads were ruining the roads; and many had come to the conclusion that neither statute duty nor turnpikes, even with the additional provisions of broad wheels and limitation of the number of horses, could effectually keep them in repair without the assistance of canals. Of course, without the adoption of improved principles of road-making no permanent results for the highways would have been gained, but, doubtless, one of the great impediments to their improvement was ameliorated by transferring a considerable portion of the heavy carriage to the canals. It is impossible to measure by any kind of calculation what share of this improvement of the roads was due to the indirect influence of the canals and what part was due to the other agencies, already considered, for securing this end directly; but it was probably true in this, as in many other instances, that results secured by indirect means were fully as significant and important as those obtained by direct methods.

---

[1] *Cambridge Chronicle and Journal*, June 26, 1812, p. 2. The writer; "A Cantab," said that there was no need to show the public utility of inland navigations generally, since they were already too well known and too gratefully felt. See also *The Times*, May 9, 1846, pp. 4–5, an editorial regarding the competition of rail and canal; and Phillips, *History of Inland Navigation* (1803), p. vi.

Another result of canal construction, concerning which we have but meagre information, was the diversion of traffic from one place to another or from one course to another. As we found, when studying the roads, that the construction of a better road would tend to turn the greater part of the trade and travel from an earlier and poorer road to the later and mechanically superior road, so was it, in all probability, with the canals: the shorter and cheaper route would tend to draw the traffic from the more circuitous and expensive route, and the waterway which was more favourable for the passage of barges would tend to prosper at the expense of its less favourably situated or constructed rival. Probably this was one of the reasons which induced the Staffordshire and Worcestershire Canal to oppose the granting of an Act to the Worcester and Birmingham, for the latter was to connect with the Severn river at a point lower down, and was to reach Birmingham by a more direct route than the earlier canal. We are informed that the construction of the Staffordshire and Worcestershire Canal had turned the trade away from Bewdley to Stourport, in consequence of which the former decayed and the latter flourished[1]; and, in the face of this reality, it would be but natural for this canal company to foresee that if another canal were made to join the Severn still farther down, at Worcester, Stourport and the Staffordshire and Worcestershire Canal traffic might in their turn decline, while Worcester and the canal from there to Birmingham might correspondingly increase in importance. We have shown that this was one of the most influential considerations with the Thames Commissioners and others who opposed the execution of one or two canals out of that river to overcome the impediments of the navigation, lest the trade might be completely diverted from the river to the canals and thus cause the partial or total decay of the former. The whole subject is one which is very elusive and would require the close examination of a great quantity of local material for its satisfactory elucidation.

Then there were advantages which canals had over navigable rivers which placed them in the forefront of the waterways. For instance, canal navigation was not dependent upon wind and tide, as was the case with the navigation of most rivers; canals were not so subject to the process of silting up as were rivers, and, therefore, there was no liability to stoppage by shoals and similar obstructions; on the canals, passage boats, where in use, had definite times of arrival and departure, which was impossible on rivers where navigation was available only by means of the tide and where it was subject to all accidents of wind and weather. And, lastly, canals could be put

[1] Felix Farley's *Bristol Journal*, Jan. 21, 1786, p. 1.

through sections of country which were not naturally provided with any waterway, and consequently the benefits spoken of above might obtain even there.

From the foregoing, we now turn to the financial results that accrued to the canal proprietors from the operation of their waterways. In many instances the canals not only furnished the public with good transportation facilities but brought to their owners ample rewards as paying investments, especially when they traversed districts that had sufficient traffic to warrant their construction. But in some sections where the amount of traffic was small in comparison with the great cost of construction, or where the expenses of operation were high because of natural obstacles or poor connexions with markets, canals earned for their proprietors but scant return upon the capital embarked in them. This was true not alone of the canals in the more remote portions of the kingdom, as, for instance, in the extreme north and the far south-west; but even in the very centre of the realm they were frequently a declining property[1]. Some were

[1] Vallance, *Sinking Capital in Railways* (1825), p. 98, quoting from Wickens, gives the following facts regarding the original worth and present value of some canal shares, showing the decline in value that had taken place:

Name of Canal	Original worth		Present value	
	£	s.	£	s.
Bolton and Bury	250		100	
Grantham	150		126	
Brecknock and Abergavenny	150		60	
Ellesmere and Chester (united)	138		75	
Oakham	180		45	
Wey and Arun	110		68	
Wisbeach	105		60	
Chelmer and Blackwater	100		90	
Leicester and Northampton	100		87	10
Montgomery	100		70	
Dudley	100		59	
Gloucester and Berkeley	100		54	
Grand Surrey	100		54	
Basingstoke	100		50	
Rochdale	100		45	
Grand Union	100		37	
Thames and Severn	100		17	
Ashby-de-la-Zouch	100		15	
Andover	100		10	
Sleaford	100		5	
Ashton and Oldham	97	18	65	
Worcester and Birmingham	79		25	
Peak Forest	78		63	
Stratford-on-Avon	75		16	10
Regent's	49		28	
Kennet and Avon	40		19	10
Wilts and Berks	20		8	
etc., etc.				

It must be said, as a word of caution, that the foregoing list should be received

begun but never finished[1]; others were never even begun after the Act was obtained[2]. Some struggled along with very indifferent results, paying little or nothing upon the capital[3]; and others got

with reservation. I cannot vouch for its strict accuracy. It seems somewhat strange, for example, that the Grantham and the Ashton and Oldham canals should appear in this list of canals whose shares were declining in value and also in Momsen's list of canals that, in 1830, were paying over six per cent. dividend (see footnote 6, page 425). It is possible that both statements were true, on account of the fluctuations in value of many canal shares. Again, by comparing the original values of some shares in this list with the corresponding figures given in *Herepath's Railway Magazine* (1838), iv, pp. 384–8, as taken from Fenn's *English and Foreign Funds*, we do not find strict accordance; for example, the original worth of Ellesmere and Chester Canal shares from these two sources are, respectively, £138 and £133; Leicester and Northampton are £100 and £83. 10s.; Rochdale are £100 and £85; Ashby-de-la-Zouch are £100 and £113; and Regent's Canal shares are £49 and £83. 17s. 6d. respectively.

[1] For instance, the St Columb Canal in Cornwall, authorized by Act 13 Geo. III, c. 93. For its history see Hitchins, *History of Cornwall*, i, p. 514. Also the Tamar Navigation and Great Western Canal (Moore, *History of Devonshire*, pp. 47, 50), and the Crediton Canal (ibid., p. 51). The Salisbury Canal, after spending the original £100 per share and an additional call of £50 per share, abandoned its works (*County Chronicle and Weekly Advertiser*, Nov. 10, 1818, p. 4; *Royal Commission on Canals and Waterways*, 1906, Vol. i, Pt. ii, Minutes of Evidence, Q. 10877). From the *Report of the Royal Commission on Canals and Waterways*, 1906, Vol. i, Pt. ii, Minutes of Evidence, Q. 10861–10883, and in Appendix No. 1, statement No. 5, and Appendix No. 9, statement No. 1, we find a list of canals which were partially or totally derelict and the history of each is given.

[2] For example, the Bude Haven Canal in Cornwall, authorized by Act 14 Geo. III, c. 53, to be constructed from some part of Bude Haven to the river Tamar in the parish of Calstock (vide Hitchins, *History of Cornwall*, i, p. 515). Another case was that of the Polbrook Canal (ibid.). For other examples, see Moore, *History of Devonshire*, p. 51.

[3] Allen, in his *History of Liskeard*, p. 382 (note), says that it is believed that most of these undertakings in Cornwall have made very poor returns to the shareholders. The Liskeard and Looe Union Canal was an instance to the contrary, but even this canal did not pay dividends of more than four to four and one-half per cent. (ibid., p. 382). The writer of *Observations on the Comparative Merits of Navigations and Railroads* (p. 47), tells us that the three canals connecting London and Bristol completely failed to remunerate their projectors. This fact is attested also by the figures showing the declining value of these canals (see page 416, footnote 1). The Ashby-de-la-Zouch Canal, in 1828, paid its first dividend, amounting to two per cent., although the canal was completed more than thirty years before. The Stratford-on-Avon Canal paid only one and one-half per cent. about 1830; and the Worcester and Birmingham about the same time paid only two per cent. (*Birmingham Journal*, May 31, 1828, p. 3; West, *History of Warwickshire*, pp. 101–102, 104, 105). The profits of the Chesterfield Canal in 1789 were merely sufficient to pay interest on the amount the proprietors had to borrow to finish their canal; so that the stockholders evidently obtained no return on their investment (Pilkington, *View of the Present State of Derbyshire*, i, p. 281). The returns to the shareholders of the Tamar Navigation, only about three miles of which were constructed, were

into the hands of men who reduced the freight rates for their own profit and cared little about a dividend for the stockholders[1].

We have admitted that if many canals failed many others succeeded[2], and it would be desirable if we could find the relative proportions of each. But this is not an easy problem to solve, because canal companies did not directly make known their rates nor their financial results to any extent. We have, however, other sources of information of a private and of a public character which give us, not accurate facts, but statements which are as nearly true as we can obtain, until such time as certain valuable private collections are made available for our inspection. Among these sources, perhaps the statistics of

very small up to 1829 (Moore, *History of Devonshire*, p. 47). The Tavistock Canal, at the same time, afforded its proprietors but a small rate of interest (ibid., pp. 47–48); and the Bude Canal with its branches, which was completed in 1826, had paid no return on the capital invested up to 1829 (ibid., p. 49). The Sheffield Canal was not built for the profit of the canal promoters; but it was clearly understood that public benefit and convenience were the main objects of its projectors. It was calculated that in time three and one-half per cent. on the outlay might possibly be obtained; but on account of the bad location of the waterway and its high cost, up to 1835 it had not paid one per cent. (*Sheffield Iris*, April 7, 1835, p. 2). The shares of the Croydon Canal, the original value of which was £100 each, paid a dividend of one per cent. for the first few years, but nothing after that; and by 1818 these shares were selling for about £4 each. It would seem, also, that the Thames and Medway Canal totally failed, and the subscribers lost all the money they had put into it (*County Chronicle and Weekly Advertiser*, Nov. 10, 1818, p. 4, and Nov. 24, 1818, p. 2). The Basingstoke Canal was unprofitable, for in 1818 its £100 shares were selling for less than £10 (ibid.). The Kennet and Avon Canal paid no dividend from 1800 to 1814; in the latter year it paid 11s. upon an original £130 share; for two succeeding years it paid 15s. each year; and the next two years furnished a total dividend of but 17s. 6d. (ibid.). Many other canals, traversing districts of large population and resources, likewise failed in the prospects held out of remunerating the proprietors, as instances of which we may mention the Huddersfield, the Rochdale, the Lancaster, the Worcester and Birmingham, and the Ellesmere Canal (*Cambridge Chronicle and Journal*, Dec. 25, 1812, p. 1; ibid., Jan. 1, 1813, p. 1). When the Bill for the Cambridge and Saffron Walden Canal was before the House of Lords, in 1814, Lord Redesdale said that after investigation he had found a few canals which were profitable to their proprietors and to the public, but that by far the larger proportion were either partly finished and then given up, or when completed had been ruinous to the subscribers. There were, according to his statement, thirty or forty which were in this condition (*Cambridge Chronicle and Journal*, June 24, 1814, p. 3). We are informed that in 1790 many canal proprietors were subject to continual calls and yet never got any dividend (*Public Advertiser*, Feb. 2, 1790, p. 1, letter from "The Inland Navigation Reformer").

[1] This was the case with the navigation made from Fisher's Cross (renamed Port Carlisle) on the Solway to the city of Carlisle (Ferguson, *History of Cumberland*, p. 279).

[2] See also *Hampshire Telegraph*, Oct. 14, 1816, p. 3, letter from "Vetus" regarding 'The Intended Canal.'

the Stock Exchange are about as useful as any that we can obtain; and from Wettenhall's List of shares, which was published by the authority of a committee of the Exchange, we learn that in 1816 there were thirty-nine canals which were wholly unprofitable. Of these, two canals paid annually £2 per share of dividend, one paid a dividend of £1 per share, one paid 17s. 6d. and all the others paid no dividend at all[1]. In accordance with this was the statement, made in the same year by a celebrated engineer, that there were some canals which paid well, but they were few in comparison with those which paid meagre returns or nothing at all[2]. In 1825 an investigation was made regarding the productiveness of the canals and from an analysis of the reports of eighty of these corporations it was ascertained that, in the aggregate, the dividends of that year amounted to five and three-fourths per cent. upon the capital. But on further analysis we find that, of the total capital of £13,205,117, only £3,200,530 paid dividends of 10 per cent. and over, while £10,004,588 paid dividends of less than 10 per cent.; and of the aggregate capital £7,808,588 paid dividends of less than two and one-half per cent. Out of the total capital £3,734,910 had not, up to that time, paid any dividends to the original subscribers[3]. If we may take the figures which were collected at that time with care as being approximately correct, it is apparent that a large part of the canal capital was very inadequately remunerated. But we must go one step further in order to safeguard our conclusion. For 1838, we have a summary of the principal English canals, the shares of which were marketable in London, and the dividends are given in most cases[4]. Upon grouping these canals according to the amount of the dividend paid, we get the following results, namely: Out of the total number of fifty-six canals, there were six which paid above twenty-five per cent., thirteen which paid between ten and twenty-four per cent. inclusive, nine which paid between five and nine per cent. inclusive,

[1] The names of these unprofitable canals, together with their market values per share and the amount of dividend paid, are transferred from Wettenhall's List to the letter from "An Enemy to Delusion," appearing in the *County Chronicle and Weekly Advertiser*, Nov. 24, 1818, p. 2. Among those which paid no dividend were such canals as the Worcester and Birmingham, the Stroudwater, the Stratford-on-Avon, the Wilts and Berks, the Ashby-de-la-Zouch, the Basingstoke, the Grand Union, the Grand Surrey, the Gloucester and Berkeley and the Leeds and Liverpool.

[2] Sutcliffe, *Treatise on Canals and Reservoirs*, p. iv. He said there were many canals that would never pay any return upon the amount subscribed toward their construction nor upon the amounts advanced as loans. He thought shareholders should receive at least five per cent. on their investment (pp. ii, vi).

[3] *Quarterly Review*, xxxii, pp. 170–71.

[4] *Herepath's Railway Magazine* (1838), iv, pp. 384–8, taken from Fenn's *English and Foreign Funds*.

nineteen which paid from nothing to four per cent. inclusive, and nine which did not report any dividend, but which certainly belonged in the list of those paying from nothing up to four per cent. Rearranging these in another form, we find that one-half the number of canals belonged to the class of those which were earning four per cent. or less, and the other half to the class earning five per cent. or over. From the list given, it is clear that there were many of the poorer canals omitted; and if they had been included the number of those earning below four per cent. would have been much augmented. We are, therefore, easily within the limit of safety when we say that at least one-half, and possibly two-thirds, of the number of canals in England were recompensed by dividends below that which was recognized as a reasonable minimum; and a considerable number of important canals were paying but one per cent.

Under these circumstances, there must have been a large amount of capital that was sunk in unproductive enterprises; and we have already learned that at the time of the canal mania of 1792–3, the funds paid over were not usually those of the rich, but of men and women living by their daily work, clerks, and many others of similar station, who embarked their all in such projects, thinking thereby to assist themselves in making ample provision for their years of age. The whole system seemed to them like a veritable *el dorado* and when once they had invested their savings in these undertakings they could not be withdrawn. It was these who suffered in the ruin and annihilation of capital, or, in the words of a discerning engineer, "a great part of the capital sunk in making modern canals in this kingdom has been found upon enquiry to belong to those who can ill spare it[1]." It will assist us to realize how great were these losses if we give a few statistics, to show that sometimes only a small amount was lost, while in other cases almost the entire amount expended was swallowed up[2]. We may refer, *en passant*, to the contrast between the canals and railways in this respect; for in the railway mania of 1844–6 it was the wealthy who were subscribers, and it was these, therefore, who had their capital tied up in ventures that were sometimes unremunerative for a considerable time.

[1] Sutcliffe, *Treatise on Canals and Reservoirs* (1816), p. vi.
[2] In Felix Farley's *Bristol Journal*, Jan. 4, 1834, p. 2, letter of "Caveat Emptor," we have some facts, taken from the daily share lists of the London brokers, and so arranged as to show in tabular form how much was lost in some particular instances through the decline in value of the canal property. The examples here set forth were all in the south of England; but it is not to be inferred from this summary that all canals were equally unprofitable, although it was apparently true that, as a rule, the canals south of the Thames and its navigable connexion with the Severn

But it not infrequently occurred that a navigation which at the beginning, or in the early days, of its existence was not a paying concern, paid well after a time. For instance, in 1774 the shares of the Bridgewater Canal would scarcely bring £20 each; but by 1804 they had advanced to over £120 per share[1]. A still more notable case is that of the Monkland Canal, near Glasgow, which was in such adverse circumstances at one time that its shares of the par value of £100 were selling for only £5 and £7, and it was seriously contemplated to fill it up. Shortly after, however, on account of the development of the mineral riches of the district, the shares rose to be worth £3200[2]. The change from an unprofitable to a remunerative undertaking was due to various causes and differed in different instances: sometimes, as above, it was the result of opening up new sources of wealth; in other cases it came through securing a connexion with a profitable canal[3], either by way of a working agreement, or by furnishing an outlet and wider market for local products[4]. Sometimes, as in the case of the Don, improvements in the navigation itself led to its

were not paying concerns. This tabular view is intended to show, in a few specific cases, the extent of subscribers' losses.

Canals	No. of shares	Average cost per share	Amount expended	Present price per share	Present value = present price per share × No. of shares	Loss = original value − present saleable value
		£	£	£	£	£
Andover	350	100	35,000	30	10,500	24,500
Basingstoke	1,260	100	126,000	5¼	6,615	119,385
Bridgewater and Taunton	712	100	71,200	70	49,840	21,360
Croydon	4,546	31	140,926	1	4,546	136,380
Grand Surrey..........	1,521	100	152,100	23	34,983	117,117
Grand Western	3,096	100	309,600	21	65,016	244,584
Kennet and Avon......	25,328	40	1,013,120	26	658,528	354,592
Portsmouth and Arundel	2,520	50	126,000	20	25,200	100,800
Regent's	21,418	34	728,212	17	364,106	364,106
Thames and Medway ...	4,805	30	144,150⎫			
Thames and Medway new shares	3,344	3¼	11,704⎭	15s.	6,111	149,743
Thames and Severn	1,300	100	130,000	28	36,400	93,600
Thames and Severn new shares	1,150	50	57,500	33	37,950	19,550
Wilts and Berks	20,000	17	340,000	5¼	110,000	230,000
Wey and Arun	905	110	99,550	23	20,815	78,735

[1] *Gentleman's Magazine*, LXXIV, Pt. II, p. 1131.

[2] Leslie, *Inclined Plane on Monkland Canal*, p. 4.

[3] This was the case with the canals which have been merged into what is now called the Shropshire Union Canal.

[4] Ibid.

earning good dividends[1]; and in other cases it was merely the development of the industry of the district through which the canal passed
that made the canal prosperous. The Ashby-de-la-Zouch Canal
whose shares, which originally cost £118 but subsequently fell to,
and remained for some years at, £10, revived in public estimation,
until in 1828 they were selling for more than £60[2], and the dividend
of two per cent. in 1828—the first dividend paid since the completion
of the canal more than thirty years before—was doubled within the
next ten years, for in 1838 it was four per cent.[3] These are only a
few of the examples that we might mention; but they are sufficient
to show us how much canal shares might fluctuate in value, and to
warn us against making hasty conclusions in regard to the desirability
or otherwise of these shares for investment purposes.

Some canals were paying investments almost from the start. The
traffic of the districts they served was large enough to ensure this
result before they were constructed. This prosperity was due in
some cases to the favourable location of the canal, as tributary to
a large manufacturing centre[4], or as an important link in a longer
through route[5]; in other cases, to an abundant supply of some local
product, which was finding its way to the great cities for consumption[6].
But, as a matter of fact, the prosperity of most canals was subject to
much variation; and since they usually refused to ally their interests
with others and were, therefore, dependent upon the local traffic for
their revenues, it is evident that the prosperity of the canals was
closely bound up with the prosperity of the immediate constituency
which they served. As the latter varied, so did the former.

We have exemplified the consequences of this fluctuation and of
other and more stable causes in bringing many canal shares to an
unremunerative basis; and it will now be in order to show the opposite
result, namely, how the prices of some shares rose to a high level and

---

[1] Leader, *History of the Cutlers' Company*, i, p. 171. Another example was the
Birmingham and Fazeley Canal, which, by 1791, was so improved that a share
which cost £140 had been sold by auction for £1080 (*Annual Register*, xxxiii, p. 47).

[2] *Birmingham Journal*, May 31, 1828, p. 3.

[3] Ibid.; also *Herepath's Railway Magazine*, N.S., iv, pp. 384–8.

[4] This was exemplified by the Birmingham Canal, which served that industrial
centre; and also by the Grand Trunk Canal, which served both the iron and pottery
districts, as well as the salt district of Cheshire.

[5] The Grand Junction Canal and the Oxford Canal are examples.

[6] For instance, the success of the Chesterfield Canal was due to the great coal-
mines near-by. The same thing is true of the Loughborough and the Leicester
canals. The success of the Weaver Navigation was due to the fact that it tapped
the immense salt wells of Cheshire.

paid enormous dividends. Perhaps we can best do this by taking a few illustrations. In 1825 two Coventry Canal shares, of the par value of £100 each, sold at the auction market in London, one for £1220 and the other for £1230[1]; and in 1828 they were still selling for £1220[2]. But by 1834 the value of these shares had decreased to one-half this amount, and then stood at £610[3]. In 1818, 1825 and 1829 these shares received an annual dividend of forty-four per cent. and a bonus[4], while in 1838 the dividend had increased to forty-six per cent.[5], although in the interim the dividend had apparently decreased to thirty-two per cent. in the year 1833[6]. The shares of the river Don Navigation, which were originally £100 each, had so greatly increased in value that in 1822 the sum offered for seven and one-half shares that were put up at public auction was £12,960 (exclusive of a quarter's dividend due), which was equivalent to about £1726 per share[7]. In 1826 a share was sold for £2160; and in 1832 a share (a freehold) brought £2420. When the first alarm was heard as to the construction of a railway along the Don river valley, some shares sold for as little as £1500 each; but when, in 1847, the navigation passed over into the hands of the Manchester, Sheffield and Lincolnshire Railway Company, the price per share was at the rate of £3000[8]. The shares of the old Birmingham Canal, which in 1792 were selling for £900 each[9], so increased in value that for convenience of the market each share was divided into eight parts, and each eighth part sold

---

[1] *Coventry Mercury*, Feb. 14, 1825.

[2] *Birmingham Journal*, Feb. 2, 1828, p. 2. If we may believe the *Manchester Gazette*, Sept. 24, 1825, p. 4, we shall have to accept the statement that the Coventry shares were at that time selling for £1340, but the fact that the £1220 mark was probably more permanent for a period of years may indicate that the £1340 was probably exceptional. Momsen, *Öffentlichen Arbeiten in England*, p. 33, gives the price of these shares in 1829 as £1080.

[3] Felix Farley's *Bristol Journal*, Jan. 11, 1834, p. 2.

[4] *County Chronicle and Weekly Advertiser*, Mar. 31, 1818, p. 3; *Manchester Gazette*, Sept. 24, 1825, p. 4; Momsen, *Öffentlichen Arbeiten in England*, p. 33.

[5] *Herepath's Railway Magazine*, N.S., iv, pp. 384–8.

[6] Shaen, *Review of Railways and Railway Legislation*, p. 30.

[7] In 1786 the shares of the Don Navigation were producing an income of fifty per cent. See the advertisement in the *Morning Chronicle and London Advertiser*, April 1, 1786, p. 4.

[8] For these particulars regarding the Don Navigation, see Holland, *Tour of the Don*, ii, pp. 373–4, and Leader, *History of the Cutlers' Company*, i, p. 171. By a close comparison of the figures in these two sources, there are one or two variations noted, but they are slight. See also *Sheffield Local Register*, pp. 65, 167.

[9] Steuart, *Account of a Plan for the better supplying the City of Edinburgh with Coal*, p. 44. Steuart was wrong in saying that the par value of these shares was £100, for it was, in fact, £140.

for twice or three times as much as the original cost of a single share. For instance, in 1825 an eighth share of this canal, of the par value of £17. 10s., was selling for £355[1], in 1828 for £300[2], and in 1834 for £286[3]. At the same time, the dividend paid on the original £140 share was 100 per cent. annually[4]. On Feb. 21, 1826, one of the original £100 shares of the Sheffield Canal was sold by public auction for £2160[5]. The Trent and Mersey Canal shares, of the original value of £200 per share, which in 1813 were paying fifty per cent.[6], were selling in 1825 for £4600, and were paying an annual dividend of £150 with a bonus[7]; by 1834 they seem to have declined in value to £2560[8], and were yielding in 1838 a dividend of £180[9], while the market value of the stock in 1836 was £2400[10]. The statement made in the House of Lords that the quarter shares of this canal had been sold for £12,000 each[11] must certainly be an error, either of fact or of the reporter; for at that rate, even assuming that the dividend of £150 per full share paid in 1825 had been continued in 1836—which it evidently was not, since the dividend in 1838 was only £130—the rate of return on the investment would have been only three-tenths of one per cent., and this would never have tempted anyone to make the investment. Another noted illustration of a financially profitable canal was that of the Loughborough; in 1792 its shares, of the original cost of £142, were selling at 324 guineas (£340)[12], in 1828 at £3850[13], in 1834 at £1760[14], and in 1836 at about £1250[15]. Before the advent of railway competition, these shares had been selling for £4800[16].

---

[1] *Manchester Gazette,* Sept. 24, 1825, p. 4. The shares had at one time sold for £3200 (vide *Birmingham Journal,* Sept. 2, 1826, p. 1).

[2] *Birmingham Journal,* Feb. 2, 1828, p. 2.

[3] Felix Farley's *Bristol Journal,* Jan. 11, 1834, p. 2.

[4] *Manchester Gazette,* Sept. 24, 1825, p. 4; *Birmingham Journal,* Sept. 2, 1826, p. 1.

[5] *Sheffield Local Register,* p. 185.

[6] *Cambridge Chronicle and Journal,* Nov. 12, 1813, p. 2, letter of "C. F."

[7] *Manchester Gazette,* Sept. 24, 1825, p. 4.

[8] Felix Farley's *Bristol Journal,* Jan. 11, 1834, p. 2. The quarter shares were selling at £640.

[9] *Herepath's Railway Magazine,* N.S., IV, pp. 384–8. The quarter shares were paying £32. 10s.

[10] *The Times,* June 17, 1836, p. 3, statement of Lord Hatherton.   [11] Ibid.

[12] *Reading Mercury,* Nov. 12, 1792, p. 3. This was the time of the canal mania.

[13] *Birmingham Journal,* Feb. 2, 1828, p. 2. In 1829 they were £2590 (Momsen, op. cit., pp. 33–34).

[14] Felix Farley's *Bristol Journal,* Jan. 11, 1834, p. 2.

[15] *The Times,* June 22, 1836, p. 4; *A Few General Observations on the Principal Railways,* p. 20.

[16] Stretton, *Stone Roads, Canals, Edge-Rail-Ways, etc.* (Brit. Mus. 8235. aa. 76), p. 232; *A Few General Observations on the Principal Railways,* p. 20.

The dividend on the Loughborough canal shares in 1829 was £140[1] (practically 100 per cent.), in 1833 it was 124 per cent.[2], in 1836 it was £90 to £100 a year[3], in 1838 it was £110[4], or about seventy-seven per cent.; while in 1847, notwithstanding the fact that the company had reduced its tonnage charge from 2s. 6d. to 4d., it still divided seventy-four per cent.[5] These are only a few examples, which might be greatly multiplied, to show how valuable a possession some canal shares were before the advent of the railway[6]. But we must not be misled by the foregoing statements into thinking that the majority of the canals paid good returns on the amounts expended upon them;

[1] Momsen, op. cit., pp. 33–34.
[2] Shaen, *Review of Railways and Railway Legislation*, p. 30.
[3] *The Times*, June 22, 1886, p. 4.
[4] *Herepath's Railway Magazine*, N.S., IV, pp. 384–8.
[5] Shaen, *Review of Railways and Railway Legislation*, p. 30.
[6] We take the privilege of adding some further details regarding this aspect of canal finance. The Erewash Canal, which was finished in 1779, was so prosperous that within ten years its shares rose to three times their original value, and even as late as 1833 it was paying a dividend of £40 on every £100 share (Pilkington, *View of Derbyshire*, I, p. 282; *Herepath's Railway Magazine*, N.S., IV, pp. 384–8). The Staffordshire and Worcestershire Canal shares, of the par value of £140 each, were selling in 1825 at £960 and paying an annual dividend of £40 and a bonus; by 1828 they were selling for £800; but in 1838 the annual dividend was still £40 (*Manchester Gazette*, Sept. 24, 1825, p. 4; *Birmingham Journal*, Feb. 2, 1828, p. 2; *Herepath's Railway Magazine*, N.S., IV, pp. 384–8). The Birmingham and Staffordshire canals, for some years before 1800, were said to have divided annually not less than thirty per cent. on the original cost of construction (Steuart, *Account of a Plan for the better supplying the City of Edinburgh with Coal*, p. 44). The original shares of the Aire and Calder, in the twenty years preceding 1816, paid fully 100 per cent. (Sutcliffe, *Treatise on Canals and Reservoirs*, p. 127). The Grand Junction Canal shares sold for 420 guineas (£441) in 1792, for £306 in 1828, and for £240 in 1834; but its dividend in 1838 was only twelve per cent. (*Reading Mercury*, Nov. 12, 1792, p. 3; *Birmingham Journal*, Feb. 2, 1828, p. 2; Felix Farley's *Bristol Journal*, Jan. 11, 1834, p. 2; *Herepath's Railway Magazine*, N.S., IV, pp. 384–8). The Oxford Canal, the trade of which was chiefly coal and pottery, paid thirty-one per cent. to its subscribers in 1812–18 (*Cambridge Chronicle and Journal*, Sept. 11, 1812, p. 3, Nov. 19, 1813, p. 2, and *County Chronicle and Weekly Advertiser*, Mar. 31, 1818, p. 3), and the price of its shares in 1828 was £720 (*Birmingham Journal*, Feb. 2, 1828, p. 2). Even as late as 1838 it was paying thirty per cent. dividend (*Herepath's Railway Magazine*, N.S., IV, pp. 384–8). For the great earning power of the Bridgewater Canal and the Grand Trunk Canal, see pamphlet by Joseph Sandars, on the "Projected Railroad between Liverpool and Manchester," to which we shall refer more particularly when we come to discuss the early railways. The Wolverhampton Canal in 1813 was paying forty-four per cent. (*Cambridge Chronicle and Journal*, Nov. 12, 1813, p. 2, letter from "C. F."). In 1790 the papers mentioned a certain baronet who was receiving annually as profits the full amount of his investment in a navigation (*Public Advertiser*, Feb. 2, 1790, p. 1).
The following table is taken from Momsen, *Öffentlichen Arbeiten in England*,

for we have already shown that fully one-half of the number of canals and probably considerably more than one-half of the capital expenditure realized returns that were inadequate in order to maintain the canals as effective agents for the work they were intended to accomplish. So dazzled were the eyes of some by the large dividends received in certain cases that in 1793 the House of Commons granted one of its members the privilege of introducing a bill to limit the profits on canals[1]; but we do not find that anything further was done about it. The fact that this was, apparently, an agitation which had few

pp. 33–34, to show those canals which, on Aug. 7, 1829, were paying over six per cent. dividend, or whose shares were worth over £125:

Name of canal	Yearly dividend	Value of share Aug. 7, 1829
	per cent.	£
Ashton and Oldham (on Feb. 6, 1827) ..	6⅔	147
Barnsley	8¾	200
Birmingham	71 3/7	1668 4/7
Coventry	44	1080
Cromford	18	420
Derby	6	160
Erewash	70	1500
Forth and Clyde	6¾	161⅕
Grand Junction	13	295
Glamorganshire	7⅘	153 19/40
Grantham	6⅔	143⅓
Leeds and Liverpool	18	470
Leicester (on Feb. 6, 1827)	11 3/7	285 5/7
Loughborough	140	2590
Monmouthshire	12	239
Melton Mowbray	9	220
Mersey and Irwell	40	830
Neath (on Feb. 6, 1827)	15	330
Oxford	32	670
Shropshire	6⅔	116
Somerset Coal	7	109⅓
Stafford and Worcester	28 4/7	578 4/7
Shrewsbury	8	212
Stourbridge	8 8/29	151 21/29
Stroudwater	15⅓	326⅔
Swansea	15	270
Trent and Mersey	37½	790
New Thames and Medway	—	200
Warwick and Birmingham	12	270
Warwick and Napton	10½	215
Wyrley and Essington (on Feb. 6, 1827)	4⅚	140⅘

See also the prices of canal shares and the dividends paid, as given in *Wettenhall's Commercial List*, Dec. 10, 1824 (Gray, *General Iron Railway*, pp. 155–6).

[1] *Public Advertiser*, Mar. 19, 1793, p. 2, and Mar. 22, 1793, p. 1.

supporters, and that it was during the period of the canal mania, may account for the slight notice that was taken of it.

Having seen the financial results which accrued to the canals, it will now be germane to this subject to inquire into some of the reasons why canals were unsuccessful. We must, of course, recognize that the reasons for failure in one case were not necessarily the same as those in another case; for each canal had its own individual peculiarities, and not until we know the inner history and working of the organization and the external environment of the canal can we in any sense adequately determine what were the factors making for its failure or its success, as the case might be. We must here again remind the reader that the canal companies jealously guarded their own business interests and secrets, so that it is difficult for us to understand all the forces that were at work in bringing prosperity in some cases and adversity in others.

But one element in the lack of success of some canals was the fraud and misrepresentation in their promotion. The projectors of a canal were not always high-minded, unselfish men, who looked to the welfare of the community without reference to themselves and their own pecuniary interests; and this led to complications which were detrimental to the canal almost from the first. Two of the projectors of a certain canal, for example, found that the best line for the general good would not be the best line for their particular good; and, without consulting the subscribers, who had paid in their deposits, they caused a deviation from the most universally satisfactory line, thus entailing an increased cost of construction[1]. This action induced those who were honest to have no faith in the projectors as a body; it sometimes caused the withdrawal of subscriptions that had already been made and so the enterprise was branded at its inception. When once a stigma had fallen upon an undertaking or its promoters it was more difficult to get other people interested in it so as to carry the work to a conclusion. Another phase of this fraudulent promotion was the attachment of names, especially the names of influential persons, to the subscription list, without their knowledge or consent, in order to induce others to subscribe, or to make a good showing before Parliament when authority was to be asked for an Act to enable them to carry out their design. How common this was we have no means of knowing; but when the solicitor for a company would falsify the list of subscribers, or attest its correctness when he knew that it

---

[1] *The Times*, Feb. 18, 1828, p. 3. In this case some of the subscribers seceded from the company, as a result of the unwarranted actions of the promoters. See also ibid., Feb. 15, 1828, p. 4, letter from "Once a Subscriber for Ten Shares."

contained the names of men who had given no such authorization, we are justified in suspecting that such illegal acts were not infrequent[1]. Sometimes a few individuals were specially eager to have a canal constructed that would directly and immediately benefit them; and, after obtaining the aid of men whose names on the prospectus would be of considerable importance, the former pushed on the enterprise while the latter did not even attend the meetings of the committee[2]. Gross discrepancies are observed as to the amount of tonnage that would pass along a proposed canal and the cost of carrying these commodities, according to the statements of the promoters and of the opponents of the measure[3]; and it would be very charitable to attribute these to lack of knowledge of the conditions, although it would probably be nearer the truth to assign them, at least in some cases, to the motive of wilful prevarication[4].

Another reason for the failure of some canals was the high initial cost of securing their Act of Parliament and the great expense of construction. The law required that, when a canal bill was brought into Parliament, the company had also to present a plan and an estimate of the cost. The plan was to be followed minutely and the amount of the estimate was to be the utmost that would be required to carry it into execution. But, time after time, the estimated cost had been expended without the canal having been completed and the company would return to Parliament to ask for permission to increase their capital or to borrow the extra amount that they thought requisite for the satisfactory completion of their works. This, along with other details of administrative inefficiency, imposed an immediate barrier to ultimate success. For example, when the Regent's Canal Bill was brought up for the third reading, it was shown that the

---

[1] *The Times*, May 21, 1830, p. 2, shows that, for such conduct, the solicitor of the Birmingham and London Junction Canal Bill was called to the bar of the House of Commons and reprimanded by the Speaker. *Cambridge Chronicle and Journal*, Dec. 18, 1812, p. 1, shows a frank confession of this insertion of names in the Bedford Canal prospectus without the consent of the persons. See also ibid., April 22, 1814, p. 3, letter from "A. B."; and *Birmingham Journal*, April 10, 1830, p. 3, which shows that the names of men were used who were wholly unable to pay a subscription.

[2] *Cambridge Chronicle and Journal*, Dec. 11, 1812, p. 1, letter from "Verus."

[3] *Cambridge Chronicle and Journal*, Feb. 7, 1812, p. 2; June 12, 1812, p. 2; June 19, 1812, p. 2, letter from "A Friend to Public Improvement;" June 26, 1812, p. 2, letter from "X. Y."; etc.

[4] The truth of our conclusion here will be very apparent to anyone who will take the time to read the correspondence, in the issues of the *Cambridge Chronicle and Journal*, 1812–16, between the advocates and the opponents of the Cambridge and London Junction Canal. The references are too many for me to cite them in detail.

company's management had been exceedingly bad; that they had not begun at one end and finished the canal as they went along and, therefore, had lost the benefit of tolls; and that they had spent all the money originally estimated as the total cost of the canal and, consequently, were insolvent and unable to pay their obligations. Under these conditions the company returned to Parliament seeking authority a second time to raise as much money as they had originally asked[1]. When unduly large sums were expended in the construction of the works, in addition to the extravagant charges of getting a bill through Parliament and paying the amounts demanded for the land and for compensations to adjacent landowners and others, it is easily seen that some concerns would be weighed down at the first with a burden of costs that would be a certain forerunner of disaster[2]. We have previously shown that even the Coventry Canal, which was so successful in its later years that it could divide a dividend of forty-four to forty-six per cent., suffered seriously during its earlier years because of financial embarrassment; and it was not until the canal got better connexions that it was able to surmount the obstacles to its success. How far the failure of canals in various instances may have been owing to their being executed on a scale too large and expensive, or to palpable mismanagement and lack of skill in constructing them, is a question.

The defects in location, engineering and construction of some canals made it a foregone certainty that partial or complete failure would be the outcome in such cases. The Sheffield and Tinsley Canal, in 1814, apparently in order to placate the Duke of Norfolk who was the owner of the land in and around Sheffield, made an unhappy choice of line and level. Instead of adhering to the line first contemplated, the directors placed the canal on the south side of the Don river, and took it by a roundabout, instead of the straight course originally intended. The level of the canal was made too high; for while it had been expected to carry it almost on a level with the river, so that there would be an abundance of water easily procurable, it was taken uphill where there was no water except the scanty, precarious and expensive supply that was pumped up out of the coal pits. The many locks that were required wasted the time of the bargemen, wasted the water, increased the cost of operating the canal and hence the freight charge, and diminished the returns of the proprietors of the canal. Instead of being nearly without locks and requiring very little water, which would probably have been secured without charge had it followed the intended line, it had twelve locks within a distance

---

[1] *The Times*, June 27, 1816, p. 2.
[2] For other examples, see Moore, *History of Devonshire*, pp. 50–51.

of three miles, and the company paid £450 a year for pumping water into it. The cost of construction had been increased from £60,000, as at first expected, to nearly £130,000; and the damages that had to be paid to coal proprietors, in addition to the heavy debt with which they were struggling, made the company decidedly unfortunate. The canal was not expected to pay large dividends, but was intended to be of great public benefit; it was thought that, if well managed, it might in time pay three and one-half per cent.; but the results showed that it had not paid one per cent., and instead of lessening the cost of carriage, it had, in some cases, increased it, and produced public disappointment and regret[1].

Several other illustrations may serve to add emphasis to what we have just said. On the Monmouthshire Canal, one branch, nearly eight miles long, had a perpendicular fall of water of 365 feet, which was overcome by thirty-two locks; while the other branch, eleven miles long, had a perpendicular fall of water of 447 feet, which was overcome by forty-two locks. The average depth of the canal was but three and one-half feet and the greatest barge load that could be accommodated was twenty-five to twenty-eight tons. The small size of barges that could be used and the waste of time and water in surmounting a height of forty-five feet per mile of canal would seem to be disastrous to effective operation[2]. The Rochdale Canal was another which was ill-constructed and in its operation used up a vast amount of water. Its reservoirs were as tight as any others, and yet, by leakage and evaporation alone, their level in a summer day would be lowered one inch over the total area of the reservoirs, 318 acres[3]. The Leeds and Liverpool Canal was also of ill-advised construction and in dry seasons boats sometimes had to wait two or three days before they could pass certain locks[4]. The Huddersfield Canal, like those just mentioned, was greatly distressed in summer for lack of water; its reservoirs were unsatisfactory, its water was scarce, and it had too great a rise in ground that was unfavourable. "The slavery of working

---

[1] The history of the Sheffield Canal may be culled from the following sources: *Sheffield Iris*, Oct. 7, 1834, p. 1, giving the prospectus of the proposed railway from Sheffield to Rotherham; ibid., Mar. 31, 1835, p. 3, letter from "A. B."; ibid., April 7, 1835, p. 2, letters from "C. D." and "A. Z."; ibid., Sept. 15, 1835, p. 4, and Sept. 22, 1835, p. 4, letter from W. Ibbotson.

[2] Coxe, *Historical Tour through Monmouthshire*, p. 65. With such difficulties of operation, it is strange that this canal should have paid so well, for in 1829 it was paying a dividend of twelve per cent. (Momsen, *Öffentlichen Arbeiten in England*, pp. 33–34).

[3] Sutcliffe, *Treatise on Canals and Reservoirs*, pp. 77–90.

[4] Ibid., pp. 90–117.

vessels through the tunnel" 5720 yards in length, together with the fact that vessels could only enter the tunnel every twelve hours, was a great inconvenience to trade[1]. Such an expensive and narrow canal was a standing invitation to failure.

In river navigations there were corresponding barriers to the greatest usefulness and financial success. The numerous shoals in the larger rivers, especially when the water in the river was low, were a great obstacle to the conveyance of barges; and between Reading and London there were many shallows in the Thames which in summer months or in dry seasons were not more than two feet nine inches or three feet deep, so that barges were frequently detained there from twelve to fourteen days, and sometimes longer, for want of water[2]. These conditions necessitated the use of a great many horses in towing the barges[3], and transhipment from barges of larger to those of smaller capacity, or vice versa, was frequent[4]. At the same time the passing of locks and weirs was expensive, for the high tolls charged by the owners or lessees of these works, which had multiplied so greatly on the Thames, made a heavy burden upon the barge-owners[5]. When periods of drought came the conveyance of goods on the two largest rivers of the kingdom had sometimes to cease[6], even in the lower portions of the rivers where there was most water. The process of "flashing," that most primitive of all devices for permitting barges to pass over shallows in the river, was still much used; it was a baneful practice, for after the "flash" was drawn and the lock closed again it left the river almost dry for twelve to twenty-four hours, so that the barges going up-stream and the mills along the river were all stopped. The evils resulting from the application of this injudicious device evoked much hostility from those who were desirous of seeing the navigation brought to the condition of its most complete utility[7].

---

[1] Sutcliffe, op. cit., pp. 118–26.

[2] *Parl. Papers*, 1793, XIII, 'Miscellaneous Reports, No. 109,' pp. 5, 7, 25–26, 28, evidence of Mr Truss, Mr Court and Mr Mould; Sutcliffe, op. cit., p. 182; *Public Advertiser*, Oct. 11, 1786, p. 1, letter from "An Esteemer of Steam Engines." At times the barges had to be unloaded and the goods carried to their destination by broad-wheeled waggons (*Public Advertiser*, Oct. 11, 1786, p. 1).

[3] *Parl. Papers*, 1793, XIII, 'Miscellaneous Reports, No. 109,' pp. 24, 25, 26, evidence of Messrs Court and Langley.

[4] *Parl. Papers*, 1793, XIII, 'Miscellaneous Reports, No. 109,' pp. 23, 25, evidence of Mr Court.

[5] *Parl. Papers*, 1793, XIII, 'Miscellaneous Reports, No. 109,' pp. 6–7, 8, 10, 25, 34, evidence of Messrs Truss, Allnutt and Court, and Report of the Committee on the evidence.

[6] *Birmingham Journal*, Sept. 9, 1826, p. 3; *Public Advertiser*, Oct. 11, 1786, p. 1.

[7] *Parl. Papers*, 1793, XIII, 'Miscellaneous Reports, No. 109,' pp. 7, 20, 35, evidence

In other instances, millers drew off the water from the river for their own purposes, so that there was not enough left to answer the ends of navigation[1]. From the foregoing causes, and a long series of petty annoyances and frauds by the agents of the navigations[2], we can readily understand how personal self-interest would militate against the welfare of the navigations, both as to their standing in the eyes of the community and as to their revenues.

Other and probably more important reasons for the lack of success of the canals will be considered at a later stage of this work, when taking up the reasons for the decline of these waterways; and until then we shall leave the subject with these few particulars.

Of the organization of the canal companies we have but scanty information. They differed fundamentally from the turnpike trusts in that they were incorporated bodies and issued shares of stock, which were subscribed for in the same way as in the case of the railways at

of Messrs Truss and Mylne, and the Report of the Committee on the evidence; Sutcliffe, op. cit., p. 182; *Public Advertiser*, Oct. 11, 1786, p. 1, letter from "An Esteemer of Steam Engines."

[1] *Reading Mercury*, Nov. 25, 1793, p. 4, letter from "A Commissioner," on the 'Thames Navigation;' *Parl. Papers*, 1793, XIII, 'Miscellaneous Reports, No. 109,' p. 24, evidence of Mr Court. The conditions are aptly described in a letter, addressed to the corn factors of London, in the *Public Advertiser*, Oct. 11, 1786, p. 1. The writer told them that, if they wanted to buy up the grain of the western counties of England, it would have to be sent to London coastwise, or in broad-wheeled waggons. The millers along the Thames had forbidden the barge-masters to carry grain unless the whole of it were ground at the millers' mills. Heavy tolls were imposed for passing through every lock and even then the barges would stick on shoals for months together. The millers could also fix their own price for grinding the grain. After all this, the corn factors would have to pray and pay for continual flashes of water to carry them over the shoals on their way down the river, and even these flashes would not always do what was expected of them. The writer urged the abolition of this monopoly of the Thames; he would overthrow all locks and mills and abolish all shoals throughout the length of the river from Cricklade to Gravesend.

[2] Lock-keepers did not treat all alike, apparently, but gave to their favourites some privileges which they denied to others. It was shown that a certain sluice keeper on the River Cam would allow certain boats to pass and detain others; in fact, that he was generally troublesome and obnoxious (Brit. Mus., Add. MSS. 35,679, pp. 316, 317, 319–23). Fraud in the conveyance by canal was very extensive, and even corn, malt, salt, etc., were taken by the bargemen from the quantities of these things that were entrusted to them for carriage. In some cases the weight of the material taken out would be made up by absorption of moisture from the atmosphere. They knew to a nicety how much water that quantity of salt would contain, without changing its colour or losing in quality; and they would abstract a certain amount of salt and sell it, and put back as much water as it would bear, thus greatly increasing their remuneration by this system of pilferage (*Parl. Papers*, 1834 (517), VII, 1, 'Report from Select Committee on the Sale of Corn,' evidence of James Sutton, Q. 1948–1958).

a later time[1]. They were private corporations whose aim was private profit, while the turnpike trusts were simply associations of men to whom was entrusted the administration of their respective portions of the roads, in the public interest, but without emolument. In other words, the turnpike trustees were regarded as public officers who had charge, for the time being, of a part of the country's affairs. It was customary, when a canal was in prospect, for the promoters of it to carry on an agitation until they had what they thought was sufficient support to enable them to push it through; and then they drew up a plan of and reasons for the intended canal[2], and with this carried the case before Parliament, seeking authority to accomplish their desire. A solicitor had to be engaged to conduct the case through the Legislature, and to him usually a large fee had to be paid for expert service. After the expenditure of a large amount to secure an Act incorporating the projectors of the enterprise, they were given the right to take the necessary land for the canal, but had to make adequate remuneration for the property thus appropriated. When the canal was opened the company was allowed to take stipulated tolls from those whose boats and barges used the navigation; these tolls were to be applied, first to the maintenance of the canal works, and after that the surplus remaining might be, either wholly or partly, apportioned as dividends. It was intended that the waterway should be like an ordinary highway, open to all upon payment of the tolls; and that any person should be at liberty to put his barge on the canal and tow his goods to any place on the canal chain, provided he paid the tolls for the use of the canals along which he passed and those which had to be paid for the privilege of passing from one canal to

---

[1] One apparent exception to this statement I have found. In the *Leeds Intelligencer*, Jan. 17, 1835, p. 1, there is given an advertisement of "road shares" in the Holm Lane End and Heckmondwike Turnpike Road, amounting to £685. 7s. 6d., that were to be sold by ticket. The advertisement stated that interest was then being paid on these shares at five per cent. per annum and that they would be sold with the interest due on them. It would seem as if, in this case, the trustees, instead of borrowing money on a mortgage of the tolls, had raised the necessary funds for repairing the road by the issuance of shares.

[2] For a few of these plans, see Brit. Mus., Maps 88. d. 13, Nos. 8, 12, 25, 32, 37; also *Cambridge Chronicle and Journal*, Dec. 11, 1812, p. 1, for the prospectus of the proposed Bedford Canal; and *The Times*, Nov. 8, 1811, p. 1, for the prospectus of the London and Cambridge Junction Canal. The prospectuses of the later canals were more elaborate than those of the earlier; they frequently gave the reasons for the canal, the kind and amount of traffic that was expected to come upon the proposed canal, the assurance of financial returns that would justify its construction, sometimes answers to any objections that had been made to the canal, and other data that would probably prove serviceable in securing public and private support for the undertaking.

another. The shares were bought and sold on 'Change, just as railroad and industrial shares are bought and sold to-day. And here let us be careful to note that the canal companies were not to put barges on the canals and act as public carriers, in opposition to or rivalry with private concerns: they were merely the custodians of the water-way, receiving their recompense in the way of tolls, and maintaining a highway for traffic that could be used by all alike upon equal terms. The canals were, therefore, in every essential respect like a turnpike; but in the one case the administration was in the interest of a private corporation, while in the other it was for the public benefit. Concerning the internal organization of the canal companies we know practically nothing, for their records have been closely guarded, and in any investigation of their activities as transportation agents the inner constitution of the companies has not been considered. But incidentally we learn the importance of two or three officials of such a company. The chairman of the board of directors seems to have been the executive head of the corporation, and it was he who shaped the policy of the company in all its exterior relationships. Under him, the clerk of the navigation performed the functions of secretary and treasurer, and directed from day to day the actual working and the inner affairs of the company. He kept the records of all shipments made and received, furnished information as to charges and facilities, and was, in general, the executive officer between the company and its patrons. The chief engineer was the official upon whom devolved the responsibility for the actual upkeep of existing works, such as locks, bridges, wharfs, barges, etc., and for the extension of the canal or its additional connexion with other canals. So far as I have seen, no canal Act specifies what kind of control was to be exercised, what form the government should take, what the duties and rights of the various elements of the governing body should be, nor where should centre the ultimate authority for making final decisions. These were matters which were probably determined in private, after the Act of incorporation had been secured; and they, doubtless, varied considerably in different concerns. If the minute or record books of some of these companies should be brought to light for the benefit of the investigator, we might then be able to obtain definite information along lines that are now obscure; but until the uncovering of such treasures has placed additional material at our hand, we shall have to rest contented with the present hazy, because incomplete, view of the directive system by which the canals were managed.

We have said that the canal companies were not allowed to be public carriers; this was probably because, on account of their owning the waterway, there might be a tendency for them, since they would not have to pay tolls to themselves, to put down their tonnage rates, and by thus lowering the charges of carriage draw the traffic away from other carriers.   But the political economy of the time favoured competition and, possibly, in order to place all the carriers on an approximate equality the above prohibition was made.   But if this restriction applied to companies, it did not apply to privately-owned canals, such as those of the Duke of Bridgewater, Sir John Ramsden, Mr Templar[1], Lord Rolle[2], and others.   The Duke of Bridgewater, for example, both owned and operated his canal; he put his own barges and packet boats on the canal, and in addition to conveying his own coal to market he fulfilled the functions of a common carrier of both passengers and goods.   In other cases, also, although the owning company could not operate the canal, there seemed to be nothing to prevent any man who was a member of the company from putting his barges on the canal and thus doing the carrying for himself or others.   This was actually done by Josiah Wedgwood, whose great interests in the Potteries led him to associate with others in the construction of the Grand Trunk Canal.   Although the canal company could not put barges on the canal and do the carrying, Wedgwood could have and use his own barges for that purpose and thereby secure the same results as if the company did the carrying for him.   It was not until the year 1845 that canal companies were allowed to be public carriers on their navigations; and this privilege was given them in order that they might endeavour to compete even-handed with the railway companies, which had the privilege of being common carriers on their own lines[3].

We hesitate, however, in making the above statement, for while the carriage of goods by the canal companies was not legally allowable there is good evidence to show that it was occasionally allowed.   In 1790 the Basingstoke Canal was evidently carrying goods regularly between Basingstoke and London; and by their having the traffic by water organized there was a constant supply of goods brought from London to the Basingstoke wharf, to be forwarded every week to Winchester, Southampton, Romsey, Salisbury, Andover and adjacent parts.   But while they were carriers on the waterway, they had not provided any facilities of their own for the carriage of these goods from the canal wharf at Basingstoke back to the towns to which they were destined; and in the early spring of that year they asked that

---

[1] Moore, *History of Devonshire*, p. 46.   [2] Ibid., p. 49.
[3] This authority was conferred by Act 8 & 9 Vict., c. 42.

those who wished to contract for the carriage of these goods between Basingstoke and the places above mentioned should apply by tender to certain designated persons, who would give them particulars as to the quantity to be forwarded every week, the exact times of going and returning, and other details. In this way, we would judge, there must have been a fairly complete circulatory system for the conveyance of commodities[1]. Soon after this, one or more of the tenders must have been accepted and the prices of land carriage fixed; for the canal was then able to advertise the exact freight rate between London and Basingstoke by canal and the cost of carriage by waggons from Basingstoke to these near-by towns[2]. Now it is easily conceivable that, since this canal company had entered into the carrying business, others may have pursued the same course. We grant the possibility of this, although we have not found any other instance to establish it as a fact. But there is one thing which makes us think that the exercise of this function was closely restricted: if it had been at all general, the names of these canal companies that were acting as carriers would probably have been included along with other carriers in any list that aimed at completeness; but the fact that nothing of this kind is found in the list of carriers for 1802 and 1809 makes highly probable our conclusion that it was very rare for a canal-owning company to be, at the same time, a carrier on its canal[3]. Moreover, if it had been common, there would have been no use for the Act of 1845 sanctioning this merging of functions[4].

By reason of the impossibility of the canal companies acting as carriers and the necessity of having a well-developed carrying service this trade had been organized on the canals by special companies; and the vast network of carriers engaged on the inland waterways at the beginning of the nineteenth century attests the great value and extent of their service to the community[5]. In the organization of the

---

[1] *Reading Mercury*, April 18, 1796, p. 2.

[2] Ibid., May 30, 1796, p. 4. The freight rate on goods in barges from London to Basingstoke was to be 9d. per cwt. (of 112 lbs.); and by waggon from

Basingstoke to	Winchester	was to be	10d.	per cwt.
,,	Southampton	,,	18d.	,,
,,	Romsey	,,	18d.	,,
,,	Salisbury	,,	18d.	,,
,,	Andover	,,	9d.	,,
,,	Whitchurch	,,	6d.	,,

[3] See Holden's *Annual List of Coaches, Waggons, Carts, Vessels, etc., from London to all parts of England, etc.*, 1802 (3rd edition) and 1809.

[4] Act 8 & 9 Vict., c. 42.

[5] See Holden's *Annual List* just mentioned.

trade, two divisions must be recognized, first, the quick or fly trade, and, second, the slow or heavy trade.

The fly trade was conducted by carriers only. The boats used could travel faster than those employed in the heavy trade and usually went three to three and one-half miles per hour. It required four men to work each boat properly, since they travelled both day and night, and a steerer and a driver had to be constantly on duty. The average load did not exceed ten tons per boat[1]. The number of relays of horses, or the length of stage that any group of horses worked, varied according to the arrangements of the individual carrier. From Birmingham to the Mersey, for example, would be divided into from three to five stages, or, on an average, perhaps four stages, so that each horse worked twenty miles per day.

To exemplify the working of the system we may take the case of, perhaps, the most celebrated carrier, Messrs Pickford & Co. of London. The method of their operation was as follows: They kept teams in London collecting the goods which were to be sent (say) to Birmingham, Liverpool and Manchester. The goods were loaded into their barges on the canal in London, and when a sufficient number of loaded barges was obtained they put their horses on the towpath and set out with their "barge-train" for the above-named cities[2]. When they reached

---

[1] Skey, *Report to the Committee of the Birmingham and Liverpool Junction Canal*, p. 6. Skey was secretary of this company and his statements are authentic and his knowledge complete.

[2] The hauling of barges on the rivers and canals had long been done by human force, but after there had been a more complete and enterprising organization of the carrying trade horses were substituted for men. In some cases both were used. Pitt, *Agriculture of Worcester* (1813), p. 268, shows us that on the Severn river the barges going against the stream, when not favoured by wind, were hauled chiefly by men, with ten, twelve, or more, to a barge; but he said that just before the time of his writing horses had been introduced and then it was not uncommon to see a horse assisting a number of men in drawing barges.

Messrs Pickford & Co.'s advertisement of their 'Expeditious Canal Conveyance' between Birmingham and London by means of "Fly Boats" is given in Aris's *Birmingham Gazette*, Nov. 9, 1807, p. 1; Dec. 21, 1807; Feb. 8, 1808; etc. It shows the connexions they had, through other carriers, with the chief towns in the neighbourhood of the Coventry, Oxford and Grand Junction canals.

In Aris's *Birmingham Gazette*, Feb. 8, 1808, p. 1, is a similar advertisement of 'Expeditious Canal Conveyance' by the Wolverhampton Boat Company, between Wolverhampton and London. This company was established, like Pickfords, with their own waggons and men to collect and deliver goods.

In the same paper, of June 13, 1808, p. 1, is the advertisement of Thomas Dixon and Co., carriers between Birmingham and Bristol. But let it be noted that none of the above three carriers published their rates.

In the same paper, of April 2, 1810, p. 1, is given the advertisement of Wm. Judd & Sons' boats between Birmingham and Sheffield and Derby.

Birmingham, the barge with goods for that place was left there and the company's teams unloaded the goods and delivered them in the city to the proper consignees. The same thing was done in Liverpool and Manchester. The Pickfords paid the tolls on the various canals passed through and collected the cost of carrying from their patrons. The same teams in Birmingham, Liverpool and Manchester, which unloaded the barges, collected at the canal depots in those cities the goods which were to be sent to other places along this same route; and in this way the traffic was regularly carried on. The lighter and more fragile goods were sent in the company's vans, or by coach, on the highways, for in that way greater speed and more careful handling could be secured[1]. Of course, we must not suppose that the Pickfords were the only carrying company; for, indeed, they had much rivalry not only from the other carrying companies, but also from individuals who undertook the same work[2]. These carriers were responsible for the safe keeping and safe delivery of the goods, from the time they left their patrons' establishments in London until the time they were given to the consignees at the point of destination; and what we have said of the traffic along this line applies equally to that on every other line. The carrier's relation to the consignor was almost a personal one; the carrier felt personally responsible for the goods and articles given into his care, just as if they were his own, and this made the shipper feel secure. Usually a business house employed the same carrier year after year and the trust reposed in the latter was seldom misplaced. It will be observed from the foregoing that this carrying trade on the canals was almost identical with, and complementary to, the organization of the carrying trade on the highways[3].

[1] Then, too, such small parcels, sometimes of great value, if sent by canal would probably have been stolen from the boats and this loss would have had to be made good by the carrier.

[2] From Manchester alone, before the introduction of railways, there were about thirty carriers by water to the south; besides, there were a great many others to Liverpool, Hull, Leeds, etc. (Slugg, *Manchester Fifty Years Ago*, pp. 223–4).

The vast extent of the canal carriage that centred at Birmingham in 1830, and its wide ramifications, may be gathered from West, *History of Warwickshire*, p. 422, where a list is given of 34 firms engaged in the carrying trade to and from Birmingham. This, compared with the extent of 'Canal Conveyances' (ibid., pp. 480–83), will give us some idea of the amount of traffic carried on the waterways. The enormous amount of coaching that centred at that city is given in ibid., pp. 468–71; and the great extent of the carrying trade by road waggons and caravans is shown in ibid., pp. 472–80.

[3] This was impossible when the railway era came, and had to be greatly modified to suit the changed conditions of railway facilities.

The slow trade was carried on by three different parties: first, by the carriers who were engaged also in the fly or quick trade; second, by the so-called "iron carriers" who confined themselves entirely to the carriage of iron and other heavy articles which they collected and distributed at the works, and who were, therefore, free from the burden of wharf, warehouse, teams and other necessary expenses of the regular carrier; and, third, merchants dealing in timber, slate, corn, etc., and iron-masters and other manufacturers of heavy and bulky articles, who used their own boats for bringing raw material to their works and delivering the manufactured products[1]. It was by the two last classes that most of the heavy business was done, and the first class has so small a share in it that they may be regarded as a negligible factor. The boats used in this trade carried a much greater weight than the fly, for their average load was over twenty tons; and when loaded to their capacity they rarely maintained greater speed than two miles an hour. Each boat was drawn by one horse, which travelled about twenty-four to thirty miles a day, and was attended by two men or their equivalent[2]. At night the boat was tied up and all rested.

A comparison of the two methods of carriage may here be appropriately introduced by taking a particular example. A cargo of, say, 120 tons had to be taken from Birmingham to Preston Brook, or to Ellesmere Port on the Mersey tideway, a distance of eighty to ninety miles. By the fly boats, under the most favourable conditions, it would need twelve boats for thirty-six hours, forty-eight horses for one day each, and forty-eight men each for a day and a half of twelve hours to the day; and this would not take into account time lost at either terminus in unloading and reloading the boats, nor the expense of time in attending to so many horses at the different stages along the entire route. It is evident from this that the fly boats required a large amount of human labour, and the fact that this expense had to be borne by a comparatively small weight of goods made the per-unit expense high. By the slow boats, to convey the same amount of tonnage, it would be necessary to employ six boats for three days, six horses for three days, and twelve men or their equivalent for three days. The expense of this system of carriage would vary according to the amount of the boat's paying cargo, for the same expense divided out over a large cargo would make the per-unit expense smaller than if divided out over a small paying load. The carriers by fly boats stated

1 Skey, op. cit., p. 8.
2 The equivalent of two men (as steerer and driver) would be sometimes a man and his sons, or a man and his wife and son, for not infrequently a whole family lived on board and the wife took her turn at the helm.

their cost along this distance to range from 5*s.* to 6*s.* 6*d.* per ton; but by the slow boats the cost was generally estimated at one half-penny per ton per mile, or 3*s.* 6*d.* per ton for the whole distance[1]. By the slow boats, therefore, the cost was a little more than half, but the time required was twice, that of fly carriage.

There are other aspects of the organization of the carrying trade which we can best consider under the heading of complaints against the canal service. In the first place, the loose and irregular system pursued by the carriers was the occasion of much dissatisfaction on the part of the mercantile interests. The arrival and departure of boats on the canals took place at all hours. The fluctuations in the amount of trade which a carrier had between any given points caused many and frequent delays. The carrier's establishment was always in accordance with the average amount of his traffic; he wisely refused to keep in reserve an additional force of boats, horses and men to meet seasonal emergencies and variations in the amount of trade, and, on the contrary, when cargo was deficient he did not send as many boats as when it was abundant. In either case, therefore, whether cargo was abundant or deficient, it was inevitable that there should be more than usual delay. If it were unusually abundant, the carrier did not have the means to accommodate it, and some of the goods must wait until another time. This abundance was generally the result of an urgent and increasing demand, so that the lack of accommodation occurred at the worst time and frequently caused loss to manufacturers and merchants. On the other hand, if the amount of freight to be conveyed were scarce, the carrier would not be warranted in sending a less quantity than would pay expenses; and his awaiting the arrival of more to make up a boat-load would cause delay in carrying what had been offered some days before[2]. Then, when the boat had been dispatched, the horse, with his feeding bag attached to his mouth to save the trouble of feeding him at regular intervals, went listlessly on his way, half eating, half sleeping, half working. Delays were frequently made at the public-houses along the canal, where time was squandered in drinking; and the time of the arrival of the cargo at its destination was uncertain. Closely connected with this, there was extensive fraud practised by those who were in charge of the boats or barges, for these men systematically pillaged the goods given into their charge to such an extent that there was good reason for thinking that the families of these men were largely supported in this way. They would abstract wine and put in water, withdraw salt and make up its weight by water, and add to their income by taking

[1] Skey, op. cit., pp. 6, 8.　　　　[2] Boyle, *Hope for the Canals*, p. 24.

groceries, provisions, etc., from the cargo entrusted to their care[1]. This moral delinquency was aided and abetted by the fact that the boats were not separated into compartments which could be locked, but all kinds of cargo were loaded into the hold indiscriminately and packed generally higher than the sides of the boat; after which the whole was covered with tarpaulin as the only protection against exposure, accident or theft.

Another great cause of dissatisfaction with the carrying service was the many transhipments that had to be made if the cargo were going any distance. Here we must remember that the canal companies were not carriers; but the great number of independent carriers engaged upon the canals multiplied the number of transhipments to more than three times the number there might have been under a properly regulated system. The termini of each canal were necessarily points of transhipment and the commencement and termination of each carrier's stage involved a corresponding movement of the goods from the boat belonging to one carrier to that of the next. All these changes produced delay, inconvenience and injury to the goods, and exposed them with frequent regularity to the covetous eyes of all who were disposed to profit by their opportunity. The cargo, for example, which was forwarded from Wolverhampton to Hull for shipment passed through the hands of four distinct parties, the stages being Wolverhampton to Shardlow, Shardlow to Gainsborough, Gainsborough to Hull, and from Hull to destination[2]. With this necessity for repeated transfers of cargo from carrier to carrier, from one canal to another, and from boats of one dimension to those of another, we can readily conceive how dilatory and unsatisfactory the conveyance would be.

The competition for cargo among many carriers was a circumstance that, at that time, would not have been suspected of having anything but beneficial results; yet it was one of the fruitful causes of delay in canal traffic and in other ways was not productive of good. In the matter of freight rates, the local, competing canals, in their later years of prosperity, got together and agreed to maintain fixed charges to the various places, unless in instances, common in all trades, where an individual would secure some temporary advantage by departure from this general understanding. But, notwithstanding their adherence to these fixed rates, the carriers of a locality were still allowed to compete

---

[1] Boyle, op. cit., p. 21; *Parl. Papers*, 1834 (517), vii, 1, 'Report from Select Committee on the Sale of Corn,' evidence of James Sutton, Q. 1948–1958. There was a strong presumption that the proprietors of these public-houses received goods stolen from the canal boats.

[2] Boyle, op. cit., p. 24.

with one another for a share of the traffic of that section. In time, however, it was discovered that other advantages could be secured by the various carriers getting together and dividing up the traffic among themselves; and by acting on this tacit or expressed agreement they would "live and let live." Where competition thus led to combination, the effect of which was to divide the goods among the carriers in certain proportions, it impaired the efficiency of the system of canal carriage by relaxing the ability of any individual carrier to send boats to any place or places as often as the public interest would require. Moreover, a merchant or manufacturer usually confined himself exclusively to one carrier. The latter might be enabled to afford him an advantage on some particular route while placing him at a disadvantage on some other; but as carriers were generally situated somewhat alike in this respect, and as the shipper of goods usually had an account-current with his carrier, he could not change in a moment from one to another, according as each might offer a particular advantage, but preferred to remain with the one carrier rather than lose the advantage of his account-current and incur the trouble which each separate transaction would entail in the absence of a settled agreement. If his goods were delayed for a few days or a week occasionally on account of shipping conditions he would rather brook this annoyance than be subjected to the annoyance of perpetual change of carriers[1].

The organization of the service was productive of other causes of complaint. The lack of system in the official conduct of the affairs of the company caused great difficulty in procuring rates and general information, because there were no published lists of rates or classes of goods. The uncertainty, partiality and inconsistency of the carriers' charges, because of their being made without reference to any fixed rule, tended to arouse resentment on the part of aggrieved shippers[2]. The lack of promptness in settling claims and rectifying errors, served to increase the patron's discontent, by allowing it to rankle for a time and assume larger proportions. These, along with the absence of unanimity or understanding among the various departments so that they failed to work together harmoniously, produced conditions that did not make for the well-being of the canal interests.

This study of the carrying service would not be complete without some reference to a possible improvement that seemed likely to be made in it, particularly through the introduction of steam. The use of steam power for hauling boats on canals was tried on the Forth and Clyde Canal in Scotland in 1802, but was abandoned because of the

[1] Boyle, op. cit., p. 25.   [2] Boyle, op. cit., p. 19.

surge arising from the paddles washing away the banks of the canal. In 1812 new trials were made, while the canal banks were protected by coarse stone pitching. Passenger boats, drawn by post-horses, were also tried in England and Scotland at the beginning of the nineteenth century, but were soon afterwards abandoned[1]. With the coming of the railway, canal companies had to look to their perpetuation, and the speed of conveyance by the railway induced canal proprietors to try to make such improvements as would enable them to compete with the railways. This revived, or else intensified, the desire to utilize the power of steam for the more speedy carriage of goods on the canals; and experiments were conducted, both in England and Scotland, which showed some remarkable results that were greatly at variance with the common opinions regarding inland navigation. It was found by experiments on the Grand Junction Canal that all speed from four to eight miles per hour was attended by a considerable wave; but when the speed of the horses was above that, the wave went on diminishing[2]. The experiments on the Forth and Clyde Canal demonstrated practically the same facts, that when the speed of the boat approached nine miles per hour the heavy surge, that preceded and followed a boat going more slowly and that in many cases ruined the banks, almost entirely ceased. At the same time, the power required to move the boat was very little more at nine miles than at seven miles per hour[3]. These experiments were continued in 1831–3 to try the effect of steam navigation, and two steamboats were constructed for the proprietors of the Forth and Clyde Canal, one to carry passengers and the other merchandise. The results of these trials showed that such steamboats could be economically employed on canals[4]. Similar success was being attained contemporaneously in England, where, for example, a boat constructed of sheet iron, seventy feet long and five and one-half feet wide, was drawn by horses on the

---

[1] The complete details of all this experimental work in the application of steam are given in Fairbairn, *Remarks on Canal Navigation, illustrative of the Advantages of the Use of Steam as a Moving Power on Canals.* See also O'Brien, *Prize Essay on Canals and Canal Conveyance*, p. 10.

[2] *Birmingham Journal*, May 11, 1833, p. 4. Rennie, the celebrated engineer, found by experiment that the diminished resistance at the higher speed was due to the fact that the boat rose out of the water. These boats were constructed of iron plates and were seventy feet long and six feet wide.

[3] *Manchester Guardian*, Feb. 26, 1831, p. 3, 'Improvements in Internal Steam Navigation;' also Fairbairn, *Remarks on Canal Navigation.*

[4] *Manchester Guardian*, Feb. 26, 1831, p. 3, gives the results in detail. Later experiments and their results are given in ibid., April 2, 1831, p. 3, and May 5, 1832, p. 2.

Paddington Canal at the rate of ten miles per hour or more[1]. The conclusion arrived at, as the outcome of all this practical work, was that on canals a speed of ten to eleven miles per hour, which was equal to the speed of the best coaches, could be attained and maintained; and we are informed that, in 1833, boats were running constantly on the Lancaster and the Edinburgh and Glasgow canals, with goods and passengers, at a rate of ten miles per hour, and at one-half the usual cost[2]. It is probable that there may have been some exaggeration in certain of the reports of that day as to the speed that could be developed on canals; but we have eliminated those which did not seem to square with the facts as given by such a careful engineer as Fairbairn, and the evidence shows that the foregoing conclusion was just and sane. Then why was not the canal traffic immeasurably increased, both as to the number of fast boats and the amount of goods sent by that means? The answer to this question will be taken up more in detail when we come to the consideration of the reasons for the decline of the canals.

We come, finally, to consider the effect of canals upon the cost of road transportation. Here we encounter several difficulties. Neither the land nor water routes formed long connecting systems, over each of which one price for carriage prevailed. The rates were local, both for the roads and the canals. Each canal had its own rates made without regard to any other and scrupulously guarded all details of its financial operations; and while this was done by the individual companies there could not be any possibility of through rates. The canal carrying companies also were private corporations; their rates were not made public, as railway rates have been; and, in fact, their rates were not the same to all shippers, but, on the contrary, they gave preferential treatment to some favoured patrons. The relation of the carrier to the shipper was a personal relation, lasting often over a period of years; and personal considerations entered into their mutual rights and obligations, even those which were financial. Moreover, even for the best-known districts or highways of trade we have a paucity of material. In all the above-mentioned ways, therefore, the few statistics we have are likely to be vitiated.

Canal transportation is different from highway or railway transportation in that a mileage rate, either per hundredweight or per ton, has rarely, if ever, been established. It was difficult to adhere to

[1] *Birmingham Journal,* April 13, 1833, p. 4. The speed of the horses was given as ten to thirteen miles per hour.

[2] *Birmingham Journal,* May 11, 1833, p. 4; Fairbairn, op. cit., shows that a few steam vessels were regularly employed on the Scotch canals.

such a rate on road or rail, because of the many variable elements which had to be taken into account; but, so far as our information goes, it was never attempted with waterways. Rates were made as so much per ton or per hundredweight, and were graded roughly according to distance; but neither over the country as a whole nor yet for any small section of the country were rates fixed as so much per ton (or per hundredweight) per mile. If, however, we are going to get any reasonable comparison of the costs of transport by road and by canal, we shall have to reduce each to the same basis, and as the ton (or hundredweight) mile basis is that which is used for highway and railway carriage, we prefer to bring our figures for the canals to this basis. It will be remembered by the reader that this can give only a rough approximation to the truth and is not intended to be strictly accurate; but we are willing to say that our results are as close to the truth as is possible with the present scattered pieces of information that can be brought together. In discussing the cost of carriage by waggon, we endeavoured to bring together facts from so many and diverse sources that the possibilities of error would be reduced to the lowest minimum. The same course has been pursued here, only that the amount of our information is less, and the sources fewer, than in the former case. But a general mileage rate for road and for canal carriage will not give us the most precise view as to the effect of canals in reducing the cost of conveyance; in order to secure this, we must see what influence the canals exerted in particular cases, and with this object we proceed to the consideration of some special instances.

When the Bridgewater Canal was projected, the cost of land carriage between Liverpool and Manchester was 40s. a ton, and the expense by river navigation was 6s. 8d. per ton one way and 10s. per ton the other way[1]. The charge on the canal was limited by the Act to 6s. per ton[2], which was less than one-sixth of the cost of land carriage. The amount of reduction in this case appears to have been greater than in most other cases of which we have thus far obtained statistics. For example, before the Basingstoke Canal was constructed the expense of carriage by waggons from Basingstoke to London was £2 per ton, while by the canal this was to be reduced to 11s. 7¾d. per ton, which was but little more than one-fourth of the cost of land carriage[3]. The

---

[1] Brit. Mus. B. 504 (4), 'Advantages of Inland Navigation,' p. 18.

[2] Act 2 Geo. III, c. 11; Aikin, *Description of the Country around Manchester*, p. 116. This 6s. per ton (tonnage included) was to be the highest amount he could charge for carrying goods by his own vessels. The Act limited him to an amount not exceeding 2s. 6d. per ton for canal dues alone.

[3] Brit. Mus. K. 6. 58 (c), 'Basingstoke Canal Navigation.'

effect of the opening of the Grand Trunk Canal was to reduce the cost of carriage to an amount only one-third to one-fourth of that previously paid for land carriage[1]. In 1777, two years after the opening of the Trent and Chesterfield Canal, the cost of carriage on this canal, for lime, coal and other heavy articles, was asserted to be about one-fifth of the expense of the usual land carriage[2]. In 1793 the cost of land carriage between London and Reading was 33s. 4d. per ton, while that by river was 10s. per ton; so that, even with the crude condition of that navigation, the expense of water carriage was scarcely one-third of that by land[3]. Near the Grand Trunk Canal, the cost of land carriage was about 9s. per ton for ten miles, but this was reduced to 2s. 6d. on the canals, thus showing that the expense of canal carriage was but little more than one-fourth of that by land[4]. In 1835 the freight on tallow from London to Reading by the river was 15s. per ton, but when brought by waggon the freight was 32s. to 35s. per ton[5]. If the cost of water carriage, with the river in its bad condition, were less than half the expense of land carriage, we can readily conceive that, with a good canal, the cost of carriage would be but one-fourth to one-third of the cost of land carriage. In the same year, the cost of moving goods by waggons and vans on the roads was repeatedly given as 4d. per ton per mile, while the cost along the canal was stated as a little over 2d. per ton per mile[6]. For the year 1792, we have the prices charged for the carriage of goods by land and by canal between Manchester, Liverpool and Chester and the important centres in the Midlands[7]; from which we learn that the cost of canal carriage was usually from one-third to one-fourth, although occasionally one-half, of that for land carriage. About the same time, the charge for bringing cheese· from Lechlade to London by the defective navigation of the

---

[1] For the statistics to prove this, see Appendix 8.

[2] *Gentleman's Magazine*, XLVII, p. 124. When this canal was opened, the price of coal at Retford was reduced from 15s. 6d. to 10s. 6d. per ton, and lime from 16s. to 9s. a chaldron, despite the fact that there was the expense of land carriage for four miles from the nearest collieries to the navigation. *Annual Register*, XVIII, p. 116.

[3] *Parl. Papers*, 1793, XIII, 'Miscellaneous Reports, No. 109,' p. 7, evidence of Mr Truss.

[4] Steuart, *Account of a Plan for the better supplying the City of Edinburgh with Coal*, p. 33 (footnote); Phillips, *Plan for a Navigable Canal*, p. 21.

[5] *Great Western Railway Bill. Minutes of Evidence taken before the Lords' Committees to whom the Bill was committed*, pp. 395, 396, 407.

[6] Ibid., pp. 417, 418, evidence of R. J. Venables.

[7] Salt, *Statistics and Calculations*, p. 71, gives us the following table, the material of which he says he obtained by his own experience—information that had been carefully withheld from the public:

Thames, during the months from the beginning of August to May when the cheese would not be likely to spoil on account of the delays on the river, was 18*d.* per hundredweight; the other months of the year it came by land carriage at an expense of 2*s.* 6*d.* to 2*s.* 9*d.* per hundredweight[1]; and from this we note that here the cost of river conveyance was only about two-fifths of that by land.

But we have given enough detail for our purpose. On river navigations the cost of conveyance varied in some cases according to the direction in which the freight was moving, being higher if going up-stream than down[2]. Some change was occasionally made according to whether the goods were going over several navigations or were merely local[3]; but this was rarely a determining factor in fixing the freight rate, since canals were usually independent of each other and did not work together. Frequently the charge varied according to the nature of the goods, and heavy commodities, like coal, ore, iron

Prices of Carriage of Goods by Land and Canal, 1792.

	By canal per ton			By land per ton		
	£	*s.*	*d.*	£	*s.*	*d.*
Between Gainsborough and Birmingham	1	10	0			
„ Manchester and Etruria		15	0	2	15	0
„ Manchester and Bromley Common (three miles from Lichfield)	1	0	0	4	0	0
Land carriage, Bromley Common to Lichfield					2	6
Between Manchester and Shardlow (six miles from Derby)	1	10	0	3	0	0
Land carriage, Shardlow to Derby					5	0
Between Manchester and Newark	2	0	0	5	6	8
„ „ Wolverhampton	1	5	0	4	13	4
„ „ Birmingham	1	10	0	4	0	0
„ „ Stourport	1	10	0	4	13	4
„ Liverpool and Etruria		13	4	2	10	0
„ „ Bromley Common	1	0	0			
„ „ Shardlow	1	10	0			
„ „ Nottingham and Newark	2	0	0			
„ „ Wolverhampton	1	5	0	5	0	0
„ „ Birmingham	1	10	0	5	0	0
„ „ Stourport	1	10	0	5	0	0
„ Chester and Wolverhampton	1	15	0	3	10	0
„ „ Birmingham	2	0	0	3	10	0
„ „ Stourport	2	0	0	3	10	0

These prices were for perishable goods. Non-perishables were carried at lower prices.

[1] *Parl. Papers*, 1793, xiii, 'Miscellaneous Reports, No. 109,' p. 27, evidence of Mr Mould.

[2] Ibid., pp. 11, 24, 25; *Leeds Intelligencer*, April 21, 1794, p. 2.

[3] *Leeds Intelligencer*, April 21, 1794, p. 2.

and stone, were carried at lower rates than lighter articles of greater value per unit of weight[1]; but in other cases the cost was the same for all goods, according to their weight[2]. We must expect, therefore, that comparisons of rates in one case with rates in another would not give us definite data and even relative rates would be subject to certain deviations from precision.

When we look at the results effected by canals in the cost of transport, the statements of some contemporaries of that era would lead us to surmise that the change was almost incalculable. In the early years of the nineteenth century, a very careful historian referred to the fact that where, forty years before, a single horse toiled along the road from Knaresborough to Skipton with a sack of wheat upon his back, now a horse would draw, with equal or greater ease, a canal boat loaded with forty tons of wheat[3]. Another writer, a few years earlier, stated that one horse would draw upon a canal as much as thirty horses on ordinary turnpike roads, and that on the canal one man alone would transport as great a quantity of goods as three men and eighteen horses usually did on common roads. In order to add a still greater inducement for multiplying the number of canals, he asserted that a mile of canal was often made at less expense than a mile of turnpike[4]. In 1791 a German traveller in England noted the fact that one horse could draw as much on the canals as forty horses on the high-road[5]. These facts were indisputable; but we must not conclude that the cost of carriage by canal was lessened in the same proportion as the amount that one horse could convey was increased. In the prospectus of the proposed London and Cambridge Junction Canal, it was declared that a forty-ton barge such as would pass upon the canal, together with tackle and other equipment, would cost at the utmost £300; while the expense of eight waggons and sixty-four horses, which would be required to convey an equal amount of tonnage on the road, would not be less than £4,000[6]. With such advantages as these put

---

[1] Brit. Mus., Add. MSS. 35,649, pp. 222, 224; *Manchester Collectanea*, in Chetham Society Publications, LXVIII, p. 136; Allnutt, *Rivers and Canals West of London*, pp. 3, 5, 6, 8; etc.

[2] *Jackson's Oxford Journal*, May 27, 1780, p. 2; *Reading Mercury*, Sept. 8, 1794, p. 4; Brit. Mus. K. 6. 58 (c), 'Basingstoke Canal Navigation;' *Cambridge Chronicle and Journal*, Nov. 23, 1810, p. 2; etc.

[3] Whitaker, *History of Craven*, p. 187.

[4] Phillips, *History of Inland Navigation* (1803), p. xii.

[5] Wendeborn, *A View of England towards the Close of the Eighteenth Century*, I. pp. 191, 237.

[6] The prospectus is given in full in *The Times*, Nov. 8, 1811, p. 1; see also *Cambridge Chronicle and Journal*, Oct. 8, 1813, p. 3, where it is stated that the forty-ton barge would cost complete £550.

before the public, it is little wonder that canal enterprise attracted public attention. But we should be in error to assume that the cost of carriage decreased, on account of the canals, in the same ratio as this decrease of the cost of equipment.

What, then, were the probable facts as to the change that was caused by the canals? We have already adduced evidence in particular cases to show what was the outcome; and we have elsewhere brought together a still greater amount of testimony upon this subject[1]. Our conclusion is that the cost of canal carriage normally did not exceed one-half, and in most cases was from one-fourth to one-third, of the cost of land carriage. One who was conversant with the conditions said, in 1813, that there was no instance in the kingdom of conveyance by canal being above one-half of the price of land carriage[2], and to his statement our own results give confirmation. But the consensus of contemporary opinion and the facts that have come down to us from that time point unmistakably to the inference that more commonly the expense of canal carriage was but one-third or one-fourth of the cost of waggon carriage[3].

Of course, the difference in the expense of conveyance is not the only thing to be taken into consideration in forming an estimate of the relative services performed by the canals and roads. The speed and the regularity of the service of the two agencies would also have a strongly determining influence in favour of the one or the other, according to which offered the greater facilities in these respects. Some canals, as we have seen, endeavoured to adhere to a regular schedule for the times of arrival and departure of their boats; others were unrestricted so that whenever the carrier had his barges sufficiently filled he made his irregular journeys. But the same thing was true of road

---

[1] See Appendix 8.

[2] *Cambridge Chronicle and Journal*, Sept. 10, 1813, p. 1.

[3] In Brit. Mus., Add. MSS. 35,649, p. 224, we have an account of the tolls paid on a series of canals in England and the expenses of towing, barge-hire, etc., from which we deduce the statement that, on the average, the expense of conveying all kinds of heavy goods on canals was about $2\frac{1}{4}d$. per ton per mile, and all kinds of light goods about $3\frac{1}{4}d$. per ton per mile. But as the average expense of conveying all kinds of heavy goods on the roads was about 1s.—the MS. erroneously gives it as 1d.—per ton per mile, and light goods 1s. 4d. per ton per mile, it is evident that canal carriage cost only one-fourth as much as road carriage.

Steuart, *Account of a Plan for the better supplying the City of Edinburgh with Coal*, pp. 33 (footnote), 72, says that in England it was usually estimated that transportation by a canal saved two-thirds to three-fourths of the cost of land transportation. See also *Cambridge Chronicle and Journal*, Oct. 29, 1813, p. 2, letter from " C. F." In this same letter, " C. F." says that the cost of sending goods from London to Bristol was £10 by land and £2 by inland navigation; so that if this were true water carriage there would have been but one-fifth of that by land.

traffic, for we have noted that some carriers were trying to keep to fixed times of setting out and return, in the same way as the coaches; while many, although possibly advertising a certain time of departure, never kept to any well-defined plan of this kind. Perhaps it is because of these circumstances that Sutcliffe, in 1816, made the statement that the competition between public roads and canals was great and that, in many instances, the difference in the expense of carrying upon them was very trifling[1].

In order that we may not get too exalted an opinion of the canals, it will be well for us to look for a moment at the rate of speed at which goods were conveyed on the navigations. In considering the organization of the canal carrying trade we mentioned that the fly boats went on the average about three and one-half miles an hour and the heavy barges about two miles an hour. A few more details will help us to appreciate the circumstances of the time. In 1793 barges required two to two and one-half days, and at flash times three days, to go from Lechlade to Oxford. With favourable water it took one day to go from Wallingford to Oxford and two and one-half days to go from Oxford to Wallingford; but sometimes it had taken three weeks and from London to Oxford eight weeks[2]. From London to Marlow required three days at the least; but usually four, and, when the days were short, five days were necessary[3]. The usual time consumed in going by canal from London to Liverpool in 1831, with heavy goods, was seven or eight days and nights[4]. In 1832 it required four days for the "fly boats" to bring glass from Birmingham to London[5]. On the navigations from London to Bristol, i.e., rivers Thames and Kennet, Kennet and Avon Canal, and river Avon, the small boats with a full load of twenty-seven tons of merchandise, in 1825, took five days in summer and six days in winter[6]. By 1835 the average time required for this same journey was given as from seven to ten days, but on occasions it had taken three or four weeks when the traffic was impeded[7]. In this latter year, during fair weather, it took,

---

[1] Sutcliffe, *Treatise on Canals and Reservoirs*, p. viii.

[2] *Parl. Papers*, 1793, XIII, 'Miscellaneous Reports, No. 109,' p. 26, evidence of Mr Court.

[3] Ibid., p. 27.   [4] *Birmingham Journal*, Mar. 5, 1831, p. 3.

[5] *London and Birmingham Railway Bill. Extracts from the Minutes of Evidence given before the Committee of the Lords on this Bill*, p. 4, evidence of Mr Hemsley.

[6] Brit. Mus. 08,235. f. 77, 'Observations on the General Comparative Merits of Inland Communication by Navigations or Railroads,' (1825), pp. 45–46.

[7] *Great Western Railway Bill. Minutes of Evidence taken before the Lords' Committees*, p. 11, evidence of C. L. Walker. With him agreed Mr John Harley (ibid., p. 14).

on the average, three days to go the eighty-nine miles from London to Reading, but sometimes it had taken a month[1]. While, therefore, the great changes in industry, agriculture and commerce were being facilitated and promoted through lowering the charges of carrying by means of canals, there was yet much to be done in improving the speed of conveyance.

The immediate effect of the canals was, then, to reduce the cost of transportation; but, in some cases, the ultimate results were somewhat different[2]. As the expense of land carriage had been increased in certain places before the introduction of canals, due to the great increase of the carrying trade and the carriers' inability to meet all requirements[3], so it was with the canals when these two means of conveyance were no longer sufficient for the volume of traffic. It would appear that, before the advent of the railway, both land and water carriers had sometimes put up their rates to a high level, so high as to be a serious burden on industry[4]. This will be more fully considered in a later chapter.

---

[1] *Great Western Railway Bill. Minutes of Evidence taken before the Lords' Committees*, pp. 398, 408, evidence of Mr Davis and Mr Morris.

[2] It would seem that, in certain instances, the reduction of the cost of transportation, if the charge had been actually reduced at all, was of short duration. Blackner, *History of Nottingham* (1815), p. 15 n., tells us, in reference to the Nottingham Canal, that the people of that city had been miserably deceived regarding the price of coal; for instead of having that article cheaper, as was expected, through the facilities of the canal, the price had been considerably advanced. Of course, the higher price of the coal might not have been due to higher carriage charges, although Blackner gives a very decided intimation that this was the reason.

[3] In addition to the statements of fact given when considering this subject under the general topic of highways, see also *J., H. of C.*, Mar. 14, 1758, xxviii, pp. 183–44, where it is stated that the cost of carriage had been increased, and that even then wool had been known to lie at Leeds and Wakefield three or four months, waiting for conveyance. See also *Local Notes and Queries*, No. 1648 (Birmingham Free Reference Library, No. 144,053), showing how the canal connecting Birmingham with the collieries reduced the "very exorbitant price" of coal to a "very moderate" price.

[4] A noted instance of this was that of the waterways connecting Liverpool and Manchester, on which the rates of carriage in 1810 were nearly three times what they were in 1795 (Sandars on the *Liverpool and Manchester Railway*, p. 11).

CHAPTER VI

STEAM NAVIGATION

At first it might seem as if the subject of steam navigation had little, if any, connexion with the development of traffic on the interior highways by land and water, but were more definitely related to the transmarine trade. When we keep in mind, however, that the application of steam to the propulsion of boats on canals and rivers was closely correlated with its use in boats on the navigable tideways of rivers and along the coasts, we shall see that the development of steam navigation had an intimate bearing upon the circumstances connected with the growth of the internal trade of the kingdom. In fact, the increasing use of steam on some of the larger estuaries was just as much a factor in the furtherance of internal trade as was any other improvement in river or canal navigation. The effects may not have been so marked in the former as in the latter, but they were none the less vital and tangible.

The first instance of applying steam to vessels in English waters was that which occurred in 1736, when Jonathan Hulls obtained from the King a patent, granting him for fourteen years the sole use of his invention of a "machine for carrying vessels or ships out of or into any harbour, port or river, against wind and tide, or in a calm[1]." He discussed, in the pamphlet describing his "machine," whether the machine should be placed in the vessel that was to be towed along, or whether it should be fitted into a boat which, by being attached to the vessel, might draw it along. For several reasons that were sufficient for him, he preferred the latter method; and his description

[1] Hulls, *A Description and Draught of a new-invented Machine for carrying Vessels or Ships out of, or into any Harbour, Port, or River, against Wind and Tide, or in a Calm*. In this pamphlet, after describing some physical principles, Hulls gives a short account of the mechanics of his vessel and a cut to illustrate it. For the earlier development of the use of steam as a motive power in other countries, dating from 120 B.C., see Boyman, *Steam Navigation, Its Rise and Progress*, pp. 69–74; Woodcroft, *A Sketch of the Origin and Progress of Steam Navigation from Authentic Documents*, pp. iv–19.

of the mechanism shows us a regular steamboat in its essentials. The application of the paddle-wheel appears to have been originally suggested by this patent. Next in succession were the experiments of the Duke of Bridgewater to use steamboats for towing barges on canals. In 1781, the Marquis of Jouffroy constructed a steamboat at Lyons, one hundred and forty feet in length; and with this he made several successful experiments on the river Saône[1]; but these successes did not lead to any immediate efforts to perpetuate them in France and so that nation lost the honour which might have been theirs had they made effective this achievement.

In Scotland the year 1788 marked an epoch in the application of steam to navigation purposes. For some time Mr Miller of Dalswinton had been working on the construction of ships, not only to improve their security, but to make them independent of the wind. This latter was secured by the use of paddle-wheels turned by human strength[2]. After the arrival of James Taylor, in 1785, as a member of Miller's family, engaged to teach his children, the two men had much discussion as to the best means of getting rid of the necessity of using human strength to drive the paddle-wheels; and they finally came to the conclusion that the steam-engine was the only thing that could supersede human muscle. Taylor drew up plans to show that the steam-engine could be used for this purpose; and at last, in 1787, Miller was satisfied and agreed to have a small engine constructed and put to a test on the lake at Dalswinton. Taylor applied to his old school friend, William Symington, and Miller gave Symington the order for the engine[3]. In 1788, in the presence of hundreds of spectators, the success of the engine was shown by the vessel moving at the rate of five miles per hour[4]. The projectors wishing to repeat this success on a larger scale in the Forth and Clyde Canal, Symington and Taylor had another engine constructed, and trial was made in the last days of the year 1789. This time the vessel moved freely as

[1] Boyman, *Steam Navigation*, p. 94.

[2] Woodcroft, *A Sketch of the Origin and Progress of Steam Navigation*, pp. 21–29, gives a pamphlet published by Miller in 1787, with drawings of the vessel which he propelled by paddle-wheels turned by men; and in the pamphlet Miller stated: "I have also reason to believe that the power of the steam-engine may be applied to work the wheels." The vessel used on Dalswinton lake was like two boats set by side, with a deck fitted on top of them.

[3] *A Brief Account of the Rise and Early Progress of Steam Navigation: intended to demonstrate that it originated in the Suggestions and Experiments of the late Mr James Taylor of Cumnock, in connection with the late Mr Miller of Dalswinton*, pp. 1–3.

[4] Ibid., p. 4; Woodcroft, *Steam Navigation*, pp. 32–39. An account of this was sent to and published by the Dumfries newspaper about the middle of October, 1788.

before, at a rate of nearly seven miles an hour, and these results were published in all the Edinburgh newspapers[1]. As the outcome of all these experiments, Miller became dissatisfied, Taylor was too poor to go on, and so Symington was the only one of the three who persevered and whose name has been immortalized.

In 1795 Lord Stanhope constructed a vessel to be propelled by steam; but its success led him no farther and Symington was left as the only important worker in this field of investigation. In January, 1801, the latter, under the patronage of Lord Dundas, began work upon a steamboat, called the "Charlotte Dundas," to be used upon the Forth and Clyde Canal. This boat was tried in March, 1803, and completely demonstrated the practicability of steam navigation; for it towed two other boats of seventy tons each, well loaded, nineteen miles, to Port Dundas, Glasgow, against a strong head wind. But a committee of the canal managers were opposed to its use, lest it should wear away the canal banks, and it was tied up and left lying at its moorings[2]. Unable to procure patrons to provide assistance in his work, and with his own resources too much diminished to be able to continue his experiments, Symington was compelled to forego the satisfaction of seeing steam navigation develop under his skill.

But two others profited by his success. Henry Bell, of Glasgow, and Robert Fulton went to Falkirk and were shown Symington's vessel in all its details; then each proceeded to use the results in his own way[3]. Fulton, who applied steam navigation on the Hudson river in the United States, seems to have obtained his first model engine from Henry Bell, and the engine he used on his first steamboat, the "Clermont," was made by and obtained from Boulton and Watt, of Birmingham[4]. The next steamboat built in British waters, after the "Charlotte Dundas," was the "Comet," constructed in 1811 by Henry Bell, to whom a worthy monument has been erected on the Clyde. Bell, apparently, took his drawings from Symington's

---

[1] *A Brief Account of the Rise and Early Progress of Steam Navigation*, pp. 4–5. Miller declined to have his steamboat patented; and there has been much discussion as to whether it was Miller or Taylor who first made the steamboat practicable in Great Britain. The details of that discussion are found in Woodcroft, *A Sketch of the Origin and Progress of Steam Navigation*, and in *A Brief Account of the Rise and Early Progress of Steam Navigation*. See also Boyman, *Steam Navigation*, p. 95.

[2] *A Brief Account of the Rise and Early Progress of Steam Navigation*, p. 5 ; Boyman, op. cit., p. 95; Kennedy, *Steam Navigation*, pp. 4–5; Woodcroft, op. cit., pp. 52–58.

[3] Kennedy, *Steam Navigation*, pp. 6, 11; *A Brief Account of the Rise and Early Progress of Steam Navigation*, p. 5.

[4] Kennedy, op. cit., p. 7.

engine[1]. The "Comet" was a vessel of twenty-five tons, with a four horse-power engine; and its success in navigating the Clyde in 1812 led to the construction of several other steamboats, of larger dimensions and greater steam power, by other persons. While this first vessel on the Clyde showed that she could successfully navigate the river, the number of passengers she carried was so small that she could hardly clear her expenses; but the success of the vessel aroused public confidence and the popular fear of the bursting of the engine was soon dissipated[2]. Experience gradually showed that, in order to secure economy, vessels should be built larger and with more powerful engines; and, consequently, the three additional steamers that were constantly plying on the Clyde between Glasgow and Greenock in 1813 were faster than the "Comet" and were twice as large[3]. In the same year the "Comet" went through the Forth and Clyde Canal, and thence on the Forth to Leith, from which she returned by the same route. In 1814 the "Stirling," with a twelve horse-power engine, began to ply between Stirling and Leith; and in 1815 two other vessels were added on the Forth. In 1814 a vessel was run on the Tay between Perth and Dundee[4].

In England, in 1813, one steamboat was plying on the Avon between Bristol and Bath, and another was introduced on the Yare between Norwich and Yarmouth. From that time the number gradually increased, until by 1816 there were also vessels on the Trent, Tyne, Ouse and Humber, Orwell, Mersey and Thames[5]. In the case of the Thames, the first steamer to ply on its waters left the Clyde in the winter of 1814–15, went through the Forth and Clyde Canal and then down the east coast to the Thames, where it was run during the season of 1815 between London and Gravesend. This was the first steam vessel ever seen on the Thames. Another Clyde steamer went, via Land's End, to the Thames, and plied between London and Margate[6].

---

[1] Boyman, op. cit., p. 96. On all this historical development, see the elaborate *Report of the Select Committee of the House of Commons on Steam Navigation*, 1817; also Baines, *History of Liverpool*, Chap. XVII.

[2] Buchanan, *Treatise on Steamboats*, p. 12.

[3] Ibid., p. 29; Kennedy, op. cit., p. 12. These larger vessels were 75 feet long and 14 feet wide. *Annual Register*, LVII, p. 504, shows the increase of passenger traffic on the Clyde due to the cheapness and facility of conveyance; and also the increase in power and tonnage of the vessels between 1801 and 1812.

[4] Buchanan, op. cit., pp. 61–62. [5] Ibid., pp. 62–64, 171–5.

[6] Buchanan, op. cit., pp. 15, 173. In the Thames, in 1816, there were three steamboats, of twelve, fourteen and sixteen horse-power respectively, plying between London and places as far out as Margate; another of twelve horse-power was soon to start and another of six horse-power was being built. See also Cruden, *History of Gravesend*, p. 484.

In the same year there was the arrival from the Clyde of the first steamer ever seen on the Mersey[1]. On the Humber, the first steamboat was introduced in October, 1814; and when the weather was favourable a speed of fourteen miles an hour was said to have been attained, which was much more than any previous rate[2]. This, however, was probably exaggerated, for in 1816 the best rate of a vessel on the Trent river between Hull and Gainsborough was given as six miles per hour[3].

The successful steam voyages from the Clyde to the Thames and Mersey gave an impetus to steam-packet construction and created active opposition, especially on the London to Margate service[4]. But it was not till 1816 that a steamboat was used to perform regular voyages at sea. In that year, the first steamer plied regularly between Greenock and Belfast[5]. In 1816 also a vessel of one hundred and fifty tons went daily between Holyhead and Dublin[6]; and in the autumn of the same year there was the first actual use of steam in the port of Liverpool for towing vessels out to sea[7]. The steam ferry service between Liverpool and the Cheshire shore began in July, 1816, when the "Princess Charlotte" sailed twice each way between Liverpool and Eastham, where the steamer connected with coaches to and from Chester, Shrewsbury and many other places; and the ferry between Liverpool and Tranmere was established in the same year[8]. The steam-packets on the Clyde increased with great rapidity; for while in 1812 there was only the "Comet" plying between Glasgow, Greenock and Helensburgh, in 1815 there was a fleet of seven steamers sailing regularly from Glasgow, southward to Largs, Ardrossan, Troon and Ayr, and westward to Rothesay, Tarbert, Lochgilphead and Inverary[9]. In 1816, two steamers on this river advertised the granting of season tickets for families; and on August 7, 1816, the "Dumbarton Castle"

---

[1] Buchanan, op. cit., pp. 15, 64.

[2] *The Hull Rockingham*, Oct. 15, 1814, said of this vessel "that, with both wind and tide against her, her speed is very considerable...travelling at the rate of fourteen miles an hour." This was during favourable weather.

[3] Buchanan, op. cit., p. 174. This vessel was said to go the 50 miles between Hull and Gainsborough in eight hours.

[4] Kennedy, op. cit., p. 23.

[5] *Liverpool Mercury*, April 5, 1816. This service was established by Napier, who put on the "Rob Roy," a vessel of 90 tons and thirty horse-power.

[6] Billinge's *Liverpool Advertiser*, Nov. 4, 1816. Kennedy, op. cit., p. 26, gives the capacity of this steam-packet as 112 tons. He says that her average time between Holyhead and Howth was about seven hours and that she was frequently faster than the mail packets.

[7] Ibid., Oct. 21, 1816.

[8] Kennedy, op. cit., p. 24 f.   [9] Ibid., p. 27.

was advertised to take passengers for a trip from Glasgow around Ailsa Craig. She was the first British vessel, except the one that first reached the Thames from the Clyde, to take passengers on a deep-sea trip[1]. From that time on, sea-going steamers began to appear in increasing numbers in British waters; and in October, 1817, a new steamer began to run between Hull and London. In 1818, after a service of two years which it initiated between Greenock and Belfast, Napier's vessel, the "Rob Roy," was transferred to the English Channel, as a packet between Dover and Calais[2]. In 1819 there was the commencement of the service between Belfast and Liverpool and soon the addition of other vessels provided a regular fleet on that route[3]. The first steamer that traded between Liverpool and Glasgow was advertised to sail from the former place on August 2, 1819; and it was expected that she would perform the journey in thirty hours[4]. From 1819 there was a rapid expansion of the British steam coasting trade. During the years 1819–21, seven steamboats, of large tonnage and great power, conveyed passengers between Greenock and Belfast and Liverpool, between Liverpool and Dublin, and between Liverpool and Bagillt in Flintshire. Most of these vessels were constructed on the Clyde. From these beginnings, it was only a few years until the chief ports of the three kingdoms were linked up by regular communication[5]. Regular services were advertised between Liverpool and the

---

[1] Kennedy, op. cit., pp. 29, 30.   [2] Ibid., p. 26.   [3] Ibid., p. 33.   [4] Ibid., p. 34.

[5] From Brit. Doc. 1822 (417), vi, 115, 'Fifth Report of the Select Committee on the Holyhead Roads,' much of our information on this subject has been obtained. For the number of steamboats built since 1811, and used in the United Kingdom, see this Fifth Report, Appendix No. 1; also *British Almanac and Companion*, pp. 112–13.

In 1826 there were 24 steamships sailing along the east coast from Hull in the summer months; and London was the extreme limit to which any of them ran. In 1835 the number had increased to about 40, some of which were in use between England and Holland (Sheahan, *History of Kingston-upon-Hull*, p. 363).

A good indication of the great increase in the use of steam vessels is furnished by the following abstract showing the number of vessels belonging to the port of London, from 1814 to 1830. The increase in the number of vessels was accompanied by a great increase in the tonnage of each vessel:

In 1814 there were	0 vessels.	In 1823 there were	26 vessels.
„ 1815 there was	1 vessel.	„ 1824 „	29 „
„ 1816 there were	2 vessels.	„ 1825 „	44 „
„ 1817 „	3 „	„ 1826 „	59 „
„ 1818 „	6 „	„ 1827 „	59 „
„ 1819 „	7 „	„ 1828 „	57 „
„ 1820 „	9 „	„ 1829 „	61 „
„ 1821 „	16 „	„ 1830 „	57 „
„ 1822 „	22 „		

Of the 57 vessels in 1830, 42 navigated the river Thames; the remainder were either laid up or employed at other ports. Brit. Doc. 1831 (335), viii, 1.

Isle of Man, Whitehaven, Dumfries, the Clyde ports and the Irish cities of Belfast and Dublin; and often two or three companies advertised sailings for the same ports[1]. In the spring of 1821 a new steamboat, at that time the largest in the United Kingdom, was launched at Perth; but she was later placed on the Newhaven (Edinburgh) to London service. In May, 1821, two steamboats, of over 400 tons each, were built for the Leith to London passenger service. They were not intended to carry cargo, and each had sleeping accommodation for one hundred passengers. They had engines of one hundred horse-power and were expected to make the voyage in about sixty hours[2]. The peculiarity of steamboats which recommended them was that their sailings were not dependent upon favourable conditions of weather, wind and tide, but were made with regularity.

In 1821 there was the establishment of two steam-packets, instead of sailing vessels, for carrying the mails from and to Holyhead and Dublin[3]; and, later, others were built on the Thames and established at various English ports to facilitate the carrying of mails and passengers. As "an amazing instance of facility in travelling by steam vessels," mention is made of a journey from London to Leith in fifty hours. The same vessel had regularly plied between Leith and Aberdeen, 108 miles, one of the most exposed coasts of Great Britain, and was never more than fourteen hours in making the passage[4]. In 1822 the Plymouth, Devonport, Portsmouth and Falmouth Steam-Packet Company was formed and in the next year regular communication was maintained with each of these places. In 1836 there was one line of seven vessels and another of eight between London and Plymouth[5]. And thus connexion among all the principal ports of the kingdom was speedily obtained by steamboat: a connexion which was much favoured by those who wanted a cheaper and safer mode of travel than the stage coach. We shall not follow out any further the detailed growth of these facilities[6].

The progress of steam navigation was very rapid after 1814[7]; but its great increase was chiefly confined to the large rivers, to the coast

[1] Kennedy, op. cit., p. 34.   [2] Kennedy, op. cit., pp. 35–36.

[3] Brit. Doc. 1822 (180), vi, 9, 'Second Report on Holyhead Roads.' For the rates for passengers, etc., on these packets, see Brit. Doc. 1822 (417), vi, 115, 'Fifth Report on Holyhead Roads,' p. 130; also Appendix No. 8 to same.

[4] Aris's *Birmingham Gazette*, Nov. 5, 1821, p. 1.

[5] Worth, *History of Plymouth*, p. 341.

[6] Kennedy gives a fine treatment of the *History of Steam Navigation*, and to his work the reader is referred for the further details of the subject.

[7] The statistics of this growth are given in Brit. Doc. 1822 (417), vi, 115, 'Fifth Report on Holyhead Roads,' Appendix No. 1.

trade, and to the near-by sea traffic. On the rest of the inland water-ways, other than the great river estuaries, there was but little use of steam. By 1822, when engines had been perfected, it was recognized that the safety of steam vessels, even in the most tempestuous weather, had been proved beyond all doubt[1]. They therefore came into favour as a means of furnishing a pleasant outing along the rivers and coasts for the public, who were eager to have them. In this way, the Thames became crowded with all kinds of these vessels and accidents were not infrequent. A committee of Parliament was constituted to examine into these matters and from their report we gather that steam vessels had not always been built strong enough for their work; that they had been carrying too many passengers; and that their speed had sometimes exceeded the limit of safety[2]. This, of course, was largely due to their continuously increasing competition in the conveyance of passengers[3]. To remedy these abuses and others, vessels had to receive a license; the speed of vessels in the crowded parts of the river was reduced and regulated; and limitations were placed on the number of passengers that each could carry[4].

We have just said that great increase in the amount of steam navigation was not found upon the canals. On September 19, 1828, a small steamboat of very moderate power passed through the canals from London to Manchester, with the object of seeing whether steam might be used on the canals without injury to them[5]. Various desultory attempts were made to introduce steam power on them, but without practical results[6], probably because it was thought that the action imparted thereby to the water would tend to break down the banks of the canals, which were often unprotected by any walls, and also because it was not long until the railway came to completely over-shadow the canal as a means for the conveyance of goods. Only a very few of the canals of England have been sufficiently improved to warrant the use of steam as an economical motive power[7].

[1] See 'Report of Select Committee on Holyhead Roads,' 1822.
[2] Brit. Doc. 1831 (335), viii, 1, 'Report of Select Committee on Steam Navigation.'
[3] *Gentleman's Magazine*, c, Pt. 1, p. 552.
[4] The steam vessels along the coasts and on the rivers, through their com-petition, took away much traffic from the roads. This phase of the subject will be dealt with in a later chapter.
[5] *Gentleman's Magazine*, xcviii, Pt. 2, p. 265.
[6] *The Palatine Note Book*, iii, pp. 42-43.
[7] The Aire and Calder and the Weaver Navigations are probably in the front rank at the present time.

CHAPTER VII

DEVELOPMENT OF RAILWAYS

It is not easy to trace the origin of railways, but the earliest approximation to the modern railway was, doubtless, the wooden tramroad, the existence and use of which dates far anterior to the modern railway era. The earliest system for the conveyance of coal from inland collieries was by the use of pack-horses, mules or asses, over the backs of which were slung the bags filled with coal; and this method prevailed down to the close of the eighteenth century[1]. Of course, with the gradual improvement of the roads, some carts had come into use; and the amount of load that could be drawn upon these roads had increased. There still remained, however, the difficulty of bringing the coal from the pit's mouth down to the river or to the road; and to effect this end, wooden tramroads came in time to be laid down.

We are informed that, as early as 1555, there was a tram from the west end of the Bridge Gate in Barnard Castle, for the repairing of which the proprietor of the castle left the sum of 20s. The word "tram," at that time, seems to have been used in the north of England and the south of Scotland to describe the special track or road and the truck that ran on it. The truck was drawn along this way by men or horses[2]. The use of the tramroad in the coal districts, however, for facilitating the conveyance of that heavy commodity, does not seem to have come into public attention until half a century or a century after that time; for a record in the books of one of the free companies in Newcastle, dated 1602, states that from time immemorial the coal carts had been accustomed to carrying eight "baulls" of coal from the pits to the river[3], but recently that amount had been reduced

---

[1] Jeans, *Jubilee Memorial of the Railway System*, p. 5. Jeans says of his work that the facts and figures were all "compiled from official and accredited sources, so that their accuracy may be accepted as unimpeachable."

[2] Gordon, *Our Home Railways: How they began and How they are worked*, I, p. 4. Gordon says that his work has been drawn from the original sources, and has been officially approved by the railway authorities as authentic.

[3] Eight bolls of coal were equal to 17 cwt.

to seven "baulls[1]." The expense of carrying such heavy loads on poor roads would naturally cause them to seek some other méans of conveyance than by cart; but it would seem as if no great change had been made before the middle of the seventeenth century, for a gentleman, writing in 1649, said that many thousand people were employed in the coal trade, some by working in the pit, and others by carrying in waggons and wains to the river Tyne[2]. Some change, however, had been made, for shortly before that time, perhaps about 1630, a man by the name of Beaumont went north to Newcastle with new kinds of implements for mining the coal, and he it was who introduced the "wooden way" and waggons for carrying the coal from the pits down to the river. He, apparently, had thirty thousand pounds in money with which to begin his improved system of mining the coal and sending it on its way toward the market; but in a few years he had used up all his money and "rode home upon his light horse," having lost all his capital[3]. By the old system, it was not uncommon for these northern mine-owners to employ five or six hundred horses and carts in this traffic; and hence it was of vast importance to reduce the great expense incurred in keeping so many horses and drivers, in the wear and tear of carts, and in the making and repairing of roads[4]. It was recognized that the difficulties of the soft roads would be overcome by the adoption and use of the wooden rails upon which to draw the loaded waggons; so that, although Beaumont lost all that he had, others took up his ideas and put them successfully into operation. About 1670 the use of wooden ways seemed to be a common method for conveying the coal from the pits to the river, and those who had lands between the collieries and the river would lease or sell strips of these lands to the mine-owners, upon which the latter would lay their rails from the mines to the bank of the river. Rails of timber were laid down and bulky four-wheeled waggons were made to fit these timbers; so that a keen observer, in 1676, asserted that by this means the carriage was made so easy that one horse would draw down as much as four or five chaldrons of coal at one time, which was an immense benefit to the coal merchants[5].

---

[1] Wood, *Practical Treatise on Railroads* (1825), p. 34.

[2] Gray, *Chorographia, or a Survey of Newcastle upon Tine* (1649), pp. 24–25.

[3] Ibid., pp. 24–25; Wood, *Practical Treatise on Railroads* (1825), p. 35, quoting from Gray; Stretton, *A Few Notes on Early Railway History*, p. 3.

[4] Jeans, op. cit., p. 5.

[5] North, *Life of the Right Honourable Francis North, Baron of Guilford, Lord Keeper of the Great Seal*, pp. 136–7. See also Wood, *Practical Treatise on Railroads* (1825), p. 36, and Cumming, *Rail and Tram Roads* (1824), p. 7. In *Transactions of the Highland Society*, vi, p. 6 et seq., Scott gives an account of the origin and

In constructing such a road an effort was usually made to have it on a slight decline from the pit's mouth to the place where the coal was to be discharged from the waggons, so that heavy loads might be easily conveyed without a great expenditure of energy by the horses employed in hauling. The rails were not always laid so as to give a uniform declivity throughout the whole length; but they followed more or less the surface of the ground. Where, on part of the road, there was a steep declivity, the speed of the waggon was regulated by a brake attached to the vehicle and managed by the driver. The waggons used had low wheels, for the smoothness of the rails made high wheels unnecessary; and upon the roads of ordinary declination it was easy for a horse to draw three tons of coal from the pit to the river[1], although the ordinary load for one horse was nineteen bolls or about forty-two hundredweight[2]. The economy of the waggon-way over the old way of carrying coal may be noted from the fact that, upon the common roads, the regular load for a horse with a cart was only about seventeen hundredweight[3]. Of course, the empty waggons had to be drawn back up the incline and the road was made so that horses could meet and pass at certain places.

In time the wooden rail had its upper surface worn away; and it is probable that at first such repairs were made by fastening another rail or plank upon the top of the one that was worn[4]. But on some parts of the road where occasional acclivities occurred which could not be levelled, or where sudden windings of the road had to be made, and where, therefore, there would be an unusual amount of friction with the wheels, thin plates of wrought-iron were laid on the wooden rails and fastened to them. The advantages secured by this means in diminishing friction and keeping the rails from wearing would suggest the obvious advantage of having the wooden rails plated throughout with sheet-iron, or covered with iron plates or bars nailed on them. These were called "plateways." When these rails were first faced with iron, we do not know; but the use of such plated rails

development of railways, and says that from the records of Ravensworth Castle it would appear that railways came into use there in 1671. See also the petition of Charles Brandling, Lord of the manor of Middleton, owner of coal mines there, and several other owners and occupiers of grounds in Leeds, asking Parliament for permission to lay a waggon-way, from the coal mines, through the grounds of the other petitioners, to Leeds, where Brandling agreed to deliver coal at reduced prices (*J., H. of C.*, xxviii, p. 57).

[1] Cumming, op. cit., p. 9, and Wood, op. cit., pp. 36–41, give full details as to the construction of the road.

[2] Cumming, op. cit., p. 8; Wood, op. cit., p. 41.

[3] Wood, op. cit., p. 41.

[4] Stretton, *A Few Notes on Early Railway History*, p. 3.

soon caused much wear to the wooden wheels of the waggons, and the next improvement was to replace the wooden wheels by those made of cast-iron[1]. From one source we are informed that as early as 1734 cast-iron wheels with an inner flange were in use near Bath[2]; and another who has looked carefully into the history of some early railways gives the introduction of cast-iron wheels as having taken place about 1753[3]. Which of these dates is more nearly correct we shall not undertake to say.

The use of the plated rail did not seem to be very much extended. Probably the chief reason for this was that the nails, which were intended to securely hold the plates on the rails, would be continually working loose and demanding constant expense in effecting repairs. Instead of this method being continued, the plated rails were displaced and cast-iron rails were adopted. At what time the introduction of the latter occurred we are unable to definitely ascertain; the year 1738 has been given as the time when cast-iron took the place of wooden rails[4], but we have not found anything to substantiate this, and regard the date given as too early. Probably the year 1767 more nearly marks the date when the cast-iron rail superseded the old plated rail; for the books of the great Coalbrookdale Iron Works in Shropshire show that on November 13, 1767, there were between five and six tons of such rails cast there and tried as an experiment[5]. These rails were made with a flange on the inside, and they were so long and without any support in the centre that the heavy waggons frequently caused them to break; but, later, that was remedied by making the waggons smaller and the loads lighter, and coupling the cars together so as to distribute the weight over a greater length of road[6]. In 1776 a similar cast-iron railway was laid down by John

---

[1] Stretton, op. cit., p. 3; Gordon, *Our Home Railways*, p. 4.

[2] Gordon, op. cit., p. 4.   [3] Stretton, op. cit., p. 3.

[4] Wood, op. cit., p. 44, quoting from an anonymous writer. See also Cumming, *Rail and Tram Roads*, p. 10, and Francis, *History of the English Railway*, I, p. 45.

[5] Stretton, op. cit., p. 4; also Wood, op. cit., pp. 44–45, who quotes the statement of Robert Stephenson to substantiate this assertion. See also Gordon, op. cit., p. 4. Scott, in *Transactions of the Highland Society*, VI, p. 7, said that below ground, in the pits, cast-iron rails did not begin to replace wooden ones till 1776.

[6] Stretton, op. cit., p. 4; Wood, op. cit., p. 44. In *Communications to the Board of Agriculture*, I, p. 203, there is given the description of a tramroad from the coal-pit to Alloa, Scotland. The sleepers were eighteen inches apart; the wooden rails were covered by iron on top; and the waggons would each hold one and one-half ton of coal. Two, and sometimes three, waggons were linked together; so that, under the latter circumstances, one horse would draw four and one-half tons of coal and three tons weight of carriages. The first cost of construction was £900 to £1000 per mile.

Curr at the Nunnery Colliery, near Sheffield[1]; but it was not until about 1794 that cast-iron rails were first used in the collieries of Durham and Northumberland[2].

One of the greatest improvements was made in 1789, when, at the suggestion of Smeaton, William Jessop, in constructing a railway from Nanpantan to the Loughborough Canal, used narrow, cast-iron "edge rails," three feet long, and removed the flange from the rail to the inner side of the wheel[3]. This form of rail and of wheel has been the model upon which the construction of rails and wheels has proceeded during most of the time since that day.

Up to this time wooden sleepers had been in use and the rails had been bolted or pinned to them. But in 1797, when laying a railroad at the Lawson Colliery, near Newcastle-upon-Tyne, Barns introduced stone blocks instead of wooden sleepers, the inducement, doubtless, being that the stone supports would be more suitable for a road which had to carry loads of such heavy material as coal[4]. This was also applied by Benjamin Outram, in 1799, in the construction of a line from Ticknall to the Ashby Canal; the rails used here were of the same form as those used in 1776 by Curr on the Sheffield line, namely, with the flange on the outside of the rails, and the latter had both ends fastened securely to the stone supports. It is evident, therefore, that Outram was not the first to employ stones as the foundation of the railway, although some have supposed that his name, shortened by the maintenance of only the last part of it, has been perpetuated by calling these "tram-ways[5]."

In 1799 there was a proposal to build a line from London to Portsmouth and in 1801 the Surrey Iron Railway Company obtained an Act for accomplishing the first part of this road, from Wandsworth to Croydon. This was the first railway company, the first public railway, and the first so-called railway Act, although it was not the first Act in which a railway was authorized. The line was opened in 1804 between these two places and traction was effected by horse-power; and the rails, resting on stone block sleepers, were four inches

[1] Wood, op. cit., p. 45, quoting from Carr's *Coal Viewer and Engine Builder*; Stretton, op. cit., p. 4; Gordon, op. cit., p. 4.

[2] Jeans, *Jubilee Memorial of the Railway System*, p. 6.

[3] Wood, op. cit., p. 48; Stretton, op. cit., p. 4; Gordon, op. cit., p. 5. The Jessop rail may now be seen in the South Kensington Museum, London.

[4] Wood, op. cit., p. 46; Stretton, op. cit., p. 5.

[5] Wood, op. cit., p. 46; Stretton, op. cit., p. 5; Gordon, op. cit., p. 5. The name "tram" was in use much earlier than this, for in 1555 Ambrose Middleton bequeathed 20s. "to the amendinge of the highwaye *or tram* from the weste ende of Bridgegait, in Barnard Castle." (*Surtees Society Publications*, xxxviii, p. 37 note.)

wide, one inch thick, and with an arched flange one inch thick and three and one-half inches high. The delight of a certain nobleman in witnessing the economy of horse-power on this railway caused him to think that such lines should be extended from London to Edinburgh, Glasgow, Holyhead, Milford, Falmouth, Yarmouth, Dover and Portsmouth; but the idea of the general extension of railways over the country was at that time considered as absurd[1].

Between this Act of 1801 and that of 1821 sanctioning the construction of the Stockton and Darlington Railway, there were not less than nineteen railway Acts passed, five of which were allowed to lapse. Among the plate-ways which were constructed at this time were some of great significance, such as the Peak Forest line over the Derbyshire hills[2]; a line at Ashby-de-la-Zouch in Leicestershire[2]; the Forest of Dean line; the Gloucester and Cheltenham; the Dartmoor Railway to connect with the port of Plymouth[3]; and others, such as the proposed railways from Stortford to Cambridge and from Liverpool to Manchester, had been under consideration but had not been authorized[4]. As early as 1810 there was a movement for the construction of a railway or canal from Stockton, via Darlington, to Winston, in order to provide an outlet for the mineral wealth of that district[5]. We are justified in saying that the first quarter of the nineteenth century was a time when, gradually, the claim of the railways upon public attention was becoming accentuated and the mechanical advances were showing that this was to be the coming means of locomotion.

The constant breaking of the cast-iron rails induced interested individuals to attempt to find a better substitute for them; and the results of experiment seemed to indicate that malleable iron rails, if properly supported so that they could not bend too much in the centre, would be more durable than cast-iron. Nicholas Wood, whose knowledge of early railway development was unsurpassed, informs us that malleable iron rails were tried at the Wallbottle Colliery, near Newcastle-upon-Tyne, about 1805, but because their narrow surface cut the periphery of the wheels they were superseded by cast-iron rails of wider surface[6]. Robert Stephenson's assertion was that this kind of

[1] Stretton, op. cit., p. 5; Gordon, op. cit., p. 6.

[2] Stretton, op. cit., p. 5.

[3] *The Times*, Nov. 21, 1823, p. 4. The road was just completed at this time after four years of construction work. For others that were constructed at this time see Francis, *History of the English Railway*.

[4] *Cambridge Chronicle and Journal*, Aug. 30, 1811, p. 2; *The Times*, July 29, 1822, p. 3.

[5] Jeans, *Jubilee Memorial of the Railway System*, p. 14.

[6] Wood, *Practical Treatise on Railroads*, pp. 60–61; Stretton, op. cit., p. 5.

rail was first introduced about 1815, at Lord Carlisle's coal works, on Tindale Fell, in Cumberland[1], and Jeans seems to have followed Stephenson in his acceptance of this date[2]; but Wood shows that, by the statement of the agent of these coal works, the date given by these two men was erroneous, for malleable iron rails were laid down on this tramroad in 1808[3]. We may say, apparently with truthfulness, therefore, that cast-iron rails began to be replaced by those of malleable iron in the first years of the nineteenth century; and although the wooden plated rails and the cast-iron rails were not immediately displaced by the malleable iron rails, the results obtained from the gradual adoption of the latter showed the wisdom of their increasing use[4].

The construction of the tramroads, or iron railways[5], varied according to the nature of the ground and the traffic to be carried upon them. If most of the traffic went in one direction, as in the carriage of coal from the mines, the road was made to slope slightly in that direction; and the degree of declivity was determined by the traffic and its extent, the object being to equalize the draught each way as nearly as possible. The road was sometimes made single and sometimes double, according to the expected amount of carrying upon the line and the way in which this was to be done[6]. As a rule, the track was not double, but the single track was made more serviceable by having turnouts, where waggons might meet and pass. The width of the rails apart was largely decided by the shape of the waggons and by the physical characteristics of the country through which the road passed; it usually varied between three and four and one-half feet. Where stone sleepers were in use, they were generally embedded about three feet apart from centre to centre, so as to accommodate the fastening of the rails which were generally three feet in length; and the space between the sleepers was filled up with gravel or other material to make a good hard road[7]. The rails were then laid end to end and fixed in position by having an iron spike driven through

---

[1] Wood, op. cit., p. 61.   [2] Jeans, op. cit., p. 12.

[3] Wood, op. cit., p. 61.

[4] Wood, op. cit., pp. 61–70, enters into a long discussion of the advantages and disadvantages of the malleable iron rail.

[5] In the early Acts, no difference is made between the terms tramroad and railway in regard to their meaning.

[6] If the descent of loaded waggons by gravity were to be used to draw up the waggons when unloaded, it was, of course, necessary to have a double track. So too, when one track was not sufficient to accommodate the amount of the carrying.

[7] On the tramroad from the coal-pits to Alloa (Scotland) the sleepers were only eighteen inches apart. See description of this road in *Communications to the Board of Agriculture*, i, p. 203.

each end into a wooden plug in the centre of the stone sleeper, or else by driving the spike through the rail into a cavity in the stone sleeper and fixing this securely in position by means of molten lead[1]. Of course, after malleable iron rails were substituted for those of cast-iron, it was not necessary to have the sleepers set so close together. We must not form the picture of these tramroads from what we know of our modern railways; it was rare, indeed, that the former demanded the cuttings and the embankments that we see on railway lines to-day, for instead of cutting through a hill in order to get a more or less straight course, they easily wound around the sides of the hills, preferring the circuitous rather than the straighter but more expensive road. In the same way, they avoided the filling up of large concavities along the line, and sought a more devious route in order to avoid the heavy costs of filling and embanking. The line of rail followed the great inequalities of the surface of the route chosen; and was not, as the present railways, laid upon as nearly a uniform, straight and level road as possible[2].

---

[1] Scott, in *Transactions of the Highland Society*, VI, pp. 8–10, gives the method of constructing the railway between Kilmarnock and Troon harbour. It was a double-track line, laid with flat or plate rails, although he says that at that time the edge-rail was generally introduced.

[2] Wood, op. cit., pp. 36–37, quotes a description of these tramways as given in Jaa's *Voyages Métallurgiques*, I, p. 199, in 1765; and he amplifies this description of the early railways on pp. 38–40 of his book. I have thought that the account of such a tramway, as given by Coxe, in his *Historical Tour through Monmouthshire* (1904), p. 202, would be illuminating, and would show what they were like in the year 1801, when his book was first published. As much as possible his own words are preserved in the narration of the process of making such a railroad: "The ground being excavated about six feet in breadth, and two in depth, is strewed over with broken pieces of stone, and the frame laid down. It is composed of rails, sleepers, or cross bars, and under sleepers. The rail is a bar of cast-iron, four feet in length, three inches thick, and one and a half broad; its extremities are respectively concave and convex, or in other words are morticed and tenanted into each other and fastened at the ends by two wooden pegs to a cross bar called the sleeper. This sleeper was originally of iron, but experience having shown that iron was liable to snap or bend, it is now made of wood, which is considerably cheaper, and requires less repair. Under each extremity of the sleeper is a square piece of wood, called the under sleeper, to which it is attached by a peg. The frame being thus laid down and filled with stones, gravel, and earth, the iron rails form a ridge above the surface, over which the wheels of the cars glide by means of iron grooved rims three inches and a half broad." At the junction of two roads, and to facilitate the passage of two cars in opposite directions, movable rails, called turn rails, are occasionally used, which are fastened with screws instead of pegs, and may be pushed sideways. "The declivity is in general so gentle as to be almost imperceptible: the road, sometimes conveyed in a straight line, sometimes winding round the sides of precipices, is a picturesque object, and the cars filled with coals or iron, and gliding along occasionally without horses, impress the traveller, who is unaccustomed to such spectacles, with

Another particular in which the tramroads differed from the railways of to-day is that the latter are the principal means for the land conveyance of goods, whereas the tramroads were regarded not as principal but as auxiliary agencies for transportation. In the later years of the eighteenth century, when the early tramroads were coming into notice in certain localities, the canals and the ordinary roads were claiming more and more attention, since these were considered as the permanent routes along which the conveyance of goods should be effected. It was but natural, therefore, that the new facilities of transport should be thought of, not as displacing in any sense the existing modes of conveyance, but as accessories to them. Canal companies considered the tramroads as valuable additions to their facilities, for by means of them trade and communications could be effected with districts that would otherwise be inaccessible. There were canals that were constructed through territory to tap the coal resources of that region; but either because they could not economically reach the source of the coal on account of its height above a water supply, or because, if they did reach it, the operation of the canal was too expensive to be remunerative, these canals were impotent to supply the necessary facilities for the development of this mineral wealth. In such places, the use of a tramway would enable such materials as coal, stone and iron-ore to be brought down to a lower level, where the canals could do the effective service that they were designed to fulfil in carrying these things thence to the places of manufacture. In fact, at the end of the eighteenth century, and even during the early part of the nineteenth century, tramways were regarded as strictly tributary to the canals; for in many petitions to Parliament, asking for authority to construct canals, there were also requests for the privilege of making collateral cuts, "with proper railways and other roads to communicate with these canals[1]." In most of these cases, the tramway was to reach some

pleasing astonishment." The expense of construction of these roads varies according to circumstances. It is seldom less than one thousand pounds per mile, and sometimes exceeds that sum. The cars weigh not less than three and a half tons. They are drawn by a single horse, and the driver stands on a kind of footboard behind, and can instantaneously stop the car by means of a lever and a drop, which falls between the wheels, and suspends their motion. In places where the declivity is more rapid than usual, the horse is taken out, and the car impelled forward by its own weight. For description of the waggons used, see Wood, op. cit., pp. 76–80; and other data regarding the formation and operation of these roads are given in Cumming, *Rail and Tram Roads.*

[1] See, for example, *J., H. of C.*, Mar. 11, 1789, regarding railways and roads to lead to the Cromford Canal; *J., H. of C.*, Feb. 9, 1791, regarding railways and roads to connect with the Hereford and Gloucester Canal; *J., H. of C.*, Dec. 20, 1792, for railways and roads to connect with the proposed Grand Junction Canal; *J., H. of C.*,

high and broken land where there were considerable amounts of mineral resources and where a canal would have been impossible or unprofitable[1]. From the above we can see that the tramways and roads were closely associated in their relation with the canals; but the fact that tramways were occasionally constructed to terminate at a certain bridge or a certain turnpike road is still stronger evidence that the iron roads were closely associated with the ordinary roads and subsidiary to them[2],—in other words, the tramways were collecting and distributing agencies for products carried along the great highways of the kingdom.

In England, Wales and Scotland, these tramroads were in some parts fairly numerous, and most of them were only short lines, branching off from the navigable rivers and canals to the different mines. The majority of those in the United Kingdom were in the extensive mining districts south of the Severn, including South Wales; in the coal districts near Newcastle and Sunderland along the rivers Tyne and Wear; in the coal and other mining areas of Lancashire and Yorkshire, as well as of Derbyshire and Staffordshire; in the mining regions of the county of Salop and adjacent parts of the Severn valley; in the mining sections near Glasgow, and in the coalfields of Midlothian and

Jan. 31, 1793, regarding railways and roads to connect with the Stratford-on-Avon Canal; *J., H. of C.*, Feb. 11, 1793, for railways and roads to connect with the Birmingham and Fazeley Canal. See also Pitt, *Agriculture of Leicester* (1809), p. 313, and Dickson, *Agriculture of Lancashire* (1815), p. 613, both of which show that by means of these iron roads coal and iron were brought down from the pits to the canals by a cheap and very convenient way, and that the tramways were primarily regarded as subservient to the canals, even down to the first quarter of the nineteenth century. That railways were not much thought of apart from canals, is shown also by *Communications to the Board of Agriculture*, II, p. 478, and *Transactions of the Highland Society*, VI, pp. 10–11. For details of this close relation between tramroads and canals, see the pamphlets of C. E. Stretton given in bibliography.

[1] There were many of these early tramroads in southern Wales, where there are mines of coal and iron; also in Lancashire, Derbyshire and the Newcastle region, as well as in Scotland. Anderson, *Recreations in Agriculture*, IV, p. 198, urged the construction of railways where canals were not possible, and showed (ibid., IV, pp. 199–201) to what extent railways had already been constructed in the Midlands of England and what a great increase in carriage had been effected by them. *Communications to the Board of Agriculture*, II, p. 477, shows the utility of the railways in extending the influence of canals for ten to twenty miles on each side of the latter, and also into the mountainous sections where canals were almost impracticable. See also Hassall, *Agriculture of Monmouth* (1812), p. 105, containing an account of the iron railways of that county and their effects.

[2] *J., H. of C.*, Feb. 15, 1826, petition for a railway or tramroad from the Grosmont railway at Llangua (co. Monmouth) to Wye Bridge, in the city of Hereford; *J., H. of C.*, lxxxv, p. 59, petition for the Leicester and Swannington railway or tramroad.

Fife, where they were found in great numbers but on a small scale[1]. These roads in South Wales, and in the counties along the Severn valley, were chiefly inclined planes with various slopes, on which one horse could easily take down thirty to forty tons together with the weight of the waggons, but it required three or four horses to bring the empty waggons up again, and even then the up-grade work was the heavier. There were, however, a few inclined planes on which the loaded waggons in descending brought up the empty ones, but this method was employed only in cases where the declivities were very great[2]. In the county of Salop and adjoining mineral areas of the west of England, and in Wales, these iron roads increased considerably in number in the first quarter of the nineteenth century[3]; and it can be justly claimed that this district may boast of being the place where the inclined plane was first used to introduce railways in aid of inland navigation and for the development of the wealth of the country[4]. The tramroads in the coalfields of Northumberland and

[1] Scott, in *Transactions of the Highland Society*, VI, pp. 11–15, gives the names of the many tramways or railways, in 1824, in the Severn valley, in Yorkshire, Derbyshire, Leicestershire, Staffordshire, Lancashire, and in the coal counties of Northumberland and Durham, as well as those in Scotland. Regarding the difficulties and dangers connected with carriages moving up and down these inclined planes, see Wood, *Practical Treatise on Railroads*, pp. 86–103.

[2] That is, from six to eighteen inches in the yard. The lengths of these inclined planes varied from 100 to 600 yards at one place. On inclined planes, see Scott, in *Transactions of the Highland Society*, VI, pp. 15–30, who goes into this subject very fully. This double railway was in use in Shropshire, for instance, in the railway connexions of the Ketley and Shropshire Canals (Plymley, *Agriculture of Shropshire* (1803), pp. 291 ff.). See also Scott, op. cit., VI, pp. 8–9.

[3] As late as 1790, there was hardly a single railway in all South Wales; while in 1824 the aggregate extent of rail and tramroads in the counties of Monmouth, Glamorgan and Carmarthen alone was thought to exceed 400 miles (Cumming, *Rail and Tram Roads*, p. 27).

[4] The history of the great Sirhowy tramway, in the county of Monmouth, may enable us to see more clearly the relation of these early roads to the development of the wealth of the country through which they passed; and we have chosen this one because, in point of magnitude, it was one of the greatest. It extended from Pilgwelly, near Newport, to the Sirhowy and Tredegar Iron Works (24 miles), whence it was continued five miles further to the Trevil Lime Works, in Brecknockshire, along with a branch westward to the Rumney and Union iron works. This railway was constructed at the suggestion of Mr Outram. On being consulted by the Monmouthshire Canal Company, as to the best means of supplying that canal with water, of which there was such a great scarcity that trade was suffering severely, Outram recommended a few reservoirs to be made, but more particularly a tramroad, to run parallel with the Crumlin line for eight or nine miles out from the town of Newport. In order to ease or take away part of the trade from the canal, this line was to pass through Tredegar park, the property of Sir Charles Morgan; and it was finally arranged between Sir Charles, the Monmouthshire Canal Co., and the

Durham were many and important, and were used not only in carrying the coal from the mouth of the mine to the river, but in bringing it from the interior of the mine to the entrance. It is in connexion with these colliery roads that we get some very important advances in the practical application of steam to locomotion on rails. Apart from these tramroads leading to coal and other mineral supplies, the only important tramroad made during these first two decades of the nineteenth century was the Surrey Iron Railway, from Croydon to Wandsworth[1], which was authorized in 1801. It was to be of advantage to a very populous agricultural country through which it was to be built, by opening up cheap and easy communication for carrying coal, corn, merchandise, and, in fact, commodities of all kinds; in other words, as we have already noted, this was doubtless the first attempt to construct a public railway for the carrying of miscellaneous products.

In all these cases the introduction of the tramway was for the purpose of facilitating the carriage of commodities, especially of heavy commodities like coal, and thereby reducing the cost of carrying these along the highways that were then and there available. Experiments

Tredegar Iron Works Co., that he should make one mile, which was in his park, the Monmouthshire Canal Company to make eight miles, and the Tredegar Iron Works Company to make the remaining fifteen miles, each to take tonnage on its respective part of the road. The road was completed about 1804, and also a turnpike by the side of it for about seventeen miles, at a total cost of about £74,000, or about £3000 per mile. About £40,000 of this sum was spent by the Canal Company in building a bridge and making some very deep and expensive cuttings; while the Tredegar Iron Co. completed nearly double the distance at a cost of £30,000. Sir Charles Morgan spent £4000 upon one mile, but he too had some deep cuttings and a double road to make. Notwithstanding the expense, this road, in 1824, paid the proprietors thirty per cent., by reason of having a considerable trade upon it in coal and iron, which paid the same tonnage as upon the canal. For the first nine miles out of Newport (the parts made by the Canal Company and Sir Charles Morgan) it was a double road: one for the loaded waggons to come down, and the other for the empty ones to return; and on the Tredegar Iron Company's part (fifteen miles) it was a single road, with frequent places for teams to turn out and pass. The whole length of the road for twenty-four miles was an inclined plane, averaging about one-eighth of an inch in the yard, or a little more; but the Tredegar Iron Company's part was of somewhat greater declivity than the rest. The coal and iron were conveyed on it in waggons, each carrying about forty-five to fifty hundredweight, exclusive of the waggon; and a team of four or five horses would draw about fifteen of these waggons down with ease. The waggons were variously constructed, according to the fancy of the parties, some of wood, some wholly of iron. The width of the road was four feet two inches, and it was laid down with cast-iron plates, three feet long, fastened to the sleepers by a pin passing through the rail, and into a hole bored in the stone block four to five inches deep, and there secured with lead (Cumming, *Rail and Tram Roads*, pp. 25, 26, 28–30).

[1] *J., H. of C.*, Feb. 27 and Mar. 5, 1801, LVI, pp. 112–13.

had been conducted to show how much more effective was the work of a horse when drawing upon a railway than upon the ordinary roads, and important results had been obtained. For example, in 1799, on a railway at Measham, the declivity of which was five-sixteenths of an inch in the yard, one horse drew nineteen waggons, which with their loading amounted to thirty tons, and was not subjected to extraordinary work in doing so. At a later time, on the same road, one horse drew down a load amounting in all to thirty-five tons; while up the grade or ascent he drew five tons with ease[1]. From the facts which were being demonstrated, it was becoming more evident that there were possibilities in this method of conveyance which were not fully realized; that, instead of being confined to the operations of mining, it was also fitted to take an important place in the conveyance of all kinds of products and merchandise, and to facilitate the interchange of traffic from one centre to another. But it was recognized that if tramways were to be used for general traffic, where there was carriage of goods each way, the more the line approximated to a perfect level the better it would serve the purposes for which it was intended[2]. While most people regarded the railways as useful in the limited sphere in which they had been employed, there was but an occasional individual, at the beginning of the nineteenth century, who contemplated a broader field of service for them. There were at that time at least two who foresaw the general extension of railways over England; and one of these proposed that all the railroads constructed should be owned by the state and free to all so that each could use his own waggons upon them[3].

But tramroads or railways for general purposes could have but partial success until some other than horse-power could be employed[4];

[1] *Communications to the Board of Agriculture*, II, pp. 475–6.

[2] Statement of Robert Stephenson, in *Transactions of the Highland Society*, VI, p. 186.

[3] Sir Richard Phillips, after witnessing the economy of horse-labour on the Surrey Iron Railway, thought that such lines should be extended from London to the principal places of the kingdom (Stretton, *A Few Notes on Early Railway History*, p. 5). Dr James Anderson, in his *Recreations in Agriculture*, IV, pp. 204 ff., 214, pointed out the advantages of carrying railroads from London to every other part of the country and recommended that they be owned by the public. In 1818, the scientific men of the country were offered a reward for the advancement of the railway system; and a piece of plate of fifty guineas value was to be given for the best essay on the construction of railroads for the conveyance of ordinary commodities. See this advertisement in *Transactions of the Highland Society*, VI, pp. 3–4; and the essays sent in are printed immediately following these pages.

[4] Cumming, *Rail and Tram Roads*, p. 33, in speaking of the Surrey Iron Railway, says: "But it must be observed, that rail-ways, as hitherto worked by horses, generally speaking possess very little, if any, advantage over canals." The fact is,

and the perfecting of the steam-engine by Watt turned the attention of many to the application of steam-power to locomotion on common roads, and of a few others to the possibility of its use on the tramways. We have already seen that the limited use of steam for navigation was a practical reality before the beginning of the nineteenth century; and we have also noted the introduction of the steam-carriage in the early years of that century, to take the place, to some extent, of the stage coach. While experiments were being conducted with the steam-engine, with a view to its use on the common roads, the possible application of steam for traction purposes on railway lines was also a subject of interest; and in 1804, for the first time, a steam-engine, constructed by Richard Trevithick, was employed on the railroad at the Merthyr Tydvil coal mines in South Wales. It was very imperfect but was used for a short time.

As early as 1800 the possibilities of the use of steam on railways were foreseen by some and were loudly proclaimed[1]; but the public mind failed to grasp the real importance of this new power in its wider applications. Many became engaged in its investigation whose names and results have not come much into public notice; but of these we do not propose to treat in detail here[2]. One of the most important

that railways were not constructed as a rival conveyance to the canals, but merely as supplementary to them. Yet railways certainly had advantages that were not possessed by canals, else there would not have been so many of them authorized during the first quarter of the nineteenth century. Note the perfect rage for railways, in 1825, when horse-power alone was in general use in connexion with them. For example, the prospectus of the Surrey, Sussex, Hants, Wilts and Somerset Rail-Road Company [Brit. Mus. 8223. e. 10 (148)], issued in 1825, says: "The necessity of using locomotive engines is not contemplated, every calculation being made on the use of horses only, although scientific improvement, when fully confirmed, will be availed of."

[1] Anderson, *Recreations in Agriculture*, iv, pp. 198–214.

[2] On the historical development of the steam-engine and the locomotive, see Gordon, *A Historical and Practical Treatise upon Elemental Locomotion*; Archer, *William Hedley, Inventor of Railway Locomotion*; Gordon, *Our Home Railways*, pp. 7–19. A few facts regarding one of the most ingenious inventors, William Murdock, whose name has been largely obscured by the glory attached to others, may help to place this man in his true light. As a Scotch boy, he came down to Boulton and Watt's works at Soho to secure employment and after some hesitation Watt engaged him. The boy soon showed his ability and began working during his spare time to produce an engine that could be used for locomotion. Watt discouraged this and the firm, in order to withdraw him from his purpose, sent him to Redruth, in Cornwall, about 1780, as engineer, to look after some of their engines that were in use in the mines there. Away from Watt, Murdock had a freer hand, and he again took up the problem of making a locomotive carriage, which he brought to a successful issue. In 1786 Murdock was on his way to London to take out a patent on his steam-carriage when he was met by Boulton who prevailed on him to

men to devote time and ability to the study of this new power was Richard Trevithick, and we are disposed to mention him in particular, not only because of the good results which he secured, but also because some have arrogated to themselves what was really accomplished by Trevithick. We have already learned of his success with the steam-carriage in the first three years of the century; but he was convinced that better results could be obtained on a smooth than on a rough road and he lost no time in applying his locomotive engine to tramways. In February, 1804, his locomotive was working on a tramroad at Penderyn, near Merthyr Tydvil in Wales, and running with facility up and down inclines of one in fifty[1]. The ten tons which the locomotive drew were soon increased to twenty-five tons, on this same road, with its unevenness and sharp curves[2], and this load was drawn at the rate of four miles per hour[3]. After Trevithick had made some further improvements in his engine and had constructed several of them for various purposes[4], he demonstrated in London for several months of the year 1808 that the locomotive with smooth wheels on smooth rails could draw heavy loads with no other assistance than the force of adhesion or gravity[5]. This is a fact which is supported by such apparently incontrovertible evidence that we wonder that any subsequent worker in this field should have attempted to take away the honour which belongs to Trevithick as the "father of the locomotive engine[6]." For some reason, which is not very clear, Trevithick's results were obscured by the partial success of Blenkinsop, who in 1811 patented his device of the rackrail and cog-wheel which was in use in his colliery. Under this arrangement the wheels were prevented from slipping on the rails by having the toothed wheels of the locomotive fit into the corresponding notches of the rails. But in 1813 and 1814 both William Hedley and George Stephenson again demonstrated the possibility of drawing loads by using locomotive engines with smooth wheels on smooth rails by the mere action of the friction of the wheels

come back, which he did. The Soho firm did not want to lose Murdock, and, loyal to them, he was deprived of the honour of introducing the locomotive. On Murdock's work, see Samuel Timmins's essay on him and his accomplishments, in the Birmingham Free Reference Library; also Gordon, *Our Home Railways*, pp. 7–9; Smiles, *Lives of the Engineers*; Wood, op. cit., pp. 123–57.

[1] Trevithick, *Life of Trevithick*, I, p. 160.

[2] Ibid., I, pp. 164, 167, 182. [3] Ibid., I, p. 182.

[4] Ibid., I, pp. 191–2. [5] Ibid., I, pp. 192, 201.

[6] Ibid., I, pp. 193–4; ibid., I, p. 206, testimony of Luke Hebert, in his *Railroads and Locomotion*, p. 30; ibid., I, pp. 201–3. To this we may add the confirmation given by Wood, *Practical Treatise on Railroads* (1825), p. 124; Stretton, *A Few Notes on Early Railway History*, p. 6; Gordon, *Our Home Railways*, pp. 11–16.

on the rails[1]. From that time on, there was a gradual increase in the employment of the locomotive, at first on colliery railroads, and, later, on the railways built for general purposes.

It is not our province to enter into details concerning the development of the locomotive, nor to trace the successive changes in the application of this power on the colliery roads in the north of England. Some good results had been secured by such men as Hedley, Stephenson and Wood, proving conclusively the great superiority of the locomotive engine over all other kinds of power. For example, on the Killingworth Colliery railroad, in 1814, an engine was tried on a line the steepest gradient of which was one in four hundred and fifty; and the locomotive ascended this with eight loaded waggons, weighing altogether about thirty tons, at the rate of four miles per hour[2]. In the years which followed these initial successes, improvements were made by Stephenson both in the locomotive itself and in the mode of constructing and laying down the rails; and these results were attracting attention all over the country.

[1] Both Hedley and Stephenson claimed the priority of this discovery; but, as we have seen above, Trevithick was some years ahead of either of them. In a letter written by William Hedley, he says: "I beg to say that I am the individual who established the principle of locomotion by the friction or adhesion of the wheels upon the rails." Trevithick, *Life of Trevithick*, I, p. 203; Archer, *William Hedley, Inventor of Railway Locomotion*, pp. 4–6. How false this statement is, we have already noted. But as between Hedley and Stephenson, a dispute has arisen as to their claim for precedence. Smiles, in his *Lives of the Engineers*, III, p. 142, clearly admits that Hedley discovered and demonstrated the sufficiency for traction of the smooth wheel and rail, but he fails to give him very ample credit; on the contrary, he reserves this for Stephenson. Archer, op. cit., pp. 4–6, in taking up the case for Hedley, gives a letter from the latter to Dr Lardner, to show that Hedley was really earlier than Stephenson in the application of this principle; and Archer says that this letter and the complete absence of denial from any source whatever is the clearest possible proof that Hedley's claim was considered incontrovertible. He says, moreover, that this fact has never been challenged nor answered by Stephenson or anyone else. This last statement of his does not seem to accord with what we find in the report of the *Proceedings of the Great Western Railway Company*, p. 27, for in this case when George Stephenson was asked: "You are the first person who suggested the using of locomotive engines, and applying them to the purposes to which they are now applied?" his answer was, "Yes." Evidently, then, Stephenson did claim priority in this matter. While there is a decided probability that Hedley's claim to priority is the stronger, we have not sufficient data to prove it conclusively. But the point to be emphasized is that Trevithick was ahead of either of them. The claim of the latter inventor is substantiated also by Sir John Rennie, in his *Autobiography*, pp. 230, 232. He says (ibid., p. 233) that Trevithick's principle had been forgotten by later experimenters. Stretton says that Trevithick's principle of the adhesion of the wheels to the rails was apparently not understood at that time (*A Few Notes on Early Railway History*, p. 6). See also Gordon, *Our Home Railways*, p. 18.   [2] Jeans, *Jubilee Memorial of the Railway System*, p. 10.

One of the most important of the coal-roads, which was constructed after the traction power of the locomotive had been demonstrated, was the Stockton and Darlington; and as the history of this enterprise is instructive from several standpoints it is desirable that we should consider it minutely.

A protracted controversy had been taking place as to the easiest and most advantageous way of improving the carrying facilities from the Durham coalfield. In those times, Stockton was the port of the river Tees, but the winding of the river from its mouth up to that town made the time required for sailing this distance sometimes as long as that occupied in the journey from London to the Tees[1]. In the year 1805 it was decided to shorten the channel of the river by a "cut" at Portrack, near Stockton; the Act for this purpose was passed in 1808, and by 1810 the work was completed by which that part of the river was shortened two miles[2]. In the same year began a movement for constructing a railway or canal from Stockton, by way of Darlington, to Winston, in order to provide a better outlet for the mineral and other traffic of South Durham and North Yorkshire. A committee in 1811 confirmed the great advantage of such a railway or canal, and reported this to a meeting of those desirous of promoting this undertaking, held at Darlington in the beginning of 1812. Those who were present at this meeting resolved to engage Rennie to make a survey before any further measures were taken[3].

For some years there was diversity of opinion, some favouring a railway and some a canal; and this is not to be wondered at, for up to that time no locomotive had been made that could attain a greater speed than four or five miles per hour, whereas steam navigation had many years before reached the rate of seven miles per hour[4]. Apparently, therefore, the railway offered no advantage over a canal in the matter of speed. Nor was there yet any widespread or generally accepted idea in favour of making railways take the place of the stage coach for passenger travel. In the public mind, railways seemed to be designed chiefly for the better and faster carriage of minerals and goods, and only a few saw the latent possibilities in the locomotive engine. Whatever the cause, this project was allowed to rest until in 1818 it was actively revived by the advocates of the canal. These

[1] Pease, *Diaries of Edward Pease*, p. 83.
[2] Ibid., p. 83; Jeans, op. cit., p. 14.
[3] Jeans, op. cit., pp. 14–15; Pease, *Diaries of Edward Pease*, pp. 83–84. From Jeans has been obtained much of the historical account of this railway that is here given.
[4] Jeans, op. cit., p. 15, quoting from John Willox, *The Steam Fleet of Liverpool* (1865).

recommended that the contemplated canal, according to the suggestion of Rennie in 1813 and Whitworth in 1768, should begin at Stockton and take its course by way of Darlington to Winston, where, perhaps by the aid of a railway, it could secure a rich harvest from the coal-field. Or, if that were not deemed advisable, the end sought might be accomplished by the construction of a railway at one-half the expense of a canal; and according to Rennie's opinion the railway would be satisfactory in cases like this where the principal carriage must proceed from one end only[1]. The only result of this meeting was the appointment of a committee to investigate the comparative merits of the two schemes; but later in the year another meeting was held to consider the whole subject and at that time most of those who were present advocated the adoption of a railway in preference to a canal. That meeting decided in favour of a railway, and drew up a prospectus showing the estimated cost and anticipated revenue from the railway, as well as its advantages to the country[2].

The road had been surveyed by Overton, but as there was much doubt as to the best route and the probable cost Robert Stephenson was called in to report on the proposed line. His survey was not very satisfactory to the committee that had the work in charge; and the latter, retaining Overton as engineer, prosecuted their work according to his directions. A Bill was brought into Parliament to secure the required authority for the construction of the road; but the opposition offered, especially by some of the landowners, was so formidable that it became necessary to employ every means to conciliate them, by the promoters using all authority and influence they could command in Parliament; and even then some of the most pertinacious opponents, like Lord Darlington, remained implacable and the Bill was lost[3].

But the committee that had the work in charge were undaunted by this defeat and it was determined to bring the measure before Parliament again in the next session. The former route lay through one of the Duke of Cleveland's fox-covers, which, to the nobility of those days, were of greater importance than public highways; so it was agreed that a new survey should be made to get another route and the committee wisely decided to lose no time in conciliating opposition. After making this survey, Overton made a report to the directors on September 29, 1820. This report gave little that was new[4]; and on the basis of that report the committee, in November,

---

[1] Jeans, op. cit., pp. 16–17.

[2] Pease, *Diaries of Edward Pease*, p. 84; Jeans, op. cit., pp. 23–24. Jeans gives the prospectus on pp. 24–26.

[3] Jeans, op. cit., pp. 28–31. [4] Jeans, op. cit., pp. 32–34.

1820, issued a manifesto showing the advantages of the railway in the conveyance of coal. They declared that everything had been done to conciliate those who hitherto opposed the railway and to avoid any injury to private property; that, because one horse on the railway could draw as much as ten on the common road, a vast reduction in the price of carriage would take place; that easier access to markets would be of great benefit to the farmers in enabling them to procure coal, as well as lime and manure for their land, while permitting them more conveniently to dispose of their surplus produce; that the commercial, mining and manufacturing interests would secure important benefits from the reduced rate of carriage for their respective products; and that the population at large would partake of beneficent results in the reduced price of fuel. In the matter of revenue it was shown that, from data already presented, there was reasonable expectation of the subscribers receiving fifteen per cent. a year, without anticipating any increased consumption, which was invariably the consequence of a reduced cost of conveyance. A very significant statement of the committee was to the effect that public opinion toward the railway had changed, as shown by the fact that there were very few who objected to the railway crossing their property[1]. Under these conditions application was again made to Parliament for a Bill in 1820; but on account of the circumstances due to the death of the King it was determined to defer proceedings until the session of 1820–21. For this second Bill, as for the first, they had to make a great fight, in which they were led by their Quaker promoter, Edward Pease, whose name is indissolubly associated with the Stockton and Darlington railway. "Every member of Parliament that could be influenced, directly or indirectly, was pressed into the service of the promoters. Every peer that was known to have any doubt or hesitation was seized upon and interviewed until he became a convert, while those who looked upon the measure with favour were confirmed in the faith. Nay, more, the promoters and their friends even carried their influence as far as the hustings, and spared neither trouble nor expense in endeavouring to secure—especially in the north of England—the return of candidates known to be partial to their cause[2]." This second Bill was passed in April, 1821[3].

After legislation had been secured, George Stephenson was appointed engineer of the Stockton and Darlington railway. This first Act comprises sixty-seven closely printed pages, embodying the whole of the law relating to railways; it was the earliest and probably the longest

---

[1] Jeans, op. cit., pp. 34–35.   [2] Ibid., pp. 35–36.
[3] Act 1 & 2 Geo. IV, c. 44.

railway Act that received the sanction of Parliament. No mention was made of the employment of engines, for it was intended to work the line entirely by horse-power; although a general provision was made that the company should "appoint their roads and ways convenient for the hauling or drawing of waggons and other carriages passing upon the said railways or tramroads, with men or horses, or otherwise." The adoption of steam-power was, apparently, not seriously considered until the construction of the roadway was far advanced. Then Edward Pease went to Killingworth Colliery to see Stephenson's engine working, and from that time he had implicit faith in the locomotive engine. Through his influence the amended Stockton and Darlington Railway Act of 1823 gave the company authority to erect one stationary steam-engine in a suitable position and to make and use locomotives or movable engines for the conveyance of goods and passengers along the line[1]. In this there was a wide departure from the first Act, which said nothing about passenger traffic and made no mention of locomotives. According to the statute, the road was to be free to all persons who chose to place their waggons and horses upon it for the hauling of coal and merchandise, provided they paid the tolls fixed by the Act; and the gauge of the railway, four feet eight and one-half inches, was taken from the width of the road waggons.

On the success or failure of Stephenson's locomotive engines on this "Quaker line" very much depended; if failure, a check would be given to railway enterprise; if success, a new era would dawn which would show a complete revolution in the means of communication. The first engine used on this railway was built by Stephenson; and in comparison with later results its performance was very modest. The best it could do was to travel at the rate of four to six miles per hour; and an engine and tender of fifteen tons could draw on a level nearly forty-eight tons gross load at the rate of five miles per hour[2]. Stationary engines were used for drawing the waggons up the incline. But even this result was enough to cause many a speculative mind to become enthusiastic over the prospects and to predict the time when high rates of speed would be attained. To them it seemed as if the vision were already within their grasp as a reality and they lost no occasion to communicate to the public, in glowing terms, the picture of the

---

[1] Jeans, op. cit., p. 43; Pease, *Diaries of Edward Pease*, pp. 85–87. This second Act was 4 Geo. IV, c. 33.

[2] Jeans, op. cit., pp. 53–54. On the early life and training of Stephenson, see *Autobiography of Sir John Rennie*, p. 235 et seq.; also the life of Stephenson in Smiles' *Lives of the Engineers*, which gives full details.

near future. Concerning railroads and other speculative schemes of
that day, Lord Eldon said that Englishmen, who were wont to be
sober, had grown mad; and to aid in forming a more reasonable view,
Nicholas Wood, who was recognized as an expert in railway affairs,
declared: "It is far from my wish to promulgate to the world that
the ridiculous expectations, or rather professions, of the enthusiastic
speculist will be realized, and that we shall see them travelling at the
rate of twelve, sixteen, eighteen, or twenty miles an hour. Nothing
could do more harm towards their adoption or general improvement
than the promulgation of such nonsense[1]."

But people did not have to wait long before they found that some
of the dreams of the enthusiasts were already accomplished facts. On
Sept. 27, 1825, when the railway was opened amid great demonstration
of splendour[2], it was shown that, on an incline, one engine could draw
a whole train, with a weight of at least eighty tons, at the rate of ten
to fifteen miles per hour[3]. The success of the railway was immediate
but not startling[4]; and soon the line was extended back to Witton
Park Colliery, about 125 miles from Stockton, so that Darlington was
just about half-way along the line. In 1827, the first year in which the
coal and merchandise traffic was fully worked, the revenue from coal
was £14,455, while the receipts from lime, merchandise and sundries
was only £3285. The chief source of revenue was the coal, the tolls on
which in 1830 were six or seven times the amount of revenue derived
from all other sources combined[5]. Both in the amount of revenue that
accrued to the company and the speed at which the traffic was carried,
it was evident that this line of road was a paying investment.

The Stockton and Darlington promoters did not at first count upon
any revenue worth speaking of from passengers. Between these two
places there was only one coach, which went three or four times a week,

---

[1] Jeans, op. cit., p. 66.

[2] Concerning the opening, see Pease, *Diaries of Edward Pease*, p. 88; *Newcastle
Courant*, Oct. 1, 1825, which gives an account of that great occasion. Smiles, *Lives
of the Engineers*, III, pt. 2, ch. viii, gives an extremely interesting account of
the arrangements for this railway. Tweddell's *History of the Stockton and Darlington
Railway* was well intended, but it does not get far enough to even touch the subject
of which it proposed to treat.

[3] Jeans, op. cit., p. 70.

[4] The success of the railway is shown in a statement signed by S. P. (probably
Samuel Pease, one of the directors of the railway), showing the facts for the railway
as on Mar. 23, 1829 (*Collection of Prospectuses*, etc., pp. 121–4). Note also *Remarks
upon Pamphlet by Investigator on the Proposed Birmingham and London Railway*,
p. 4, showing that by 1831 the shares of the company had risen in value from £100
to £200 each.

[5] Booth, *History of the Liverpool and Manchester Railway*, p. 2.

on the turnpike road; and the amount of passenger travel scarcely paid a reasonable profit to the coach proprietor. Nor was there much likelihood that there would be any increase of passenger traffic on the rail unless greater speed could be developed in order to encourage the desire to travel. The railway made no special provision for this aspect of the business. It was originally intended to allow proprietors of stage coaches or other vehicles to use the line under certain specified conditions for the conveyance of passengers, and on similar terms to allow carriers to make use of the line for the carriage of goods, so that both these phases of enterprise might be carried on independently of the railway company. After the railway had been opened two weeks, the company put on a coach of their own for the conveyance of passengers[1]; but shortly after, a contract was made with Pickersgill, who leased the railway company's coach and operated it on the railway. Up to 1830 the two or three coach proprietors on the line carried on the passenger and merchandise business; they used horses to draw the coaches along the line, paying the tolls for the use of the line and receiving the amounts paid for these services[2]. They seem to have had their own way, in large measure, as to regulating hours and traffic; and it appears certain that their arrangements must have clashed, for on Jan. 22, 1830, the company began to regulate the times of arrival and departure at each end so as to give them equality of advantages[3]. The early organization of the passenger and goods traffic on the line shows us, then, several coach proprietors each of whom took out a licence for himself and paid his tax to the state, but gave no account to the railway company except the total number of journeys each coach had made per month, on the basis of which they paid the company for the use of the line. Anyone was at liberty to put his horse and carriage on the railway and draw for himself or others, provided he complied with the company's by-laws. The growth of the passenger traffic was slow, for before 1832 the number of passengers travelling between Stockton and Darlington did not average more than 520 per week[4], although the number of coaches had increased from two or three in 1830 to seven in 1832[5]. About 1833, the company found that, instead of having so many different interests

---

[1] Jeans, op. cit., p. 81, gives in full their hand-bill concerning the passenger service. This is also given in Layson, *Life of George Stephenson*, p. 93, advertising the passenger coach between Stockton and Darlington. It gives the times of departure and arrival at each place along the line for each day of the week. It is interesting as the first railway passenger time-table.

[2] Jeans, op. cit., pp. 81–82; Booth, *History of the Liverpool and Manchester Railway*, p. 2.

[3] Jeans, op. cit., p. 84.

[4] Details are given in Jeans, op. cit., pp. 85–86.   [5] Ibid., pp. 84, 86.

represented on their line, it would be more convenient and advantageous if they should take the whole carrying trade into their own hands and displace horses by steam-power[1]. The respective interests of the different proprietors were acquired by purchase and arrangements were made for more comfortable and speedy carriage of passengers; and on April 7, 1834, the company announced that they had commenced to run coaches and carriages by locomotives for the conveyance of passengers and goods between Stockton and Middlesborough "six times per day at present fares, thus forming a regular line of communication via Stockton and Darlington with Shildon, Auckland," etc.[2]

We have given somewhat fully the conditions regarding the operation of this railway, to show the way in which the carrying trade was organized on it, for, since this line was an intermediate between the colliery roads and the modern railway designed for both passenger and freight carriage, it is instructive to see the steps through which the orderly process of development has taken place. It will help us to appreciate the circumstances under which the enterprise was carried on if we picture to ourselves what two noted railway engineers observed on this road in 1829; between Stockton and Darlington there were several locomotive engines of different forms and power and horses also were employed upon the same part of the line; while, toward the upper end of it, there were two inclined planes with stationary engines[3]. When the declivity was such that the waggons would run down without the horse drawing, the animal was detached and took his place in his own carriage behind the train of waggons until his services were again required[4]. With this aggregation of the different kinds of power in

---

[1] Competition among the rival coach proprietors using the single line of roadway led to confusion and collisions among them; their merchandise trains sometimes got so heavily loaded that they had to be helped by the locomotive engine in order that other traffic might not be held up or delayed. This method, of course, was simply carrying out the same conditions that existed in the carrying trade on the canals. Even before the railway was opened, the committee in charge of the work, after careful investigation, had declared that it would "greatly conduce to the interest of the company that they should become the principal carriers on the line." They had been asked by a certain individual for permission to use his locomotive engine on the railway, but the committee thought that it would be improper to grant this application (Jeans, op. cit., p. 63).

[2] Jeans, op. cit., pp. 87–89.

[3] Walker and Rastrick, *Report to the Directors of the Liverpool and Manchester Railway, on the Comparative Merits of Loco-motive and Fixed Engines, as a Moving Power*, p. 3. On the application of stationary engines on some railroads, see Wood, *Practical Treatise on Railroads*, pp. 110–23.

[4] Macturk, *History of Railways into Hull*, p. 29, quoting from Walker's observations concerning the operation of the Stockton and Darlington Railway.

use upon the line, together with the facts already noted in regard to the diversity of interests in the passenger and merchandise traffic, we can see how difficult it would be to maintain harmony among the different carriers and to fix the responsibility for any breach of the company's regulations or any misuse of its property.

From our present-day standpoint, we would imagine that the question as to whether locomotive engines should be employed, or whether horse power should be used for traction purposes, could be easily settled; but it does not seem that the former was immediately accepted as the motive power that was soon to displace all other. Even after the Liverpool and Manchester Railway had been in full operation for some years and the utility of the locomotive engine had been completely demonstrated, there were still those, and some of them engineers, who clung tenaciously to the idea that, under certain conditions, horses or stationary engines might be profitably employed. In 1833 when the agitation was going on for a railway to connect London with the west, one writer urged the employment of horses because they would be more economical than steam power[1]. In 1825 when George Stephenson had surveyed the line of the proposed Leeds and Selby Railway, he recommended three inclined planes for part of the line and locomotive engines for the rest[2]; but as the committee in charge of the work did not agree with him they asked James Walker, another famous railway engineer, to make another survey. He opposed the use of stationary engines recommended by Stephenson; but said that, if the road as then designed were to be constructed, he would favour the employment of horses, as on the Stockton and Darlington, instead of stationary engines. If they were used, the inclination from Leeds toward Selby was such that the horse might ride six to seven miles, in the proposed distance of about thirty miles, and in the opposite direction it might ride about three miles. Walker's report seems to imply that the committee had decided to use horses on the railway[3]; to this he agreed if speed were not desired; but, taking everything into consideration, he strongly favoured the uniform level, without inclined planes, and the employment of locomotive engines upon it[4]. His

---

[1] *Bristol Mercury*, Oct. 5, 1833, p. 4, letter of "A Well Wisher."

[2] Macturk, *History of the Hull Railways*, pp. 18–32.

[3] Ibid., pp. 18–32, gives in full Walker's report to the committee of the proposed Leeds and Selby Railway Co.

[4] Walker said that on the Darlington line the horse-power amounted to about one halfpenny per ton per mile on the coal conveyed down to Stockton, and, all things considered, the cost of hauling by locomotives was not less; but at the rate of eight or ten miles per hour the engine-power would be very much cheaper if there were enough traffic to form full loads for the engines. Macturk, op. cit., p. 32.

calculations, however, were made with a view to the use of rails that would be strong enough to support locomotive engines, although he expected that at first horse-power chiefly would be used. It is evident, therefore, that public attention did not turn away immediately and entirely from the time-honoured motive power as soon as the locomotive engine had demonstrated its possibilities.

From the results that had been accomplished on the coal-roads, it was apparent that the ultimate triumph of steam locomotion on rails was certain; but the efforts toward its actual realization did not wait until the success of the Stockton and Darlington had been shown. In the meantime, other roads were in process of construction, such as the Moreton and the Liverpool and Manchester; and here, too, the decision had been made in favour of the employment of locomotive engines[1]. When, therefore, it was obvious, beyond all doubt, that it was practicable to use the locomotive engine for hauling heavy loads on rails, the canal proprietors found that their waterways had a powerful rival in bidding for traffic; and in the decade beginning with 1820, when the railway propaganda was being vigorously pushed, a very active discussion was going on as to the relative merits of steam railways, canals and turnpike roads. Such periods of change, when the social and industrial world must be adapted to some new development in commercial life, necessarily elicit much controversy and bring to light the underlying current of public thought in regard to existing conditions and proposed improvements. Under the circumstances, during the above decade, one of the great questions was as to the relative importance of the railways and the canals, since it was the competition between these two agencies that was likely to produce the most pronounced effects. It may help us to understand the situation more perfectly, therefore, if we can see the way in which the friends of each regarded them just at the time before the railway came to assume such great importance.

One of the great reasons put forward for the construction of railways was the reduced cost of carriage that would thereby ensue. In many cases a considerable part of the prices of articles of necessity consisted of the cost of transporting them from producer to consumer, and therefore it was recognized that every saving in this cost would produce a corresponding reduction in the prices of the articles. The decreased cost of commodities would redound to the benefit of the consumer, by giving him command over a larger supply of the necessaries of life, and this larger supply at lower prices would stimulate consumption, both at home and abroad. The increased consumption would, in turn, call for

[1] Cumming, *Rail and Tram Roads*, p. 33.

increased production of both manufactured and agricultural products; and so the whole fabric of rural and urban industry would be strengthened and developed[1]. In addition to securing their food supplies cheaper, they would also be able to obtain a cheaper and more regular supply of coal. During severe frosts, when the canals were frozen for some weeks, the price of coal sometimes went up to exorbitant figures and even the supply was inadequate to the need; but during the most inclement weather the railroad would be able to continue the bringing of the usual amount of this much needed article the same as at other times, so that the possibility of scarcity or high prices would not cause any alarm[2]. The same thing applies with regard to the provision of abundant supplies of coal and other raw materials for manufacturing; and the cities that could draw upon wider and wider areas for these necessaries of manufacture would flourish all the more abundantly[3]. What was true of the necessity of regularity and certainty, as well as cheapness, in supplying consumption goods was equally true in regard to goods intended for export; if the goods did not reach port from the interior in time for the sailing dates of the vessels the shipper lost the orders and the shipowner the amount of the freight. This was no infrequent occurrence[4]. But cheapness of carriage, in addition to benefiting consumers, would be equally profitable to producers, both in industry and agriculture. By reducing the cost of transport there would remain to the producer a greater surplus to reward his labour; lands more distant from markets could be cultivated because of being more nearly on a parity with those nearer the markets, and in this way also the margin of cultivation could be extended; land that had hitherto lain waste could now become productive, and, therefore, while there would be much increase in the food supplies of the country, there would also be a larger return to the landlords as well as to the farmers. By making possible the application of more capital to tracts already under cultivation and increasing the

---

[1] Cundy, *Inland Transit*, 2nd ed. (1834), pp. 19–21; Godwin, *An Appeal to the Public on the subject of Railways*, pp. 8–18; *The Times*, Mar. 16, 1836, p. 7.

[2] *Manchester Guardian*, Jan. 1, 1831, p. 4, letter from " W. N. R." on the "Liverpool and Leeds Railway;" ibid., Jan. 29, 1831, p. 1, prospectus or "report" on the Manchester and Leeds Railway; Mudge, *Observations on Railways*, p. 2.

[3] *Manchester Guardian*, Jan. 29, 1831, p. 1; *The Times*, Oct. 28, 1837, p. 3, on the first general meeting of the Sheffield and Manchester Railway; *Sheffield Iris*, July 29, 1834, p. 2, on the "New Railroad;" ibid., Oct. 7, 1834, p. 1, prospectus of the proposed railway from Sheffield to Rotherham.

[4] *London and Birmingham Railway Bill. Extracts from Minutes of Evidence given before the Committee of the Lords on this Bill*, pp. 1–12, evidence of Messrs Barry, Hemsley, Barnes, Dillon, Mason, Moore; Mudge, *Observations on Railways*, p. 3.

extent of tillable land, the population of the kingdom would be provided with a more ample and less expensive food supply and the amount of labour would be increased, thus reducing the poor rates[1]. The extension of the markets for the product of any section would tend to maintain uniformity of price, so that the farmer would not be subjected to the occasional alternations of over-abundance and scarcity and the price fluctuations which accompanied such changes. The more equal distribution of goods throughout the country would result in benefit to the consumer also, by making his food cheaper and less precarious. What we have said as to the prospective advantages to the consumers and producers of agricultural produce applied with equal force in the case of manufactured products; to maintain and enlarge both the home and the foreign market, the articles supplied must be cheaper and better than could be produced elsewhere, and that necessitated cheaper communication and facility in executing orders. The opening up of new and larger markets would infuse a new spirit into industry as well as agriculture, and the material resources of the realm would no longer lie waste[2].

Increased speed in the conveyance of passengers and goods was another great desideratum which was anticipated as the outcome of the introduction of railways. As a consequence of this, both producers and consumers expected that new and more distant markets would be opened for commodities of a perishable nature, such as vegetables, dairy produce and meat. In order that these should be most successfully marketed they would have to be sent as quickly as possible from the producer to the consumer; and as the railway speed would be six or seven times as great as that of the carts or waggons the railway would cause the area of production of these things to be thirty-six to forty-nine times greater than its present extent[3]. Butter, cream, vegetables and similar commodities would not stand transportation by the slow-going canal or road-waggon, and, therefore, were confined to the markets at a very limited distance from the grower or feeder;

---

[1] Cundy, *Inland Transit*, 2nd ed. (1834), pp. 19–20; Cundy, *Observations on Railways*, 2nd ed. (1835), pp. 23–24; *Manchester Guardian*, Jan. 1, 1831, p. 4, letter from " W. N. R.;" Mudge, *Observations on Railways*, p. 3.

[2] *Herepath's Railway Magazine*, N.S., I, pp. 96–100, "A Few Words on Railways," by " Delta." He said that the cost of conveyance in the case of coal was a large element of the price; that coal was sold at the pit's mouth for five to ten shillings per ton, and at the distance of fourteen miles it cost double that amount, so that for lack of cheap transportation facilities the natural resources of the country were lying waste. He regarded this as an unanswerable argument for railroads.

[3] Cundy, *Observations on Railways*, 2nd ed. (1835), pp. 21–23; Cundy, *Inland Transit*, 2nd ed. (1834), pp. 22–23.

but with the faster conveyance by rail they would secure an enlarged market that would make them profitable for production. Similar conditions would enlarge the area for the remunerative rearing of animals to provide the meat supply of the large centres. Before the railway, if animals were transported alive from the growers to the consumers' market, the market was limited by the power of the animals to travel and the cost of their support on the road; or if they had to be carried by waggons the cost was still greater by reason of the added expense of horses and waggons. But, by the railway, lambs, calves and other animals could be sent easily and cheaply to the metropolitan markets to meet the requirements of these large centres. In this way the urban population would be able to draw upon wider and wider sources of supply and thus eliminate any peculiarities of local conditions; while the rural producers would find a more extended market for their surplus and a more stable equilibrium of prices and of demand[1]. The improvement of the marketing would enhance the value of the land which produced these things, and so both the tenant and the owner would derive advantage from being made independent of merely local circumstances. Along with accelerated speed in the conveyance of agricultural and industrial products, there would be a similar advance in the rate of passenger travel. Towns under existing conditions some stages distant from London or other large city would become its suburbs; men doing business in the greater centres would be able to reside at considerable distances from the places of their employment, and thus not only enjoy a more healthful environment for themselves but also help to prevent the overcrowding of population within confined areas[2]; and time that had been spent on the slow journeys of the coaches could now be saved, in great measure, and devoted to remunerative employment. This saving of time that would accompany the frequency of communication between places of great commercial intercourse would be a considerable economy in enabling them to expedite the transaction

---

[1] Cundy, *Inland Transit*, p. 23; *London and Birmingham Railway Bill. Extracts from the Minutes of Evidence given before the Lords Committee*, pp. 18–21, evidence of Messrs Warner, Whitworth, Sharp, Attenborough, and Kay; *The Times*, Mar. 16, 1836, p. 7, concerning the South Eastern Railway; Macturk, *History of Railways into Hull*, p. 44, original prospectus of the Hull and Selby Railway; Boyle, *Hope for the Canals*, p. 19; *Hampshire Advertiser and Salisbury Guardian*, Mar. 29, 1834, p. 2, on the "Southampton Railway;" Parkes, *Claim of the Subscribers*, pp. 4–17.

[2] Cundy, *Inland Transit*, p. 24. A notable instance of this is observed to-day when we see business men, who carry on their occupation in smoke-begrimed Glasgow, going to and from their residences in the royal city of Edinburgh. On the saving of time and expense in travelling, see *London and Birmingham Railway Bill. Extracts from the Minutes of Evidence given before the Lords Committee*, pp. 22–24, evidence of Messrs Mason and Cheetham.

of business both by personal visits and through the medium of corre-spondence[1]. Then, too, in the transportation of troops and of military and naval supplies the railway would be of vast importance over the former slow and expensive means of conveyance[2].

Railways were desired also because they would bring increased facilities and introduce certainty and regularity of conveyance. The lack of accommodation and equipment on the part of the canal companies was, in some instances, notorious, especially on the routes connecting the great industrial and distributing markets, like Manchester, Liverpool, Birmingham and London[3]. The inadequacy of the canals between Manchester and Leeds for the conveyance of general merchandise was so strongly felt that even with the carriage of a large part of the traffic by waggons the need for a railway was keenly appreciated[4]. But, supposing the carrying facilities of a canal chain to be sufficient for all ordinary purposes, there were other elements which affected the desirable uniformity of the flow of traffic. The droughts of summer or the frosts of winter frequently caused delays of several weeks; and these were attended with serious results to those who were affected to the greatest extent by them[5]. The cessation of trade on a canal that served a particular town would, at times, cause the price of coal to increase as much as 100 per cent., on account of the scarcity of that commodity at that special time[6]. Exporters who were depending upon goods from the interior reaching the port by a certain sailing day were sometimes disappointed, and the goods, when delivered, were rejected because out of time. Orders were frequently lost because the goods could not be

[1] Cundy, *Observations on Railways*, 2nd ed. (1835), pp. 29–32.

[2] This was especially emphasized in the case of the London and Southampton Railway. See summary of evidence on this railway Bill in *Hampshire Advertiser and Salisbury Guardian*, Mar. 29, 1834, p. 2.

[3] To each of these we shall refer in more detail when we come to consider the railways connecting these places.

[4] *Manchester Guardian*, Jan. 29, 1831, p. 1, prospectus of the Manchester and Leeds Railway.

[5] *Manchester Guardian*, Jan. 29, 1831, p. 1, prospectus of the Manchester and Leeds Railway; Sandars' pamphlet on the *Liverpool and Manchester Railway*; Parkes, *Claim of the Subscribers to the Birmingham and Liverpool Railroad*, pp. 46–51; Brit. Mus. 8223. e. 10 (70), 'Prospectus of Kentish Railway Company,' 'Prospectus of the Birmingham and Liverpool Rail Road Company;' Brit. Mus. 8223. e. 10 (149), 'Prospectus of the Taunton Great Western Railroad;' Vallance, *Sinking Capital in Railways*, p. 9; Macturk, *History of Railways into Hull*, pp. 43–44.

[6] *Manchester Gazette*, Jan. 1, 1831, p. 4, letter from "W. N. R." refers to the fact that in January and February 1830, the canals were frozen for some weeks, and during that time "the price of coals in Liverpool rose, in many instances, upwards of one hundred per cent.," and the daily demand for the town was more than the supply.

got to the seaport in time for shipment by certain vessels[1]. All such vexations would be avoided by having railways upon which to carry the goods and the public thought turned to the desirability of this new accession to the agencies of conveyance. Furthermore, many of the canals took a circuitous route, which added greatly to the length of the journey; and while this enabled them to get access to all those places that would be likely to have most traffic to offer, it was very inconvenient for those who wished the transport of their goods with the least possible delay.

Another prospective advantage of railways was that there would be a saving in the amount of capital required to be invested in business. It was expected that capital would not need to be locked up in warehouses where individual merchants had to keep on hand large supplies of stock because of the uncertainty and difficulty of renewing their supply. Under the existing circumstances, for example, coal was unable to stand the expense of land carriage, and so every dealer had to lay in an immense stock before winter, lest the canals should freeze, and before summer, lest they should be deficient in water supply. To remunerate the extra capital that was thus unproductively tied up something had to be added to the price of the coal. But it was a foregone conclusion that the railway would be able to operate without reference to these accidents of time or season, so as to bring a uniform supply throughout the year; and, therefore, dealers would not need to have a large capital lying barren for months at a time. The retail merchants of the country could go or send to London in the morning and have their purchases in the evening; this would obviate the necessity of their keeping an expensive and redundant stock, and in their country establishments, which would cost less than in the town, they would be enabled to enter advantageously into competition with the London dealers[2].

The coming of the railway was eagerly awaited in other cases because it was thought that this would emancipate the people from the thraldom of a canal monopoly that had become oppressive, sometimes almost beyond endurance. The canals between London and the Midland metropolis long enjoyed a monopoly and reimbursed their capital with

---

[1] *London and Birmingham Railway Bill. Extracts from the Minutes of Evidence given before the Lords Committee*, pp. 6–11, evidence of Messrs Barnes, Dillon, Mason, Moore and Westall. Sometimes when there was an insistent demand for the goods by a certain time the dealer paid the heavy cost of land carriage, rather than depend upon the slow movement of freight by water.

[2] Cundy, *Observations on Railways*, 2nd ed. (1835), p. 50; Mudge, *Observations on Railways*, p. 2; Shaen, *Review of Railways and Railway Legislation*, p. 33; *The Times*, Mar. 16, 1836, p. 7, on the "South Eastern Railway."

great profit; and one of the objects of projecting a railway to connect these two places was to get rid of the high charges which the canals continued to impose[1]. Among the chief reasons for the construction of the proposed railway between Sheffield and Rotherham, in 1834, was the desire to break down the monopoly in coal that then existed, and to supply these cities, especially the former, with coal brought from greater distance. The unfortunate condition of the navigation along this six and one-half miles, particularly of the Sheffield and Tinsley Canal, precluded all hope of reduction of charges on this waterway; and the people turned to the proposed railway as the only means of affording relief[2]. The conditions in the county of Monmouth at the time the plan was formed for a railway from Newport, through Pontypool, to Blaen Avon and Nant-y-glo, exemplify a situation which called loudly for and warranted the construction of the railway along this valley. The carrying traffic of the extensive mineral country there was almost monopolized by "The Company of Proprietors of the Monmouthshire Canal Navigation," which had been incorporated in 1792[3] and had been given most arbitrary powers for making canals and railways. Its Act exempted the tolls receivable by the company from the payment of any rates, and the lands purchased by them were to be rated at their original, not their improved value. Within a few years the company had completed two lines of canal, one from Newport to a little above Pontypool, and the other from Newport through another valley to Crumlin. From these canals a variety of ill-constructed railways and tramways had been made in order to open communication with new works. For lack of water the Crumlin branch could not be operated, and necessity compelled the construction of a tramroad which almost superseded the canal. The other branch of the canal, because of the numerous locks and continual impediments and cessation from one cause and another, was almost useless[4]. Yet notwithstanding the unsatisfactory state of their works, the company charged such exorbitant rates and provided such poor facilities that many of the iron-masters of that section preferred to send their products

[1] *Birmingham Journal*, Feb. 5, 1831, p. 3, letter from "A Subscriber to the London and Birmingham Railway."

[2] *Sheffield Iris*, Oct. 7, 1834, p. 1, 'Prospectus of the Proposed Railway from Sheffield to Rotherham.'

[3] Act 32 Geo. III, c. 102.

[4] Blewitt, *New Monmouthshire Railway*, p. 6, informs us that, in 1844, a boat from the Pontypool works could make only five journeys in a fortnight between that town and Newport, a distance of not over ten miles; and coal taken down the canal about five and one-half miles did not, on the average, reach Newport in less than eight hours.

around by a more circuitous route, sometimes at great expense, than to utilize this shorter waterway[1]. Then, too, the iron-masters of Merthyr, by means of the Taff Vale Railway, were able to send down and ship their iron from Cardiff, in first-class condition, without rust, just as it came from the rolling mills. This led to the desire for a better quality of material among the iron merchants of Liverpool and elsewhere; and they demurred to receive iron from the Monmouthshire district which, on account of having been so long on the canal and waiting so long at the docks at Newport, exposed to air, had become badly rusted[2]. Cardiff, the adjoining port and great rival of Newport in the iron trade, had always had the advantage of lower rates by canal, and, after the construction of the Taff Vale Railway, its trade had markedly increased[3]. From the foregoing facts we can judge how oppressive was the Monmouthshire Canal Company's monopoly upon the iron-masters of that portion of the country, and with what eagerness the industrial community contemplated, and actively set to work to secure, the advantages of a railway[4]. But the greatest canal monopolies, from which release was earnestly sought by the projecting of railways, were probably those between Liverpool and Manchester and between Liverpool and Birmingham; these we shall consider in detail when we come to discuss the formation of railway lines along these routes[5].

---

[1] Blewitt, *New Monmouthshire Railway*, pp. 7–9. The Ebbw Vale Iron Works had made, at their own expense, a tunnel a mile long, to communicate with the Sirhowy Tramroad, by which their iron reached Newport much cheaper, although the route was four miles longer. The Bailey Iron Works brought their iron from Beaufort to Nant-y-glo by a tunnel about a mile long, at the end of which it was raised by a water balance, and then sent via Abergavenny to Newport, 31 miles, though the canal company's road was convenient to their works and the distance by that route to Newport was only 22 miles. The greater part of the Blaen Avon iron was sent at heavy cost, via Abergavenny, to Newport, about 28 miles, when by the canal route it was only about 16 miles.

[2] Ibid., pp. 7–8.

[3] Ibid., p. 10. In 1829 the amount of coal sent to Cardiff was only about one-sixth of that sent to Newport; while in 1840 the amount sent to Cardiff was more than one-half of that sent to Newport.

[4] The canal company had turned a deaf ear to all complaints of the traders, as to the excessive charges, bad construction and indifferent repair of their canals and tramroads, until the wholesome fear of threatened competition compelled them to take the first step toward amelioration by reducing their tonnage rates on iron and coal (ibid., pp. 12–14).

[5] On the canal monopoly between Liverpool and Manchester, see Sandars' pamphlet on the Liverpool and Manchester Railway, and *The Times*, April 7, 1826, p. 2, on the Liverpool and Manchester Railway Bill in the House of Commons. On the conditions that prevailed between Liverpool and Birmingham, see Parkes, *Claim of the Subscribers to the Birmingham and Liverpool Railroad*.

Of the other reasons which were given as incentives to railway construction we shall give but brief mention, because they did not assume such importance in the estimation of the public. Railways were urged by some because it was thought that they would reduce the number of horses required for the transportation service, and that land which had been devoted to the keeping of horses for the carriage of freight and passengers along the highways could now be used for growing food supplies for the families of the kingdom. Adam Smith had said that to support each horse required, on the average, as much land as would support eight men; and if there were, as was estimated, over 1,000,000 horses engaged on the roads, the land required to provide for them would be able to support an additional 8,000,000 people, or else it would largely increase the means of subsistence of the existing population[1]. This same argument was used, as we have seen, in favour of the canals when their introduction was the subject of public interest; but in neither case did the contemplated decrease of horses employed on the highways materialize, for increasing facilities of conveyance brought an increasing demand for horses in collecting and distributing traffic[2]. Another circumstance favourable to the new means of conveyance was that the introduction of the railway would furnish a more efficient method of handling large quantities of freight than was possible on the canals. In the ports of the north of England from which coal was shipped machinery was used for lifting a loaded car and suspending it over the hold of the vessel, after which the bottom of the car was displaced and the coal was allowed to fall easily into the vessel. But no such service was possible with canal barges and so the cost of unloading them was much greater. Of course, this system was not known until after the tramways were found in effective operation[3]. Of other inducements for the formation of railways, there was the expectation that thereby the pilfering from canal barges, which in some cases seems to have been an organized business systematically pursued, would be for ever abolished[4]; the ordinary roads would be greatly relieved of the transport of coal, lime and other heavy articles, so that the expense of their maintenance would be much less than under

---

[1] Cundy, *Inland Transit*, 2nd ed. (1834), pp. 20–21; *Bristol Mercury*, Sept. 1, 1832, p. 2, letter from "J. O." Suppose the coaches on the road from London to Edinburgh changed horses twenty-five times; that would require one hundred horses for one journey of each coach, besides the supernumerary ones kept in case of accident. But the work of a great many coaches might be performed by the expense of one steam-engine and this would result in great economy.

[2] Cundy, *Inland Transit*, 2nd ed. (1834), p. 21; Cundy, *Observations on Railways*, 2nd ed. (1835), p. 47.

[3] Blewitt, *New Monmouthshire Railway*, p. 11.   [4] Ibid., pp. 9–10.

existing conditions; and, in fact, no limit could be assigned to the wealth that would be saved and the increase of wealth that would be produced by this change in inland conveyance.

In contrast to the claims made for the railways, those made for the canals seem decidedly lacking in many features of economic significance. The two enterprises were entirely different in character. The railway had an air of parade and display that dazzled and tended to deceive the superficial observer. Its general aspect was that of vitality, energy and efficiency: the large trains, their promptitude of arrival and departure and the speed of the engines were all subjects of admiration, and stood out in great relief when viewed alongside the quiet, unseen canal and its slowly plodding barges. In consequence of this there were few who ventured to lift up their voices in favour of the canals as an effective competitor of the railways. There were, indeed, some who, despite the unfortunate system of construction and maintenance of the canals, argued in favour of them and urged their claim from the stand-point of cheapness and facility of carriage. For example, a writer, in 1825, after showing the relative advantages of rivers and canals in the matter of ease and speed of carriage, finally concluded that, at a given rate of speed, a horse could move four times the weight on a canal or river that he could on a railroad[1]. Others who looked into this subject carefully and with scientific precision were convinced that, up to a certain low rate of speed, a horse could draw more on a canal than on the railway; but this rate of traction was so much less than what was possible on the rails that the waterway would be thought of only in connexion with the conveyance of commodities for the carriage of which speed was of little account[2]. Occasionally, other reasons were adduced to prove that railways were inferior to canals as means for the carriage of freight[3];

[1] Brit. Mus. 08,235. f. 77, 'Observations on the Comparative Merits of Inland Navigations and Railroads,' pp. 22–23.

[2] Wood, *Practical Treatise on Railroads* (1825), p. 157 et seq., and Thomas Tredgold on *Railroads*.

[3] Brit. Mus. 08,235. f. 77, 'Observations on the Comparative Merits of Inland Navigations and Railroads,' pp. 28–38, gives arguments against the locomotive and in favour of canals; and the writer finally says: "But I certainly think sufficient proofs appear, that in competition with a long line of canal or river navigation, enjoying a general trade, and affording the means of free and open competition, any project of a railroad would prove ruinous to the adventurers, and useless to the public" (p. 43). To the same effect were the words of another in 1832, after the Liverpool and Manchester Railway had been in operation for two years. He showed that for passenger traffic the railway was superior to any other mode of travelling; but, in regard to the freight traffic, his conclusion was that, mile for mile, goods were not carried so cheaply on the Liverpool and Manchester Railway as on the great lines of canals, and could neither remunerate the carrier as to his

but the number of people who laid any emphasis upon this possible outcome of the competition seemed to be very small.

The facts appear to indicate that the canal companies, instead of becoming more active and endeavouring to secure more of the traffic in the field which was now being invaded by a rival, usually acted on the defensive in trying to protect their alleged rights[1]. The great argument put forth by those who favoured the canals was the constant plea of vested interests: that Parliament had, by statutory provision, authorized the construction of and investment in canals, and, therefore, nothing should be done to destroy such facilities, under which the trade of the country was said to be flourishing[2]. Canal property, in many cases. was the only basis of security for wills, settlements and family incomes; and to destroy them would ruin thousands of families. If the canal were not sufficient for the increased traffic, why should the canal companies not be allowed to enlarge their works to meet the needs of an expanding commerce?[3] To interfere with private property was to overthrow the stability which. was fundamental to social life and the protection which the individual might justly claim from the government; and this appeal of protection to individual rights seldom failed of response when addressed to any class of the English people[4]. In

freight nor the proprietor as to his tolls in the same manner as canals did (P., *Letter to a Friend, containing Observations on the Comparative Merits of Canals and Railways*, pp. 2, 3, 8). His inference was "that the level railway 30 miles long between Liverpool and Manchester cannot put down two navigations, between the same points the first of which is fifty and the other forty-five miles long" (ibid., pp. 12–13); and in this opinion he seems to have been entirely sincere, since his pamphlet shows the utmost candour in giving the railway its full share of praise. See also *Birmingham Journal*, Mar. 5, 1831, p. 3, "On Railways."

[1] There were, of course, some exceptions to this. At Manchester, for example, in 1825, the activity of the railroads in carrying on their plans excited the feeling of competition among the proprietors of inland navigation; and the latter (called the fourth estate of the realm, because of their immense parliamentary interest) determined to prove that the speed on inland navigation was much greater than it had been represented. To establish their point, a flat left Manchester on the Mersey and Irwell Navigation in the morning and reached Liverpool by one o'clock. There she loaded a full load of cotton and started back for Manchester which was reached at 10.30 that night. This was repeated the following day (*Manchester Gazette*, Jan. 15, 1825, p. 3, on "Effect of Competition"). It is possible that such sporadic attempts as this were found elsewhere; but that does not disprove the statement we have made.

[2] *Hansard's Parliamentary Debates*, N.S., xii (1825), pp. 845–9; ibid., xv (1826), pp. 89 ff.; Brit. Mus. T. 1371 (18), pp. 9–11.

[3] "J. C." in *Gentleman's Magazine*, xcv, Pt. 1, pp. 113–15, showing the "advantages of canals over railways."

[4] *Hansard's Parliamentary Debates*, N.S., xii (1825), p. 847, debate on the Liverpool and Manchester Railway Bill. Jeaffreson, *Life of Robert Stephenson*, i, p. 268,

answer to this plea of vested rights it was said that Parliament, by sanctioning the building of railways, would not be breaking faith with the canal proprietors, for it was never contemplated that monopolies should be protected nor that further impetus to the development of commerce should be denied. The carriers by waggon had, in vain, urged the same consideration against the development of water conveyance, when the latter had become absolutely essential to the material advance of the country[1]. The canals had been given an opportunity to meet the increasing demands of commerce, and even those which were best situated had not done so, but had raised their charges and treated the demand with insolence[2]. It had never been the function of government to protect such injustice at the expense of the public good.

To recapitulate: the chief arguments advanced in favour of the railways were their speed and cheapness of carriage as contrasted with the canals, and the insufficiency of water carriage to serve the necessary purposes of the rapidly growing trade and industry of the country; the fact that they did provide good investments in some cases was not one of the primary inducements to their formation[3]. Additional facilities for the carriage of goods were essential, a *sine qua non* for the material advance of the country's interests. Some of the canals had not been improved at all, others very little, since their construction;

tells us that among all classes of society so universal was the antagonism to railways, from a fear that they would be injurious to vested interests, that gentle and simple with equal complacency viewed the constitution of tribunals which necessarily sympathised in a very high degree with the prevailing prejudice. See also Parkes, *Claim of the Subscribers to the Birmingham and Liverpool Railroad*, pp. iv–v, 3–4, 63–66, 72–74.

[1] *Gentleman's Magazine*, xcv, Pt. 1, pp. 199–200.

[2] See Sandars' pamphlet on the *Liverpool and Manchester Railway*, pp. 3–9.

[3] *Gentleman's Magazine*, xcv, Pt. 1, pp. 199–200; *Remarks upon Pamphlet by "Investigator" on the Proposed Birmingham and London Railway*, pp. 10–12, 24 ff.; Brit. Mus. 8223. e. 10 (70), 'Prospectus of Kentish Railway Company;' Brit. Mus. 8223. e. 10 (148), 'Prospectus of the Surrey, Sussex, Hants, Wilts, and Somerset Rail-Road Company;' 'Collection of Prospectuses, Maps, etc., of Railways and Canals,' pp. 13–14, giving the prospectus (dated May 10, 1830) of the two companies, the Liverpool and Birmingham Railway Company and the Birmingham and Liverpool Railway Company, whose interests were identical for the construction of a railway between these two cities; ibid., p. 65, giving the announcement of the Directors of the London and Birmingham Railway, 1833, and showing that the railway would more than double the speed of the stage coach, that the cost of passenger travel would be less than half that by coach, and that merchandise would be carried for only two-thirds of what was formerly paid on the canals. See also ibid., p. 74, 'Statement of the Case of the Liverpool and Birmingham Railway Bill (1831),' and pp. 139–40, 159–60, 169–70, 176, which give other prospectuses of railways; also *Railway Times*, v, pp. 639, 711, 973.

and this in the face of an incessant demand from industrial and commercial interests.   In some cases, where the trade was heavy, navigation companies had refused to incur the extra expense of maintaining a sufficient number of barges to provide for emergencies, but had made only meagre provision for even the usual requirements[1].   Then, too, on some canals the charges had been raised by the exorbitant demands of the proprietors[2], until this increased cost, along with the inadequacy of the service[3], led to the promotion of railway enterprise.   Private interests and individual advantage had already too long dominated in matters of transportation; necessity required added equipment for, and new life imparted to, the service; the public must not be sacrificed to the individual benefit; and the railway system was the result of the operation of these imperative calls for the national advance along this line.

It will be fitting at this point to consider the nature of the opposition which was encountered by the railways in their efforts to become established along the highways of trade.   In the first place, opposition arose from many of the landowners, who stubbornly resisted the encroachment upon their domains of these black monsters, the locomotive engines, with their trailing clouds of smoke, disfiguring the landscape, destroying the privacy and seclusion of their estates, and causing a great decrease in the value of their lands.   As a rule, the landlords thought much more of the peacefulness of their own estates and mansions than of the public good; and the mental picture of a railway with its tail of smoke curling across the country, blackening everything even to the fleeces of wool on the sheep, reckless of the aesthetic rural conditions and of the security of individual or public property, was to them the symbol of all that was disagreeable, vulgarizing and mercenary[4].   The introduction of such "infernal machines," as the

---

[1] Sandars on the *Liverpool and Manchester Railway*, p. 16.

[2] Ibid.; Cumming, *Rail and Tram Roads*, p. 33; *Gentleman's Magazine*, xciv, Pt. 1, pp. 415–17 ; 'Prospectus of the Birmingham and Liverpool Railroad Company.'

[3] On account of delays on the waterways from Liverpool to Manchester, more time had sometimes been consumed in the carriage of goods that short distance than in the transatlantic voyage (*Gentleman's Magazine*, xciv, Pt. 2, p. 556).

[4] See leter from "No Railer at the Present System," in Aris's *Birmingham Gazette*, Jan. 10, 1825, p. 1, acknowledging that railroads were superior to the canals in the matter of speed, but opposing them chiefly from the aesthetic standpoint: he would not like to see the country disfigured by the clouds of smoke.   His letter ends by saying: "Do, good Mr Editor, lend your potent aid, at the commencement of the coming year, to avert this mass of evils, and help by advice, by entreaty, by warnings, by ridicule, by anything, to thwart the designs of these iron-hearted speculators, who would take from the people of this free country all hopes of another

locomotives were sometimes called, must be stoutly resisted. The destruction of the unity of the farm by having part of it cut off from the homestead; the dividing of closes that were convenient in form and size into "ill-shaped fragments;" the formation of deep cuttings across the hills and of large embankments across the low lands, thus preventing the natural flow and drainage of water; the inconvenience and danger to the public on account of the railway crossing the highways on the same level; these, along with the declaration that there was no necessity for greater speed of travelling nor facilities for conveyance, added pretext upon pretext for the opposition of the landholding classes[1]. Others were aroused to hostility lest a projected railway might pass through their fox-covers, or in some other way interfere

merry Christmas. If we must be slaves let it not be to iron-masters—let us open our eyes before the accumulation of smoke renders it impossible for us to see, and let us, above all things, beware lest Rail-roads, like party, prove 'the madness of many for the gain of few.'" A similar letter is found in ibid., Jan. 17, 1825, p. 2, from one who subscribes himself as "Common Sense." See also *Hansard's Parliamentary Debates*, N.S., xii (1825), pp. 845–9; ibid., xv (1826), p. 89 et seq.; Booth, *Liverpool and Manchester Railway*, p. 31 et seq.; Brit. Mus. T. 1371 (18), pp. 5–9; *The Times*, May 4, 1824, p. 2 and Mar. 5, 1825, p. 5, on the "Tees and Weardale Railroad;" ibid., Nov. 18, 1830, p. 3, letter from "A Landowner;" ibid., July 17, 1832, p. 3. As an example of utter lack of sanity in the treatment of such a subject, perhaps the letter of George Jones in *The Times*, May 3, 1834, p. 6, stands unrivalled. See also his petition against the London and Southampton Railway, in *Hampshire Advertiser*, Mar. 22, 1834, p. 2. A farmer in Northamptonshire refused his assent to the proposed London and Birmingham Railway on the ground that the smoke would injure the fleeces of his sheep (*Birmingham Journal*, Jan. 29, 1831, p. 2). As to the landlords' opposition to the Stockton and Darlington Railway, see Jeans, *Jubilee Memorial of the Railway System*, pp. 28, 29, 32. A writer, signing himself "Ebenezer," evidently a Quaker, in a letter to the *Leeds Intelligencer*, Jan. 13, 1825, p. 3, while admitting that the engines on the rail were much faster than the canal barges, expresses the evil connected with the railway as follows: "On the very line of this railway, I have built a comfortable house; it enjoys a pleasing view of the country. Now judge, my friend, of my mortification, whilst I am sitting comfortably at breakfast with my family, enjoying the purity of the summer air, in a moment my dwelling, once consecrated to peace and retirement, is filled with dense smoke or foetid gas; my homely, though cleanly, table covered with dirt; and the features of my wife and family almost obscured by a polluted atmosphere. Nothing is heard but the clanking iron, the blasphemous song, or the appalling curses of the directors of these infernal machines." This was not the sentiment of one but of a multitude, and references could be added at great length. See also *Leeds Intelligencer*, Mar. 4, 1830, p. 3, and Mar. 11, 1830, p. 3, on the Leeds and Selby Railway; *Sheffield Iris*, Mar. 3, 1835, p. 2, and Sept. 22, 1835, p. 4, on a landlord's opposition to the proposed Sheffield and Rotherham Railway; *Birmingham Journal*, Mar. 11, 1826, p. 1; *Manchester Courier*, Feb. 4, 1832, p. 3, on the London and Birmingham Railway.

[1] *Birmingham Journal*, Jan. 22, 1831, p. 1, on the London and Birmingham Railway.

with their amusement of hunting; and the great landholders were not to be expected to make any concession, or to be coerced into anything, even although their estates would thereby become more valuable and great benefit result to the public[1]. Many were averse because it seemed to them that a railway, with its force of men who were by no means scrupulous of others' property and property-rights, would be an unmitigated evil; for it would permit the passage along the line of men of the worst class who would be ready to cause much annoyance to landowners on account of the nuisances which they would commit upon private property adjacent to the railway. Vast sums of money were required at first, under the plea of "compensation," to buy off the opposition of property holders and to pay for the strips of land that were necessary for these public enterprises; and when it was found that the money would be forthcoming for this purpose, some impecunious peers enriched themselves by demanding exorbitant prices for their land, under the specious pretence of injury to their estates[2]. "*Any* amount that could by *any* means be squeezed from the funds of a railway company under the name of compensation, public opinion decided to

---

[1] Jeans, *Jubilee Memorial of the Railway System*, p. 32; *Sheffield Iris*, Sept. 22, 1835, p. 4.

[2] Jeaffreson, *Life of Robert Stephenson*, I, chap. ix, tells about the opposition to railways and shows that "in some cases enormous sums of money were paid for the acres of obstinate landowners" (p. 180). See also, *Remarks upon Pamphlet by* "*Investigator*" *on the Proposed Birmingham and London Railway*, p. 6; 'Great Western Railway. Evidence on the London and Birmingham Railway Bill,' pp. 12, 14, 34–37; *Railway and Canal Cases*, I, pp. 326 ff., 347 ff., 416 ff., 462 ff., show how often it was necessary to buy off the opposition of landowners. Parkes, *Claim of the Subscribers to the Birmingham and Liverpool Railroad*, p. vi, shows that certain landowners were incited by the canal companies to oppose railways. See also ibid., p. 70. Jeaffreson, op. cit., I, pp. 268–70, gives us an instance: "An impoverished nobleman, owning a house and park of the value of £30,000, in a county through which one of the earliest railways was carried, for a small strip of his park, occupied by the railway, which ran quite beyond the sight range of his windows, obtained no less a sum than £30,000—or the entire value of the estate which the line was supposed only to depreciate. A few years afterwards this same peer sold another corner of the same park for another line for a second £30,000, and when he had thus extracted from two companies £60,000 as compensation for damage done to his estate, the original property was greatly augmented in value by the lines which, it was represented, would inflict upon it serious injury." He tells us that similar cases were of constant occurrence; and far from rousing public indignation, they met with public approval. The way in which compensation for lands was determined is fully set forth in *Parl. Papers*, 1845 (420), X, 417, 'Report of Select Committee of the House of Lords on Compensation for Lands taken for or injured by Railways;' see especially the evidence of Messrs Duncan, Clutton, Driver and Cramp. They show the way in which "extravagant sums," often far greater than the lands were worth, were paid in order to get rid of opposition.

be legally and honourably acquired[1]." Later, when the public benefits of the railway were known, this was changed; and those who had protested against this means of carriage became strong supporters of it[2]. Parliament went even so far as to pass a resolution excluding from the committee sitting on any railway bill any member who either held land through which the line was to run or was otherwise commercially interested in the rejection or passing of the bill[3].

Other great opponents of the railways were the canal companies, which in some instances had become strongly entrenched in the commercial life and activities of the kingdom. Their hostility was naturally to be looked for where the railway was to be constructed parallel to the canal, for in that case it was possible that the revenue of the canal company would be decreased on account of the railway competition[4]. The motives of opposition were sometimes concealed or thinly veiled; but underneath them all there was the one pervading object, namely, to keep the monetary returns from the canals as high as possible. Sometimes it was said that the railway could not carry heavy goods as

---

[1] Jeaffreson, op. cit., I, pp. 269–70. As compensation for "severance" of his estate, a proprietor, after requiring that bridges should be built at so many points along the line that the "severance" would practically cease to exist, would demand two, three, or four thousand pounds, in addition to the extortionate price already paid for the land actually given up to the line. It was useless for the agents of the railway company to show that this "severance" was merely an imaginary grievance, which effected no real injury to the estate. Refusing to see it in this light, the owner remained steady in his demand and gained his "severance" compensation. Having thus sold a strip of land at four, five, or six times its value, as recompense for a purely imaginary damage, the owner would then candidly avow that this severance caused him so little discomfort, that he could do with only half or quarter of the stipulated bridges, and that, for a further sum, he would free the company from the obligation to build the unnecessary bridges. In these early days railway companies were powerless to resist such extortions. They had to buy the goodwill of the community by hard cash. See also *The Economist, Weekly Commercial Times, and Bankers' Gazette*, 1845, p. 758.

[2] In 1844, Mr Croker wrote: "I know persons who were adverse to railroads, and who would now give £500 a mile to have them nearer their residences." (*Croker Papers*, III, p. 25.) Mr Earle, before the Committee on the London and Birmingham Railway Bill, testified that he would no longer oppose any railway, as he had determinedly opposed the construction of the Liverpool and Manchester Railway ('Great Western Railway. Evidence on the London and Birmingham Railway Bill,' p. 12). Other great estate-owners had also changed front in this way (ibid., pp. 34–37).

[3] Jeaffreson, op. cit., I, p. 271.

[4] Teisserenc, *Études sur les voies de communication*, pp. 21, 31; Whishaw, *Analysis of Railways*, p. 164, showing that the Manchester and Leeds Railway was opposed by the Rochdale Canal, Calder and Hebble and Aire and Calder Navigations until restrictive clauses in their favour were inserted in the railway Act; *Leeds Intelligencer*, Mar. 11, 1830, p. 3, Jan. 23, 1836, p. 3, and April 23, 1836, p. 3.

economically as the canals could[1]; in other cases the canal companies declared that there was not enough traffic to warrant the additional facilities of carriage[2]. The commissioners of river navigations opposed railways that would probably take away some of their traffic, for the same reasons as they opposed the construction of canals parallel to or out of the river over which they had control, namely, that the river would rapidly fill up on account of disuse and, therefore, the lands and towns along the river would be materially injured and in danger of inundation; that the tolls from the traffic remaining on the river would not be sufficient to defray the interest of the debt and expenses of maintenance; and that the value of adjacent estates and of the mercantile property connected with the river and its trade would seriously decline[3]. Perhaps the canals and canal carriers were the most indefatigable of all the opponents of the railway. They would inevitably lose more than most other interested parties by the entrance of this new and effective rival into a realm which they had thought was pre-empted by themselves; and in proportion to their probable loss was their effort to save themselves from the impending disaster[4].

But if the freight carriers by water were vigorous in their animosity to railways, we should expect that the coaching establishments would also be hostile to them, because of invading their field for the conveyance of passengers. And, of course, ranged with the coach proprietors we should expect to find others whose interests were closely bound up with the prosperity of the coaching and the carrying trade on the roads. Whether it is because our information along this line is not so complete, or because there was less capital embarked in road carriage than in canal carriage—from whatever cause, we do not find the same volume of complaint and the same keenness of antagonism from the representatives of the carriers along the highways that we find from those interested in the waterways. It would be wholly unnatural for those large concerns that had from 700 to 1800 horses, or even those which had a much smaller business, to allow their enterprise to be disintegrated without making efforts to save it; but while they petitioned Parliament to care for their interests, they did not, apparently, endeavour to arouse such a storm of opposition as did the canal forces. It must not be understood,

[1] *A Few General Observations on the Principal Railways*, pp. ix, x, 9.

[2] *Leeds Intelligencer*, Mar. 4, 1830, p. 3, Mar. 11, 1830, p. 3, and Mar. 18, 1830, p. 3; *Sheffield Iris*, April 7, 1835, p. 4, on the Sheffield and Rotherham Railway Bill.

[3] See, for example, Felix Farley's *Bristol Journal*, Feb. 22, 1834, p. 1, and Mar. 15, 1834, p. 1.

[4] See also *Railway Chronicle*, Sept. 20, 1845, pp. 1299–1300; *Sheffield Iris*, April 7, 1835, p. 4, and June 2, 1835, p. 2; *Manchester Guardian*, April 2, 1831, p. 3, on Manchester and Leeds Railway Bill.

however, that their claims upon public consideration were not strongly presented; but, in seeking favourable action, they did not stir up disaffection among other classes of the community in order to secure their co-operation. The character of their opposition, in trying to uphold their own stability and permanence, is well exemplified by a petition to the House of Lords from coach proprietors, post-masters and waggon-masters on the lines of road between London, Worcester, Hereford and Gloucester, asking for protection of their interest by the rejection of all applications for railways in general, and, particularly, for the Liverpool and Birmingham, and Birmingham and London railways[1].

Of a similar nature to the foregoing was the opposition of trustees of turnpike roads and of those who had loaned them money on the security of the tolls. They were averse to the building of a railway which would take traffic from that highway and thereby reduce the amount of toll that would be collected at the gates; for if the tolls were to become lessened, the security for the money loaned would be correspondingly lessened[2], while the revenues for the maintenance of the road and the payment of interest on the debt would be depleted. When it was known in advance that a railway was to be constructed which would probably have this effect upon a certain turnpike, very few persons were ready to bid for the gates at the time they were put up at auction; and, reasoning from analogy with those cases where railways had already made themselves felt, the few bidders who did come forward would not assume the risk of taking the gates except at greatly reduced rentals. When trustees found such difficulties in the financing of their roads after the completion of a railway, it is little wonder that they objected to the introduction of the latter knowing that their difficulties would increase with the passage of the years.

Some towns rejected the boon that was offered them, and opposed the railways so strongly that they would not allow the company to build their line within the city limits. For instance, to satisfy the people of Northampton and to meet their objections, the London and Birmingham Railway Company carried their roadway a considerable distance from the town, and built their works and shops at Wolverton, instead of, as originally intended, at Northampton[3]. The town of

[1] *Hampshire Advertiser and Salisbury Guardian*, May 11, 1833, p. 2; see also *Proceedings of the Great Western Railway Company*, p. 10.

[2] Brit. Mus. T. 1371 (18), p. 14; 'Great Western Railway. Evidence on the London and Birmingham Railway Bill,' p. 15, evidence of Joseph Pease; *Birmingham Journal*, Feb. 12, 1831, p. 2, on the London and Birmingham Railway; Cundy, *Observations on Railways*, p. 15.

[3] Stretton, *History of the London and Birmingham Railway*, p. 1. See also Markham and Cox, *Northampton Borough Records*, II, pp. 543–4.

Maidstone in Kent assailed the South Eastern Railway so vigorously that the Dover line was carried far away from them[1]. Owing to representation from Windsor, a clause was inserted in the Great Western Railway Act forbidding any station at that important town[2]. It was not till after repeated applications that a branch of the Great Western was allowed to be constructed to Oxford[3]; and then the authorities of the university had a clause inserted by which the station at Oxford was to be erected at a spot as remote as possible from the colleges[4]. Perhaps the opposition of Oxford University and of Eton College was the most vehement that the Great Western Railway encountered. The latter institution refused to allow the railway to come within three miles of the school; and in the railway Act Eton College obtained the insertion of a clause forbidding the erection of a station at Slough and requiring the company to provide policemen to patrol the line for a certain distance on each side of Slough so as to safeguard the Eton boys from danger. But although the Act forbade the building of a station at Slough, yet the railway trains, from the very first, stopped there to set down and take up passengers, and as an office the company used two rooms in an adjoining public-house. On account of this supposed breach of the law, the Eton College officials entered suit against the railway company; but it was shown that the latter had observed the terms of the Act of incorporation and consequently the suit was dismissed[5].

Local jealousies of one kind and another aroused opposition to railways that were highly advantageous from the point of view of public benefit. Farmers near a large centre of population were found to object to the construction of a railway back from that centre into the more distant country, lest their monopoly in a lucrative market would thereby be broken up, because produce grown at a much greater distance from that market could come into competition with that grown in the nearer areas. As an illustration of this, we note that the

---

[1] *Railway Times*, IX, p. 961.

[2] Shaen, *Review of Railways and Railway Legislation*, p. 29.

[3] Ibid., p. 29.

[4] Sekon, *History of the Great Western Railway*, p. 8.

[5] *Railway and Canal Cases*, I, pp. 200–10, gives this case in full. See also Sekon, *History of the Great Western Railway*, pp. 6–8; Markham and Cox, *Northampton Borough Records*, II, pp. 543–4; *Felix Farley's Bristol Journal*, Mar. 15, 1834, p. 4. The Great Western Railway was a rival scheme of the London and Southampton Railway; and the latter joined with Eton College and with many landowners in opposing the Great Western (Fay, *A Royal Road*, p. 20). Brit. Mus. 8235. bb. 87 (1), 'Speech of Counsel on the 30th May, 1848, before a Select Committee of the House of Commons, on behalf of the Head, Lower and Assistant Masters of Eton against the Great Western Railway Extension from Slough to Windsor,' shows how strong was Eton's opposition to the railway coming within easy access to that college.

Great Western Railway was opposed by the Middlesex, Berkshire and Buckinghamshire farmers because they feared that London would be able by this means to get food supplies from a distance at cheaper prices than those at which they had been accustomed to selling[1].   Of the same nature, apparently, was the opposition to the proposed Tees and Weardale Railway, in 1824, the opponents of the line asserting that the outlet for North Durham coal by the rivers Tyne and Wear was sufficient, while its advocates desired additional facilities of outlet by the river Tees[2].   Some were averse to railways because they feared that thereby trade would be transferred from one place to another.   For example, the contemplated Great Western Railway stirred up some of the people of Bristol because of their alarm lest a large part of the trade then transacted at that city should be afterward centred at London. If the products which were brought into Bristol from Ireland were to be taken to London immediately upon their arrival at the quay, it was thought that the mercantile interests of the former city would be injuriously affected; and if the shipping and the West India trade should subsequently locate at London, instead of remaining at Bristol, this western emporium of commerce would be sacrificed and the "ancient and once-flourishing city of the splendid name" would probably be dismantled[3].   Sometimes the owners of coal mines at a certain place opposed the introduction of a railway that would enable other coal mines to compete with theirs, and this local monopolistic spirit characterized much of the antagonism that railway promoters realized[4].

[1] *Proceedings of the Great Western Railway Company*, pp. 7, 13.   We observed the same objection to the extension of the turnpike system from London to the remote parts of the kingdom.   See *Railway Times*, VI, p. 242, article entitled "Railways and the Agricultural Interests," showing that the prices formerly received by the farmers of Southall and Perivale, co. Middlesex, for their cattle when sold in the London market had been forced down by reason of the great numbers of cattle and sheep that had been brought by the Great Western Railway from the West of England to London.   But, of course, there were compensating advantages that the farmers enjoyed, even if they had forgotten them.

[2] *The Times*, May 4, 1824, p. 2.

[3] *Bristol Mercury*, Feb. 16, 1833, p. 2, and Mar. 2, 1833, p. 2, letters from "Scrutator."   It seems strange that this man should have been showing how Bristol would decline after the railway was constructed, when another man, signing himself "A Burgess," was, at the same time, writing a series of thirty letters on the trade of Bristol, showing the causes which had brought about its decline and the means necessary for its revival (*Bristol Mercury*, Feb. 2, 1833, p. 2; Feb. 9, 1833, p. 2; Feb. 16, 1833, p. 2; Feb. 23, 1833, p. 2; etc.).

[4] *Manchester Guardian*, Feb. 26, 1831, p. 3, concerning the Oldham Railway; *Sheffield Iris*, Oct. 7, 1834, p. 1, prospectus of the proposed railway from Sheffield to Rotherham; ibid., Mar. 31, 1835, p. 2, editorial on the Sheffield and Rotherham

Upon the other forms of, or reasons for, opposition to railways we shall dwell but briefly. Some turned against them as enterprises in which money would be sunk without any adequate return, considering them as a new and dangerous form of speculation[1]. It was contended that the actual amount invested in them would greatly exceed the estimates that had been made for the purpose of inducing capital to embark therein[1], and hence there would soon be financial embarrassment when interest could not be paid on the great investment. Moreover, the absorption of the national capital to such a vast extent would divert it from more legitimate channels; and the fact that railways were not regarded by some as legitimate enterprises would seem to have been chiefly due to branding the whole system with the same characteristics that had been displayed in the case of some crude and other dishonest ventures[2]. When many plans were being formed for railways, it was inevitable that some ill-devised and delusive schemes were encouraged, which ended in the ruin or injury of those concerned in them; and also some dealers in shares who profited in one case but lost in another used unworthy means of accrediting or disparaging particular undertakings. Thus, no matter how good and substantial the enterprise, it might be given a wrong character, at least for a time, and in this way be subjected to popular condemnation.

Railway; ibid., April 7, 1835, p. 2, on the Sheffield and Rotherham Railway; ibid., Sept. 15, 1835, p. 4, letter from W. Ibbotson. See also *The Times*, June 17, 1836, p. 3.

[1] Aris's *Birmingham Gazette*, Oct. 4, 1824, p. 3, letter signed "Common Sense;" *Remarks upon Pamphlet by "Investigator" on the Proposed Birmingham and London Railway*, pp. 4, 6; Parkes, *Claim of the Subscribers to the Birmingham and Liverpool Railroad*, pp. 3–4, 64–67. Vallance, *Sinking Capital in Railways*, pp. 6–23, warned against investing in railways the £30,000,000 which at that time (1825) it was proposed to lay out upon them. His opinion was that it was doubtful whether the railway could travel with safety at more than six miles an hour (p. 18), i.e., about half the rate of the coach (p. 30); that, at that rate, the railway could not travel regularly and uniformly from the beginning to the end of the journey (pp. 19–20); and that, because it had to carry so much coal and water, the locomotive would not be able to do its work at the least possible expense. Since railway operation was therefore defective in all these essentials, the people should beware of sinking capital in this new mode of transit.

[2] *Herepath's Railway Magazine*, N.S., III, pp. 24–27, "On Railways as Investments." This writer says that it was very obvious that the greater number of capitalists were not friendly to railways and generally stood aloof until the profit of the work was known. The great fundholders and the landed proprietors, with few exceptions, hung back from enterprises which were ultimately very successful; while the commercial classes, who were accustomed to tracing out results from the operation of certain principles, were the men who had been chiefly responsible for the development of national improvements.

Railways were at times prevented from, or delayed in, obtaining legislative sanction on account of the political expedients that were resorted to. Not infrequently the House of Lords blocked measures that had passed the House of Commons after careful inquiry and close scrutiny. Some members of a committee to which a particular bill had been referred, were known to absent themselves from all hearings upon the bill because they either had no interest in or were opposed to it, and to walk into the committee room just before the time for the decision in order to give their vote adversely[1]. In other cases, some of the supporters of a bill, from motives of delicacy, abstained from attending at all and this left the measure in the hands of an opposing majority, so that after all the expense of time and money involved in hearing witnesses and paying parliamentary agents, the measure was foredoomed to rejection. With the committees upon private bills constituted as they were, it was not always the best line that secured favourable consideration; but that line was likely to be carried which could exert the greatest influence in commanding public attention and obtaining the predominance in the committee[2].

Some very trivial reasons were occasionally given for the opposition that was manifested to railway projects. Sometimes they were opposed for the same reason that stage coaches in early days were opposed, namely, because they would induce people to go flying about the country, instead of attending to their work at home[3]. Some were afraid that the velocity at which the trains would travel would occasion great accidents and the suggestion was made that it might be desirable to establish every five or six miles along the line what would be practically well-equipped hospitals to take care of the injured[4]. At other times railway bills failed to pass because of insignificant breaches of the Standing Orders when the measure was brought before the committee[5]. Even where a noble landowner knew that the projected railway would not injure, but rather immensely benefit, his property, he still opposed it, without being able to assign any valid reason for

---

[1] *Sheffield Iris*, Sept. 22, 1835, p. 4, letter from W. Ibbotson.

[2] *Parl. Papers*, 1836 (0.96), xxi, 235, 'Minutes of Evidence before the Select Committee on Railway Bills,' evidence of James Walker, C.E., Q. 178.

[3] Fay, *A Royal Road*, pp. 23–24, gives some of the contemporary statements; for example, one of the great canal proprietors said: "I foresee what the effect will be—it will set all the world a-gadding. Twenty miles an hour! why, you will not be able to keep an apprentice boy at work....Grave plodding citizens will be flying about like comets."

[4] *Leeds Intelligencer*, Jan. 13, 1825, p. 3, letter on "Railways."

[5] Shaen, *Review of Railways and Railway Legislation*, pp. 36–40. He cites his cases from the parliamentary reports.

this decision[1].   In fact, until the railways had fully demonstrated their
utility there was active opposition to every scheme that was brought
forward; railways, apparently, were treated as nuisances and every
impediment was thrown in their way to cause the promoters to desist
from such activity[2]. We do not wonder, however, at the rooted
prejudice to the railways, when such a great engineer as Thomas Telford
was strongly opposed to them[3].

Among those who took a prominent part in the discussion in favour
of railways were George Stephenson and his son Robert, Nicholas Wood
(who was intimately associated with George Stephenson), William James
and Thomas Gray.   Perhaps the last-named, more than any other, kept
this subject before the public, not only by his contributions to the
current press, but by a work of considerable magnitude on what he
called a "general iron railway[4]."   His mind was full of this one idea,
and he gave it expression on all occasions.   The locomotive engine was
sure to supersede all other kinds of conveyance, and even to do away
with the necessity of horses.   He would, therefore, leave the turnpike
roads as they were, and perfect steam railways as a system more in
keeping with the time and with the increasing traffic of a commercial
nation[5].   And, as for canals, he deplored the fact of engineers still
wasting, as he thought, their time and the public money in these delusive
speculations[6].   He warned the public against subscribing to canal
schemes, "for the time is fast approaching when railways must, from
their manifest superiority in every respect, supersede the necessity both
of canals and turnpike-roads, so far as the general commerce of the

---

[1] *Sheffield Iris*, Sept. 22, 1835, p. 4, letter of W. Ibbotson.

[2] *Observations on the Comparative Merits of Navigations and Railroads*, pp. 43 ff.;
Shaen, op. cit., pp. 29–30.

[3] *Autobiography of Sir John Rennie*, p. 244.

[4] The complete title is *Observations on a General Iron Rail-way, or Land Steam
Conveyance; to supersede the Necessity of Horses in all Public Vehicles; showing its
vast Superiority in every respect, over all the present Pitiful Methods of Conveyance by
Turnpike Roads, Canals, and Coasting-Traders. Containing every Species of Information relative to Rail-roads and Loco-motive Engines.* The first edition was published
in 1821 and the fifth in 1825.

[5] See letter of T. Gray in *Gentleman's Magazine*, xciv, Pt. 2, pp. 313–16; also
Gray, *General Iron Railway*, pp. vii, xx–xxi, 2, 6, etc. He speaks of the many
complaints as to the state of the roads and of the impossibility of finding an effectual
remedy: of the accumulating debt on the turnpike roads, as shown by their annual
statements; of the waste in trying to keep them up, for they had "nothing, save
folly and extravagance, to recommend them." He favoured the "general introduction of mechanic power, so as completely to supersede the necessity of
horse power in all public waggons, stage and mail-coaches, and post-chaises"
(p. xi).

[6] *Gentleman's Magazine*, xciv, Pt. 2, pp. 313–16.

country is concerned[1]." The expense of making a canal and canal boats, the expense of men's wages, of horses' keep, etc., he thought, must render the transport of merchandise much dearer by canal than by an improved railway which combined economy of time and of labour.

Gray's scheme is an interesting one. He would have the railway system undertaken as a national work; for unless the nation took up the matter it would not be carried out on proper principles[2]. He would have a national board appointed to introduce the most simple and general principle of uniform connexion throughout the country; there should be facility of national communication by having uniformity of rails and vehicles and provision should be made for the easy interchange of traffic. The central feature of his plan was to have a general iron railway centring at London, with one main trunk line running from London to Edinburgh and another trunk line from London to Falmouth[3]. From these, branches should be constructed to run to all the important places in the kingdom, so that London might be connected with all the industrial and commercial centres. By having these roads laid out in straight lines and on perfect levels, the distances between the chief places would be greatly reduced and thus the time and the cost of carriage and travelling would be much lessened[4]. On these great through routes there should be different roadways for trains going to and those departing from London; and as London was approached these should be increased in number[5].

By such a plan, Gray thought to see extended into every part of the country the advantages which would lead to permanent prosperity; and so confident was he of the ultimate triumph of the steam railway, that he used every possible endeavour to secure its accomplishment. In 1820 and 1821 he submitted two addresses to His Majesty's Ministers

---

[1] *Gentleman's Magazine*, xciv, Pt. 2, pp. 313–16. He says: "Were canal proprietors sensible how much their respective shares would be improved in value, by converting all the canals into rail-ways, there would not, perhaps, in the space of ten or twenty years, remain a single canal in the country." (Gray, *General Iron Railway*, p. 9.)

[2] Letter of T. Gray on "Railway Advantages," in *Gentleman's Magazine*, xcvi, Pt. 1, pp. 126–8; also ibid., xcv, Pt. 2, pp. 310–12.

[3] *Gentleman's Magazine*, xcv, Pt. 1, p. 205. For his general plan for the railways of Great Britain, see his diagram in *General Iron Railway*.

[4] *Gentleman's Magazine*, xcv, Pt. 1, p. 205; also ibid., xcvi, Pt. 1, pp. 126–8; Gray, *General Iron Railway*, p. 10.

[5] Gray, *General Iron Railway*, p. 12. He says: "In order to establish a general iron railway, it will be necessary to lay down two or three railways for the ascending and an equal number for the descending vehicles. In the immediate neighbourhood of London, the traffic might demand six railways." With the constantly increasing traffic between Liverpool and Manchester, he would lay down between these two places also six lines. Similarly for other towns, according to their importance.

of State, showing the great national importance of his scheme; and again, in 1822, he urged its importance upon them by giving a detailed account of its advantages[1]. In 1823 he renewed his petition to the Ministers of State and asked for the appointment of a Select Committee of the House of Commons to investigate his plan[2]. He likewise petitioned the Board of Agriculture and tried to show them that the many important advantages which his proposed system of railways would afford to the public must overcome every prejudice and finally prevail over every other means of conveyance[3]. In a petition to the Lord Mayor and Corporation of the city of London, he reiterated the advantages that would result from the adoption of this new system of carriage; and asked them, in the interests of the whole country, to favour the establishment of railways[4]. But, whatever the reason may have been, Gray's national railway project was not taken seriously, for nothing was done towards its accomplishment[5].

The name of William James seems to have been given a place secondary in importance to that of Thomas Gray and the two Stephensons; and yet he was among the earliest, if not *the* earliest, of the originators and promoters of the system of passenger transit on railways. It appears that, as early as 1799, he was engaged in laying out plans for railways, some details of which he gives in his memoranda[6]. This work was continued at least down to 1808 when his diary ends; and during that time he surveyed and completed many sections of railroad that were to be used for the conveyance of coal to navigable waterways[7]. In a paper addressed to the Grand Surrey Canal Company,

---

[1] Gray, *General Iron Railway*, pp. xvii–xviii.

[2] Ibid., p. xix.   [3] Ibid., pp. xx–xxi.

[4] Ibid., pp. xxi–xxiii. He shows that by the railway the people of London might be regularly supplied with coal on comparatively reasonable terms, instead of "suffering under abominable extortion," under the existing conditions.

[5] Anyone who will read Gray's book through will find some things which are visionary and even the more serious part of the book contains much that would antagonize the public and turn them away from the writer of it. In the *Railway Record*, ii, pp. 401, 563, 595, 628, 658, 692, there are a series of articles dealing with "The Railway System and its Author," giving the chief facts in connexion with the work of Thomas Gray for the introduction of railways into England. He was at that time (1845) in destitute circumstances; and there was an agitation in favour of raising a sum of money that would put him beyond the necessity of hard manual work in those days of his old age. See also Wilson, *The Railway System and its Author, Thomas Gray, now of Exeter. A Letter to the Right Honourable Sir Robert Peel, Bart.* This was an appeal to Peel, by a friend of Gray's, that the latter might be relieved, and that his name might have the honourable place it deserved among England's great men in connexion with the railways.

[6] P., *The Two James's and Two Stephensons*, p. 18; *Mining Journal*, Dec. 5, 1857, in which is found James's diary down to 1808.   [7] Ibid.

he spoke of himself as an experienced engineer, "in railroads especially;" and other expressions of similar import are given in this same document[1]. But the most remarkable part of his diary refers to his plan for the formation of a general railroad company, with a capital of £1,000,000, "to take lands for ever to form railroads," and to fulfil other designated purposes[2]. We need not here follow his career and the testimony which was borne to his accomplishments; it will be sufficient to say that his wide experience and his ability were recognized at that time, but we have been unable to ascertain why his work has been overshadowed by that of his compeers[3].

At this formative stage in the history of railways it was to be expected that a considerable variety of plans would be suggested for their construction and operation; they were an entirely new feature in the industrial and commercial world, and those who were most interested in them were groping their way in the endeavour to ascertain the conditions of the greatest economy for this new instrument of conveyance. It will, therefore, not be amiss to note some of the proposals that were made, with the intention of securing this object, in the early years of the railway development. Before the success of the locomotive engine had been fully assured, the use of the inclined plane on the coal-carrying railways was, as we have seen, a feature of common occurrence; and even after the tractive power of steam had been demonstrated there were still some roads which were planned by engineers of repute, partly as inclined planes to be worked by stationary engines and partly level to be worked by locomotive engines[4]. Instead of steam power, it was

---

[1] P., *The Two James's and Two Stephensons*, p. 22. For instance, he says in that paper: "...and that the said railroads and all person or persons, and their servants, carriages, and cattle passing thereon, shall be under the control and management of the said William James and his co-partners...."

[2] Ibid., p. 23; also his diary referred to above.

[3] For the rest of his work, and his connexion with George Stephenson, see ibid., pp. 23–105. He advocated the possibility of attaining on railways a speed of twenty or thirty miles per hour—contrary to the opinions of George Stephenson and Nicholas Wood, who thought that railway travelling could never exceed eight or ten miles an hour (pp. 40 ff.). Even Robert Stephenson acknowledged that it was not his father, but William James, who was the "Father of Railways" (ibid., p. 105). See also *Autobiography of Sir John Rennie*, pp. 234–6.

[4] The Cromford and High Peak Railway, from the Cromford Canal to the Peak Forest Canal in Derbyshire, was made to rise by inclined planes to the summit level of one thousand feet above the former canal and then descend seven hundred and sixty feet to the latter canal. The rough country there made it necessary to propose the construction of both level parts and inclined planes, on the former of which locomotive engines were to be used and on the latter stationary engines (*Manchester Gazette*, Nov. 13, 1824, p. 3). On the Stockton and Darlington Railway they had both levels and inclined planes and both kinds of engine. They also used horse-power

not infrequently planned to use horse power, either on the incline or on the level[1]. As late as 1829, when the Leeds and Selby Railway was in prospect, it was decided to make the operation of the line possible by either horse power or locomotive engines, or to permit the company to use locomotive carriages if this were thought desirable[2]. George Stephenson, who had surveyed that line in 1825, had recommended for part of the line three inclined planes which could be worked by horse power or stationary engines, and for the remainder of it level reaches upon which locomotive engines could be employed. But in 1829, after Stephenson's suggestions had been rejected by the committee that had the work in charge, James Walker, who had also come into great prominence as a railway engineer, was asked to re-survey the line; and his opinion was decidedly in favour of the uniform system without inclined planes. Under these circumstances each shipper could utilize the line most favourably; and he calculated the strength of the rails, so that although at first horses would, in all probability, be the principal power used, yet locomotive engines might be used then or at a later time[3]. With accumulated experience of the results secured on railways, it became evident that for all ordinary purposes, where there would probably be traffic in both directions, the more nearly the line approximated to a perfect level the more economically could it be operated and the more efficient would be its service.

Another suggestion that seemed to find some favour was that railways might very acceptably be laid down at the sides of the ordinary highways and might be worked by either steam or horse-power. By this plan, the cost of the roadway would be greatly reduced, for the utilization of the land at the sides of the public roads for such a public purpose would not call for the enormous expenditures that were made by existing railways for the right of eminent domain. The carrying of

(Jeans, *Jubilee Memorial of the Railway System*, pp. 33–35, 43, 53–54). See also Walker and Rastrick, *Liverpool and Manchester Railway. Report to the Directors on the Comparative Merits of Locomotive and Fixed Engines*, pp. 3, 4, showing that there were inclined planes and stationary engines on other colliery roads, such as those of the Hetton Colliery and the Brunton and Shields. See also Wood, *Practical Treatise on Railroads* (1825), pp. 93–123, and *A Few General Observations on the Principal Railways*, pp. ix, 19–20, showing that in 1838, on the Stockton and Darlington and on the Leicester and Swannington, there were both self-acting and stationary-engine inclined planes, and these abrupt inclines were great drawbacks on all railways.

[1] Walker and Rastrick, op. cit., p. 49. Rastrick here shows the great advantage of the locomotive over horse-power.

[2] *Leeds Intelligencer*, Nov. 5, 1829, p. 3, on the Leeds and Selby Railway.

[3] Macturk, *History of the Hull Railways*, pp. 18–32, gives Walker's report to the committee of the proposed Leeds and Selby Railway Company.

railways along the course of the highways would not cause the dislocation of the usual currents of trade; the inns along the roads would not suffer, the various establishments that had grown up as links in the customary trade circulation would not be endangered, the diverse interests that had grown up around the system of road carriage would not be threatened with annihilation, and consequently the change from the old régime to the new would be accomplished with as little adverse effect as possible. So hopeful were some in regard to the application of this method that a writer in 1829 observed that "it is therefore nothing problematical to expect in the course of the next ten years to see railways by the road-sides extending from London, Liverpool, Hull, Edinburgh," etc.[1] He asserted that by having railroads laid down on the high road from London to Liverpool, the mails drawn by a light locomotive engine might go this distance, 204 miles, easily in twelve hours, carrying twice their usual number of passengers and at much lower cost[2]. Another, in 1833, considered horse-power more economical than steam, and he would have this applied on tram or railways, constructed as nearly as possible along the sides of the turnpike roads[3]. This suggestion was not the product of visionary minds, for even such a competent engineer as Fairbairn advocated the plan[4]. In addition to securing the advantages already mentioned, of reducing the cost of construction and perpetuating the existing trade routes with all their appointments for commercial purposes, railways located in this way would cause no invasion of estates, against which there was much complaint at that time. The decreased cost of construction would result in lower freight rates; and the increased traffic along the roads would ensure the receipt of tolls sufficient to repay the debts upon the various turnpike trusts. If the railways were built and owned by the state, as was suggested by Fairbairn, all revenues therefrom would accrue to the state. We see, therefore, that there were several reasons why this would appeal to the public; but when we remember that the locomotive engine works most economically on long lines of straight road we can see one physical reason why this method was not adopted[5]. The fact, too, that some

---

[1] *The Times*, Oct. 19, 1829, p. 3, on "Locomotive Carriages."

[2] *The Times*, Oct. 19, 1829, p. 3.

[3] *Bristol Mercury*, Oct. 5, 1833, p. 4, letter from "A Well Wisher," on the comparative advantages of horse and steam-power on railways.

[4] Henry Fairbairn's *Treatise on the Political Economy of Railroads* (1836).

[5] Ibid. He gives a full description of this plan and the benefits that would result from it. In his chap. iv, he shows that steam-power is too expensive for use in conveying merchandise; horse-power is best for that purpose. This sounds grotesque in view of the present circumstances. Many other of his statements are ludicrous; for instance, he says that all the great navigable rivers, like the Shannon,

apparently impractical conceptions were associated with this scheme, must have militated against its serious consideration.

Another plan for the improvement of railways has the name of Henry R. Palmer associated with it[1]. He proposed that where substances were likely to get on the rails, as was customary when they were so close to the surface of the ground, the rails should be elevated; but to elevate two lines of rail would cost too much, and, therefore, he would endeavour to arrange the form of a carriage so that it would travel upon a single line of rail without overturning. His method was to have the carriage so constructed that the two parts of it would balance upon the rail, irrespective of whether the number of passengers or the amount of freight were the same in each compartment[2]. A line of railway on this suspension principle was constructed for practical purposes of demonstration at Cheshunt, in Hertfordshire; apparently, it did its work successfully and answered the design in every respect[3], but it was intended more to exemplify the principle upon which it worked than to actually engage in the general carriage of all kinds of traffic. Why it was not employed as a regular means of conveyance, we have not the means of determining, although it was probably because of mechanical defects; and from that time onward all efforts at securing a workable monorail system were unsuccessful, until within the last few years when the gyroscope seems to have exhibited its practicability for the carriage of passengers at a high rate of speed.

We must now return from this digression as to the attitude of the public at this early time toward the railways, and the consideration of some of the proposals for securing their greatest effectiveness, to resume the historical development of the network of lines which was soon spread over the country. Through the discussion which was going on among engineers and the public generally, it was becoming evident that, not only from a mechanical standpoint[4], but also economically, the railway

Mississippi, St Lawrence, Thames, etc., will now be deserted for land conveyance, when his system is put into effect. See his chapters vi and viii for such ethereal projects.

[1] Palmer, *Description of a Railway on a New Principle.*

[2] The details of the plan may be found in the work last referred to.

[3] A full description of this railway and its method of operation is very clearly given in *The Times*, June 27, 1825, p. 3.

[4] On the relative mechanical advantages of railways, canals and turnpike roads, especially the two former, see Sylvester's *Report on Railroads*; Maclaren, *Railways as compared with Canals and Common Roads*, p. 58 et seq.; Tredgold on *Railroads*; Nicholas Wood, *Practical Treatise on Railroads*; and contrast these with Gordon, *Observations on Railways and Turnpike Roads*, pp. 4–11, who thought that the mechanical advantage of an edge railway was small when compared with a good turnpike road. Of these mechanical features we shall not treat here.

was to largely supersede both the canal and the highway as a means for the facilitation of the carriage of goods and passengers. Besides, the success of the coal-roads, and especially of the Stockton and Darlington, on which locomotive engines were being used with admirable results, made widely known the benefits to be obtained from the new means of locomotion. But we must not suppose that the success of the Stockton and Darlington was the reason for the construction of the other roads which were opened a few years afterward; on the contrary, at least two of the most important roads were projected before the Stockton and Darlington line was opened, namely, the Liverpool and Manchester and the Liverpool and Birmingham. As we have already seen, the chief reason why the railways came into existence was because of the need of more adequate facilities for conveyance than the canals could give. The enormous profits which some canals were making were also an inducement for railways to come in and secure a share of these benefits[1], and the success of existing railroads, giving additional encouragement to the projectors of new lines, had an important effect in initiating these enterprises along routes where they were much needed.

What we have just said applies with special force to the transportation conditions and requirements between Liverpool and Manchester. Under the stimulus of the Industrial Revolution, which assumed its greatest prominence in the cotton industry of Lancashire, villages had grown into towns and towns into large cities. Since the year 1760 Manchester and Salford, which are separated by only a small river, and which are considered as one, had probably increased in population at least eight-fold before 1830[2]. The increase in the amount and value of

---

[1] The great profits of the navigations between Liverpool and Manchester are considered when we come to take up the necessity for the Liverpool and Manchester Railway. Regarding the profits of the Bridgewater Canal, see also Parkes, *Claim of the Subscribers*, p. 24. As to the amount of the profits of the Birmingham, Grand Trunk and other canal companies, see Parkes, op. cit., pp. 16–20, 24, 43–44, 61. See also the examples given in 'Prospectus of the Birmingham and Liverpool Rail Road Company.' In Aris's *Birmingham Gazette*, Dec. 13, 1824, p. 1, a letter from F. Finch regarding the proposed Birmingham and Liverpool Railway speaks of the "inordinate profits" that the canals had enjoyed long enough.

[2] Sandars' pamphlet on the *Liverpool and Manchester Railway* gives us the following figures for the population of Manchester and Salford:

$$
\begin{aligned}
\text{in } 1757 &= 19{,}837 \text{ (estimated)} \\
1773 &= 27{,}246 \text{ (estimated)} \\
1821 &= 133{,}788 \text{ (census figures)} \\
1824 &= 163{,}888 \text{ (estimated).}
\end{aligned}
$$

Booth's figures were:

in 1760 about 22,000 ⎫
„ 1824 „ 150,000 ⎭ Booth, *Liverpool and Manchester Railway*, p. 6.

the cotton manufactured there had been very great, amounting to fifty per cent. in the eight years following the close of the Napoleonic war[1]. So great had been the change, that, while in 1814 there was not one power loom in Manchester, in 1824 there were nearly 30,000 of them[2]. Manchester had become the focus of a large manufacturing population, from which large quantities of cotton goods were sent to Liverpool and thence to all parts of the world. Liverpool also was rapidly attaining commercial importance and as a seaport was second only to London. Her population had almost doubled between 1800 and 1825[3]; and her colonial and foreign trade had been making great progress, as is shown by the tonnage and customs statistics[4]. Foreign produce of all kinds

[1] The following figures show this fact:

	Cotton manufactured at Manchester	
Year	(a) Quantity	(b) Value
1815	110,000,000 lbs.	£7,487,562.
1823	160,000,000 lbs.	£10,875,000.

[2] Booth, *Liverpool and Manchester Railway*, p. 6. In 1790 there was only one steam-engine in use in Manchester while in 1824 there were over two hundred.

[3] Sandars gives the following figures for the population of Liverpool:

in 1720 = 11,833 (estimated)
„ 1760 = 25,787 (estimated). Same figure given by [Corry], *History of Liverpool*, p. 119.
„ 1801 = 77,708 (census)
„ 1811 = 94,376 (census)
„ 1821 = 118,972 (census)
„ 1824 = 135,000 (estimated).

[4] On the imports, exports and shipping of Liverpool, see Brit. Doc. 1825 (182), II, 409, and 1825 (206), II, 413.

According to Sandars (p. 44), the statistics of tonnage and dock duties at Liverpool were as follows:

Year	No. of ships	Tonnage	Dock duties £	s.	d.
1752	—	—	1,776	8	2
1760	1,245	—	2,330	6	7
1770	2,073	—	4,142	17	2
1780	2,261	—	3,528	7	9
1790	4,223	—	10,037	6	2½
1800	4,746	450,060	23,379	13	6
1805	4,618	463,482	33,364	13	1
1815	6,440	709,849	76,915	8	8
1822	8,136	892.902	102,403	17	4
1824	10,001	1,180,914	130,911	11	6

From these figures it is evident that the tonnage of this port had much more than doubled between 1800 and 1824, while the dock duties in 1824 were almost six times as much as in 1800. Sandars gives these figures for each year, but his figures

passed daily from Liverpool to Manchester and manufactured goods went from Manchester to Liverpool, whence they reached the world's markets. The amount of this interchange of commodities between these two cities was conservatively estimated at 1000 tons a day and it was constantly increasing[1]. This great advance in population and in industry meant a greatly increased demand in the facilities for handling both in-coming and out-going freight.

How had this increasing demand for carrying facilities been met by the existing transportation agencies? The cost of carriage by land was 40s. per ton, which was so high as to be almost prohibitory for all goods except those of the finest quality and highest value. In reality, land carriage was more largely concerned with the carrying of passengers than of goods, although, on account of delays by water carriage, it frequently occurred that waggons and carts had to be resorted to for taking cotton up to Manchester and manufactured goods back to Liverpool[2]. This was done in the face of a freight rate that was three times that on the canal, in order to secure speedy and certain delivery of goods that were required for immediate shipment from Liverpool[3]. But the carriage of most of the heavy goods was done by the navigation companies, which felt themselves secure in the possession of a monopoly that they had long enjoyed to the public detriment. To the consideration of this monopoly we now turn our attention in order that we may see how their work had been carried on.

By various devices, both the Mersey and Irwell (or "Old Quay") Navigation Company and the Bridgewater Canal Trustees had contrived to raise their rates above what they were legally allowed to charge[4].

for the earlier years do not exactly correspond with those of Enfield, *Essay towards History of Leverpool*, pp. 67–69.

In 1770 the customs receipts at the port of Liverpool were £231,994. 12s. 5d. In 1822 they were £1,591,123, and in 1823, £1,808,402, which shows how rapidly they were increasing. (In addition to above references, see Baines, *History of Liverpool*, p. 492.)

In 1784 there were eight bags of American cotton imported into Liverpool; in 1824 there were 409,670 bags. Of this, the great bulk went to Manchester (Booth, *Liverpool and Manchester Railway*, p. 5). In the footnote to pages 6 and 7, Booth shows the progress of the port of Liverpool from 1824 to 1830.

[1] Booth, *Liverpool and Manchester Railway*, p. 4; Sandars, *Liverpool and Manchester Railway*, p. 13; also 'Prospectus of the Liverpool and Manchester Railway Company,' as given in Booth, p. 11.

[2] Sandars, op. cit., p 17.

[3] Ibid., p. 17. He says that goods going from Manchester for immediate shipment from Liverpool, often paid £2 or £3 per ton for carriage.

[4] Concerning the conditions of carriage by these two routes between Liverpool and Manchester, we shall follow Sandars, *Liverpool and Manchester Railway*, p. 4 et seq., in his description of their methods. His statements were uncontradicted, were based upon documentary evidence and bore the sanction of authority.

The Old Quay Navigation Company, by their Act of 1733, were allowed to levy a tonnage duty of 3s. 4d. per ton, but were not restricted• as to the rate of charge for freight. They adhered faithfully to this rate of tonnage; but as they owned nearly all the warehouses in Manchester on the banks of their navigation they were able to make much more revenue by freight, since for the use of these warehouses they could charge what they pleased and without the warehouses the navigation would be useless. In this way they were able to put their charges up so high as to drive all the other carriers off the navigation, and thus almost monopolize the carrying trade on their route. It will be observed that, in increasing their charges, this company did not violate their own laws. But this cannot be said of the Bridgewater Trustees, who, apparently, transgressed in several ways the statutory authority under which they were expected to operate, as we shall now show. The proprietor of the Bridgewater Canal was bound by his Acts not to charge more than 2s. 6d. per ton for canal dues; and for this charge the Duke was required to provide, for all persons carrying goods on his canal, wharfage or warehouse room for a certain period of time. He also bound himself not to charge more than 6s. a ton (tonnage included) for any goods which he might carry by his own vessels[1]. How was this fulfilled?

On the suggestion of Brindley, who surveyed the Trent and Mersey Canal, the Duke arranged with the Trent and Mersey Canal Company to unite the two canals at Preston Brook in order to faciitate the transfer of goods from one canal to the other; and His Grace contracted to cut the canal from there to Runcorn at his own expense. By this means the two canals would have a common outlet to the Mersey tideway and thus the communication would be more convenient and complete[2]. In consideration of his expenses in making and maintaining the canal from Preston Brook to Runcorn and the necessary locks and other works to accompany the canal, the Duke was empowered to receive a tonnage duty of 6d. per ton on all goods destined to enter the Trent and Mersey Canal[3]. This sum was exacted by the Duke on *all* goods that were conveyed between Liverpool and Manchester, in addition to the 2s. 6d. allowed by his own Acts, although he would have had to bring his canal to Runcorn and charge no more than the 2s. 6d. if he had not arranged with the Trent and Mersey Canal Company. This made the tonnage duty between Liverpool and Manchester 3s.

---

[1] Acts 32 Geo. II, c. 2 (1759), sec. 29; 2 Geo. III, c. 11 (1761), sec. 11.

[2] Act 6 Geo. III, c. 96 (1765), sec. 84.

[3] Trent and Mersey Canal Act, pp. 100–4; Act 6 Geo. III, c. 96 (1765), secs. 86–87.

But this was not all. His Grace bound himself to the Trent and Mersey Canal Company that if he ever found it necessary to make increased accommodation at Runcorn, the tonnage charge should still not be more than 6*d*. He soon found it necessary to construct a large reservoir there into which vessels destined to enter his canal were admitted at tide time. He pretended to construct this for his own vessels, but he kindly permitted those of other carriers to enter on condition that they paid 1*s*. per ton for the privilege, which was almost a necessity. This amount he collected on all goods passing along his own canal, as well as on those destined for places along the Trent and Mersey. This raised the tonnage to 4*s*. per ton.

A third means of increasing this tonnage rate remains to be pointed out. When the Rochdale Canal Company obtained its Act for cutting a line from Rochdale to Manchester, the Duke of Bridgewater obtained permission to make the lock to connect his canal with the Rochdale; and for this he was empowered to levy 1*s*. 2*d*. per ton on all goods which passed his lock, as indemnification for the loss which his warehouse property might sustain by this junction. In return for this payment he was bound to find warehouse room, gratis, for the goods for a certain limited time. Instead of this legal charge, he exacted the 1*s*. 2*d*. per ton on all goods that were carried on his canal between Liverpool and Manchester, whether they passed the junction lock or not. In this way he managed to secure 5*s*. 2*d*. per ton on all goods carried on his line, while the Legislature never intended him to have more than 2*s*. 6*d*.

In addition to this unduly high charge, another extortion of the Bridgewater Trustees was that all goods which passed from Liverpool to Runcorn to enter the Trent and Mersey Canal had to pay about twice the amount of freight which they should have paid, owing to the fact that these Trustees had monopolized nearly the whole of the land and warehouses at Runcorn. They would not allow the goods to be landed at all without paying what they asked.

It is evident, therefore, that each of the existing navigations between these two great cities was acting so as to get the greatest possible amount from the service rendered; each was operating as a virtual monopoly. The Duke was strongly advised to buy the Mersey and Irwell Navigation at the price for which it was offered, about £10,000; but he was confident of the superiority of his canal and rejected the offer[1]. But

---

[1] On his refusal, it was bought by some Manchester merchants and in the years preceding 1825 it was producing an average annual revenue of about £15,000; in other words, the yearly receipts were one and one-half times the total cost. **Brit. Mus. 08,235. f. 77,** 'Observations on the General Comparative Merits of Inland Communication by Navigations or Railroads,' pp. 6–7.

although the Duke declined to purchase this Old River Navigation, it seems that the two companies found it advantageous to share the monopoly with each other.   The alliance of their interests was effected by an agreement made in 1810; and in that year the two concerns publicly advertised that they had mutually agreed upon an advance of freight rates[1].   The rates of 1810 were nearly three times those of 1795 and about one-third more than those of 1824[2].   Of course, by the latter year it was becoming evident that there was the possibility of putting down a railway between these two places; and, apparently to placate those who wanted the railway, the navigation companies reduced their rates, but even the reduced rates were twice as much as those of 1795[3]. Until this possibility came before them there was an unqualified refusal to make any reduction[4], and any objections made by shippers were met with insolence on the part of the navigation companies[5].

Not only were the charges for transportation high, but the delays in the carriage of goods were often long and vexatious.   These were occasioned, sometimes because of the entire stoppage of the waterways by frost or drought, and at other times by their being blocked up on account of the pressure of traffic.   At times, storms and adverse winds prevented the navigation of the tideway of the Mersey, for it frequently occurred that when the wind blew very strong either south or north, the vessels could not move against it.   Merchandise was often brought across the Atlantic to Liverpool in twenty-one days; while, owing to the various causes of delay above mentioned, goods were in some instances longer than this on their passage from Liverpool to Manchester[6].

[1] Brit. Mus. 08,235. f. 77, 'Observations on the General Comparative Merits of ...Navigations or Railroads,' p. 7.   The advertisement of this change is given in *Liverpool Advertiser*, Sept. 29, 1810.   Each company gave public notice of the change of rates over the signature of its own agent and the two advertisements are exactly alike.

[2] Sandars, op. cit., pp. 11–13.

[3] Ibid., p. 13.   Their rates in that year (1824) were: on heavy goods, such as corn, 12s. 6d. per ton, and on light goods, like cotton, 15s. per ton.

[4] See letter of Captain Bradshaw, who had charge of the Bridgewater Canal interests, in reply to a Memorial from the Corn Merchants of Liverpool asking for a reduction of freight.   He refused to make any move toward such lowering of rates. Bradshaw's letter is given in full by Sandars, p. 12.

[5] If the merchant complained of delay, he was told to do better if he could.   If he objected to the rates, he was warned that if he did not pay promptly his goods might not be carried at all.

[6] See 'Prospectus of the Liverpool and Manchester Railway,' as given in Booth, op. cit., p. 13.   In regard to this lack of carrying facilities and the delays, see also 'Collection of Prospectuses, Maps, etc., of Railways and Canals,' p. 14, and Brit. Mus. 08,235. f. 77, 'Observations on the General Comparative Merits of...Navigations or Railroads,' p. 7.

Even the opponents of the railway did not deny that this had occasionally been the case[1]. But the causes of delay were mostly of such a nature that the navigation companies were powerless to effect much change in them, for the forces of nature were beyond their control; and, therefore, in the complaints as to the inadequate service, more emphasis was laid upon the extortionate charges which were voluntarily imposed than upon the impediments which could not be avoided[2].

The results of this monopolistic policy pursued by the two navigations were highly satisfactory to them, but not to the public generally. For nearly half a century, the thirty-nine original proprietors of the Mersey and Irwell Navigation had been paid every other year the total amount of their investment[3]; and shares in that navigation company, the original cost of which was £70 each, had been sold in 1824 for £1250 each[4]. In the case of the Bridgewater Canal, the results were similar; and one who knew the financial position of that concern as fully as anyone could know it without being a trustee had good reason to believe that, since about 1800, the net income of this canal had averaged nearly £100,000 a year[5]. Remembering that the cost of this canal was £200,000 to £220,000, we see that, at the above rate, the whole cost of the undertaking would be repaid every two years or a little more. These statements are in accord with that made by another in 1826, who, speaking upon this point, said that because the canals had recently raised their rates they were then making more than 100 per cent. profit[6].

---

[1] Booth, op. cit., p. 16.

[2] It was "the enormous charge for the freight of goods" between Liverpool and Manchester that had "become quite insufferable" (Sandars, p. 3). Of course, some delays were directly traceable to the navigation companies, for it was a well-known fact, especially among corn and timber dealers, that great difficulty had been found in getting vessels or barges to convey these things to Manchester. Timber had frequently been detained in Liverpool a month for want of barges to carry it inland, and corn and other commodities had been delayed eight or ten days for lack of a vessel for their conveyance. This was a serious evil, for men would not go to a market from which they had such difficulty in getting their goods (Sandars, p. 16). The delays on the canals also made it possible for much pilfering of goods to be carried on. On this monopolistic policy, see a.so Brit. Mus. 08,235. f. 77, 'Observations on the General Comparative Merits of...Navigations or Railroads,' pp. 6–7; Aris's *Birmingham Gazette*, Dec. 13, 1824, p. 1, letter from F. Finch; P., *Letter to a Friend on the Comparative Merits of Canals and Railways*, pp. 29–30, speaking of the overgrown monopolies between Liverpool and Manchester.

[3] Sandars, op. cit., p. 21.

[4] Ibid., p. 21; also 'Prospectus of the Liverpool and Manchester Railway,' as given in Booth, p. 12.

[5] Ibid., p. 21. The fact that no one ventured to deny what Sandars said seems to point pretty conclusively to the accuracy of it. See also Parkes, op. cit., p. 24.

[6] *Hansard's Parl. Debates*, xv (1826), p. 93.

From these facts, it is apparent that the monopolizing policy of the two navigation companies was, for the time being, highly advantageous to them, although their benefit was secured by means that were derogatory to the best interests of industry[1].

In the light of what we have here presented it is clear that a new line of conveyance was essential if adequate provision were to be made for the growing needs of that district. The proprietors of the navigations said that, by allowing time for increasing the number of their boats and the facilities for loading and unloading, they would be able to take care of the increase of trade; but this would not put an end to the delays or reduce the expenses of transport, against which there were such persistent complaints. Another canal was out of the question for the existing navigation had possession of all the water that was available; and it never seemed to occur to them that by lowering their rates they might perpetuate their business and also their profits[2]. Canal navigation had failed to meet the conditions of an expanding trade and a developing industry; and therefore the only thing to do was to obtain parliamentary authority for laying down a railway, which would combine the requisites of speed[3], economy[4], and safety[5].

In 1822 a project was formed for constructing a railway between these two cities, on which carriages driven by steam should carry both merchandise and passengers at the rate of ten miles per hour. The expenses of a survey were contributed, and in the autumn of that year

[1] It was not the desire of the Duke of Bridgewater that his canal should thus be used for the personal enrichment of the one individual who controlled it. On the contrary, his will (which gave R. H. Bradshaw the position of "superintendent" of the Duke's possessions) showed that he intended the canal for the public good, for it says that the almost unlimited authority conferred on the superintendent was "to the intent that the public may reap from the same those advantages which I hope and trust the plan adopted in this my will is calculated to produce for their benefit" (Brit. Mus. 10,815. c. 35, 'Will of the Duke of Bridgewater,' p. 50).

[2] Brit. Mus. 08,235. f. 77, 'Observations on the General Comparative Merits of... Navigations or Railroads,' p. 7.

[3] In the passage from Liverpool to Manchester, goods going by canal took, on the average, about 30 hours (Sandars, p. 17). But by a railway it would not take more than one-sixth of that time. The railway would not be hindered by drought or frost, or any of the other impediments and dangers of water carriage. The railway would have extra carriages ready to meet any emergencies of business and thus prevent delay from that source.

[4] The rates by the railway would be greatly reduced and competition would prevent their becoming exorbitant. Thus coal and other necessaries would be procured cheaper than at present. Goods shipped by railway from either terminus for the other would not have to break bulk and be transhipped at the tideway.

[5] When goods were sent by railway there would be no losses in the Mersey tideway due to storms. There would be no breaking open of packages and pillaging of contents if the goods were in the railway car (Sandars, p. 17).

William James completed the survey and suggested a line of road. Public notices were given of the intention to apply to Parliament for authority to execute this line; but, probably owing to the fear of opposition from the whole body of inland navigation proprietors throughout the kingdom and for other causes, the measure was not followed up[1]. The enterprise, however, was not allowed to sleep; men were sent to investigate the Stockton and Darlington and other coal-roads in the north, especially near Newcastle and Sunderland, where both locomotive and stationary engines were in use for the conveyance of coal, and after their return it was decided to form a company for building a double railway between Liverpool and Manchester. The promoters were men of the highest standing and influence in these two cities. On Oct. 29, 1824, the company issued its prospectus, which detailed the reasons why the railway was desired and the benefits to be secured by it[2]. In the early part of 1825 application was made to Parliament for an Act to authorize the construction of this road.

Strong opposition was aroused against this Bill. The proprietors of the three navigations which connected Liverpool and Manchester forgot their former jealousy and disagreements and made common cause against the proposed railway. Their chief argument was that of vested interests: that their canals had been brought into existence under the authority of a former Act of Parliament and that now Parliament could not consistently pass a Bill which would destroy that property[3]. But this pretext was taken away when it was shown that there was a great difference between superseding an old machine that had paid its owners thirty times over, and superseding one that had not paid its owners the amount of its first cost. As these navigations belonged to the first class, there could be nothing against their being displaced by more advanced means of carriage. In league with the navigation companies were the large landholders, like the Earls of Derby and Sefton, a part of whose estates would be crossed by the railroad. They opposed the railway because they believed the sanctity of their domains would be invaded and the privacy of their residences destroyed by thus bringing into their neighbourhood a public highway, with its varied traffic of coal, merchandise and passengers[4]. The canal companies that were interested in the traffic of this region issued circulars calling upon "every

[1] *Manchester Gazette*, Oct. 16, 1824, p. 3.

[2] The full text of the prospectus is given in Booth, *Liverpool and Manchester Railway*, p. 9 et seq. This first railway prospectus is an interesting document.

[3] See report of the committee of the House of Commons on the Liverpool and Manchester Railway Bill, 1825, which gives full details as to the character of the opposition. See also Sandars, op. cit., p. 21.

[4] *Hansard's Parl. Debates*, N.S., XII, p. 848; Booth, op. cit., p. 15.

canal and navigation company in the kingdom" to oppose to the utmost, and by a united effort, the establishment of railways wherever contemplated[1]; and these must have had great influence when the cause was aided by Bradshaw, the superintendent of the Bridgewater Canal, whose authority was almost as good as law[2]. So intense was this opposition of the canal interests that, in their opinion, it was impossible for a man to hold any of these railway shares and still be loyal to the canal company of whose shares he held any considerable amount[3]. The hostility of the estate-owners was also vigorous: they had used every means to prevent the making of a survey for the proposed railroad. They had blockaded their grounds on every side and had men employed to watch them. Bradshaw even fired guns through his grounds in the course of the night to prevent the surveyor coming on in the dark[4]. Both the navigation companies and the large landowners employed parliamentary representatives to work in their behalf, so as to put down such an intolerable innovation in established modes and vested rights[5]. The railway company likewise sent down a committee

---

[1] See postscript attached to the prospectus that was distributed to members of Parliament and others, as given verbatim in Baines, *History of Liverpool*, p. 603.

[2] According to Sandars (pp. 31, 32, 34) no bill could be brought forward in Parliament for a canal in any part of the kingdom but Bradshaw interfered to give full directions. He made the trade of the country tributary to him in all directions. Sandars, pp. 31–33, gives examples of this: "Every man, every Corporate Body, seems spell-bound the moment Mr Bradshaw interposes his authority."

[3] At a meeting of delegates from different parts of the kingdom, to consult as to canal interests in general, one of the delegates was turned out because he had five shares in the railway. The fact that he held canal property of the value of £40,000 was no protection to him. Sandars, p. 34.

[4] See letter of George Stephenson to Joseph Pease, dated Oct. 19, 1824, giving details of this opposition. This letter is reproduced in Jeans, *Jubilee Memorial of the Railway System*, pp. 55–56, to show that the railway promoters had "sad work with Lord Derby, Lord Sefton, and Bradshaw."

[5] Of course, those members of Parliament who acted in this way were not acting in any judicial frame of mind, but as those who were biased in favour of their friends. Mr Creevey, who represented Lords Sefton and Derby, was a member of the parliamentary committee to deal with the Liverpool and Manchester Railway Bill; and the attitude of some is well exemplified in Mr Creevey's letters (Maxwell, *Creevey Papers*, ii, pp. 87–88). Under date of Mar. 16, 1825, he writes: "...Sefton and I have come to the conclusion that our Ferguson is insane. He quite foamed at the mouth with rage in our Railway Committee in support of this infernal nuisance—the locomotive Monster, carrying 80 tons of goods, and navigated by a tail of smoke and sulphur, coming through every man's grounds between Manchester and Liverpool." On Mar. 25, 1825, he writes: "...I get daily more interested about this railroad—on its own grounds, to begin with, and the infernal, impudent, lying jobbing by its promoters." See also under dates of May 31 and June 1, of the same year.

to London to watch and aid the progress of the Bill through the House of Commons[1]. After a contest of about three months, during which the necessity of additional means of conveyance was emphasized and thoroughly acceded to, some errors were discovered in the survey that had been made and this created so unfavourable an impression on the committee that the Bill was withdrawn[2]. Before the next year an accurate survey had been made; the line of way was changed so as to be less objectionable to the Earls of Sefton and Derby; the Marquis of Stafford, representing the Bridgewater Canal, had been induced to subscribe for 1000 shares of stock in the railway; and a new prospectus was issued, explaining the causes of the former unsuccessful application, how these had been overcome, and the benefits that would accrue from the railway[3]. Early in 1826 the Bill was introduced a second time, and in that session it passed both Houses[4]. Various estimates are given as to the cost of obtaining the Act, varying from £40,000[5] to £70,000[6], but, of course, either of these estimates may be far from the actual amount. George Stephenson was then appointed resident engineer, and under his direction the work was pushed to completion as rapidly as possible.

Of the difficulties connected with the construction of the line we shall not speak[7]. The means adopted to overcome the immense bog called Chat Moss, which the railway crossed as if it were dry and firm land, when at any point a piece of metal would sink out of sight by its own weight, forms a chapter in engineering which is of great interest. But while we shall not discuss the physical and mechanical features connected with the formation of the road, there is one aspect of its development which we may profitably refer to, namely, the choice of motive power. The line was nearing completion at the end of the year 1828, but no agreement had been reached as to whether stationary or locomotive engines should be employed[8]. In order to settle this

---

[1] Booth, p. 14; Baines, *History of Liverpool*, p. 603.

[2] Booth, p. 18.

[3] Booth, pp. 25–31, gives this prospectus also in full.

[4] The new survey put the line so that it did not touch the Earl of Sefton's estate and crossed only a few detached fields of the Earl of Derby's estate. The opposition of the Bridgewater Navigation, the most powerful of the two direct routes, was disarmed by the Marquis of Stafford taking such a large interest in the railway (see second prospectus as given by Booth).

[5] *Birmingham Journal*, May 27, 1826, p. 3.

[6] Ibid., Feb. 5, 1831, p. 3, letter from "A Subscriber to the London and Birmingham Railway." Compare these estimates with that of Booth in the Appendix of his work.

[7] On this aspect of the work, see Smiles, *Lives of the Engineers*, George Stephenson.

[8] Chattaway, *Railways*, p. 2, tells us that even horse-power was considered.

important matter, two celebrated engineers, James Walker and John Urpeth Rastrick, were asked to investigate this question and report their results to the directors of the railway. They visited the important places where steam-engines were used, notably the Stockton and Darlington and other coal roads in the north; and afterward each made out his own report showing his conclusions, in which there was almost entire harmony between the two engineers. In their reports, they were agreed that, having regard for the present and prospective interests of the company, locomotive engines would be found the more satisfactory. These should travel at the rate of ten to fifteen miles per hour[1]. In addition, they would employ two stationary engines upon the Rainhill and Sutton inclined planes to draw up the locomotive engines along with the carriages and goods. Their view, that on the line as a whole locomotive engines should be used, found acceptance with the directors; but the locomotive engines that had been used for some years in connexion with a few of the large collieries for the conveyance of coal were utterly unsuited to the requirements of passenger traffic. Knowing the vital importance of the character of the motive power, the directors offered a premium of £500 for the best locomotive adapted to the purposes of their line, two of the conditions being that it should be capable of drawing at least three times its own weight, at a speed of not less than ten miles per hour, and that it should consume its own smoke[2]. Several competitors entered this contest, and in October, 1829, the various designers of the engines brought their locomotives for trial on the railway. On the first day, the engine made by Braithwaite and Erickson, of London, exceeded all others in speed; but when the competition had continued for some days, in order to have a good test of all the engines, the prize was finally awarded to George Stephenson's engine, the "Rocket[3]." After the expiration of almost another year,

---

[1] Walker and Rastrick, *Liverpool and Manchester Railway, Report to the Directors on the Comparative Merits of Loco-motive and Fixed Engines, as a Moving Power.* Nicholas Wood thought that the locomotive engine ought not to travel more than eight miles an hour; but these two engineers believed it could go at the rate of ten miles per hour with perfect safety, provided it did not exceed eight tons gross weight, exclusive of the tender (ibid., pp. 49, 76).

[2] Chattaway, *Railways*, p. 2. For the conditions of this competition, see Jeaffreson, *Life of Robert Stephenson*, I, pp. 124–5. The Liverpool and Manchester Railway Act, 7 Geo. IV, c. 49, required the engine to "effectually consume its own smoke."

[3] Full details of the trial of the engines are given in *The Times*, Oct. 8, 1829, p. 3; Oct. 9, 1829, p. 3; Oct. 12, 1829, p. 3; Oct. 16, 1829, p. 3; Oct. 24, 1829, p. 4; Oct. 31, 1829, p. 2. See also Smiles, *Life of George Stephenson*. These experiments, and others later, showed that the locomotive engine could easily attain a speed of 24 to 30 miles an hour. An account of this trial of the engines is given also in

during which the construction of the roadway and its accessories proceeded toward completion, the line was formally opened with great. éclat and enthusiasm, on September 15, 1830[1].

With the Liverpool and Manchester line, the railway era really began. It was the first railway that was constructed for the express purpose of carrying passengers as well as freight; and no other power was ever used on it but that of locomotive engines. Up to this time, all others, except the Surrey Iron Railway, had contemplated the carriage of one commodity (usually coal, iron, or stone) and were operated as adjuncts to a colliery, quarry, or the like; while the Surrey Iron Railway employed only horse-power in the work of conveyance. The Liverpool and Manchester, on the contrary, was constructed for the public welfare, rather than for private profit, as we can readily judge by the fact that no person could subscribe for more than ten shares, and the profit on these would not aggregate very much for any individual[2]. Indeed, under the Act of Parliament by which it was authorized[3], the profits or dividends were limited to ten per cent.[4]; and the undertakers were so anxious to encourage industry and commerce that they declared they would be satisfied with even five per cent.[5] It is very evident, then, that there was a wide difference between the Liverpool and Manchester Railway and any of those which had preceded it.

The immediate success of the Liverpool and Manchester Railway was the occasion of universal admiration and satisfaction. The rate

Liverpool Times, Oct. 13, 1829, p. 328, and Oct. 20, 1829, p. 333. The facts connected with this contest are also given in Jeaffreson, *Life of Robert Stephenson*, I, chap. ix.

[1] A full account of the opening is given in *Liverpool Times*, Sept. 21, 1830, p. 298; *Manchester Guardian*, Sept. 18, 1830, p. 3; and Smiles' *Life of George Stephenson*. See also the history of the Liverpool and Manchester Railway written by Walker, pp. 42–48, for a description of the "Grand Opening of the Railway." The rejoicing of the day was saddened by the death of Mr Huskisson, M.P., which occurred because of an accident on the line. Booth, who was treasurer of the company, gives us an account of the construction of the line and the expenditures connected therewith in his second chapter. His third chapter is an account of the railway itself. His fourth chapter shows the mechanical principles applicable to railways, and how the directors finally decided to adopt the locomotive engine. In the Appendix he gives the details of the cost of the railway, which, including stations, warehouses, etc., amounted to £820,000.

[2] *Hansard's Parliamentary Debates*, N.S., XII (1825), p. 848; *The Times*, Mar. 3, 1825, p. 2, statement of Mr Huskisson in the debate in the House of Commons on this Bill.

[3] Act 7 Geo. IV, c. 49.

[4] See also *The Times*, Mar. 3, 1825, p. 2, and April 7, 1826, p. 2, statements of Mr Huskisson.

[5] *The Times*, Mar. 3, 1825, p. 2.

of speed on passenger trains was twice that of the fastest stage coaches and the cost of travelling was reduced about one-half[1], while the amount of travelling increased fourfold[2]. Under these circumstances many of the old stage coaches ran almost empty for a short time and several were immediately withdrawn. Soon all the stage coaches disappeared from regular service along this route and the railway absorbed all the passenger traffic[3]. The freight rates also were reduced by the railway by about one-third; and in order to enable the carriers on the navigations to meet this reduction the tolls on the Bridgewater Canal and on the Mersey and Irwell were reduced by about thirty per cent.[4] The effect of the railway, therefore, was beneficial to the public by reducing overgrown monopolies within reasonable bounds, and it also stimulated these opulent canal companies to think of something else than their own pecuniary interests[5]. The value of land along the line of railway invariably increased, which was advantageous both to landowners and tenants, for the tenants had wider and better markets opened up to receive their produce and because of this enhancement of the value of the land the landowners could receive higher rents. This was observable also in cases where the railway company wanted to buy land in addition to that which they already held; their second purchase was invariably

---

[1] *Proceedings of the Great Western Railway Company*, p. 6; *Annual Register*, 1832, p. 445; 'Great Western Railway. Evidence on the London and Birmingham Railway Bill,' testimony of Henry Booth, p. 8. According to Mr Booth's statement, the fare between Liverpool and Manchester, by stage coach, had varied a good deal, but was about 10s. inside and 6s. outside. On the railway, first class fare was 5s. and second class 3s. 6d. The statement of a writer in the *Manchester Guardian*, Sept. 25, 1830, p. 2, makes the railway fares a little higher than those given by Booth, placing first class at 7s. and second class at 4s. On the reduction of rates see also 'Collection of Prospectuses, etc.,' p. 65, which is in close accord with Booth's assertion.

[2] Before the railway, there were about twenty coaches per day between Liverpool and Manchester. Supposing these to be full every trip, carrying eighteen passengers each and pursuing their daily rounds for three hundred days in the year, there would be 108,000 people carried between these places in the course of the year. But in the twelve months after the opening of the railway about 460,000 persons were carried between these two termini (*The Times*, Oct. 19, 1831, p. 4).

[3] *Manchester Guardian*, Sept. 25, 1830, p. 2, on "Railway Coaches."

[4] P., *Letter to a Friend, containing Observations on the Comparative Merits of Canals and Railways*, p. 12. The freight rate between Liverpool and Manchester was reduced from 15s. to 10s. The tolls on the Bridgewater Canal were reduced from 3s. 8d. to 2s. 8d., and on the Mersey and Irwell from 3s. 4d. to 2s. 4d. See also 'Collection of Prospectuses, Maps, etc., of Railways and Canals,' pp. 13, 65.

[5] P., op. cit., pp. 29–30; *The Times*, April 7, 1826, p. 2. For canals which were paying one hundred per cent. every year or every two years there was need of some new factor to reduce their charges.

made at a higher price than that paid for the first[1]. Not only did the
public benefit from the railway, but the company itself also realized
that the enterprise was a corporate success. In the first half of the
year 1831 the net receipts were such that, after large expenditures for
warehouses, carriages, etc., the company was able to declare a half-
yearly dividend of £4. 10s. per share[2]; and the annual rate of dividend
continued to range between eight and ten per cent. during the years
following[3]. The value of the shares in the market may also be

[1] See evidence before the committee on the London and Birmingham Railway
Bill, as summarised in *Birmingham Journal*, May 19, 1832, p. 3, e.g., evidence of
Messrs Earle, Lee, Unsworth, Pease. See also *Proceedings of the Great Western
Railway Company*, p. 6. The prospectus of the Liverpool and Birmingham Railway
Company and the Birmingham and Liverpool Railway Company, whose interests
were practically identical, showed that land which, from its vicinity to the Liverpool
and Manchester Railway had been expected to deteriorate in value, and the owners
of which had consequently claimed compensation, had, on the contrary, become
more valuable than before. See especially the testimony of Mr Lee before the
committee on the London and Birmingham Railway Bill, to the effect that some
property along the line of the Liverpool and Manchester Railway had been sold
for building purposes at three to five times the sum it would have brought before the
establishment of the railway. The almost universal testimony of those who gave
evidence before the committee on the London and Birmingham Railway Bill in
1832 was that lands along the route of the L. & M. Ry had increased in value. Even
land formerly waste had been brought into cultivation and yielded a good rent.
See also *Annual Register*, 1832, p. 445; 'Collection of Prospectuses, Maps, etc. of
Railways and Canals,' p. 65; 'Great Western Railway. Evidence on the London
and Birmingham Railway Bill,' pp. 34–37; and the notable case of increased land
values given in *Railway Times*, IV, p. 215.

[2] The receipts from Jan. 1 to June 30, 1831, as given by the *Annual Register*,
1831, p. 169, were:

	£	s.	d.
From conveyance of passengers	43,600	7	5
,,　　　　,,　　merchandise	21,875	0	0
,,　　　　,,　　coal	218	16	2
Gross receipts	65,693	13	7
Expenses for repairs, salaries, etc.	35,379	0	0
Net receipts	30,314	13	7

The net receipts divided among 7012 shares allowed a dividend of £4. 10s. per
share for the half year. (It will be noted that there is a slight error here in summing
up the gross receipts.) See also Brit. Mus. 8235. ee. 12 (1), p. 2.

[3] In 'Collection of Prospectuses, Maps, etc., of Railways and Canals,' p. 173,
there is given an 'Extract of the Report of the Liverpool and Manchester Railway,'
for the half year ending Dec. 31, 1833, with a comparison of the results for this six
months with the results for previous half years since the railway began operation.
The half-yearly dividend thus far had ranged from four to four and one-half per cent.
This financial statement was also attached to the prospectus of the Great Western
Railway, 1834 (ibid. p. 176). In Brit. Mus. 8235. ee. 12 (1), 'Reasons in favour of
a Direct Line of Railroad from London to Manchester,' p. 2, we have a comparison

taken as an index of the measure of the success of the railway. Even before the experiments of October, 1829, to find the best engine for use on the line, the railway shares had been selling at a premium; but after that time their value rose very rapidly, until, within a month after the success of the locomotive engine had been demonstrated the shares were selling at £175 when the original value was only £100[1]. So great was the demand for these shares, and so highly were they valued, that it was difficult to procure them on any terms. By 1832 the value of the shares had risen 100 per cent.[2], and by 1836 almost 200 per cent.[3], above their original value. All the important railways that were taken up immediately after 1830 put forth the success of the Liverpool and Manchester Railway as an attestation and guarantee of the success of their own enterprises[4].

But we must follow the finances of this company a little farther if we would obtain a correct idea as to its operations. Let it be said, first of all, that by its Act the company was limited in the payment of dividends to a maximum of ten per cent. a year; and it was the only railway company that was restricted in this way. As we have seen, the company early paid the full amount of the dividend that was allowed and continued to pay this for many years. When the company sought authority from Parliament to construct the road it was declared that £510,000 would be ample for all purposes[5]; and according to their Act of incorporation the capital was fixed at that amount. But it would appear that this amount proved insufficient to complete the road and its equipment, and by the Acts of 1829 and 1830 the company was allowed to increase its capital by the issuance of shares to the amount of £127,500 and £159,375 respectively, all of which was said to have

of the above receipts of 1831 with those of the year 1836, showing that the dividend in the latter year was ten per cent. In 1834 it was paying nine per cent. (*Proceedings of the Great Western Railway Company*, p. 52), and in 1842 it was paying ten per cent. (*Railways: Their Uses and Management*, p. 7).

[1] *The Liverpool Times*, Nov. 24, 1829, p. 376, informs us that before these experiments the shares were selling for £118 each, but at this date they were now selling for £75 premium, and could scarcely be had even at that price.

[2] *Proceedings of the Great Western Railway Company*, p. 52; also evidence on the London and Birmingham Railway Bill, 1832, testimony of Henry Booth. The shares in 1831 were selling for £196 (*Remarks upon Pamphlet by Investigator on the Proposed Birmingham and London Railway*, p. 4; 'Collection of Prospectuses, etc.,' p. 65).

[3] *Gentleman's Magazine*, 1836, VI, p. 421. The Liverpool and Manchester shares, the par value of which was £100, were selling in 1836 for £280.

[4] See prospectuses of the Birmingham and Liverpool, London and Birmingham, and Great Western railways.

[5] See the company's prospectus, as given in Booth's history of the railway.

been expended and yet the works were not completed. Under subsequent Acts, they were allowed to raise by sale of shares or to obtain on loan a further sum amounting to £427,500, thus bringing the total capital up to £1,224,375, of which the share capital was £808,025 and the loan capital £416,350[1]. In 1837 the company presented a bill to Parliament, stating that although the above amount had been spent "on or about the undertaking," its works had not yet been completed; and accordingly it was desired to obtain authority to borrow an additional sum of £400,000, which, if authorized, would raise the capital to £1,624,375. This extra amount was to be asked as a loan from the Government, that is, from the Exchequer Loan Commissioners, and in case the Government advanced the money it was to have the prior claim upon the revenues of the company[2]. In the six years up to 1837, the company had paid in dividends £442,504. 7s. 6d.; but during the same time the amount obtained on loan and by the sale of shares was much in excess of this amount, and, therefore, the company would seem to be obtaining money from others to pay dividends, while all the time becoming more embarrassed[3]. But when seen in another light, these several accessions to capital presented facts which led to an entirely different conclusion. Their expenditure upon additional works was said to have brought additional revenue; so that after paying the interest on these increasing amounts obtained from creditors the company was still able to pay the maximum dividend of ten per cent.[4] Instead, therefore, of the company becoming more hopelessly embarrassed financially, it was ostensibly getting upon a more secure foundation. We prefer to think that this was the explanation of the above-mentioned great increase of capital. But there is another way in which it can be, and was, accounted for, namely, as a device for overcoming the restriction of their profits to ten per cent. It was held by some that if Parliament had rigidly enforced this provision of the Act and steadfastly refused to allow the distribution of additional profits under any other guise, the company would have been compelled time and again to reduce the fares and charges to the public; but since this provision was not enforced the railway company, under the semblance of increasing the "public accommodation," created a pretext for the issuance of new shares, and thus extra profits were divided out in

---

[1] These facts appear in the Bill presented to Parliament in 1837, asking for further authority (*The Times*, May 9, 1837, p. 6, letter from "T.G."), and are confirmed by W. S. Moorsom, C.E., in ibid., May 23, 1837, p. 6.

[2] *The Times*, May 9, 1837, p. 6, letter of "T. G."

[3] Ibid., May 9, 1837, p. 6. This was the contention of "T. G."

[4] Ibid., May 23, 1837, p. 6, letter of W. S. Moorsom giving quotations from the company's semi-annual financial statements.

the form of new stock[1]. If this were the explanation of the great increase of capital from time to time, the road must have been sufficiently profitable to pay at least forty to fifty per cent. It seems to be more consonant with the facts to accept the first solution of this problem; for if the company's business were so flourishing that surplus profits could be divided out in this way, there would have been no need of applying to Parliament for a Government loan. And yet, in the face of these facts, several persons, by their publications, attempted to prove to the public that this railway was nothing but an unprofitable speculation[2].

We have now brought our subject down to the time of the initiation of the modern railway; but in order that we may consider in detail the effect of this new means of transportation we must see it in a more advanced stage of development, for it is impossible to form any correct estimate of its value and influence from a single example apart from a system. It will, therefore, be necessary for us to outline the history of railways to about the middle of the century in order to see the forces which were at work throughout this early period when the railway was attaining a position of importance as a public carrier.

During the third decade, when the railway had not yet demonstrated its great superiority, but was in the tentative evolutionary stage, and when the locomotive engine was still in the experimental period of its development, there was uncertainty and instability of the public mind concerning the utility of this newcomer in the field of transportation. Some expected that the railway would only add another means of conveyance to those already existing, in the same way as the introduction of canals had done sixty years before, but that every facility given to the carriage of materials, while adding to the general carrying trade, would cause no injury to canal property[3]. It seems, however,

---

[1] *The Times*, Oct. 1, 1846, p. 5, letter from "Cato."

[2] Gordon, *Treatise upon Elemental Locomotion*, 2nd ed., p. 225 et seq.; Gordon, *The Fitness of Turnpike Roads and Highways*, p. 28; Cort, *Railroad Impositions Detected, or Facts and Arguments to prove that the Liverpool and Manchester Railway has not paid One per cent. Nett Profit*, etc. These based their opinion upon the probability that nothing had been set aside for depreciation. See also *Remarks upon Pamphlet by Investigator on the Proposed Birmingham and London Railway*, p. 4. Grahame, *Treatise on Internal Intercourse and Communication* (1834), p. 159, in summing up his statistics and arguments regarding the Liverpool and Manchester Railway, said: "No one, who fairly considers these results, but must acknowledge that the whole is a failure, at least, as presently conducted. The expenses are so enormous, as completely to absorb every advantage of speed, and each year these expenses increase." He would have the road open to all, so that, upon payment of the tolls, anyone could use the road as freely as they did the turnpikes.

[3] *Manchester Gazette*, Jan. 15, 1825, p. 3, editorial comment.

that there were few who regarded railways in this way. Many people foresaw in them very decided advantages, and, while fairly assured in their own minds that a new era was dawning in the transport service, they had not yet received complete proof that its successful establishment was at hand. But whatever were the prospects of the railways, whether favourable or unfavourable, there was a large amount of capital in the country seeking investment and this superabundance of capital introduced the rage for speculation[1], in which the railways shared. The years 1825 and 1826 seem to have been the climax of this speculative fever. All kinds of projects were promoted by men who were eager to take advantage of the circumstances of the time to reap large returns from credulous and unsophisticated prospective investors. Men were induced to believe that they had only to embark in one of these schemes to ensure themselves a life of affluence and ease; labour and care were to be at an end and the golden harvest would soon appear. In February, 1825, there were at least five railway companies and thirty dock companies, loan companies, insurance companies, and other kinds of undertaking, that were being floated[2]. Railways were being planned to connect the most important mercantile and manufacturing towns in the kingdom, and the success of the Stockton and Darlington line gave added impetus to this movement[3], notwithstanding the secret opposition which was very active on behalf of interested bodies for their own private good. This fever was instituted mostly for purely speculative purposes, in order that projectors and their attorneys and other assistants might profit to a large extent through trafficking in shares[4]. The latter were brought into the market at a premium and pushed to as high a price as possible; then they were unloaded upon unsuspecting and unfortunate individuals who were duped and left stranded "after the waters of delusion had ebbed away[5]." Of the great number of these schemes that were brought forward, but few ever came to completion; of the others, no vestige remained except in the disaster which

[1] Brit. Mus. 08,235. f. 77, 'Observations on the Comparative Merits of Inland Navigations and Railroads,' pp. 8, 10.

[2] *County Chronicle and Weekly Advertiser*, Feb. 1, 1825, p. 2, gives a list of thirty-five such companies then afloat.

[3] *The London Magazine*, I, N.S. (1825), p. 33, on "Railways;" Grinling, *The History of the Great Northern Railway*, p. 1. Among these railways may be mentioned the London and Birmingham, the Great Northern from London to Cambridge, and the Liverpool and Birmingham. Aris's *Birmingham Gazette* for the year 1825 (note, for example, the issue of Jan. 31, 1825) shows a great many projects for railroads that were then occupying public attention.

[4] *The Times*, July 17, 1832, p. 3, statement of Lord Wharncliffe; Mudge, *Observations on Railways*, p. 35.

[5] Investigator, *Beware the Bubbles*, p. 10; Mudge, op. cit., p. 35.

overtook those who had been deceived by the wiles of the mercenary speculators. How much capital was lost from legitimate productive industry we have no means of ascertaining; but if we were to receive the statements of contemporaries[1], and then make much allowance for exaggeration, we should still be required to believe that this panic assumed proportions of considerable magnitude. Fortunately, however, only a few of these projects which were brought forward were authorized by Act of Parliament to proceed to execution, for most of them were ventures of such a nature that their success could not be definitely foretold[2]. But when the success of the Stockton and Darlington was assured the year 1826 saw the authorization of eighteen new railways, among them the Liverpool and Manchester.

Following the policy that had been pursued with great benefit to the country for three-quarters of a century, in allowing private enterprise to develop and manage inland communication, the Legislature considered each of the schemes brought forward according to its own merits; and for each one that met with approval a private Act was passed, which contained the entire statutory provisions applicable to the undertaking.

After the utility of the locomotive had been shown on the Stockton and Darlington line, and especially after the results of the trials of the locomotives on the Liverpool and Manchester, in the autumn of the year 1829, had been made known, interest was aroused anew in the prospects of railways. The vast range of possibility which opened up when it was seen that locomotive engines could travel at rates of speed from twenty-five to thirty miles an hour, seemed to fire the imagination of many. By this means, places then considerable distances apart would be brought very close to one another; the capitals of Scotland and Ireland would be within twenty-four hours' journey of London; facility in the communication of intelligence would enable the people in all corners of three kingdoms to keep in direct touch with the measures

[1] Investigator, *Beware the Bubbles*, p. 1, speaks of the "uncontrollable exercise of the spirit of speculation, which, in 1825 and 1826, brought about so fatal a crisis, involved so many in ruin," etc.; and again (p. 10) he refers to the "melancholy wrecks of men of important station." In Felix Farley's *Bristol Journal*, Oct. 5, 1833, p. 2, a letter from John Weedon speaks of the "rash and improvident speculations which led to the frightful commercial catastrophe of 1826." Mudge, *Observations on Railways*, p. 35, deplores allowing the "delusive and ruinous speculations" of 1825 to go on unchecked, and says that this "injury to the wealth and prosperity of the country" was felt for nearly ten years.

[2] On the details of this panic, see Francis, *History of the English Railway*. Jeaffreson, *Life of Robert Stephenson*, I, pp. 272 et seq., shows the difference between the railway crises of 1825 and 1836 and the railway mania of 1844–6.

that were before the Government for consideration, and public opinion would acquire a strength and concentration that it never possessed before. By the rapidity and cheapness of travel, workers in any part of the country could readily go to any other part, and the inevitable consequence would be that sooner or later there would be only one rate of wages throughout the United Kingdom. The ease and celerity with which markets could be reached would cause land to be brought into cultivation that had hitherto been required to lie waste because of the expense of transporting the produce to a suitable market. By means of steam, it was thought, the produce of land twenty or thirty miles from the market would be brought to the place of sale in as short a time, and at as small a cost, as the produce of land five or six miles distant had been by waggon; and, therefore, while the consuming public would profit by this increased supply, the landlords would also derive advantage because of the increased value and rentals of their lands and the farmers would receive greater net returns from the sale of their surplus[1]. With such vast national benefits as these and many others presented to an admiring world, it would have been strange, indeed, if there had not been an outburst of sentiment in favour of an expansion of railway construction; and the statement in 1829 that within ten or twenty years the whole country would be united by railways which would convey passengers and goods at twice the speed and one-third of the expense that then prevailed[2], was abundantly fulfilled in strict literalness of detail. With the accustomed tendency to exaggeration, people had been talking of travelling in the near future at fifty or sixty miles per hour[3]; but more conservative minds were counselling moderation. It was thought that the rate of thirty miles an hour of actual progress would be as great a velocity as would be compatible with safety. In any case, railway promoters should wait until the Liverpool and Manchester Railway was opened before making a survey, since a few months' operation of that line would teach many things of which people were then ignorant[4]. Even as late as 1831 there were some who, after the Liverpool and Manchester Railway had been in operation for a half year, still advised to go slowly in the further establishment of railways; it was said that the greater economy of this new means of carriage had

---

[1] See, for example, *Liverpool Times*, Nov. 17, 1829, p. 362, on "Future Changes." So great would be the advantages secured in the way of linking up closely the great towns of the kingdom, that the writer thought the country would become like Sir Thomas More's Utopia, where "*tota insula velut una familia est.*" He was particularly interested in the great benefits which would accrue to Liverpool in making it the most important port of England.   [2] Ibid.

[3] *Birmingham Journal*, Dec. 5, 1829, p. 4, on "Steam Coaches and Locomotive Engines."   [4] Ibid., Dec. 12, 1829, p. 4, on "Steam Travelling" (editorial).

not yet been proved, and that experience alone would show whether railways could carry cheaper than canals[1]. But, while urging the necessity of caution and the desirability of avoiding undue haste, it was felt that the locomotive on the rails was to be the coming means of transportation, and, therefore, consideration should be given to making the road as nearly level as possible and to preventing all chances of obstruction, so that the engines might develop the greatest power and the highest rate of speed[2].

The first lines that were actually constructed after the opening of the Liverpool and Manchester were in connexion with it, and chiefly in Lancashire. A branch was formed from Bolton to Leigh, and another from Leigh to Kenyon, where it formed a junction with the main line. Other branches were made from Newton, on the main line, to Wigan on the north and Warrington on the south, and still another from the main line, near St Helens, to Runcorn. It is not our purpose, however, to enter into details as to the filling in of the shorter lines; rather do we consider it as consonant with our object to describe the laying down of only the foundations of the railway system, and to the greater lines only shall we devote our attention.

Following the year 1824, an active campaign had been pursued to secure a railway between Liverpool and Birmingham. A large traffic was carried on along this route between the midland metropolis and the great port on the Irish Sea; yet the carrying facilities of these two places, like those of Liverpool and Manchester, were uncertain, expensive and totally inadequate to their necessities. We have already detailed the conditions which existed between the two latter places for the carriage of goods before the railway was constructed; but, according to the prospectus of the two companies which were desirous of having the railway between Liverpool and Birmingham, the conditions attending water carriage between these two cities were "infinitely worse" in regard to delays, charges and impediments[3]. It would almost seem as if this statement were exaggerated; and in order that the reader may see the relative conditions along the two routes we shall present a few facts concerning the conveyance of merchandise by the waterways between Liverpool and the Midlands[4].

As in the case between Liverpool and Manchester, so also between

---

[1] *Birmingham Journal*, Mar. 5, 1831, p. 3, on "Railways."

[2] Ibid., Dec. 12, 1829, p. 4.

[3] 'Collection of Prospectuses, Maps, etc., of Railways and Canals,' pp. 13–14, gives the prospectus in full. It is also given in *Birmingham Journal*, Mar. 11, 1826, p. 1, and in *Liverpool Times*, May 11, 1830, p. 149.

[4] For the facts pertaining to water carriage between Liverpool and Birmingham, we shall refer much to Parkes, *Claim of the Subscribers to the Birmingham and Liverpool*

Liverpool and Birmingham, the canal companies constituted probably the strongest opposition to the construction of a railway. Their monopolistic policy was not to be overthrown without a struggle to save it. For many years the canals along this route had made inordinate profits: one of the canals connecting with Birmingham paid an annual dividend of £100 on the original cost of £140 per share, so that the annual profits divided among the shareholders closely approximated the first cost of the canal[1]. It would appear that some canals profited still more largely from their trade, for we learn that one of them passing through this midland district paid an annual dividend of £140 upon an original share of £140, and the value of such shares had been increased from £140 to £3200; while another in the same district had paid an annual dividend of £160 upon the original shares of £200, and the shares had been enhanced in value until they had reached £4600 each[2]. Impediments of one kind and another caused delays to the transit of merchandise; for example, all goods that arrived at Runcorn had taken three or four hours, and occasionally as many days, in the Mersey estuary; then at Runcorn every ton of goods had to be transhipped and the loaded barges had to be elevated through the locks to a height of seventy-five to ninety feet before they could proceed on their way; after that they set out along the canal for Birmingham, which they reached four to six days after leaving Liverpool[3]. Similar delays and barriers had to be endured by the finished products of the Midlands on their way to the port, whence they could be shipped to the great markets. In addition, the cost of conveyance along the canals, together with the great amount of compensation tolls imposed when

Railway. His statements were not contradicted, and may therefore be taken as accurate.

[1] Parkes, *Claim of the Subscribers*, pp. 19–20; West, *History of Warwickshire* (1830), p. 100.

[2] See Prospectus of the Birmingham and Liverpool Railroad Company. This statement was not controverted by any other evidence, not even before the parliamentary committee to which the Bill was referred, and it may therefore be considered as true. Had it been possible to contradict it, the denial would certainly have been made, since every effort was being used at this time to discredit the railways. See confirmation in Parkes, op. cit., p. 24; also Aris's *Birmingham Gazette*, Dec. 13, 1824, p. 1, letter from F. Finch, in which he speaks of the "inordinate profits" which these canals had long enough enjoyed. Cumming, *Rail and Tram Roads* (1824), p. 47, evidently quotes from the Prospectus of the Birmingham and Liverpool Railroad Company. See also *Birmingham Journal*, Sept. 2, 1826, p. 1, Prospectus of the Proposed Railway from Birmingham to Wolverhampton.

[3] Prospectus of the Liverpool and Birmingham Railway, as given in the *Liverpool Times*, May 11, 1830, p. 149. The delays, charges and impediments of the navigation were felt as early as 1771 by Brindley, who projected an aqueduct bridge over the Mersey as a remedy.

a barge passed from one canal into another, were serious obstacles to the development of traffic[1]. But, despite these restrictions upon the system of transit in that part of the country, there had been a great increase in the amount of, and revenue from, the tonnage which centred in the Midlands[2]; so great, in fact, that the existing facilities could not accommodate the traffic, and, therefore, there was a persistent demand for new means of conveyance. To rely upon water carriage for auxiliary facilities would be to invite disaster, and this for several reasons. In the first place, no more water was available for an extra canal throughout that section[3]; and even had there been abundance of water for an additional canal, there were circumstances that were decidedly opposed to the formation of such a waterway. The conveyance by canal, under the best conditions, was altogether too slow for the carriage of meat, butter and other agricultural produce, since these might be spoiled before they reached the market for which they were intended[4]; and the transport of manufactured commodities from the interior to the coast was frequently so uncertain that shippers sometimes suffered considerable loss through their inability to ship goods by a pre-arranged vessel. During the drought of summer and early autumn, the boats often had to go with a light load and wait their turns in passing the locks, so as to economize in the use of water[5]. Moreover, some of the castings and apparatus, then sent at great expense by land carriage, could not be sent by canal, because their size would not permit them to pass through canal locks, and, occasionally, because their weight exceeded the tonnage of a single barge[6]. The stoppages of traffic on the canals along this course were frequent and sometimes prolonged, for floods damaged the navigation works, repairs consumed much time, and frost was sometimes a still more serious barrier. All these suspensions deranged the accustomed production, distribution and consumption of products, and consequently the price, so that both producer and consumer suffered thereby[7]. Considerable delays occurred also from the lack of a sufficient number of boats to convey the accumulation

---

[1] Parkes, op. cit., pp. 42–43.

[2] Ibid., p. 44.   [3] Ibid., p. 45.

[4] Aris's *Birmingham Gazette*, Feb. 2, 1829, p. 2, letter of "A Looker-On," says that the average speed of a boat passing along a canal, with a full load, and without the interruption of locks, was two and one-half miles per hour. In Mr Lee's letter (ibid., Feb. 9, 1829, p. 1), replying to the foregoing, while he contradicts some things mentioned by "A Looker-On," he does not deny this statement as to the rate of speed. For further confirmation of this, see also ibid., Dec. 17, 1827, p. 3, letter from Mr Lee containing some facts as to canal traffic rates.

[5] Parkes, op. cit., p. 47.

[6] Ibid., p. 45.   [7] Ibid., pp. 47–52.

of goods[1]; and on various occasions goods had been in transit from four to six weeks[2]. Because of these circumstances, namely, the vast increase of trade in the districts between Birmingham and Liverpool, the increasing importance of cheap and rapid transport, and the glaring fact that, although the profits of water carriage had increased beyond those of any other branch of enterprise, no reduction had been made in the charges of conveyance, the commercial and industrial classes proposed to construct a railway that would furnish adequate facilities to meet the enlarged needs of trade.

The first efforts toward securing such a line were made in the year 1824, when two companies were formed to undertake this work: one, the Liverpool and Birmingham Railway Company to construct the portion of the line from Liverpool, and the other the Birmingham and Liverpool Railway Company to construct the portion of the line from Birmingham. Their interests were identical, and they worked together to secure separate acts of incorporation and to frame regulations under which they could collaborate for their mutual good and the public advantage. Each company issued its prospectus in 1824, showing the existing conditions and the changes which would be effected should they be successful in securing parliamentary authorization to construct their line; and in this announcement they promised to carry "by day and night, at all times of the year, in periods of frost or of drought, at the rate of at least eight miles an hour," and at an expense "less by one-third, probably by one-half," than the existing rate by canal[3]. Immediately the opposition was aroused, led by those who were interested in maintaining the monstrous monopoly of the navigation interests; and owing to the hostility of the combined opponents, and to some inexplicable causes[4], the railway companies were powerless to secure the passage of their Bill. The advantages to be obtained from the proposed railway, however, were too great to allow the project to permanently fail; and a quiet agitation was continued in its favour. The companies endeavoured to placate opposition and to solicit the concurrence of those whom they had been able to win over from their attitude of dissent[5]. Early in the year 1826, the Birmingham and Liverpool Railway Company issued an address and appeal to the public, explaining their motives and the principal grounds upon which

[1] Parkes, op. cit., p. 53.   [2] Ibid., p. 54.
[3] Parkes, op. cit., pp. 57–60, gives the prospectus.
[4] *Birmingham Journal*, Dec. 12, 1829, p. 3.
[5] See, for example, *Birmingham Journal*, Jan. 21, 1826, p. 3, showing that the railway companies sent representatives to attend at the meetings of the commissioners of the turnpike roads along and near the route, in order to get them to favour the railway. In this, they seemed to have poor results.

they based their case[1]. They showed that because there was only one canal connecting Birmingham with Liverpool there was no competition in the conveyance of goods; hence the need of a railway to introduce that element in the life of trade. The distance between Birmingham and Liverpool by canal and the Mersey river was approximately 120 miles, but by the proposed railway it would be only ninety miles. Moreover, the time required by fly boats to follow this waterway between the two termini would be at least sixty hours, but by the proposed railway it would not exceed fifteen hours. The freight rate for merchandise would be reduced from 45*s*. per ton by the above waterway to not more than 30*s*. per ton on the rails. The stoppages on the canals, due to frost, drought and other causes, which occasioned great inconveniences and frequent losses to shippers, would be unknown with the railway; and the injury to corn, merchandise and other goods, on account of the leakage and sinking of boats, could not occur on the railway. In the previous session of Parliament the canal proprietors had strenuously denied that there was any need for additional means of conveyance; but in the session following there was a Bill before Parliament for a canal from Autherley to Nantwich, along the line of the intended railway, which was a virtual admission that there was need of greater accommodation for the public. This being granted, it was a question whether parliamentary sanction should be given to a railway or to a second canal. The railway line had been lately re-surveyed by Jessop and Rennie and the greatest care had been taken to render the route as satisfactory as possible to the largest number of the landed proprietors, although there were some whom they had not been able to conciliate. Having detailed some of the local advantages that would result from the railway, its promoters also showed its importance from the national point of view, as forming part of one great line of direct communication between London and Ireland, and they concluded by requesting Parliament and the public to consider it impartially and to obtain for the country the benefits it held out to agriculture, commerce, manufactures and the political security of the realm. This appeal was signed by Robert Peel, the chairman of the company, and doubtless carried much weight except with those who were personally interested in opposing the railway.

For some months we are unable to follow the company's history, but in August of that year (1826) there was held at Birmingham a general meeting of the subscribers to the railway, at which an unexpected turn was given to their affairs. Those present at this meeting, while fully satisfied regarding the advantages of a railway between Birmingham

---

[1] *Birmingham Journal*, Mar. 11, 1826, p. 1, gives this address in full.

and Liverpool, yet decided that, taking into consideration "the existing pecuniary embarrassments of the country," and the present difficulties in the way of obtaining an Act for a line upon the extensive scale originally proposed, it would be prudent to confine their efforts to the establishment of a railway between Birmingham and Wolverhampton, with such branches as might be thought necessary to the neighbouring towns and works. This line would not exceed fourteen miles in length and could be executed for £150,000; it would pass through the heart of the mining district of Staffordshire, and thus provide another means of supplying Birmingham, Wolverhampton and intermediate places with coal, iron, lime and other materials for their manufacturing industries. Those subscribers who preferred to retire rather than co-operate in this limited enterprise would be allowed to do so under reasonable terms, and those who wished to continue as subscribers would be retained, but in no case could a subscriber hold more than fifty shares of fifty pounds each[1]. At the same meeting there was read a prospectus of this proposed shorter line of railway, showing the large population and business interests of this locality, the need of additional means of conveyance to compete with the monopolistic canals and reduce the freight charges, and the desirability of eliminating long-existing grievances[2]. It would seem that this project did not materialize, probably on account of opposition that was aroused through party squabbling; for at a meeting of the subscribers to the undertaking in the early part of the year 1831 it was agreed, with only one dissenting voice, to suspend further prosecution of the work for a year[3]. Evidently they had not got much, if any, nearer to the execution of the proposed undertaking.

But although this partial enterprise was devoid of results, the earlier plan for a line between Birmingham and Liverpool had, in the meantime, been resumed. Toward the end of the year 1829, when the people had seen the probable, if not the positive, success of the Liverpool and Manchester Railway, a large and enthusiastic meeting of the wealthy merchants and manufacturers of Birmingham was held, to promote the construction of the line to Liverpool[4]. About the same time, a meeting was held in Liverpool, at which it was determined to form a company for constructing a railway from that city along the same course as that projected by the Birmingham people; and, in order to further this plan,

---

[1] *Birmingham Journal*, Sept. 2, 1826, p. 1, gives the exact words of the resolutions accepted at that general meeting.

[2] Ibid. The prospectus is here given verbatim.

[3] Ibid., Feb. 12, 1831, p. 2, on the "Wolverhampton and Birmingham Railroad."

[4] Ibid., Dec. 12, 1829, p. 3, on the "Birmingham and Liverpool Railway."

and to reduce the opposition as much as possible, the interests of the navigation companies were to be given attention. To remunerate canal proprietors for the losses which they would probably sustain from impending competition, extraordinary inducements and privileges were held out for them to become shareholders in the railway[1]. Before the middle of the year 1830 this line had been surveyed by Stephenson and Rastrick on the same double-track plan as that of the Liverpool and Manchester[2]; and a new prospectus had been issued detailing the reasons for the proposed railway and the advantages which would be secured by it[3]. As in 1824, the work was to be carried out under the superintendence of two companies, one beginning at each end of the line; their interests were to be identical, in making application to Parliament for separate Acts of incorporation, in framing their laws and regulations, and in fixing their tolls, the object being to secure unity of design and harmony of operation. In the session of 1831 application was made to Parliament for a Bill to authorize the construction of this road; but after a little time it was decided not to proceed with the measure in that session of Parliment and consequently the Bill was withdrawn. But although the Bill for the whole line was withdrawn, the Birmingham committee resolved to apply for the part of the intended line between Birmingham and Wolverhampton. The House of Commons, however, refused to entertain their application under such circumstances and the committee abandoned, for the time being, their legislative activity[4]. In November of that year, the subscribers to the proposed Liverpool and Birmingham Railway met to consider a report from the committee which had been appointed to determine the best course to be pursued to further their object. This report showed what had already been done, the opposition encountered and the difficulties overcome; and recommended that, instead of beginning at Liverpool, the railway should join with the Warrington and Newton line at Warrington and proceed southward from there, thus forming a

---

[1] *Birmingham Journal*, Dec. 26, 1829, p. 2.

[2] Ibid., June 5, 1830, p. 2; *Manchester Gazette*, June 5, 1830, p. 3. The line was to be 100 miles long.

[3] *Liverpool Times*, May 11, 1830, p. 149, gives the prospectus in full. In brief, the advantages of the proposed railway, as given in the prospectus, were as follows: (1) avoiding the dangerous and uncertain navigation of the Mersey; (2) much greater speed in the carriage of goods; (3) reduction of the cost of carriage; (4) passengers would then be conveyed in one-half the time and at one-half the cost; (5) hence, great saving to the agricultural, commercial and manufacturing classes; (6) Ireland would be benefited by a wider market for her produce. The prospectus is also given in 'Collection of Prospectuses, Maps, etc., of Railways and Canals,' pp. 13–14.

[4] *Manchester Guardian*, July 16, 1831, p. 3.

continuation of that short road. The committee also recommended that the line stop at Wolverhampton, instead of being carried all the way to Birmingham, in order to placate the strong opposition among the canal proprietors who were antagonistic to the formation of a railway between Wolverhampton and Birmingham; but a deputation from Birmingham showed that if this course were followed most of the subscribers of that city would withdraw their names[1]. Early in 1832 the Liverpool and Birmingham Railway Company, with which the Birmingham and Liverpool Railway Company had consolidated[2], decided that the railway should commence at Warrington, where it would virtually join the Liverpool and Manchester, and terminate at Birmingham, the distance between these two points being seventy-four miles[3].

In the latter part of the year 1832, it was found expedient for these two consolidated companies to unite under one head, forming the Grand Junction Railway Company. This company revived the undertaking which had been previously postponed, and planned to connect Birmingham with Warrington, whence connexion would be secured through Newton with Liverpool and Manchester by means of the lines that were already in operation. The road was to be made through the mining and manufacturing sections of Warwickshire and Staffordshire, with branches finally to the Pottery districts; and the prospectus which the company issued expressed the conviction that the traffic would be sufficiently great to yield a net return of fifteen per cent. upon capital[4]. Application was made to Parliament for authority to give effect to the company's purposes[5], and on May 3, 1833, the Grand Junction Railway Act received the sanction of the House of Lords[6]. In 1835, these two

---

[1] *Manchester Guardian*, Nov. 19, 1831, p. 3.

[2] *Birmingham Journal*, Dec. 10, 1831, p. 3. It would seem that the Birmingham and Liverpool Railway Company came in for some scathing censure, because that after acting as a self-constituted body and "fattening from the deep subscription purse," they had incurred expenses of about £17,000 in connexion with their three-fold application to Parliament, and yet had accomplished nothing. Ibid., Dec. 24, 1831, p. 3, letter from "A Sufferer."

[3] Ibid., Jan. 14, 1832, p. 2; *Manchester Guardian*, Jan. 14, 1832, p. 2; *The Times*, Jan. 13, 1832, p. 4. It was expected that this road would pay a clear profit of about fourteen per cent. on the capital expended.

[4] The Grand Junction Railway prospectus is given in full in the *Manchester Guardian*, Oct. 27, 1832, p. 1. It was dated Oct. 15, 1832. The estimated cost of the road between Birmingham and Warrington, 75 miles, was £1,000,000, which was thought to be in excess of the amount that would be actually needed. By using the Liverpool and Manchester and Warrington and Newton connexions, the distance between Liverpool and Birmingham would be 95 miles, and between Manchester and Birmingham 96 miles.

[5] *Birmingham Journal*, Nov. 17, 1832, p. 1.   [6] Ibid., May 4, 1833, p. 3.

divisions, from Newton to Warrington and from Warrington to Birmingham, were incorporated into one line, thus forming continuous rail communication for over eighty miles. This was a very important road, not only because it was the longest line at that time, but because it brought the Midlands and the intervening commercial and manufacturing district into close connexion with Liverpool, the second largest port of the kingdom.

Contemporaneous with the efforts to secure a railway between Liverpool and Birmingham, there was corresponding activity to obtain rail connexion between Birmingham and London. The need of this was greatly felt at that time. The commercial and industrial classes were prevented from reaping the full reward of their activities, because of the impediments to the carriage of goods on the canals. The latter were frequently stopped, sometimes for considerable periods, on account of frost, drought, or the necessity of repairs[1]; and at such times shippers, who had arranged to send goods on a particular vessel, were unable to fulfil their orders, while the shipowner also lost by being deprived of the revenue from this freight[2]. Even if the canals were not stopped, the rates of conveyance were so slow that merchants lost orders because they could not get their goods in time to ship by a certain sailing. By "fly boats" on the canal, the fastest water conveyance of the time, it took four days to bring such products as glass from Birmingham to London[3], and then there were losses to be borne on account of breakage and pilferage, which, on the great amount of traffic along that route, amounted to a large toll annually[4]. In addition to these barriers, the freight rates were so high that the monopoly of the canal companies had long been recompensed by a great profit on their capital[5]. Some commodities, such as linens, silks and others, for the carriage of which speed was a desideratum, had to be brought by coach and pay charges which were two, three, or four times as much

---

[1] 'London and Birmingham Railway Bill. Extracts from the Minutes of Evidence given before the Committee of the Lords on this Bill,' pp. 3, 8, 10, etc., evidence of Messrs Barry, Dillon, Moore. The latter, who was a Birmingham merchant, said that some of his goods had been delayed in transit on account of the canal being frozen from Dec. 24 till Feb. 20, and then part of the goods were rejected because out of time. Mr Barry had known the canals closed by frost for six or seven weeks.

[2] Ibid., pp. 1, 2, 3, 4, 6; etc., evidence of Messrs Barry, Hemsley, Barnes, etc.

[3] Ibid., p. 4, evidence of Mr Hemsley; *Birmingham Journal*, Dec. 1, 1832, p. 3, "Advantages of a London and Birmingham Railway."

[4] 'London and Birmingham Railway Bill. Extracts from Minutes of Evidence before Lords Committee,' p. 5, evidence of Mr Hemsley; *Birmingham Journal*, Dec. 1, 1832, p. 3.

[5] *Birmingham Journal*, Feb. 5, 1831, p. 3, letter from "A Subscriber to the London and Birmingham Railway."

as canal carriage would have cost[1]; and as the mercantile classes were conducting their business more and more from-hand-to-mouth, they were feeling the necessity of having some means by which rapidity of communication could be effected. If this could be secured they would be able to carry on their enterprises without having so much capital locked up in unproductive forms[2].

The necessity of a railway was felt also by the agricultural interests, especially the farmers. The supplying of the London market with vegetables, dairy produce, etc., from the country demanded accelerated transit; otherwise these perishable commodities could not be carried any great distance. To meet the requirements of that market for meat, cattle and sheep in vast numbers were taken from the country; but the road expenses connected with taking these animals on the hoof were considerable and the cattle were much injured by the long journey. At times they travelled till their feet were sore, and they had to be sold at the towns along the road for what they would bring. This was true also of the sheep. If the animals were slaughtered in the country and the meat carried to the metropolis, it would sometimes be spoiled before it reached its destination. So that, whether the one method or the other were adopted, it was uneconomical, for both the weight was lessened and the quality deteriorated[3]. But by means of a railway, animals, meat, dairy products, vegetables and all other necessaries of life could be sent to the London market and be received there almost as fresh as when they left the country; and, at the same time, the expense of reaching the metropolis and the loss or injury suffered on the way would be either lessened or prevented[4].

While these classes were not being provided with suitable facilities for the transportation of their products, the inevitable tendency was to retard all the best interests of the community. Deficiencies in the means of conveyance reacted upon the cost of the goods and commodities to the consumers, and the burden which was felt by the producing classes pressed with equal or greater weight upon those who had to

---

[1] 'London and Birmingham Railway Bill. Extracts from Minutes of Evidence before Lords Committee,' pp. 6, 7–8, 11, evidence of Messrs Barnes, Dillon and Westall.

[2] Ibid., pp. 5, 12, evidence of Messrs Hemsley and Westall.

[3] Ibid., pp. 13–20, evidence of Messrs Warner, Whitworth, Sharp and Attenborough.

[4] Ibid., evidence of Messrs Warner and Attenborough. It was estimated that cattle going from Braybrooke to London, 80 miles, lost 10s. a head in walking that distance upon the common roads; and the cost of driving them that distance was about 7s. a head in summer and 8s. in winter. See also *Birmingham Journal*, Dec. 1, 1832, p. 3, "Advantages of a London and Birmingham Railway."

purchase these things. Not only was the expense of the carriage of freight unduly high, but the cost of travelling was also felt to be too great for the advantages in regard to speed and comfort that were offered by the coaches; and this barrier to trade, it was certain, would be removed by a railway, which would save both time and expense[1].

Perhaps the only interest to profit from the existing conditions was the body of canal proprietors, who reaped large returns from their high charges and impositions. The increasing traffic on the canals in this chain, especially on the Grand Junction and Oxford canals, was not accompanied by any disposition on their part to reduce their rates; but they exacted all that they could lawfully charge and endeavoured to swell their receipts to the utmost extent[2].

During the railway fever of 1825, among many projects that were brought forward was one for the connecting of Birmingham and London by a railway; but this scheme, like several others, failed to materialize at that time[3]. It was revived in 1827, but without any success; and in 1829, when the results of the trials of the locomotive engines on the Liverpool and Manchester line had demonstrated the vast possibilities of mechanical traction, the plan was taken up with greater vigour. The possibility of a railway as a competitor of the canals induced those interested in the waterways to get together in 1827 and propose the formation of a new canal, to be called the London and Birmingham Canal, which, it was hoped, would render such great aid in the carriage of the traffic that there would be no need for a railway. It appears from contemporary evidence that the promoters of this canal, or one of their officials, notoriously falsified the subscription list; but there were other reasons also which helped to decide the issue against the proposed canal[4], especially the necessity of increased speed in the

---

[1] 'London and Birmingham Railway Bill. Extracts from Minutes of Evidence before Lords Committee,' pp. 22–24, evidence of Messrs Mason and Cheetham.

[2] Parkes, *Claim of the Subscribers*, pp. 42–43. Parkes gives (op. cit., p. 44) a table of the "Tonnage Receipts on the Grand Junction Canal" from 1795–1824, which shows that in little more than twenty years (1800–1823) there had been a more than tenfold increase in the tonnage receipts on this canal. See also 'Collection of Prospectuses, etc.,' p. 19, which gives the distance and tonnage rates on the canals between Birmingham and London.

[3] Grinling, *History of the Great Northern Railway*, p. 1.

[4] The agitation for this canal is given in Aris's *Birmingham Gazette*; see, for example, the issues of Dec. 17, 1827, and Feb. 2, 1829. 'Collection of Prospectuses, etc.,' pp. 31–33, gives the complete prospectus of this proposed canal. See also *Birmingham Journal*, Nov. 17, 1827, p. 1; Dec. 15, 1827, p. 3; Dec. 19, 1829, p. 2; Dec. 26, 1829, p. 2; April 10, 1830, p. 3. There was much fraud connected with this project, and the unabashed jobbing is shown also in *Parl. Papers*, 1830 (251), x, 719, 'Report from the Committee on the Birmingham and London Junction Canal Petitions.'

conveyance of both goods and passengers. But the agitation for the railway continued and its advocates used the results obtained from the locomotive tests on the Liverpool and Manchester, in October, 1829, as an inducement to secure support for this new enterprise. It was asserted that this railway would be of great national benefit for forwarding troops and military stores; that by it the manufactures of Birmingham and its neighbourhood would be conveyed to London in much less time and at less expense than by canal; that the agricultural produce of all the intervening section would be able, on account of the faster speed and reduced cost of conveyance, to find wider markets and better prices; that the coal of Staffordshire could be taken to satisfy the needs of the consumers at the metropolis and along the railway; and that the expense of maintaining the turnpike roads adjacent to the proposed railway would be almost all saved, amounting on the average to about £250 per mile annually[1]. Throughout the year 1830 there was a deeper interest manifested in the project and this continued to intensify until application was made to Parliament to secure authority to carry out this enterprise.

It will be appropriate here to examine the nature of the opposition to the proposed railway, and, first, we shall consider that of the landed interest. The landlords feared that the railway would injure the property through which it would pass, by destroying the privacy and unity of the farms; that the closes which were now convenient in form and size might be divided into ill-shaped fragments; that the deep cuttings across the slopes of the hills might intercept the supply of water to the wells and grounds below; that the large embankments across the low lands would interfere with the natural drainage of the parts above them; and that, where the railway crossed the highways on the same level, it would be inconvenient and dangerous to the public. It was said that the existing means of land and water carriage were greater than had ever been required; that no necessity had been shown for accelerated communication; and that the absence of the support of the landowners was undeniable proof that the undertaking was uncalled for

---

[1] *Birmingham Journal*, Nov. 28, 1829, p. 3, letter from "T. B.," entitled "Observations on the Advantages of a Railway Communication between Birmingham and London." The writer shows that at a very moderate calculation the returns from passengers who now patronized the existing seventeen coaches each way daily between London and Birmingham, would be enough to pay large profits on the cost of the railway. As a matter of fact, his calculation of the cost of construction was much too low, being only about one-fifth of the actual cost. His figure for the expense of building the road was £963,000, whereas the actual cost was over £4,500,000. See *Herepath's Railway Magazine*, N.S., vi, pp. 16–17. On the advantages of this railway, see also *Birmingham Journal*, July 17, 1830, p. 3, on "Railroads."

by the wants or wishes of the country[1]. The promoters of the railway answered these objections by an array of facts which, to an unbiased mind, should have been convincing. Regarding the plea that the privacy of the estates would be destroyed and the homesteads severed, the answer was made that privacy was one of the worst features of a farm; that a farm on a great public thoroughfare was worth much more than one in a country lane; and that the construction of bridges over and arches under the railway would give facility of communication between the divided portions of the farms. The objection that the railway cuttings would prevent the circulation of water to the lower grounds was answered by showing that, if the water collected in the ditches on each side of the railway, it could be drawn off and used for irrigation and that the railway would act as a drain to those lands that had too much water, and therefore would do for the farmer what he had long wanted but had not the money to do for himself. The claim that large embankments across the low lands would interfere with the natural drainage was met by the assertion that the railway would cross streams and watercourses by means of viaducts so as to leave these outlets as open as before. The supposed interference with traffic on the highways was shown to be without foundation, because the railway would pass either over or under all great thoroughfares and every precaution would be taken to protect the public from risk. The alleged sufficiency of the existing facilities of carriage provoked acrimonious reply: it was true that the roads and canals could convey more goods and passengers than had passed on them, and for the obvious reason that a narrow limit was imposed by the expense and delay in each case; but by lowering the charge of conveyance and by quickening the return on capital through increased speed and regularity the amount of the traffic would be greatly augmented. It was very clear that, when the canals were frozen, and the people, especially the labouring classes, of the Midlands could not get coal, there was much need of some additional means of conveyance. The wealth of some mineral districts was, to a great extent, excluded from the London markets on account of the heavy expense of canal transport. Moreover, every man's time was part of his capital: it made considerable difference whether a person had to spend six hours or twelve hours upon the road, for in the former case there was a greater use of time and less expense involved than in the latter. A few landowners might not be put to any inconvenience by reason of the slow and expensive transit of passengers and goods,

[1] *Birmingham Journal*, Jan. 22, 1831, p. 1, on the "London and Birmingham Railway;" *The Times*, Nov. 18, 1830, p. 3, letter from "A Landowner," concerning the London and Birmingham Railway.

but the public were the best judges as to the loss occasioned by the present impediments to everyday business. And, finally, the declaration that the absence of the support of the landowners was undeniable proof that the railway was uncalled for by the wants or wishes of the country, was of absolutely no validity. The wants of the country gentlemen were no index of the wants of the country generally. The fact that the Liverpool and Manchester line had benefited both landowners and tenants was conclusive evidence that corresponding benefits would accrue to the landed classes in this case[1].

The other great class that were opposed to the formation of the railway included the canal interests between the Midlands and London. It was indubitable that the railway would take part of the traffic which had been accustomed to going on the canals; and the revenues of the canal companies would probably be reduced, at least relatively if not absolutely. There was no doubt but that some of the canals along this route, like the Grand Junction and the Oxford, had remunerated their proprietors handsomely and it was but natural that they should seek to perpetuate these conditions of their own prosperity; but, on the other hand, the freight charges were high, and, judging from the results which had been attained by the Liverpool and Manchester Railway in reducing the rates along its line, there was every reason to believe that comparable results would be secured in this case. There was, apparently, no remedy for the injury which canal proprietors would sustain, unless they would join and make common cause with the railway company. Persevering hostility, conducted at great sacrifice of property, might delay the railway, but could not prevent it, since it was for the public benefit. The canal companies should not expect the progress of improvement to be halted to secure the continuance of their enjoyment of monopoly. They had remunerated their capital for a long time with immense profits; and they should not now complain at the introduction of a cheaper and faster means of conveyance[2]. The Marquis of Stafford, the greatest canal proprietor in the world, had formerly opposed the Liverpool and Manchester Railway, but later he became convinced of its usefulness and in 1831 he owned 1000 shares of its stock[3]. But all the owners of canal shares were not so readily convinced

[1] *Birmingham Journal*, Jan. 22, 1831, p. 1, on the "London and Birmingham Railway;" ibid., Feb. 5, 1831, p. 3, letter from "A Subscriber to the London and Birmingham Railway."

[2] *Birmingham Journal*, Jan. 22, 1831, p. 1, on the "London and Birmingham Railway;" ibid., Feb. 5, 1831, p. 3, letter from "A Subscriber to the London and Birmingham Railway;" ibid., Mar. 5, 1831, p. 3, "Public Meeting to Support Railways."

[3] *Birmingham Journal*, Feb. 5, 1831, p. 3.

that the greater economy of railways had been proved. Because the Liverpool and Manchester Railway was considered as a brilliant success was no reason to conclude that experience would confirm this result in every other instance; and, partly in support of this vague hope of being able to compete with the railways, and partly in the expectation that Parliament would protect them from ruin, the canals offered strenuous resistance to the authorization of the railway[1].

The third class from which opposition was encountered by the railway included the coach proprietors, waggon masters and postmasters, the amount of whose business was likely to be seriously reduced by the new means of conveyance. Before the line had been put in operation between Liverpool and Manchester, there were about twenty-two regular coaches on that road; but, by the beginning of the year 1831, almost all these stage coaches had been laid aside, and soon the railway was carrying about three times as many passengers as had formerly patronized the coaches[2]. This apparently inevitable decline of road carriage of passengers and goods induced the proprietors engaged in this business to resist the establishment of the new enterprise which was destined to destroy their means of support; but, probably because they were not backed by the large amount of wealth that was available for the landowners and the canal proprietors, their claims seem to have commanded but little public attention. An interesting case of such opposition, in 1833, comes to us in the form of a petition to the House of Lords from those who were carrying on these undertakings on the lines of road between London, Worcester, Hereford and Gloucester; they requested the Lords to protect their interests by rejecting all applications for railroads in general, and particularly the Liverpool and Birmingham and Birmingham and London railways[3]. It would be but natural that the owners of these vehicles along the same or parallel lines of road should oppose the formation of a railway which would take away their business; but why those should oppose it whose line of activity was more or less in the opposite direction, is by no means so clear.

In addition to neutralizing the arguments of their enemies, the railway company put forward some other strong reasons in favour of their

[1] *Birmingham Journal*; Mar. 5, 1831, p. 3, "On Railways."

[2] *Birmingham Journal*, Feb. 5, 1831, p. 2, letter from "A Railway Subscriber," on the London and Birmingham Railway; 'London and Birmingham Railway Bill. Extracts from the Minutes of Evidence given before the Committee of the Lords on this Bill,' evidence of Henry Booth (treasurer of the Liverpool and Manchester Railway Company), pp. 54–55.

[3] *Hampshire Advertiser and Salisbury Guardian*, May 11, 1833, p. 2, "Coaches v. Railway."

project. As the landowners had profited from the construction of the Liverpool and Manchester line, and some who formerly were very active against that enterprise were now as strongly in favour of it, so it would be to the advantage of estate owners in this other section to have the midland metropolis connected with London. The hostility of the landlords was, therefore, ill-advised. Not only would the railway add to the value of their property, but the proximity of larger and better markets for farm produce would give the tenants higher prices for what they had to sell, and thus render them more prosperous. A few hours at the most would suffice to carry fatted animals from their pastures to Smithfield, without their losing in weight or being injured, as at present, by drovers. The railway would supply the metropolis market better, and with more facility and regularity, from a distance of eighty miles, than at present from the neighbouring districts; and the steady market would be a boon for agriculture, while providing steadier employment for labour. The expenditure of millions upon this work would lighten the burden of poor rates and prove beneficial to the country through which the railway would be carried[1]. The passenger fares would be reduced from the coach fares of 4d. per mile inside and 2$\frac{1}{4}d$. per mile outside, to 2d. and 1$\frac{1}{4}d$. per mile respectively on the railway; and this would be the accompaniment of a rate of speed double that of the average speed of coaches. Corresponding reduction in the time and expense of the carriage of goods was anticipated[2]. To placate the owners of coaching establishments, it was shown that, instead of there being less work for coaches, there would be more after the railway were put in operation. Doubtless, the construction of the railway would cause the coaches along that line to be set aside; but throughout a belt of many miles in width on each side of it, numerous cross coaches would be immediately established to meet the railway at important stations according to the convenience of passengers. For example, the many steamboats connecting London, Dover and Calais had increased, rather than diminished, the number of post-horses on the Dover road[3]; and evidence was given before a committee of the House of Commons to the effect that while, before the opening of the Liverpool and Manchester Railway, the coaching business on the main

[1] *Birmingham Journal*, Feb. 5, 1831, p. 3, letter from "A Subscriber to the London and Birmingham Railway."

[2] *Birmingham Journal*, Jan. 22, 1831, p. 1, on the London and Birmingham Railway. According to this writer the carrying of goods by the fly vans was done at the rate of five miles per hour and at a cost of at least 9d. per ton per mile, while the railway would carry them at fifteen or more miles per hour.

[3] *Birmingham Journal*, Feb. 5, 1831, p. 2, letter from "A Railway Subscriber," on the London and Birmingham Railway.

road between these two termini was carried on by 400 horses belonging to Liverpool proprietors and 400 to Manchester proprietors, after the opening of that line, although coaches soon ceased to run on the direct road, there was such a great increase on the cross roads that the proprietors at Manchester, by 1834, had 800 horses employed and the demand was still increasing[1]. The safety, certainty and rapidity of conveyance were of themselves sufficient to recommend the railway in preference to any other means of carriage[2].

In 1830 a Bill was introduced into Parliament seeking authority to construct a railway between these two termini; but the strong opposition which was manifested against this measure, especially by several of the great landowners[3] and the canal companies along the route[4], but also by proprietors of coaching establishments and turnpike trustees[5], caused the failure of the Bill to pass in the session of 1831–2. As was the case in the promotion of the railway from Birmingham to Liverpool, so also in this case, there was the existence of two separate companies which were later merged into one before the Act of Parliament was passed to sanction the undertaking[6]. The defeat in 1832 was made the occasion of greater earnestness and the supporters of this scheme got together to inquire into the reasons for their failure and to devise more effective measures for securing their ends. In addition to foes without, the company had to meet and harmonize internal dissension. Some of the subscribers to the undertaking had made their subscriptions and signed the contract deed, in 1830, on the assumption that this line would connect with the projected

[1] See summary of this evidence in *Hampshire Advertiser and Salisbury Guardian*, Mar. 29, 1834, p. 2.

[2] *Birmingham Journal*, Mar. 5, 1831, p. 3, "Public Meeting to Support Railways."

[3] In addition to foregoing references, see also Brit. Mus. 1890. c. 9 (5). Stretton, *History of the London and Birmingham Railway*, shows the opposition of the Earl of Clarendon and the Earl of Essex.

[4] In addition to previously-mentioned references, see *Remarks upon Pamphlet by Investigator on the Proposed Birmingham and London Railway*. "Investigator" evidently represented the canal interests and he had tried to show the evils that would be caused by the railway. See also Brit. Mus. T. 1371. (18), the writer of which was apparently a canal proprietor, since the pamphlet presents that side of the case. It was entitled, 'The Probable Effects of the London and Birmingham Railway.' Since it is impossible to get too clear a view of the way in which railways were regarded at the time of their introduction, the contents of this pamphlet are worthy of perusal.

[5] *Birmingham Journal*, Feb. 12, 1831, p. 2, for example, shows the opposition of the trustees of the Dunchurch and Stonebridge road to the proposed railway.

[6] *Birmingham Journal*, Sept. 21, 1833, p. 3, on "London and Birmingham Railway;" *Manchester Guardian*, April 7, 1832, p. 1, on "London and Birmingham Railway."

Liverpool and Birmingham Railway; but the latter had been abandoned, and, therefore, the whole situation was changed. It would seem, too, that the directors of the company that was formed by a union of the former two concerns had changed the plans for the railway and increased the estimated cost of the line without submitting these plans to a general meeting for approval. Many of the subscribers had requested the directors to publish a full and authentic report of the condition and prospects of the company, and afterward to convene a general meeting of the shareholders to take action as to what should be done; but all they had received was a circular giving a few loose details. Because of these conditions, those who were dissatisfied, including a number of the great landlords, sent a petition to the House of Commons requesting that they might be released from their obligations and not be considered as subscribers to the present undertaking[1]. Earlier in the year 1832 there had been a meeting of the owners and occupiers of land along the proposed course of the railway, at which there appeared to be agreement among those present that the railway as planned would depreciate the value of their property, and they, therefore, decided to protest against the granting of an Act of Parliament[2]. What was the outcome of this discord we need not trace; suffice it to say that by midsummer of that year the company issued its new prospectus, showing the public the advantages to be gained by this proposed railway, in opening up new sources of supplies of provisions for the metropolis, in facilitating and cheapening travel, in providing rapid and economical interchange of the great articles of consumption, and in connecting London with Liverpool and the great manufacturing sections of Lancashire and the Midlands[3]. Once more application was made to Parliament and Lord Wharncliffe, the chairman of the parliamentary committee to which this measure was submitted for examination and report, asserted that in his long experience in Parliament he had never seen a measure passed by either House that was supported by evidence of a more decisive character. But,

---

[1] This petition is given in full in *Manchester Guardian*, April 7, 1832, p. 1. It shows that the estimated expense of the railway was at first £1,500,000, but in 1832 it was £2,500,000.

[2] *Manchester Courier*, Feb. 4, 1832, p. 3, on "London and Birmingham Railway."

[3] This announcement or prospectus is given in full in *Manchester Guardian*, July 7, 1832, p. 1, and also in 'Collection of Prospectuses, etc.,' p. 61. It refers to the great success of the Liverpool and Manchester Railway. This line would connect with that to be constructed from Birmingham to Liverpool, and through the port of Liverpool it would furnish rapid connexion between Ireland and London. See also Brit. Mus. 1890. c. 9 (5), and "Statement of the Case in support of the London and Birmingham Railway Bill," as given in 'Collection of Prospectuses, etc.,' p. 74.

notwithstanding this, the Bill was thrown out, owing chiefly to the opposition of the landowners, who feared that their estates would be prejudiced or injured by the railway. This failure had resulted after £32,000 had been expended on the application, but still the company was not deterred in their efforts. Subsequent changes were made in the line in order to avoid the properties of two of the nobility who had strongly opposed it and to keep at a considerable distance from the town of Northampton[1]. With these alterations, the line was regarded favourably at the company's third application and the Act was passed in 1833. In 1837, the first section of the line was opened between London and Tring[2], but it was not until the following year that the whole line was opened[3]. By this railway and what were later its north-western connexions, there was established a complete communication from London to Birmingham and from Birmingham to Liverpool; but we must remember that, at this time, these were entirely separate roads, not working in harmony, and, therefore, there was no through rate nor through traffic.

A few facts regarding the finances of this railway may be appropriately given. In 1830, when the line was being agitated, it was computed that the amount paid by passengers and parcels conveyed by coaches between London and Birmingham exceeded £300,000 a year, and that paid for the carriage of goods between the same places exceeded £500,000 a year. The expense of building the railway upon the best possible plan was estimated not to exceed £1,500,000; so that one-fourth of the amount paid for the conveyance of passengers and goods would be ample remuneration on the capital to be spent on the construction of the railway[4]. But soon the plans were changed and the estimated cost was augmented, so that in the company's original Act of incorporation the capital was stated at £2,500,000. By later Acts, the company was empowered to raise a capital in shares and on loans amounting to £4,500,000; but by 1839 even this sum had been

---

[1] Stretton, *History of the London and Birmingham Railway.* The announcement of the directors of the railway in 1833 is given in 'Collection of Prospectuses, etc.,' p. 65. It showed the advantage of the railway in regard to safety, expedition and economy, and the benefits that would accrue to London and the public generally. The change of front of many landlords is apparent in the testimony that was given before the committee that had this Bill for consideration, and those who had formerly opposed railways were now favourable to them (see 'Great Western Railway. Evidence on the London and Birmingham Railway Bill,' especially the evidence of Mr Earle, Mr Joseph Pease, and Mr J. Moss).

[2] *The Times*, Oct. 21, 1837, p. 2, on the London and Birmingham Railway Company.

[3] Stretton, op. cit.   [4] Brit. Mus. 1890. c. 9 (5).

exceeded by £500,000 on account of calls and loans, and the company proposed to go to Parliament again for authority to raise another £1,000,000, making the total capital £5,500,000[1], for it was admitted that the road would cost at least that sum. These vast amounts in excess of the estimated expenditures for the road caused disappointment and called forth some sharp criticism; it was thought that the revenues of the company would not be sufficient to pay a reasonable return upon the great outlay[2]. But when it was shown that a large part of this increased expenditure was for the construction of additional lines of railway, so that new sources of income, which had developed subsequently to the origin of the railway, might yield to the company a good return after paying the interest on the capital embarked in these accessories[3], the sting was taken out of the adverse comment, and it was seen that the company was working with ultimate, rather than proximate, issues in view. The enormous amounts that were wasted in proceedings before Parliament and the extraordinary sums that were demanded to make complete settlement for their right of way will be apparent from the figures for this railway company, which show that the cost of obtaining the original Act of incorporation was £72,868. 18s. 10d., and the payments made for "land and compensation" were £622,507. 3s. 10d.[4]

When London had been connected with the great centres in the Midlands and the north-west, the next project of most importance was to secure connexion between the capital and Bristol, so as to give facility of access to the immense trade of the Severn valley. In reality, the agitation for this line did not wait even for the authorization of the London and Birmingham, but began after the success of the Liverpool and Manchester had been assured. Bristol had formerly been second only to London in its importance as a port, but Liverpool had risen into such prominence that it assumed the position which had been held so proudly by Bristol. As a consequence, the latter city had declined to third place, and its trade was languishing in the competition with its north-western competitor[5]. To some, it seemed as if this were

[1] *Herepath's Railway Magazine*, N.S., VI, pp. 16–17, letter of "A Friend to Railways and Truth," in regard to the London and Birmingham Railway.

[2] Ibid., VI, pp. 17–18, 113–18, 235–6; *The Times*, May 9, 1837, p. 6.

[3] *The Times*, May 23, 1837, p. 6, letter from W. S. Moorsom.

[4] Brit. Mus. 1890. c. 9 (21), 'Plans, Prospectus, Reports, and Minutes of Evidence, in reference to the London and Birmingham Railway.'

[5] A series of thirty letters from "A Burgess," relating to the trade of Bristol, showing the causes of its decline and the means by which its revival could be effected, appeared in the *Bristol Mercury*, beginning with the issue of Feb. 2, 1833, p. 2, and ending with that of Jan. 4, 1834, p. 4.

the acceptable time to restore the old commercial prosperity and prestige of this ancient city, and the railway question formed the nucleus of a conflict which helped to arouse Bristol from her lethargy. It will help us to understand the issue which confronted the people at this time if we look more closely at the conditions of transportation by land and water in 1832, when the problem as to the construction of a railway came into public attention.

The goods traffic along this route, especially the carriage of heavy commodities, was largely confined to the canals and connecting waterways, namely, the River Avon Navigation, from Bristol to Bath, the Kennet and Avon Canal and River Kennet Navigation, from Bath to Reading, and the Thames Navigation, from Reading to London. The delays and uncertainty of water carriage were becoming unbearable to the commercial interests, at a time when the mercantile practice was undergoing revision and the old system of keeping a large stock on hand was giving way to the method of keeping less stock but more frequently replenished. The average time occupied in the traffic by water from London to Bristol was from seven to ten days, but barges had been detained, on account of drought, flood, frost, or other stoppage, for weeks and even months on their journeys, and during these delays there was a great amount of pilferage carried on[1]. Such interruptions on account of natural conditions occurred several times a year, and the time when the canals were not in working order was increased by the necessity of stopping them for repairs. The vast volume of complaint concerning these obstacles to trade was persistent, prolonged and almost universal[2]. In the case of articles of constant consumption, such as coal, groceries and other food-stuffs, the hindrance of the

---

[1] *Proceedings of the Great Western Railway Company*, pp. 9, 28; 'Great Western Railway Bill. Minutes of Evidence before the Lords Committees,' evidence of Messrs Walker, Harley, Wilkins, Davis, Morris; Felix Farley's *Bristol Journal*, April 19, 1834, pp. 3, 4, evidence of Messrs Hire and Stone; ibid., April 26, 1834, p. 2, evidence of Messrs Keys, Sheppard, Luscombe, Provis, Walker, Taylor, Moline, Wilson, Kendall, et alii.

[2] It will help us to realize the situation more fully if we give a few instances of what actually took place, as taken from the evidence before the committee on this railway Bill (Felix Farley's *Bristol Journal*, April 19, 1834, p. 3, and April 26, 1834, p. 2). Mr Hire, of Bristol, asserted that in one case several hogsheads of sugar were sent to him from London; but, instead of arriving in 13 days, they did not arrive for two months, so that he lost about £300 by this delay. Mr Davis, of Reading, in January, 1834, had his goods coming from London delayed a month all but two days, which prevented him from executing his orders and thus caused him great loss. His goods, especially tobacco and sugar, were much injured by exposure to moisture. Butter was injured in hot weather by these delays and on one occasion he was glad to sell £200 worth at half-price.

regular supply often produced most serious inconveniences; and it not infrequently happened that, by the stoppage of the canal, waggons had to be sent miles to procure necessaries from the barges which were unable to proceed on their voyages. This, of course, greatly increased the cost to the consumer, and, on such a necessary article as coal, was quite a burden[1]. Another inconvenience on the Thames was that vessels were of large capacity and would not set out on their voyage up the river until they had a full load. A merchant might, therefore, have ten tons of goods to be sent as quickly as possible by water in fulfilment of an order; but if the vessel that was to carry these goods was of eighty tons burden, she would not start until her cargo was complete. This compelled the tradesman to wait for the goods[2]. So absolutely uncertain was the conveyance that not even an approximate calculation could be formed by the most experienced traders as to when their goods would arrive at the point of destination[3]; and merchants and manufacturers frequently received or sent their goods all the way by land carriage at twice the cost, or more, rather than send them by the navigations and not know that they would be certain to reach their destination at the required time[4]. In addition to the uncertainty of the navigation, its expense and the injuries which the commodities were likely to sustain, there was much annoyance on account of the losses by pilferage, which were considerable under ordinary conditions, but were very heavy when delays occurred to cause the barges to stand still.

---

[1] *Proceedings of the Great Western Railway Company*, pp. 9, 28; 'Great Western Railway Bill. Minutes of Evidence before the Lords Committees,' evidence of Mr Davis; Felix Farley's *Bristol Journal*, April 26, 1834, p. 2, evidence before the committee on this railway Bill, given by Mr Ogden, Mr Ray, and others.

[2] *Proceedings of the Great Western Railway Company*, p. 28; 'Great Western Railway Bill. Minutes of Evidence before the Lords Committees,' evidence of Mr Davis, p. 396.

[3] Felix Farley's *Bristol Journal*, April 19, 1834, p. 4, "Committee on the Great Western Railway Bill," statement of Mr Harrison. Sometimes on the Thames there would not be more than two "flash days" a week, and often barges were stranded on the shallows and could not move (ibid., April 28, 1834, p. 2, evidence of Robert Ray and others).

[4] Evidence of Messrs Walker, Stone, Shepherd, Provis, Ogden, Wilson, Harris, Mills, Davis, Pearman, et alii, given to Committee on the Great Western Railway Bill; also 'Great Western Railway Bill. Minutes of Evidence before the Lords Committees,' evidence of Mr Wilkins, Mr Marling, Mr Morris, and Mr Venables. Saxony wools, which were brought into the eastern ports of England, were carried west principally by waggons to avoid the delay on the canal. Woollen goods manufactured in the west of England were sent to London by waggon paying 5s. per cwt., or sometimes by coach at 1d. per lb., rather than by canal barge at 2s. 9d. per cwt. The goods were too valuable to risk sending them by canal, with the necessary transhipment, for they would become crushed and often wet before they were delivered.

The passenger traffic, too, was not carried on as expeditiously as was desired, and the gross abuses, the inconveniences and the cost connected with coaching were impediments for which no adequate remedy had been devised. The insecurity of life had been the cause of continual complaint, and, as we have seen, measures had been taken to prevent the perpetual recurrence of those things which endangered the lives of travellers, but still the evils went on without serious check. Twenty-two coaches went up and down every day, and there were also four mail coaches a day, two up and two down. The great number of passengers who were carried by the coaches may be readily calculated from their returns, which showed that the average number of passengers by a four-horse coach was nine, by the mails five, and by a pair-horse coach six[1]. The average time taken by the stage coaches from London to Bristol, including stoppages, was fourteen hours, and by the mail thirteen hours. This was an average rate of speed of about nine miles per hour. But those who knew the speed attained on the Liverpool and Manchester Railway were eager to see the same twenty to twenty-five miles per hour accomplished on the way between Bristol and London, particularly when it could be secured at lower cost and with greater safety and comfort than by the coaches.

The above-mentioned reasons were by no means all that were adduced in support of the plan for a railway along this course. Much emphasis was laid upon the fact that Bristol, being the natural entrepôt for Ireland, Wales and the West of England, would attract the trade from these sections, and their products could then be sent to London on a shorter haul and at a cheaper rate than if they were sent via Liverpool. In this way the metropolis would be furnished with quantities of food supplies from Ireland, with which Bristol had regular communication, and these could be greatly increased. Fish, also, instead of coming from the north, could be supplied in great abundance from that island[2]. The immense quantities of coal and iron in South

---

[1] Felix Farley's *Bristol Journal*, May 3, 1834, p. 2, evidence of Thomas Cooper, coachmaster at Bath and Bristol, before the Committee of the House of Commons on the Great Western Railway Bill. See the returns of the passenger traffic as given by the records of the Stamp Office, a table of which is given in 'Great Western Railway Bill. Minutes of Evidence before the Lords Committees,' evidence of R. J. Venables, p. 416. The same table is inserted in the *Proceedings of the Great Western Railway Company*, and in the evidence given before the committee of the House of Commons on this Bill.

[2] Felix Farley's *Bristol Journal*, April 19, 1834, p. 4, "Committee on the Great Western Railway Bill," statement of Mr Harrison; ibid., Sept. 28, 1833, p. 3, on the Great Western Railway; ibid., Oct. 5, 1833, p. 2, letter from John Weedon, on the Great Western Railway.

Wales, with which Bristol had immediate connexion, and in the vicinity of the city of Bristol, would provide fuel for the increasing demands of London and the intervening places, at a reduced cost that would soon greatly increase the consumption of that commodity and the revenue to be derived from it[1]. The agricultural interests were appealed to by the possibility of opening up wider markets for their surplus produce, thus tending toward higher prices for everything they had to sell, and by the inevitable enhancement of the value of their lands should the railway be put into operation. For these claims they had the utmost justification from the results which had accrued along the two lines which were already carrying on their work as general carriers, namely, the Stockton and Darlington and the Liverpool and Manchester. The farmer could get supplies of manure from greater distances and at a cheaper rate than before, so that the land would be brought into a higher state of cultivation and the fertility of the soil improved. By the greater productiveness of the land and the better marketing facilities the rental value of the land would be increased, and consequently both owner and occupier would receive the benefit[2]. Farmers would also be able to send their cattle, sheep, etc., to the London markets at diminished expense, and at the same time the better means of conveyance would prevent any deterioration in the quality of the meat. This would make it possible for the butcher to pay the farmer higher prices for this meat supply, and to give the consumer a better quality of product. Encouragement was also given that, when the heavy road traffic had been transferred to the rails, the highways would be greatly improved and their maintenance would not involve such a heavy burden of expense. Lastly, the construction of such a great public work would give employment to a large number of men, and this would be a significant relief at the time when the pressure of distress was severely felt and the obligation of poor rates was being increasingly realized[3].

Now, let us consider the nature of the opposition which was aroused against this scheme. As in the case of the other railways which were authorized before this, so in the case of the Great Western, the opposition of the landlords and of the inland waterway interests was the most powerful. On Nov. 19 and Dec. 9, 1833, there were

[1] Felix Farley's *Bristol Journal*, Nov. 17, 1832, p. 3, "Railway from Bristol to London;" ibid., Oct. 12, 1833, p. 3, editorial under the caption "Great Western Railway."

[2] Felix Farley's *Bristol Journal*, April 26, 1834, p. 2, evidence of Mr Geo. W. Hall and Mr Joseph Pease, before the Commons Committee on the Great Western Railway Bill; also ibid., May 3, 1834, p. 2, evidence of Thomas Pearman and others.

[3] *Bristol Mercury*, Aug. 18, 1832, p. 3, letter from "Ignotus."

meetings of noblemen and gentlemen, owners and occupiers of lands through or near which it was proposed to make this railway, declaring that no case of public utility had been made out to justify or palliate such an uncalled-for encroachment upon the rights of private property; and that the projected railway would be repugnant to the feelings and injurious to the interests of the landed classes. They decided in each case to enter into a subscription and appoint a committee who were to see that all possible legal measures were taken to counteract the activity of the promoters of the railway, and were to bring pressure upon their members of Parliament to induce the latter to oppose the sanctioning of such a baneful innovation[1]. But it must not be inferred from what we have said that all the landowners were opposed, for there were some who were sufficiently open-minded and public-spirited to see that their own personal predilections should be subordinated to the general good; and there were others, whose property would not be crossed by the railway, who were convinced that the proximity of that convenience would be of great value in the marketing of their products[2]. In the inland counties there were some who recognized that in sending their products to London by railway at a lower expense they would come into competition with the south of Ireland, which would also be afforded great inducement to place its products on the same market; and if the market were thus taken away from the home producer the agriculture of these southern counties would suffer[3]. On the other hand, there were certain who could foresee that, with the lowering of the prices of food supplies, there would be a greater demand for them on account of greater consumption, and, consequently, there was little fear that Irish competition would be injurious to English interests. It is clear, however, that landlord opposition was active in preventing the favourable consideration of the Great Western Railway Bill.

The animosity of the waterway interests was likewise vigorous.

---

[1] Felix Farley's *Bristol Journal*, Feb. 22, 1834, p. 1, gives the resolutions which were adopted at each of these meetings. Ibid., Mar. 15, 1834, p. 4, "Proceedings in the House of Commons on the Great Western Railway Bill," showed many petitions for the measure, and some against, with very strong opposition from the landowners. Countess Berkeley petitioned against the Bill and said that her residence would be uninhabitable if the line marked out by the company was selected.

[2] Felix Farley's *Bristol Journal*, May 31, 1834, p. 3, evidence of Lord Kensington on the Great Western Railway Bill.

[3] Felix Farley's *Bristol Journal*, Oct. 5, 1833, p. 2, letter from John Weedon. Middlesex landowners and farmers opposed it because they thought it would bring produce to London from a distance as cheaply as they could send it there, and thus destroy their monopoly in that market. Buckinghamshire and Berkshire farmers opposed it because they feared Irish competition. (*Proceedings of the Great Western Railway Company*, pp. 10–11.)

The Kennet and Avon Canal Company and the Commissioners of the River Thames Navigation were loud in their denunciation of a scheme which would take away their trade and nullify all that they had endeavoured to do. The canal company, through its special committee, unanimously resolved to oppose the railway[1]. They thought, from what they had already seen in other instances, that most of the traffic would leave the canal and go on the rails; that, therefore, the money invested in the canal would be largely lost, and, as a result, great numbers who were depending upon this undertaking for their income would be deprived of their maintenance[2]. But it was shown to them that the opposition evoked against other railways had been powerless to stem the tide of progress; that the principle of public good must prevail over that of private advantage, here as well as in the other cases, by the construction of a superior means of conveyance; and that the canal company should not blind themselves to the evidence of experience and throw away their money in useless legal contests[3]. The general committee of the Thames Navigation formed a more potent antagonist to the proposed railway than was the Kennet and Avon Canal Company, because they represented also the great majority of the owners of land adjacent to the river. In order to prevent the authorization of the railway, they endeavoured to enlist "the active assistance of the various interests, threatened by this widely destructive speculation with inevitable ruin[4]." They sought to rouse public support to their side, by showing that the great body of commissioners, acting gratuitously, had, by judicious expenditure of over £250,000, made that navigation one of the most perfect in the kingdom; that anything which would lessen the amount of tolls they received would prevent the meeting of their obligations to their creditors and the maintenance of the navigation; and that, if the proposed railway were constructed, the river would fall into disuse and become silted up, the floods would increase in height and duration, many towns on the river would have their trade injured, and the lands along the river would deteriorate in value. The wide range and the nature of their appeal included the bondholders, whose security would be endangered by the railway; the landholder, the value of whose property would be affected; the great trading towns along the river, whose commercial prosperity was threatened; the owners of mills, wharfs, and other mercantile establishments, whose trade would be

[1] Felix Farley's *Bristol Journal*, Mar. 1, 1834, p. 2, letter from "Aequus."

[2] *Bristol Mercury*, Mar. 2, 1833, p. 2, letter from "Scrutator."

[3] Felix Farley's *Bristol Journal*, Mar. 1, 1834, p. 2, letter entitled "Canals versus Railways."

[4] Ibid., Feb. 22, 1834, p. 1, "Thames and Isis Navigation in opposition to Great Western Railway."

annihilated; and the owners of old locks whose revenues would be destroyed. To those who presided over, and those who were educated at, Eton College and Oxford University, appeal was made by the sanctity of their present trust and their former recollections and associations; and, lastly, it was requested that all those who resided upon the banks of this river, whether attracted there by its beauty, its salubrity, or its utility, would lend their aid to prevent the sanction of Parliament being given to "so useless a scheme" as that of the Great Western Railway[1]. With the great influence which the inland navigation companies exerted, it is little wonder that they were called the "fourth estate of the realm[2]."

Of the vehement opposition of the authorities of Eton College and the University of Oxford, we have already spoken in a former connexion. It was not until after repeated applications had been made that a branch line of the railway was sanctioned to Oxford, and then it was stipulated that the station should be built as far away from the city as it could conveniently be placed[3]. In the Act as first passed, there was also a clause forbidding the erection of any station at the important town of Windsor[4].

While we have been impressed by the fact that the commercial classes, generally, were strongly in favour of the railway, we note in this case, what we have not observed in any of the foregoing, that some of the mercantile elements were averse to this railway. Some feared lest Bristol might become merely a way station between London and Wales and Ireland, and as such would be overshadowed by the metropolis to such an extent that it would cease to grow. Moreover, since there was always a prejudice in favour of the London market, the rapid transit by rail would enable purchasers in South Wales and the west of England to go directly to London for their supplies, and thereby Bristol's importance as a great entrepôt would probably decline[5]. It would seem as if there were not a few people in Bristol who shared this apprehension that the railway, if constructed, might transfer part of the Bristol trade to London, and that the shipping and West India trade might also leave Bristol and follow the domestic trade to the metropolis[6].

Of the other sources of opposition we shall merely mention a few; and of these the most important was the rivalry of other railways which

---

[1] Felix Farley's *Bristol Journal*, Feb. 22, 1834, p. 1, "Thames and Isis Navigation in opposition to Great Western Railway."

[2] *Manchester Gazette*, Jan. 15, 1825, p. 3, "Effect of Competition."

[3] Shaen, *Review of Railways and Railway Legislation*, p. 29; Sekon, *History of the Great Western Railway*, p. 8.   [4] Shaen, op. cit., p. 29.

[5] Felix Farley's *Bristol Journal*, Jan. 19, 1833, p. 4, letter from "S. T. C."

[6] *Bristol Mercury*, Mar. 2, 1833, p. 2, letter from "Scrutator."

were projected at the time the Great Western was seeking incorporation, such as the line from Windsor to London[1], the London and Southampton, and several others. Some trustees of turnpike roads did not favour the railway, on the ground that their revenues would be depleted because of the transference of the traffic from the roads to the rails[2]. Of a similar character was the opposition of the town of Maidenhead, on the plea that all the existing traffic which paid toll on the bridge over the Thames at that place would be diverted to the railway[3]. As in other instances, coachmasters and the representatives of the carrying trade on the highways presented feeble resistance to the movement in favour of the railway. But it would seem, from the records of the time, that one of the greatest factors with which the advocates of the line had to reckon was the inactivity of Bristol and its people to rouse themselves for the accomplishment of a great future good. We have not found the manifestation of any such sluggish, self-satisfied spirit in the promotion of any other line. The Bristolians of that day, unlike those of the present, seemed to be in favour of the quiet enjoyment of the old, rather than of the reaching out after the new; they seemed to be rejoicing in the peaceful returns from their investments, rather than utilizing their wealth in channels which might greatly aid in restoring their former commercial ascendancy[4]. While the probability of good returns from the railway was inducing capitalists in Liverpool, Manchester, Birmingham, and other important towns in the north to subscribe largely for its shares, the wealthy classes in Bristol were, apparently, indifferent to the opportunity before them. Even after all the facts had been gathered and made public, and it had been conclusively proved before the parliamentary committee that the road would well repay the subscribers, it was with much difficulty and persuasion that they could be induced to support the railway by taking stock in it[5].

[1] *The Times*, Jan. 18, 1834, p. 3, and Jan. 20, 1834, p. 3.

[2] Felix Farley's *Bristol Journal*, Mar. 15, 1834, p. 4; "Proceedings in the House of Commons on the Great Western Railway Bill;" ibid., Mar. 15, 1834, p. 1, report of the meeting at Reading, statements of Mr Harris and Mr Law.

[3] *Proceedings of the Great Western Railway Company*, pp. 10–11.

[4] Felix Farley's *Bristol Journal*, Nov. 16, 1833, p. 4, letter from "R. R.;" ibid., Sept. 28, 1833, p. 3, on "Great Western Railway;" *Bristol Mercury*, June 30, 1832, p. 4, letter from John Ham; ibid., Aug. 11, 1832, p. 3, address of "G. R. C." to rouse the Bristolians from their apathy; ibid., Sept. 1, 1832, p. 2, emphasizing the same thing, and bemoaning the curse of "party spirit;" ibid., Sept. 29, 1832, p. 2, letter from "A Well-Wisher."

[5] Felix Farley's *Bristol Journal*, Oct. 11, 1834, p. 3, letter from "R. R.;" ibid., Oct. 18, 1834, p. 4, and Nov. 8, 1834, p. 2, letters from Thomas Motley, "Good Speed," and E. Jones, urging the necessity of support.

After more than a year had been spent by a committee of citizens of Bristol in an elaborate investigation of the prospects for the railway and in ascertaining minute and accurate information regarding the sources of revenue and the amount of the returns from each source, and after the survey by two engineers had shown that the line was very favourable[1], the matter was brought before the public with the object of enlisting popular support. Subscriptions did not come in very rapidly, so that the company did not secure enough money to warrant their applying to Parliament for permission to construct the complete line between Bristol and London; but in the latter part of 1833 they gave notice that they intended to make application in the ensuing session for authority to construct the two end sections of the line, that from London to Reading, with a branch to Windsor, and that between Bath and Bristol[2]. It was thought that the company was acting wisely in their determination to secure the two ends of their line, first, because if they had applied to Parliament for the whole line, and if for any cause they had failed to obtain their Act, it was highly probable, they thought, that the Windsor Railway Company might obtain the Act they sought, to enable them to build a railway from Windsor to London. In that event, the most profitable part of the whole undertaking would have been lost to the Great Western Railway and the latter would have been at the mercy of its fortunate rival as to the terms of transit on that part of the line. The Bill was read in Parliament for the first time on Feb. 26, 1834[3], and with the great support given it by London merchants it passed rapidly from stage to stage. At its second reading the vote stood 182 for and 92 against the measure, and the advantages of the railway as a national undertaking were becoming firmly established[4]. After a debate of fifty-seven days in the committee of the House of Commons, during which there was strenuous exertion by the contending parties—the one to preserve monopoly, the other to throw open the resources of the kingdom for the general benefit

[1] Felix Farley's *Bristol Journal*, Aug. 3, 1833, p. 4, gives the report of this committee. It is evident from the report that all possible care was taken to secure facts that could be relied upon and to avoid any kind of exaggeration or false security.

[2] Ibid., Nov. 2, 1833, p. 2, Great Western Railway Notice. The committee of promoters decided that for the completion of the whole line £3,000,000 would be needed; but they could not raise this amount in the two months that were left; and the Standing Orders of the House of Lords required that four-fifths of the proposed capital should be actually subscribed before any railway Bill could be read a third time. Hence the decision to get the two most important parts of the line first (*Proceedings of the Great Western Railway Company*, p. 7).

[3] Felix Farley's *Bristol Journal*, Mar. 1, 1834, p. 3.

[4] Ibid., Mar. 15, 1834, p. 3.

—the Bill went to the House of Lords, but that body threw it out without even a hearing[1]. The reason for this failure was probably the fact that there was no security given for the completion of the whole line between these terminal sections[2]. A great public dinner was held by the opposition to celebrate the defeat, to secure which they had diligently and systematically arrayed all possible influence against the measure[3]. But the promoters of the railway set to work more vigorously than before to obtain the necessary amount of subscription to enable them to apply at the next session for authority to construct the whole line. The facts regarding the need and the advantage of such a railway were kept before the public, a new prospectus was issued[4], opposition was allayed in some cases by seeing the real situation in a new light; and in the session of 1835, despite much hostility which could not be placated, the Great Western Railway Act was passed. The road was opened in 1889–41.

In 1824 began the agitation for a railway to connect Newcastle and Carlisle, but it was not until 1829 that this line was authorized, and not before 1835 was it all open for traffic[5]. In 1834, the London and Southampton Railway Act was passed, with almost unanimous support, and the line was opened in 1838–40[6]. In 1825 the surveys for a railway between Leeds and Hull had been made and the work begun; but in 1826 the work was stopped on account of commercial difficulties, and also because of the increased water accommodation due to the opening of the new port of Goole, while at the same time many wanted, first of all, to know what would be the success of the railways then being formed before they should go on with additional construction[7]. In 1829 it was thought wise to construct only the part of the line between Leeds and Selby in the hope that the possible use of steam tugs on the

---

[1] Shaen, *Review of Railways and Railway Legislation*, p. 29.

[2] *Proceedings of the Great Western Railway Company*, p. 7.

[3] For particular instances, see Felix Farley's *Bristol Journal*, Mar. 15, 1834, p. 8, editorial.

[4] 'Collection of Prospectuses, Maps, etc.,' p. 176, gives this prospectus in full (1834).

[5] Cumming, *Rail and Tram Roads*, p. 33.

[6] It would seem that few landowners petitioned against the formation of this railway (*Hampshire Advertiser and Salisbury Guardian*, Mar. 22, 1834, p. 2, petition of George Jones to the House of Commons) and that the amount of opposition to it was very insignificant. See the summary of the evidence upon this Bill as given in ibid., Mar. 29, 1834, p. 2; also Fay, *A Royal Road : being the History of the London and South Western Railway*, pp. 1–28. Yet, notwithstanding the slight opposition, the cost of obtaining the Act was £31,000 (ibid., p. 17).

[7] *Leeds Intelligencer*, Feb. 3, 1825, p. 3, on "Leeds and Hull Railway;" ibid., Feb. 10, 1825, p. 3; ibid., Jan. 29, 1829, p. 3, on "Railroad from Leeds to Hull."

river from Selby to Hull might furnish an acceptable continuation of the railway service. The subscribers to the original undertaking were organized as the Leeds and Selby Railroad Company and application was made to Parliament for an Act to carry out their purpose. It was decided to make the railway available for either horse-power or locomotive engines, or, if thought desirable, to enable the company to use locomotive carriages[1]. The Act was passed in 1830[2], and the work completed in 1834. It was the current testimony that, during the first year of operation, the increased speed and reduced expense had brought about an almost ninefold increase in the number of passengers travelling between these two centres[3]. In all probability, it was this rapidity, cheapness and safety 'of railway carriage, in contrast to the delay, uncertainty and danger of river navigation, that led to the design of a railway between Selby and Hull in 1835[4], but as there was an insufficient response for subscriptions to the latter railway at this time, the project could not be brought before Parliament for another year, and so the Act was not passed for the Hull and Selby Railway until 1836[5].

Although direct rail connexion was thus secured between the manufacturing section of Yorkshire and the port of Hull, there was need of extending these facilities through the industrial sections of Yorkshire and Lancashire, so as to join this great seaport on the east with Manchester and Liverpool on the western sea. The authorization of the Leeds and Selby line was the signal for activity looking toward the junction of Leeds with the large centres of Lancashire. Here, two different routes were suggested: one following the general direction of the Leeds and Liverpool Canal, and the other a more southerly course,

[1] *Leeds Intelligencer*, Jan. 29, 1829, p. 3; ibid., Mar. 26, 1829, p. 3; ibid., Nov. 5, 1829, p. 3. See report of James Walker, the engineer, concerning this line, as given in Macturk, *History of Railways into Hull*, pp. 18–32. He was decidedly in favour of the uniform line, without inclined planes, because the public could then, upon payment of the tolls, freely use the line to convey their own goods in either direction, and because there would be greater simplicity and certainty in its operation.

[2] *Leeds Intelligencer*, May 20, 1830, p. 3. Opposition was encountered from the Marchioness of Hertford and other landowners on the ground of the railway's interference with private property (ibid., Mar. 11, 1830, p. 3; April 1, 1830, p. 2; and May 13, 1830, p. 3). The Aire and Calder Navigation Company at first opposed it, but afterwards withdrew their opposition (ibid., Mar. 18, 1830, p. 3; April 1, 1830, p. 2; and May 13, 1830, p. 3). Slight opposition was also made by the watermen along the river and by the captains and owners of vessels there (ibid., Mar. 4, 1830, p. 3).

[3] *Sheffield Iris*, Sept. 29, 1835, p. 3, on Leeds and Selby Railway.

[4] Macturk, op. cit., pp. 42–46, gives the prospectus of this railway, showing the reasons for its proposed construction.

[5] Ibid., p. 46.

from Manchester to Leeds. In the case of the proposed Liverpool and Leeds railway there were alternative routes suggested, but either of them would take the line through a region of productive industry of manufacturing and mining, and through a series of flourishing towns. So important did the railway appear that along its route some of the occupiers of land and the workers of the mines offered to pay an increase of rent in the event of its being established[1]. Application was made in 1831 for an Act to permit the construction of this railway but it was denied. The promoters, however, immediately set to work to remove the obstacles which had caused their defeat. In the agitation for the other railway between Manchester and Leeds, which also actively began in 1830 by a survey of the line[2], there seemed to be more vigour than in the case of its rival. The citizens of both Manchester and Liverpool, as well as those of Leeds, were eager to see the line constructed[3]; and a report or prospectus of the undertaking was issued, detailing the necessity for the line and the objects to be secured by it[4]. The chief purposes to be served were the accelerating and cheapening of the transport of passengers and commodities and the opening up of wider markets for the productions of the section through which it passed. In 1831 the measure was first brought before Parliament; but the Rochdale Canal Company and other opposing interests gave evidence to show that the existing means of conveyance were ample for all the traffic of the country, and the Bill failed, or else was abandoned for that session because it was too late to get it through[5]. For five years the project

[1] *Manchester Guardian*, Dec. 11, 1830, p. 3, on "Liverpool and Leeds Railway." See also ibid., Jan. 1, 1831, p. 4, letter from " W. N. R.;" *Liverpool Times*, Nov. 16, 1830, p. 365, letter from "Observer;" ibid., Nov. 23, 1830, p. 373, letter from a correspondent; *Leeds Intelligencer*, Nov. 25, 1830, p. 3, on "Liverpool and Leeds Railway."

[2] *Manchester Guardian*, Sept. 18, 1830, p. 2, on Railways; *Leeds Intelligencer*, Sept. 23, 1830, p. 2. A company had been formed in 1825 for making this railway, but at that time of universal depression it was deemed advisable to postpone the measure (*Leeds Intelligencer*, Oct. 21, 1830, p. 2).

[3] *Manchester Guardian*, Oct. 16, 1830, p. 2.

[4] This report is given in ibid., Jan. 29, 1831, p. 1. The delays and inadequacy of the canals along this route were so strongly felt, according to this report, that by far the largest proportion of the merchandise was, with some difficulty, conveyed over the mountainous district by waggons and carts, at great expense and with the squandering of much time. The average time taken by the stage coaches between Manchester and Leeds was seven to eight hours and the time required for the carriage of goods was about twenty-four hours. But by the proposed railway there would be considerable reduction of expense and the time required for the carriage of goods or passengers would be three to four hours.

[5] *Manchester Guardian*, April 2, 1831, p. 3; ibid., July 16, 1831, p. 3; ibid., July 30, 1831, p. 3

slumbered, and then the promoters, with more spirit than before, began a campaign which, in spite of the opposition of the Aire and Calder Navigation, secured favourable consideration by Parliament and the passage of the Act authorizing the construction of the railway[1]. Another line connecting two important centres in this northern manufacturing area was the Manchester and Sheffield. An Act had early been obtained to make a railway here[2]; but, apparently, it was designed as a purely speculative scheme, and when the shares would not bring a premium in the money market the whole thing was given up. But in 1835 the project was revived by those who were vitally interested in securing better facilities of carriage[3]. At that time the only means of conveyance between these places was by waggon over a rough country, and the time occupied in performing the journey was about forty hours. There was neither existing nor prospective water carriage, so that nearly all the traffic would go by the railway, if constructed[4]. No opposition was encountered from the landowners along the route, nor from any other interests; its advantages were indubitable; and the Act was passed for giving effect to the line in 1836[5].

It is not our purpose to give a complete account of each railway that was formed; and we have traced in sufficient detail a few of the most important of the early undertakings, in order to see the various influences pro and con which were operative in laying down these roads. Many of the other lines are equally instructive, but we cannot follow their history here[6]. Among the railways for which surveys were made during the railway fever of 1825, was one from London to

[1] *Leeds Intelligencer*, Jan. 23, 1836, p. 3; ibid., April 23, 1836, p. 3.

[2] *Manchester Guardian*, Aug. 28, 1830, p. 2. The construction of the railway appeared to be considered as certain at that time.

[3] *Sheffield Iris*, Oct. 13, 1835, p. 3, and Jan. 5, 1836, p. 3, on the Sheffield and Manchester Railway. When the measure was given up in 1830, there seemed to be nothing more done about it until 1832, when at a meeting of the subscribers the whole undertaking was discussed. Some regarded it as useless and impracticable, and wanted it abandoned. Others thought nothing further should be done about it for three years. Finally, it was agreed that those who were friendly to it should try to take up the shares of the dissentients, and if they were unsuccessful the concern should be abandoned. Evidently they were unable to meet the last condition.

[4] See prospectus as given in *Sheffield Iris*, May 10, 1836, p. 2; also *The Times*, Oct. 28, 1837, p. 3, report of first general meeting of the Sheffield and Manchester Railway.   [5] *Sheffield Iris*, Oct. 10, 1836, p. 2.

[6] We would mention, among the shorter lines, the Sheffield and Rotherham Railway, the history of which is intensely interesting. The necessity of this line for the industrial development of Sheffield and its environs, the antagonism of a strong but unserviceable navigation monopoly, the hostility of the landlords, two of whom were implacable, are detailed in the columns of the *Sheffield Iris*, especially the following issues: July 29, 1834, p. 2; Oct. 7, 1834, p. 1; Oct. 14, 1834, p. 3;

Cambridge. In 1827 the survey was extended north through Lincoln to York, but by that time the fever had stopped and nothing further was then done toward constructing this railway. In 1833 this Great Northern Railway line was again surveyed from London, via Cambridge, Lincoln and Gainsborough, to York, with several branches; but before the building of the road was authorized many years intervened, during which George Hudson, the "Railway Napoleon," was manipulating the railways of England through his control over the North Midland and the York and North Midland lines. It was not until the year 1845 that an Act was passed to construct the Great Northern from London to a little north of Doncaster[1]. In 1834, the prospectus was issued for the Eastern Counties Railway, which was to run from London, via Colchester, to Norwich and Yarmouth[2], and very glowing accounts were given of the great things which were to be accomplished by this railway. It was sanctioned by an Act passed in 1836; and at the company's first general meeting in that year, the Chairman showed what an "ample return" the stockholders would receive on their capital, and that the enterprise rested on "the broad and stable basis of national utility." But his optimism was eclipsed by the extravagant statements of some of the shareholders who thought that a dividend of at least twenty-two per cent. would be paid, and that this railway and other similar undertakings would provide such a social amelioration as to almost banish misery from the earth. But the perfidy of the Eastern Counties Railway Company, which, instead of building the road through to Yarmouth, stopped short at Colchester, and wanted to leave to another company the construction of the rest of the line, which would not pay so well but which would afterwards be used as a feeder for their more important part of the road, is a chapter upon which we shall not enter[3]. Notwithstanding the troublous days of its early history, the Eastern Counties Railway became an important

Mar. 17, 1835, p. 2; Mar. 31, 1835, pp. 2, 3, 4; April 7, 1835, pp. 2, 4; June 2, 1835, p. 2. Ibid., Sept. 15, 1835, p. 4, and Sept. 22, 1835, p. 4, gives a letter from W. Ibbotson which is very important.

[1] Probably the best account of this railway is Grinling, *History of the Great Northern Railway*, which gives much detail also of Hudson's career. Acworth, *The Railways of England*, ch. v, may also be consulted. On Hudson's career, see also *Railway Times*, VI, pp. 1058, 1084, 1095–6, 1122, 1312–13; VII, pp. 62, 131, 173–4, 327–8; VIII, p. 2127.

[2] Grinling, op. cit., p. 2. An earlier project had been brought forward during the railway fever of 1826, for the construction of a line from Norwich to London; but it was apparently intended to be a speculative venture and not to materialize (*The Times*, April 8, 1826, p. 3, letter from "A Shareholder").

[3] On the Eastern Counties Railway see Acworth, *The Railways of England*, ch. x. Concerning the administrative fraud and financial corruption which made the name

constituent in the Great Eastern, when this latter, in 1862, was formed by the amalgamation of five small lines.

Two other lines running out of London remain to be mentioned. From early days, Brighton had been noted as a fashionable resort, and along the three branches of this road there was a perpetual succession of coaches, each one vying with the others in speed and comfort. Along these lines of travel, too, large sums of money had been spent in cutting off curves, reducing or cutting through hills, and straightening, shortening and improving the road to the greatest extent, so that the numerous coaches which travelled it at all times of the day might not be impeded in their journeys. As soon as railways had demonstrated their many points of superiority over former means of communication, there was a movement for a line between London and Brighton, to provide for the constantly increasing passenger traffic which was overtaxing the coaches. The Bill was brought into the House of Commons in the early part of the year 1836, and ere long there were no fewer than five lines seeking authority to connect these termini, each line being the result of a survey by a different engineer. Then began the parliamentary contest, in which immense sums were spent, varying from £16,500 for the least expensive, to £72,000 for the most expensive. The fortunate line was completed and in operation before the critical period of 1843[1]. The movement for the railway between London and Dover, afterwards called the London and South Eastern, also began in the early months of 1836[2]. This road would be beneficial to the farmers, as, for example, in the quick conveyance of their stock to market; it would enable traders to carry on business with much less capital when they had easy access to London; and it would facilitate and encourage the passenger traffic between London and the Continent[3]. Authority was granted to construct the line, and it was in active use before the middle of the next decade[4]. With the completion of the above-mentioned lines, the chief arteries from the metropolis to the

of this railway a by-word for treachery and deceit, see *Railway Times*, iv, pp. 63–64, and *Herepath's Railway Magazine*, N.S., iii, pp. 92–94, letter from "A Suffering Shareholder."

[1] *Railways as they Really Are : or Facts for the Serious Consideration of Railway Proprietors. No. 1, London, Brighton and South Coast Railway.* This gives the history and finances of this company in brief form, using almost exclusively the parliamentary documents, and citing minutely the references. The writer exposes the fraud practised by the company upon the public, showing the way in which the dividends paid were added to capital, etc. See also Acworth, *The Railways of England*, ch. viii.

[2] *The Times*, Mar. 16, 1836, p. 7.   [3] Ibid.

[4] Consult Acworth, *The Railways of England*, ch. ix, for some interesting details not of an economic character.

different parts of the kingdom were laid down in outline. It is beyond the scope of this work to enter into the minutiae of the construction of the railway net; we merely wish to present its general features in a series of great roads leading out from London[1], with transverse roads where they were most required[2].

The railway fever of 1825–6, as already noted, brought forward many projects which never materialized, and others which took form at a later time. But when the success of the Liverpool and Manchester was demonstrated, there were many who were eager to embark their capital in similar enterprises with the object of reaping corresponding rewards from analogous public services. Public attention was centred upon railways, and with the prospects that were held out by sanguine investors many were induced to put their earnings or capital where they would secure the largest returns. Since there was a disposition to readily devote funds to these particular channels, there came to be a prevailing mania in regard to railroads. Schemes were brought forward which were mere speculations, undertaken for purposes of individual profit and without any thought that they would ever be carried through to completion. Every day new companies were announced, some of them very visionary and destined to end in ruin to those who put their money into them; but as the prices of the shares were advanced, speculation became rampant, and this in turn reacted to push the prices of shares still higher. Railway lines were planned along routes which could barely support a coach. Newspapers contained numerous prospectuses; and, on the basis of the statements made in these, millions were subscribed with eagerness and zeal. Railway Bills were coming before Parliament in great numbers, and in 1836 alone there were presented fifty-seven petitions involving an estimated outlay of over twenty-eight million pounds[3]. Many of these Bills were, of course, left without any action having been taken upon them. The great number of enterprises that were sanctioned during

[1] For full statistics as to railway construction up to 1844, see Brit. Doc., 1844 (318), xi, 17, Appendix No. 2, pp. 4–5; also 'Report of Royal Commission of 1867,' pp. xxxiii–xxxiv. For descriptions of the various lines, see Francis, *History of the English Railway*, i, chs. vii–ix, also Smiles, *Lives of the Engineers*, iii, pp. 346–96.

[2] Such as the Newcastle and Carlisle, Manchester and Leeds, Leeds and Selby, Hull and Selby, Manchester and Sheffield, Leeds and Liverpool, Whitstable and Canterbury, etc.

[3] *Sheffield Iris*, Mar. 22, 1836, p. 3, on "Railways." See also, in regard to this mania, ibid., Oct. 13, 1835, p. 3, on "Railway Speculations;" *The Times*, Feb. 13, 1836, p. 3, April 1, 1836, p. 3, and June 17, 1836, p. 3; Whishaw, *Analysis of Railways*, p. v; Grinling, *History of the Great Northern Railway*, p. 3; *Leeds Intelligencer*, Oct. 31, 1835, p. 4.

this mania, from 1835 to 1837, absorbed so much money that in the years from 1838 to 1844 very few new lines were authorized[1].

By the middle of this fourth decade of the century, it was obvious that railroads were no longer to be regarded as mere private enterprises, but as great public concerns, forming a new but most material element in the development of commerce, national wealth and national resources. Since they were in future to constitute the regular and established modes of communication between the different parts of the kingdom, and by their more rapid speed the value of time would be relatively enhanced, it became a matter of expediency that the lines should be planned according to some well-devised system, and that care be taken not to sacrifice public good to private advantage. If no supervision were to be exercised over the formation of these lines, they would be constructed in the same piecemeal fashion as the canal network, in consequence of which local and individual, rather than national and public, benefit would be considered. The railway mania of 1835–7 seems to have brought the issue more prominently before those who were looking beyond the temporary adjustment; and to them it was clear that to leave the railways to speculators, to be decided according to their judgment and interest, would be the greatest folly. The lines should be made to dovetail into one another; and to have such a preconcerted plan as a basis of action for the Legislature in sanctioning these undertakings, the country ought to be thoroughly examined and studied as to its needs and obstacles. One prime essential was that there should be ready communication between the capital and all parts of the kingdom; London was regarded as the heart from which, by the system of arteries and veins, the life of the whole organism should be maintained. How such a system was to be established and adjusted gave rise to differences of opinion. Some were agreed that the best plan would be to have a survey of the country made under the direction of a Government commission, with a view to laying down the great trunk lines in the most favourable situations, from which branches might be made according to the wants of different sections. In this way the country would avoid the evils of the parliamentary committee system of handling these Bills, under which it was not the best line, but the line whose personnel could exert the greatest influence in the committee, that received the recognition

---

[1] Jeans, *Jubilee Memorial of the Railway System*, p. 141, says that up to and including 1836, Parliament had sanctioned 34 lines of railway, of a length of 994 miles, at an estimated cost of £17,595,000; and that in 1837 there were fourteen new companies incorporated, with power to construct 464 miles of railway at a cost of £8,087,000. Teisserenc, *Études sur les voies de communication*, p. 19, says that in 1838–41 only 200 kilometres were authorized.

sought[1]. Another advocated that each of the great towns, like Manchester, Birmingham, Sheffield, etc., should, as far as possible, have its own direct railway connecting with London, so as to maintain the natural healthy condition of direct communication between the heart and the extremities[2]. It will be noted that this movement in the direction of systematization and correlation in the railway structure of the country was in harmony with the plan of Thomas Gray, more than ten years before, to have a consistent and effective development of the railway facilities; .but in neither case did the proposals meet with favourable action from Parliament[3], and lines continued to be treated as separate entities without regard to any organized relations with others.

Railway enterprise was something wholly new in the history of the world, and Parliament did not know what legislative principles to adopt so as not to stifle their development, but at the same time to safeguard the public interests. As *laissez-faire* doctrines were so predominant in every other aspect of the national life, and had proved to be productive of good in the case of the canals, the same policy was adopted at first regarding the railways. Each project was considered on its own merits; the conditions in that particular locality were expected to be carefully investigated by a parliamentary committee in regard to the need for the proposed line; and by the Act that was passed the railway company was allowed to charge a certain specified maximum of rates for different classes of goods, but otherwise it could conduct its business as it thought best[4]. This was the only restriction imposed upon the company in the operation of its road, for it was thought that other matters would be regulated by competition. The aim of the Legislature, at the outset, was to maintain the same freedom on the railways as on the old roads.

It was the avowed purpose, in the construction of the railway lines, that they should be open for the public use, on the payment of the tolls. This was enacted by Parliament to prevent monopoly, that is, to prevent the railway companies from getting exclusive control over

---

[1] *Parl. Papers*, 1836 (0.96), xxi, 285, 'Minutes of Evidence before Select Committee on Railway Bills,' evidence of James Walker, C.E., Q. 177–212; Mudge, *Observations on Railways*, pp. 30–67.

[2] Brit. Mus. 8235. ee. 12 (1), 'Reasons in favour of a Direct Line of Railroad from London to Manchester,' pp. 1–5.

[3] On this whole subject, in addition to the above references, see also *The Imperial Railway of Great Britain*, by M. A., and *Hansard's Parliamentary Debates*, 1836, xxxiv, pp. 984–8.

[4] This maximum of rates was practically inoperative, for the companies found it convenient to lower their rates, in most cases, below this maximum.

the conveyance of passengers and goods along their respective lines; and even railway proprietors said that they wanted no monopoly: that they were merely toll-takers, and that it was neither their wish nor their interest to undertake the work of a public carrier upon their own lines[1]. It was expected that merchants and others would put their own carriages on the line, and either furnish their own horse or steam-power, or pay the railway company for the use of their power. Even after the introduction of steam-power this system in part prevailed in the case of goods traffic; for we find that in 1838 "engines belonging to different parties, coach proprietors, and others," were running upon the Liverpool and Manchester line[2], and so closely associated was the railway with the ordinary highways, in the public mind, that a select committee of the House of Commons, in 1837–8, recommended that the right enjoyed by private persons of running their own engines and trains upon any railway, should be extended to the Post Office[3].

This system, of having divided responsibility on the same line, was not found to work well. In the first place, there was great danger in the running of rival trains over the same rails, on account of the struggle for the greatest possible use of the railway facilities. In the second place, no provision had been made to ensure, for private trains and engines, access to stations, watering places and other equipment along the line. In the third place, the rate of toll limited by Act of Parliament was almost always so high as to make it impossible for other parties than the railway company to work at a profit, even if

---

[1] 'Report on Railway Communication,' 1837–8, Q. 428, 495.

[2] 'Report of Select Committee on Railroads, 1837–8, Minutes of Evidence,' p. 133. See also 'Report of James Walker to the Committee of the Proposed Leeds and Selby Railway Company,' given in Macturk, *History of the Hull Railways*, pp. 18–32.

[3] Brit. Doc. 1837–8 (257), xvi, 341, 'Report of Select Committee on Railroads,' p. iv. The Post Office had already been forced to put the mail on the Manchester and Birmingham Railway, because, since the introduction of the railway, the passenger traffic had left the mail coaches for a more speedy and economical conveyance, and therefore there was no one who was willing to contract for carrying the mail by mail coaches (ibid., 'Minutes of Evidence,' p. 1).

The reason why it was recommended that the Post Office should run its own cars, was because the railways carrying the mails were often late and usually very irregular (ibid., 'Minutes of Evidence,' pp. 12–17, 61–62). In 1837 the Post Office had entered into agreement with the Grand Junction Railway to carry the mails regularly between Birmingham and Liverpool and Manchester. The regularity, however, was often affected by temporary imperfections in the machinery, breaking down of waggons, taking too heavy traffic, station delays, etc. (ibid. 'Minutes of Evidence,' p. 98).

the other obstacles were removed[1]. Then, too, great difficulty arose from the fact that private parties were not willing to build engines and carriages under such regulations as were necessary to work well on the road[2]. Soon it became evident to the railway companies that, with due regard to the efficiency of their line and to the public convenience and safety, they could not allow rival parties to run engines and carriages on the same line; and it was eventually acknowledged that these lines of communication must be placed under undivided control and authority. Accordingly, a Parliamentary Committee of 1839 urged the necessity of prohibiting, as far as locomotive power was concerned, the rivalry of competing parties on the same line of railway[3]; and the Committee of 1840 decided that railway companies using locomotive power possessed a practical monopoly for the conveyance of passengers, and that under existing circumstances this monopoly was inseparable from the nature of their business[4]. It became imperative, therefore, that each railway company should take over the working of its own line. This difference between railway and other kinds of business was early recognized: that competition of rival interests on the same railway line is impracticable, and that the railway company is in essence a monopoly[5].

But although the practice of traders or independent carriers running their own trains fell early into disuse, the theory of the railways being public highways is found in all the early Acts, and even in a great part of the modern railway legislation[6]. This privilege is preserved, indeed, to the present time, since it is conferred by the Railways Clauses

---

[1] 'Fifth Report of Select Committee of 1844 on Railways,' Appendix 2, p. 22.

[2] Brit. Doc. 1840 (299), xiii, 167, 'Third Report of Select Committee on Railway Communication.'

[3] 'Second Report of the Committee of 1839 on Railways;' also 'Third Report of Select Committee on Railway Communication,' 1840.

[4] Brit. Doc. 1840 (299), xiii, 167, 'Third Report of Select Committee on Railway Communication,' under heading "The Conveyance of Passengers by Railway."

[5] As we have seen, this fact was fully recognized at least as early as 1839 (v. 'Second Report of Select Committee on Railways,' 1839).

As owners of the roads, railway companies were not intended by Parliament to have any monopoly or preferential use of the means of communication on their lines; on the contrary, provision was made in all or most of the Acts of incorporation, to enable all persons to use the road on payment of certain tolls to the company, under such regulations as the company might make to secure the proper and convenient use of the railway. But when railways began to be worked on a large scale with locomotive power, it was found that the necessities of the case demanded the non-recognition of this Parliamentary safeguard.

[6] See remarks of Wills, J., in Hall *vs.* London and Brighton Railway Company, 90, in 15 Queen's Bench Decisions, p. 536.

Consolidation Act of 1845[1]; but the right is one to which it would be impossible to give practical effect, except in a very limited way. Almost the only remaining trace of the theory is found in the "running powers" exercised by one company over the lines of another; but these are usually arranged by agreement or by special statutory provision in each case.

Each railway Act, therefore, provided for the use of the railway by the public, subject to the company's approval of the engines and carriages to be used on it and to the payment of tolls not to exceed the maximum amounts stipulated in the Act. These tolls, in the case of animals and passengers, were on a mileage basis, and in the case of minerals and goods on a tonnage basis. The latter, of course, were divided into different classes[2], according to the nature, bulk and value of the articles and their liability to damage. These tolls were payable merely for the right of passage along the railway. But after 1833 it became the practice to insert in railway Acts a clause allowing the company to charge for supplying the traction power also[3]. Here, then, were two tolls, the "road toll," for the use of the roadway, and the "locomotive toll," paid when the company supplied haulage.

It was not long before the companies took a third step. In two or three cases railway companies were required by their Acts to be carriers[4], but these were very exceptional. It soon became necessary for railways to provide the whole equipment of rolling stock and a staff of officials for doing the carrying themselves, and from 1833–40 we find, in consequence, that the railway Acts contained not only toll clauses, but another clause authorizing the company, "if they shall think proper," not only to provide engines for use by other persons, but also to use and employ them themselves, in carrying the goods and passengers that might require that service performed[5]. The charges authorized by the Acts of that period, therefore, as pertaining to goods, fall into three classes: first, the road toll, for the use of the roadway; second, the locomotive toll (without any specified limit) for the use of the engine; third, a "reasonable charge" for conveyance, in addition to the above tolls, when the company provided everything

---

[1] 8 Vict., c. 20, sec. 92.

[2] The classification of goods for the railway traffic was borrowed directly from that of the canal Acts.

[3] v. Great Western Railway Act, 1835, sec. 166.

[4] Liverpool and Manchester Railway Act, 1827 (7 Geo. IV, c. 49, sec. 138); Newport and Pontypool Railway Act, 1845 (8 & 9 Vict., c. 159, sec. 128); Monmouthshire Railway and Canal Act, 1852 (15 & 16 Vict., c. 126, sec. 128).

[5] See, for instance, the Great Western Railway Act of 1835 (5 & 6 W. IV, c. 107, sec. 167).

and conveyed the traffic along their line. All three of these charges were paid by those who were engaged as carriers on the railways[1]. The reason for these payments is probably to be found in the traffic conditions of the railways at that time. As to the road-bed, the company had the monopoly and therefore Parliament thought best to limit and fix the rates of toll that might be taken for its use; but in regard to the other two charges, it was expected that they would be determined by competition, since the carriers might legally employ their own engines and do their own carrying.

But experience soon taught that competitive carriers on the same line were an anomaly; that the work of conveyance had to be undertaken by the company; and from 1841 on, further restrictions were placed upon the charges of the new companies that were authorized. A new form of clause began to prevail by which an increased toll, of *specified* amount, was authorized when the company had to provide the rolling stock and power and also had to do the actual work of carrying. Under this form of Act, which includes most of the railway Acts from 1841–4, the charges were: first, the road tolls, which even in early Acts had been of *fixed* amount; second, increased tolls of *fixed* amount for the use of the company's carriages; and, third, a *fixed* additional charge for locomotive power[2]. The fixing of the charges for rolling stock probably shows that Parliament recognized the futility of trying to regulate these charges by competition. It may be noted that neither railway company nor independent carrier was allowed to charge more than the aggregate of these three tolls; for a clause in each Act provided that "neither the company nor any other person using the railway as a carrier shall demand or take a greater amount of toll, or make any greater charge, for the carriage of passengers or goods than the company are by this Act authorized to demand[3]."

When conveyance by the railway company had become the usual mode, another change was introduced, in the "Maximum Rates Clause," which limited a company's total charge for conveyance to *something less* than the aggregate of the three tolls; in other words, if the company

---

[1] These three features of the charges that railways were allowed to make may be noted in the Acts of several large railway companies, e.g., Grand Junction Railway Act, 1833 (3 W. IV, c. 34); London and Birmingham Railway Act, 1833 (3 W. IV, c. 36); Great Western Railway Act, 1835 (5 & 6 W. IV, c. 107); Bristol and Exeter Railway Act, 1836 (6 & 7 W. IV, c. 36).

[2] This form of charging clause may be seen in the Oxford Railway Act, 1843 (6 Vict., c. 10, secs. 281. 284); the Warwick and Leamington Union Railway Act, 1842 (5 Vict., c. 81); the Yarmouth and Norwich Railway Act, 1842 (5 Vict., c. 82).

[3] See Oxford Railway Act, 1843 (6 Vict., c. 10, sec. 288).

had the advantage, as conveyers of traffic, of performing all three services, they were to be content with something less than the aggregate of the three sums which, as toll-takers, they were authorized to charge for each service separately. This gives us for the present-day railway Acts two sets of charging clauses: first, the toll clauses, including the three charges spoken of above, and, second, the maximum rates clause, limiting the total charge for carrying[1]. The maximum rates clause insured to the public cheap conveyance, while the toll clauses protected the companies against rival conveyers on their own lines, whether private carriers or other railway companies with running powers, by enabling them to levy tolls upon persons using the railway to such an amount as would prevent competition[2].

When the railway companies had taken over the working of their lines and undivided control was accorded to each over its own line, competition became active between the different railways, and also between the railway companies and the canal companies, in the same territory. The natural effect of this competition was to cause the rates of carriage to be put down, sometimes to ruinously low figures, and when this could not be continued any longer, working agreements were entered into or amalgamations effected, without any Parliamentary sanction[3]. Under these private arrangements, made for the mutual profit of the formerly competing companies, a higher scale of tolls and charges was usually established, sometimes in excess of even the original rates[4]. As soon as Parliament was aware that secret agreements were being made, it endeavoured to encourage those companies that wished to consolidate

---

[1] The earliest Act in which this Maximum Rates Clause was inserted was probably the Kendal and Windermere Railway Act, 1845 (8 & 9 Vict., c. 32).

[2] In regard to these statutory provisions of railway Acts, see Butterworth, *Railway Rates and Traffic*, p. 3 et seq.

[3] Brit. Doc. 1846 (200), XIII, 85, 'First Report of Select Committee on Railways and Canals Amalgamations.' This called attention to the legislative amalgamations, and also to the fact that some important lines of railway, originally formed by independent companies, and which had not proposed any legislative amalgamation, were at that time practically under the same control and management; and so long as these parties felt it to be to their interest to combine, all the evils to be feared from amalgamation might be produced by private arrangements between them. Ibid., 'Minutes of Evidence,' p. 7, shows a list of the railways and canals that proposed amalgamation at this time.

[4] Brit. Doc. 1846 (275), XIII, 93, 'Second Report of Select Committee on Railways and Canals Amalgamations;' also 'Fifth Report of the Select Committee on Railways, 1844, Minutes of Evidence,' p. 200 et seq. Evidence showed that several railway lines had formed working agreements, and had raised their charges to keep up dividends as high as eight to eleven per cent. See also Brit. Doc. 1872 (364), XIII, 1, 'Minutes of Evidence,' p. 332.

to come forward and obtain an Act authorizing this, for, by so doing, some method of general superintendence and control might be adopted, so that competition among lines might not be obliterated. In some cases, amalgamations had been sanctioned by Parliament from the first[1].

While the most competent witnesses favoured amalgamation of competing lines, either of railways or canals, where competition might be destructive, they almost invariably favoured also the amalgamation of closely related lines which were not rivals. It was recognized that where two roads competed for the same traffic they had everything to gain and nothing to lose by amalgamation, or by an arrangement under which the traffic was divided. But the interests of the public must also be looked after, as well as those of the railways and canals. As early as the panic year of 1836, when so many railway bills were being brought before Parliament, attention was called again and again to the fact that railway competition could not be relied upon to ensure the protection of the public from unjust charges[2]. The railway was essentially monopolistic, and even if another railway were formed as a rival it would be to their ultimate advantage to make some under-standing to work together, and thus the possibility of competition would be further removed than ever. But there were a few who saw that it was not economical, nor would it prove effective, to construct two or three lines along a certain route, with the object of securing competition, when one company could carry all the traffic that was likely to be offered[3]. Even for the purpose of making competition effective, this would be a flagrant waste of capital; and the Legislature ought to prevent unnecessary waste of funds by seeing that lines were built only for necessities. But this cry for protection of the public, at the time of the panic, was different from that which came a few years later, after railways became more aggressive and formed closer working relations with one another. At the earlier time it was more spasmodic and individual; at the later time it was prolonged, profoundly and universally felt, and officially recognized. As early

[1] Brit. Doc. 1846 (200), xiii, 85, 'First Report on Railways and Canals Amalgamations.'

[2] *The Times*, June 17, 1836, p. 3, statement of the Duke of Wellington; ibid., June 22, 1836, p. 4, editorial, showing how the various concerns established to provide water for the city of London and its suburbs had finally combined and parcelled out the city for their own profit. This is also referred to by Mr Morrison in ibid., June 22, 1836, p. 4. See also *Hansard's Parliamentary Debates*, 1836, xxxiii, pp. 977–94, and xxxiv, pp. 1–4.

[3] *The Times*, June 17, 1836, p. 3, statement of the Duke of Wellington; Whishaw, *Analysis of Railways*, p. v.

as 1844, Parliament was strongly urged to retain within its power sufficient authority to curb the railways, should these tend to unduly increase their influence.   It was impossible to foresee what turn affairs might take in the following years, and the public must be protected should the railways try to deal illiberally[1].   In the reports of various committees with reference to the railways and canals, we are impressed by the fact that the advantage to the public from competition between these two instrumentalities was fully recognized; but how to maintain that competition for the future was a subject which was constantly pressing for attention, and yet wholly unsolved[2].   They recognized that it would not be a wise policy to always refuse to sanction the amalgamation of railways and canals, for this was frequently for the public good; and the most fruitful suggestion they could make was that a searching inquiry, should be made into the merits of each case, and that Parliament should permit only those amalgamations which could be effected without prejudice to the public[3].   In the light of

[1] Brit. Doc. 1844 (166), xi, 5, 'Third Report of Select Committee on Railways;' also the 'Fifth Report of Select Committee on Railways,' 1844, p. 82.   The same caution was urged by the Board of Trade in the following year, v. Brit. Doc. 1845 (279), xxxix, 153, 'Report of the Railway Department of the Board of Trade on the Proposed Amalgamations of Railways.' They say: "If these extensive powers are to be granted to private companies, it becomes most important that they should be so controlled as to secure the public, so far as possible, from any abuse which might arise under this irresponsible authority."   Then, after showing the complications that had recently arisen in railway operation, and the advantages to some railway companies of amalgamating with others, they say: "Accordingly we suggest for the consideration of Parliament that general and unlimited powers of granting or accepting a sale or lease of a railway or canal by another railway or canal company, or of otherwise merging the independence of one company in another, should not be allowed to be inserted in any Bill; and that when such powers are applied for in any specific instance, they should only be granted after a full consideration of the probable results as regards the interests of the public as well as of the parties."

[2] Brit. Doc. 1844 (166), xi, 5, 'Third Report of Select Committee on Railways;' 1845 (279), xxxix, 153, 'Report of the Railway Department of the Board of Trade,' pp. 3–4; also 1846 (275), xiii, 93, 'Second Report of Select Committee on Railways and Canals Amalgamations;' etc.   In the latter we find it stated that, "There are now few parts of the country which have not derived material advantage from the competition between railways and canals.   It is obviously important that Parliament should not sanction lightly any arrangements which would tend to deprive the public of this advantage; and it has been a subject of consideration with your Committee whether, in order to maintain future competition, it might not be the duty of Parliament to refuse its assent to all bills uniting the interests of the railways and canals."

[3] Brit. Doc. 1846 (275), xiii, 93, 'Second Report of the Select Committee on Railways and Canals Amalgamations,' under heading "Conclusion." This Committee recommended that, since the system of railways and canals had become so complicated, some department of the Executive Government should be given full supervision over them, with power to enforce such regulations as were indispensable

these facts, it is clear that in the various investigations of the railways and canals at that time, the interests of the people as a whole were regarded as paramount; and if these were not conserved it was not because they were not urged upon Parliament, but chiefly because of the lack of knowledge in that body of how to deal with the situation[1].

While Parliament, without experience as a guide, was busy examining the conditions under which railways were operating and endeavouring to secure adequate legislation for their proper regulation as agents of the public service, the companies themselves were active in maturing plans for working agreements or consolidations. These were at first among lines that might be connected into a longer line of communication, and afterwards with parallel and competing roads. Experience showed that lines of short length were generally worked at great disadvantage; and the saving of expense that would result from the consolidation of establishments was another reason why amalgamation was sought by those companies that wanted to add to their pecuniary prosperity[2]. For this reason, the amalgamation of short independent links or branches, and of unprofitable lines, with others of larger extent and in more prosperous circumstances, was eagerly sought from purely economical considerations. Another, and even more important, factor was that the full development of traffic upon a system of railways often depended very materially upon the existence of a uniform system of management and unity of interest over a considerable extent of line[3]. In regard to passengers also, serious inconveniences often resulted from the conflict of interests and lack of uniformity of system among independent companies. The more

for the interest of the public. See also Brit. Doc. 1845 (279), xxxix, 153, 'Report of the Railway Department of the Board of Trade on Proposed Amalgamations of Railways,' p. 4.

[1] Even as early as 1840 (v. Brit. Doc. 1840 (299), xiii, 167, 'Third Report of Select Committee on Railways'), a Committee of Parliament recommended that an authority be appointed to watch the carrying systems practised on differen: lines of railway, with a view to obtaining the best system "for the public welfare." See also the references under footnote 2, p. 579; Brit. Doc. 1844 (166), xi. 5, 'Third Report of Select Committee on Railways;' also 'Fifth Report' of same year, 'Minutes of Evidence,' p. 82; *The Economist, Weekly Commercial Times, and Bankers' Gazette,* 1845, p. 1078, letter from Lawrence Heyworth, urging Parliament to insist that railways be undertaken on such principles of economy as to secure the greatest possible benefits to the public.

[2] Brit. Doc. 1845 (279), xxxix, 153, 'Report of Railway Department of Board of Trade on Proposed Railway Amalgamations.' It was to put an end to the costly warfare of the London and Birmingham, the Grand Junction, and the Manchester and Birmingham, that they were amalgamated to form the London and North Western.

[3] Ibid., pp. 2–3, gives an example of this.

obvious of these evils were those in which attempts were made, by companies holding one portion of a great line of communication, to extort an undue charge by compelling passengers who had arrived at a terminus of one road in second or third-class carriages, either to wait, or to proceed on the adjoining road in carriages of a more expensive class[1]. Even first-class passengers were often subjected to delay and inconvenience in changing carriages and luggage upon a journey, owing to the same cause[2]. From such conditions, it would naturally be assumed that the more complete the unity of interest and management throughout the more satisfactory and efficient would be the arrangements for traffic that had to pass over more than one line. It was these considerations of inter-railway operating economy, then, that led to the early working arrangements and consolidations[3].

These began at an early stage in the history of the railways. By 1844, a number of Bills were being introduced into Parliament to secure authority for consolidating certain lines, but we may be certain that this was by no means the beginning of such things[4]. Private working agreements were, doubtless, in existence for several years before this; for the railways had increased their power so much that in that year a Committee of Parliament urged upon the House the necessity of seeing that the railways did not unduly extend their influence by destroying competition[5]. If there had not been such working agreements in force, there would have been no need for the strong appeal that was thus made to the Government, for there would have been no "illiberal" dealings of the railways toward the public to be guarded against. But we are not left in doubt upon this subject, for the evidence of witnesses is too conclusive to be discredited[6].

[1] See 'Fifth Report of Select Committee on Railways, 1844,' Appendix 2, pp. 20–21. Here is given a good account of the "Nullity of Parliamentary Provisions for the Protection of the Public," and several "Instances of inconvenience to the public from the existence of so many independent railway companies."

[2] Ibid., pp. 20–21; Brit. Doc. 1845 (279), xxxix, 153, 'Report of Railway Department of the Board of Trade on Proposed Railway Amalgamations,' p. 3.

[3] These, of course, were not the only reasons why railways sought consolidation. For instance, the Liverpool and Manchester wanted amalgamation with the Grand Junction Railway so that the two companies together might provide sufficient means to make their station commodious (v. Brit. Doc. 1846 (275), xiii, 93, 'Second Report of the Select Committee on Railways and Canals Amalgamations,' p. 18).

[4] 'Fifth Report of Select Committee on Railways, 1844, Minutes of Evidence,' p. 82, where one witness said: "Now is the time for Parliament to protect the public, when these Amalgamation Bills are being brought in."

[5] Brit. Doc. 1844 (166), xi, 5, 'Third Report of Select Committee on Railways.' This Report is very explicit upon this point.

[6] 'Fifth Report of the Select Committee on Railways, 1844, Minutes of Evidence,' p. 81. Here we are told that the London and Birmingham, and Birmingham and

Whether we call these arrangements railway pools or not may be simply a matter of nomenclature; but the fact is that before 1844 there were quite a number of such agreements for division of traffic, or for adjustment and maintenance of rates. In addition to the amalgamations that were in force before 1845, many others were proposed in that year[1]; and in 1846 there seems to have been a great number of such proposed mergers[2]. While many of these were not in that year sanctioned by Parliament, yet a considerable extent of both railways and canals came into the control of the powerful railway companies[3];

Derby railways contemplated amalgamation, and they were to be amalgamated with the North Midland Company. It would seem, therefore, as if working arrangements must have been in force for these roads before this, else such an extensive amalgamation would not have been projected, without knowing the benefits that would accrue from it. Further, the Birmingham and Derby and the Midland railways, after running a short time, made an arrangement that the Midland Company should take all passengers coming by certain trains by the North Midland line to London, and that the Birmingham and Derby should take all the passengers coming by other trains. This agreement was broken and the two companies quarrelled, after which they carried for almost nothing. Then a second agreement was made (ibid., p. 82).

The Bolton and Preston and the North Union railways which were competing lines for traffic between Preston and Manchester, after a short contest, amalgamated, and in 1844 were applying to Parliament for this amalgamation to be confirmed.

The York and North Midland Railway and the Leeds and Selby Railway were competing lines for part of the traffic between Leeds and York and Hull. But the Leeds and Selby had been leased to, and later bought by, the York and North Midland Company (ibid., p. 83). See also the arrangement of the Manchester and Leeds Railway Company with the Calder and Hebble Navigation (ibid., p. 140). Other agreements are mentioned on p. 169 et seq., pp. 384, 488; see also pp. 20–21 of Appendix No. 2 to this 'Fifth Report of the Select Committee on Railways, 1844.'

On pp. 20–21 of Appendix No. 2 of this 'Fifth Report of the Select Committee on Railways, 1844,' we learn that this movement of amalgamation or consolidation had "made rapid progress of late," and seven instances of this are there given.

[1] For these, see Brit. Doc. 1845 (279), xxxix, 153, 'Report of Railway Department of Board of Trade,' p. 4.

[2] Brit. Doc. 1846 (275), xiii, 93, 'Second Report of Select Committee on Railways and Canals Amalgamations, Minutes of Evidence,' p. 3, shows proposed amalgamations among existing railways, as stated in the titles of the Bills; pp. 3–6 show proposed amalgamations of "new with existing railway companies," about 105 of which sought power to lease or sell to some other railway; p. 7 shows a list of the railways and canals that proposed amalgamation. Some of these amalgamations were for filling up of old canals and building new railways instead, and often these railways were to be united with other railways.

[3] For the canals and navigations acquired by the railway companies, by amalgamation, purchase, or lease, from 1846–72, see Brit. Doc. 1872 (364), xiii, 1, Part ii, pp. 755–6. This is given in Appendix 9. See ibid., pp. 966–71 (Appendix T), for 'Returns from Each Railway Company of the Names, Number, and Extent of the Canals and Navigations under their Control, and How Held.' This does not give the railway amalgamation that had occurred.

and we may be certain that much of the amalgamation that failed to obtain the consent of Parliament became effective by secret agreements between the companies interested[1].

When the time came that railways were allowed to take over canals, and to consolidate with other railways for the formation of great systems, we have a new epoch in the history of railway transportation. Instead of small, detached roads, having poor, if any, connexions with the next adjoining roads, long lines were formed and worked with a degree of economy and efficiency that was hitherto unknown. The times of arrival and departure of trains, instead of being a matter of caprice, and not made to suit the public convenience, were made to dovetail into a general scheme that grew to meet the needs of the public. Lines already constructed, by getting together, could save in the number of officers that were necessary to man them. The public also gained, because by uniting their interests the railways were better managed, their finances were put in better condition, and by thus putting an end to the wastes of competition the roads were able to deal more liberally with the public in the way of supplying conveniences[2].

The completion of these great systems was not effected until after the railway mania of 1844–6 had done its work, and to that subject we must now give brief consideration. What was the cause of this third and greatest railway mania, we may be unable to determine, but it seems pretty certain that it was not brought on by the universal success of the railways which were then in existence. From the list of important railways which was published in 1841, we see that only eight out of twenty-two had their shares selling in the market above cost, while many were selling for prices that were much lower than the paid-up values of the shares[3]. The more probable cause was speculation.

[1] Brit. Doc. 1852–3 (736), xxxviii, 447, 'Fifth Report of Select Committee on Railway and Canal Bills, Minutes of Evidence,' p. 187, where we are told that there were vast amounts of amalgamation that were not sanctioned by Parliament.

President Hadley, in his excellent work on *Railroad Transportation*, p. 159, after mentioning that the early history of English railway pools is obscure, says: "They first assumed importance some thirty years ago," which would make it about 1855. We have shown in the foregoing that there were many working agreements before 1844, under which there was division of traffic among the lines interested; and from a careful study of this period, I would place the time when they became important at least ten years earlier than the date given by President Hadley. The editor of the *Railway Times* characterized the year 1843 as the "year of amalgamation," and said that amalgamation was the order of the day (*Railway Times*, VI, pp. 1128, 1387).

[2] Brit. Doc. 1852–3 (736), xxxviii, 447, 'Fifth Report of Select Committee on Railway and Canal Bills, Minutes of Evidence,' p. 1.

[3] *Railway Times*, IV (1841), p. 107, gives statistics of the more important railways. Those whose shares were selling above cost were the Liverpool and Manchester,

The work of such a man as George Hudson, who rose from a position of obscurity until he could command the policy of several railroads, simply because of his gambling in railway shares and his ability to exercise undue influence over railway directors, was, doubtless, an incentive to others to try the same method of piling up wealth[1]. The names he received as the "Railway King" and the "Railway Napoleon" are typical of his shrewd, grasping policy, his work as a stock-jobber, and his ability to lord it over railway officials for his own material ends. Others were, doubtless, imitating his example; and the rage for speculation was fostered by the weekly reports and circulars of the many brokers. In the latter part of the year 1844 railway projects were numerous, money was abundant, and its investment in railways was encouraged by the prospects of profit held out by scheming designers as bait to the unwary[2]. The editor of the most important railway journal of the time informs us that the fashionable phrase regarding the numerous railway undertakings that were daily making their appearance, was to say that a railway fever was raging[3]; and the editor of the London *Times* was urged to raise his voice in warning against the mania which was then spreading rapidly over the land, and which promised a severe financial crisis that would shake the country[4]. By November 1844, a list of projected lines involving the aggregate capital of £563,203,000 was published, but many of these were abandoned[5]. The rage for shares continued and increased in intensity in 1845, until it infected all classes from peer to peasant and from private individual to government officials[6]. The press was full of

Grand Junction, London and Birmingham, Great Western, Birmingham and Gloucester, London and South Western, Manchester and Leeds, and the York and North Midland.

[1] For his career, see the files of the *Railway Times*, vi (1843), pp. 1058, 1084, 1095–6, 1122, 1312–13; vii (1844), pp. 62, 131, 173–4, 327–8 (in which he is called the "Prince of Premium Hunters"), etc. Also the great detail given by Grinling, *History of the Great Northern Railway*.

[2] Brit. Mus. 1396. g. 21, 'Railways and the Board of Trade,' p. 7.

[3] *Railway Times*, vii (1844), p. 485.

[4] *The Times*, Nov. 20, 1844, p. 7, letter from "John Trot."

[5] Jeans, *Jubilee Memorial of the Railway System*, p. 142, quoting from Spackman's published list.

[6] Brit. Mus. 1396. e. 22 (4), 'Ruminations on Railways,' No. i, on "Railway Speculation," discussed the mania with sanity. The writer said (p. 6): "Such is the delirium in the share market, that many an honest, industrious tradesman withdraws from his more sober pursuits behind the counter, and dubs himself that delver into the mines of Golconda, a share-broker." In subsequent pages he goes on to describe how vehement was the fever for railroads. See also Brit. Mus. 8235. d. 27, 'Railways and Shareholders,' p. 3.

railway prospectuses[1]; and a large amount of the shares were in the hands of persons who were holding them, not for investment, but merely for speculation[2]. Share jobbing was rife[3]. Even such journals as the *Railway Times* and *The Economist* were encouraging this vast expenditure of money and declaring that railway securities would constitute important means of investing capital[4]. On the other hand, the editor of the London *Times* was giving words of warning to the public against the time when blind confidence would be displaced by doubt and when the inevitable collapse would come[5]. All kinds of fraudulent methods were employed to delude the public and secure their funds[6]. The significance of the mania may be judged by the fact that, during the three years 1844–6, Parliament sanctioned Bills for the construction of 8470 miles of railway, which was just about three times the mileage then constructed; and the amount of capital required for them, £180,138,901, was so great that the further growth of the railway system was checked for some years[7]. The financial panic which followed the railway mania, and which was probably in large measure due to the locking up of so much money in these temporarily

---

[1] See, for example, the *Railway Times, The Economist, The Times,* in their advertising columns.

[2] *The Times,* Aug. 9, 1845, p. 6, on "Railway Speculation."

[3] Ibid., July 11, 1845, p. 5; July 12, 1845, p. 5; July 14, 1845, p. 5; July 21, 1845, p. 7; July 25, 1845, p. 8; etc.

[4] *The Economist,* 1845, p. 1013; *Railway Times,* VII, p. 485.

[5] *The Times,* Aug. 9, 1845, p. 6. It is interesting to contrast the editorial opinion of the *Railway Times* in 1844, when it was said that there was "neither fever nor lunacy in forming new railway schemes without end" (VII, p. 485), with that in 1845, when his verdict was that those embarking in new railway schemes ought to exercise more caution (VIII, Pt. I, p. 569).

[6] *Railway Times,* VIII, Pt. I, p. 1013, letter from "Expositor;" 'Railways as they Really Are,' No. I, on the 'London, Brighton and South Coast Railway,' which exposes the frauds of this company, also Nos. II and VII; Brit. Mus. 8235. d. 27, 'Railways and Shareholders,' pp. 3–4; *The Times,* Oct. 18, 1845, p. 5, editorial; ibid., Oct. 23, 1845, p. 7, letter on "Railway Speculation." This subject was continued in ibid., Oct. 25, 1845, p. 4; Oct. 27, 1845, p. 4; Nov. 8, 1845, p. 4; Nov. 14, 1845, p. 4; Dec. 2, 1845, p. 4; in which the editor wrote strongly against the tide of speculation that was flooding the country. See also Brit. Mus. 1396. e. 22 (4), 'Ruminations on Railways,' No. II, 'The Railway Board of Trade,' and Brit. Mus. 1396. g. 21, 'Railways and the Board of Trade,' 3rd ed., pp. 20–28, showing the evils that attended the work of this body.

[7] Jeans, op. cit., p. 142. In the year 1846 alone the length of railway authorized was almost double the total length of line authorized up to the end of 1843. Brit. Doc. 1854–5 [1965], XLVIII, 1, 'Report of the Railway Department of the Board of Trade' for 1854, p. xi. The amount of money authorized to be raised for the railways that were sanctioned in 1846 was £132,617,368; for those of 1847, £39,460,128; for those of 1848, £15,274,237; for those of 1849, £3,911,331 (ibid., p. vii).

unproductive enterprises, was severely felt in the spring of the year 1847, but we shall avoid any further reference to that subject.   One outcome of the mania to which we may here allude was the great number of suits that were brought before the courts; some of these were instituted by railway companies against shareholders, because the latter refused to pay up the calls that were made upon them in connexion with their subscriptions; others were brought by individuals to recover deposits of money that they had advanced for the construction of railways which had not materialized.   Some suits were started as a consequence of the winding up of undertakings that had proved abortive; and others were due to a variety of causes, which we need not enumerate[1].

We have already noted that up to and including the year 1843 there had been considerable amalgamation of railways[2]; in fact, it was asserted in 1843 that "amalgamation is the order of the day[3]."   But after the cessation of the mania in 1846 there was a still greater agitation for amalgamation[4].   With the great amount of construction and reorganization which took place immediately following the mania, the weaker roads found it necessary to ally with the stronger, not only to reduce the expenses of management and operation, but to produce peaceful relations among the companies.   Beginning with this epoch-making time in the history of the railways, a vast amount of consolidation was effected[5], and the railways, instead of being left as independent units, were gradually becoming organized into a system which was beginning to take on its permanent form[6].   We may say that, by 1850,

[1] *Railway and Canal Cases*, Vols. IV and V, give many of these.

[2] *Railway Times*, Vol. VI, gives much material on this subject, in addition to what we have already given.

[3] Ibid., VI, p. 1128.   [4] Ibid., IX, p. 316.

[5] Brit. Doc. 1847–8 (510), LXIII, 449, gives very complete returns of all existing railway amalgamations in Great Britain and Ireland, accompanying which is a map showing the amalgamation of railways that had taken place.

[6] The tendency in 1844 for railways to consolidate into a few great systems was becoming daily more manifest.   The results that had already been realized showed conclusively that the probability was that the principal lines would be grouped into six or eight leading divisions.   For the consolidations that had been made by 1844, see Brit. Doc. 1844 (318), XI, 17, 'Fifth Report of Select Committee on Railways,' Appendix No. 2, p. 21.   For the probable results of the tendency toward amalgamation, see ibid., Appendix No. 2, p. 21.

The leading systems, as developed in outline, by 1844 were as follows:

First.   The Great North Western artery, extending for 238¼ miles in a direct line from London to Lancaster and connecting Birmingham, Manchester, Liverpool, and the manufacturing districts of Lancashire with the metropolis.

Second.   The Great Midland and North Eastern system, beginning at Darlington, passing through York, within a few miles of Leeds and Sheffield, through Derby

the present-day grouping of lines into the great arteries of communication had been effected; and the changes since then have been the filling in of the network.

In connexion with the subject of amalgamation, there are one or two other features which require mention. The earlier railways had been formed by companies owning comparatively short lines; for example, the line from London to Liverpool belonged to three companies; and great loss of time and inconvenience arose from the want of unity of management and from disputes between the companies. Therefore, partly for economy of management, and partly for the convenience of the traffic, some of these companies whose lines formed links in a through route obtained powers to amalgamate. But as time went on a further increase in the number of railways led to competition of rival lines at many more points. This resulted in further amalgamation and buying up of rivals. Thus, amalgamation, which at first was a question of economy of management and public convenience, became later a matter of offensive and

and Leicester, and meeting the Great North Western artery at Rugby. The length of this line was 201¾ miles, and it was soon to be extended to Newcastle.

Third. The Great Western system, from London to Bath, Bristol and Exeter. This when completed would give 194¼ miles in a continuous line.

Fourth. The great transverse system, formed by the Liverpool and Manchester, the Manchester and Leeds, the Leeds and Selby, and the Hull and Selby railways, connecting the two leading ports of the east and west coasts, by a line of communication 132 miles long, and passing through the heart of the great manufacturing districts of Lancashire and Yorkshire.

Fifth. The South Western system, from London to Southampton and Portsmouth.

Sixth. The South Eastern system, consisting of the Dover and Brighton Railways, which diverged from a common trunk a little south of Croydon.

Seventh. The Eastern Counties system, intended to have connected Norwich and the Eastern counties with the metropolis, but only finished at that time from London to Colchester, 51 miles.

Eighth. The Northern and Eastern system, intended to connect London with York by a line passing through Cambridge and Lincoln, and completed for only 32 miles out of London.

Among the minor and subsidiary lines may be mentioned:

First. The Newcastle and Carlisle Railway, connecting the North Eastern and North Western arteries.

Second. The Birmingham and Gloucester, Bristol and Gloucester, and Cheltenham and Great Western railways, connecting the North Western and Great Western arteries.

Third. The Birmingham and Derby Railway, connecting the Midland and the North Western arteries.

(v. Brit. Doc. 1844 (318), xi, 17, 'Report of Railway Department of Board of Trade,' Appendix No. 2, p. 6.)

defensive policy, to enable the companies to fight one another more successfully.

The benefits from amalgamation were so apparent that many could foresee its continuance until all the railways of the country were united under the control of a few large corporations. Some went even further than that, and advocated a general amalgamation of all the railways, not only from the standpoint of economy of operation, but to prevent a great deal of the jobbing and indiscretion that existed in some boards of directors[1]. As early as 1846 a scheme was proposed for merging the shares of all railway companies into one common stock, under the management of a general proprietary board[2]; and this idea so occupied public attention that in 1852 the issue was investigated by a committee of Parliament[3]. That committee, however, reported adversely upon the plan, and it was never attempted. Another proposed solution of the railway problem, after the principle of monopoly had been recognized, was that certain districts should be assigned to particular railway companies, and that in those districts each should be protected from competition, in exchange for certain advantages that they were to give the public[4]. No definite plan was brought forward to carry this into execution and its futility soon became evident.

Another factor tending to the harmonious operation of railways, after the principle of amalgamation had been quite largely followed out, was the establishment in 1847 of the Railway Clearing House. In the earlier period of railways, the rolling stock of one company did not generally pass from one line to another and the inconvenience and expense due to change of vehicles or transhipment were very great. But when the railway system had been developed to a considerable extent, it was necessary for the companies to have a mutual understanding in regard to the sending of traffic over one another's lines. For this purpose, the chief railway companies formed from among themselves an association, with a central office in London, to regulate certain questions of interchange of traffic as between the several companies, and to adjust the accounts arising out of the united action of the companies: to settle disputes as to the division of, and to

---

[1] Brit. Doc. 1846 (489), XIII, 217, 'Report of Select Committee of House of Lords on Railways,' evidence of Mr W. Cubitt, p. 101, Q. 898; also 1852–3 (736), XXXVIII, 447, 'Minutes of Evidence,' p. 32.   [2] Ibid.

[3] Brit. Doc. 1852–3 (736), XXXVIII, 447, 'Fifth Report of Select Committee on Railway and Canal Bills.' It shows that such a general amalgamation would be undesirable, and why.

[4] This view was taken by Gladstone's Committee of 1844 and by Lord Dalhousie's Railway Commission of 1845. See also Brit. Doc. 1852–3 (170), XXXVIII, 5, 'Second Report of Select Committee on Railway and Canal Bills, Minutes of Evidence,' p. 30.

apportion the receipts from, the traffic that might pass over more than one line, under agreements made by the several companies; and to keep the records of the movements of waggons and carriages when these might pass off the lines of the company to which they belonged, to the lines of other companies. This Railway Clearing House was a purely voluntary association at first, but in 1850 it was incorporated by Act of Parliament and had become a very important feature in preserving amicable relations among the various roads.

The width of gauge was another important problem which came up for consideration in 1846. When the Great Western Railway was constructed, the engineer, I. K. Brunel, constructed the road with the rails seven feet apart, while other roads generally had the rails only four feet eight and one-half inches apart[1]. This diversity of gauge was a serious barrier to interchange of traffic, and in the above year, before the Committee of the House of Lords, railway engineers and others were in perfect agreement that the width of gauge should be uniform[2]. The settlement of this was important on account of the enormous number of lines that were then in progress and in prospect. In the session of 1846 the Gauge Act was passed, which enacted that unless it should be otherwise specified in the special Acts all future railways in Great Britain should be constructed upon the gauge of four feet eight and one-half inches, with the exceptions of railways forming branches of the Great Western, or those situated in the counties of Somerset, Dorset, Devon and Cornwall[3].

In a former connexion we spoke of the advantages which were anticipated from the development of railways, and we have referred to some of the results which were actually obtained in the cases of the Stockton and Darlington and the Liverpool and Manchester railways. A few words more as to the benefits that were definitely conferred by railways may not be out of place. Of course, the greatest results came

[1] There were also mixed gauge lines, that is, roads with part of one gauge and part of another. In 1854, out of a total of 6114 miles of railway in England, there were 206 miles of mixed gauge, 647 miles of broad gauge, and 5261 miles of narrow gauge (i.e., 4 ft. 8½ in.). See Brit. Doc. 1854–5 [1965], XLVIII, 1, 'Report of Railway Department of Board of Trade' for 1854, p. xii.

[2] Brit. Doc. 1846 (489), XIII, 217, 'Report of Select Committee of the House of Lords on Railways, Minutes of Evidence,' pp. 106–7. See also Brit. Doc. 1846 (353), XXXVIII, 371, 'Copy of Minute of the Lords of the Committee of the Privy Council for Trade, on the Report of the Commissioners for inquiring into the Gauge of Railways, June 6, 1846.' This has some good things on the subject of gauge.

[3] Brit. Doc. 1854 (139), LXII, 441, 'Report of the Board of Trade to the General Committee on Railway and Canal Bills, on the Railway Bills of 1854,' p. 26; also Brit. Doc. 1867 [3844], XXXVIII, 1, 'Report of the Royal Commission on the Railways of the United Kingdom,' Part I, p. xv.

through the development of traffic, consequent upon the reduced cost and the increased speed of conveyance. Before the Stockton and Darlington was constructed, the number of passengers travelling between these two places was scarcely sufficient to pay for the running of a coach three or four times a week[1]. Between 1825 and 1832, when there were separate coaches running on the line, belonging to different individuals, the average number of passengers did not exceed 520 per week; so that the growth of the passenger traffic was slow but steady[2]. After that, the company took over the passenger business and so greatly increased the comfort and speed of their trains that, according to the statement of F. W. Cundy, a celebrated engineer, in 1834, 600 passengers per day were frequently conveyed along this line, where, formerly, by the coach, there were not more than ten passengers per day[3]. On the Liverpool and Manchester railway, in 1832, according to the evidence of the treasurer of the company, there were almost three times as many passengers conveyed as had been carried by the twenty-two regular coaches before the railway was opened[4]. In the case of the Leeds and Selby line, the number of passengers who travelled between these places during the first year of the operation of the railway increased, we are informed, from about 400 to about 3500 per week[5]. It is difficult to believe that there could have been as much as a nine-fold increase here in that short time, and yet we must remember that Leeds was flourishing as an industrial centre and Selby as a shipping centre. Perhaps some of this increase may have been merely experimental, indicative of the popular curiosity to try this new agency of travel, and may not have represented anything like as great a gain in the substantial, permanent increase of the business. But if passenger traffic increased so much there was a corresponding gain in the freight traffic and many a place was galvanized into new life by the advent of the railway. For example, during the first quarter of the nineteenth century the port of Stockton seemed to be subject to a gradual decline, but after the railway was built to connect with that port there was almost immediate reversal to a condition of steady progress. At the Tees ports the number of ships which cleared outwards in 1830 were three British ships of 262 tons and four foreign ships of 318 tons; but in 1841

[1] Jeans, op. cit., p. 79.

[2] Ibid., p. 86. The details are given in pp. 85–86. See also *Birmingham Journal*, July 8, 1826, p. 2.

[3] *Sheffield Iris*, Oct. 14, 1834, p. 3.

[4] 'London and Birmingham Railway Bill. Extracts from Minutes of Evidence given before the Committee of the Lords,' evidence of Henry Booth, pp. 53–55.

[5] *Sheffield Iris*, Sept. 29, 1835, p. 3, editorial comment on the Leeds and Selby Railway.

there were 454 British ships of 80,139 tons and 596 foreign ships of 44,392 tons[1]. The export figures for the coal trade are also instructive, since this railway was designed to tap the great coalfield behind Darlington. The total shipments of coal from Stockton, both coastwise and foreign, amounted to 1224 tons in 1822; 10,754 tons in 1826; 66,051 tons in 1828; 704,781 tons in 1835; and 1,500,374 tons in 1840[2]. Of course, it is possible that the railway was not the only cause of the great development of this traffic; but the fact that the great upward trend synchronized with the opening of the railway furnishes a strong presumption that the railway was the chief cause of this development. The Liverpool and Manchester, in addition to saving cotton manufacturers and others large amounts on the conveyance of goods[3], and increasing greatly the amount of business carried on in this locality, increased also the value and the extent of the traffic of the Leeds and Liverpool Canal, although the latter, fearing injury to their property, had opposed the railway at a very large expense[4]. The influence of the Manchester and Birmingham and the London and Birmingham lines in the development of traffic along that route was such that, in 1846, these lines were regarded as no longer capable of handling the immense amount of freight that was offered to them; and the manufacturers of both Manchester and Birmingham wanted to see a direct line constructed to connect them with London[5]. The great development of industry and agriculture which gave rise to the above-mentioned increase of traffic was one of the accompaniments of the railways.

Another of the immediate effects of the railway was the enhanced value that was given to land adjacent to it. The fact that by this means good markets could be brought nearer to the farmer made the land more valuable; and since the cultivator could secure a larger net return from

---

[1] Jeans, op. cit., pp. 173–4, gives the statistics for each intervening year.

[2] Ibid., pp. 174–5. Ibid., p. 176, gives comparative statistics of coal exports from Newcastle, Sunderland and Stockton, in the period before 1850, showing the extremely rapid growth of the exports from Stockton from 1821 to 1850, as compared with the exports from the other two ports.

[3] See evidence of many witnesses on the London and Birmingham Railway Bill before both Commons and Lords, 1832.

[4] 'London and Birmingham Railway Bill. Extracts from Minutes of Evidence given before the Committee of the Lords,' evidence of James Forster, p. 44.

[5] Brit. Mus. 8235. ee. 12 (1), 'Reasons in favour of a Direct Line of Railroad from London to Manchester,' pp. 8–11. The writer of this pamphlet says: "The trains are now frequently of such vast size as to render it impossible for the Company to keep time. These facts can be abundantly established, even by the testimony of their own admissions and declarations."

it he could pay a larger amount for his use of the land[1]. Land which formerly had been of little or no value, such as Chat Moss along the Liverpool and Manchester line, soon became veritable garden spots, and the proximity to large consuming centres, effected by the railway, made the land valuable for gardening and other agricultural purposes[2]. Its value for building and industrial purposes also was soon recognized, and if the railway company, after its line was constructed, wished to purchase more land adjoining what they already had, they had to pay twice to five times as much for this subsequent purchase as for the first[3]. When land was advertised as being for sale or to let, if it were at all possible the advertisement would stipulate that the railway either passed through the estate or near to it, for under these conditions a higher price would be paid[4]. After railways had been carrying on their work for a few years, and it became known that they had paid at times large sums for the real estate they required, the contemplated formation of railways in different parts was the signal to put up the price of land. In some instances exorbitant prices were asked by landowners, and, as the railway companies were not willing to accede to these prices, juries were summoned to assess the value and decide between the two parties. In these cases the almost invariable result was that the jury assessed the value of the land at a lower figure than that offered by the railway

---

[1] *Manchester Guardian*, Dec. 11, 1830, p. 3, showing that occupiers of land and mines volunteered to pay higher rentals if the railway were put within easy reach of them. *The Times*, Sept. 4, 1835, p. 2. This was also in accord with the testimony of Mr Pease, a director of the Stockton and Darlington Railway, who said that not only had the value of his land along the line been increased, but his rentals had likewise increased, and that amid falling prices (*The Times*, Feb. 13, 1836, p. 3).

[2] *Birmingham Journal*, May 19, 1832, p. 3, evidence of Messrs Moss, Earle, Lee and Pease; *Sheffield Iris*, Oct. 7, 1834, p. 1; Cundy, *Observations on Railways*, 2nd ed., pp. 11–15, 17–24.

[3] *Sheffield Iris*, Oct. 7, 1834, p. 1. Mr J. Moss, Deputy Chairman of the Liverpool and Manchester Railway Company, said that for the first eight miles outside of Liverpool his company paid 5s. 8d. a yard for the land they needed; but land all around the railway was sold before 1832 at 22s. a yard. The company also bought land at another part of their line for 7s. a yard, but in 1831, when more was wanted, 10s. 8d. a yard had to be paid. At another part of the line the company's subsequent purchase of land had to be made at double the price of the original purchase. Thomas Lee's testimony was that after the construction of the railway, land had been sold for building purposes at from three to five times the sum it would have brought before the establishment of the railway (*Birmingham Journal*, May 19, 1832, p. 3, evidence of Messrs Moss and Lee).

[4] 'London and Birmingham Railway Bill. Evidence before the Lords Committee,' p. 46, evidence of Joseph Pease; *Birmingham Journal*, May 19, 1832, p. 3, evidence of Messrs Moss, Earle and Pease.

company[1]. What we wish to impress is that either the prospective or the actual construction of a railway was accompanied by a movement toward higher prices for the land in the circumjacent territory.

Of the other immediate benefits secured by railways, we might enumerate a long list. Sometimes they conferred public benefits by reducing overgrown monopolies within reasonable bounds, as was done by the Liverpool and Manchester Railway when it entered the contest against the three navigation companies that operated between these two cities[2]. Sometimes they stimulated the more opulent canal companies to make improvements in their canals and thus contribute to the public welfare instead of dividing among the proprietors the enormous profits that had been made by some of them[3]. They created an immediate and great demand for labour and thereby eased the burden of the labourers and of the community; they furnished in some cases good investments for English capital, and thus kept these funds within the country for the development of the kingdom, rather than having them seek employment in foreign countries. But why need we go any further, for the history of the remainder of the nineteenth century is the record, in part, of the achievement of the railway.

With all the benefits which accrued from the construction of railways, there were also some evils which were a natural accompaniment of such a great change. In the first place, in railway initiation there were features which were decidedly objectionable. Some lines were formed for no other purpose than pure speculation; their promoters wanted to influence the market in such a way that the prices of their shares would reach a high figure, and then they would unload their holdings upon others who were innocent of the game that was being played. Values were given to shares purely on account of market manipulation, without any reference to the intrinsic value of the property upon which they were based, for in not a few instances they

[1] *The Times*, Oct. 28, 1837, p. 3, gives a number of instances to show the relation of railways to the price of land. For example, in Bath a gentleman claimed £6780 for land taken by the Great Western Railway Company; the company offered him £4500 merely to save litigation, but this would not satisfy him, and the jury awarded him only £4223. Under similar circumstances, Lord Manvers, in Bath, on his claim of £9000, received from the company an offer of £4500; but the jury awarded him only £3375. Many other cases are given in this reference. On five claims of £16,067, the jury gave only £2053. 7s.

[2] The same outcome was the result of the Grand Junction and of the London and Birmingham Railway competition with the canals.

[3] P., *A Letter to a Friend, containing Observations on the Comparative Merits of Canals and Railways*, pp. 12–13, 29–30. Examples are given to confirm this fact.

had no material basis at all[1]. In other cases railways were projected, not with any idea of construction, but to induce existing railways to buy them off rather than have to meet the threatened competition[2]. Wilful misstatements of fact, in order to induce the public to come forward and invest in these undertakings, were not at all uncommon; and mere probable estimates were put forth with an assumption of confidence and reliability of accuracy that were intended to deceive the unwary[3]. In some cases, impecunious individuals were able to reap considerable fortunes by bringing forward schemes for railways and having wealthy landowners along the proposed line pay large sums to cause the promoters to desist from what they regarded as a possible disfigurement of their estates. When we remember that many schemes projected at the times of the three manias were unworthy of being entertained, but were started for gambling purposes, we can faintly discern the expense and magnitude of such an evil[4]. Sometimes subscription lists were swollen by using the names of persons who had never given their consent, especially of those who would be influential in inducing others to signify their allegiance to the proposed scheme; and this moral turpitude must have been quite prevalent since there was the passage of an Act in 1844 for punishment of this offence[5].

[1] *Sheffield Iris*, Oct. 13, 1835, p. 3, on "Railway Speculations," showing that the Manchester and Sheffield Railway, as first planned by some Liverpool people, was got up for this purpose, and that this view was probably present in the minds of those who went in for the North Midland Railway. See also Whishaw, *Analysis of Railways*, p. v, who speaks of these schemes as noxious weeds to be eradicated; Investigator (pseud.), *Beware the Bubbles!!!*, 2nd ed., pp. 1, 8; *Herepath's Railway Magazine*, N.S., I, pp. 32–35, where the editor shows up these fraudulent methods that he had known: ibid., I, pp. 72–78, letter of "Detector;" *The Times*, Oct. 18. 1845, p. 5, editorial; Morrison, *Defects of English Railway Legislation* (1846), p 5.

[2] See editorial in *The Times*, April 8, 1826, p. 3, expressing deep regret at the spirit of speculation that had broken out among all ranks; also, on same page, letter from "A Shareholder" of the Norfolk, Suffolk and Essex Rail-Road Company, showing this nefarious scheme to have been only on paper, and not intended to be realized.

[3] *Herepath's Railway Magazine*, N.S., I, pp. 72–78, letter of "Detector," concerning the South Eastern and Dover Railway; ibid., II, pp. 114–17, false statements from the promoters of the Cheltenham. Oxford and Tring Railway, which they did not attempt to deny.

[4] Martin, *Railways*, p. 33, quoting from 'Report of Parliamentary Committee on Railway Acts Enactments,' Aug. 25, 1846. See also letter in *Morning Chronicle*, Nov. 17, 1848, on the results of competition.

[5] *The Times*, Oct. 18, 1845, p. 5, editorial; *Hansard's Parliamentary Debates*, 1837, XXXVI pp. 855–63. See Act 7 & 8 Vict., c. 110, sec. 65, which punished this offence by a fine not exceeding £10. Compare the padding of the subscription list of the proposed London and Birmingham Canal, as given in *Parl. Papers*, 1830 (251), X, 719.

The construction of too many lines along certain routes was another initial detriment. Following the accustomed policy, Parliament, thinking that competition would be desirable also in railways, sanctioned many competing lines, which swallowed up capital and seemed to waste the national wealth. It was not foreseen that competition, instead of causing low rates, might operate just the reverse. While this employment of the country's capital did produce results which were immediately injurious, it is open to question whether this supposed excess has not, in the long run, been of substantial benefit to the country. But even with all this expenditure, it seems clear that it was not always the best line, but the one which could command the greatest influence in Parliament, that secured recognition[1]; and this opened the way for political corruption and jobbing.

In the second place, in railway finance there were some things which were not for private welfare or public good. The enormous sums paid by the railway companies for lands and for compensation constituted a heavy preliminary obligation. An investigation of this matter in 1845 showed that landowners frequently obtained for their lands a much larger amount than the land was really worth; and a landowner who was a member of Parliament, and who would otherwise be likely to oppose the Bill in Parliament, was sometimes given a higher price for his land than another who could not wield that influence. The companies recognized that it was often better to spend money in this way and stop opposition at the beginning, than to pay the higher expense of getting Bills through Parliament in the face of such opposition. In some cases, extravagant sums were paid in order to get rid of opposition, not only from landlords[2], but from rival railway companies and other interests[3]. These expenses, along with the legal fees that

---

[1] *Parl. Papers,* 1836 (0.96), xxi, 235, 'Minutes of Evidence before Select Committee on Railway Bills,' evidence of James Walker, C.E., Q. 178.

[2] *Parl. Papers,* 1845 (420), x, 417, 'Report of Select Committee of the House of Lords on Compensation for Lands taken for or injured by Railways,' evidence of John Duncan, John Clutton, Edward Driver and John Cramp. Q. 289 shows that the gentleman who bought the land for the London and Brighton line testified that he paid fully ten times what the land was worth, simply to get rid of opposition.

[3] Young, *Steam on Common Roads,* pp. 67–68. Shaen. *Review of Railways and Railway Legislation,* pp. 36–40, cites cases from the parliamentary reports showing that large sums had to be squandered to buy off opposition to certain Bills. Ibid., pp. 43–46, shows how much time and money was wasted before railway committees of Parliament. Chattaway, *Railways,* pp. 23–24, said that "the sums paid by many of the railway companies for land and compensation are almost fabulous;" and as he was an officia of the North British Railway, he should have known the facts. He referred to one property valued at £5000 that had been sold to a railway company for £120,000. See also Marshall, *Railway Legislation,* p. 30.

had to be paid for competent solicitors who were experienced in the work of guiding measures through Parliament, and which were also excessive and sometimes extortionate[1], made the cost of obtaining an Act of Parliament very burdensome. Another element in the situation that was to be deprecated was that rival roads endeavoured to outdo each other in the fineness of their equipment; and engineers, anxious to make a name for themselves, put into the construction of bridges and other works a large amount of money, which added to the aesthetic value, but not to the traffic value of the road. By these and other means the costs of the railways were often increased two or three times as much as the original estimate[2].

Moreover, there was much financial manipulation that was derogatory to the welfare of the companies. When the shareholders elected their directors to look after the affairs of the company, they allowed them, in too many cases, to have full authority over the property and policy of the company, without keeping any oversight of the way in which the directors fulfilled the trust that was reposed in them. If one or two of the directors were particularly aggressive they sometimes got too large a share of the control of the railway, and used it for their personal interests rather than for the benefit of the owners of the

[1] Young. op. cit., pp. 70–71. He gives statistics as to these costs in particular cases. In one instance the expense was £146,000, and then the Bill was defeated; in another the solicitor's fee was £240,000. See also Fay, *A Royal Road*, pp. 14, 17. Chattaway, *Railways*, p. 23 says that the parliamentary and legal expenses of the Great Northern were £2400 per mile. or a total of £683 053; of the Cornwall, £129,147; of the Eastern Union, £242,385; of the South Western, £279,500; and of the Shropshire Union, £111,855. It is impossible for me to know how much confidence to place in the reliability of these figures. His figures for the London and South Western are widely different from those given in Fay, *A Royal Road*, p. 17, where the cost of obtaining the Act is given as £31,000. Fay was a traffic official of the railway. Pratt, *History of Inland Transport and Communication in England*, pp. 255–6, among other figures, gives £41,467 for the London and South Western, but he is quoting from Porter's *Progress of the Nation*, and Porter gives no authority for his statement, but says that his figures do not include the same items of expense in each case. It was, doubtless, true that many of the companies did not know exactly how much these expenses were, but, in any case, they were high; and the wide differences here noted may be explained by the inclusion of different elements of cost in each case. See also Martin, *Railways, Past, Present and Prospective*, p. 32; *The Times*, Nov. 16, 1848, which shows that at a meeting of the shareholders of the Liverpool, Manchester and Newcastle Junction Railway, to dissolve the company, it was shown that they had already spent £100,000 in law expenses.

[2] Marshall, *Railway Legislation*, p. 29; Chattaway, *Railways*, p. 24; *Herepath's Railway Magazine*, N.S., iii, pp. 92–94, letter from "A Suffering Shareholder," to the shareholders of the Eastern Counties Railway; *The Times*, May 9, 1837, p. 6, letter from "T. G." on "Railways;" "A Few General Observations on the Principal Railways... ," p. vi.

property[1]. Wasteful expenditures were allowed to go on unnoticed and without any accounting on the part of the officials[2]. The most reckless extravagance had, in many instances, been shown, not alone in the actual construction of the main lines, but in the formation of secondary lines at a cost that was unjustifiably high; and while all this was going on, shareholders remained singularly apathetic and only a small fraction of the total proprietary of a railway even attended the annual meeting of their company[3]. Sometimes a series of transactions were carried through that were injurious to the revenues of the railways, such as leases, purchases and other contracts that were paid for at too high a price[4]. Some railways sacrificed other considerations of great importance to the payment of dividends, and revenues which should have been put back into the property, or used for the liquidation of

[1] As in the case of George Hudson and the York and North Midland Railway. Hudson had bought shares in the Hull and Selby Railway to the amount of £35,646, and immediately sold these shares to the York and North Midland Railway Company for £38,842, thus netting himself £3196. He was able to do this because of the influence he had acquired over the directors of the York and North Midland. At a later time, after an investigation of the affairs of the York and North Midland by a committee of its directors, these shares were given back to Hudson, and he was required to pay back to the company the amount he had received through the sale of the shares ('York and North Midland Railway, First Report of the Committee of Investigation (1849),' pp. 6–7; ibid., 'Second Report,' p. 3, in which this committee reported that Hudson had become "almost sole and absolute manager" of the railroad, and that he had "abused the confidence which was placed in him, by wielding the power he obtained to forward his own interest." He had "lost his better judgement and moral rectitude when left with the entire control." See also *Railway Times*, IV, p. 85; ibid., V, p. 1268; ibid., V, pp. 1309 (letter of Charles Penfold), 1315–16; and ibid., VI, pp. 83, 84.

[2] Marshall, *Railway Legislation*, 2nd ed., p. 15; *Herepath's Railway Magazine*, N.S., III, pp. 92–94, letter from "A Suffering Shareholder;" *Railway Times*, IV, p. 38, in which the editor says, in regard to railway management, that "extravagance is the rule, economy the exception." See also ibid., IV, pp. 38, 39, 42, 43, 61 et seq. ('Proceedings of the Meeting of the Manchester and Birmingham Railway Extension'); ibid., V, p. 1268.

[3] *Railway Times*, IV, pp. 13, 14, 38, 85 (editorial on the "necessity of observing the most rigid economy in the future management of railways"); ibid., V, p. 1220; Marshall, *Railway Legislation*, p. 18, said that the extraordinary disclosures of the affairs of some of the English companies had created so much suspicion among shareholders, that nothing short of a searching inquiry into the condition of every company would allay the prevailing alarm.

[4] Handyside, *Review of the Manchester, Sheffield and Lincolnshire Railway*, pp. 5–26. For example, this railway, he says, bought for £21,000 the Dearne and Dove Navigation, which cost but £6000; and the Don Navigation which cost £15,000 they bought for £450,000. He gives many other examples. See also Chattaway, *Railways*, pp. 25–26, showing that branch lines which were intended to be feeders to the main line had often sucked the company dry, through guarantees, leases, etc.

debt, were used for paying dividends of six, seven, or eight per cent. Had the directors furnished full statements of their affairs, there would have been sufficient light thrown upon the condition of the companies' affairs that it would be seen that dividends were not warranted. The payment of dividends out of capital; the charging of other expenditures improperly to capital, rather than to revenue; the neglect to provide properly for repairs, depreciation, renewals of permanent way and other essentials of good financing; these and similar methods enabled companies to pay good dividends and thus have their shares command a high price in the share market[1]. In one instance, at least, and probably in several others, the accounts were manipulated and falsified by those who were in charge of the road; one station agent was securing large amounts of money, through representing it as wages to be paid to the men; false statements were made wittingly; a general manager whose delusive methods and irregularities were known by the directors was kept at his post because he was capable; and all these things were going on while shareholders were ignorant or indifferent, more usually the former, in regard to their property[2]. It was not until well on in the fifth decade of the century that public opinion began to be aroused to really see what had been taking place; and the owners of the various properties were urged to take active interest thenceforward in the management of the companies' affairs, and to put in directors who would administer their trust for the public well-being[3]. Closely connected with the foregoing were the wide fluctuations in the prices of railway shares, by which some became wealthy and others impoverished. Many causes may be assigned for this, but the more important were the instability of the whole system of railways, the lack

[1] Marshall, *Railway Legislation*, pp. 12–16. With the kind of statements that were issued, it was frequently impossible to know how much had been spent on rolling stock, how much on permanent way, how much on stations, etc. Langley, *The Dangers of the North British Railway Policy*, 2nd ed., pp. 5–6, shows that the North British admitted in their reports that they were sacrificing other considerations, like the upkeep of rolling stock and permanent way, to the payment of dividends. In contrast with this, the North Eastern spent large sums on maintenance. See also 'Railways as they Really Are: or Facts for the Serious Consideration of Railway Proprietors,' Nos. I and II.

[2] A good illustration is furnished by the case of the York and North Midland under Hudson's régime (v. 'York and North Midland Railway, Report of the Committee of Investigation,' first, second and third reports). This was, apparently, a one-man power, and the results of the investigation were terribly damaging to Hudson.

[3] Ibid.; Marshall, *Railway Legislation*, pp. 19–20; *The Times*, Jan. 16, 1843, p. 5; ibid., Jan. 23, 1843, p. 3, report of the committee on the Midland Counties Railway Company; ibid., Feb. 20, 1843, p. 5.

of adequate reports as to the condition of the various companies, the decisions of the Railway Department of the Board of Trade, which, after 1844, had to sanction every new railway project before it was authorized by Parliament, and the gambling spirit which was prevalent in the early years of the establishment of railways[1].

In the third place, there were some phases of railway operation which were objectionable from the standpoint of the companies and of the public. In the internal organization of the companies, there was for more than twenty years an imperfect definition of the authority of the various officials, so that responsibility for errors could not be fixed; reports were not rendered to the ·higher officials frequently enough for their guidance; there was lack of discipline in carrying out regulations and orders; the desire for economy went so far that the road was under-manned, or else the repairs and alterations were deficient in amount or defective in the quality of materials used; and in cases of accident or irregularity there was a lack of individual responsibility, since the heads of departments did not know to whom to look for instructions[2]. Competition brought some of the railways almost to the verge of ruin and entailed much loss to those whose funds were invested in these enterprises[3]; and even after different lines had entered into working agreements with one another, these agreements became so intricate and chaotic that constant disputes were inevitable. Nothing was more common than to see a company eagerly seeking authority to make a branch which could only bring it loss, but which, it was feared, would cause still greater loss if it fell into the hands of a rival[4]. In some cases the companies ran a greater number of trains than the traffic warranted, or carried traffic, for the time being, at unremunerative rates in order to take it away from their rivals. The time-tables show that, on the great routes, passenger trains moved about as regularly as the coaches had done formerly on the roads, the aim being to provide such conveniences of travel as would take the trade from the coaches. This excess of accommodation was neither warranted by public necessities nor remunerative to the railways, and through the

---

[1] On the subject of price fluctuations of railway shares, see Marshall, *Railway Legislation*, pp. 10–12, 34–36; Brit. Mus. 1396. g. 21, 'Railways and the Board of Trade,' 3rd ed., pp. 24–40; *The Times*, Jan. 16, 1843, p. 5, giving a table of the fluctuations of railway shares during the year 1842, which showed wide variations of prices.

[2] McDonnell, *Railway Management*, pp. 1–23.

[3] Cotterill, *The Past, Present and Future Position of the London and North Western, and Great Western Railway Companies*, p. 31; Civis (pseud.), *The Railway Question*, p. 11.

[4] Ibid., p. 11.

formation of working agreements of one kind or another they gradually learned to reduce the unnecessary expenditure incurred in this way[1]. At first, too, there was the desire on the part of the companies to cater more carefully to the first-class passengers and to neglect to some extent the third-class passengers. The former were provided with good coaches; their trains were run at good speed, with as few delays as possible, and with the best connexions. On the other hand, the third-class passengers were poorly provided with coach accommodation; during the early years, the third-class coaches were open to all changes of weather; they were attached close to the engine, and the smoke and cinders from the engine were a source of great annoyance to the passengers; these cars were not run nearly as often as those of the higher classes along the same line and they were run at inconvenient hours; they were subjected to frequent and sometimes long delays, and it was aggravating for these passengers to lie on sidings while the first-class trains went speeding by. Not uncommonly the third and second-class passengers reached a junction point and then found that they would either have to stay there for some time or else pay the higher fares in order to proceed immediately in first-class coaches to their destination[2]. It would seem as if the object of the railway companies was to compel passengers to give up third-class and go first-class; for even second-class passengers received but meagre consideration on some lines. That the monopoly of the railway company was used to the detriment of the public is evident from the current testimony of the time[3], and from the fact that Parliament was desirous of having working agreements, amalgamations, leases, etc., sanctioned by the authority of the legislature and subject to their jurisdiction and control. Many were in favour of giving up the principle of competition

---

[1] Marshall, *Railway Legislation*, pp. 15, 50–51.

[2] In addition to the references given on this subject when we were considering railway amalgamations and working agreements, see Galt, *Railway Reform* (1844), pp. 14–18; ibid. (1865), pp. xvii, 34–35; Brit. Mus. 8235.c. 72, 'Railway Management,' pp. 8–9; Young, *Steam on Common Roads*, pp. 71–84, showing the discomforts and indignities of third-class passengers; *The Times*, Nov. 3, 1837, p. 5, letter from "A Passenger;" ibid., Sept. 22, 1843, p. 6, letter from "A Second-class Traveller;" ibid., July 2, 1844, p. 6, letter from "A Commercial Man;" also other letters to the same effect in ibid., Aug. 22, 1844, p. 6; Aug. 27, 1844, p. 6; Aug. 28, 1844, p. 6; Sept. 26, 1844, p. 7. In ibid., Sept. 20, 1844, p. 6, a letter from J. L. Ramsden, F.R.S., who held a great number of shares in the London and South Western, showed how little attention was paid to second and third-class passengers on his line.

[3] Galt, *Railway Reform* (1865), pp. 34–36. On the abuse of railway monopoly, see also *Parl. Papers*, 1857–8 (0.77), xv, 11, 'Minutes of Evidence taken before the Select Committee on the Manchester, Sheffield and Lincolnshire, and Great Northern Railway Companies.'

as applied to railways and of having agreements entered into wherever possible; but after seeing the early results of monopoly those who advocated working agreements did so because they wanted to see greater unity of action, greater economy and improved accommodation, under some parliamentary supervision which would guard the public interests[1]. These, and other accompaniments of the extension of the railways, together with the political effects in securing what was called a "Railroad Parliament," were certainly fraught with a power for evil[2]; and yet many of them were imputable to the newness of the system, to the universal ignorance of its tendencies, and to the wonderful suddenness of its growth. Looked at from the distant point of view which the present affords, we can see that these evils were but incidents in the rapid expansion that was taking place.

From the foregoing, it is not difficult to decide why many railways were unprofitable enterprises, so far as their owners were concerned. With the payment of very high charges in order to secure the act of incorporation, and often exorbitant prices for land and compensation, followed by the great extravagance in the management of the companies' funds; the costly construction and equipment which greatly exceeded the needs of traffic; the extraction of funds by dishonest officials, and the expenditures for ostentation rather than utility— these, and the disastrous results of early competition, must have proved to be a burden, for some of the railways, that was hard to endure. For example, we learn that the Manchester, Sheffield and Lincolnshire Railway, during the first fourteen years of its activity, scarcely paid a fraction of a dividend on the amount of the original stock[3]. Even some of the main lines paid but low rates of interest upon the capital expended, for, in addition to the above-mentioned disabilities, they frequently had, in self defence, to link up with themselves certain branch lines of railway or canal, which, instead of being feeders to the

---

[1] Marshall, *Railway Legislation*, pp. 52–60; Galt, *Railway Reform* (1865), p. 36, showing that, under existing conditions, occasionally one company paid another a large sum of money if the latter would refrain from competition. Civis (pseud.), *The Railway Question*, pp. 15–17, said that there was a feeling gaining ground that the roads of the country were properly the dominion of the state; and that to secure harmony of interests for the public good, the companies should admit a more regular and extensive exercise of state control.

[2] Whitmore, *Letter to Lord John Russell on Railways* (1847), pp. 9–10. We have not given a full account of all the evils that accompanied this period of great advance, but have mentioned the most significant of them. Some others were almost entirely imaginary (P., *Descant on Railroads*); and others still were more or less unsubstantial (Jeans, op. cit., Ch. XII). See also Gordon, *Observations on Railway Monopolies*, pp. 1–55.

[3] Handyside, *Review of the Manchester, Sheffield and Lincolnshire Railway*, p. 4.

main lines, proved rather to be suckers, withdrawing funds from the treasury of the main lines in order to pay interest to the claimants upon the branch lines[1]. Among the early railways in the north of England, there seems to have been quite a number that paid either no dividend at all or else but a very small return, and many of these were in the very centre of the industrial and mining section, especially in the county of Durham[2]. A writer, with intimate knowledge of the railways, has given us in 1854 the dividends paid by the fifty-nine railway companies of England and Wales for the first half of that year; and working this out upon the annual basis, we find that, of this number, fifteen paid no dividend, thirty-four paid dividends from less than one per cent. to less than five per cent., five paid dividends of five to six per cent. inclusive, and five paid dividends of seven to ten per cent. inclusive[3]. There is no reason to suppose that the year 1854 was anything but an ordinary year, for by that time the country had recovered from the mania of 1844–6 and was again going on its normal way. With forty-nine out of fifty-nine railways, or eighty-three per cent., paying less than five per cent. per annum, it would appear that the number of companies which secured reasonable remuneration on invested capital was small in comparison with the number of those which fell below the reasonable minimum. But although many of the railways were not profitable to their owners in yielding large financial returns they may still have been beneficial to the public in providing for the necessities and conveniences of traffic.

When considering the subject of roads, we noted the fact that, before 1830, the consolidation that had taken place in the turnpike trusts was not the consolidation of those which formed continuous lines of road, but of those that were found in particular counties, or parts of counties, such as those in the vicinity of London or Bristol. But in the case of canals, the small amount of consolidation that had taken place was the merging of those which were parts of a continuous line of navigation[4]. The amalgamation of the railways followed closely the type of that of the canals, not of the roads, by the formation of

---

[1] Young, *Steam on Common Roads*, p. 67; Galt, *Railway Reform* (1865), p. 36; Civis (pseud.), *The Railway Question*, p. 11.

[2] Jeans, op. cit., p. 171, mentions some of them.

[3] Chattaway, op. cit., p. 20. Here, also, he gives the average dividends that were paid during the period 1850–4.

[4] Such as the consolidation of the Birmingham and Bilston Canal with the Birmingham and Tamworth, in 1783; the Trent and Mersey with the Oxford and Coventry canals, in 1785; the Chester and Ellesmere canals, in 1813; the Grand Junction and Regent's canals, in 1819; and the North Wilts Canal with the Be.ks and Wilts, in 1821.

adjoining railway lines into great through routes. Why should the roads have been different in this respect from the canals and railways? The answer would seem to lie in the fact that the control was different. The canals and railways were in the hands of private companies, which, under their several Acts, were given authority to construct and operate their transportation facilities in the several counties or districts through which they passed; but the turnpikes were under the authority of the Justices of the counties, and it would have been almost impossible to get several successive counties to work harmoniously in the proper maintenance of great through roads, such as that from London to Holyhead, or that from London to York.

From the foregoing outline of the railway systems of England we are impressed by the similarity of their position with that of the ancient Roman roads, and also with that of the turnpike and canal systems which played so important a rôle. It was because of this juxtaposition of the turnpikes, canals and railways that the subject of competition between them assumed so conspicuous a place in the public mind, and to this we now turn.

CHAPTER VIII

EFFECTS OF STEAM UPON ROAD TRANSPORTATION

FOLLOWING, probably, the example of the coach proprietors, who provided for both outside and inside passengers, the railway companies which undertook the conveyance of passengers provided two, and often three, different kinds of accommodation, at different prices[1]. The first-class were covered carriages, intended only for the well-to-do; the third-class carriages were at first open and exposed to all the changes of the atmosphere, and were for the poor; while the second-class accommodation was intermediate in quality and cost, and was for the great middle class. The people who were expected to travel third-class were those who belonged to the working group; and it was thought that by providing this cheap means of conveyance the poor would be able to live out in the country where they could have agriculture or gardening as a by-employment, and have also better sanitary conditions, while they could go to and from their work every day. In making any comparison, therefore, between railway and stage coach charges, we must keep in mind this difference between first, second, and third-class rates on the railway, and must draw our analogies between first-class rate and inside coach fare, and between second and third-class rates and outside coach fare.

Before the opening of the great trunk lines, about 1838–40, the coach fares on some roads were very high[2], while on others they were kept moderate by the influence of the competition of several coaches[3].

---

[1] Brit. Doc. 1844 (318), XI, 17, 'Fifth Report of Select Committee on Railways,' Appendix No. 2, pp. 12–13, showing that some railways rejected altogether or limited the third-class accommodation.

[2] See Appendix 7.

[3] See the great number of coaches licensed to run in 1837 between London and other important places in the kingdom, as given in 'Collection of Prospectuses, Maps, etc., of Railways and Canals,' p. 80. Between London and Birmingham, for example, there were 122 journeys weekly and 1098 passengers carried; between London and Liverpool there were 68 journeys weekly and 612 passengers carried; between London and Manchester there were 119 journeys weekly and 1071 passengers carried; etc. Of course these coaches were not all engaged in competition; for all those which were under one management would not be rivals among themselves.

In our discussion of the cost of travel by coach during this period we found that, as a general thing, the inside fare was from two and one-half to four pence per mile; and more commonly the latter would come nearer to the actual truth than the former. This figure seems to be slightly lower than that given by one of the great coach proprietors, who said that before the introduction of railways the fares were about four and one-half pence per mile inside and two and one-half pence per mile outside[1]. But the Committee of 1844, in speaking of this, said that upon most of the leading roads, where competition was effective, this rate of four and one-half pence per mile was somewhat higher than was customary. Probably, therefore, the coach rate which was prevalent on the great roads was three and one-half pence to four pence per mile, or in some cases a little more. The fares adopted by the leading railway companies were about three pence per mile for first-class passengers, two pence per mile for second-class, and one to one and one-half pence per mile for third-class[2]. These, it will be observed, were but little lower than the coach fares, probably just enough to be an additional inducement for passengers to travel by rail; but when we supplement the reduced rate by the combined incentives of greater comfort and speed[3] of railway trains we can easily see why the railway would attract the passenger traffic away from the stage coaches.

After making the foregoing general statement, let us examine some particular instances of the results of this competition. In the investigation of 1844, as to the effect of railways on the interests of the poorer classes, we have some very definite information given as to the relative cost of travelling by canal, stage coach and railway. The cost of passage from Manchester to London, for an ordinary family consisting of two adults and three children, was, by canal boat £3. 14s., by coach

---

[1] Testimony of Mr Chaplin before the Committee of the House of Commons on Railways, 1808. See also Brit. Doc. 1844 (318), XI, 17, 'Fifth Report of Select Committee on Railways,' Appendix No. 2, p. 9. Galt, *Railway Reform* (1865), p. 71, said that about 1830 the general average fares by mail coaches were 5*d*. per mile inside and 3*d*. per mile outside, and by the stage coaches 3*d*. per mile inside and 2*d*. per mile outside. But we have formerly shown that the facilities by mail coaches were not better than those of the stage coaches at that time, and, therefore, the fares by the former could not be much, if any, in excess of those by the latter means. It is clear from the evidence we have previously given that Galt's figures for stage coaches were too low.

[2] Brit. Doc. 1844 (318), XI, 17, 'Fifth Report of Select Committee on Railways,' Appendix No. 2, pp. 10–11, gives table of fares on leading roads. The average fares charged per mile were, for first-class 2·727*d*., for second-class 1·745*d*., and for third-class 1·151*d*.

[3] The average rate of travelling, stoppages included, on the principal passenger railways was about twenty-four miles per hour (ibid., Appendix No. 2, p. 11).

£6. 2s., and by railway £4. 15s.[1] Thus, taking into consideration the necessary expenses incident to such a journey, we judge that travelling by railway cost only about three-fourths of that by coach. From London to Coventry, before the railway was opened, there was one stage waggon, charging nine shillings fare and taking thirty-six hours, and several stage coaches charging for outside fare seventeen shillings by night and twenty shillings by day, which took from ten to eleven hours; but in 1844, when the railway was in operation, there was no stage waggon on this route, and only one night stage coach, charging ten shillings and taking twelve hours, while by the railway the third-class fare was twelve shillings and the time occupied six and one-half hours[2]. That is, the introduction of the railway brought a reduction of the fare and of the time required for this journey amounting to almost one-half. When we consider the greater expense for fees and meals when travelling by the stage coach, the cost of travelling by the latter vehicle must have been fully twice as much as by the railway. Again, before the establishment of the Liverpool and Manchester railway the coaches between these two places, at full capacity, could not carry more than 688 persons per day, and, on the average, probably

---

[1] The following figures were given in detail as to this journey (Brit. Doc. **1844** (318), xi, 17, ' Fifth Report of Select Committee on Railways,' Appendix No. 4):

By *Canal* boat (Manchester to London).

	£	s.	d.
2 adults' passage, 14s. each	1	8	0
3 children's passage, 7s. each	1	1	0
Provisions, etc., for 5 days' passage, 5s. each	1	5	0
Total	3	14	0

By *Coach*, Manchester to London, 186 miles.

	£	s.	d.
2 adults' passage, 30s. each	3	0	0
3 children's passage, 15s. each	2	5	0
Coachmen and Guard		7	0
Food, etc.		10	0
Total	6	2	0

By *Railway*, Manchester to London, 212 miles.

	£	s.	d.
Third-class, Manchester to Birmingham,			
2 adults' passage, 11s. each	1	2	0
3 children's passage, 5s. 6d. each		16	6
Third-class, Birmingham to London,			
2 adults' passage, 14s. each	1	8	0
3 children's passage, 7s. each	1	1	0
Food, etc., 1s. 6d. each		7	6
Total	4	15	0

[2] Brit. Doc. 1844 (318), xi, 17, 'Fifth Report of Committee on Railways,' Appendix No. 4, p. 63.

carried not more than 450 or 500. The railway at its commencement carried an average of 1070 per day[1]. The fare by coach varied according to the season and the amount of travel, but on the average it was ten shillings inside and five to six shillings outside; the fare by the railway in 1832 was five shillings for first-class and three shillings and six pence for third-class[2]. The time occupied in making the journey by coach was four hours; the time occupied by the railway was but one and three-fourths hour[3]. It is evident that here, too, the establishment of the railway reduced by one-half the cost and the time of travelling. Before the advent of the railway, the twenty-four-hour journey by coach between London and Liverpool cost £4. 4s., but, including the fees and the meals, the cost would approximate £5. After the railway had been opened some time, the cost by rail first-class was 37s. and second-class 27s.[4], showing the expense of travelling by railway to have been less than half that by road. While, therefore, our general conclusion, above stated, that the railway fares, on the whole, were not much lower than those of the coaches, is probably close to accuracy, we must, nevertheless, realize that, in some instances, there had been a reduction of as much as fifty per cent. in these charges. Similar results were secured in the conveyance of commodities. For example, before the opening of the Great Western, the waggon rate from London to Oxford was £3 to £3. 10s. per ton; but the railway charged only 30s. per ton, which was practically one-half the former charge[5]. The latter rate included rail carriage from London to Steventon and then waggon carriage for ten miles between Steventon and Oxford. Had there been rail carriage all the way, so as to avoid the necessity of a waggon haul and its attendant loading and unloading, the cost would have been still less than that mentioned, and probably would not have exceeded 25s.[6] Before the opening of the railway to

---

[1] In 1836 the average was 1200 daily. See *Advantages of the Progressive Formation of Railways*, p. 23; 'London and Birmingham Railway Bill. Extracts from Minutes of Evidence before Lords Committee,' evidence of Henry Booth, treasurer of this railway company, pp. 53–55.

[2] Ibid.

[3] *Annual Register*, 1832, p. 445; also 'London and Birmingham Railway Bill. Extracts from the Minutes of Evidence before Lords Committee,' evidence of Henry Booth, pp. 53–55.

[4] Shaen, *Review of Railways and Railway Legislation*, p. 32.

[5] Brit. Mus. 8235. ee. 4 (1), 'Oxford and Didcot Railway Bill. Evidence taken before Commons Committee,' evidence of Mr Sadler, p. 8, and of Mr Sheard, p. 10.

[6] Ibid., evidence of Mr Clarke, p. 27. The evidence of Messrs Sadler, Sheard, Underhill and Clarke gives much detail by way of comparison of rail with road carriage, both as to passengers and goods, showing the decreased cost and reduced time by the former.

connect London and Manchester, the cost of carrying general goods by road was 70*s*. to 80*s*. per ton, but after the railway had been in operation for some time the charge by this faster conveyance was only 30*s*. to 40*s*.[1], or one-half the former charge. We must not be understood, however, as implying that railway rates in general were only one-half as much as those charged by waggon.

In like manner, the change in the amount of coaching and posting, after the coming of the railway, was almost immediate. Along the line of the Liverpool and Manchester Railway there had been each day twenty-two regular and seven occasional coaches for carrying passengers, but, within five months after the opening of the railway all these, with the exception of four, had disappeared. By 1832 all but one of these coaches had ceased running and that one was chiefly for carrying parcels[2]. On the road from London to Birmingham, before the railway was opened, one of the chief London coach proprietors had nine coaches; but after the opening of that line this number was gradually reduced until in 1839 he was working only two coaches and had difficulty in keeping them on. The fares charged by coach were only one pound inside and twelve shillings outside; yet he got no inside passengers, because people could go by the railway for the same fare and they preferred that means of travelling[3]. The great number of coaches that travelled the road from London to the west of England was soon reduced after the railway was established in operation[4]; but it was not until after some years of rivalry, namely, about 1843, that

[1] Shaen, op. cit., p. 33.

[2] *Birmingham Journal*, Feb. 5, 1831, p. 3, letter from "A Subscriber to the London and Birmingham Railway;" ibid., May 19, 1832, p. 3, evidence on the London and Birmingham Railway Bill; *Annual Register*, 1832, p. 445. See also Shaw, *Liverpool's First Directory*, p. 19.

[3] Brit. Doc. 1839 (295), ix, 369, 'Evidence of Mr Sherman,' p. 8. Others testified to like results. See also Brit. Doc. 1837 (456), xx, 291, 'Report of Committee on the Taxation of Internal Communication,' p. iv, and evidence of Messrs Horne (pp. 1–51), Gray (pp. 5–10), Fagg (p. 15), Wimberley (pp. 35–38), Kemplay (pp. 38–39), etc. Slugg, *Manchester Fifty Years Ago*, p. 221, says that scores of posting houses were ruined by the introduction of railways; but he was speaking from memory. Stretton, in his *History of the London and Birmingham Railway*, says that the result of the first run over this line was that the stage coach proprietors at once decided to raise their rates, and the following quotation appeared in several newspapers: "Coach Fare from Birmingham to London.—The coach proprietors on this line of road, aware that on even the partial opening of the London and Birmingham Railway, they may cry 'Othelo's occupation gone,' are making hay while sunshine is left them. The fare from Birmingham to London, which of late years has averaged from 18*s*. to 25*s*., has recently been trebled, the proprietors now modestly ask £3. 12*s*. fare from Birmingham to the metropolis."

[4] To give some idea of the amount of coaching on certain roads at the time railways were introduced, we give the following statistics from the records of the

the last coach was driven off this road[1]. Before the railways could cause the coaches to give up the struggle they had to reduce their charges to a point almost equal to the fares of the coaches; and any slight excess of railway fares above that point was sufficient to bring back the coaches on some of the roads[2]. If railway companies had charged as much as the law allowed, their lines would have been comparatively deserted in most cases, for they would have been used almost exclusively by the opulent classes; but by putting down their charges to an approximate equality with those of the coaches they diverted to the rail all through traffic and most of the local coaching business along lines of road which were near to and parallel with the railways[3]. The decrease in the amount of coaching was accompanied by a corresponding reduction in the amount of posting along these roads.

Stage Coach Office (v. *Proceedings of the Great Western Railway*, evidence of Mr Sutherland, p. 39).

The number of coaches licensed and the number of journeys performed along the main western highway, in 1834, before the Great Western Railway was built, were as follows:

Number of coaches	From	To	No. of journeys per week
6	London	Bath	40
20	,,	Bristol	136
4	,,	Cheltenham	24
3	,,	Devonport	14
19	,,	Exeter	81
1	,,	Farringdon	6
1	,,	Great Marlow	12
6	,,	Gloucester	38
1	,,	High Wycombe	12
1	,,	Henley	12
1	,,	Harlington	16
1	,,	Maidenhead	12
1	,,	Marlborough	6
4	,,	Newbury	24
10	,,	Oxford	62
11	,,	Reading	80
8	,,	Stroudwater	20
5	,,	Taunton	24
7	,,	Uxbridge	82
7	,,	Windsor	96
3	,,	Wallingford	24
1	,,	Wantage	6

[1] Galt, *Railway Reform* (1844), p. 7; Brit. Mus. 8235. ee. 4 (1), 'Oxford and Didcot Railway Bill,' pp. 5–6.

[2] Galt, *Railway Reform* (1844), p. 7; Young, *Steam on Common Roads*, p. 84; *The Times*, Dec. 7, 1843, p. 6.

[3] In *Railway Times*, v (1842), pp. 639–40, 711, 973, we find a comparison of travel by railways and coaches showing the vast change that had been effected by the

We must bear in mind that this was a period of transition, and like all other similar periods was fraught with disaster to those upon whom the burden rested most heavily. Perhaps the classes that suffered most were the proprietors of coaching establishments and the innkeepers along the great roads. The latter class found their inn and posting business rapidly declining[1]; and the papers of the time contained the advertisements of whole coaching and carrying establishments that were selling out[2]. Through many years the coach-masters had endeavoured to provide facilities for a greatly increasing amount of travel and had, in some cases, many hundreds of horses; but when the railway came and took the passenger traffic from these great roads we can easily understand that ruin seemed to stare them in the face[3].

introduction of the railways. The last of the coaches between London and Cambridge made its final journey on Oct. 25, 1845 (*The Times*, Oct. 29, 1845, p. 5). The opening of the railway from Salisbury to Bishopstoke was the signal for the withdrawal of the coaches which went through Andover, which, but a few years before, numbered about forty daily (*The Times*, Mar. 12, 1847, p. 6).

[1] *The Times*, Dec. 14, 1843, p. 5, "Turnpike v. Railway."

[2] *The Times*, Sept. 28, 1837, p. 1, gives three such advertisements; ibid., Oct. 21, 1837, p. 1, gives two advertisements; etc.

[3] Some idea of the great traffic that centred in London may be gathered from a table (v. 'Collection of Prospectuses, Maps, etc., of Railways and Canals,' p. 80) showing the number of coaches licensed in 1837 to run between London and many other places, the number of passengers carried, and the weekly receipts from these licenses. From the table we take the following data to show the extent of the coaching business between London and the north:

Places	No. of journeys weekly	No. of passengers weekly
London to Birmingham	122	1098
,, Liverpool	68	612
,, Manchester	119	1071
,, Glasgow	14	70
,, Holyhead	14	70
,, Shrewsbury	40	360
,, Woodside	14	126
,, Worcester	51	459
,, Edinburgh	14	70
,, Halifax	28	252
,, Leeds	70	630
,, Leamington	12	96
,, Leicester	14	126
,, Newcastle-on-Tyne	28	252
,, Nottingham	28	252
,, York	14	126
,, Barton	12	96
,, Lincoln	18	162
,, Northampton	14	126
,, Aylesbury	38	304
,, Luton	18	180
,, Watford	14	140
,, Pinner	26	260

The traffic from London to the Eastern Counties is well represented by a diagram

This was particularly true, of course, concerning those in the great centres, notably London. It would be erroneous, however, to suppose that the decline or the disappearance of the country innkeepers on the important thoroughfares was due entirely to the advent of the railway; as a matter of fact, many of them were gradually eliminated before this time, on account of the necessity of the coaches making fewer stops as they developed greater and greater speed[1]. Nor was the picture of the disappearance of the coaches entirely unrelieved by a brighter aspect. It is, doubtless, true that along the main roads, where they were in competition with the railways, the coaches were soon taken off; yet the increase of business brought by the railroads, not only at their stations but also on the country roads leading to the stations, caused a greater demand for the labour of horses in the carriage of passengers and goods[2]. We have ample proof of this from the increase in the number of such vehicles that were licensed[3]. We may, therefore, say that a decrease of coach traffic along roads that were adjacent and more or less parallel to the railway, which meant almost annihilation to some proprietors of coaches, was only the forerunner of greater business of this kind when once the readjustment

in the Appendix to Vol. IX of the Brit. Doc. for 1839, showing the gradual diminution of the amount of passenger travel from London to the towns farther east:

> The number of stage coaches from London to West Ham and Stratford was 62.
> The number of stage coaches from West Ham and Stratford to Romford was 41 and 2 mails.
> The number of stage coaches from Romford to Brentwood was 36 and 2 mails.
> The number of stage coaches from Brentwood to Chelmsford was 32 and 2 mails.

Bearing in mind that some coach proprietors conducted the traffic on several of the chief roads where the density of travel was fully as great as the aforementioned, it is not hard to see what the sweeping away of all this business would mean to such establishments.

[1] *Herepath's Railway Magazine*, N.S., VI, p. 463, letter of Joseph Lockwood.

[2] Brit. Doc. 1839 (295), IX, 369, 'Report of Select Committee on Railroads;' also ibid., Minutes of Evidence of Mr Macadam. Both of these references give us to understand that the increase on the lateral lines was not at all commensurate with the loss on the principal lines. This may have been the immediate effect in some cases, but it certainly was not the ulterior effect. On the increase in the number of horses and vehicles that were used on the cross roads tributary to the Liverpool and Manchester Railway, see Godwin, *Appeal to the Public on Railways*, p. 40; *Hampshire Advertiser and Salisbury Guardian*, Mar. 29, 1834, p. 2, evidence of Mr Langston, of Manchester; Felix Farley's *Bristol Journal*, April 19, 1834, p. 4, Committee on the Great Western Railway Bill. See also Brit. Mus. 8235. ee. 4 (1), 'Oxford and Didcot Railway Bill,' evidence of Mr Sadler, p. 6, and *Herepath's Railway Magazine*, N.S., VI, p. 461.

[3] *Railway Times*, VI (1843), p. 443, statement of the Earl of Hardwicke, on the "Effect of Railways." He gives accurate statistics to substantiate this fact.

was effected[1]. Mention must also be made of the fact that one of the largest coaching establishments in London, and we cannot say how many more, became an ally of the railway to act as collectors and distributors of goods at the terminus[2].

As a result, it was said, of the competition between the railways and the turnpike roads for the traffic of the country, which, in many cases, was accompanied by a great decrease or total decline of traffic on the turnpikes parallel with the railways, we find constant complaints from the turnpike trusts that their tolls were diminishing because of the diminution of posting and stage coach business[3]. It must be borne in mind that the trusts depended mainly upon the passenger traffic for their revenues. On account of this decrease of revenues, the debts of the trusts were constantly increasing, for it was the prevailing practice to convert the unpaid interest into principal, by the trustees giving interest-bearing bonds to cover the full amount. That there was a great increase in the debts of the trusts is beyond dispute[4], as is also the fact that the debt was increased through the consolidation with it of interest that was in arrears[5]. The continuation of this practice, of

[1] This is well exemplified in the Earl of Hardwicke's statement in the preceding reference.

[2] *Railway Times*, IV (1841), p. 209, showing that the Grand Junction Railway Co. had employed Chaplin and Horne for some time as their agents in London to unload and deliver goods. It is probable that few of the coaching firms were fortunate enough to attach themselves to the railways in this way.

[3] Brit. Doc. 1837 (456), XX, 291, Minutes of Evidence of Mr Hall. Also 1839 (295), IX, 369, 'Report of Select Committee on Railroads,' and Minutes of Evidence of Messrs Bicknell, Levy and Macadam. On the reduction of traffic and tolls on particular roads, see *Railway Times*, V (1842), pp. 18, 21; but on the roads as a whole the tolls had apparently increased (ibid., VI, p. 443).

[4] Brit. Doc. 1833 (24), XV, 409, 'Second Report of Select Committee on Turnpike Trusts.' The Committee "contemplate with alarm the results of the great and increasing debt on many roads." Statistics to prove this are given in ibid., 'Minutes of Evidence,' pp. 174–5.

From Brit. Doc. 1836 (547), XIX, 335, 'Report of Select Committee on Turnpike Tolls and Trusts,' we learn that the trusts' debts at that time amounted to nearly £9,000,000, and that the probability was that they would constantly increase as in the past. See also 'Report of Select Committee of 1839' on the influence of railways on turnpike trusts, Brit. Doc. 1839 (295), IX, 369, with evidence of Messrs Bicknell, Levy, and Macadam.

[5] Brit. Doc. 1836 (547), XIX, 335, 'Report of Select Committee on Turnpike Tolls and Trusts,' states that several trusts were at that time insolvent because the amount of interest due annually was more than the amount of the annual income. Also Brit. Doc. 1839 (295), IX, 369, 'Report of Select Committee on Railroads and Turnpike Trusts,' tells us that the debt then exceeded £9,000,000. Also ibid., evidence of Sir Jas. Macadam, who said that this policy prevailed very generally throughout the trusts of the kingdom, where interest payments could not be made. He said this was the chief cause for the increase of the bonded debt for some years

course, increased both the principal and the interest of the debt. But when we come to consider the additional reasons why the trusts' revenues were insufficient to keep up their interest payments we meet with a problem which involves several other factors than that of railroad competition.

In a former chapter we have seen that, in the management of many of these trusts, the funds were squandered by injudicious expenditures and keeping up official parasites who were incapable of accomplishing anything for the good of the roads from which they drew their salaries. The same thing was, doubtless, still prevalent, although the accounts of the trusts did not show it[1]. Then, too, the statute duty, or statute labour, was abolished in 1835[2], and also the composition in its place. This loss was estimated by Sir James Macadam, who had an intimate knowledge of the condition of the turnpikes, to amount to £200,000 a year[3]. But, notwithstanding the abolition of the statute labour, in the administration of which there was much fraud, the financial condition of many trusts became worse and worse[4]; and as a means toward obtaining economical and efficient management the consolidation of small trusts into larger trusts and of the larger trusts into unions of trusts was urged upon Parliament[5]. This suggestion, of course, came

before that. This was very acceptable to creditors, in that it gave them additional security by a bond for the payment of their interest. His statement was that he knew some roads upon which there were *sixty years' arrears of interest due.*

[1] Brit. Doc. 1836 (547), xix, 335, 'Report of Select Committee on Turnpike Tolls and Trusts.' The Committee put forth a plan that would be "useful in preventing any wasteful expenditure of funds in some trusts;" and although they do not expressly mention this form of extravagance, we are warranted, from what we have found hitherto, in saying that it still existed.

[2] Act 5 & 6 William IV, c. 50.

[3] Brit. Doc. 1839 (295), ix, 369, 'Report of Select Committee on Railroads and Turnpikes;' also ibid., evidence of Mr Macadam. See also Brit. Doc. 1837 (457), xx, 343, 'Minutes of Evidence,' p. 9 et seq.

[4] Brit. Doc. 1839 (295), ix, 369, 'Report of Committee on Railways and Turnpikes,' and evidence of Sir James Macadam.

[5] The evidence was nearly unanimous that such consolidations would be desirable from many points of view: It would save the amounts now spent in salaries to officers of small trusts; it would give cheaper road materials by purchasing them in larger quantities; it would abolish the competition for such materials that was common among small trusts; and it would obtain an improved system of management by merging the small trusts into large trusts.

But there were also objections raised to the consolidation of such interests. Some trusts were in good circumstances, and were opposed to allying with those that were in debt. Then, the creditors of those trusts that had given good security were averse to the adoption of any measure likely to lessen their security, by uniting the solvent trust, to which they had advanced money, with one or more that were financially embarrassed. See Brit. Doc. 1836 (547), xix, 335, 'Report of Select

from the good results which were secured by the consolidation of the metropolitan turnpike trusts. It is not within the present plan to follow out the subsequent history of these roads[1]; but the point to be noted is that such recommendations were the outcome of the reduction of revenues which accrued to the trusts.

Again, the competition of steam vessels on the rivers, and,. more important still, in the coasting trade, drew away traffic, and consequently revenues, from the turnpike trusts. Wherever there was a route for steam vessels near the coast, and a more or less parallel coaching route on the land, whether near to or somewhat distant from the shore, the vessels almost invariably took the greater part of the passenger traffic, especially during the warmer part of the year. This rivalry of the two means of conveyance was most noticeable along the east coast, and from London, around Kent, to the south-east coast. The preference that was shown by the public for the steam vessels was chiefly due to the fact that they were much more comfortable, and at the same time cheaper, than the coaches. For instance, a passenger

Committee on Turnpike Tolls and Trusts;' also 1839 (295), IX, 369, 'Report of Select Committee on Railroads and Turnpike Trusts.'

[1] A brief summary of the later history of the turnpike trusts will be *à propos* here. After 1830 many of the trusts were unable to maintain their roads in reasonable condition, and, according to the common law, the burden of maintenance devolved upon the parish. This liability was not enforced after the Highway Act of 1835, but in 1841 an effort was made, in Act 4 & 5 Vict., c. 59, to restore it, by authorizing the Justices to demand a payment out of the highway rates toward the repair of turnpike roads where the tolls were insufficient. In this way, the parish had a double burden to bear, the payment of the tolls and the cost of repairing the road; and, while ineffective to improve the finances of the trusts, this system aroused hostility. The "Rebecca Riots" in 1842–3 were the outcome; and the conditions in South Wales, where these riots were particularly vigorous, are told in the 'Report of the Royal Commission of Inquiry of 1844.' Finally, as a result of this inquiry, an Act was passed to put an end to the administration of the trustees in South. Wales, by merging all the trusts under "County Roads Boards," composed of Justices of the Peace. In England, no such centralization of control was possible, on account of the opposition of rival interests of one kind and another. But soon after the middle of the century there came to be a growing sentiment in favour of the abolition of the turnpikes and toll-gates. The committee of the House of Commons which, in 1864, investigated the subject reported that the tolls were "unequal in pressure, costly in collection, inconvenient to the public, and injurious as causing a serious impediment to intercourse and traffic," and advocated the union of the trusts in some such way as had been carried out twenty years before in South Wales. Still the whole matter was left in abeyance, so far as any general public policy was concerned; but from this time onward successive committees of the House of Commons began the gradual dissolution of the trusts, and their administration was handed over to the highway districts, or to the highway parishes, in which they were located. By 1887 only 15 trusts remained; by 1890 these had been reduced to two; and in 1895 the toll system ceased.

could get by packet from London to Gravesend for 1*s*. 6*d*., and from Gravesend to Maidstone for 2*s*. 6*d*., making a total of 4*s*. from London to Maidstone; but the coaches charged 6*s*. for this distance[1]. From London to Newcastle the fares by coach were £4. 10*s*. inside and £2. 5*s*. outside; while by steamer the fares, including provisions and all expenses, were only £3 for the best cabin and £2 for the fore cabin[2]. Between London and Hull the fares by steamship were, for the best cabin £1. 1*s*. and for the fore cabin 15*s*.[3]; and the fares between Hull and York, at these rates, could not exceed 5*s*. and 4*s*. respectively; so that the steamer fares between London and York could not have been more than £1. 6*s*. and 19*s*. respectively. The coach fares, on the other hand, were £3. 5*s*. inside and £1. 14*s*. outside[4]. There was, therefore, a decided advantage in travelling, where possible, by steamer. From the point of view of the coach proprietors, one vital element in their higher rates was that they had to pay duties and taxes from which steam navigation was free on account of the sea being an open highway that required nothing for its maintenance[5]. It was recognized by the owners of coaches that they could not maintain their position in the face of this competition, and they were compelled to take off many of their coaches during the summer months, when the traffic was most

---

[1] Brit. Doc. 1837 (456), xx, 291, 'Report of Committee on the Taxation of Internal Communication,' evidence of Mr Horne; also 'Report,' p. v.

[2] Harris, *The Coaching Age*, p. 194.

[3] Macturk, *History of the Railways into Hull*, p. 11, advertisement of the "Enterprise" steamship. Harris, *The Coaching Age*, p. 194, gives the fares between London and York, not including expenses, as 8*s*. and 4*s*. 6*d*. for best cabin and fore cabin; but it is pretty certain that his figures are altogether too low, when we compare them with those which we have just given.

[4] Harris, *The Coaching Age*, p. 194.

[5] The conveyance of passengers by water was free of duty; but on making as close an approximation as possible to the truth, through a comparison of the rate of duty and the average number of passengers conveyed it was found that the duty paid by the stage coach was ½*d*. per passenger per mile (Brit. Doc. 1837 (456), xx, 291, 'Report of Committee'). On the road from London to Dover one coach proprietor had five coaches, on which he paid (in 1836) a mileage duty of £2273. 16*s*. 6*d*., from which his competitors by steam power were wholly free. The same conditions were found on the great north road (ibid., evidence of Mr Horne). In Harris, *The Coaching Age*, p. 193, we are given a statement of the duties and other expenditures of the "Wellington" coach from London to Newcastle for a year, drawn up by one who was thoroughly familiar with the accounts, which shows that the taxes paid by this coach to the Government were £2568. 18*s*. 6*d*. The tolls paid were extra, over and above this amount, and were annually over £2500. From all of these expenses the vessels were free. Steamship proprietors had another advantage over coaches in being allowed to retail wines and spirits without paying an excise license (Brit. Doc. 1837 (456), xx, 291, 'Report of Committee on Taxation of Internal Communication,' p. v).

profitable, because there was not enough business for both rivals[1]. In the case of passengers who wished to reach any of the places on or near the great north road it was more congenial to them to take the vessel to the nearest point on the coast, and then reach their inland destination by coach, than to take the coach all the way[2]. Thus, people flocked to the steamboats and left the long coach roads wherever it was possible to do so conveniently. In some instances, the travelling on the lateral lines leading to these main roads was much increased, so that although steam navigation might interfere with the business on the roads that were parallel with it, it produced a considerable increase in the collateral trade[3]. Whether this increase of transverse trade made up immediately for the loss of trade on the longer through routes it is impossible for us to determine.

But, in addition to the effect on the revenues of the roads due to the abolition of statute labour (or composition therefor) and to the competition of steam navigation, it is certain that the railways directly attracted the traffic from the adjacent thoroughfares that were parallel with them. This reduction of road carriage, by diminishing the tolls on the turnpikes, made it more difficult for the trusts to pay their interest obligations and maintain the roads. As it is impossible to make general statements on this subject with great accuracy, it will serve our purpose better to note the results in particular instances in which the great roads are involved.

One of the clearest cases of the influence of the railway in curtailing the revenues of the roads was that of the line from London to Birmingham. This road was practically parallel with the London and Birmingham Railway, and was one of the best constructed and managed roads in England. The railway was fully opened in 1888, and the tolls received on the various road trusts between these two cities, for the half-year ending Mar. 29, 1889, amounted to £7899, which when doubled would make £15,798 as the approximate amount of the tolls

---

[1] Brit. Doc. 1837 (456), xx, 291, 'Report of Committee;' also evidence of Messrs Horne, Wheatley, Wimberley and Kemplay. These facts were evidenced by witnesses from almost every district of the kingdom. Baines, *History of Liverpool*, pp. 564–5, shows that when Bell put his first useful steamboat on the Clyde, plying between Glasgow and Greenock, four coaches between these two places were immediately discontinued, on account of the transfer of the passenger travel to the vessel, although the ordinary speed of the vessel was only four to four and one-half miles an hour, and less than that when the wind and tide were unfavourable. Buchanan, *Practical Treatise on Propelling Vessels by Steam*, p. 13, says that in 1816, when he was writing, the vessels along the Clyde had largely superseded the coaches and that the steamers had greatly increased the amount of travelling.

[2] Brit. Doc. 1837 (456), xx, 291, evidence of Messrs Wimberley and Collins.

[3] Ibid., evidence of Mr Wimberley.

for the first year after the opening of the railway. The tolls for the year 1836, the last year before the opening of the railway[1], were £28,525. This shows a decrease of approximately £12,727 a year, or almost fifty per cent., on the tolls of 1836[2]. This could not have been due to canal competition, for the road traffic was derived largely from passengers and parcels while the canal traffic was that of heavy articles. It is evident, therefore, that the decrease of the road tolls was a direct accompaniment of, and caused by, the operation of the railway. A similar result may be noted in connexion with the Liverpool and Manchester Railway, which was opened in the latter part of the year 1830. Soon after its operation began, an attempt was made to let the tolls at two bars near the Manchester end of the turnpike road which followed the same direction as this railway. The Eccles bar, which had been let in 1829 for £1575, and in 1830 for £1700, was offered for the next year at £800; and the Irlam bar which had brought in 1829 a rental of £1335, and in 1830 of £1300, was offered for £500; but because of the reduction of revenue anticipated or experienced as a result of the railway, no one was found who wanted to farm these tolls, even at the immense reduction for which they were offered[3].

At a ganglion like London, where great roads converge, the effect of a railway would necessarily be felt with great intensity. We would also expect considerable reductions of the tolls on roads that were parallel to railways but at short distances removed on either side. Taking those trusts which were parallel and close to, but not adjoining,

---

[1] This railway was partly opened in 1837 and completely in 1838.

[2] Brit. Doc. 1839 (295), IX, 869, 'Minutes of Evidence,' p. 66. The details of this are as follows:

Turnpike Trusts between London and Birmingham				Tolls received for half-year ending Mar. 29, 1839	Tolls received in year 1836
Whetstone Trust	..	..	..	£2,207	£5,365
St Albans Trust	..	..	..	1,063	3,821
Dunstable Trust	..	..	..	720	2,770
Puddlehill Trust	..	..	..	656	2,525
Hockliffe and Stratford Trust	..	..	735	3,507
Stratford and Dunchurch Trust	..	..	1,324	6,335
Dunchurch and Stonebridge Trust ..	..	466	1,707
Stonebridge and Birmingham Trust	..	728	2,495

∴ Tolls for half-year after the opening of the London and Birmingham railway were 7,899

2

∴ Estimated toll for 1839 would be .... £15,798 .... £28,525

[3] *Manchester Guardian*, Feb. 12, 1831, p. 3.

the London and Birmingham Railway, we find that the tolls in 1839 were only from one-half to two-thirds of what they were in 1834[1]; and, of course, those roads that were the more distant from the railway were less affected than those that were nearer[2]. So great were the reductions of the tolls in some cases, and the uncertainty of their amount, that it was becoming increasingly difficult to get anyone to farm them[3]. Sir James Macadam, who was the General Superintendent of the Metropolitan Roads, after speaking in 1839 of other factors which had caused some roads to go from bad to worse[4], added the significant statement: "The calamity of railways has also fallen upon us, which, of course, has aggravated the evil[5]." It must not be understood from what we have said, and from the instances we have adduced, that wherever the railway went the roads fell into decay. The reduction of turnpike revenues noted above was an inevitable concomitant of the introduction. of such a novel and effective instrument of transport as the railway; and there is no doubt that in some cases the financial embarrassment

[1] The *British Almanac and Companion* for 1842, p. 119, gives us the following information regarding these roads:

Names of the Trusts				Amount of tolls	
				1834	1839
Metropolis Roads, North	..	..	..	£86,676	£77,944
St Albans and Barnet	..	..	..	3,472	1,896
Dunstable	..	..	..	2,680	1,011
Sparrows Herne	..	..	..	3,458	2,613
Hockliffe and Woburn	..	..		2,519	1,230
Holyhead Road, Hockliffe district		..		3,250	1,198
Old Stratford and Dunchurch		..		5,894	2,702
Northampton and Newport Pagnell		..		2,260	1,505
Market Harborough and Welford	..		..	3,847	2,562
Dunchurch and Stonebridge		..	..	1,525	1,027
Market Harborough and Loughborough		..		6,591	5,646
Stone, Stafford and Penkridge		..	..	1,536	901

[2] See above table. For other instances of similar reductions of tolls, see Brit. Doc. 1839 (295), IX, 369, evidence of Mr Levy and others; also *Railway Times*, V (1842), pp. 18, 21.

[3] Ibid.

[4] In a communication from the Grand Junction Canal Company to the Board of Trade, in 1846, as to the desirability of keeping the Regent's Canal open and free from railway control, as a means of outlet for the inland canals, there are these words: "Now it is at once admitted that if this new power [i.e., locomotives and railways] can prove itself competent to under-carry canals, the Proprietors of the latter cannot reasonably expect to be shielded, either by Parliament or by Her Majesty's Government, from the ruin which has already befallen a considerable portion of our macadamized roads, with the various establishments...which are dependent thereon." This would seem to be good (because disinterested) testimony in regard to the decline of some of the best constructed roads.

[5] Brit. Doc. 1839 (295), IX, 369, evidence of Sir James Macadam.

of the trusts was directly traceable to the railway. But we have already shown that before the iron road came into active use there were other factors which were causing many of the trusts to be pecuniarily involved, and these, doubtless, still continued after the railway came into operation. If the railway did, along certain routes, exercise the most potent influence in effecting an immediate dislocation of business from the road to the rail, it is evident that this transfer would leave the roads subject to less injury, and therefore they would not need so much expended upon them for maintenance and repair. It would seem, then, that if some other things had not been detrimental to the finances of the roads, the railways alone would not have brought about all the evils that were attributed to them.

Having now considered the effect of the railways in causing a decrease of the traffic and revenue of many of the turnpike trusts, we next inquire as to the reasons why the railways attained the ascendancy over the stage coaches. In the light of past experience, we to-day can see many reasons why they should have gained the pre-eminence, such as, their greater speed and comfort, greater accommodation and cheaper rates, to say nothing of the mental stimulus from railway travel[1]. But, beside these, at this early time there were some additional reasons for railway predominance which are not apparent to us to-day.

Railway companies had a much lighter burden of taxation than those who carried on the ordinary highways. In addition to the turnpike tolls, the chief taxes paid by regular stage coaches were: the license duty of five pounds (£5) on each coach kept to run, and one shilling on each supplementary license; the assessed tax on coachmen and guards, which was £1. 5s. for each[2]; the stage coach duty, which was levied on a graduated scale according to the number of passengers which the coach would carry, but irrespective of the number of passengers actually carried or the number of horses used; and, finally, the assessed tax on all draught horses[3]. The license tax had to

---

[1] See *Railway Times*, v (1842), pp. 639–40, 711, 973.

[2] Brit. Doc. 1837 (456), xx, 291, 'Report of Committee ón Taxation of Internal Communication, Minutes of Evidence,' p. 3, Q. 31. Harris, *The Coaching Age*, p. 195, says that £5 was assessed for every coachman and guard.

[3] Brit. Doc. 1837 (456), xx, 291, 'Report of Committee on Taxation of Internal Communication.' The stage coach, or passenger, duties were changed from time to time, but they were at best so burdensome that coach proprietors wanted them reduced to the minimum; and in order to accomplish this it was customary, with the approach of winter, to lessen the number for which the license was taken out at the Stamp Office. For example, a coach which was licensed to carry six inside and twelve outside during the summer when business was active might be licensed for the winter to carry only four inside and eight outside. In this way, the stage coach duty would be reduced about one penny a mile per single mile, or about thirty

be paid whether the coach were run only a few days or for the whole year, and the same regulation was enforced concerning the assessed taxes. The stage coach rates were paid each way by the coach proprietors, the lowest amount being for a coach with a capacity of four passengers, which paid one penny per mile each way[1]; in other words, the lowest duty was one-fourth penny per passenger per mile. Beside the foregoing taxes, from which the railway company was exempted, the coaches had to pay a mileage duty, on the basis of the number of miles the coach travelled but without any reference to the number of passengers the coach was licensed to carry. Coach proprietors, in a few cases, bought their coaches outright; but in most instances they made an arrangement with the coach builder to pay him, for the use of the coaches that were required, a certain mileage rate, which varied from two to three pence per mile according to the contract they were able to make with the owner of the vehicles[2].

On the other hand, the taxation of the railway, as an operating agent, consisted merely of a mileage duty of one-half penny per mile on every four passengers, that is, one-eighth penny per passenger per mile[3]. It will be seen, therefore, that the mileage rate was the only one of the stage coach taxes that applied to railways, and it was only a small fraction of the amount charged on the coaches. But there was this further distinction to be carefully noted between the steam and the stage coaches, that while the railway was charged mileage rate

per cent., during the winter. It may be asked why a coach proprietor could not take out a license for a smaller number of passengers, paying therefore the lower duty, but carry the larger number of passengers on his coach. The answer is that penalties were heavy for the transgression of the law, and on the chief roads there were men who made their living by informing on persons who broke the law, since the informers got one-half of the penalty imposed on offenders. The number of passengers that a coach could carry had to be painted on it in a conspicuous place; and if a coach were found with more than its legal number of passengers the magistrate's fine made a considerable expense for the proprietor. See also Harris, *The Coaching Age*, pp. 196–8.

[1] These stage coach duties, as given in the schedule to Act 2 & 3 William IV, c. 120, were as follows:

For 4 passengers, 1*d.* per mile a single mile.
      6   ,,   1½*d.*   ,,   ,,   ,,
      9   ,,   2*d.*   ,;   ,,   ,,
     12   ,,   2½*d.*   ,,   ,,   ,,
     15   ,,   3*d.*   ,,   ,,   ,,
     18   ,,   3½*d.*   ,,   ,,   ,,

[2] Harris, *The Coaching Age*, pp. 198–9.

[3] Brit. Doc. 1837 (456), xx, 291, 'Report of Committee on Taxation of Internal Communication.' In 1842, this was changed to five per cent. of the receipts from passenger traffic.

only on the passengers actually conveyed, the stage coaches were charged their rate on the number of passengers which the coaches were licensed to carry, whether they were 'full or empty. This was a detriment to the stage coaches; for if one of them was capable of carrying twelve passengers, only an average of eight passengers could be counted on; and, therefore, in paying both ways, they paid duty for twenty-four passengers, but carried and received payment for only sixteen[1]. Some advocated repealing the duties on stage coaches to enable them to compete with the railways, and a proposal had been made to take the tax off coaches running parallel with the railway, but neither of these was carried out[2]. This manifest unfairness could not but prove prejudicial to the proprietors of stage coaches, many of whom expected that their business would be overwhelmed[3]. Some of them, however, saw clearly that, even if stage coaches were made free

---

[1] Brit. Doc. 1837 (456), xx, 291, evidence of Mr Horne. He handed in the following computation to exemplify the difference between railway and road carriage in the matter of mileage duty alone:

Coaches to Birmingham, say 108 miles,—	£	s.	d.
If licensed for 15 passengers, say average 10, at 3d. a single mile, is per journey	1	7	0
Railway at ⅛d. per head, say 10 passengers, is per journey		11	3
Difference		15	9

The mileage duty, therefore, is 2s. 8¼d. by coach, and 1s. 1¼d. by railway, for each person actually carried.

Mr Horne	had 8 Birmingham and 8 Liverpool and Manchester coaches.						
Mr Chaplin	,, 2	,,	,, 7	,,		,,	,,
Mr Sherman	,, 8	,,	,, 8	,,		,,	,,
Mr Gilbert	,, 1	,,	,, 0	,,		,,	,,
Mr Mountain	,, 1	,,	,, 1	,,		,,	,,
Mr Nelson	,, 0	,,	,, 1	,,		,,	,,
	10		15				

	£	s.	d.
On 10 Birmingham coaches, the difference between railway and coach −	7	17	6
On 15 Liverpool and Manchester coaches, the difference between railway and coach .... =	23	12	6
On 6 coaches between Liverpool, Manchester and Birmingham, the difference between railway and coach .... =	4	14	6
Difference each way =	36	4	6
Difference for journey =	72	9	0

For a comparison of the mileage duties paid by stage and railway coaches, see also *Herepath's Railway Magazine*, N.S., v, pp. 582–8, the figures of which are all right, but the editorial comment on them contains statements that cannot be accepted.

[2] *Herepath's Railway Magazine*, N.S., vi, pp. 458–64; *Hansard's Parliamentary Debates*, xiv (1832), pp. 1300–2, statement of Lord Althorp.

[3] Brit. Doc. 1837 (456), xx, 291, evidence of Mr Horne and Mr Gray.

of duty, they could not compete with the railways in the same direct line[1], on account of the many other advantages which the railway had.

Another thing which tended to defeat the coaches in their competition with the railways was that the latter were frequently permitted by the Treasury to compound for their taxes at a very low and perfectly illusory rate: a privilege that was uniformly refused to the proprietors of stage coaches[2]. In the three years 1835–7 the railways which were compounding for their mileage rate paid, in all, £1519. 10s., whereas the amount of mileage duty which would have been paid if no composition had been entered into would have been £5727. 14s. 3d.; that is, they paid about one-fourth of the statutory duty[3]. With such favouritism or protection to the younger and progressive means of communication it was inevitable that the railways should soon dominate in the carriage of passengers.

Enough has been said to show the effects of the introduction of railways, in particular cases, upon the previously existing means for the conveyance of passengers, and upon the roads. But it requires to be emphasized that the particular cases must not be taken as exemplifying or attempting to prove that the foregoing results were universally found to follow the construction of railways. On the other hand, we have the statements of some that the revenues of certain trusts which

---

[1] Brit. Doc. 1837 (456), xx, 291, evidence of Messrs Horne, Gray and Collins.

[2] Ibid., 'Report of Committee,' p. iii.

[3] Brit. Doc. 1839 (517), x, 127, 'Appendix, No. 23,' pp. 406–7; also Brit. Doc. 1837 (456), xx, 291, 'Minutes of Evidence,' p. 23.

This fact is more fully exemplified if we take the individual cases of those railways which paid composition during the three years 1835–7, as follows:

Railways	Total amount of composition paid			Amount of mileage duty that would have been paid, if not compounding		
	£	s.	d.	£	s.	d.
Bolton and Leigh	30	0	0	877	4	2
Canterbury and Whitstable	51	0	0	239	8	9
Hartleyburn and Brampton	6	0	0	23	5	0
Leicester and Swannington	60	0	0	257	11	6
Newcastle and Carlisle	255	0	0	1294	3	7
North Union (Wigan and Preston) .	360	0	0	1112	7	10
St Helens and Runcorn Gap	25	0	0	99	9	3
Stanhope and Tyne	2	10	0	46	13	2
Stockton and Darlington	600	0	0	1801	5	11
Stratford and Moreton	20	0	0	51	13	8
Warrington and Newton	110	0	0	424	11	5
Total	1519	10	0	5727	14	3

were paralleled by railways had increased[1].   Whatever may have been
the immediate results upon the revenues of the turnpike trusts, it is
almost certain that the roads did not suffer any permanent set-back,
nor did the number of coaches decrease; on the contrary, the number
of licenses for coaches increased because of the necessities that accompanied the great stimulus given to travel[2].

[1] Statements of Mr Pease and others before the Parliamentary Committee, as
quoted in *Advantages of the Progressive Formation of Railways*, pp. 16–21; also
Proceedings of the Great Western Railway Company, 'Minutes of Evidence,' p. 49.

[2] See *Railway Times*, VI (1843), p. 443, giving statistics to prove these statements.

CHAPTER IX

COMPETITION OF RAILWAYS AND CANALS

In a former chapter we have described the manner in which the carrying trade was effected on the canals; for before 1845 the canal companies themselves were not authorized to carry, but this work was done by private carriers and regularly chartered companies, who placed their own barges on the canals and furnished the traction power, paying only the tolls demanded by each canal company. We have also outlined the changes that were made in the organization of the carrying trade on the railways; but as it was in connexion with this that the first great railway struggle was precipitated we venture, even at the possible risk of repetition, to consider the three chief systems of railway operation with reference to the way in which goods were transported.

The system adopted by the London and Birmingham Railway Company allowed the carriers who chose to avail themselves of the terms offered by the railway company the opportunity of sending goods to any amount, the waggons and the locomotive power being provided by the railway company[1]. The carrier collected and delivered

---

[1] Brit. Doc. 1839 (517), x, 1, 'Second Report on Railways,' pp. viii–ix; also Brit. Doc. 1840 (487), xiii, 181, 'Fourth Report of Select Committee on Railway Communication, Minutes of Evidence,' p. 110, showing that the London and Birmingham Railway Company believed it was more advantageous to them to allow carriers to come on their line and pay the railway company tolls for the use of their road, than to be carriers themselves. The company thought that the public would also be better served, because the individual carrier who had charge of their goods would be more responsible. The railway company charged by weight only. They claimed all small parcels under 100 pounds, and transported them themselves (v. Brit. Doc. 1840 (487), xiii, 181, 'Minutes of Evidence,' p. 98). That the legislature intended to ensure the right of carriers to use the railways, upon payment of the tolls to the railway companies, is evident from the fact that every original railway Act, except that for the Liverpool and Manchester Railway, contained a clause to that effect. But since this right could not be exercised without great danger to the public, on account of the admission to the railway of steam-power that was not under the immediate control of the company, the provision was made in one of the later Acts of the London and Birmingham Railway Company that the latter should provide the carriers with both waggons and power at a fair and reasonable charge (*Railway Times*, iv, p. 866). See also Whitehead, *Railway Management*, 2nd ed., p. 6.

the goods, took all risks, and paid the tolls and haulage charges, which were so regulated as to yield good profits to the railway company and a reasonable return to the carriers. This arrangement could hardly be said to offer such competition as to secure the public against exorbitant charges, because the demand for carriage depended upon the terms and rates fixed by the company, and, consequently, the rivalry between the carriers was, in a great measure, restricted to the collection and delivery of the goods with which they were entrusted. This competition, therefore, afforded no guarantee that the service would be performed at the lowest remunerative charge. The profits of each carrier depended on the amount of his business, and this could only be maintained and increased be incessant attention to the wishes of his employers. This open system pursued by the London and Birmingham ceased when that line became merged with others in the London and North Western.

An entirely opposite system was that in force on the Liverpool and Manchester Railway. That company was required by its Act to undertake the carriage of any goods that might be brought to its representatives for conveyance along the line; and thus private carriers were excluded from a share in the goods traffic. The company was limited as to the amount of charge which it might demand for the carriage; but, in reality, it did not charge the maximum rates specified in the Act. On the contrary, the rates were fixed with reference to the cost of water carriage between these places[1].

The third system was a combination of the other two, and was found in operation on the Grand Junction Railway, from Birmingham to Manchester and Liverpool. The company could not prevent any private carrier using their line, but, at the same time, they themselves undertook the carriage of goods, and therefore competed at every point with the private carriers[2]. They retained for themselves the conveyance of all Birmingham and Lancashire goods coming from or going to London. Before admitting any carrier on their line, the railway

---

[1] Brit. Doc. 1839 (517), x, 1, 'Second Report on Railways,' pp. viii–ix. It would seem, however, that the Liverpool and Manchester, at a later date, found it wise to adopt a different plan from this which they had pursued for many years (*Railway Times*, vi, p. 152). This railway company was different from any other in that it was made a carrier by its Act of incorporation. The Stockton and Darlington, before that line was opened, had received application requesting the privilege of carriage by locomotive engine on its roadway, but had refused this; for the committee in charge, after careful inquiries, were convinced that the company's welfare would be best served by being the principal carriers on its own line. Jeans, *Jubilee Memorial of the Railway System*, p. 63.

[2] Brit. Doc. 1839 (517), x, 1, 'Second Report on Railways,' pp. viii–ix; also Brit. Doc. 1844 (318), xi, 17, 'Fifth Report of Select Committee on Railways,' Appendix No. 2, pp. 22–23.

company bound him by agreement not to charge his patrons less for the carriage of goods than the rate demanded by the company for the same service[1]. Hence, the public could derive little or no benefit from this kind of competition. . Moreover, we can readily see that when many carriers were allowed on the one line of railway, using the same track, stations, terminal facilities and other equipment, much confusion and no little strife would ensue, not only among the carriers themselves but also between the railway company and the carriers, for the carriers were not always careful in their use of the appurtenances of the road. Then, when anything went wrong, or any injury was done, it was almost impossible for the company to know who had been the cause of the trouble. Besides, the safety and the convenience of passengers were endangered by the presence of so many rivals on the line. The jealousies and complaints that arose from such a confusion of interests on the same line had shown the railway company, as early as 1840, the necessity of excluding private carriers altogether, and undertaking all the carrying trade themselves[2].

The decision arrived at by the Grand Junction Railway Company was being reached by other railway companies also, as the only solution for the ills of the existing situation in regard to the goods traffic on railways[3]. In support of the contention that the railway companies should be the only carriers on their lines, it was urged that, as it was necessary for them to perform so much of the carriage as was equivalent

---

[1] See Brit. Doc. 1840 (437), XIII, 181, 'Minutes of Evidence,' pp. 88–89, for "Copy of an Agreement between the Grand Junction Railway Company and Messrs Robins & Co. (carriers) of Liverpool." The Grand Junction Railway Company charged by the parcels, and these could not be boxed together by putting small ones inside of large ones (ibid., p. 98).

The Bolton and Leigh Railway, communicating with the Liverpool and Manchester, was let to a single carrier. The North Union Railway (from Wigan to Preston) professed to follow the example of the London and Birmingham, but up to 1839 only one carrier had established himself upon the line, and as he was the lessee of the Bolton and Leigh it was thought that his wealth and influence might exclude all other competition from these two lines. The Newcastle and Carlisle Railway Company was the only carrier on its line. The Stockton and Darlington was the principal carrier on this line; although there were also other parties, using horse-power, who were engaged in carrying goods. The Leeds and Selby Company was the sole carrier on that line. Brit. Doc. 1839 (517), X, 1, 'Second Report of Committee on Railways.'

[2] Brit. Doc. 1840 (299), XIII, 167, 'Third Report of Select Committee on Railway Communication.' See also *Railway and Canal Cases*, I, p. 592 et seq., Pickford et al. *vs.* The Grand Junction Railway Company. This case is fully discussed in Appendix 14.

[3] *Railway and Canal Cases*, III, p. 568 et seq., Parker *vs.* Great Western Railway Company (1844), shows that this railway company was following the Grand Junction in trying to exclude the carriers from their line.

to at least eighty per cent. of the whole cost, namely, the transport of the goods along the railway, the conditions were not such as to enable the public to benefit by the competition of private carriers, and the companies could perform the remaining twenty per cent. of the work more economically. Then, by the companies taking the carrying trade completely into their own hands, the shippers served by each line would be assured the advantage of uniformity of charge. Finally, it was said that, in order to bring railway conveyance fairly into competition with the old canal monopoly that existed in many cases, it was essential that the railway companies should become carriers, since the great private carrying firms were generally interested in continuing the canal[1].

On the other hand, it was asserted that it was neither so economical nor so convenient for the public that the railway company should step out of its legitimate sphere, by becoming collectors and distributors of goods; and that the competition of the private carriers, though confined to twenty per cent. of the total charge, was sure to reduce this portion of it to a minimum, and hence was worthy of a place in the public economy. Moreover, if the railway companies should become carriers, and the private carriers should be driven off the rails, the railways would then be in a position to combine with the canals and force the public to pay monopoly prices[2].

[1] Brit. Doc. 1844 (318), xi, 17, 'Fifth Report of Select Committee on Railways,' Appendix No. 2, also 'Minutes of Evidence,' Q. 3933–6.   Ibid., p. 290 et seq., gives much discussion as to whether it was best to have private carriers on the railway lines or to have the railways act as carriers on their own lines.

The last argument, of course, has no weight. The railway could compete with the canal, by its having passenger traffic to add to its income, while the canal had none; so that, in order to meet the canal monopoly, it was not necessary to drive the private carriers off the railway.

[2] Brit. Doc. 1844 (318), xi, 17, 'Fifth Report of Select Committee on Railways,' also 'Minutes of Evidence,' p. 290 et seq.

In the *Railway Times*, iv, p. 366, the editor of that journal in discussing the pamphlet of Henry Booth on the "Carrying Question," shows that Booth's arguments against admitting private carriers on the railway, though applicable to the Liverpool and Manchester Railway, did not bear upon the general question as it related to the vast and complicated interests over the whole kingdom. Then, when he had thus dismissed the consideration of that pamphlet, he reiterated his own views; and after showing that the carriers had a legal right to the use of the railways on payment of the tolls, he went on to exemplify how it was to the advantage of all parties that such competition should be encouraged. In addition to the reasons here suggested in the text, the editor says that the private carriers would compete also in the matter of attention and civility to the public, which was scarcely less important than the economy of charges; whereas the railway companies as carriers would show but slight moral responsibility. The carriers had also well-established collecting and distributing facilities in all the important towns and cities, and were therefore prepared to look after the goods traffic at all places remote from the

There certainly was a good deal of weight in some of the arguments advanced on each side. The effect of railway companies becoming carriers was undoubtedly, in many cases, beneficial, and led to a material reduction of the existing charges[1]; and, under certain circumstances, it might be as convenient for the public to employ the railway companies as carriers. This would be the case where the railway had a station at the point of destination of the shipment; for example, if a merchant in London were sending goods to Birmingham it would be just as convenient to have the railway take them there as to send them by a carrier on the canal, for as soon as they were unloaded at that railway station the Birmingham merchant would have no trouble in getting his goods. But, in regard to much of the traffic of the country, it would certainly be felt as an inconvenience to be obliged to employ a railway company as the sole carrier. For example, if goods were to be sent from London to some place near Birmingham, and they were given into the hands of a private carrier in London, they would be taken by him or his agent from the station at Birmingham, when they had reached there, and delivered to the consignee; but, if the railway company were the only carrier, the freight would be left at Birmingham until removed by the consignee. To cite the instance of Coventry: if goods were sent by private carrier from London to Coventry, they would, if suitable, be taken by him on the railway to Birmingham (there being no station at Coventry), and then carted back by the carrier for the seventeen miles to Coventry and there delivered at the consignee's door. Even this method of getting the goods to Coventry did not cost as much as to take them from London to Coventry by canal. But if the shipment were given to the railway company as the carrier it would be taken to Birmingham and left there to await the coming of the consignee or his agent who would ship the goods back to Coventry by road or by canal as seemed most

---

railway termini. He regarded it as in the interest of the railways as well as the public to continue the carriers on their lines. For further discussion of the carrying question, see Appendix 14.

[1] Brit. Doc. 1840 (299), xiii, 167, 'Third Report of Select Committee on Railways,' under heading II, "Carriage of Cattle and Goods by Railways."

Note what we have formerly said regarding the effect of the Liverpool and Manchester Railway in reducing the charges made by navigations between these places. Also Brit. Doc. 1840 (437), xiii, 181, 'Fourth Report on Railways, Minutes of Evidence,' pp. 82, 110 et seq.; and Brit. Doc. 1844 (318), xi, 17, 'Minutes of Evidence,' p. 76 et seq., where we learn that when the Leeds and Selby Railway Company opened their line and did their carrying the rates on the Aire and Calder Navigation were very materially reduced. See also ibid., 'Minutes of Evidence,' p. 527 et seq.

desirable[1]. The carriers who used both railway and canal conveniences carried the more valuable articles on the railway, and the cheaper freight, i.e., the more bulky and heavy commodities, by the canals, for the latter class generally did not require rapid transport. The private carriers who did nothing else than that work, and who had well-established facilities for the economical collection and distribution of traffic, asserted that it was not possible for the railway company to perform these services as acceptably and cheaply as themselves; but, of course, the railway company could develop just as good facilities if it were thought best to put the performance of these duties in their hands. There was another advantage in employing private carriers, in that they were responsible for the goods from the time they left the consignor until they reached the consignee; but the railway assumed no such responsibility[2].

As we have seen, it was the intention of Parliament that railways should be on the same footing as canals and that railway proprietors should have similar rights to those of canal proprietors, that is, receiving tolls, but not carrying at all[3]. When this matter was under public discussion many of the carriers said that the only thing to do was to carry out the intention of Parliament and preserve competition by excluding the railway companies from carrying on their own lines. Others were in favour of suppressing the private carriers and giving all the work into the hands of the railways[4]. Out of the mass of

---

[1] Brit. Doc. 1840 (437), xiii, 181, 'Fourth Report on Railways,' evidence of Messrs Tibbits, Derham and Harnett'(p. 24). The London and Birmingham Railway carried only through traffic.

[2] Brit. Doc. 1844 (318), xi, 17, 'Fifth Report on Railways, Minutes of Evidence,' Q. 3941. The personal relations of the private carriers with their patrons were a valued element in the conduct of business. Carriers allowed their customers from three to six months' credit; and permitted them to warehouse their goods, without charge, till they could conveniently send them to their destination. If any inconvenience were suffered and a complaint made to the carrier, the latter was always amenable and an answer was obtained. Even though this redress was sometimes tardy and not entirely satisfactory to the shipper, it was better than to be treated with indifference. On the contrary, it seems to have been the policy of the railways to be more overbearing. They required monthly settlements of bills. They did not allow goods to be left in their warehouses without the payment of storage charges. When complaints were made, the responsible railway official was so far removed from the complainant, and the company was so unresponsive, that unless the complainant had enough influence to enforce attention to his claim he could not depend upon receiving justice. Boyle, *Hope for the Canals*, pp. 17–18.

[3] Brit. Doc. 1840 (299), xiii, 167, 'Third Report of Select Committee on Railways;' also Brit. Doc. 1840 (474), xiii, 189, 'Fifth Report on Railways, Minutes of Evidence,' p. 40, Q. 959; etc.

[4] On this whole question, see Brit. Doc. 1840 (299), xiii, 167, 'Report and Evidence,' which deals very fully with it; also Brit. Doc. 1844 (318), xi, 17, 'Minutes of Evidence,' p. 106 et seq. See also Appendix 14.

conflicting testimony, the Parliamentary Committee of 1839 came to the conclusion that the intention of the Legislature in this respect could not be carried into effect in the way contemplated; for it was obvious that the payment of legal tolls was only a very small part of the arrangements that were necessary to open railroads to public competition, and the rest of the arrangements were wholly disadvantageous to the private carriers on the line. They decided that, upon grounds of safety and economy, there should be upon every railway one system of management, under one superintending authority, which should have the power of making and enforcing all regulations necessary to the proper conduct and maintenance of the traffic. Because of this, it was essential that the railway company should possess a complete control over their line of road, even though they should thereby acquire an entire monopoly of the means of conveyance[1].

We have entered thus fully into this subject because it is one of the pivotal points in the competition of railways and canals, and because it is interesting to see how early, after the introduction of railways, it was recognized and settled that they were unlike most other enterprises in being essentially monopolistic. Later reports from parliamentary and other public bodies reiterated and emphasized this characteristic feature, and also the need for some general supervision and control so that the public might derive the utmost benefit from this natural monopoly[2]; but into this latter phase, that of railway control, it is not our purpose to enter.

By what means did this monopoly power actually realize its monopoly, or, in other words, how did it drive the private carriers off the railway? In many cases the railway company gave no better terms to the carriers than to the occasional shipper, and so the carrying

---

[1] Brit. Doc. 1839 (517), x, 1, 'Second Report of Select Committee on Railways,' pp. vi–vii, xiii.

Another factor which contributed to the taking over of the carrying trade by the railway company was the systematic efforts of the carriers to secure advantage over the railway company by making false declarations as to the weights and descriptions of the goods that they loaded for carriage on the company's waggons. The London and North Western had to appoint a detective to see that their interests with reference to this were protected. In 1847, the next year after that company was formed, the system of toll carrying was abolished, and the railway company gradually began to carry directly for the public (v. Stevenson, *Fifty Years on the London and North Western Railway*, p. 17 et seq.). On this subject, see also *Railway Times*, iv, pp. 208–9, the affidavit of John Moss, and ibid., vii, p. 217, on "Railway Companies and Railway Carriers." Refer also to Appendix 14.

[2] See, for instance, Brit. Doc. 1840 (299), xiii, 167; ibid. 1844 (166), xi, 5; ibid. 1845 (279), xxxix, 153; ibid. 1846 (200), xiii, 85.

trade became unremunerative[1]. For example, the company charged the carriers for the mere transport of a certain weight of goods over the line, independently of the collection and distribution of these goods, the same rates as were charged the public for the carriage and the additional services of collection and distribution[2]. This was sometimes put into effect against all the carriers on the line at once, and in other cases the carriers, one at a time, were compelled to suspend operations on account of the imposition of these practically prohibitive rates. Sometimes lower rates were quoted to some carriers than to others[3], and in at least one case the railway company absolutely refused the use of its carriages to a certain carrier. It appeared that the company had made arrangements to carry goods for another firm of carriers only, by which that firm obtained a monopoly of the conveyance of goods along that line of road[4]. A few years later, when this railway company allied its interests with others in the formation of the London and North Western, the latter company adopted the policy of being themselves the exclusive carriers on their line; but they retained an arrangement with Chaplin and Horne, who were probably the largest carriers into and out of London, to collect and distribute in London the goods going from and coming to that city by this railway[5]. During the tentative stages of

[1] Brit. Doc. 1840 (437), xiii, 181, 'Fourth Report on Railways, Minutes of Evidence,' p. 37, Q. 918. This was done by the Grand Junction Railway Company, which was engaged in carrying on the London and Birmingham Railway, as well as on their own, and on the Liverpool and Manchester Railway.

[2] Brit. Doc. 1844 (318), xi, 17, 'Fifth Report of Select Committee on Railways, Minutes of Evidence,' pp. 138–9. Also Brit. Doc. 1852–3 (170), xxxviii, 5, 'Second Report on Railway and Canal Bills, Minutes of Evidence,' pp. 35–38, shows the means by which Kenworthy & Co., carriers, were driven off the canals and railways by the railway companies that got control of these canals.

[3] Brit. Doc. 1844 (318), xi, 17, 'Fifth Report on Railways, Minutes of Evidence,' pp. 384 ff.; ibid. 1852–3 (170), xxxviii, 5, 'Second Report of Select Committee on Railway and Canal Bills, Minutes of Evidence' of Mr Pixton.

[4] Willmore, Wollaston and Hodges, *Reports of Cases argued and determined in the Court of Queen's Bench, and upon Writs of Error from that Court to the Exchequer Chamber, and in the Bail Court,* i, pp. 578 ff., ex parte Robins and others. Messrs Robins, general carriers, made application in 1838 for a mandamus to compel the London and Birmingham Railway Company to carry the goods of the applicants; but the Court decided that, under the Act of incorporation, the company could not be compelled to carry all goods sent for conveyance and the application was refused.

[5] *Parl. Papers,* 1857–8 (0.77), xv, 11, 'Minutes of Evidence taken before the Select Committee on the Manchester, Sheffield and Lincolnshire, and Great Northern Railway Companies Bill,' Q. 4683–4, 4901–16. Chaplin and Horne would not state exactly what their relation was with the London and North Western Railway. They had also close business relations with the London and South Western, and finally invested a considerable sum in that railway. Fay, *A Royal Road,* p. 28. It would

the development of the carrying trade, sometimes the railway company bought out the business of respectable carriers by payments that were much in excess of the real value; but even after the purchase was made some railways did not exclusively collect goods for themselves, but gave discounts and allowances for the collection and delivery of goods to and from the stations, and allowed a certain percentage for loading, unloading and invoicing, until it was discovered that frauds were being practised which tended to destroy the company's own carrying trade[1]. Under these conditions the tendency was for the company to eliminate the carriers entirely. But amid the variety of causes tending to take traffic from the carriers and give it to the railways was the growing conviction among the commercial classes that, because the railway company did not have to pay tolls on its own line, and the cost of locomotive power to it would be no greater than if furnished to a private carrier, therefore the railway company could do the carrying cheaper than any private carrier; and if either were to be stopped they would prefer to see the company left as carrier[2]. The railway company had so many advantages over any other carrier fulfilling this office along their line, that gradually it became the universal practice for them to do all this work, including the collection and distribution of the goods at their starting-point and destination[3].

It must not be concluded from what we have said that railway companies (except as regards passengers) superseded the old carriers at

seem that Chaplin and Horne were retained for this service on condition that they would cease carrying on the canals. Boyle, *Hope for the Canals*, p. 5. See also Whitehead, *Railway Management*, p. 7.

[1] Nash, *Railway Carrying and Carriers' Law*, p. 75; also *Railway Times*, IV, pp. 208–9, and ibid., VII, p. 217.

[2] Brit. Doc. 1844 (318), XI, 17, 'Minutes of Evidence,' p. 527 et seq.; also Brit. Mus. 8235. b. 57 (1), 'The Carriers' Case considered,' pp. 8–9. This was especially the case with the Grand Junction Railway Company, whose highest rate for the lightest articles of merchandise, up to April 1844, had been 5s. a cwt., but after that was 4s. a cwt., from London to Liverpool. Other carriers charged up to 6s. and 7s. a cwt. Note the two examples given in footnote 3, p. 631. See also Brit. Doc. 1840 (437), XIII, 181, 'Fourth Report on Railways, Minutes of Evidence,' p. 37.

This conviction, however, was long in being established, and we find strong opposition, up to the middle of the century, against the oppressive and unjust conduct of the railway companies toward the carriers. Petitions were sent in by large and influential bodies of traders against the monopolistic policy of the railways to defeat fair competition. See *Herepath's Railway and Commercial Journal*, XI, pp. 585, 599. Also Appendix 14.

[3] In all probability it is because the railways at this early time took over all the work of the carriers that to-day the English railways, unlike those of the United States and some other countries, do the collecting and distributing of the goods carried on their lines. Because of this, there is no need for such secondary concerns as Express Companies which we find in the United States.

once. Few of the carriers tried the experiment of running their own trains along the railway, and these few, for reasons already given, were forced to give up; but for some years a considerable part of the carrying business remained in the hands of the old firms[1], who continued to collect goods from the public and to arrange for their safe delivery, employing the railway companies, which would give them access, to convey them along their lines. On certain railways, as we have already shown, this practice prevailed for some years exclusively; on others, from the first, the companies seem to have undertaken the business of general carriers for the public, as well as conveying for the carriers[2].

Now that we have considered the organization of the carrying trade on the canals and on the railways, we are able to appreciate more fully the effects of the competition which occurred between these two rivals. In an earlier chapter of this work it was shown that before the introduction of the railways many canals had put up their rates, until, with their monopoly, some of them were making enormous profits. This fact is attested by the high market value of some shares, and by the large dividends obtained by the shareholders of certain canals[3]. In some cases, as soon as a railway was threatened and action taken toward that end, the adjacent canal, which had been deaf to all complaints, found it desirable to reduce its tonnage rates and to think

---

[1] Pickford, Parker, Robins, Chaplin and Horne, etc. See the advertisement of Chaplin and Horne in *Railway Times*, VII (1844), p. 1447, showing that they forwarded goods by the various railways, "on their own account or as Agents of the Companies." Then they mentioned the different railway lines they used and the places in England to which they shipped.

[2] Brit. Doc. 1840 (299), XIII, 167, 'Third Report on Railways,' p. 3; ibid. 1844 (318), XI, 17, 'Fifth Report on Railways,' Appendix No. 2, p. 22; ibid. 1881 (374), XIII, 1, 'Report of Select Committee on Railway Rates and Fares, Minutes of Evidence,' p. 573.

[3] These large profits are reflected in the prices of some of the canal companies' shares; for example, the Staffordshire and Worcestershire Canal shares (of the par value of £100) sold in 1810 for £735–50, and in 1829 for £810; the Grand Junction Canal shares sold in 1810 for £260–86, in 1825 for £330, and in 1828 for £315; the Trent and Mersey shares sold in June 1825 for £2150, in June 1828 for £3280, and in June 1829 for £3160. See the quotations of the share market in the *Gentleman's Magazine* for these various dates.

The rates of dividend paid are also a good indication of the profits reaped by some canals. In addition to those we have formerly noted, we may mention that in 1833:

The annual dividend of the Coventry Canal was	82 %
,,　　,,	Oxford Canal was	34 %
,,　　,,	Stafford and Worcester Canal was	34 %
,,　　,,	Trent and Mersey Canal was	37 %
,,　　,,	Erewash Canal was	47 %
,,　　,,	Loughborough Canal was	134 %

(v. Martin, *Railways—Past, Present, and Prospective*, p. 27.)

of the necessity of making improvements in its waterway in order to maintain the traffic[1]. But when a railway was actually constructed the first effect was to cause a reduction in the freight rates that had been in existence on the more or less parallel canals; and this cut in rates was almost immediate, for when the railway put a low rate into force the canals had to meet it or lose the traffic. With this diminution of freight rate, and the accompanying decrease of traffic due to a portion of the traffic being turned to the rails, it was inevitable that the railway should cause a decline in the revenues of the canals[2]. The amount of

[1] Blewitt, *New Monmouthshire Railway*, pp. 11–15.

[2] Teisserenc, *Voies de communication*, pp. 571–4, shows that, on account of the railway competition, the revenue of the Wilts and Berks Canal was reduced from 482,500 fr. in 1839 to 212,500 fr. in 1843, and the revenue of the Kennet and Avon Canal during the same period declined from 1,150,000 fr. to 800,000 fr. The opening of the London and Birmingham Railway in 1837–8 caused a reduction of the revenues of the Grand Junction Canal from 4,957,500 fr. in 1838 to 2,700,000 fr. in 1844, and a corresponding reduction in the case of the Coventry Canal. The effect of railway competition between Manchester and Leeds is seen by the fact that the gross revenue of the Rochdale Canal from bulky commodities declined from 1,473,250 fr. in 1840, when the railway was opened, to 680,000 fr. in 1841 and 485,000 fr. in 1844. These statistics are corroborated by those given in the Appendix to a statement issued on behalf of the Grand Canal Co. of Ireland, as printed in *The Times*, July 20, 1844, p. 6.

On the route from Manchester to Hull, 99 miles by canal, the rates per ton before and after the opening of the railway (1840) were as follows:

	Before			After		
For corn, flour, etc........	£1	4s.	0d.	£0	18s.	0d.
For cotton twist	1	12	6	1	0	0
For manufactured goods	2	5	0	1	4	0

Brit. Doc. 1845 (61), xxxix, 293, p. 13. Because of railway competition, the Calder and Hebble Navigation, which was part of the through water-route between Manchester and Hull, reduced their dividend from 18 % in 1848 to 8 % in 1849. *Herepath's Railway and Commercial Journal*, xi (1849), p. 1241.

The effect of the opening of the Great Western Railway was also to reduce the charges of carriage on the Thames, as follows:

	Date	Cost by water	Cost by railway
London—Windsor......	1829	9s. per ton	———
	1846	———	5½–6s. per ton
London—Reading	1829	15s. per ton	———
	1846	———	7–8s. per ton
London—Oxford	1829	£1. 2s. per ton	———
	1846	———	10–12½s. per ton

As soon as the Liverpool and Manchester Railway was opened, the former insolence of the navigations connecting these two cities was immediately abandoned, and under competition their rates had to be cut down. The rate on light goods carried on the canal was 15s. per ton; the railway reduced this to 10s. *Annual Register*, 1832, p. 445. See also Boyle, *Hope for the Canals*, pp. 5–6; Shaen, *Review of Railways and Railway Legislation*, pp. 33–34.

the reduction of the charge depended, of course, partly upon the conditions which prevailed before the railway came in; for if the canals had been charging unduly high rates the decrease was the greater, but if they had been contented with ample but not exorbitant profits the cut made in their rates was not so excessive. From these circumstances it will be seen that it is wholly impossible to make any explicit general statement that will be a close approximation to the truth; but from material collected elsewhere in this volume, we may say that, putting it at the minimum, the reduction in the rates was from one-third to one-half of the rates previously in effect on the navigations[1]. On those waterways which were adjacent to the railways the reduction would, of course, be greater than on those more remote. On the basis of the diminution of the freight rates alone, however, we would not get an adequate conception of the influence of the railways; we must take into account also the entire change in the method of conducting business as a result of the more speedy conveyance. Orders given a long time in advance became more rare; retailers kept smaller stocks of goods; less capital was, therefore, tied up in unproductive forms; and, taking all things into consideration, there was probably a saving of at least seventy-five or eighty per cent. in the conduct of business.

The decreased revenues of the canals were reflected in the lower market values of some of the canal shares which had previously brought high prices; and the prices which ruled on the Exchange will be a corrective, if necessary, of the above-mentioned conclusion. For example, before the opening of the London and Birmingham Railway, the shares of the Grand Junction Canal were selling in 1838 for £250 and two or three years later for £303 to £330; but after the railway was in operation the shares of this canal fell to £155 in 1844, to £100 in 1846, and to £60 to £70 in 1853[2]. These were on the par value of £100. By 1844 the shares of the Warwick and Birmingham Canal had fallen from £330 to £180, the shares of the Worcester and Birmingham from

---

[1] In Appendix 10 there have been brought together some tables which will illustrate the reduction of rates that was brought about by competition, and for detailed information the material there collated may for the present suffice. In making the general statement that railway competition caused a reduction of at least one-third to one-half of the previous navigation rates, we have endeavoured to keep well within the limits of accuracy, as revealed by the statistics given in Appendix 10. This conclusion is authenticated by the statement of a writer in the *Railway Times*, vii (1844), p. 217, who said that railways had caused a reduction of over 50 % in the cost of carriage of goods, and also by Teisserenc, op. cit., pp. 34–38, 571–4.

[2] Teisserenc, op. cit., pp. 34–35; Brit. Doc. 1852–3 (246), xxxviii, 175, 'Third Report on Railway and Canal Bills, Minutes of Evidence,' p. 14.

£84 to £55, the Rochdale Canal shares from £150 to £61. 10*s*., and those of the Kennet and Avon Canal from £25 to £9 per share[1]. We seem to be standing on firm ground, therefore, in saying that the revenues of canals which were parallel with railways were reduced from one-third to one-half. In some instances we see still greater changes; for instance, Coventry Canal shares, which at one time were as high as £1200, fell as low as £315[2]; and the shares of the Loughborough Canal, which before the opening of the railway sold as high as £4300 or £4400 each, had fallen to £1200 to £1500 in 1838, and to £180 to £200 in 1872[3].

Usually, however, when competition between a railway and a chain of canals had gone on for a little time, so that the profits of each had been considerably decreased by the reduction in the charges for conveyance, the competing concerns made a working agreement, which put an end to the competitive efforts of the canals. In all cases, the railway company was the aggressive rival of the canals. Sometimes these agreements were made secretly; at other times they were initiated in secret but afterwards ratified by Parliament; and there were other instances where they were entered into at first by consent of Parliament. The nature of these arrangements varied in different cases; some were really pooling agreements, others were simply a tacit understanding in regard to rates, while many forms of leasing the canal tolls to the railway were also found[4]. These working agreements were first formed in

[1] *The Times*, July 20, 1844, p. 6. It would appear that the Rochdale Canal shares had been as high as five hundred guineas (£525). Brit. Doc. 1844 (318), xi, 17, 'Minutes of Evidence,' p. 488.

[2] *The Times*, July 20, 1844, p. 6.

[3] Brit. Doc. 1872 (364), xiii, 1, 'Report of Committee on Railway Amalgamations, Minutes of Evidence' of Mr Allport, Q. 4348, and *A Few General Observations on Railways*, p. 20. See also Brit. Doc. 1844 (318), xi, 17, 'Fifth Report on Railways, Minutes of Evidence,' p. 76.

[4] As examples of these early working agreements, we give the following:

The Manchester and Leeds Railway had for a long time been competing with the Calder and Hebble Navigation, both charging very low rates; then they made an agreement that the rates should be raised to a certain point, in consideration of which the railway company was to guarantee that the canal company's traffic should amount to a certain sum, and any excess beyond that sum was to be shared between them, the railway company having the right of putting inspectors on the canal to watch that the traffic that they had thus guaranteed was fairly conducted. This was done by consent of Parliament. Brit. Doc. 1844 (318), xi, 17, 'Fifth Report on Railways, Minutes of Evidence,' p. 140.

After the joining of Manchester and Leeds by railway, the canal route along this course came into conflict with a powerful rival. The railway company, however, had a difficulty to meet, in that they did not know what the canal charged. They said they charged certain rates, but they used to let 50 tons go as 30 tons. By the competition, the revenue of the canal was reduced from £70,000 to about £28,000 a year; and this induced the canal company to come to terms. In order to put a

the last years of the decade 1830–40, but they became much more
numerous in the fifth decade, during and after the railway mania.
Sometimes the entire length of a canal, or some important link in it,
was leased to, or purchased by, or otherwise amalgamated with the
railway which was its strong competitor. In some cases arrangements
were made for the conversion of canals into railways; and the initiative
for this sometimes came from the side of the railway and sometimes
from the canal company. In certain instances the canal companies,
in their opposition to railways, and with the concurrence of their
engineers, promoted Bills to convert their canals into railways, or to
construct lines of railway parallel to or in connexion with their waterways. But as it was to the advantage of the railways to bring all
conveyance under their control, they considered it necessary to prevent
canal companies from obtaining powers to make railways. On the
other hand, the canal companies probably exaggerated the power of
the railways to destroy their profits, and opposed the railways in order
to get the latter to come to some favourable terms for the protection of
the canal shareholders. At times a company organized to construct
a railway found a canal which followed the direction of the line they
had projected, and negotiated for the acquisition of it, in order to be
able to utilize its channel, lands and other equipment to save money
and economize time[1]. In most cases, however, it was the canal

stop to such gross frauds and misrepresentation in regard to weights and rates,
the Manchester and Leeds Railway Company, the Rochdale Canal Company, and
the Calder and Hebble Navigation Company agreed that they should be fully
informed of each other's rates, that these should not be changed without conference
among themselves, and that the collection of dues should be more strictly attended
to. Brit. Doc. 1844 (318), xi, 17, 'Minutes of Evidence,' p. 488.

In 1846, the London and North Western Railway Company made an agreement
with the Birmingham Canal Company, consequent upon the following conditions:
The Birmingham Canal Company were not only the owners of an important canal,
but also of a good deal of adjacent land; and they were proposing to make a railway
of their own very much in the course of the Stour Valley branch of the London and
North Western Railway. That led to negotiations between the two concerns, and
afterwards it was felt that if a railway were to be made and if the canal company
were not to make it, but an independent company were to make it, the canal company
ought to be guaranteed from loss. This guarantee was dated 1846, and assured
four per cent. to the canal company if the canal did not earn that much. Brit. Doc.
1883 (252), xiii, 1, 'Minutes of Evidence' of Mr Evans, Q. 1493.

The North Staffordshire Railway, in applying for their Act, proposed to amalgamate with the Trent and Mersey Navigation. The railway company was to
guarantee a certain percentage on the capital of the canal, on condition of their
giving up the management of the canal to the railway company. Brit. Doc. 1846
(275), xiii, 93, 'Minutes of Evidence,' p. 57.

[1] On the whole subject of the conversion of canals into railways, see Teisserenc,
Voies de communication, pp. 29–30, 477–86. He gives examples of canal companies

companies, apparently, which were eager to have the railways take them over, either by purchase or by some form of working agreement.

By 1845 some of the possible evils of allowing railways to acquire too much control over canals had become evident; the railways had grown to be the predominant party in the contest, completely over-shadowing most of the canals; and it was thought advisable that Parliament should give some encouragement to canals, as the weaker party in the competition. An Act was passed[1], therefore, in that year, giving to canals a similar power to that possessed by railway companies, of varying their tolls or of leasing their tolls to each other[2]. By having this privilege canal companies might be enabled to work together and quote through rates on the long lines of canals—rates that would be less than the aggregate of the rates charged by each canal individually; or, one canal might take over the management of several adjoining canals, and, by reducing the rates of toll, make competition with the railways possible. This Act was passed for the purpose of obtaining "greater competition for the public advantage[3]." In the Act passed in the same session to enable canal companies to become carriers of goods upon their canals, and to make working arrangements with, and to lease their canals to, other canal companies, we see the same object kept in view, namely, to place the canals more nearly on an equality with the railways, so as to permit even-handed competition[4].

which were thinking of transforming their works into railways, and of railway companies that were planning to take over and utilize the equipment of canals. He shows that when the canal companies turned to the best engineers for guidance the advice given was usually favourable to the alteration of the canals into railways. See also the examples given in *Leeds Intelligencer*, Nov. 25, 1830, p. 3 ; ibid., July 15, 1830, p. 4, and Oct. 7, 1830, p. 3 : ibid., Oct. 21, 1830, p. 3, letter from "A Constant Reader," and note by editor; ibid., Nov. 4, 1830, p. 3 ; *Railway Chronicle*, Aug. 30, 1845, p. 1115, editorial; ibid., Aug. 2, 1845, pp. 931-2, on 'Railway and Canal Amalgamation;' *The Economist*, 1845, pp. 985, 994, 1015, and 1081. Sometimes canal proprietors were induced to convert their canals into railways because of lack of water to operate the canals. Sutcliffe, *Treatise on Canals*, p. 73. For other examples of railway companies becoming owners of canals, and for two instances of canals that were controlled, but not absolutely owned by railways, see *Report of Royal Commission on Canals and Waterways*, VII (1909), pp. 9–11.

[1] Act 8 & 9 Vict., c. 14.

[2] By the Railway Clauses Consolidation Act of 1845, railways were allowed to vary their rates, so as to work together with other railways in harmonious agreement, especially as to through rates.

[3] Brit. Doc. 1852-3 (736), xxxviii, 447, 'Fifth Report of Committee on Railway and Canal Bills, Minutes of Evidence,' p. 69. Act 8 & 9 Vict., c. 14.

[4] Act 8 & 9 Vict., c. 42. In 1840 a Bill had been introduced into the House for this same purpose, of allowing canal companies to be carriers and to make traffic arrangements with other canal companies. Brit. Doc. 1840 (405), I, 287.

Canal companies were not commonly carriers before this time, although a few

How was this new legislation received by the railways? It was not long before the railway companies saw that the aim of the Canal Carriers' Act was to keep them from securing monopoly, by allowing the canals to collaborate and thus obtain harmonious action in the contest against their rival. But acute minds soon recognized also that this Act gave power to railway companies that had become owners of canals to obtain a control over other canals, without coming under the notice of Parliament; and under such a plan no opportunity would be afforded to Parliament of taking the course usually taken when sanctioning arrangements between railway companies, of investigating the terms of the proposed arrangement before confirming it, or of subjecting it to the approval of the Board of Trade. If a railway company could obtain a controlling interest in a canal it would then be entitled to rank as a canal or navigation company, and claim the privileges of traffic arrangements that were allowed by this Act[1]. Accordingly, railways set to work to secure this standing, and thus make the statute that was intended for the benefit of their rivals, contribute to their own advancement. Having become in effect canal companies, through acquiring control over navigations, the railway companies were then

had been carriers for some time. The Bridgewater Trustees had been carriers on their canal, but, of course, it had been constructed and operated under the control of a private individual. The Trent and Mersey Canal Company had also been carrying for the public on their line. Other canal companies had been engaged in this carrying trade, but not under their own names. Even where the work was done by the canal company there were always other carriers who were doing the same work, upon payment of the tolls; and on the Bridgewater Canal a small part of the traffic was carried by the Trustees, while the larger part was taken by other carriers. Brit. Doc. 1844 (318), xi, 17, 'Fifth Report on Railways, Minutes of Evidence,' p. 169 et seq.; Brit. Doc. 1840 (437), xiii, 181, 'Fourth Report on Railways, Minutes of Evidence,' Q. 960. But it was a very rare thing that the canal companies did the actual work of carrying, either before or after the passage of the Act of 1845. As late as the year 1883, several witnesses advocated the carrying business being taken up by the canal companies, as well as by the private carriers: and it was said that at that time the system of carrying goods on the inland waterways was almost exclusively in the hands of the traffic senders, who put their own boats on the canals and paid the toll to the canal company (v. for example, Brit. Doc. 1883 (252), xiii, 1, evidence of Mr Lloyd, p. 23; also Brit. Doc. 1867 [3844], xxxviii, 1, evidence of Mr. Wilson, Q. 10,021, p. 433).

[1] Brit. Doc. 1857–8 (411), xiv, 1, 'Report of Select Committee on Railway and Canal Legislation,' p. 40. The Act authorized the owners of canals and navigations to carry as common carriers on their own canals and navigations; to enter into arrangements with each other in the way that railway companies were authorized to do, so as to avoid the delays incident to a diversity of interests; to enter into agreements for the division and apportionment of tolls and charges; and to let the tolls and duties to be levied on any canal or navigation, or any railways or tramways belonging to them, to any other canal or navigation companies for a period not exceeding 21 years.

ready to enter into negotiations with other canal companies which were powerful rivals, and to make such agreements with them as would prevent their competing with the railways, so that the latter would have the whole field to themselves[1].   In this way, the acumen of the railway managers or directors proved more than a match for the legislators, and the more powerful transportation rival was able to still further obtain the predominance.   So great was the influence that might be acquired by railway companies which were in a position to make use of the powers conferred by the Act of 1845, that the Board of Trade suggested whether it might not be proper to place some restriction on the exercise by these companies of the power of entering into traffic arrangements with canal companies[2].

The impetus given to the amalgamation of railways and canals before the beginning of the railway mania continued in the following years, and in 1846 there were over 200 Bills presented to Parliament

[1] As an example of this strategy, we give some facts in the history of the Leeds and Liverpool Canal.   Before the commencement of railway competition, the tolls on this canal for general merchandise varied from 1*d*. to 1½*d*. per ton per mile.   To meet railway competition, the canal tolls were reduced to ¾*d*. to ½*d*. per ton per mile. As competitors for the traffic of the district traversed by this canal, there were three lines of railway, the London and North Western, the Midland, and the Lancashire and Yorkshire.   These railways, having under authority of Parliament secured the property of certain navigations, and desiring to put down all competition for traffic in this district, engaged the Leeds and Liverpool Canal, in 1851, in consideration of an annuity of £41,860, to give up all competition and to practically close up their navigation by raising their tolls to a prohibitory figure, obtaining thereby for the united railways a complete monopoly of the traffic of that district.   The arrangement was made to assume the appearance of a lease of the canal tolls, under the powers of the Act of 1845.   The lease, however, was a fiction; the £41,860 yearly was paid, not as a rent, but in consideration of a rise in the canal tolls, which shut up the navigation and compelled the traffic to go by rail.   While the canal was charging the aforementioned reduced rates, these three railways, together with the East Lancashire Railway, offered the Leeds and Liverpool Canal this annuity, the counter condition being an increase of all the canal tolls to 1½*d*. per ton per mile, which was an advance of 100 % to 200 % on the existing tolls.   The canal accepted the annuity offered, but refused to allow the East Lancashire Railway to appear as a party to the transaction, since the latter did not have any canal whereby to legalize the agreement.   The arrangement was therefore completed under the pretence of a lease of the Leeds and Liverpool Canal tolls, by the London and North Western Railway, as proprietors of the Huddersfield Canal, the Lancashire and Yorkshire Railway, as proprietors of the Bolton and Bury Canal, and the Midland Railway, as proprietors of the Ashby-de-la-Zouch Canal.   The proportions in which the £41,860 was divided among the four railway companies were not publicly known.   Brit. Doc. 1852–3 (736), xxxviii, 447, 'Fifth Report of Select Committee on Railway and Canal Bills, Minutes of Evidence' of Thomas Grahame, p. 69.   For other instances, see Brit. Doc. 1857–8 (117), xxxi, 335, 'Report of Board of Trade on the Railway and Canal Bills of that Session,' p. 40.

[2] Brit. Doc. 1857–8 (117), xxxi, 335, p. 40.

containing provisions for uniting canals with railways[1].   The committee
that was appointed to look into this subject recognized the growing
tendency to union and extension, with its advantages of harmonious
management and its accompanying evils of monopoly; and they
recommended the appointment of a department of the Government
to provide more effective supervision of railways and canals[2].   Still
the amalgamations went on, with some effects that were detrimental
to the public; and the committee of 1853, that was appointed to
report on the railway and canal bills of that year, urged that working
agreements between different companies, for the regulation of traffic
and division of profits, should be sanctioned under proper conditions
and for limited periods, but that amalgamation of companies should
not be sanctioned except in special cases, where its object was to secure
public benefit through economy of management[3].   They also recommended that the good results of such merging of interests should be
retained, and the evils arising from them should be done away, by
compelling every railway company to afford to the public, in regard to
both goods and passengers, the full advantage of convenient interchange
from one system to another[4].   Since competitors were able, in a great

---

[1] Brit. Doc. 1872 (364), xiii, 1, 'Report of Select Committee on Railway Amalgamations,' under heading No. 8.

[2] Brit. Doc. 1846 (275), xiii, 93, 'Second Report of Select Committee on Railways
and Canals Amalgamations.'

The recommendations of this, the first committee on railways and canals amalgamations, are important, and we give them as follows:

(1)   The imposition of a low scale of tolls and charges upon all parties to the
amalgamation.   In the case of canals, the scales of tolls were of much greater
importance than in that of railways, for, in most instances, the public were the
carriers upon the canals.

(2)   Strict regulations should be made for keeping the canals in effectual repair
and with a proper supply of water.

(3)   The public must have the right of carrying passengers and goods on the
canals.

(4)   The privilege of making by-laws should be subjected to careful revision.
By this means, many of the canal companies exercised much power and could
prevent fair competition.

(5)   Where a canal was converted into a railway, care should be taken that no
district would be deprived of efficient means of communication.

[3] Brit. Doc. 1852–3 (736), xxxviii, 447, 'Fifth Report of Select Committee on
Railway and Canal Bills,' pp. 20–21.   If working agreements were entered into and
found to be injurious, they could easily be dissolved at any time; whereas if amalgamations were allowed they would be permanent and could not be subsequently
broken.   Brit. Doc. 1865 (3), xlix, 219, p. 23.

[4] Brit. Doc. 1852–3 (736), xxxviii, 447, 'Fifth Report of Select Committee on
Railway and Canal Bills,' pp. 20–21.   Running powers were generally discouraged
on the score of danger, and were to be conceded only in cases where free transit

measure, to secure the benefits of combination by agreements with each other, without authority of Parliament, and there were many such private agreements[1], it became necessary for Parliament to adopt some means of protecting the public by compelling proper arrangements for traffic between the companies. For this reason, the Legislature acted in accordance with the recommendation of the above-mentioned committee, and in the following year passed the "Railway and Canal Traffic Act, 1854." This Act enunciated two principles: that every company should afford, both for passengers and goods, proper facilities for forwarding traffic, and that no preferences should be given[2]. It was the first really important step in the direction of solving the difficulties that had arisen in connexion with the conduct of the traffic of railways whose interests were at variance with one another or with the interests of the public[3]. The Act also provided a summary remedy against

from one system to another could not be adequately ensured by other means (ibid., pp. 20–21, No. 6). The Board of Trade in 1865 also opposed the granting of running powers, that is, conceding to one company power to pass over the lines of another company without the consent of the latter, on the ground of its being questionable from considerations of public safety (ibid., p. 24; also 'Fourth Report of Select Committee of 1853,' p. 6).

[1] Brit. Doc. 1852–3 (736), xxxviii, 447, 'Fifth Report of Select Committee on Railway and Canal Bills,' p. 6. Here it is stated that such combinations of interests under private agreements were a matter of constant occurrence.

[2] Act 17 & 18 Vict., c. 31. Under this Act, "every railway company, canal company, and railway and canal company, shall afford all *reasonable* facilities for the receiving and forwarding and delivering of traffic upon and from the several railways and canals belonging to or worked by such companies respectively, and for the return of carriages, trucks, boats, and other vehicles, and no such company shall make or give any undue or unreasonable preference or advantage to or in favour of any particular person or company, or any particular description of traffic, to any undue or unreasonable prejudice or disadvantage in any respect whatsoever." The rest of the Act gives provisions for its enforcement. Brit. Doc. 1854 (87), vi, 19; also Brit. Doc. 1854–5 [1965], xlviii, 1, 'Report of Railway Department of the Board of Trade for 1854,' pp. x, xi give the provisions of this Act.

The necessity for this Act may be further illustrated by the following instance: In 1853 there was a complaint sent to Parliament by the coal-owners in Lancashire, that the railway company did not provide locomotive power to meet their needs, and that their coal had been forwarded at the company's convenience, rather than their own. The company took higher class traffic, which paid higher rates, and left the coal, which paid lower rates. Then, too, the railway left the coal-owners' rolling stock and coal on sidings along the line, which required the maintenance of a larger amount of rolling stock. The complaint also alleged that there was much delay in sending back the empty waggons from London. Brit. Doc. 1852–3 (736), xxxviii, 447, 'Fifth Report of Select Committee on Railway and Canal Bills, Minutes of Evidence,' p. 4.

[3] The Railway Department of the Board of Trade, in 1865, observed that the necessity there might formerly have been for allowing running powers were, to some extent, obviated by the passage of the Railway and Canal Traffic Act, 1854, and

railway companies for any violation of its enactments, by an application to the Court of Common Pleas[1]; but despite this it remained for many years practically a dead letter.

As soon as the Traffic Act of 1854 had been passed, large numbers of Bills were laid before the House by railway companies, asking that authority be given to enter into various descriptions of agreements for working in connexion with other companies, or for forwarding or inter-changing traffic with other companies. Out of a total number of 138 Bills introduced in 1854, seventy-five were for making working arrangements and this movement for working agreements increased in importance during subsequent years[2]. Most of those that were authorized were for ten years, but the power of renewal at the expiration of that period was generally granted, subject, of course, to the approval of the Lords[3]. The reason why there were so many of these agreements consummated about this time was because the trunk lines had been laid out, and the many short lines that were being constructed had to be merged with them in order to acquire any stability of operation[4]. To have attempted to remain apart from one of the main lines would have been to invite ruinous competition from the other roads in the same district; and, on the other hand, it was for the public good that new lines, which were extensions of, or feeders to, existing lines, should form part of one or other of the great systems and thus facilitate intercommunication.

The amount of amalgamation that was effected between railways and canals we are unable to trace with minuteness through successive stages in the growth of the transportation system. Some had been accomplished before the railway mania of 1844-6; much more was

that they were necessary only where a company required to pass for a short distance over the line of another company to reach a station at which to deposit and receive traffic, or when such short piece of line was a link necessary for the completion of a special railway system. Brit. Doc. 1865 (3), XLIX, 219, p. 24.

[1] See also Brit. Doc. 1867 [3844], XXXVIII, 1, 'Report of Royal Commission,' p. xxi.

[2] Out of 71 Bills introduced in the Session of 1858, there were 46 seeking sanction for working and traffic agreements. Brit. Doc. 1857-8 (117), XXXI, 335, 'General Report of the Board of Trade upon the Railway and Canal Bills of the Session of 1858,' p. 11.

[3] Brit. Doc. 1854 (139), LXII, 441, 'Report of the Board of Trade on Railway Bills of 1854,' p. 14; also 1854-5 [1965], XLVIII, 1, 'Report of the Railway Department of the Board of Trade for 1854,' p. viii.

[4] For the full text of the English and Scotch Traffic Agreement, among seven great railway companies, for apportioning the receipts from the Scotch traffic, see Brit. Doc. 1856 [2114], LIV, 1, 'Report of the Railway Department of the Board of Trade for 1855,' Appendix No. 4. Some other traffic agreements are given in ibid., Appendix No. 5.

completed during those years; and the subjecting of canals to railway control went on more gradually subsequent to that time[1]. By 1850 a considerable proportion of the canals had passed into the hands of the railway companies[2]; and by about 1865 that proportion had been increased, until nearly one-third of the total length of the canals and navigations of Great Britain had gone over to the railways[3]. From the report of 1872 we learn that there was a still greater extent of navigable waterways under railway control, amounting to about three-eighths of the whole[4]; and in 1883, in England and Wales alone, one-half of the total mileage of navigations had become allied with the railways and was no longer independent[5].

[1] The statistics of such amalgamations from 1846 to 1872 are given in Brit. Doc. 1872 (364), xiii, 1, pp. 755–6, and are found in Appendix 9.

[2] Brit. Doc. 1851 [1332], xxx, 1, 'Report of the Commissioners of Railways for the year 1850,' p. xix et seq.

[3] Brit. Doc. 1867 [3844], xxxviii, 1, 'Report of the Royal Commission, Minutes of Evidence' of Mr Thomas Wilson, p. 428 et seq. Mr Wilson was hon. sec. of the Canal Association of Great Britain. The following summary is given (ibid., Q. 9902–4):

Extent of navigations in England and Scotland, in 1865:

109 canals, total length	2552 miles
49 improved rivers, total length	1339 ,,
158 navigations, of a total length of	3891 ,,

Of these 3891 miles of navigation,

5 navigations have been converted into railways	68 miles
37 navigations have been amalgamated with railways	1026 ,,
2 navigations were wholly or partly leased to railways and virtually amalgamated with them	177 ,,
Total	1271 ,,

Therefore, about one-third of all the mileage of navigations had gone into railway hands.

The particulars in regard to this 1026 miles of amalgamated canals and the railway companies that had absorbed them are given in Brit. Doc. 1867 [3844], xxxviii, 1, 'Report of Royal Commission, Minutes of Evidence,' Q. 9906, pp. 428–9.

The extent to which the canals had passed under railway control, by 1872, is shown on the map given in Appendix R of the 'Report of the Select Committee (of 1872) on Railway Amalgamations,' Brit. Doc. 1872 (364), xiii, 1.

[4] Brit. Doc. 1872 (364), xiii, 1, 'Report of the Select Committee on Railway Amalgamations,' p. xx; also ibid., pp. 755–6. We may take the total length of navigable waterways of Great Britain in 1872 to be the same as that of 1865, namely, 3891 miles. According to the returns of that year (1872), there were then 1544 miles of canal in Great Britain held by railway companies, of which 1300 miles were held in perpetuity and the remaining 244 miles under temporary tenure. Therefore, at that time, there was about three-eighths of the total length of canals under railway control.

[5] Brit. Doc. 1883 (252), xiii, 1, 'Minutes of Evidence' of Mr Calcraft, Q. 3–6.

This subjecting of the inland waterways to the railways had its counterpart

We have already observed that the first effect of the introduction of a railway, as a competitor to a canal, was to cause the rates on the latter to be lowered; and by thus reducing the business and profits of the canal the railway company hoped to bring the canal proprietors to terms[1]. But it sometimes occurred that a canal was able to maintain competition with the railway; and where this was the case, the railway was compelled to charge lower rates at competitive points, while it recouped itself by imposing higher rates at non-competitive points[2]. On the passenger traffic the railways generally charged their maximum rates, because in that they had no competition; but on the goods traffic they charged much less than their maximum rates. In addition, the higher class goods, for conveying which the railways offered specially good facilities as compared with canals, were charged rates very much

in the railways obtaining a strong foothold in the external trade. By 1847 the London and South Western Railway Company had made a deed of settlement with a Steam Navigation and Packet Company connecting the channel ports with ports of the Continent, which gave the railway control of much of the trade between England and Europe. Brit. Doc. 1847 (164. IV), XXXI, 33 and 1847–8 (148 (30)), XXXI, 399, 'Reports of the Commissioners of Railways on the London and South Western Railway.' By 1858, the South Eastern Railway Company had obtained power to build, hire and work vessels for the purpose of affording communication between the ports of Folkestone, Dover, Hastings, Ramsgate, Margate, Rye, Whitstable, or Gravesend, and any port in France or Belgium. Brit. Doc. 1857–8 (117), XXXI, 335, 'General Report of Board of Trade upon the Railway and Canal Bills of that Session,' p. 37.

[1] See also Skey, *Report to the Committee of the Birmingham and Liverpool Junction Canal, on the Present State of the Competition between the Canal Carriers using that Line and the Grand Junction Railway Company*, p. 4. He shows how the railways lowered freight rates to a point which was disastrous to the canals, while at the same time keeping up their passenger rates, so that no individual canal carrier could long compete against a rival armed with such powers. Refer also to Boyle, *Hope for the Canals*, pp. 5–7, and Palmer, *British Canals*, pp. 19–20.

[2] In 1853 the rate on second class goods between Liverpool and Birmingham was 15s., but the rate between Manchester and Birmingham was 17s. 6d. The distance was about the same in both cases; but between Liverpool and Birmingham there was the competition between the canal and the railways [the Grand Junction Railway and the recently opened Shrewsbury line of railway], while between Manchester and Birmingham there was no such competition.

The following table of charges on the Midland Railway between Birmingham and intermediate places to Gloucester, and between Birmingham and several other points, will illustrate still more fully the difference in the railway rates where canal competition existed and where it did not.

Note that the charge between Birmingham and Gloucester, 53 miles, was 7s. 6d. per ton, whereas the charge between Birmingham and Cheltenham, 46 miles, was 10s. At Gloucester, the competition of the waterways kept down the rate, but at Cheltenham there was no such competition. Similarly in other cases. Note also that the rates on first and second class articles were the same to Bromsgrove and to Gloucester, although the distance in the former case was hardly one-third of that

higher than lower class goods, for which the canals could enter into fair competition with them[1].

After competition had proceeded to a certain length, and canals found it advisable to merge their interests with the railways, this step was usually accompanied by an increase of rates, especially on the more valuable goods[2], to a point higher than the competitive rates, and frequently higher than those which existed before competition became

in the latter (Brit. Doc. 1852–3 (246), xxxviii, 175, 'Third Report on Railway and Canal Bills,' p. 32).

Midland Railway rates between Birmingham and the following places:

	No. of miles	1st class		2nd class		3rd class		4th class		5th class		Smalls	
		s.	*d.*	*s.*	*d.*	*s.*	*d.*	*s.*	*d.*	*s.*	*d.*	*s.*	*d.*
Gloucester, in competition with Birmingham and Worcester Canal and Severn Navigation	53	7	6	8	4	12	6	20	0	40	0		9
Cheltenham, no competition	46	10	0	15	0	20	0	30	0	40	0	1	0
Droitwich and Bromsgrove } no competition	20} 15}	7	6	8	4	10	0	15	0	20	0		9
Worcester	26½	7	6	8	4	12	6	20	0	30	0		9
Hull, in competition with canals and Trent Navigation	134	20	0	20	0	25	0	30	0	40	0	1	3
Sheffield, no competition	86	20	0	25	0	35	0	40	0	60	0	1	0
York, no competition	129	26	8	30	0	33	4	40	0	60	0	1	6
Newcastle, in competition with navigations to Hull, and coasters to Newcastle	216	25	0	30	0	35	0	45	0	‹60	0	1	9

This same thing was observed by the Select Committee of 1872, from the testimony of several witnesses who appeared before that body. Brit. Doc. 1872 (364), xiii, 1, 'Report of Select Committee on Railway Amalgamations,' p. xxii; also 'Minutes of Evidence' of Messrs Nicks and Clegram, Q. 2919, 2987–8. See also ibid., p. xxi, and 'Evidence' of Mr Wilson, p. 233 et seq.

[1] In 1853, from Birmingham to Liverpool and Manchester, the rate on the lowest class of goods was 11*s.* a ton, but on the highest class it was 30*s.* a ton. Brit. Doc. 1852–3 (246), xxxviii, 175, p. 32.

[2] This was done when the Manchester and Leeds Railway made an agreement with the Rochdale Canal and the Calder and Hebble Navigation. Brit. Doc. 1844 (318), xi, 17, 'Fifth Report on Railways, Minutes of Evidence,' p. 488. When the Birmingham Canal came under the control of the London and North Western Railway Company its tolls were raised; and the rate on iron going along that canal was 1½*d.* per ton per mile, while the rate on the Trent and Mersey, Bridgewater and Staffordshire and Worcestershire Canals was only ½*d.* per ton per mile. Brit. Doc. 1852–3 (170), xxxviii, 5, 'Second Report of Committee on Railway and Canal Bills,' p. 70.

operative. In this way, independent canals began to find that their traffic was inconvenienced and injured by the high rates on the canals that were joined to railways, because by raising their rates to a point that was almost prohibitory to private carriers the railway-controlled canals not only drove the carriers off their own waterways but also materially aided in driving them off the independent canals. If carriers could not secure sufficiently favourable terms from all the canals in the chain, it was frequently useless for them to make any attempt at carrying; for to carry on a short stretch of canal, and then be compelled to transfer to a railway, or pay the high charges of the latter's canal, was wholly destructive of any advantages from water carriage[1]. This increase of railway and canal·rates was but another phase of the general policy of the railways to realize the utmost results from their monopoly, and to secure ample returns for losses sustained during what were sometimes prolonged periods of competitive rate cutting.

The policy of the railways in regard to the canals was, in all cases, to drive the traffic from the water to the rails. From the earliest days of the competitive period this tendency was manifested and its dangers recognized; but the problem was, how to secure the canals from the interference and control of the railways, and to afford the former a good opportunity of testing their capabilities as a rival system[2]. It would have been an easy matter for Parliament, had its members foreseen the outcome, to have passed legislation forbidding the railways doing anything that might prejudice canal interests, but that would not prevent *private* negotiations which looked toward a settlement of difficulties that would be acceptable to both the parties concerned. In contrast to this lack of knowledge on the part of the Legislature, there was the enterprise of the railway companies, which pursued their aim with steady and determined zeal. There were constant warnings given to Parliament against Bills which, if passed, would involve new or increased inducements to divert traffic from canals to railways; but in the face of these, the latter kept up the contest with their rivals.

---

[1] Brit. Doc. 1852-3 (246), xxxviii, 175, 'Third Report of Select Committee on Railway and Canal Bills,' evidence of Mr Mellish and Mr Loch, p. 26. The committee of 1872 reported that "Where Railway Companies amalgamate, or where Railway Companies acquire a navigation, the result is usually an increase of rates." Brit. Doc. 1872 (364), xiii, 1, 'Minutes of Evidence,' p. 332. See also Brit. Doc. 1881 (374), xiii, 1, 'Report of Select Committee on Railway Rates and Fares,' evidence of Mr Hingley, Q. 5489, 5659. In Appendix 11, we have brought together some tabular statements of freight rates, showing how much they were raised by the amalgamation of railways with canals.

[2] Brit. Doc. 1846 (275), xiii, 93, 'Second Report of Select Committee on Railways and Canals Amalgamations.'

Sometimes canals were purchased or leased by the railways, frequently at a loss so far as the revenue from the waterways was concerned, and then their free use was forbidden to the public, through the imposition of prohibitory tolls[1]. In other cases the railway companies used their passenger traffic as a means by which they could put down their freight rates and thus appeal to shippers, from the standpoint of economy, to patronize the railway. Then, when the canal companies or other carriers on the waterways had found it impossible to compete for traffic, and had sold their stock of horses, it would be hard for them ever again to get back their traffic, since the public had become accustomed to having their goods carried by the faster conveyance of the railway[2]. In some instances, railways neglected or refused to repair the canals they held; and although the necessity for keeping them in good condition had been early shown to Parliament[3], yet they

---

[1] Brit. Doc. 1881 (374), xiii, 1, 'Report of Committee on Railway Rates and Fares, Evidence' of Mr Lloyd, Q. 10,181–2. In that year the Great Western had practically £1,000,000 invested in canals, and the net revenue was only £276 (in 1880). For some of these canals they had to pay rent charges of £8243, so that on the canals the company lost £7967; but this closing of the canals was to bring the traffic on to the rails. Some of these canals they were forced by Parliament to purchase when they obtained power to construct their railways (v. Brit. Doc. 1881 (374), xiii, 1, 'Minutes of Evidence,' Q. 13,720). See also Brit. Doc. 1846 (275), xiii, 93, 'Second Report of Committee on Railways and Canals Amalgamations, Evidence,' p. 47. The oppressive policy of the Birmingham Canal Navigation, controlled by the London and North Western Railway Company, was notorious (see example given in Brit. Doc. 1881 (374), xiii. 1, 'Report of Committee on Railway Rates and Fares, Evidence' of Mr Spence).

As showing to what extent the Great Western Railway Company diverted the traffic from three of the most important of the canals of which it got control, note that on the Hereford and Gloucester Canal, in the thirty years following 1848, the gross receipts had decreased seventy per cent. During the same period, the receipts on the Stratford-upon-Avon Canal decreased seventy-seven per cent.; and in the same time the receipts of the Kennet and Avon Canal decreased eighty-seven per cent. That is, during that time, in the case of these three canals, seventy, seventy-seven and eighty-seven per cent. respectively of the traffic had been shunted on to the rails. Brit. Doc. 1881 (374), xiii, 1, evidence of Mr Spence.

The tolls on the Leeds and Liverpool Canal, during the time it was under the control of the London and North Western Railway, were the maximum rates, and were as much as the through freight rate on the railway. This, of course, prevented the use of the canal, and it was in reality closed up. Brit. Doc. 1872 (364), xiii, 1, 'Report of Select Committee on Railway Amalgamations,' Q. 5772.

[2] Brit. Doc. 1846 (275), xiii, 93, 'Second Report of Committee on Railways and Canals Amalgamations, Evidence,' p. 35; also 1852–3 (246), xxxviii, 175, 'Third Report of Committee on Railway and Canal Bills, Evidence,' p. 16.

[3] Brit. Doc. 1846 (275), xiii, 93, 'Second Report of Committee on Railways and Canals Amalgamations; Recommendations of the Committee,' among others that "strict regulations should be made for maintaining the canals in an efficient state of repair, and for securing a proper supply of water."

were allowed to decline[1]. Notwithstanding the passage of the Act of 1873, requiring railway-controlled canals to be kept open and navigable for the public without interruption and delay[2], and that Parliament in many cases tried to annex conditions to the amalgamation, compelling the companies to maintain the canals in an efficient working state[3], many of these canals went from bad to worse; they became silted up, the locks became broken, and the navigation fell into disuse[4]. From what we have just said, we can easily see how the railways could draw to themselves the traffic formerly carried on competing canals, and leave the latter in a state of hopeless decay.

An examination of the English canals to-day reveals the fact that the amount of traffic carried on them, *tout ensemble*, is comparatively insignificant. By way of summary, we shall now note some reasons for their failure to compete successfully with the railways. To discuss this fully would require a more minute investigation of the policy and management of each than we have the space here to describe; and so we shall endeavour to give only the salient factors which bear upon the problem.

In the first place, the disjointed state of the canals prevented their being used to advantage. Very few of them had the same dimensions[5]. They were constructed usually as short independent canals, and not as long through routes. Their dimensions were made to accord partly with the amount of money that had been subscribed or contributed

---

[1] Brit. Doc. 1872 (364), xiii, 1, 'Report of Select Committee on Railway Amalgamations,' p. xxii; also ibid., 'Minutes of Evidence' of Messrs Clegram (Q. 2936) and Lloyd (Q. 5041). See also Brit. Doc. 1881 (374), xiii, 1, 'Report of Committee on Railway Rates and Fares, Evidence,' Q. 10,184–8; and Brit. Doc. 1883 (252), xiii, 1, 'Report of Select Committee on Canals, Evidence,' Q. 564, 630, 632–3.

[2] Brit. Doc. 1881 (374), xiii, 1, 'Report of Committee on Railway Rates and Fares, Evidence' of Mr Lloyd, Q. 10,194.

[3] Brit. Doc. 1872 (364), xiii, 1, 'Report of Select Committee on Railway Amalgamations,' p. xxii; also 'Evidence' of Mr Bartholomew, Q. 5779.

[4] The Act of 1873 was not enforced, because it would have cost the public too much to enforce it on account of the legal complications involved. Brit. Doc. 1883 (252), xiii, 1, 'Report of Select Committee on Canals, Evidence' of Mr Lloyd, Q. 564. As to the manner in which that Act was evaded, see Brit. Doc. 1881 (374), xiii, 1, 'Report of Committee on Railway Rates and Fares, Evidence' of Mr Spence, Q. 10,438 et seq.

[5] See Brit. Doc. 1872 (364), xiii, 1, 'Report of Select Committee on Railway Amalgamations,' Part ii, Appendix X, which gives in detail the dimensions of all the navigations. With depths of water varying from $4\frac{1}{2}$ feet to $14\frac{1}{2}$ feet, widths varying from 7 to 22 feet, and corresponding variations in length of locks, it would be difficult to get any boats that could be used to good effect on a through route. See also Palmer, *British Canals*, pp. 19, 22; Boyle, *Hope for the Canals*, pp. 29–30.

by the stockholders of the individual companies for the completion of their works, and partly with the difficulties that had to be overcome in the location of the canal, or the soil through which it had to pass. When a Bill was presented before Parliament, the proposed canal was considered solely on its own merits, and not in regard to any connexion that it might have in future with any other. Not only did the original dimensions of the canals show wide diversity, but changes were some-times made in these, at times when improvements were subsequently carried out[1]. Even on the same canal, there were sometimes differences in the size of the locks which had been constructed[2]. This lack of uniform gauge was utterly destructive of any economy of operation. If a barge were required to go along a through route, its carrying-power and dimensions had to be limited to suit the smallest locks on the route. If one boat were not to be used throughout the course, there had to be frequent loading and unloading from one barge into another. Both these methods of carrying were wasteful: the former in the utilizing of the capacity of the boats and canals, and the latter in the employment of time and labour. So also, the canals with large locks often con-sumed a large portion of their water inefficiently without passing an effective cargo, while on the narrow canals the carriers were greatly restricted as to the weight they could take. These discrepancies of gauge were wholly subversive of the greatest usefulness of the canals[3].

[1] Brit. Doc. 1883 (252), xiii, 1, 'Report of Select Committee on Canals, Minutes of Evidence,' p. 38, Q. 785–92.

On the Aire and Calder, for example, the locks were originally 60 feet × 15 feet, with a depth of water of 3 feet 6 inches. Under the Act of 1776 the locks were made 66 feet × 15 feet and the depth of water 5 feet. Under the Act of 1828, the locks were made 72 feet × 18 feet and 7 feet depth of water. After 1860, the locks were made 215 feet × 22 feet and 9 feet depth of water.

[2] Brit. Doc. 1883 (252), xiii, 1, 'Evidence' of Mr Bartholomew, Q. 804. On the Leeds and Liverpool Canal, the locks on the Yorkshire side were 66 feet × 15 feet 2 inches, and were capable of admitting boats 60 feet × 14 feet 6 inches; but on the Lancashire side the locks were 76 feet × 15 feet 2 inches, and they would receive boats 70 feet × 14 feet 6 inches.

On the canal route connecting the river Severn at Saul with the Thames at Abingdon, the Stroudwater Navigation gauge was 75 feet × 15 feet, the Thames and Severn Canal gauge 75 feet × 12 feet 6 inches, and the Wilts and Berks Canal gauge 80 feet × 7 feet. The Thames alone had three gauges upon it, the gauge in every case being regulated by the size of the locks (ibid., 'Evidence,' Q. 107).

[3] Canals could not now be economically widened so as to make a uniform gauge, because of the fact that tunnels, stone bridges, etc., along the routes could not be widened except at vast expense. In one case, we are told, the canal runs under the houses in Manchester. This certainly could not be made wider. Brit. Doc. 1883 (252), xiii, 1, 'Report of Select Committee on Canals, Evidence,' Q. 1700–1. See also Boyle, *Hope for the Canals*, p. 23.

Another reason for the failure of the canals was the lack of unity of management, due to the great number of companies which controlled them. In 1883, between London and Liverpool there were three distinct routes: on the first there were nine different canals and navigations, on the second route also there were nine different companies, and on the third there were ten separate companies[1]. From London to Bristol there were four routes: on the first, via the Kennet and Avon, there were three companies; on the second, via the Wilts and Berks Canal, there were five companies; on the third route, via the Stroudwater Canal, there were three companies; and on the fourth route, via the Warwick Canal, there were nine companies, and this was the only one in practical use[2]. From Birmingham to Bristol there were three routes[3]. Between Hull and Liverpool there were four ways: on the first route, via the Leeds and Liverpool Canal, there were four separate companies; on the second, via the Rochdale Canal, there were seven companies; on the third, via the Huddersfield Canal, there were nine companies; and on the fourth route, via the Trent and Mersey Canal, there were at least five navigations[4]. If, in 1883, there was such lack of unity, it could not have been less, but, possibly, more diverse in the period before the middle of the century; and these conditions have not been improved since 1883[5].

[1] Brit. Doc. 1883 (252), xiii, 1, 'Report of Select Committee on Canals, Evidence,' Q. 231.

[2] Ibid., 'Evidence,' Q. 232.

[3] Ibid., 'Evidence,' Q. 233.

[4] Ibid., 'Evidence,' Q. 234. In all these cases, the navigable tideways, such as the Mersey, Severn, Ouse, Humber, etc., are included as separate jurisdictions. Compare ibid., 'Evidence,' Q. 783–4.

[5] The 'Final Report of the Royal Commission on Canals and Waterways, 1909,' vii, p. 16, gives a few illustrations of this diversified control.--For example: "Taking Birmingham as a centre, we will assume that it is proposed to despatch thence three cargo boats, one to the port of London, one to that of Liverpool, and one to Hull, by the most direct routes. The boat which went to London would have to traverse some portion of the Birmingham Canal system, next 22 miles of the Warwick and Birmingham Canal, next 14 miles of the Warwick and Napton Canal, then 5 miles of the Oxford Canal, then either 98¼ miles of the Grand Junction Canal to Brentford, and finally the Thames,—or else 100¼ miles of the Grand Junction Canal to Paddington, and finally 8¼ miles of the Regent's Canal to the Thames at Limehouse. All these waterways belong to different authorities. A cargo proceeding to the port of Liverpool would traverse first some part of the Birmingham Canal, then 2¼ miles of the Birmingham and Warwick Junction Canal, then 17 miles of the Birmingham and Fazeley Canal, then 5¼ miles of the Coventry Canal, then 60 miles of the Trent and Mersey Canal, and would then go, probably not without transhipment, by 12 miles of the Weaver Navigation and then by the Mersey to Liverpool, or, without going down the Weaver, proceed by the Trent and Mersey to its junction with the Bridgewater Canal at Preston Brook, and by that canal to the Manchester Ship

In connexion with this want of unity of management along all the great through routes, some of the canal companies, whose waterways formed central links in a longer chain, took advantage of their peculiar position to raise their rates so as to secure for themselves the largest possible return on their investment, even upon a small amount of traffic[1]. When the different canals along a through route would not work in harmony, it was impossible to get a through rate that might enable the carrying to be conducted at a profit to all, for the other companies that were not so advantageously situated would be obliged to reduce their rates below a reasonable minimum if the amount of the through rate were to be made acceptable to the carrier. If the canal companies, therefore, would not adopt concerted action, there certainly could not be any fair competition with the railways[2]. The jealousy that existed between adjoining canals during the time preceding and immediately succeeding the introduction of railways is well exemplified in the junction or bar tolls. They were a sort of protective system, originally granted to the existing canals, so that whenever any new canal formed a junction with them the older canal could charge the amount of the bar toll merely as a gratuity for allowing traffic to enter

Canal, and thus to the Mersey; but as the narrow boat could not navigate the estuary, transhipment would be necessary. A cargo going to Hull would pass over some miles of the Birmingham Canal, 5¼ miles of the Coventry Canal, 26 miles of the Trent and Mersey Canal, 9¼ miles of the Trent Navigation, 2¼ miles of the Nottingham Canal, 21 miles of the Trent Navigation, 4 miles of the Newark Navigation, 30 miles more of the Trent Navigation, 26 miles of the open Trent River, and then 18 miles of the Humber. Transhipment, probably at Nottingham, would be necessary." On this subject, for the earlier period before 1850, see Boyle, *Hope for the Canals*, pp. 23–24; Palmer, *British Canals*, pp. 19–23.

[1] Brit. Doc. 1846 (275), xiii, 93, 'Second Report on Railways and Canals Amalgamations,' p. iv. As a special instance of a canal taking advantage of its position to raise its rates, we may mention the Grand Junction Canal, which extends from Paddington to Braunston where it joins the Oxford Canal. The Grand Junction Canal was an important link between London and the great mining and manufacturing sections of Warwickshire, Cheshire, Staffordshire, etc. It was a monopoly without competitor; its exactions, excessive rates, discriminatory rates, and its supercilious conduct caused loud and general complaints even as late as 1836. Mercator, *Tonnage Rates on Grand Junction Canal*, pp. 8–24. The Oxford Canal had pursued a similar policy. See also Palmer, *British Canals*, p. 19.

[2] About 1847 the Aire and Calder Navigation offered to lease the Calder and Hebble Navigation at a guaranteed net dividend of sixteen per cent., but the offer was refused. This high rate of dividend was surely a tempting offer, and why it was rejected we do not know; but it was not more than two years before railway competition had caused such a decrease of the revenues of this navigation that they were able to divide but eight per cent., or one-half of the dividend that had been guaranteed by the Aire and Calder. *Herepath's Railway and Commercial Journal*, xi, p. 1241.

their canal at will[1]. These bar and compensation tolls were sometimes so extraordinarily high that they alone amounted to a sufficient income to pay a large dividend on the canal capital[2]. With such onerous charges upon the carriage of goods on the canals, it is no wonder that the through traffic declined and that the railways came to have the upper hand. Of course, when such canals passed into the control of the railways, these tolls were still continued; for it was the railway policy to divert all the trade from the canals to the rails, and this formed a useful auxiliary agency in the carrying out of this plan[3].

[1] Brit. Doc. 1881 (374), XIII, 1, 'Report of Committee on Railway Rates and Fares, Evidence' of Mr Lloyd, Q. 10,174. We must not confuse these bar tolls with the bars which were occasionally allowed to remain between two canals that almost formed a junction end to end. For example, by Act of 1791 for making the canal from near Worcester to Birmingham it was provided that this canal should not come within seven feet from the end of the Birmingham Canal without the consent of the proprietors of the Birmingham Canal in writing under their common seal. Up to 1815 this bar, of the width of seven feet, still remained to prevent any passage from one canal into the other or to prevent any waste of water out of one canal into .the other. Over that bar all the traffic between the two canals had been conveyed out of boats upon one canal into boats upon the other. In that year, the Worcester and Birmingham Canal obtained the consent of Parliament to remove this bar, upon condition that the Birmingham Canal might not be injured in any way, and this canal agreed to the removal of the bar upon these conditions (v. *Case in Support of the Bill for removing the Bar between the Birmingham, and the Worcester and Birmingham Canals*, Birmingham Free Reference Library, No. 87,368). See also *Herepath's Railway Magazine*, N.S., IV, p. 378, address by "A Canal Proprietor."

[2] Brit. Doc. 1881 (374), XIII, 1, 'Report of Committee on Railway Rates and Fares, Evidence' of Mr Lloyd, Q. 10,174–6. See also the examples given in Shaen, *Review of Railways and Railway Legislation*, p. 31.

The Oxford Canal Company, at the junction of this canal with the Warwick and Birmingham Canal at Napton, by authority of its Act, was allowed to take a toll of 2s. 9d. per ton upon coal, and 4s. 4d. per ton upon all other articles. This was not for any service rendered, but merely for allowing traffic to pass from the other canal into the Oxford. In the first twenty years from the opening of the Warwick and Birmingham Canal, the Oxford Canal received a quarter of a million in bar tolls at that junction, which was enough to pay a ten per cent. dividend on the entire cost of construction.

The Grand Junction was a still more extraordinary case. This canal joined the Oxford Canal, seven miles from the junction of the latter with the Warwick and Birmingham Canal at Napton, and here there was a compensation toll of 6d. a ton leviable on all coal turning towards Oxford, which, in reality, never went within seven miles of the Grand Junction Canal, and in which they never could have any possible interest. Mercator, *Tonnage Rates on Grand Junction Canal*, p. 24, says: "The trade of the country at the present day (1836) groans under the excessive imposts and complicated system of the Grand Junction Canal and the abominable tolls." called compensation tolls, paid to the Oxford Company.

[3] That the railways took the full amount of these bar and compensation tolls on their canals, is evident from Brit. Doc. 1872 (364), XIII, 1, 'Report of Select Committee on Railway Amalgamations,' p. xxi; also 'Evidence' of Mr Lloyd,

The railway companies, seizing upon this diversity of management in the canals, purchased or leased the important links of through routes, raised the tolls on these divisions to the utmost limit allowed by law, and thus made it impossible for the companies that owned the remainder of the lines of canal and for the common carriers on a through route to maintain their traffic in competition with the railways. This was one of the earliest and most effective ways which the railways used for breaking up and paralyzing whole chains of waterways[1]. The railway companies seldom, if ever, had possession of the whole of such a canal route, for that would require too much capital to be tied up in un-productive business, especially when the control of a few miles would answer the purpose they had in view just as effectively. But besides putting up the tolls on their canals, the railway companies, in some cases, contrary to statute, neglected or refused to keep their canals in repair[2],

Q. 5031–32; also Brit. Doc. 1881 (374), xiii, 1, 'Report of Select Committee on Railway Rates and Fares, Evidence,' Q. 10,180, 10,209–10. The independent canals abolished their bar tolls (ibid., Q. 10,177).

[1] Brit. Doc. 1846 (275), xiii, 93, 'Second Report of Committee on Railways and Canals Amalgamations, Evidence,' p. 42; 1852–3 (170), xxxviii, 5, 'Second Report of Committee on Railway and Canal Bills, Evidence' of Mr Pixton; also Brit. Doc. 1872 (364), xiii, 1, 'Report of Committee on Railway Amalgamations,' pp. xxi–xxii. Mr Acworth, who is, doubtless, the greatest railway economist of England, in an article in the *Economic Journal*, June 1905, pp. 149–55, takes issue with the state-ment that railway companies "strangled" some of the canals which came into their possession or control. Whether we use the word "strangled" or not is a mere question of nomenclature; but it is undoubtedly true that when some of the important canal links came under railway dominance their day as free-acting agents ceased. From that time onward, their policy was dictated by the railway companies into whose hands they had passed. This will be evident to those who consult the references I have here given for this paragraph. See also Palmer, *British Canals*, pp. 76–77. We are perfectly willing to admit that in many cases the canals were desirous of selling out to the railways, and in doing this they were acting from the motive of self-interest. We may as well admit the application of the same legitimate principle on the part of railways, which negotiated for the taking over of certain canals in order to further their own economic interests.

[2] Brit. Doc. 1872 (364), xiii, 1, p. xxii; Brit. Doc. 1881 (374), xiii, 1, 'Report of Committee on Railway Rates and Fares, Evidence' of Mr Lloyd, Q. 10,162. On the Stratford-upon-Avon Canal, owned by the Great Western Railway Company, Mr Lloyd said that a boat would make very good speed if it went one-and-a-quarter to one-and-a-half miles per hour; and similarly, for the trade on the Hereford and Gloucester Canal, owned by the same Company. See also *London and Birmingham Railway Bill. Extracts from the Minutes of Evidence given before the Committee of the Lords on this Bill*, pp. 3, 10; Palmer, *British Canals*, p. 28. Many of the canals had remained in nearly the same condition as when they were first put into operation, their course was needlessly circuitous, their tunnels were small and inconvenient, they were inadequately supplied with water, and in the case of most companies no effort had been made to progressively improve the canals so as to keep them abreast of the development of the country and its expanding trade. Palmer, op. cit.,

and in other cases closed them at nights or stopped them for repairs just at the times when there would have been most traffic for conveyance[1]. Notwithstanding the reiterated recommendation by successive Parliamentary committees that every means should be adopted for the maintenance of effective competition by the canals against the railways[2], the latter acquired possession of more and more of the strategic canal links, until the competition of the canals was stifled; or, in the words of Mr Conder, the canals were "struck with creeping paralysis with all those obstructions[3]." The Joint Select Committee of 1872, which investigated the railway and canal amalgamations, considered that "the most important method by which the railway companies have defeated the competition of canals has been the purchase of important links in the system of navigation and the discouragement of through traffic," and the great complaint against the railways still is that they tend to discourage traffic from going on the water routes[4]. Now it may be strictly true, although even this is open to question, that the railways did not acquire the canals with the deliberate intention of throwing obstacles in the way of their traffic development. It may not be just

pp. 20–29; *Herepath's Railway Magazine*, N.S., IV, p. 373, address to the canal navigation proprietors in Great Britain; Boyle, *Hope for the Canals*, p. 22; *Reading Mercury*, Nov. 25, 1793, p. 4, letter from "A Commissioner."

[1] As much of the canal traffic was customarily carried at night, the railways would close their canals at night. Brit. Doc. 1872 (364), XIII, 1, 'Report of Committee on Railway Amalgamations,' p. xxii, also evidence of Mr Clegram, Q. 2936–8, and Mr Lloyd, Q. 5041. In other cases, there would be a failure of the water supply, or the necessity of stopping for repairs at certain seasons when the canal would have been most used, and this would go on for weeks at a time.

[2] See, for example, Brit. Doc. 1846 (275), XIII, 93, 'Second Report of Committee on Railways and Canals Amalgamations,' under heading "Conclusion;" Brit. Doc. 1872 (364), XIII, 1, 'Report of Committee on Railway Amalgamations,' p. xxiii, "Resolutions of the Committee."

[3] Brit. Doc. 1883 (252), XIII, 1, 'Evidence before Select Committee on Canals,' p. 128, Q, 2447.

Mr Grierson, the General Manager of the Great Western Railway Company, in 1881, before a Committee of Parliament, testified that in many cases the railways were forced to purchase the canals; that his company were possessors of several canals which Parliament forced them to purchase when they obtained the Act giving them power to construct their railway (v. Brit. Doc. 1881 (374), XIII, 1, 'Report of Committee on Railway Rates and Fares, Evidence,' Q. 13,720). Mr Farrer, Secretary to the Board of Trade, also said that the purchase of certain canals was made compulsory, in consequence of the terms Parliament imposed upon the railway companies when applying for their Acts. Furthermore, the railway companies sometimes found it a matter of policy to buy off the opposition of the canal interests through the purchase of the canals. In the case of the Stratford-upon-Avon Canal, the committee would not allow the Bill to pass unless the railway company did absorb the canal (Brit. Doc. 1881 (374), XIII, 1, 'Evidence,' Q. 16,466–7, 16,488–9).

[4] 'Final Report of Royal Commission on Canals and Waterways,' 1909, VII, p. 70, paragraph 381, and p. 77, paragraph 412.

to say with some that railway companies (except perhaps in a few instances) acquired canals in order to strangle them. But it certainly is true to say that railway companies which have, in various ways, come into possession of canals feel, in most cases, little desire to do more than their barest legal duty in maintaining them. There seems no doubt but that they favour what they consider to be by far their most important business, that of placing all possible traffic on their lines of railway[1]. Where railway companies find it to their interest to maintain and improve their canals so as to promote trade on them they do so, though perhaps not always with successful results; but in the larger number of cases the railway companies seem to have neglected to promote, if not actually to have impeded by high tolls and otherwise, the traffic on the canals which they have acquired. The cases in which railway companies have a more or less strong interest in developing the trade upon canals which belong to them are exceptions to the rule[2]. But we must not suppose that it was the railway-owned canals alone which were allowed to fall into partial or total decay; as a matter of fact, many of the independent canals were fully as bad as those which were controlled by railways[3].

Along with the foregoing physical factors leading to the relative decline of the canal traffic, we must include another element which has been alluded to incidentally in several cases, namely, the fact that the canals were easily stopped or injured by frost, drought, and occasionally flood. This was one of the strong reasons put forward by the advocates of the railways, and there was ample justification for the desire to get rid of a system which entailed so much uncertainty and delay. To have goods stopped for weeks and sometimes months on account of the inability to navigate the canals, was subversive of all system in commercial life[4]. The business community was coming to depend more and more upon regularity in the transportation of commodities, and as the waterways could not assure this they were gradually abandoned in favour of an improved means of conveyance which could provide this desideratum.

[1] 'Final Report of Royal Commission on Canals and Waterways,' 1909, VII, p. 77.

[2] Ibid., pp. 74–76.

[3] Brit. Doc. 1883 (252), XIII, 1, 'Evidence before Select Committee on Canals,' Q. 1343.

[4] *London and Birmingham Railway Bill. Extracts from the Minutes of Evidence given before the Committee of the Lords on this Bill*, pp. 3, 6, 9, 10; *Great Western Railway Bill. Minutes of Evidence taken before the Lords Committee to whom the Bill was committed*, pp. 7, 8, 11, 408; *Manchester Guardian*, Jan. 29, 1831, p. 1, Report of the Manchester and Leeds Railway; *Birmingham Journal*, Sept. 9, 1826, p. 3, and Mar. 5, 1831, p. 3, statement of E. T. Moore; *The Times*, Jan. 30, 1802, p. 3; May 16, 1826, p. 2; Nov. 7, 1826, p. 2.

The next adverse feature was the lack of tone and spirit in the system itself, and the failure, due partly to inability and partly to indifference, to adapt themselves to the changing circumstances of the times. Barges were started on their journeys, not at definite times, but whenever the carriers had enough cargo to warrant their putting their horses on the tow-path. The arrival and departure of boats took place at all hours, and the horse with his feeding can attached to his mouth sauntered listlessly on his way, while those in charge of the boats systematically pillaged the goods and made frequent stops at the public-houses along the canals where time was squandered in drinking. In addition to this lack of punctuality in the delivery of goods and the frequent losses and delays, there was much difficulty in procuring rates and general information, since these were not publicly announced. The uncertainty, partiality and inconsistency of the carriers' charges, because of their being made without reference to any fixed rule; the want of promptness in rectifying errors and settling claims; and the general looseness of system and absence of unanimity or understanding among the various departments of the canal companies, conspired to perpetuate complaints against the whole system. The great number of independent carriers engaged upon the canals multiplied the number of tranship-ments, for besides having to tranship at the termini of the various canals, a similar transfer had to be made at each commencement and termination of a carrier's stage. In this way a loose, disjointed and uneconomical method of forwarding goods was imposed upon the shipper, and the number of hands through which the goods passed added needlessly to the cost and often precluded the fixing of responsi-bility for injury or loss. Moreover, the carrier, instead of having his business divided naturally into three departments for attending to the receiving, the conveyance and the delivering of the goods, each of which should have been in charge of a separate official and all joined under the supervision of one general head, sometimes required one person to attend to two or more things, in different places, at the same time. While the man was doing his work in one capacity, such as attending to the loading or unloading of a barge, he must have been neglecting it in another, such as receiving goods and making out an invoice for them; and it is no wonder, therefore, that disputes arose in regard to goods that went astray or that were not delivered. This lack of method, of system, of business acumen, in the conduct of the carriers' affairs must have been a potent reason of the decay of canal traffic when railway activity began[1].

[1] Boyle, *Hope for the Canals*, pp. 19–29; Grahame, *Treatise on Internal Inter-course and Communication* (1834), pp. 28–29.

Another reason for the failure of the canals to successfully compete with the railways was that they were at a great disadvantage on account of being unable to carry passengers.   The large revenues from passenger traffic enabled the railways to lower their charges for the carriage of freight to such an extent that the canals, in meeting these lower freight rates, failed to make sufficient profit, and were, therefore, compelled to relinquish their hold on the carrying trade, or else to amalgamate with the railways[1].   The latter, by charging the maximum fares for passengers, could make the receipts from this traffic pay all the fixed charges of the road, and allow the goods to be carried at so low a charge that the canals could not meet this rate for any length of time.   The hope early expressed, that the canals could compete with the railways in the carriage of heavy freight, was not long in being deposed from the public mind; and the railways assumed the place of carriers *par excellence*[2].

Finally, canal traffic declined because of a psychological reason.   The discovery of railways as a means of transport, surpassing both in speed and economy any that were already in existence, so took the civilized world by surprise that the public were carried away with the thought of its possibilities.   The canal traffic was carried on comparatively quietly and unseen, there was nothing fast about it.   The sight of an occasional horse passing through the country, mounted or driven by a boy, and hauling an insignificant looking barge which was managed by one or two persons, excited no surprise on the part of anyone.   In reality, most of the canal conveyance was effected at night, when it would be recognized by very few.   On the other hand, the railway had an appearance of grandeur and ostentation that charmed the public.   It seemed the embodiment of enterprise and boundless capabilities.   The enormous trains conveyed across the country at a speed of twenty to thirty miles an hour contrasted strongly with even the best speed of the fly boats on the canals, going two and one-half to four miles an hour.   The effectiveness of the engine and the substantial road-bed and rolling stock were all matters of wonder.   The promptitude of train schedules was a radical reversal of the policy of the canal carriers, who, in the conduct of their business, had no schedule to which they adhered, but set out with their load whenever it was ready.   When confronted with these

---

[1] Brit. Doc. 1846 (275), xiii, 93, 'Second Report of Committee on Railways and Canals Amalgamations,' p. iv; and 'Evidence' of Mr R. Scott, p. 60, Q. 555.   See also Brit. Doc. 1852–3 (246), xxxviii, 175, 'Third Report on Railway and Canal Bills, Evidence,' p. 32, Q. 1700.

[2] In treating of the effects of railway competition upon canals, we have touched upon the influence of the passenger traffic, but only slightly, because the central fact in that competition was the relation of the carriers of goods to the canals and the railways.

new conditions, the few who ventured to deprecate the total abandonment of all earlier means of transport, and especially those who advocated the retention and upkeep of waterways, found themselves a powerless minority. The reduction of freight rates elicited the support of the public, and writers for the railways industriously circulated the opinion that canals must ultimately give place to railways. As a result of these conditions the prices of canal shares wént down[1]. In the fascination that the railways exercised from the outset, the possibility of materially and effectively improving the waterways was lost sight of, while the economy which the former effected in the carriage of goods seemed so great as to lead to the idea that the limit of cheapness had been reached, and that it would be vain to suppose that the expense of carriage could be further reduced. Consequently any amount of money was placed at the disposal of railway schemes[2]: while the canals, occupying the background of the public consciousness, were easily let go because they were thought to be a declining property[3]. The railways lowered the rates of carriage on the canals to such an extent that the receipts and dividends of the canal proprietors were greatly reduced; and the business of the railways, both in passenger and freight traffic, was on such a gigantic scale that the canals thought it would not be long before they would be driven out of business unless they could make an alliance or agreement with their over-powering rivals. This attitude of many canal proprietors toward their property was in great contrast to the enterprise exhibited by those younger men, usually of the trading and industrial classes, who were actively pushing the construction of railways[4]. The former often gave up the battle with the railway companies in despair, and perhaps at too early a period, before they had learned what strength they really had and how largely the traffic of the country would increase[5].

---

[1] Boyle, *Hope for the Canals*, pp. 5–6, 20; Palmer, *British Canals*, pp. 23, 25–26; Teisserenc, *Voies de communication*, pp. 23–30.

[2] Brit. Doc. 1883 (252), xiii, 1, 'Evidence before Select Committee on Canals,' Appendix No. 18, pp. 257–61, statement of Lieutenant-General Rundall, R.E.

[3] Brit. Doc. 1883 (252), xiii, 1, 'Evidence' of Mr Calcraft, Q. 61–62.

[4] Ibid., 'Evidence' of Mr Abernethy, President of the Institution of Civil Engineers, Q. 1356–8.

[5] Brit. Doc. 1872 (364), xiii, 1, 'Report of Select Committee òn Railway Amalgamations, Evidence,' Q. 5814.

In a recent work by Forbes and Ashford, entitled *Our Waterways*, p. 228, the authors say: "If, however, the canal companies must be regarded as in a great measure responsible for the rapid supersession of their undertakings by those of the railway companies, the predominant position of the latter is equally attributable to the failure of the Legislature to recognize the value of our waterways." From what we have already shown, when considering the subject of railways, it seems clear

We have thus outlined the chief elements which entered into the decline of the canals; but we must not assume that this decline was always immediate, nor that all of them shared alike in the process of decay which we have just traced. We have seen that the carrying trade on the canals was the chief feature which gave them vitality in resisting the encroachments of the railways; and that the driving of the carriers off the canals was among the first of the railway tactics. But, in a few cases, the carriers were not to be so easily disposed of; they made arrangements with the independent canals as to the rates that would assuredly remain in force for some time, and by securing favourable rates they were able to compete with the railways in the matter of local traffic. The through traffic, however, was dependent upon through rates, but as the railways had got control of the important canal links and had raised the tolls on these to a point that was usually prohibitory, the amount of goods carried on long through routes of waterway was comparatively insignificant. We can, therefore, say that, for long distances, the competition of the canal with the railway was

that this statement is at least greatly exaggerated, if not wholly unfounded. In the ' Report of the Royal Commission on Canals and Waterways,' 1909, VII, p. 82, a similar implication is given, although not in the extreme form just noted; for it says: "But waterways in this country are also at a disadvantage, due not to the nature of things, but to a state of things which, in our opinion, has been to some extent brought about by errors in legislation, and by neglect on the part of Government and the Legislature." We question even this mild statement of so high an authority. From a close examination of the reports of committees in the decade 1840–50, we see that they invariably recommended that the canal competition should be maintained, thus showing that they recognized the value of the canals, but they were at a loss how to accomplish this; and if those who had so fully investigated the subject could not devise some suitable means of regulation, can we wonder that nothing was done by Parliament as a whole? Then, Parliament was dominated by *laissez faire* principles; it was under this régime that the canals had brought so many benefits to England, and it was but natural that the same policy should be allowed with the railways until it was clearly seen how to change it for the better. Even the Committee of 1844 in their 'Third Report,' Brit. Doc. 1844 (166), XI, 5, and the Committee of 1846 in their 'First and Second Reports,' 1846 (200), XIII, 85, and 1846 (275), XIII, 93, showed that great advantages had come from the competition of the railways with the canals; they said it was impossible for the Legislature to impose proper restrictions on the railway companies in this early stage; they showed that the public had derived great benefit from the cheaper carriage of goods, and urged that Parliament should not lightly sanction any arrangements that would tend to deprive the public of this advantage. See also Brit. Doc. 1851 [1332], XXX, 1, 'Report of the Commissioners of Railways for the year 1850,' p. xix, on "Railway Tolls," third paragraph. The fact would seem to be that Parliament did not know what course to pursue, other than that taken, to regulate this new power; so far as Parliament was concerned, it was lack of knowledge, rather than lack of good intention, that allowed the canal competition to go on as it did until its elimination was assured.

practically at an end by the middle of the century[1]. For short distances, however, and especially where at least one terminus of the navigation communicates with the sea, canals have frequently held their own in competition with the railways[2], and have in some cases paid good dividends to their proprietors[3]. We are, therefore, forced to conclude, as we have already said, that the canals were probably handed over to the railways before their capabilities as a rival system were fully known.

Confronted with the fact that the railways were gradually abstracting the business from the canals, an occasional advocate ventured to devise a plan for keeping the waterways competing with their formidable antagonist. One who had at first regarded the railway between Liverpool and Manchester as invincible was led soon after to a different conclusion; and, provided competition were properly conducted, he thought it possible for the private and independent canal carriers to not only recover all the carrying business that they had lost, but also to draw to themselves the carriage of passengers and light goods, which the railway had taken from the coachmasters and carriers on the turnpike roads. In order to accomplish these results, he proposed to get improved vessels to fit the navigation, with almost four times the carrying capacity of those then in use, and employ steam haulage, so that, by the cooperation of the carriers and the navigation companies, the freight on goods between these cities might be made as low as, or lower than, the actual cost incurred by the railway in carrying these goods. To get the passenger trade, he would put packet boats on

---

[1] In Appendix 12 will be found one or two illustrations showing how the canals succeeded in holding their own against the railways.

Other examples of navigations which were successful in their competition against the railways were the Aire and Calder and the Weaver. These have been constantly improved, both as to the waterway and the equipment for handling the traffic, and are even now active competitors of the railway for the carrying trade of their respective sections. Brit. Doc. 1872 (364), xiii, 1, 'Report of Select Committee on Railway Amalgamations, Evidence,' Q. 3598 et seq.; Brit. Doc. 1883 (252), xiii, 1, 'Evidence before Select Committee on Canals,' evidence of Mr Bartholomew, Q. 776 et seq.

[2] Note, for instance, the Aire and Calder Navigation and the Weaver Navigation, above noted. See also Brit. Doc. 1872 (364), xiii, 1, 'Evidence,' Q. 3604 et seq. Other examples are the Gloucester and Berkeley Canal, the Severn Navigation, the Regent's Canal, and the Birmingham Canal Navigations.

The Leeds and Liverpool Canal, after getting free from railway control in 1874, reduced its tolls by one-half and yet paid dividends of twenty-one per cent. Brit. Doc. 1883 (252), xiii, 1, 'Evidence' of Mr Bartholomew, Q. 827–32; 'Final Report of Royal Commission on Canals and Waterways,' 1909, vii, p. 57.

[3] Brit. Doc. 1872 (364), xiii, 1, Appendix X, which gives full particulars of all the canals, including the dividends paid.

the canal, each drawn by two horses and suitably built 'so as to attain a speed of ten miles per hour, the practicability of which, he said, had been established by more than two years' experience on the Paisley Canal in Scotland. By furnishing such facilities at less than half the fares charged by the railway, the canals and their carriers would be again favoured with public support, and would be able to retain their place as public servants[1]. The difficulties in trying to put such a plan into operation would have been insuperable at that time, on account of the fact that the various elements of the canal interest would not work together; neither do they operate in harmony to any extent even at the present day.

In 1841, after the Birmingham and Liverpool Junction Canal had lost half its tonnage and had been compelled to lower its rates by one-half, its secretary brought forward a method by which he hoped to save the remainder of the tonnage from also going to the railway. His plan had much in common with that mentioned above. It was realized that, even if the canal companies gave up charging the full amount of their tonnage rates, this would not begin to make up for the amount by which the railway had reduced its freight rates; and, consequently, the essential thing was to save in the expense of transit. To do this, he would fasten a train of six boats closely together, one following the other, and draw the train by three horses in order to increase the speed of conveyance. The fast or "fly" trade he would treat in like manner, since it was in this that the opposition was severely felt; and it was by concentrating the traffic in large quantities, through having the carriers work together to make up full cargoes rather than a large number of boats with only a partial cargo, that the cost of conveyance would be reduced. In order to prevent the railway company from continuing to use the passenger and parcel trade as a weapon against the canal, he would introduce on the latter fast packet boats, like those which had been in successful use on the Scotch canals, and, by granting decreased fares and rates, would take much of the passenger and light goods traffic from the railway[2]. By disarming the railway of its most potent instruments of attack, it was hoped to place the canals on a more even footing with their adversary. But here too, as in the former

---

[1] Grahame, *A Letter to the Traders and Carriers on the Navigations connecting Liverpool and Manchester*, 2nd ed. (1834), pp. 6–36. A considerable amount of error is found in this pamphlet. See also his *Treatise on Internal Intercourse and Communication*, p. 159.

[2] Skey, *Report to the Committee of the Birmingham and Liverpool Junction Canal, on the Present State of the Competition between the Canal Carriers using that Line and the Grand Junction Railway Company*, pp. 9–23. See also O'Brien, *Prize Essay on Canals*, pp. 15–21.

case, the difficulty of getting the carriers to collaborate in carrying out such a plan would have been an almost insurmountable obstacle.

Another suggested method of enabling canals to withstand the opposition of the railways differed from the foregoing in detail, but fundamentally it involved the application of the same principle of pulling together. In the first place, systematic management should displace the existing confusion. Canal offices should be organized, where possible, with a responsible head for each branch of the work, the receiving, the transporting and the distributing, and each of these heads should be acting under the supervision of one higher up. In this way, all immediate causes of inefficiency and error would be abolished, so far as the internal management was concerned. Then the various carrying establishments should be brought into accord with one another so as to work upon an intelligible principle for the general good. The carriers alone could not work together in such a way as to adhere consistently to any comprehensive plan; and even if they could, the tendency would be toward a monopoly, from which the public interests would probably suffer. The canal companies had no authority to enforce general regulations among the carriers in regard to cooperation; and, moreover, they greatly needed a much closer understanding among themselves. It seemed, therefore, as if the owners and shippers of goods were the only parties which could establish some body that would harmonize all interests; and it was therefore proposed that these should unite and appoint an agent to act for them, one who would hand over their products for transportation to that carrier who offered the greatest advantages[1]. The same barrier would have been found in any attempt to put this plan into effect as was noted in connexion with the other two suggested remedies, that is, the practical impossibility of securing sufficient united action to carry out such a contemplated project.

It would take too long to consider all the plans which have been brought forward to place the canals in a position to compete with railways and to be effective agents in the transportation of commodities. The Act of 1872, requiring railways to maintain their canals in working order, did something to arrest the decline of these waterways, although it was so meagrely obeyed that it had little constructive effect. In the last two decades of the century, partly as a result of the agitation for lower freight rates, further efforts were made to work out a solution of the canal problem and these appear to have culminated, for the time being, in the labours of the recent Royal Commission of 1906. After

---

[1] Boyle, *Hope for the Canals*, pp. 29–43. See also *Herepath's Railway Magazine*, N.S., IV, pp. 373–4, address by "A Canal Proprietor."

a thorough investigation of the entire subject, that body decided that if waterways, or certain main routes of waterways, were placed under a uniform administration and so improved as to provide the best system of mechanical traction, of transport, and of loading and unloading, the trade on these waterways would be largely increased, provided that carriage upon them were substantially cheaper than that by railway[1]. They recommended that, as the first step in any comprehensive scheme of waterway development, it would be desirable to take in hand four main routes for amalgamation and gradual but continuous improvement, namely, those which radiate as trunk lines from Birmingham, the canal centre of the Midlands, to the estuaries of the Humber, Thames, Severn and Mersey, and which have been called "the Cross[2]." These would tap the great mineral and manufacturing sections of the kingdom and give direct outlet to the four great ports of Liverpool, Bristol, London and Hull. These four routes should be amalgamated under a single control and should be so improved as to permit the use of larger barges for carrying an immense volume of long-distance traffic which did not require the highest speed. Since private capital had not sufficient inducement to embark in this enterprise of improvement, the canals along "the Cross" should be taken over by the Government and paid for by the issuance of "waterway stock;" and the development of this four-branched water route should be effected by public funds. The final control of the system should be put in the hands of a Waterway Board, created by the Government[3]. There are so many reasons why the Government should not subsidize inland waterways, that we think the half-hearted recommendation of the Majority Report should be adopted and acted upon only after much more convincing argument has been adduced in favour of it[4].

But, to return to the period before the middle of the nineteenth century: to all the other carrying agencies of that time it seemed as if the railway would inevitably abstract their business from them. They seemed to be waging an unequal contest with a powerful antagonist. The proprietors of coaches, waggons and vans realized at the outset that the increased speed and better facilities of the railway would soon take most of the traffic off the road, where the two systems came into

---

[1] 'Final Report of Royal Commission on Canals and Waterways,' 1909, VII, p. 84.

[2] Ibid., VII, pp. 93–94, where the details are given, and pp. 188–9, where a summary of their recommendations is given.

[3] Ibid., VII, pp. 165–75.

[4] Ibid., VII, pp. 84–85, 174–5. In the *Traffic World* (Chicago), XII (1918), pp. 420–4, 449–53, I have dealt more fully with the present-day conditions and the recommendations of the Royal Commission of 1906, to which article the reader is referred for more detailed consideration of this question.

competition; and most of the canals likewise soon found that their day of prosperity and independence was hastening to its close.   Closely connected with the conditions of the internal trade was the welfare of the coasting trade; and it would be strange indeed if this too were not influenced by the activity of the railways.   We have formerly observed that, when the Oxford Canal, for example, was being agitated, and a Bill therefor was before Parliament, the eastern coasting trade petitioned against it on the ground that when the metropolis received coal by means of this and other internal water connexions, the amount of coal that came from the north to London by the sea route would be greatly decreased, and this, in turn, would be detrimental to the maritime interests of the kingdom.   In the same way, it was thought by some that the development of the railways would be prejudicial to the well-being of the marine; and in 1846 memorials of the shipping interests of Sunderland, Shields and the Tyne were presented to the Treasury, requesting that efficient measures might be devised for preserving the coal coasting trade from ruin through the conveyance of this northern coal southward by railway.   It was admitted that coal could be carried by railway from the Durham and Northumberland collieries to London at charges lower than those for which ships could be navigated; and representation was made that to jeopardize or destroy this northern marine, while developing the railways, would be contrary to the best good of the kingdom, and would, in effect, be crippling "the right arm of England's strength[1]."   But it is evident that if railways were allowed to carry coal from the mines to the interior portions of the country, they could not be prevented from carrying it to London.   The point which we wish to emphasize, as a concluding thought, is that, within the first twenty years of the railway era, this young giant had overshadowed all other systems of carrying, some of which had taken centuries for development.

---

[1] *Railway Chronicle*, April 25, 1846, pp. 418–19, and June 13, 1846, p. 582.

APPENDIX 1

RIVER WEAVER NAVIGATION

AT the end of the seventeenth century and in the first two decades of the eighteenth century, the salt industry of Cheshire was coming to occupy an important place. Before 1699, most of the coal that was used in the refining of the salt was brought from Staffordshire by land carriage, and the salt was taken, also by land carriage, to Frodsham Bridge near the mouth of the Weaver, and to Worcester and Bristol in the south, where it was loaded into vessels which carried it to other parts of England, to Ireland, and some to Northern Europe.

The cost of carriage of salt seems to have varied greatly at different times of the year, and in some cases excessive charges were made for this service. This is evident from the following letter contained in the Brit. Mus., Add. MSS. 36,914, p. 3:

"Sir, 'tis very observable how the rock-men have over-acted their part in conveying their rock (salt) from their pits...by giveing excessive rates, as some days 20*s*. per ton to Frodsham Bridge—the like for seven miles has not been known—and other days their wages were so great that people were so blinded with it, that they neglected their necessary duties at home, in plowing, sowing, etc. This hurry and charge is...vain, and labour and money near lost, for by it, they too greedily presumed to have the advantage of the surplusage weight, but they are nickt, as you'l finde in the Act, for all salt...after the 15th day of May (1699) shall be weighed wherever its landed at 75 lbs. the bu., which is a subject of lament among themselves...."

The great demand for horses to carry salt to Frodsham Bridge and Worcester was supposed to be the reason why the horses with which strangers came to Droitwich and other salt towns were taken from their pastures, were used for carrying salt to Worcester, and were then found near this latter place when they had been unloaded. So often was this the case, that hotel landlords commonly advised their guests not to put their horses out in the pasture, but to keep them in the stables (v. Brit. Mus., Add. MSS. 36,914, p. 9).

On account of the difficulty in the matter of transportation of coal and salt, those manufacturers of salt which were more distant than others from the coal supplies and from the markets for the finished product found it difficult to compete with their rivals who enjoyed greater advantages than they in these respects. For example, the cost of coal at Northwich was greater than at Middlewich, because it had to be brought a longer distance by land carriage from the Staffordshire coal mines. On the other hand, those manufacturers who were nearer the supply of Staffordshire coal had a longer haul before they could bring their salt either to the Mersey or to the port of Worcester. A combination of the manufacturers had been formed at Droitwich (v. Brit. Mus., Add. MSS. 36,914, p. 10), and probably also at Middlewich and Nantwich, to control the price of salt, and an attempt made by a private manufacturer to break up this monopoly had ended in failure. But a certain Mr Slyfford, and one or two associates, who owned salt deposits at Winnington Bridge on the Weaver, just below Northwich, saw that if they could get coal from Lancashire brought up the Weaver to their works, at a price that was lower than that for which their rivals could get it, and if they could have their finished product carried down from their works to Frodsham Bridge at a lower rate than their rivals, the trade in salt would be largely in their hands. For this reason, they proposed that the river carriage should be utilized rather than land carriage, and in order to make the Weaver, which in its original state was navigable only at high tides, an effective agent for their carrying business, a Bill was brought into Parliament asking authority to make this improvement.

Immediately the opposition was aroused. The other manufacturers said that if the Weaver were made navigable only to Northwich it would advance the interests of only three or four proprietors of salt and salt rock, and would certainly ruin the estates of several proprietors in other places, as at Middlewich, as well as some thousands of people adjacent, whose livelihood depended on the carrying of salt and coal (v. Brit. Mus., Add. MSS. 36,914, p. 10). The opposition of the carriers found vent in many petitions to Parliament, of which the following is a fair sample (v. ibid., p. 16):

"The Humble Petition of several Farmers & Freeholders in Bucklow Hundred in the County Palatine of Chester, in behalf of themselves and neighbors

Sheweth

That your petitioners having heard that there is a Bill presented to this Honorable House for making the River Weever navigable some few miles, which Bill should it pass would extremely impoverish your Petitioners by depriving them of the Benefit they receive by carrying of coals to their own Houses at spare times in summer, and from thence to the Wiches in Winter, whereby they are the better enabled to pay their rents and provide for the comfortable support of their families.

Wherefore

Your petitioners make it their Humble Request that before the Bill be suffered to pass their objections against it may be heard from their Counsel or otherwise as this Honorable House shall be pleased to direct.

And your petitioners shall ever pray, etc."

Similar petitions to Parliament were sent by "the poorer sort of inhabitants of Bucklow Hundred" (ibid., p. 18), "the poorer sort of inhabitants of Northwich" (ibid., p. 18), "the poorer sort of inhabitants of Edesbury Hundred" (ibid., p. 19), "the farmers and freeholders in Northwich Hundred" (ibid., p. 20). These all presented the "ruin" which would ensue to the carriers should the Bill pass. But it is doubtful if these would have been effective in defeating the Bill, had not the "prominent landlords and gentlemen of rank" taken the matter up, among whom were Lord Gerard, Thomas Cholmondeley, G. Warburton, and Sir Willoughby Aston. These men were presenting a petition to Parliament against the proposed improvement, and in order to give it more weight they sent around the following letter to get signatures to be attached to the petition (ibid., p. 20):

"Gentlemen

By intelligence from London and some practices in the country we find that the projectors concerning whom we have formerly troubled you, have renewed their design and prepared a Bill now ready to be presented in Parliament, for making the River Weever navigable from Frodsham Bridge towards Northwich. We think it needless to represent to you how injurious this Bill would prove to those who have lands lying near the river; and Destructive to the trade of this County, especially of Middlewich and Namptwich and all the adjacent salt works; since the easy import of coals from Lancashire to Northwich and export of their salt would certainly enable the proprietors there to undersell and ruin all the other salt works which are supplied with coals from Staffordshire or Wales; whereby about four thousand families, now subsisting by the land carriage of those coals, salt, and malt, would be utterly ruined and left to be maintained at the charges of their respective parishes; and the Rents of those lands which they inhabit, and of those near their Roads would be impaired. After which some few proprietors of Salt Rock and Brine in and near to Northwich (who alone can be enriched by this project) having engrossed the trade, would impose the price of salt at their own pleasure, and raise their fortunes on the ruin of the country. We have prepared a petition to be heard by our counsel against the said Bill; and if ye approve it, we desire your concurrence with us, believing your subscription will be as serviceable to the country, as obliging to

$\left\{\begin{array}{l}\text{We send the like petition}\\ \text{to other hundreds for expedition,}\\ \text{intending to unite them all in one Roll.}\end{array}\right.$ Gentlemen
    Yr humble servants."
    (Here follow their names.)

Similar petitions were sent in by the "High Sheriff, Deputy Lieutenants, Justices of the Peace, Gentlemen, and other inhabitants of the County Palatine of Chester" (ibid., p. 24); by "the Inhabitants of Warrington, in the County of Lancaster" (ibid., p. 25), who protested because if the Bill should pass it "would subject the salt of Cheshire to a monopoly;" and by "several gentlemen and others in that part of Staffordshire adjacent to Cheshire," who said that if the Bill should pass, it "would ruin most of the salt works in Cheshire; it would also greatly impoverish that part of Staffordshire which the petitioners inhabit, by stopping the great vent of coals thence to the Wiches and by destroying that commerce and carriage whereby the farmers are enabled to pay the greater rents and many of the poorer people wholly subsisted" (ibid., p. 28).

In Brit. Mus., Add. MSS. 36,914, p. 29, we have "A Short Account of a Design for making the river Weever in the County of Chester Navigable, from Frodsham-Bridge to Winnington-Bridge, being about five or six Miles only." This is really a series of reasons *against* the project. It was intended for circulation among the members of the House of Commons, had the promoters proceeded in their purpose. Its substance follows:

This navigation is a design projected for engrossing the trade of selling salt and rock-salt into the hands of two persons only, whereby a great many families would be ruined and undone. As the trade now stands, all the proprietors of salt are upon equal terms throughout the whole county of Chester. (This, as we have already seen, was entirely wrong.)

Should the river Weaver be made navigable from Frodsham Bridge to Winnington Bridge, there will be the following evil consequences:

The two persons in this combination have salt works and rock salt adjoining Winnington Bridge where they intend to end the navigation. To there they can get coal cheap from Lancashire by water; hence, with no land carriage to trouble them, either for coal or salt, they will undersell all other salt works that have land carriage for both coal and salt. These two persons will drive out rivals, and therefore will be able to make their own prices for salt, "as formerly Northwitche did, till the erection of new salt-works in the county reduced the price of salt from above 4*s*. to 2*s*. 6*d*. per barrel." Nothing but an abundance of salt will keep down its price.

The ruin of other salt works will be followed by additional evil results:

1. The collieries in Staffordshire and some in Wales will have their market reduced.

2. Those poor people who live by carriage of coal and salt must starve or be a charge on the parishes. The same thing will happen to those who now carry malt to the Wiches from Derbyshire and Nottinghamshire. The trade of malt for salt will be at an end by the destruction of the salt-works, which will effectually and speedily be accomplished, "should the monopolizing project of this self-ended navigation take effect."

3. Continual overflowing and spoiling of the meadow grounds, which cannot be avoided because the river banks are low, and the water will be raised in the river by means of locks. The grazing lands along the river will also be injured; and this will cause reduction of rents of these lower lands.

Many more mischiefs are so obvious, "that 'tis hoped this self-designed project shall never be countenanced by Parliament, to the great prejudice and injury of the publick, for the sake of a private interest." (The feeling against this navigation was very strong; and the fact that we find no petitions in favour of it would seem to indicate that it had a selfish end.)

In 1709, the question was revived apparently with more seriousness, and the people became alarmed at this pernicious self-ended project. Every man who had any influence with the members of Parliament used his position to show them, by letters, the terrible evils that would result from making the Weaver navigable. The great hostility against the Bill came from all sources; and the vigorous opposition to it may be gathered from the letters of 1709, given in Brit. Mus., Add. MSS. 36,914, pp. 34, 40–45.

This enterprise, though held off, could not be completely turned down, and by 1715 the promoters had "a great many friends" who were "very industrious" in behalf of the scheme. It had been so much talked about, that instead of losing ground it had gained increased support (ibid., p. 54). A letter of May 16, 1715, shows the way in which the opponents of the Bill regarded it (ibid., p. 40):

"I was in hopes this ruinous project had been so often battled that we might have lived secure from any further attempts of strangers to bring sure destruction upon so many poor families in this county, and so great damage to many other; but now to help forward their designs, they have got some assistants from Liverpool who no doubt have either our interest or their own very much at heart...." Then the writer goes on to give reasons against the measure.

In that year, on June 14, the Bill was read the second time, but it was finally accorded the same treatment as in former years, and was in effect thrown out (ibid., p. 58).

In order to make their case stronger, the promoters of this navigation were supposed to have ordered to Frodsham Bridge such great numbers of ships that it was impossible to get enough white salt and rock salt to give them a full load, without keeping them lying there for many weeks, and some for months (ibid., p. 66). This would tend to show the need for vastly increased facilities for transport. But, however this might be, in accordance with another petition to them, the House of Commons, in January 1719, ordered a Bill to be brought in for making the Weaver navigable (ibid., p. 68).

In carrying on their campaign in 1719, those who favoured the navigation issued a pamphlet showing the 'Reasons for Making the River Weaver in the County of Chester Navigable;' and demonstrating that it would be of "very considerable national advantage," as well as of local benefit (ibid., pp. 86–90):

The pamphlet begins by giving some account of the trade of those parts that will be affected by making this river navigable. The salt trade is the most important of all.

The *mines of rock salt* which supply all the salt refineries erected in Ireland, and in several parts of Great Britain, viz., Cheshire, Lancashire, North and South Wales, Bristol, etc., lie about one-quarter mile from Northwich. And the *salt springs* and *salt works*, which supply Ireland, Wales and several counties of England, with great quantities of white salt, are at Northwich and within three or four miles of that town.

This white salt and rock salt is brought by land carriage to the ships at Frodsham Bridge, and is *mostly carried on horses' backs, by reason of the badness of the roads*. The ships usually come in fleets, and hence the men and horses kept for carrying this salt have sometimes more than they can do, and at other times have scarcely any work. This difficulty is tending to ruin the salt trade. This winter, some ships have lain there three months before they could get fully laden; and such a thing spoils the trade and will eventually drive away the trade entirely, if not prevented.

The best, and perhaps the only, expedient that can preserve the trade is to make the river Weaver navigable. This would make carriage considerably cheaper, and give greater dispatch to the shipping.

Other Advantages of this Navigation :

It would allow England and Ireland to get their salt at home, and thus save buying foreign salt. But it might also enable Cheshire to supply salt to the northern parts of Europe.

Opening this navigation would make a way for attaining a good share of this trade with Northern Europe. Hence it would benefit both the kingdom and the county of Chester.

It will create new employment for much shipping, breed a great many seamen, employ many of our poor, and bring in considerable sums annually to our kingdom.

It would necessitate fewer officers, and less charge and trouble of frequent weighing of the salt.

Then the pamphlet takes up what it calls some "weak but obstinate objections," namely :

1. That it would ruin the salt works at Middlewich and Nantwich.

This, if it were true, is not a reason why such a great public good should be declined, for fear of interfering with the private interests of a particular place or two. For by this navigation all those places that get salt from Cheshire would get it much cheaper than at present. But this will not ruin these two places, for as they are four miles nearer the coal supply of Staffordshire they will be able to get their coal cheaper, and this will offset the extra four miles of land carriage necessary in bringing their salt to Northwich.

2. That it will overflow and spoil the adjacent lands.

This is refuted by the experience of other rivers; such lands are less liable to be overflowed than before the navigation was made. But, on the other hand, the adjacent lands are increased in value, because of the power of overflowing them on occasions of great drought or dry seasons.

3. That it will take away the livelihood of those who were formerly maintained by land carriage.

But particular employment must give way to the public good. Then, too, these carriers, because of the uncertain arrival of ships, have only a sorry livelihood, notwithstanding the great prices they get for carriage. Their horses, etc., have to be kept, whether there is work or not. Their present business is precarious and they would make a better living by going into dairying. Moreover, the increased trade brought by the navigation would give increased work for all classes of people.

By 1719 the opposition to the navigation had assumed a somewhat different character from what it had been shortly before. Formerly the question was, whether Northwich should be allowed the advantage of this navigation, by which that town might be enabled to undersell, and consequently to ruin, all the other salt works, with the trade depending upon them, and then impose its own price of salt upon the nation. But since the discovery of the rock salt, the projector of this design, being a proprietor of the rock, a considerable merchant, and naturally qualified for a great undertaking, might, by his rock salt, not only ruin all the brine trade of Northwich and the rest of Cheshire, and then impose his own price of salt upon the nation, but might also draw all the money in specie to himself at London. So that, by 1719, the question was, whether all the ancient Wiches and other brine salt works in Cheshire, and the trade depending on them, together with the landowners adjacent to the river, and the greater value of all the other lands in Cheshire by the land carriage, should be destroyed for the interest of a few men, without any public advantage (ibid., p. 95).

Other objections to the navigation were brought forward, as the project seemed nearer to obtaining Parliamentary sanction:

1. The boats would have to be towed up the river by horses, and as this country was enclosed landowners would be obliged to make gates 'at their own expense. The negligence óf boatmen in closing the gates would cause trouble in keeping each man's animals from his neighbour's fields. Hence there would arise quarrels, breaches of the peace, etc.

2. Several landowners had large estates along the river, "where their deer, sheep, rabbits, and other household provisions are kept." Boatmen were ill-disposed persons, and as they would be compelled to pass through the grounds of these estate owners they would feed their horses on the landowners' hay and corn, at times when they would be waiting for the violence of the tides or floods to subside, or when awaiting the raising of the water by the locks. Or the boatmen might steal their deer, sheep, wood, corn, fish, etc., and perhaps break open the houses of these gentlemen.

3. The farmers who lived along the river would not be able to use their fords to cross the river after it were made navigable, without wooden bridges that would obstruct the passage of the boats. Then, too, these bridges would be destroyed by the ice, as "often happens to the danger of the stone county bridges."

4. Higher than where the tides flowed, which was about three miles above Frodsham Bridge, the river was very narrow, full of roots of trees, and in many places, both above and where the tides flowed, so shallow that half the channel lay dry several months together in summer time when the tide was out, and in the other half the water would scarcely cover the stones in the rocky fords. This would prevent the river being made navigable, except at great expense. In cutting it *wider*, which would be made difficult by the roots of trees, there would be great loss and damage to the landowners, because some of their land would have to be used in making the towing paths also. In making it *deeper*, which could not be done but by cutting the bottom lower or by raising the water higher, there would be other difficulties; for if the river-bed were cut lower the rocky shallows must be cut at great expense, and the foundations of two stone county bridges across the river would be undermined; and if the water were raised higher, locks would have to be made, and these would cause the water to overflow the banks, especially in floods. They would also cause the water to be longer in passing off the ground, and thus this rich land would become bog or marsh land.

5. This navigation would take away the living of those who were then carriers by land. The landowners were accustomed to employ the tenants' teams in carrying, and so the tenants were enabled the better to pay their rents. If this carriage should cease, rents would fall, the tenants that have leases would be ruined, the landlords impoverished, the taxes on their lands would have to be abated, and hence the nation's revenue would be lessened. In reply to the question as to why these carriers could not turn to dairying, it was pointed out that some of them were very poor people, and kept or hired horses for the purpose of carrying.

6. Northwich would, by this navigation, get such an advantage over all other salt works that she would be able to dictate the price for salt. Against this, it was shown that the Justices of the Peace had power to fix the price of salt; to which it was answered that if this Northwich projector should ruin the other salt works he would then be able to set his own price for salt.

In Brit. Mus. 357. b. 9 (76), we are shown the desirability of the proposed navigation for the benefit of the salt trade of Cheshire, the opposition of Liverpool to the Bill, and the trouble with Mr Vernon, who had large salt works at Winsford. This difficulty is very clearly stated in Brit. Mus. 357. b. 9 (78), 'Reasons Humbly Offered by the Trustees of Richard Vernon...against the Bill for Repealing Act 7 Geo. I for making River Weaver Navigable.' Other 'Proposals Humbly Offered for making River Weever Navigable from Frodsham Bridge to Northwich' are given in Brit. Mus. 357. b. 9 (75). In spite of the endless repetition of the above-mentioned objections (for which see also Brit. Mus., Add. MSS. 36,914, pp. 117–23), and the strong opposition encountered, the Bill passed into an Act in 1720, under which this river was made navigable for nearly twenty miles from its mouth, that is, to Winsford Bridge. By this Act, the £9000 subscribed to carry on and perfect the navigation could be increased by as much more, if necessary. In order to repay the cost of the improvement, a duty of 15d. per ton was to be taken on all goods carried on the river; and after payment of the cost the tonnage duty should be but 12d. per ton, the whole of which was to be applied for the public purposes of the county of Chester for ever. It was expected that, after the cost of the navigation had been defrayed, the county would get not less than £1500 a year (Brit. Mus. 357. b. 9 (72), 'Reasons Humbly Offered against Bill for Repealing Act 7 Geo. I for making River Weaver Navigable'). The merchants of Liverpool opposed this measure. Since the river was to be made navigable by three private undertakers at their own cost without any contribution from the county, it was thought unreasonable as well as unjust that the county should benefit financially from it, and that any money which would go to the county would be an overcharge on the navigation which would impede and burden trade (Brit. Mus. 357. b. 9 (73), 'Reasons Humbly Offered against allowing County of Chester any part of the Tonnage Duty for making River Weaver Navigable'). An agitation was, therefore, started to repeal the Act, but it was unsuccessful, and the work of canalization began in 1721. All revenues from tolls, in excess of the amount required to pay the cost of construction and maintenance, were to be devoted to repairing the roads and bridges of the county, and for any other purposes determined by the Justices.

In 1759 the administration of the Weaver was entrusted to a body of self-perpetuating trustees, under whom the navigation works were extended and improved and made more enduring. About 1807, the navigation was completed by a canal of four miles in length from Weston Point, where it joins the Mersey, to Sutton lock; this was intended as a surer course than the lower part of the Weaver, so that boats could enter or leave at all conditions of the tide. For further improvements, see *Ministère des travaux publics: Quatrième Congrès International de Navigation Intérieure, tenu à Manchester en 1890, Rapports des délégués français sur les travaux du congrès*, pp. 39–55. See also Hanshall, *History of Cheshire*, p. 84.

APPENDIX 2

SHAPLEIGH ON HIGHWAYS (1749)

HE says (p. 4), "For, it must be granted, that there has been always, and now is, great reason to complain of the neglect of the repair of most roads within this kingdom; and that it has always been found by experience, that the many laws, which have hitherto been made concerning their repairs, have never met with the desired success. Hence there must be some fundamental error in these laws, and there is need of further regulation."

He thinks the fundamental defect is in permitting parishes, towns, etc., to be presented or indicted for not repairing their roads (p. 5).

(p. 6) His method is:

1. To prove that the presenting or indicting of parishes, towns, etc., for not repairing their public roads is generally found to be hard and injurious to particular persons; that it seldom, if ever, answers the intended design; and that it causes the laws relating to the surveyors of the highways to be greatly neglected. Consequently, both these prosecutions should be entirely done away by law.

2. To prove that the most just and most effectual way to have the public roads kept in good and sufficient repair is to oblige the surveyors to do their duty.

3. To offer some amendments and additions to existing laws, which will more easily and more effectually oblige the surveyors to perform their offices, and the parishioners to do their six days' work.

To prove the first.

Act 3 & 4 W. & M., c. 12, sec. 3, allows the surveyors to be men in mean circumstances, men who have no property in the parish chargeable to the repair of the highways; since their qualification is £10 per annum of real estate, or £100 personal estate. But by the same Act a tenant of £30 a year may be appointed surveyor. If none so qualified can be found, then the Act directs that the most responsible persons within the parish are to be chosen. Under such mean persons for surveyors, it is not prudent nor safe for the rest of the parishioners to repair the roads, under their own directions, lest their officiousness should hereafter be used as an argument of their obligation to repair them in their own right.

And, as laws now are, the parishioners have no coercive power to oblige these mean persons to discharge their official duties. Suppose the surveyor were too idle or obstinate to call out the parishioners to do their six days' work, or to oblige them to work honestly; would it not be unjust that the whole parish should be presented for neglect? The innocent would be punished with the guilty. Some parishioners, of course, would, and others would not, work voluntarily; hence the need of the surveyor to force them (p. 9).

It is unjust to punish the innocent with the guilty. But Shapleigh says that the innocent bear the burden of the surveyor's transgressions, while he goes free. (Evidently, therefore, the law which imposed a penalty upon neglect of either surveyor or parishioner to do his duty was found to be unenforced.) For in many places, especially in the Western counties, the landlords of such tenants as are at rack-rent pay the rates, taxes, etc., for the tenants; and hence no process that can issue upon any such presentment or indictment of the parish can in any way affect the surveyor's property.

(p. 10) But suppose, again, that the parish surveyors should contribute some small matter only out of their own property toward the parish levies, or should happen to be such tenants as have all their rates, taxes, etc., paid by their landlords, and live in a parish where one or two persons own most of the property; suppose also that the landlords live at a great distance from the parish, and that the surveyors should be prejudiced against the stewards of the landlords (which is quite probable from their insolence and imperiousness) and should refuse to do anything for repair of the parish roads; would it not be unjust that the innocent landlord should be punished by such expensive proceedings as presentments or indictments for an offence which he could neither remedy nor prevent? This is no mere supposition, but actual fact.

(p. 12) Take the common case: suppose the surveyors call out the parishioners to their statute duty, and only two or three of the best householders and landholders obey the call. Suppose that (because of the surveyors' laziness, unwillingness, or probably from bribery or corruption) the surveyors do not force the rest to come out to work, and thereby the parish is presented or indicted for not having its roads properly repaired; would it not be very unjust that those parishioners who have done the work required by law should still be punished ten or twenty times as much as the other householders, etc., who ought to have done their work, and as the surveyors who should have forced the work to be done?

(p. 13) This method is wholly inconsistent with justice and reason. As the law now stands, every person having an estate within the parish is liable to be punished for not repairing the highways; and yet the law has not given him proper or sufficient power to compel the surveyors and defaulters to do their duty and contribute their part towards the repair of the public roads. So that the law in this particular instance punished persons for not doing that which it was not in their power to do.

Next, to show that notwithstanding these presentments and indictments are too often made against parishes, towns, etc., for not repairing their roads, *yet such proceedings seldom, if ever, answer the end intended by them.*

The fact is shown by experience, for everyone observes that there are some parishes which have presentments or indictments almost perpetually hanging over their heads. This could not possibly be the case if these prosecutions were so effectual for the repair of the roads, as some persons erroneously insist upon. The fact is beyond power of contradiction.

(p. 15) From the nature and reason of such presentments and indictments, no good effect can possibly be expected from them, either to the public or to individuals (except lawyers and others who attend the courts of justice). For since the surveyors are mean persons, their payment toward the fine imposed on the parish is so small as to have very little effect upon them; especially if it be considered that they are sure to have the laying out of this money—which they often do, more to their own benefit than to the improvement of the roads.

(p. 16) Besides, the presentment or indictment specifies particular parts of the road (those that are worse than the rest) to be repaired. This does not effect a thorough reformation of *all* the roads of the parish.

(p. 17) Again, these prosecutions are often made against parishes, towns, etc., in the winter; and it costs the parish at that time of the year far more to repair its roads than if it were at a seasonable time of the year. If the road is complained of in the winter, and presented then, it has to be repaired then.

No result can be derived from such prosecutions, but the expensive repair of such places as happen just then to be out of repair; for these grievous prosecutions cannot reform the inclinations of the parishioners in general, nor make them more willing for the future to repair their roads. The power of repairing rests with the surveyors; and, therefore, as the parishioners, before the prosecution, could not safely repair the roads without the concurrence of the surveyors, so neither can they do it after the prosecution is begun or ended. Can anyone think that such prosecutions will make the surveyors more diligent than they were before; when, as we have shown, they are seldom one penny out of pocket by all the presentments or indictments which the law can throw upon the parishes, towns, etc., for which they are surveyors? They would rather be pleased than displeased with such prosecutions.

(p. 19) "This argument, I must own, carries great weight with me against the allowance of any such prosecutions; and I verily believe that all considerate and disinterested persons must entertain the same opinion of it with myself."

But the vexation, oppression, expense and uselessness of these prosecutions are not the only reasons for laying them aside; Justices of the Peace would be more willing, earnest and ready to hear such complaints as are to be made against the surveyors and defaulters, and to enforce the laws against them, if they were restrained from exercising this favourite power of punishing all the parishioners promiscuously, by way of presentment, which most of them now are apt to think their safest and easiest remedy. Those interested would be more diligent in making such complaints of the surveyors and defaulters to the Justices of the Peace, in case they found they had no other redress; and the Justices, in turn, would be more ready to give an attentive ear to such complaints.

To prove the second, viz., that the best way to have the roads repaired effectively is to oblige surveyors to do their duty.

(p. 20) If the six days' work were done faithfully, it would be sufficient in most cases for repairing the public roads. Wherever it were otherwise, Acts 3 & 4 W. & M. and 1 Geo. I have given the General Sessions power to make a rate for that purpose, not exceeding 6*d.* in the pound for any year. If, then, the due execution of the above methods would effectually repair most of our public roads, whenever they are out of repair, it must always happen (unless in case of floods, great frosts, etc.) through the surveyors' default, or that of the parish, or of particular individuals in the parish. And the laws have given Justices of the Peace and surveyors power to punish delinquents.

(p. 22) Under Act 5 Eliz., c. 13, sec. 8, it is enacted that surveyors, under pain of 40*s.*, shall within one month after any person has omitted to do his statute duty, present the offence to the next Justice of the Peace, who shall certify that present-ment at the next General Quarter Sessions, which Sessions shall immediately inquire of such default and assess such fine for it as they, or any two of them, shall think fit. With such authority, it is easy for Justices of the Peace to perform their duty; and when defaulters are found guilty, Act 2 & 3 P. & M., c. 8, inflicts a penalty on them that is double the value of their neglected labour. Under this Justices can safely act, for they are simply carrying out their duties in certifying the surveyors' presentment to the next General Quarter Sessions. The punishment is inflicted by the Court of Sessions upon the defaulter.

We have now shown that the laws have given power to the Justices of the Peace to punish defaulters for non-repair of roads, and the surveyors for neglecting to present such offenders. Act 5 Eliz., c. 13, has given the Justices an easy method of procedure.

Next, we show that *this method of punishing the surveyors for neglecting to present defaulters for omitting to do their six days' work is just and equitable and the most effectual to obtain the end desired.* For by this each delinquent is punished according to the assistance which he ought, but neglected, to give towards repair of the roads. By this way of proceeding, the landowners and the rest of the parishioners are not all promiscuously punished, without making any distinction between those who either have obeyed or else were willing to obey the directions of the law, from those who either have neglected or else refused to observe its orders. By this way, the surveyors are not capable of triumphing, nor of benefiting themselves by their own neglect or open defiance of the law; but the innocent and faithful observers of the law are exempted from that punishment, which is, with a just and distinguishing hand, inflicted on each of the offenders with equality and prudent distinction.

(p. 26) But we have to show that this is not only the most just but also the most effectual way to obtain the desired end.

Since the law has appointed proper officers to take care of road repairing, that method must certainly be the most effectual which has the greatest power and influence, and is the most capable of compelling these officers to do their duty, and effectually care for and sufficiently repair the roads. But presentments affect very few of such surveyors; and in general such prosecutions are rather beneficial than otherwise to them. Whereas, on the other hand, by Act 5 Eliz., c. 13, they are liable to be punished 40s. for not presenting defaulters; and, by Act 1 Geo. I, they are, for most cases of neglect of duty, liable to pay £5. As this last method, therefore, is the most coercive, and indeed the only effectual one to force the surveyors to do their duty, I think there cannot be the least doubt but that it is by much the most effectual way to have the roads repaired and kept in repair. (Apparently, Shapleigh thought that the law which imposed the greatest punishment for neglect of duty was the most effectual for repairing the roads.)

Regarding the third—some amendments of and additions to the laws, in order the more effectually and easily to oblige surveyors to perform their duty, and the parishioners their six days' work.

(p. 28) We have said that Act 5 Eliz., c. 13, sec. 8, gives a safe and easy method for Justices of the Peace to proceed in punishing surveyors or parishioners for neglect of duty; yet it is not so easy for there are *later contradictions.* Act 22 Car. II, c. 12, sec. 9, says that complaint of defaulters to the public roads is to be made to the next Justices of the Peace, who are required, on the oath of one witness, to levy the penalties. By Act 22 Car. II, c. 12, sec. 12, the method authorized is the same as that of Act 5 Eliz., c. 13; and as Act 22 Car. II is later than Act 5 Eliz., and these two clauses seem diametrically opposite to each other, it is hard to tell which is the best and safest method to be followed. Justices are rather unwilling to proceed.

(p. 29) Again, by Act 3 & 4 W. & M., c. 12, sec. 9, all offences and neglects respecting the public roads are to be presented by the surveyor, on his oath, to the Justice of the Peace. But this Act does not direct what the Justice shall do with the presentment; it is inferred, however, that he ought to certify it to the next General Quarter Sessions, because in the second section of the Act it says that all former laws regarding the highways shall remain in effect. Besides (p. 30), Act 1 Geo. I is so worded that it may seem doubtful whether the directions given by the said former Acts, concerning these points, are not thereby repealed; and jurisdiction vested in the Special Sessions. Such vague laws weaken the hands of the Justices of the Peace, and because the Justices do not care to act under them the surveyors and defaulters often go unpunished.

(p. 32) To proceed on either of these statutes, however, is slow, for Special Sessions are only held every four months; but either method is preferable to presentments or indictments, for the latter are unjust and oppressive, as we have shown.

Then Shapleigh proceeds (p. 33 et seq.) to outline in full a law which he would recommend to the Legislature, to take the place of the old laws, so as to be clear and easily executed, and to give his reasons for particular clauses in the wording of his proposed new law.

(p. 56) "The six days' work have hitherto in most parishes been so much neglected, and so slightly performed, that I believe very few parishes can truly say, from their own experience, that the six days' work, duly and properly attended to, and performed by all the parishioners liable by law, to work in the amendment of the highways with due care and diligence, are not sufficient." [This seems to be strong testimony of the adequacy of the statute duty, if satisfactorily performed, to effect the improvement of the roads; but it also shows how poorly this work must have been done when the roads were in such a bad state.]

(p. 60) In speaking of the imposition of the assessment of 6*d.* in the pound, as authorized by Acts 3 & 4 W. & M., c. 12, sec. 17, and 1 Geo. I, when the six days of statute labour were not sufficient to repair the roads, he says: "For such a rate does, in some parishes, raise by much too large a sum to be trusted in the hands of such persons as are generally chose surveyors." (The character of the surveyors, if they were at all like what is here implied, must have been such as would not command the respect and confidence of the parishioners.)

APPENDIX 3

HAWKINS ON THE LAWS OF HIGHWAYS (1763)

(p. ix) "But this the Public may be assured of, that every attempt to amend the Highway-Laws by additional or explanatory Acts, will produce great Confusion among those whose duty it is to execute them; and that nothing can remedy the evils at present complained of, but the consolidation into one Act of the most efficacious clauses contained in those now subsisting." (It was Hawkins' chief purpose to get a consolidation of such Acts into one general Act.)

(p. 2) "It is too obvious to need insisting on that very little of the concern which has of late been shown about the roads in general has been directed to those that lead from parish to parish, and are not the ordinary channels of conveyance to cities and towns of great trade. The invention of turnpikes is manifestly calculated for great roads, which, as they are made in favour of commerce, produce a revenue sufficient to keep them in repair; but the former have been left to the care of the surveyors of the highways in their respective parishes, subject to the direction and controul of the Justices of the Peace."

Since the framing of Act 2 & 3 P. & M., c. 8, and its successors under Elizabeth, coaches, chaises and post chaises had come in; and gentlemen who had these and drove on the road constantly got off with the same road work as the poor cottager who had no such things. These were not included under the term "draught" of the Act, and hence these nobles were merely householders like the cottagers, so far as the statute labour was concerned. Hence there was great need for a change of the law.

(pp. 24–25) The statute 2 & 3 P. & M., c. 8, is also indefensible. A law without a sanction is but a dead letter, and this is the case with this statute. Suppose a farmer, who occupies a plough-land or keeps a draught, is required by the law to send a team to work six days on the highways, and that he is averse to this duty.

He does it only through fear of incurring the penalty. [Nelson, *Justice of Peace*, 6th ed., p. 332 n., says that a ploughland was formerly 100 acres, but at his time (1718) only 80 acres. By Act 7 & 8 W. III, c. 29, £50 per annum was a ploughland. See also Burn, *Justice of the Peace* (1755), I, p. 512.] But it may be to his interest to incur the penalty and not do his statute labour. For example, the labour of a cart, a team and a man is valued at 10s. a day in most places. He can get this if he lets them out to a neighbour. But if he sends his cart and team and two men to work on the roads, the labour of all will excuse him from the payment of no larger a sum. So if he lets out his team and incurs the forfeiture he saves the labour of one man. *Hence the statute is pregnant with a motive for disobedience.*

(pp. 25–26) In like manner the day labourer may argue that if he must either actually perform or forfeit the price of six days' labour, it is as well to choose the latter as the former. This defect, it must be confessed, is owing solely to the diminution in the value of money since this statute was enacted. Taking Bishop Fleetwood as authority, he says:

"In 1514, not long before the statute was made, the wages of a labourer, from Easter to Michaelmas, except in harvest, were 4d. per day, and from Michaelmas to Easter, 3d. per day. The labour of a waggon, team, and two men, amounted to 2s. 8d. per day. But by the statute the penalty for default to send a team was 10s. and for default to send two men it was 1s. Therefore the penalty was 11s. for default in sending two men and a team. If these forfeitures be compared with the respective duties they were intended to enforce, we find them to be sufficiently penal at the time the statute was made. But this is not so now."

But it is urged that whether the person charged does actually perform or pay the price of his duty, the case is the same to the public; inasmuch as the forfeiture will purchase just as much labour and assistance as was originally required of him; and if that is done, it is nothing to the public what hands were employed in it. *But is it of no consequence to a state whether the laws are obeyed or not?*

(p. 27) Let us see how the law is observed in those few parishes where the people are disposed to yield obedience to the letter of it. *The days for doing statute duty have long been looked upon as holidays,* as a kind of recess from the accustomed labour, and have been devoted to idleness, and its concomitant indulgences of riot and drunkenness.

(p. 28) Further, those doing statute duty (which is to some extent voluntary) are less obedient to the directions of those whom the law has appointed to superintend it, than is consistent with the due discharge of their duty. The men are working at four or five different places in the parish and not under the oversight of an officer, and not executing a well thought out plan.

Again, even if a surveyor were a good judge of roads, and of how to make and repair them, he has only six days to carry out his plan, and even then some may refuse to do their statute duty. How is the way to be amended under these conditions?

Some will say, let the surveyor apply the forfeitures incurred by the several defaults in the hire of teams and men and go on with his work. Very true, but first he is to get them. In order to do this, he is to enter on a new work, viz., to bring the defaulters to justice. And first he is to make out a list of their names, which, when completed, is to be returned to the Sessions, which may possibly be held either in a week or in four months after the offence. The Justices upon this return, of course, issue summonses for the defaulters to show cause in a reasonable time why they will not pay; after this, if they do not comply, distress-warrants are issued, before the execution of which the wet weather sets in, and there is an end of road work for that year. The surveyor is then busied in making up his accounts against the January Sessions, or perhaps in defending actions grounded on some irregularity in the notice, the due publication of which, or of the respective defaults, not one in fifty of them is ever prepared to prove. When January comes, his account is passed and he pays the balance to the new surveyor, who will have just the same difficulties to encounter as his predecessor.

(p. 30) Everyone knows that the highways shall be kept in repair by the several parishes of which they are part. Act 2 & 3 P. & M., c. 8, has established a form of proceeding by way of indictment against the parishioners, upon which, if the defendants are found guilty, they shall not be discharged by submitting to a fine, but a distraint shall go *in infinitum* till they repair.

(p. 31) Act 5 Eliz., c. 13, prescribes another method of proceeding—in effect much the same as that of an indictment—which is by a presentment of the surveyors to the next Justice, who is to certify the same at the next General Sessions, and the Sessions is *immediately* to inquire of the defaulters. But, despite the word "immediately," the general opinion on that clause of the statute is, that the certificate of the Justice in this case has not the effect of a presentment, but must be turned into an indictment, to which, by the Rules of Law, the offender may enter his traverse, and no trial can be had till the Sessions after.

Both these methods are objectionable. In the *first*, the law does not distinguish those who have done from those who have refused to do their statute work; but gives its judgment indiscriminately against the whole parish, and hence the innocent and the guilty are involved in the same punishment. By the *second*, there is not that expeditious justice which the statute gives reason to expect.

(p. 33) The delay and expense of these methods of proceeding are objections common to them both; and because of these reasons, as well as their inefficacy, they should either be abolished or so regulated as to be more effectual. Other complications have been introduced by Acts 22 Car. II, c. 12 (secs. 9 and 12) and 3 W. & M., c. 12 (sec. 9) in regard to what should be the mode of procedure for offences and neglects respecting the public roads.

(p. 34) Besides all this, Act 1 Geo. I, stat. 2, c. 52, is so worded that it seems doubtful whether all the authority given by former Acts as to these matters is not taken away, and the jurisdiction vested in the Special Sessions.

(p. 36) A surveyor, if he is a farmer, or engaged in some other like occupation, is very often ignorant of how the roads should be amended. What effects can we expect to follow from ignorance combined with authority on the one side (i.e., the surveyor) and invincible obstinacy on the other (i.e., the parishioners)?

(p. 37) *But the surveyors in general are not disposed to follow the law in executing their office.* One error they usually make is to consider the respective forfeitures for every day's default as a tax or rate; in consequence of which their practice is, as soon as they enter office, to assess (ex officio) every inhabitant a sum proportional to the labour required of him, which they proceed to collect as soon as possible—like the proper officers do the poor rates. But these several sums are not due until there has been a default to perform the statute labour. In this way the surveyor is open to an action at law for the sums thus collected from the parishioners; and, further, when the notice has been so negligently given, as that its publication cannot be proved by an uninterested witness (i.e., one not liable for statute work in the parish), who can swear to the reading of it by the Clerk, it is no blame if the parishioners do not go to the roads to work.

(p. 38) *Surveyors also are corrupt*, in commuting with parishioners for different amounts, receiving from some 5s., from others 4s., from others half-a-crown, etc., or what many of them like better, a bowl of punch. These things are punishable by a fine of £5, which the Justices in their Sessions have power to impose. Such evils and many others of like nature are largely owing to the practice of electing tradesmen, and persons in a situation necessarily dependent and subject to influence, into parochial offices. Inferior inhabitants get the offices; while the gentlemen, perhaps from contempt of an employment which requires little more than to be able to write and keep a year's accounts, or for other reasons, sit by and see the public defrauded and the law evaded.

Then Hawkins quotes from Burn on *Justice* (under title "Highways") saying that "Most of the books are remarkably confused under this title, occasioned by a multiplicity of statutes standing unrepealed, and yet altered perhaps five or six times or oftener, by succeeding statutes." Later, Burn says that there is no uniformity of action among the surveyors, and because each has the roads under his charge for at most six days, and his successor has other schemes and notions, the roads are never the better. Hence, it is but natural that the *people have a picnic on statute labour days.* Why should they not, when their work would be to no purpose?

District surveyors, he thinks, should be appointed, with salaries, to lay out the roads and attend and direct the work, and see that it is well done. He thinks this could be done with half the present legal maximum assessment of 6*d.* in the pound.

Burn's objections are two in number: first, the multiplicity of the laws, and second, the ill direction of the power given by them. The latter objection can be overcome by giving that power to those who have no temptation to abuse it; but the first objection he considers very serious.

(p. 43) Existing statutes relating to the amendment and repair of the highways are not fewer than twelve in number, made at different times as need required, and abounding in clauses which legal skill cannot reconcile. Clauses in older statutes have been left unrepealed, though such clauses were altered and amended by subsequent Acts; different penalties have been inflicted for the same transgression by different statutes. Thus the highway laws have so accumulated as to be a subject of universal complaint.

(p. 47) Hawkins then pleads for modifying the highway laws so as to remove inconsistencies. He would reduce all into *one* law, so as to be effective and easy in execution. In making such a law the following points at least should be looked to:

1. The burden on the public should be proportional to their circumstances and abilities to bear it.

2. Those who use the highways most should pay most.

3. The tax on gentlemen of large personal estates, who keep coaches, chariots, etc., should be adjusted in a compound ratio of their wealth and the use they are supposed to make of the highways.

4. Surveyors should be persons of greater property than are usually appointed to that office.

(p. 52) The advantages from the use of broad wheels, he says, are so apparent that it is needless to insist on them; "this is certain, that by means of them the price of carriage from York to London has been reduced forty per cent." He acknowledges that they do not succeed so well on cross-roads as on the great roads, because the former are usually so narrow as to admit of only one track. These ways should be widened, and the use of broad wheels made universal.

(pp. 61–143) In these pages, he draws up a Bill to suit the ends he has in view; and if it were passed all the old laws would be repealed and their useful provisions alone embodied in the new law.

APPENDIX 4

ON LETTING THE TOLLS

USUALLY the tolls were farmed out, and not managed by the trustees of the road on which they were to be collected. Parliament laid down the conditions for letting the tolls, which included the following: " To prevent fraud or any undue preference in the letting thereof, the Trustees are hereby required to provide a Glass with so much sand in it as will run from One End of it to the other in One Minute; which Glass, at the Time of letting the said Tolls, shall be set upon a Table, and immediately after every Bidding the Glass shall be turned, and as soon as the Sand is run out it shall be turned again, and so for Three Times, unless some other Bidding intervenes: And if no other Person shall bid until the Sand shall have run through the Glass for Three Times, the last Bidder shall be the Farmer or Renter of the said Tolls " (Hills, *History of East Grinstead*, p. 158, quoting from Act 3 Geo. IV, c. 126, sec. 55).

When the term for which the tolls had previously been let was near its expiration, the trustees of the turnpike trust usually announced in the newspapers of that locality that they would meet at a certain place, on a certain date, to again let the tolls of the turnpike gates which they controlled. This gave notice to those who wanted to bid for them to appear at that time. The advertisement would read something like the following, which appeared in the *Shrewsbury Chronicle*, Feb. 20, 1773, p. 2:

Notice is hereby given that "at a meeting of the Trustees, to be held at the Guildhall (Shrewsbury), on Tuesday the 23rd instant,...the Tolls arising on the Roads leading from Shrewsbury to Preston, Brockhurst, Shawbury, and Shreyhill, in the county of Salop, will be *let to the best bidder, for the term of three years,* commencing the second day of March next.

John Warren, Clerk to the Trustees."

It was the usual rule that the man to whom the tolls were leased had to pay for the first month in advance, as an evidence of good faith on his part (v. Act 3 Geo. IV, c. 126, sec. 56). But the following advertisement shows that this was not always required, if the lessee could give other satisfactory security. In the *Hereford Journal*, April 20, 1803, p. 1, we find:

"Hereford Turnpike Trust.

Notice is hereby given, that the next Meeting of the Trustees will be held, at the City-Arms Hotel, in the City of Hereford, on Tuesday, the Third day of May next, when the tolls arising from the several Turnpike-Gates belonging to this Trust will be Let by Auction to the best Bidder, who will be required to give security, to the satisfaction of the Trustees then present, for the performance of his or their contracts.

Particulars by applying to Mr. J. Coren, Clerk to the Trustees.

N.B. New Trustees will then be appointed.

Apr. 2, 1803."

Instead of leaving the notice in the above indefinite terms as to the price, it was common to stipulate what was the lowest amount for which the gates would be put up at auction (v. *Hereford Journal*, Dec. 4, 1805, p. 2; *Norfolk Chronicle and Norwich Gazette*, Aug. 27, 1814, p. 3); or else to give the amount for which they were let the preceding year (v. *Newcastle Courant*, April 16, 1774, pp. 2, 3); or, in some cases, in addition to the gross revenue, to give the amount which the toll realized, clear of the expenses of collection, during the last year or term of years (v. *British Volunteer and Manchester Weekly Express*, April 27, 1822, p. 1; Felix Farley's *Bristol Journal*, Feb. 10, 1787, p. 3). It is evident that, by these means, the tolls would tend to progressively increase from year to year. Sometimes the lease would be made for only one year, or two years, or three years, or the trustees might give the lessee an option on this (v. *Newcastle Courant*, April 16, 1774, p. 3).

The method of letting the tolls, as told by one who had seen the process (v. Fowler, *Records of Old Times*, pp. 18–20), well illustrates why it was that the tolls did not continuously increase. The trustees, who were mostly country gentlemen of the district, gathered at the principal inn at the county town in considerable numbers. Usually the trustees gave a guinea for each gate let, to be expended in refreshments; and as there were generally six or eight gates, there would be that number of guineas spent among about forty or fifty "pikers," as they were called, who attended, but only about six or seven of these would be bidders and lessees; *these were men of capital*, who rarely collected their own tolls. Those who witnessed these meetings called them the "Whispering Society," as the company scattered about the inn yard in small groups were in full conclave, all in whispers; one would run off and whisper to another group and return again, when they would be approached by another envoy. while circulating rapidly among them was one of the *bona fide* bidders, evidently making terms with several threatening opponents and promising from one to five pounds to the recipient who kept from bidding. At the appointed time, a rush was made to the auction room, where the trustees, with their clerks, treasurers. surveyors and other officers were assembled. After the conditions were read the letting commenced, but it sometimes happened that the whispering had been so effective, that not a single offer was made, to the astonishment of the trustees, who had not seen the manœuvres that had been going on in the yard for more than an hour. As no biddings were made, it was then announced that the upset price was (say) £200 for each gate, and that unless that sum were obtained the gates would be withdrawn and the trustees would put in their own collectors and farm the gates themselves. When the sum was announced, a general groan of horror went round, and the trustees were told that the offer was so outrageous it could not be listened to; that the last two years the gates had not produced more than £180 to £190, and that the lessees had lost all their wages and expenses, but if they would listen to reason a tenant could be found at £150. Suddenly some stranger to the "pikers," a decoy put up by the auctioneer, would bid £180, at which there would be a burst of indignation and outbreak of insulting by-play. By a continuous series of "card-playings," the bidders would keep down the prices of the gates to about the £200; and very often the former lessee who had declared that he had lost so much by taking the gates for the last two years, was anxious to again have them since they had really been profitable to him. All the whispering that had taken place beforehand represented an endeavour to buy off every dangerous opponent. Many persons came down from London and elsewhere, under pretence of taking the gates, who earned a sovereign or even £3, as payment for the day's work, from the lessee, who had probably held the gates for the past two or three years and was reluctant to lose them. Other evils connected with letting the tolls are given in Pagan, *Road Reform*, pp. 173–6, and by James and William Macadam in *Parl. Papers*, 1833 (703), xv, 409, pp. 497, 555.

The business of contracting to take leases of turnpikes was in many instances a very expensive one. The gentleman who took most of the gates in Buckinghamshire and some adjoining counties was a Mr Tongue, living at Manchester, and it was estimated that he had over £50,000 annually embarked in gate holdings. He retained a regular staff of collectors, who moved about from one part of the country to another as his confidential servants (Fowler, op. cit., pp. 18–20). Sometimes individuals, who were in the habit of hiring the tolls to a large amount, united into a company and leased a great number of gates, until they had from £100,000 to £200,000 a year embarked in this kind of investment. Because of this monopoly of tolls, it often happened that upon two parallel lines of road in the control of the same lessee, the one paying the lower toll would be sacrificed to the other paying the higher toll; *Parl. Papers*, 1833 (703), xv, 409, 'Second Report of the Lords Committee on Turnpike Returns,' p. 497; *Hansard's Parliamentary Debates*, 1835, xxix, pp. 1183–92.

APPENDIX 5

RATE OF TRAVELLING, 1750–1830

I HAVE endeavoured to bring together in the following statistical table only such data as are most authentic, and to indicate in each case the source of the information, so that it may be easily verified. It must not be thought that the matter here presented is absolutely accurate, for the writer makes no claim to such precision; as a matter of fact, it has been impossible to secure even correct distances between places, because we have no measurements of the roads which give us this information with guaranteed accuracy. Further, the changing and straightening of the roads, accompanied sometimes by slight changes in the coaching routes, would vitiate any series of mileage figures which we might have. These things I have taken into account in the computation of the mileage and it will be seen that the distance sometimes varies; for between two places the length of the road differed, according to the route taken and the straightness or crookedness of the road. The same difficulties appear with reference to time, for a day at one part of the year or with one person, did not mean the same as at another part of the year or with another person; for example, days in summer were long, while the days in winter were short. It was not until the coaches were timed by hours and minutes that we get accuracy in this particular. In some instances, the length of time required to perform a journey included the time spent at nights in the inns along the route; but we have no knowledge of how much time was thus consumed. With all these liabilities to error, and others which we need not here mention, it will be apparent that the best we can get is an approximation to the truth. The great amount of statistical material presented is intended to avoid any errors due to paucity of data upon which conclusions might be based; and, making all due allowance for these variations, it is claimed that the statistics are as reliable as the available information will permit. The inferences drawn from them will be found in the text.

Year	Termini	No. of miles	Time required	No. of miles per hour	No. of miles per day	Source of Authority	Remarks
1748	Ipswich—London	70	one day	—	70	Defoe, *Tour through the Whole Island of Great Britain*, 1, p. 80	
1751	London—Dover	75	A little more than a day	—	about 70	*London Evening Post*, Mar. 28, 1751	
1752	London—Newbury	57	12 hours	4¾	—	Advertisement of coach given in Money, *History of Newbury*, p. 388	
	(Unknown)	over 126	2 days	—	68	Brit. Mus. 10,349. g. 11, 'A Journey through England,' p. 82	
	London—Bath	110	2 days	—	55	Ibid., p. 189	Left London 2 a.m.; stayed at Newbury next night about 8 hrs; left Newbury about 2 a.m., and reached Bath 7 p.m.—two long days of summer
	Birmingham—London	110	2 days summer	—	55	Aris's *Birmingham Gazette*, Mar. 30, 1752, p. 4	
	Birmingham—London	110	3 days summer	—	87	Ibid., April 13, 1752, p. 2	This was the "Shrewsbury and Birmingham Caravan," and was probably a heavy vehicle that carried some freight too
	Shrewsbury, via Birmingham, to London	158	4 days summer	—	40	Ibid.	
1754	London—York	200	4 days	—	50	*Archaeologia Aeliana*, N.S., III, p. 247	
	London—Newcastle	290	6 days	—	48		

Year	Route	Distance	Time		Fare	Source	Notes
1754	London—Edinburgh	400	10 days summer / 12 days winter	— —	40 summer 33 winter	Advertisement of coach given in Malet, *Annals of the Road*, p. 18, as quoted from the *Edinburgh Courant*, 1754	
	Manchester—London	200	4½ days	—	45	Malet, *Annals of the Road*, p. 12	Daily in summer
	Chelmsford—London	31	5 hours	6 +summer	—	Advertisement of coach given in Thrupp, *History of Coaches*, p. 107; taken from *Ipswich Journal*, Aug. 1754	
1756	London—Brighton	57	1 day	—	57	Hills, *History of East Grinstead*, p. 147	Went three times a week; returned alternate days
	Birmingham, via Stratford, to London	120	2½ days winter	—	48	Aris's *Birmingham Gazette*, Nov. 15, 1756, p. 4	The Birmingham and Stratford Stage Coach
	Birmingham, via Warwick, to London	110	3 days winter	—	37	Ibid.	The Birmingham and Warwick Old Stage Coach
1757	London—Lewes	56	1 day	—	56	Blew, *Brighton and Its Coaches*, p. 35	
	Warrington—London	200	3 days	—	67	Williamson's *Liverpool Advertiser*, June 9, 1757	Took from Monday morning till Wednesday evening
	Liverpool—London	210	3 days	—	70	Malet, *Annals of the Road*, p. 14, quoting from the advertisement	
	Birmingham, via Enstone and Oxford, to London	120	2 days	—	60	Brit. Mus. MSS. 23,001, 'Dr Pocoke's Travels in England,' III	
	Birmingham, via Stratford, to London	120	2 days summer	—	60	Aris's *Birmingham Gazette*, May 9, 1757	
1760	Manchester—London	200	3 days	—	67	Axon, *Annals of Manchester*, p. 93	

Year	Termini	No. of miles	Time required	No. of miles per hour	No. of miles per day	Source of Authority	Remarks
1760	London—Leeds	190	4 days	—	47	Advertisement of Glanville's Coaches in Ward's *Sheffield Public Advertiser*, Nov. 4, 1760	This was doubtless winter rate
	London—Sheffield	150	3 days	—	50		
	Birmingham—Bristol	86	2 days summer	—	43	Aris's *Birmingham Gazette*, Apr. 28, 1760, p. 2	
1761	London—Leeds	190	3 days summer	—	63	*Leeds Intelligencer*, May 19, 1761, p. 3	Coach advertisement. The passengers slept the two nights at Northampton and Nottingham
	London—York	200	3 days summer / 4 days winter	— / —	67 summer / 50 winter	*Archaeologia Aeliana*, N.S., III, p. 248	
1762	Newcastle—London	290	5 days	—	58	Ibid.	
	Plymouth—Exeter	40	12 hours	3¼	—	Worth, *History of Plymouth*, p. 340	
1763	Newcastle—Edinburgh	105	2 days	—	52	*Archaeologia Aeliana*, N.S., III, p. 249	Left Newcastle in the morning, stayed that night at Kelso; reached Edinburgh next evening
1764	York—Newcastle	90	1 day	—	90	Ibid., p. 248	
	Leeds—London	190	4 days	—	47	Coach advertisement, quoted in Parsons, *History of Leeds*, I, p. 180	
	Shrewsbury—London	158	2 days summer	—	79	Owen and Blakeway, *History of Shrewsbury*, I, p. 515	Went 3 times a week, but only in summer. All former coaches went only once a week

Year	Route	Miles	Time			Authority	Notes
1765	Newcastle—London	290	3 days summer	—	96 summer	Coach advertisement, quoted in *Archaeologia Aeliana*, N.S., III, p. 248	Slower rate in winter. Nights sometimes cut short for sleep
	Bath—London	110	ca. 28 hours	4¾ to 5	—	Brit. Mus., Add. MSS. 27,828, p. 9, gives the advertisement of this coach	Take out 10 hours for the night at Andover, and we have 28 hours left
1766	Liverpool—London	210	2 days summer 3 days winter	— —	105 summer 70 winter	Picton, *Memorials of Liverpool*, II, p. 116, quoting from first Liverpool directory	
1769	Leeds—London	190	2½ days	—	76	Parsons, *History of Leeds*, I, p. 130	
1772	Shrewsbury—London	158	1½ day	—	105	Owen and Blakeway, *History of Shrewsbury*, I, p. 515	This was during the summer only
	Manchester—London	200	2 days summer 3 days winter	— —	100 summer 67 winter	*Manchester Collectanea*, in Chetham Society Publications, LXVIII, p. 127	
1773	Manchester—Liverpool	35	12 hours	3	—	Axon, *Annals of Manchester*, p. 102	.
1774	Shrewsbury, via Birmingham, to London	158	1½ day	—	105	*Shrewsbury Chronicle*, July 9, 1774, p. 1	During the summer only
	Newcastle—York	90	ca. 16 hours	5½	—	*Newcastle Courant*, April 16, 1774, p. 2	The advertisement gives the time as 18 hours, which includes two stoppages (say 1 hour each) for meals
	London—York	200	2 days	—	100	Ibid.	
	Edinburgh—London	ca. 400	4 days	—	ca. 100	Ibid., p. 4	
	Southampton—London	80	14½ hours	5¼	—	*Southampton Guide*, 1774, p. 68	
	Liverpool—Preston and return	68	1 day	—	68	*Williamson's Liverpool Advertiser*, June 24, 1774	

Year	Termini	No. of miles	Time required	No. of miles per hour	No. of miles per day	Source of Authority	Remarks
1775	Birmingham, via Oxford, to London	120	18 hours	ca. 7	—	*Shrewsbury Chronicle,* April 15, 1775, p. 3	Both these were in summer. Latter road was stony and hilly
	London—Oxford	60	9–9¼ hours	6¼–6⅔	—	Brit. Mus., Add. MSS. 17,398, pp. 50, 102	
1776	Hereford—Monmouth	17¼	8 + hours	5¼	—	Ibid., p. 58	
	Newcastle—London	290	3 days	—	97	Armstrong, *Post Roads between London and Edinburgh,* 1776, p. 5	
	Newcastle—Edinburgh	105	1 day summer / 1¼ day winter	—	105 summer / 70 winter	Ibid.	Went 3 times a week
	London—York	200	36 hours	5⅔	—	Ibid.	Flying coach
	York—Newcastle	90	1 day	—	90	Ibid.	Went 3 times a week
	York—London	200	2 days	—	100	Ibid.	York Old Coach and York Diligence
	London—Stamford	90	1 day	—	90	Ibid., p. 6	It probably did not take a full day's time, for it left London at 10 p.m.
	London—Leeds	190	2 days	—	95	Ibid.	Night coach; left London twice a week
	London—Lincoln	131	1 day	—	131	Ibid.	Stage coach or post coach
	London—Bristol	120	2 days	—	60 winter	Felix Farley's *Bristol Journal,* Nov. 2, 1776, p. 2	
1777	Bristol—Bath	11	2 hours	5¼	—	Bonner and Middleton's *Bristol Journal,* Jan. 18, 1776, p. 1	
	Bristol—London	120	18 hours	ca. 7	—	*Bristol Gazette and Public Advertiser,* Aug. 7, 1777, p. 1	
	Exeter—Bath	85	15 hours	ca. 6	—	Ibid.	

Year	Route	Miles	Time	m.p.h.		Source	Notes
1777	Exeter—Bristol	75	18 hours	ca. 6	—	Ibid.	Coach advertisement
	Bath—Exeter, via Bridgwater and Taunton	80	15 hours	5¼	—	Ibid.	Coach advertisement
1778	Bristol—Exeter	75	13 hours	ca. 6	—	Ibid.	Coach advertisement
	Birmingham—Bristol	86	1 day or 16 hours	5¼	86	Aris's *Birmingham Gazette*, Mar. 30, 1778, p. 1	Went 3 times a week
	Birmingham—London	110	1¼ day summer	—	73	Ibid., April 20, 1778, p. 3	"The Old and Original Shrewsbury, Wolverhampton and Birmingham Fly."
	Shrewsbury, via Oxford, to London	160	1¼ day summer	—	107	Ibid.	"The Shrewsbury New Fly,"
	Birmingham—Sheffield	73	1 day summer	—	73	Ibid., May 25, 1778, p. 2	"The Birmingham and Sheffield Post Coach"
	Coventry—London	90	16 hours summer	5⅜	—	Ibid.	Post coach, daily except Saturday
	Birmingham, via Oxford, to London	120	18 hours summer	6¼	—	Ibid., Oct. 5, 1778, p. 2	
	Birmingham, via Oxford, to London	120	1 day summer	—	120	Ibid.	Same speed as the one above, but went only 3 times a week
1779	Bristol—Weymouth	72	12 hours	6	—	Felix Farley's *Bristol Journal*, Aug. 14, 1779, p. 1	
	Bristol—London	120	1 day	—	120	Ibid.	This must have meant from 22 to 24 hours
	Bristol—London	120	2 days	—	60	Ibid.	This post coach probably stopped over night on the road
	London—Glasgow or Portpatrick	400	4 days	—	100	Advertisement given in full in Malet, *Annals of the Road*, p. 24	
	Shrewsbury—London	158	31 hours summer	5 +	—	Aris's *Birmingham Gazette*, April 12, 1779, p. 1	
	London—Birmingham	110	18–19 hours summer	6	—	Ibid.	
	Shrewsbury—Birmingham	46	9 hours summer	5 +	—	Ibid.	

Year	Termini	No. of miles	Time required	No. of miles per hour	No. of miles per day	Source of Authority	Remarks
1780	Birmingham—Manchester	85	1 day summer	—	85	Aris's *Birmingham Gazette*, April 10, 1780, p. 1	3 times a week (coach)
	Birmingham—London	110	19 hours summer	6	—	Ibid.	3 times a week (diligence)
	Stroud—London	110	1 day	—	110	Jackson's *Oxford Journal*, July 20, 1780, p. 3	Coach advertisement
1781	Birmingham—Manchester	85	13 hours summer	6½	—	Aris's *Birmingham Gazette*, Aug. 20, 1781, p. 1	Post coach
	Liverpool—London	210	40 hours	5	—	Picton, *Memorials of Liverpool*, II, p. 116	Daily
	Liverpool—London	210	2 days	—	105	Ibid.	These were the slower coaches; one daily and two 3 times a week
	Stroud—London	110	2 days	—	55	Jackson's *Oxford Journal*, Jan. 13, 1781, p. 3	
	London—Wantage	68	11 hours	6	—	Ibid., April 28, 1781, p. 4	Coach advertisements
	Oxford—Bath	68	11 hours	6	—	Ibid., June 9, 1781, p. 2	
	Manchester—Birmingham	85	15 hours	ca. 6	—	*Manchester Collectanea*, in Chetham Society Publications, LXVIII, p. 135	
1782	Birmingham—Bristol	86	14 hours winter	6	—	Aris's *Birmingham Gazette*, Jan. 7, 1782, p. 1	
	Birmingham—London	110	22 hours winter	5	—	Ibid.	
	Chester—London	180	2 days summer 2½ days winter	—	90 summer 72 winter	Hall, *History of Nantwich*, p. 233, quoting from *Chester Guide Book* of 1782	
	Oxford—Birmingham	60	23 hours	3	—	Jackson's *Oxford Journal*, Nov. 12, 1782, p. 4	Coach advertisement

Year	Route		Time			Source	Remarks
1782	London—Bristol	120	1 day	—	120	*Constitutional Chronicle,* Jan. 24, 1782, p. 1	Coach advertisement
	Bristol—London	120	18 hours	6¾	—	Sarah Farley's *Bristol Journal,* Aug. 8, 1782, p. 2	Coach advertisement
1783	Bristol—Birmingham	86	1 day	—	86	Felix Farley's *Bristol Journal,* Jan. 18, 1783	This probably meant 18 to 20 hours at least, for it was before Palmer's coaches started
1784	London—Bristol	120	1 day	—	120	Bonner and Middleton's *Bristol Journal,* April 17, 1784	These were Palmer's coaches
	London—Bristol London—Bath	120 110	16 hours 14 hours	7½ ca. 8	— —	Felix Farley's *Bristol Journal,* Oct. 2, 1784, p. 1	
	London—Liverpool	210	30 hours	7	—	Gore's *Liverpool Advertiser,* July 22, 1784; Baines, *Hist. of Liverpool,* p. 468	
	Bath—London	110	1 day	—	110	*The New Bath Guide,* 1784, p. 72	So there were both one and two-day coaches at the same time
	Bath—London	110	2 days	—	55	*Ibid.,* p. 73	
1785	London—Chester Chester—Holyhead	180 60	2 days 1¼ day	— —	90 48	*Morning Chronicle and London Advertiser,* Jan. 8, 1785, p. 4	Coach advertisements
	Newcastle—London	280	less than 2 days	—	145 +	*Archaeologia Aeliana,* N.S., III, p. 249	Left Newcastle 10 p.m.; got to London at close of third day
	Bristol—Portsmouth	88	15 hours	6	—	Bonner and Middleton's *Bristol Journal,* May 7, 1785	Palmer's mail coach. On cross post roads he could not go so fast
	Birmingham—London	110	19 hours	6	—	Brit. Mus., Add. MSS. 27,828, p. 11	
	London—Norwich	108	15 hours	7 +	—	Mason, *History of Norfolk,* I, p. 10; Bayne, *History of Norwich,* p. 282	Palmer's mail coaches. They brought mail from London a day sooner than by the old means

Year	Termini	No. of miles	Time required	No. of miles per hour	No. of miles per day	Source of Authority	Remarks
1786	London—Chester	180	2 days	—	90	*General Advertiser*, Oct. 20, 1786, p. 4, and Nov. 14, 1786, p. 4	Lay at Coventry the intervening night for 9 hours. This post coach carried only 4 insiders and a servant on the box
	London—Chester	180	30 hours	6	—	Ibid., Oct. 20, 1786, p. 4	
	Chester—Holyhead	60	1 day	—	60	Ibid.	
	London—Chester	180	37 hours	5	—	Ibid., Dec. 8, 1786, p. 1	There was also a two-day coach that went very leisurely
	London, via Oxford, to Birmingham	120	16 hours	7¼	—	Ibid., Dec. 22, 1786, p. 1	Coach advertisement
	London—Plymouth	215	28 hours	7¾	—	*St James Chronicle*, Dec. 7, 1786, p. 2	
1787	London—Sheffield	150	26 hours	6	—	Leader, *Sheffield in the Eighteenth Century*, p. 100	This journey in 1760 took 3 days
	London—Manchester	195	1 day summer	—	195	*General Advertiser*, May 21, 1787, p. 4	
	Bristol—London	120	18 hours	ca. 7	—	*Felix Farley's Bristol Journal*, Jan. 6, 1787, p. 1	
	Bristol—Birmingham	86	15 hours	5¾	—	Ibid., July 28, 1787, p. 1	
1788	Manchester—London	195	28 hours	7	—	*Manchester Collectanea*, in Chetham Society Publications, LXVIII, p. 158	Palmer's mail coach
	Shrewsbury—London	158	ca. 22 hours	7 +	—	Advertisement given in *Salopian Shreds and Patches*, I, p. 55	

		Miles					Post coach
1788	Oxford—Bristol (via Farringdon, Fairford, Cirencester, Tetbury, Sodbury)	72	11½ hours	6	—	Felix Farley's *Bristol Journal*, May 17, 1788, p. 3	Post coach
	Bristol—Birmingham	86	14 hours	6	—	Ibid.	
	Birmingham—Manchester	85	1 day	—	85	*Bristol Gazette and Public Advertiser*, May 15, 1788, p. 3	Palmer's mail coach Slept at Newcastle on the way
	Birmingham—Manchester	85	27 hours (16 hours)	5¾	—	Ibid.	If we take, say, 11 hours out as the time occupied in resting at night, it will leave 16 hours on the road
1789	London—Bristol	120	16 hours	7⅛	—	*Bath Chronicle*, April 17, 1788, p. 2	Some other coaches took 1½ day
	Chester—London	180	36 hours	5	—	Hall, *History of Nantwich*, p. 233, quoting from *Chester Guide Book* of 1789	
	Bath—Exeter	80	11½ hours	7	—	*Bath Chronicle*, Jan. 8, 1789, p. 1	
	Bath—London	110	1 day	—	110	Ibid., p. 2	
	Bath—London	110	1½ day	—	73		
	Bath—London	110	2 days	—	55		
1790	Oxford—Southampton	38	12 hours	3 +	—	*Jackson's Oxford Journal*, June 5, 1790, p. 4	
	London—Brighton	57	9 hours	6½	—	Blew, *Brighton and Its Coaches*, p. 39	
1791	London—Salisbury	88	8 to 10 hours	9 to 11	—	Brit. Mus., Add. MSS. 28,570, p. 38, June 15	This was summer. Left London soon after 10 a.m.; supped and slept at Salisbury
	Salisbury—Exeter	90	8 to 10 hours	9 to 11	—	Ibid., June 16	Left Salisbury soon after 10 a.m.; supped and slept at Exeter

Year	Termini	No. of miles	Time required	No. of miles per hour	No. of miles per day	Source of Authority	Remarks
1791	Exeter—Truro	88	9 to 11 hours	8 to 9½	—	Brit. Mus., Add. MSS. 28,570, p. 38, June 17	Left Exeter soon after 9 a.m.; supped and slept at Truro
	Hull—London	185	3 days	—	62	Battle, *Hull Directory*, 1791, pp. 65–73	
	Dover—Canterbury	16	a little over 2 hours	ca. 8	—	Brit. Mus. 567. e. 7, 'Beyträge zur Kenntniss von England', 1, p. 7	
	London—Brighton	57	8 hours	7	—	Blew, op. cit., pp. 42, 60	Coach advertisement
	London—Shrewsbury	158	20 hours	5½	—	*Morning Chronicle*, April 29, 1791, p. 1	
	London—Bristol	120	17 hours	7	—	Ibid.	Advertisement said this was the only coach that ran through to Bristol in a day
	London—Bath	110	15 hours	7	—	Ibid.	
	London—Leeds	190	28½ hours	ca. 7	—	*Morning Post and Daily Advertiser*, July 20, 1791, p. 1	
1792	London—Cambridge	55	8 hours	ca. 7	—	Ibid.	
	Lichfield—Birmingham	17	3½ hours	5	—	Brit. Mus., Add. MSS. 30,173, p. 22	
1793	? —Alcester	10	2 hours	5	—	Ibid., p. 23	
	London—Southampton	80	10–12 hours	6½–8	—	Baird, *Agriculture of Middlesex*, p. 35	
	London—Cambridge	55	6½ hours	9	—	*The World*, Sept. 18, 1793, p. 4	
	London—Wisbeach	98	18 hours	7½	—	Ibid.	
	London—Bath	110	15 hours	7	—	Ibid.	
1796	London—Cambridge	55	8 hours	7	—	*Cambridge Directory*, 1796, p. 159	London and Cambridge Diligence

Year	Route	Miles	Time	Speed (mph)		Source	Royal Mail Coach
1796	London—Cambridge	55	7¼ hours	7¼	—	Ibid., p. 159	
	Wisbeach—Cambridge	34	8¼ hours	4	—	Ibid., p. 159	
	Cambridge—London	55	10 hours	5¼	—	Ibid., p. 160	
	Cambridge—Birmingham	105	2 days	—	52¼	Ibid., p. 161	"The Fly" Post coach
	York—London	200	81 hours	ca. 7	—	York Guide, 1796, p. 43	
1797	Sidmouth—Exmouth	10	7 hours	1½	—	Brit. Mus., Add. MSS. 28,798, p. 46	Sept. 29, 1797
	Oakhampton—Launceston	20	4 hours	5	—	Ibid., p. 57	
	Launceston—Bodmin	21	4 + hours	5	—	Ibid., p. 63	Nov. 1797
	Penzance—Helston	14	2 + hours	7	—	Ibid., p. 124	
1798	Edinburgh—London	400	3 nights and 2 days	7	160	Harris, The Coaching Age, p. 279	I have called this 2¼ days of 24 hours each Rates include all stoppages. Difference is due to difference in amount of time consumed in stoppages
	—	—	—	5 (heavy C.) 6 (light C.) 7 (mail C.)	—	Middleton, View of the Agriculture of Middlesex, p. 394	
1800	Brighton—London	58	9–10 hours summer	6–6¼	—	Brighthelmston Guide, 1800, p. 82	Distance is given as 57–59 miles
1805	Manchester—Kendal	75	15 hours	5	—	Brit. Mus., Add. MSS. 30,929, p. 16	Time taken was from 7 a.m. to 12 p.m., or 17 hours. But take out an hour each for dinner and supper
1808	Birmingham—Nottingham	50	9–9½ hours winter	5–5¼	—	Aris's Birmingham Gazette, Feb. 15, 1808, p. 1	
1810	—	—	—	ca. 6	—	Simond's Travels in Great Britain, 1, p. 16	"Our rate of travelling does not exceed six miles an hour, including stoppages, but we might go faster if we desired it"
	Bath—Ilfracombe	100	2 days	4–4½	—	Jackson, The Bath Archives, 1, p. 182	Hills were steep and frequent, the weather was warm, and they were heavily laden

Year	Termini	No. of miles	Time required	No. of miles per hour	No. of miles per day	Source of Authority	Remarks
1810	London—Liverpool	210	31 hours	7	—	*Manchester Guardian,* April 3, 1830, p. 3	
1812	London—Bath	110	2 days	—	55	*Diaries and Letters of George Jackson,* pp. 382, 387	This certainly was extremely easy travelling on such a road
	Hull—Manchester (via York and Leeds)	105	14 hours	7½	—	*York Herald, County and General Advertiser,* July 25, 1812, p. 1	Stayed, say, 2 hours at Leeds for dinner
1813	Cambridge—Leicester (via Huntingdon, Stilton, Stamford and Uppingham)	75	12 hours	6¼	—	*Cambridge Chronicle and Journal,* Oct. 29, 1813, p. 1	
	Leicester—Birmingham	40	7–8 hours	5¾	—	Ibid. Hills, *History of East Grinstead,* p. 152, quoting from Cary's *Itinerary of the Great Roads,* 1815	
1815	London—East Grinstead	30	5 hours	6	—		
	London—Brighton	57	10 hours	ca. 6	—┊		
	Cambridge—London	55	7 hours	8	—	*Cambridge Chronicle and Journal,* Oct. 27, 1815, p. 3	
1816	London—Brighton	57	6 hours	9½	—	Blew, *Brighton and Its Coaches,* pp. 103–4, 137, 138	
1818	Shrewsbury—London	158	16 hours	ca. 10	—	*Salopian Shreds and Patches,* I, p. 55	
	York—London	200	29¾ hours	6¼	—	Hargrove, *History of York,* 1818, Vol. II, Pt. II, pp. 671–5	Mail coaches. York to Liverpool was a cross-road, not one of the main thoroughfares
	York—Liverpool	90	18¾ hours	5	—		
	York—Edinburgh	ca. 200	29¾ hours	6¼	—		

1818	York—London	200	$30\frac{1}{2}$–31 hours	$6\frac{1}{2}$	—	Ibid.	These were the stage coaches
	York—Edinburgh	ca. 200	30 hours	$6\frac{1}{2}$	—	Ibid.	
1819	Bristol—Exeter	75	14 hours	less than $5\frac{1}{2}$	—	Latimer, *Annals of Bristol in the Nineteenth Century*, p. 84	This was not one of the great thoroughfares
	Liverpool—Nottingham	100 +	$8\frac{1}{2}$ hours	12	—	*Liverpool Mercury*, July 9, 1819	
	Nottingham—Liverpool	100	9 hours	11	—	*Hampshire Telegraph and Sussex Chronicle*, July 19, 1819, p. 3	The paper called this a rate of 14 miles per hour, but that was erroneous
1820	Bristol, via Birmingham, to Manchester	170	2 days or 24 hours	7	—	Latimer, *Annals of Bristol in the Nineteenth Century*, p. 84	Coach stopped at Birmingham over night; so that the time spent in travel would not be more than 24 hours
1821	London—Dover	75	10 hours	$7\frac{1}{2}$	—	*The Times*, July 28, 1821, p. 2	
	London—Manchester	185	19 hours	ca. 10	—	Ibid., Oct. 8, 1821, p. 3	
	London—Southampton	80	9 hours summer	9	—	*The British Traveller*, July 20, 1821, p. 1	
	London—Liverpool	210	30 ,, ,,	7	—	Ibid.	
	London—Leeds	190	26 ,, ,,	$7\frac{1}{3}$	—	Ibid.	
	London—Birmingham	110	17 ,, ,,	$6\frac{1}{2}$	—	Ibid.	
	London—Manchester	185	26 ,, ,,	7 +	—	Ibid., Oct. 12, 1821, p. 1	These were three different coaches
	London—Manchester	185	27 ,, ,,	ca. 7	—	Ibid.	
	London—Manchester	185	24 ,, ,,	ca. 8	—	Ibid.	
	London—Liverpool	210	30 ,, ,,	7	—	Ibid.	
	London—Exeter	ca. 175	26 ,, ,,	$6\frac{2}{3}$	—	Ibid.	
	London—Nottingham	120	18 ,, ,,	$6\frac{2}{3}$	—	Ibid.	
	London—Norwich	108	14 ,, ,,	$7\frac{2}{3}$	—	Ibid.	
	London, via Northampton, Market Harborough and Leicester, to Derby	130	18 ,, ,,	$7\frac{1}{4}$	—	Ibid.	

Year	Termini	No. of miles	Time required	No. of miles per hour	No. of miles per day	Source of Authority	Remarks
1821	Birmingham—London	110	15 hours winter	7⅓	—	Aris's *Birmingham Gazette*, Jan. 8, 1821, p. 1	These were all different coaches
	Birmingham—London	110	15 hours summer	7⅓	—	Ibid., April 2, 1821, p. 3	
	Birmingham—London	110	18½ hours winter	6	—	Ibid., Jan. 8, 1821, p. 1	
	Birmingham—London	110	15½ ,, ,,	7	—	Ibid.	
	Birmingham—Oxford	60	7¾ ,, ,,	ca. 8	—	Ibid.	Two different coaches
	Birmingham—Oxford	60	8½ ,, ,,	7	—	Ibid.	
	Birmingham—Liverpool	98	13 hours summer	7½	—	Ibid., Oct. 15, 1821, p. 1	Two different coaches
	Birmingham—Liverpool	98	13 ,, ,,	7½	—	Ibid.	
	London—Manchester	185	19 ,, ,,	ca. 10	—	Ibid.	
	Birmingham—Manchester	85	10½ hours winter	8	—	Ibid., Nov. 19, 1821, p. 1	
1822	Shrewsbury—London	158	18 hours	9	—	Owen and Blakeway, *History of Shrewsbury*, I, p. 519	
	Birmingham—Bath	88	12½ hours summer	7	—	Aris's *Birmingham Gazette*, April 15, 1822, p. 1	
1823	Birmingham—London	110	14 ,, ,,	8	—	Ibid., July 21, 1823, p. 1	
1824	Manchester—London	185	ca. 22 hours	ca. 9	—	*Manchester Gazette*, July 24, 1824, p. 1	
	Manchester—Birmingham	85	11¼ hours	ca. 8	—	Ibid.	
1825	Colchester—London	56	ca. 6 hours	9 +	—	Cromwell, *History of Colchester*, p. 408	
	Birmingham—London	110	13 hours winter	8½	—	Aris's *Birmingham Gazette*, Jan. 10, 1825, p. 1; Jan. 31, 1825	This was called "superior travelling"
	Birmingham—London	110	12¼ ,, ,,	9	—	Ibid., Jan. 31, 1825	
	Birmingham—London	110	13½ ,, ,,	8 +	—	Ibid.	
	Birmingham—London	110	14½ ,, ,,	7⅗	—	Ibid.	
	London—York	200	24–25 hours winter	8	—	*London Magazine*, N.S., I, 1825, p. 36	

Year	Route	Miles	Time	m.p.h.		Source	Notes
1826	Liverpool—Manchester	35	2 hrs 32 mins	14	—	Baines, *History of Liverpool*, p. 624	
	London—Liverpool	210	26 hours	8 +	—	*The Times*, April 8, 1826, p. 1	
1828	Oxford—London	60	6 hours	10	—	Coach advertisement given in Oxford Historical Society, *Collectanea*, IV, p. 278	
	London—Exeter	ca. 175	20 hours	8¾	—	Brit. Mus., Add. MSS. 27,828, p. 17	This speed was not hard upon the horses
	London—Brighton	57	7 hours	8	—	Blew, *Brighton and Its Coaches*, p. 158	It would seem that the earlier six-hour coaches were not fully successful
1829	Gloucester—London	120	12 hours	10	—	Counsel, *History of Gloucester*, p. 209	
	Birmingham—London	110	12 hours	9	—	*Birmingham Journal*, Oct. 17, 1829, p. 1	
1830	London—Dunstable	40	3 hrs 50 mins	10	—	Smith, *Dunstable*, p. 112	The straightening of this road had doubtless shortened the distance
	London—Manchester	185–200	20 hours	9¼–10	—	Advertisement of "Beehive" coach given in Harris, *Old Coaching Days*, pp. 43–45	
	London—Birmingham	109	7 hrs 35 mins	ca. 15	—	*Manchester Guardian*, May 8, 1830, p. 3	Special occasion
	Liverpool—Manchester	35	4 hours	9	—	*Annual Register*, 1832, p. 445	By furious driving it could be done in 2¼ hours. See under year 1826
	Liverpool—Manchester	35	3¼ hours	10	—	Baines, *History of Liverpool*, p. 575	
	Birmingham—London	110	12 hours summer	9 +	—	Aris's *Birmingham Gazette*, May 24, 1830, p. 1	
	Birmingham—London	110	13 " "	8½	—	Ibid.	
	Birmingham—Liverpool	98	11 " "	9	—	Ibid.	
	Liverpool—Manchester	35	4 " "	9	—	*Great Western Railway. Evidence on the London and Birmingham Railway Bill*, p. 8, testimony of Henry Booth	

Year	Termini	No. of miles	Time required	No. of miles per hour	No. of miles per day	Source of Authority	Remarks
1880	Coventry—London	90	10 hours	9	—	West, *History of Warwickshire*, p. 782	It will be noted that there is much diversity between these figures and those given by Harris below. And yet different kinds of coaches travelled at different rates which may account for the diversity
	Coventry—Manchester	100	13¼ hours	7⅜	—		
	Coventry—Cambridge	81	1 day	—	81		
1831	Manchester—Leeds	45 {	7–8 hrs, stage C.	5½–6¼	—	*Manchester Guardian*, Jan. 29, 1831, p. 1	
			6 hrs, mail C.	7⅜	—		
ca. 1832	London—Birmingham	110	12 hours	9+	—	Baines, *On the Track of the Mail Coach*, pp. 30, 32	
	London—Exeter	ca.175	17½ hours	10	—		
	London—Leicester	92	11¼ hours	8	—		
	London—Manchester	185	17¼ hours	ca. 11	—		
	London—Shrewsbury	158	16 hours	ca. 10	—		
	Gloucester—Brighton	152	15 hours	10	—		
	London—Brighton	57	4 hrs 10 mins	ca. 14	—	Blew, *Brighton and Its Coaches*, p. 182	
	Braybrooke—London and return	160	12 hours	18⅓	—	London and Birmingham Railway Bill. *Extracts from Minutes of Evidence given before the Committee of the Lords*, p. 19	
	Worcester—London	118	16¼ hours	7	—	Ibid., p. 22	
	Leicester—London	92	11 hours	8½	—	Ibid., p. 24	
	London—Brighton	50 +	less than 5 hours	10	—	Brit. Mus., Add. MSS. 27,828, p. 16	

1832	London—Manchester (Mail)	185	19 hours	9¾	—	Harris, *Old Coaching Days*, p. 96, gives these as some of the fastest day coaches and mails running out of London	The rate in miles per hour has been worked out on the supposition that Harris's figures are correct. But contrast the figures given by Baines on page 700
	London—Manchester (Day Coach)	185	18½ hours	10	—		
	London—Exeter (Mail)	176	19 hours	9¼	—		
	London—Exeter (Day Coach)	165	17 hours	9¾	—		
	London—Holyhead (Mail)	259	27 hours	9⅗	—		
	London—Devonport ,,	216	21 hours	$10\frac{2}{7}$	—		
	London—Shrewsbury (Day Coach)	158	15¾ hours	10	—		
	London—Bristol	121	11¾ hours	10¼	—		
1833	London—Brighton	57	4¾ hours	12	—	Blew, *Brighton and Its Coaches*, p. 191	
	Hull—Scarborough	50	6–6¼ hours	8	—	Macturk, *History of the Hull Railways*, pp. 8, 9	Coach advertisements
	Hull—London	185	23 hours	8	—		
	Boston—Hull	60	9–9½ hours	6½	—		
c. 1835	London—Sheffield	150	16 hours	9⅓	—	Leader, *Sheffield in the Eighteenth Century*, p. 100	This was the rate made by the last Sheffield mail coach
	Shrewsbury—London	158	12 hrs 40 mins	12½	—	*Salopian Shreds and Patches*, II, pp. 26–28	The regular Time Bill, given in Harris, *Old Coaching Days*, p. 153, allowed 15 hrs 45 mins
	Shrewsbury—London	158	15 hrs 45 mins	10	—	Time Bill given by Harris, *Old Coaching Days*, p. 153	
1836	London—Edinburgh	400	45½ hours	over 9½ mls exclusive of stoppages	—	Harris, *Old Coaching Days*, p. 98	Harris knew the coaching arrangements from first-hand knowledge
	London—York	197	20 hours	ca. 10	—	Ibid.	
	Manchester—London	185	18 + hours	10 +	—	*Herepath's Railway Magazine*, I, p. 2	

APPENDIX 6

COST OF TRAVEL, 1750–1830

THE following data upon this subject have been collected and arranged in tabular form; and it may here be said that this table is subject to the deviations from accuracy that were mentioned at the beginning of Appendix 5. For example, to those who would scrutinize these statistics closely it will be apparent that the distance between certain great termini is different in some cases from that in other cases. This is due in some instances to the fact that different routes were taken between these termini. For example, the road from London to Manchester might be through Dunstable, Northampton, Loughborough and Derby; or it might lead through Coventry, Birmingham, Newcastle and Macclesfield. The improvement of the roads usually led also to the straightening of them, and consequently to the reduction of the distances. In most of these cases we have considered the cost of travelling upon the great highways of communication of the kingdom, for the statistics of travel on the minor cross-roads have been difficult to secure, probably because the great majority of the travellers were destined for the great towns and cities on business, rather than for the smaller places on the cross-roads. Our conclusion from the statistics here presented will be found in the text.

Year	Termini	No. of miles	Total fare (inside)	Fare per mile	Source of Authority	Remarks
			£ s. d.	d.		
1752	Newbury—London	57	10 0	2 +	Advertisement of coach given in Money, *History of Newbury*, p. 338	
	Birmingham, via Warwick and Oxford, to London	120	1 5 0	$2\frac{1}{2}$	Aris's *Birmingham Gazette*, March 16, 1752, p. 2	
	London—Warwick	98	1 0 0	$2\frac{1}{2}$	Ibid.	
	Birmingham—London	110	1 1 0	$2\frac{1}{3}$	Ibid., March 30, 1752, p. 4	
	Wolverhampton—London	125	1 5 0	$2\frac{2}{5}$	Ibid.	
1753	Shrewsbury—London	158	1 1 0	$1\frac{3}{5}$	Owen and Blakeway, *History of Shrewsbury*, I, p. 515	
1756	London—Brighton	57	16 0	$3\frac{1}{3}$	Hills, *History of East Grinstead*, p. 147	
	London—East Grinstead	30	6 0	$2\frac{2}{5}$		
	Birmingham, via Stratford, to London	120	1 0 0	2	Aris's *Birmingham Gazette*, Nov. 15, 1756, p. 4	
	Birmingham, via Warwick, to London	110	1 0 0	$2\frac{1}{5}$		
1757	London—Lewes	56	13 0	ca. 3	Blew, *Brighton and Its Coaches*, p. 85	
	London—Brighton	57	16 0	$3\frac{3}{8}$		
	Warrington—London	200	2 2 0	$2\frac{1}{2}$	Williamson's *Liverpool Advertiser*, June 9, 1757, gives advertisement	
1760	Liverpool—London	210	2 10 0	ca. 3	Picton, *Memorials of Liverpool*, I, pp. 203–4, quoting from the coach advertisement	This was the first stage coach on this route
	Manchester—London	200	2 5 0	$2\frac{1}{2}$	Axon, *Annals of Manchester*, p. 93	
	Birmingham—Bristol	86	1 1 0	3	Aris's *Birmingham Gazette*, April 28, 1760, p. 2	
1761	London—Leeds	190	2 5 0	2·8	*Leeds Intelligencer*, May 19, 1761, p. 8	
	Newcastle—London	290	3 5 0	2·7	*Archaeologia Aeliana*, N.S., III, p. 248	
	Newcastle—York	90	1 0 0	2·7		

Year	Termini	No. of miles	Total fare (inside) £ s. d.	Fare per mile d.	Source of Authority	Remarks
1764	Shrewsbury—London	158	1 10 0	2¼	Owen and Blakeway, *History of Shrewsbury*, I, p. 515	
1766	Shrewsbury—London	158	1 16 0	2¼		
1769	Leeds—London	190	1 11 6	2	Parsons, *History of Leeds*, I, p.130, quoted from the coach advertisement	
1772	Shrewsbury—London	158	1 14 0	2⅜	Owen and Blakeway, *History of Shrewsbury*, I, p. 515	
1774	Liverpool—Manchester	35	8 0	2¾	Advertisement in Williamson's *Liverpool Advertiser*, April 15, 1774	
	Liverpool—Preston	34	8 6	3	Advertisement of coach in Williamson's *Liverpool Advertiser*, June 24, 1774	
	Newcastle—Edinburgh	106	1 6 6	3	Advertisement of coach in *Newcastle Courant*, April 16, 1774, p.4	
	Shrewsbury—London	158	1 10 0	2¼	Advertisement of coach in *Shrewsbury Chronicle*, July 9, 1774, p. 1	
	London—Newcastle	290	3 3 0	2·6	Advertisement of coach in *Newcastle Courant*, April 16, 1774, p. 2	
	Southampton—London	80	16 0	2¼	*Southampton Guide*, 1774, p. 68	
	Hereford—Monmouth	19	15 0	9	Brit. Mus., Add. MSS. 17,398, p. 58	The chaise cost so much because the road was stony and hilly
1775	London—Birmingham	110	1 7 0	ca. 3	*Shrewsbury Chronicle*, April 15, 1775, p. 3	
	Hereford—London	125	1 10 0	ca. 3	*Hereford Journal*, Dec. 4, 1805, p. 2	
	Hereford—Gloucester	27	7 6	3·3		
	Hereford—Bristol (via Gloucester)	63	14 0	2·7		

1776	Newcastle—London	290	3	6	0	2¾	Armstrong, *Post Roads between London and Edinburgh*, 1776, p. 5	Flying coach
	Newcastle—Edinburgh	105	1	11	6	3⅗	Ibid.	Flying coach
	London—York	200	2	2	0	2½	Ibid.	Flying coach
	York—Newcastle	90	1	1	0	2¾	Ibid.	Flying coach
	York—London	200	2	2	0	2½	Ibid.	York Old Coach—a night coach
	York—London	200	2	10	0	3	Ibid.	York Diligence—day coach
	London—Stamford	90		18	0	2⅖	Ibid., p. 6	Left London nightly, 10 p.m.
	London—Leeds	190	2	2	0	2⅘	Ibid.	
	London—Lincoln	131	1	7	0	2½	Ibid.	
	London—Bristol	120	1	3	0	2⅓	Felix Farley's *Bristol Journal*, Nov. 2, 1776, p. 2, gives advertisement	Stage coach
	London—Bath	110	1	1	0	2⅓	Ibid.	Stage coach
	London—Bristol	120	1	8	0	2⅘	Ibid.	Post coach
1777	Bristol—London	120	1	11	6	3 +	*Bristol Gazette and Public Advertiser*, Aug. 7, 1777, p. 1, gives the advertisement	
	Bath—Exeter	80	1	1	0	3	Ibid.	Short distance passengers, 3*d*. per mile
	Bristol—Exeter	75		18	0	3		
1778	London—Shrewsbury	158	1	16	0	2¾	Aris's *Birmingham Gazette*, April 20, 1778, p. 8	
	London—Wolverhampton	125	1	8	0	2¾		
	London—Birmingham	110	1	5	0	2¾		
	London—Coventry	90	1	1	0	2⅘		
	Birmingham—Sheffield	73	1	0	0	3⅓	Aris's *Birmingham Gazette*, May 25, 1778, p. 2	
	Coventry—London	90	1	1	0	2⅘		
	Birmingham, via Oxford, to London	120	1	9	0	2¾	Ibid., Oct. 5, 1778, p. 2	Carried only 4 inside passengers
	Birmingham—London	110	1	5	0	2¾	Ibid.	
1779	London—Bristol	120	1	5	0	2½	Felix Farley's *Bristol Journal*, Aug. 14, 1779, p. 1, gives advertisement	"One Day Machine"
	London—Bath	110	1	3	0	2⅘	Ibid.	"One Day Machine"
	London—Bristol	120	1	12	0	2⅗	Ibid.	Post Coach or Diligence
	London—Bath	110	1	10	0	2¾	Ibid.	Post Coach or Diligence

Year	Termini	No. of miles	Total fare (inside) £ s. d.	Fare per mile d.	Source of Authority	Remarks
1779	London—Carlisle	305	3 6 0	2·6	Advertisement of coach is given in Malet, *Annals of the Road*, p. 24	The advertisement says that passengers taken up on the road were to pay 4d. per mile, inside, for either coach or diligence
	Carlisle—Glasgow or Portpatrick	100	1 16 6	4·3		
	∴ London—Glasgow or Portpatrick	405	5 2 6	8		
	Bewdley—Birmingham	—	—	8	Aris's *Birmingham Gazette*, April 12, 1779, p. 1	The advertisement reads "each passenger to pay Three-pence per Mile", "The Old and Original Shrewsbury, Wolverhampton, Walsall, and Birmingham Fly"
1780	London—Birmingham	110	1 5 0	2¾	Ibid.	
	London—Shrewsbury	158	1 16 0	2¾		
	Birmingham—Manchester	85	1 1 0	8	Ibid., April 10, 1780, p. 1	
	London—Steyning	51	11 0	2¼	Blew, *Brighton and Its Coaches*, p. 37	
	London—Brighton	57	14 0	8		
	London—Stroud	110	1 2 0	2·4	Jackson's *Oxford Journal*, July 29, 1780, p. 3	
	London—Cirencester	98	19 0	2·3		
	Liverpool—London	210	2 15 0	8 +	Picton, *Memorials of Liverpool*, II, p. 116, quoting from advertisement	This diligence was a superior kind of vehicle
1781	Birmingham—Manchester	85	1 1 0	8	Aris's *Birmingham Gazette*, Aug. 20, 1781, p. 1	
	London—Wantage	68	18 0	2·8	Jackson's *Oxford Journal*, April 28, 1781, p. 4	Short distance passengers paid 3d. per mile
	London—Wallingford	54	10 0	2·2		
	Oxford—Bath	68	1 1 0	3·7	Ibid., June 9, 1781, p. 2	
1782	Bristol—Gloucester	36	8 0	2¼	Sarah Farley's *Bristol Journal*, April 27, 1782, p. 2	
	Chester—London	180	1 11 6	2	Hall, *History of Nantwich*, p. 288, quoting from *Chester Guide Book*, 1782	

1782	Oxford—London	60		11	0	2·2	Jackson's *Oxford Journal*, Jan. 5, 1782, p. 1	
	Banbury—London (via Oxford, Wycombe, Uxbridge)	88		17	0	2·5	Ibid., May 18, 1782, p. 1	
	Bristol—Exeter	75	1	1	0	3⅓	Sarah Farley's *Bristol Journal*, April 27, 1782, p. 2	
	Exeter—Plymouth	40		12	6	3¾		
	Bristol—Gloucester	36		8	0	2·7		
	Bristol—Plymouth	118	1	15	6	3·6	Ibid., p. 3.	
	Wootton—Bristol	—		—		—	Ibid., Aug. 3, 1782, p. 2	Short distance passengers, 3*d*. per mile
	Bristol—Weymouth	72		17	0	3	Ibid.	
	Bristol—London	120	1	1	0	2·1		
	London—Bristol	120	1	8	0	2·8	*Constitutional Chronicle*, Jan. 24, 1782, p. 1	
	London—Bath	110	1	5	0	2·7		
1783	Bath—Shrewsbury	118	1	7	0	2¾	Advertisement given in Felix Farley's *Bristol Journal*, Jan. 18, 1783, p. 1	
	Bristol—Shrewsbury	108	1	4	0	2⅔		
	London—Shrewsbury	158	1	11	6	2⅖	Aris's *Birmingham Gazette*, May 5, 1783, p. 4	The proprietors of this coach say that these rates are enhanced over those previously in force, because of excessively high prices of corn, hay, etc.
	London—Wolverhampton	125	1	8	0	2¾		
	London—Birmingham	110	1	5	0	2¾		
	Shrewsbury—London	158	2	12	6	4⅖	Ibid., Aug. 18, 1783, p. 4	These higher rates are explained in the advertisement as necessary because of "the several additional duties that have now taken place on all post-coaches, together with the very great expence unavoidably incurred in supporting the same"
	Stourbridge—London	120	1	14	0	3⅖		
	Bromsgrove—London	112	1	13	0	3½		
	Alcester—London	108	1	11	0	3½		
	Oxford—London	60		16	0	3⅕		
	Bristol—Exeter	75		16	0	2½	Bonner and Middleton's *Bristol Journal*, Sept. 27, 1783, p. 3	
	Bristol—Birmingham	86	1	4	0	3·4	Ibid., Dec. 20, 1783, p. 1	
	Birmingham—Liverpool	100	1	5	0	3		
	Birmingham—Manchester	85	1	4	6	3·4		
	Birmingham—Sheffield	73	1	0	0	3·3		

Year	Termini	No. of miles	Total fare (inside)	Fare per mile	Source of Authority	Remarks
			£ s. d.	d.		
1784	London—Bristol or Bath	120 or 110	1 8 0	2⅓-3	Advertisement given in Felix Farley's *Bristol Journal*, Oct. 2, 1784, p. 1	The first of Palmer's mail coaches
	Bristol—London	120	1 8 0	2·8	Bonner and Middleton's *Bristol Journal*, March 20, 1784, p. 1	Short distance passengers, 3*d.* per mile
	Oxford—London	60	12 6	2·5	Jackson's *Oxford Journal*, Oct. 30, 1784, p. 1	
1785	Bristol—Portsmouth	88	1 9 0	ca. 4	Advertisement in Bonner and Middleton's *Bristol Journal*, May 7, 1785, p. 2	Palmer's mail coach
	Bristol—Salisbury	53	15 0	3¾		
	Liverpool—London	210	3 18 6	4⅜	Advertisement in Gore's *Liverpool Advertiser*, July 22, 1785	First mail coach on this road
	Bristol—Oxford	78	14 0	2+	Jackson's *Oxford Journal*, April 2, 1785, p. 4	
	Oxford—Bristol	78	1 1 0	3·2	Ibid., Nov. 12, 1785, p. 3	
	Oxford—London	60	18 0	3·6	Ibid., p. 4	Those charging the lower price were licensed by the Vice-Chancellor of the University
	Oxford—London	60	13 0	2·6		
	London—Chester	180	3 3 0	4·2	*Morning Chronicle and London Advertiser*, Jan. 8, 1785, p. 4	
	Newcastle—Carlisle	55	15 0	3¾	*Archaeologia Aeliana*, N.S., III, p. 252	
1786	Bristol—London	120	1 7 0	2·7	Bonner and Middleton's *Bristol Journal*, Dec. 23, 1786, p. 2	"Balloon coach"
	Bristol—London	120	1 6 0	2·6	Ibid.	"Light post coach"
	Bristol—London	120	1 2 0	2·2	Ibid.	"Four-horse coach"
	Oxford—Birmingham	60	18 0	3·6	Jackson's *Oxford Journal*, May 27, 1786, p. 3	
	London—Dover	75	1 1 0	3¼	*St James Chronicle*, Dec. 7, 1786, p. 2	
	Plymouth—London	215	2 2 0	2¼		
	London—Lincoln	142	1 11 6	2¼		

Year	Route	Miles	Fare (£ s. d.)	Pence per mile	Source
1786	London—Chester	180	3 3 0	4·2	General Advertiser, Oct. 20, 1786, p. 4
	Chester—Holyhead	60	1 11 6	6·3	
	London—Chester	180	2 5 0	3	
	London—Nantwich	161	2 2 0	3+	
	London—Stafford	137	1 11 6	2·7	
	London—Lichfield	122	1 8 0	2·7	
	London—Northampton	63	0 17 0	3·2	
	London—Holyhead	240	3 16 6	3·8	
	London—Holyhead	240	4 14 6	4·7	Ibid., Nov. 14, 1786, p. 4
	London—Chester	180	3 3 0	4·2	
	London—Holyhead	240	3 7 0	3·3	Ibid., Dec. 8, 1786, p. 1
	London—Chester	180	1 16 0	2·4	
	London—Coventry	90	0 18 0	2·4	Ibid., Dec. 22, 1786, p. 1
	London—Nantwich	161	1 13 0	2·5	
	London—Chester (via Oxford, Birmingham, Newport and Whitchurch)	185	2 12 6	3·4	
1787	Sheffield—London	150	1 17 0	3	Leader, *Sheffield in the Eighteenth Century*, p. 100
	Sheffield—York	50	0 11 0	2⅔	
	Sheffield—Leeds	35	0 5 0	ca. 2	Ibid.
	Sheffield—Birmingham	72	0 8 0	1⅓	Ibid.
	Bath—Gloucester	38	0 12 0	ca. 4	Bath Chronicle, Jan. 11, 1787, p. 1
	Bath—Tewkesbury	49	0 16 0	4	
	Bath—Worcester	68	1 0 0	ca. 4	
	Bath—Birmingham	88	1 7 0	3·6	
	Bath—Upton-on-Severn	55	0 17 0	3·7	
	Bath—Kidderminster	78	1 5 0	4	
	Bath—Bridgnorth	94	1 10 0	3·8	
	Bath—Shrewsbury	118	1 16 0	3·7	
	Bath—London	110	1 5 0	2·7	Ibid., p. 4
	Bath—Oxford	68	1 1 0	3·7	
	Bristol—London	120	1 7 0	2·7	Ibid., Jan. 18, 1787, p. 1

Remarks (1786):

No reason can be given for the great difference between these fares from London to Chester, as the kind of coach and the speed were the same in each case

I can assign no reason for these differences of fare between London and Holyhead and Chester in the same year

Year	Termini	No. of miles	Total fare (inside) £ s. d.	Fare per mile d.	Source of Authority	Remarks
1787	London—Manchester	195	2 5 0	2·8		
	London—Congleton	174	1 18 0	2·6		
	London—Newcastle-under-Lyme	162	1 13 0	2·4		
	London—Stone	150	1 10 0	2·4	*General Advertiser*, May 21, 1787, p. 4	These were all described as "reduced fares"
	London—Lichfield	126	1 4 0	2·3		
	London—Atherstone	109	1 1 0	2·3		
	Bristol—London	120	1 10 0	3		
	Bristol—London	120	1 7 0	2·7	Felix Farley's *Bristol Journal*, Jan. 6, 1787, p. 1	These coaches were almost identical as far as the facilities they offered, and yet there was considerable difference in fares
	Bristol—London	120	1 6 0	2·6		
	Bristol—London	120	1 2 0	2·2		
	Bristol—Shrewsbury	108	1 12 0	3·6	Ibid., Feb. 10, 1787, p. 8	
	Bristol—Holyhead	211	3 14 0	4·2		
	Bristol—Birmingham	86	1 5 0	3·5		These were the fares on the mail coach. Short distance passengers were charged 4*d.* per mile
	Bristol—Worcester	61	19 0	3·7	Ibid., July 28, 1787, p. 1	
	Bristol—Gloucester	36	10 0	3·3		
	Shrewsbury—Birmingham	46	14 0	3·65	*Salopian Shreds and Patches*, I, p. 55, gives the advertisement	
	Shrewsbury—Oxford	98	1 12 0	ca. 4	Ibid.	
	Shrewsbury—London	158	2 0 0	3	Ibid.	
1788	Manchester—Macclesfield	20	10 0	6		
	Manchester—Leek	38	16 0	ca. 6		I am wholly at a loss to explain these figures, showing such high rates of fare, in view of the figures which we have already presented for the preceding years. It seems inconceivable, for instance, that the rate in 1781 from Liverpool to London should be 8 + *d.* per mile, while in 1788 the rate from Manchester to London was practically 5*d.* per mile
	Manchester—Derby	61	1 10 0	ca. 6		
	Manchester—Loughborough	78	1 16 0	5·5	*Manchester Collectanea*, in Chetham Society Publications, LXVIII, p. 153	
	Manchester—Northampton	121	3 0 0	6		
	Manchester—Dunstable	154	3 18 6	5·7		
	Manchester—London	195	3 18 6	4·7		

Year	Route	Miles	£	s.	d.	Pence per mile	Authority	Remarks
1788	Manchester, via Preston, Lancaster and Kendal to Carlisle	121	2	6	0	4¼	Ibid.	
	Manchester, via Rochdale, Halifax, Bradford, Leeds and Tadcaster, to York	69	1	8	0	ca. 5	Ibid., p. 154	
	Manchester—Liverpool	35		14	0	ca. 5	*Bath Chronicle*, April 24, 1788, p. 2	Palmer's mail coach
	Bristol—Exeter	75	1	4	6	3·8	*Bristol Gazette and Public Advertiser*, May 15, 1788, p. 3	
	Bristol—Birmingham	86		10	6	1·5	Ibid.	
	Bristol—Birmingham	86		8	0	1·1	Ibid.	Post coach. No reason given why these rates were so low
	Birmingham—Manchester	85	1	7	6	ca. 4		
	Birmingham—Manchester	85	1	5	0	3·5		
	London—Manchester	195	1	11	6	2		
	London—Hull	185	1	10	0	2		
	London—Lincoln	181	1	1	0	2		
	London—Peterborough	81		16	0	2·4		
	London—Liverpool	210	1	16	0	2	*The Gazetteer and New Daily Advertiser*, April 4, 1788, p. 1	These were advertised as "reduced fares"
	London—Birmingham	110	1	1	0	2·3		
	London—Walsall	118	1	8	0	2·4		
	London—Chester	180	1	16	0	2·4		
	London—Holyhead	240	3	7	6	3·4		
	London—Bristol	120	1	7	0	2·7		
1789	Chester—London	180	2	2	0	3	Hall, *History of Nantwich*, p. 288, quoting from *Chester Guide Book* of 1789	
	Bath—Exeter	30	1	4	0	8·6	*Bath Chronicle*, Jan. 8, 1789, p. 1	Post and mail coaches
	Bristol—Birmingham	86		8	0	1·1	Ibid.	Fare by mail coach was 10s. 6d.
	Birmingham—Holyhead	130	2	14	0	5		
	Shrewsbury—Holyhead	108	2	9	0	5·7	Ibid.	
	Bath—London	110	1	5	0	2·7	Ibid., p. 2	The one-day coaches charged no more than the two-day coaches
	Bath—London	110	1	1	0	2·3		

Year	Termini	No. of miles	Total fare (inside) £ s. d.	Fare per mile d.	Source of Authority	Remarks
1790	Sheffield—York	45	11 0	3	} *Sheffield Local Register*, p. 64	I am unable to explain these low rates. It was not due to competition, for only one coach went to Birmingham four times a week (ibid., p. 88)
	Sheffield—Leeds	38	5 0	2		
	Sheffield—Birmingham	78	8 0	1½		
	London—Brighton	57	18 0	3·8	Blew, *Brighton and Its Coaches*, p. 39	
	London—Worcester	118	1 8 0	3	Jackson's *Oxford Journal*, April 10, 1790, p. 3	
	London—Oxford	60	18 0	3·6		
	Oxford—Southampton	88	1 1 0	6·6	Ibid., June 5, 1790, p. 4	Do not know why the fare should have been so high on this cross road
1791	Hull—London	185	2 2 0	2¾	Battle, *Hull Directory*, 1791, pp. 65–78	
	London—Birmingham	110	1 8 0	3	} *Morning Post and Daily Advertiser*, July 20, 1791, p. 1	
	London—Warwick	90	1 5 0	3¾		
	London—Shiffnal	140	1 8 0	2		
	London—Wolverhampton	125	1 2 0	2 +	} *Morning Chronicle*, April 29, 1791, p. 1	
	London—Bristol	120	1 10 0	3		
	London—Bath	110	1 8 0	3	Ibid.	
	London—Birmingham or Walsall	110–118	1 1 0	2¼		
1792	London—Colchester	55	14 0	3	Ibid., April 11, 1792, p. 1	
	London—Ipswich	73	19 0	3		
	London—Woodbridge	82	1 2 0	3 +		
1798	London—Southampton	80	10 6	1⅝	Baird, *Agriculture of Middlesex*, p. 35	He says this machine was little inferior to the mail coaches in ease and speed; but that the difference in expense was considerable

Year	Route	Miles	£	s.	d.	Rate	Authority	Remarks
1795	Chester—London	180	2	2	0	2·8	Chester Guide, 1795, pp. 61-62	"Royal Mail"—only 4 passengers; night coach
	Chester—London	180	3	10	0	4¾		"London and Cambridge Diligence," for 3 passengers—day coach
	Chester—Holyhead	60	1	15	6	7		
	Chester—Shrewsbury	84		13	6	ca. 5		
	Chester—Bath	150	2	9	0	ca. 4		
	Chester—Bristol	140	2	7	6	4·2		
	Chester—Worcester	84	1	9	6	2¼		
	Chester—Liverpool	18		3	6	1·8		
	Chester—Manchester	37		4	0			
	London—Cambridge	55	1	1	0	4¾		
1796	London—Cambridge	55	1	0	0	4⅜	Cambridge Directory, 1796, p. 159	
	London—Cambridge	55	1	0	0	4⅜	Ibid.	
	Cambridge—London	55		18	0	4	Ibid., p. 160	Lower rate possibly due to taking longer time than the foregoing
	Cambridge—London	55		15	0	3¼	Ibid.	Lower rate doubtless due to much slower speed of travel
	Cambridge—Ipswich	50		16	0	3⅜	Ibid.	The high rates for the coaching centring at Cambridge were probably due to the great extent of the travel to and from that city
	Cambridge—Birmingham	105	1	11	6	3⅜	Ibid., p. 161	
1797	London—Brighton	57		19	0	4	Blew, *Brighton and Its Coaches*, p. 59	Increased fares due to extra duty on coaches
1798	Edinburgh—London	400	10	0	0	6	Harris, *The Coaching Age*, p. 279	
1802	London—Chester	180	4	14	6	6·3	*The Times*, May 31, 1802, p. 3	It seems probable that these figures should have referred to the journey from London to Holyhead, which cost four guineas, as given under date 1786
	Chester—London	180	4	4	0	5·6		
1805	Hereford—London	125	1	10	0	3	Advertisement of coach is given in *Hereford Journal*, Aug. 14, 1805, p. 2	

Year	Termini	No. of miles	Total fare (inside) £ s. d.	Fare per mile d.	Source of Authority	Remarks
1808	Manchester, via Derby and Nottingham, to London	195	3 8 6	4+	*Manchester Collectanea*, in Chetham Society Publications, LXVIII, p. 160, quoting from *Manchester Directory* of 1808–9	These were the fares by mail coach in each case. Like those given on preceding pages, under date of 1788, they seem to have been much higher than the fares on most of the other roads
	Manchester—Macclesfield	21	10 6	6	Ibid.	
	Manchester—Bolton	12	6 0	6	Ibid.	
	Manchester—Birmingham	94	1 15 0	4·5	Ibid.	
	Manchester—Leeds	45	1 3 6	6	Ibid.	
	Manchester—York	69	1 15 0	6	Ibid.	
	Manchester—Liverpool	35	14 0	ca. 5	Ibid.	
	Birmingham—Sheffield	78	8 0	1¼	Aris's *Birmingham Gazette*, Feb. 15, 1808, p. 1	This was reduced fare. Before this new coach started the fare was £1. 10s. (see next item)
	Birmingham—Sheffield	78	1 10 0	5	Ibid.	
	Birmingham—Nottingham	50	10 0	2⅝	Ibid.	This is advertised under the heading "Very Cheap Travelling"
	Birmingham, via Oxford, to London	120	1 16 0	3¾	Ibid.	
1811	Birmingham—Leicester	40	12 0	3⅝	Ibid., Feb. 4, 1811, p. 1	
	Birmingham—Liverpool	98	1 1 0	2⅝		
1818	Cambridge—Stamford	44	10 0	ca. 2·8	*Cambridge Chronicle and Journal*, Oct. 29, 1818, p. 1	
	Cambridge—Leicester	75	1 1 0	3·4		
	Leicester—Birmingham	40	8 0	2·4		
1815	London—Cambridge	55	18 0	ca. 4	Ibid., Nov. 8, 1815, p. 8	
	London—Leeds	190	8 18 6	4·6	Price, *Leeds and Its Neighbourhood*, p. 271	

Year	Route	Miles	Fare (£ s. d.)	Per mile	Source	Remarks
1816	London—Hastings	60	15 0	8	*The Times*, May 24, 1816, p. 1	Stage coach
1817	London—Hastings	60	1 0 0	4		
	Sheffield—London	150	8 5 0	5¼	*Sheffield Local Register*, p. 150	
1818	York—London	200	8 18 6	4¾	Hargrove, *History of York*, 1818, Vol. II, Pt. II, pp. 671–5	
1825	London—Bath	110	1 18 0	4·1	*The Times*, Aug. 27, 1825, p. 3	
	London—Brighton	57	1 1 0	4·5	Blew, *Brighton and Its Coaches*, p. 138	
1830	Liverpool—Manchester	35	10 0	3⅝	*Annual Register*, 1832, p. 445; also *Advantages of the Progressive Formation of Railways*, p. 23	
	Liverpool—Manchester	35	10 0	3¾	*Great Western Railway. Evidence* (of Henry Booth) on the *London and Birmingham Railway Bill*, p. 8	He says the rate varied much from 10s.
	London—Newcastle	290	4 10 0	3¾	Harris, *The Coaching Age*, p. 194	
	Bristol—Oxford	68	1 8 0	5	Brit. Mus. 8285. ee. 4 (1), 'Oxford and Didcot Railway Bill,' p. 26	Fare was 25s. and coachman's fee was 3s.
	Steventon—London	60	1 5 0	5	Ibid., p. 7	
1832	Braybrooke—London and return	160	1 12 0	2·4	*London and Birmingham Railway Bill. Extracts from Minutes of Evidence before Lords Committee*, p. 19	
	Worcester—London	118	2 8 0	4·8	Ibid., p. 22	This included the usual fees
	Manchester—London	185	4 4 0	5·4	Ibid., p. 28	
1837	Hull—Selby	32	7 0	2·6	Macturk, *History of the Hull Railways*, p. 11	
1847	Steventon—Oxford	10	3 0	3·6	Brit. Mus. 8285. ee. 4 (1), 'Oxford and Didcot Railway Bill,' p. 7	In addition, each passenger had to pay the gratuity
	Oxford—London	60	1 8 6	4·7	Ibid., p. 26	Fare includes the gratuity of 2s. 6d.

APPENDIX 7

COST OF CARRIAGE OF GOODS BY LAND, 1750–1830

THE following details of the expense of conveyance by land have been brought together in this tabular form from sources which are among the most reliable. As in the preceding tables, so here, the statistics have been made as accurate as possible; but it is inevitable that some slight errors exist, on account of our inability to know exactly the length of the road from one place to another at these earlier times. But if absolute accuracy is unattainable, we can at least say that the slight limit of error renders our figures relatively correct. The data here presented have been summarized, in order to arrive at some general conclusion as to the cost of carriage, and, at the same time, to enable the reader to see the variations from that general conclusion, which is stated in the text.

Year	Termini	No. of miles	Total cost per cwt.	Cost per cwt. per mile	Source of Authority	Remarks
1749	London—Bury	72	3*s*. (=£3 per ton)	$\frac{1}{2}$*d*.	Phillips, *Plan for a Navigable Canal*, p. 21	He says it is about 9*s*. per ton for 10 miles
	Bury—Norwich	40	2*s*. 2*d*. (=£2. 3*s*. 4*d*. per ton)	·65*d*.	Ibid.	For cumbersome articles, like packs of wool and woollen goods, groceries, pottery and other merchandise, the cost was a great deal more
1754	— —	20	8*d*.	$\frac{2}{5}$*d*.	Brit. Mus. 213. i. 3 (101), 'Reasons against a Bill for permitting only carriages with Broad Wheels' etc., p. 3	The writer says that meal was brought 20 miles to London for less than 8*d*. per cwt. Further, he says that the cost of carriage to London in winter was at least $\frac{1}{3}$ more than in the summer
	London—Derby	126	6*s*. summer 7*s*. 6*d*. winter	$\frac{4}{7}$*d*. summer $\frac{5}{7}$*d*. winter	Cox, *Three Centuries of Derbyshire Annals*, II, p. 236	These rates are quoted from the rates as fixed by the Justices of the Peace at their Easter Sessions
	London—Ashbourne	139	6*s*. summer 7*s*. 6*d*. winter	$\frac{1}{2}$*d*. summer $\frac{2}{3}$*d*. winter		
	London—Bakewell	147	6*s*. 6*d*. summer 8*s*. winter	$\frac{1}{2}$*d*. summer $\frac{2}{3}$*d*. winter		
	London—Chesterfield	147	6*s*. 6*d*. summer 8*s*. winter	$\frac{1}{2}$*d*. summer $\frac{2}{3}$*d*. winter		
	London—Wirksworth	138	6*s*. 6*d*. summer 7*s*. winter	$\frac{1}{2}$*d*. summer $\frac{2}{3}$*d*. winter		
	London—Tideswell	150	7*s*. summer 8*s*. winter	$\frac{1}{2}$*d*. summer $\frac{2}{3}$*d*. winter		
	London—Chapel-en-le-Frith	155	7*s*. 6*d*. summer 8*s*. 6*d*. winter	$\frac{3}{5}$*d*. summer $\frac{2}{3}$*d*. winter		
	London—Buxton	154	7*s*. 6*d*. summer 8*s*. 6*d*. winter	$\frac{3}{5}$*d*. summer $\frac{2}{3}$*d*. winter		

Year	Termini	No. of miles	Total cost per cwt.	Cost per cwt. per mile	Source of Authority	Remarks
1760	Manchester—Liverpool	35	2*s.*	$\frac{2}{3}$*d.*	Brit. Mus. B. 504 (4), 'Advantages of Inland Navigations,' p. 18	The rate is given as 40*s.* per ton
	Birmingham—London	108	—	$\frac{1}{2}$*d.*	Ibid., p. 35	Rate is given as about 8*s.* per ton for 10 miles
	Along Grand Trunk Canal	—	—	$\frac{1}{2}$*d.*	Ibid., p. 18; also Meteyard, *Life of Wedgwood*, I, p. 275	These references say that the average cost of land carriage here is about 9*s.* per ton for 10 miles
	Burslem—Winsford	20	$\frac{9}{10}$*s.* (=18*s.* a ton)	$\frac{1}{2}$*d.*	Meteyard, *Life of Wedgwood*, I, p. 275	
	Burslem—Willington (on Trent)	34	1·7*s.* (=34*s.* a ton)	$\frac{3}{5}$*d.*		
	Burslem—Bridgnorth	40	3*s.*	$\frac{9}{10}$*d.*	Jewitt, *The Wedgwoods*, p. 171, quoting from a letter of Richard Whitworth's of that year	This figure is higher than the others, but why I cannot tell
ca. 1760	— —	10	$\frac{2}{5}$*s.* to $\frac{9}{20}$*s.*	ca. $\frac{1}{2}$*d.*	Phillips, *Plan for a Navigable Canal*, p. 21	He says the price of land carriage along the route of the Staffordshire Canal was formerly 8–9*s.* per ton per mile
	Birmingham—Liverpool	100	4*s.* 0*d.*	$\frac{1}{2}$*d.*	*J., H. of C.*, XXX, p. 520, and XXXII, p. 725	
	Manchester—London	195	7*s.* 0*d.*	$\frac{2}{5}$*d.*		
	Blackburn—Liverpool	34	2*s.* 2·4*d.*	$\frac{3}{4}$*d.*		
	Burnley—Preston	23	1*s.* 0*d.*	$\frac{1}{2}$*d.*		
	Leeds—Liverpool	—	—	$\frac{3}{5}$*d.*		
1766	Winsford Bridge—Coalbrookdale	41	·3*s.*	$\frac{9}{10}$*d.*	Whitworth, *Advantages of Inland Navigation*, p. 39	These were all heavy goods, like iron, coal, pottery, clay, etc. He says (p. 11) that the cost of land carriage differed according to the badness of the road or the number of carriers
	Madeley—Nantwich and Winsford	33–40	$\frac{1}{2}$*s.*	$\frac{1}{8}$–$\frac{1}{6}$*d.*	Ibid., p. 41	
	Winsford—Doddington	12	·3*s.*	$\frac{3}{10}$*d.*	Ibid., p. 60	

1766	Newcastle and Burslem—Bridgnorth	38–40	3s.	ca. 1d.	Ibid.	
	Burslem and Newcastle—Bridgnorth and Bewdley	38–42	2½s.	¾–⅘d.	Ibid.	
1767	Ripon—York	22	6d.	$\frac{3}{10}$d.	*J., H. of C.*, March 2, 1767	
1770	Blackburn—Liverpool	34	2½s.	¾d.	*J., H. of C.*, XXXII, p. 725	Quoted as 5*s*. 6*d*. a pack, or 44*s*. a ton
	Burnley—Preston	23	1s.	½d.	Ibid.	Using same basis as above, 2*s*. 6*d*. a pack = 20*s*. a ton
	Leeds—Liverpool	82	—	⅗d.	Ibid.	The cost is quoted as 1*s*. per ton per mile
	Chester—London	180	8s.	½d.	Brit. Mus. 8235. h. 44, 'Remarks relating to a Canal intended to be made from Chester to Middlewich,' p. 5	
1771	Sheffield, via Middlewich, to Chester	65	3s. 6d.	⅗d.	*J., H. of C.*, XXXIII, pp. 204–7	
	Sheffield, via Manchester, to Chester	75	4s.	⅗d.	Ibid.	
	Birmingham—Chester	70	4s.	⅔ + d.	Ibid.	
	London—Chester	180	7s. summer 8s. winter	$\frac{7}{15}$d. summer $\frac{8}{15}$d. winter	Ibid.	This = ca. ½*d*. per cwt. per mile
1773	Derby—Northampton	54	3s.	⅔d.	Cox, *Derbyshire Annals*, II, p. 236	The other rates given under 1754 apply also to this year
	Derby—Leicester	24	1s. 6d.	¾d.		
1774	Cooper's Bridge—Huddersfield	4	⅛s.	⅜d.	*J., H. of C.*, XXXIV, p. 414	
	Southampton—London	80	3s. 6d.	½d.	*Southampton Guide*, 1774, p. 64	
1781	Birmingham—London	108	4s. 6d.	½d.	*Manchester Collectanea*, in Chetham Society Publications, LXVIII, p. 136	
	London—Birmingham	108	3s.	⅓d.		
1785	— —	—	—	⅗d.	Phillips, *Plan for a Navigable Canal*, p. 23	He gave the price of land carriage as, on the average, 1*s*. per ton per mile

Year	Termini	No. of miles	Total cost per cwt.	Cost per cwt. per mile	Source of Authority	Remarks
1785	London—Ongar	21	£1. 0s. per ton	$\frac{4}{7}$d.	Phillips, *Plan for a Navigable Canal*, p. 24	
	London—Chelmsford	31	£1. 5s. ,, ,,	$\frac{1}{2}$d.		
	London—Braintree	44	£1. 10s. ,, ,,	$\frac{2}{5}$d.		
	London—Bradford	53	£1. 15s. ,, ,,	$\frac{2}{5}$d.		
	London—Sudbury	75	£2. 10s. ,, ,,	$\frac{2}{5}$d.		
	London—Lavenham	80	£2. 15s. ,, ,,	$\frac{2}{5}$d.		
	London—Stowmarket	91	£4. 0s. ,, ,,	$\frac{1}{2}$ + d.		
	London—Attleborough	112	£4. 0s. ,, ,,	$\frac{3}{7}$d.		
	London—Hingham	116	£5. 0s. ,, ,,	$\frac{1}{2}$d.		
	London—Wyndham	120	£5. 10s. ,, ,,	$\frac{1}{2}$d.		
	London—Norwich	132	£5. 0s. ,, ,,	$\frac{1}{2}$d.		
	London—Swaffham	132	£5. 0s. ,, ,,	$\frac{1}{2}$d.		
ca.	Basingstoke—London	50	2s.	$\frac{1}{2}$d.	Brit. Mus. B. 263 (6), 'Basingstoke Canal Navigation'	
1786	Basingstoke—Reading	16	$\frac{5}{8}$s.	$\frac{2}{5}$d.		
1788	Lewes—Lindfield	11	$\frac{3}{8}$s.	$\frac{2}{5}$d.	Jessop, *Estimate of Expense of Making River Ouse Navigable*, p. 4	These were coal and timber rates respectively
	Lewes—Lindfield	11	1s. per ton per mile	$\frac{3}{5}$d.		
1792	Leeds—London	190	6s. 0d.	·38d.	*Leeds Intelligencer*, Jan. 9, 1792, p. 4	
	Leeds—Birmingham	106	4s. 0d.	$\frac{4}{9}$d.		
	Leeds—Newcastle	100	6s. 0d.	·72d.		
	Manchester—Etruria	34	2s. 9d.	1d.	Salt, *Statistics and Calculations*, p. 71	This information was obtained from the compiler's own experience. The prices were for perishable goods. Non-perishables would be carried at lower prices
	Bromley Common—Lichfield	3	$1\frac{1}{2}$d.	$\frac{1}{2}$d.		
	Manchester—Shardlow	62	3s. 0d.	$\frac{3}{5}$d.		
	Shardlow—Leicester	20	10d.	$\frac{1}{2}$d.		
	Manchester—Newark	72	5s. 4d.	$\frac{8}{9}$d.		
	Manchester—Wolverhampton	70	4s. 8d.	$\frac{4}{5}$d.		
	Manchester—Birmingham	85	4s. 0d.	·56d.		
	Manchester—Stourport	93	4s. 8d.	$\frac{3}{5}$d.		
	Liverpool—Etruria	42	2s. 6d.	$\frac{5}{7}$d.		
	Liverpool—Wolverhampton	85	5s. 0d.	·7d.		
	Liverpool—Birmingham	100	5s. 0d.	$\frac{3}{5}$d.		
	Liverpool—Stourport	101	5s. 0d.	$\frac{3}{5}$d.		
	Chester—Wolverhampton	55	3s. 6d.	$\frac{4}{5}$d.		
	Chester—Birmingham	70	3s. 6d.	$\frac{3}{5}$d.		
	Chester—Stourport	74	3s. 6d.	$\frac{3}{5}$d.		

Year	Route	Distance (miles)	Rate	Rate (d.)	Reference	Note
1793	London—Reading	46	1s. 8d.	$\frac{7}{8}$d.	*Parl. Papers*, 1793, XIII, Report No. 109, 'Thames and Isis Navigation,' p. 7	
	Cirencester—London	93	3s. 0d.	$\frac{5}{8}$d.		
	Tetbury—London	107	3s. 6d.	$\frac{5}{8}$d.		
	Tetbury—Lechlade	20	9d.	$\frac{1}{2}$d.		
	Cirencester—Lechlade	11	6d.	$\frac{1}{2}$d.	*Ibid*, p. 27	
1794	Manchester—Huddersfield	25	1s. 3d.	$\frac{3}{5}$d.	*Leeds Intelligencer*, April 21, 1794, p. 2	
1796	Basingstoke—Winchester	18	10d.	$\frac{1}{3}$d.	*Reading Mercury*, May 30, 1796, p. 4	
	„ —Southampton	30	1s. 6d.	$\frac{3}{5}$d.		
	„ —Romsey	28	1s. 6d.	$\frac{3}{5}$d.		
	„ —Andover	18	9d.	$\frac{1}{2}$d.		
	„ —Whitchurch	10	6d.	$\frac{3}{5}$d.		
1822	London—Leeds	190	13s. 0d.	$\frac{4}{5}$d.	Young, *The Economy of Steam Power on Common Roads*, p. 11	
1808	Basingstoke—London	50	2s. 0d.	$\frac{1}{3}$d.	Brit. Mus. K.6.58(c), 'Basingstoke Canal Navigation'	
	Basingstoke—Reading	15	10l.	$\frac{1}{3}$d.		
	London—Derby or Ashbourne	123–139	6s. 0d.	$\frac{1}{2}$–$\frac{3}{4}$d.		
	London—Bakewell, Chesterfield, or Wirksworth	147–158	6s. 6d.	·53–·56d.		These were the rates assessed by the Justices for the summer half of the year. The rates for the winter half of the year were from ½ to ⅓ more than these. They are given in Farey, III, p. 275
	London—Tideswell	150	7s. 0d.	·56d.	Farey, *Agriculture of Derby*, III, p. 275	
	London—Buxton or Chapel-en-le-Frith	155	7s. 6d.	$\frac{3}{5}$d.		
	Derby—Northampton	54	3s. 0d.	$\frac{2}{3}$d.		
	Derby—Leicester	24	1s. 6d.	$\frac{3}{4}$d.		
	Manchester—Sheffield	35	2s.–2s. 2d.	$\frac{2}{3}$d.	*Sheffield Local Register*, p. 117	This was in January, 1808
	Sheffield—Liverpool	66	2s. 10d.	1·2d.		
	London—Staines	20	£1 0 0 per ton	$\frac{3}{5}$d.		
	London—Windsor	27	£1 6 8 ,, ,,	$\frac{1}{3}$d.		
	London—Maidenhead	33	£1 10 0 ,, ,,	$\frac{1}{3}$d.		
	London—Reading	46	£2 10 0 ,, ,,	$\frac{4}{5}$d.		
	London—Wallingford	50	£3 0 0 ,, ,,	$\frac{3}{5}$d.		
	London—Abingdon	58	£3 10 0 ,, ,,	$\frac{3}{5}$d.		
	London—Oxford	80	£4 0 0 ,, ,,	$\frac{4}{5}$d.		
	London—Lechlade	87	£5 0 0 ,, ,,	$\frac{3}{5}$d.	Mavor, *Agriculture of Berkshire*, p. 581	

Year	Termini	No. of miles	Total cost per cwt.	Cost per cwt. per mile	Source of Authority	Remarks
1809	Sheffield—London	150	11*s*. 8*d*.	$\frac{9}{10}$*d*.	*Sheffield Local Register*, p. 123	
	On 16 chief roads	—	—	¾*d*.	*Parl. Papers*, 1809, III, 431, 'Third Report of Committee on Broad Wheels,' p. 461	The average cost of carriage on these roads was 1*s*. 3½*d*. per ton per mile
1810	London—Windsor	27	1*s*. 6*d*.	⅔*d*.	Allnutt, *Rivers and Canals West of London*, p. 3	
	London—Henley	40	2*s*. 8*d*.	⅘*d*.		
	London—Reading	46	2*s*. 6*d*.	⅔*d*.		
	London—Wallingford	50	2·9*s*.	$\frac{7}{10}$*d*.		
	London—Abingdon	58	3¼*s*.	⅔*d*.		
	London—Lechlade	87	4*s*.	½ + *d*.		
	London—Weybridge	20	1¼*s*.	¾*d*.	Ibid., p. 5	
	London—Guildford	30	1¾*s*.	$\frac{7}{10}$*d*.		
	London—Godalming	34	2*s*.	$\frac{7}{10}$*d*.		
	London—Basingstoke	50	1¾*s*.	⅖*d*.	Ibid., p. 6	
	London—Reading	46	2*s*.–2¼*s*.	½*d*.–⅗*d*.	Ibid., p. 8	Heavy goods went slightly cheaper than other goods (such as groceries, malt, etc.)
	London—Newbury	52	3½*s*.–3¾*s*.	⅘*d*.–$\frac{9}{10}$*d*.		
	London—Abingdon	60	3¼*s*.	⅔*d*.	Ibid., p. 9	
	London—Wantage	68	4*s*.	¾*d*.		
	London—Swindon	75	4¾*s*.	¾*d*.		
	London—Semington	96	5⅓*s*.	⅔*d*.		
	London—Bath	110	6$\frac{3}{10}$*s*.	⅔*d*.		
	London—Bristol	120	7¼*s*.	¾*d*.		
	Liverpool—Etruria	42	2½*s*.	¾*d*.	Baines, *History of Liverpool*, pp. 439–40	
	,, —Wolverhampton	85	5*s*.	¾*d*.		
	,, —Birmingham	100	5*s*.	⅗*d*.		
	Manchester—Lichfield	65	4*s*.	¾*d*.		
	,, —Derby	55	3*s*.	⅔*d*.		
	,, —Leicester	82	6*s*.	⅞*d*.		
	,, —Newark	72	5⅓*s*.	$\frac{8}{9}$*d*.		
	,, —Nottingham	65	4*s*.	¾*d*.		
	,, —Wolverhampton	70	4⅔*s*.	⅘*d*.		
	,, —Birmingham	85	4*s*.	½ + *d*.		

1810	Cambridge—Watford	50	2*s*. 6*d*.	$\frac{3}{5}$*d*.	*Cambridge Chronicle and Journal*, Nov. 9, 1810, p. 2	6*s*. per qr. for oats would be 2*s*. 6*d*. per cwt. (of 112 lbs.)
	Cambridge—Stortford	26	1*s*. 1$\frac{1}{2}$*d*.	$\frac{1}{2}$*d*.	Ibid., Nov. 23, 1810, p. 2	
1812	Cambridge—London	55	3*s*. 0*d*.	$\frac{3}{5}$*d*.	Ibid., June 19, 1812, p. 2	
	Cambridge—Stortford	26	1*s*. 6·5*d*.	·7*d*.	Ibid., Oct. 9, 1812, p. 1	
1813	London—Bristol	120	10*s*. 0*d*.	1*d*.	Ibid., Oct. 29, 1813, p. 2	
1816	Portsmouth—London	80	4*s*. 6*d*.	$\frac{2}{3}$*d*.	*Hampshire Telegraph and Sussex Chronicle*, Oct. 14, 1816, p. 3	
1829	Newcastle—Carlisle	62	1*s*. 9*d*. to 2*s*. 3*d*.	$\frac{1}{5}$ to $\frac{3}{7}$*d*.	'Evidence taken before Commons Committee on Newcastle and Carlisle Railway Bill,' pp. 2, 13, 14, 15	Cost was 35*s*.–45*s*. per ton. A little less for return carriage. Low cost of coal carriage was probably due to ocean competition
1831	— —	—	—	·45*d*.	*Birmingham Journal*, Jan. 22, 1831, p. 1	Cost by fly vans was given as 9*d*. per ton per mile
1832	London—Bristol	120	4*s*.	$\frac{2}{5}$*d*. }	*Bristol Liberal*, Jan. 7, 1832, p. 1	Advertised as reduced rates of carriage
	Bristol—Birmingham	86	3*s*.	$\frac{2}{5}$*d*. }		
1833	London—Bristol	120	4*s*.	$\frac{2}{5}$*d*. }	*Proceedings of the Great Western Railway Company*, pp. 11, 12, 19	These were the rates on groceries, teas, etc.
	London—Bath	110	6*s*.	$\frac{3}{5}$*d*. }		
	London—Reading	46	1$\frac{1}{2}$*s*.	$\frac{2}{5}$*d*. }		
	Birmingham—London	110	5*s*.	$\frac{1}{2}$*d*.	*Great Western Railway. Evidence on the London and Birmingham Railway Bill*, pp. 32–33, testimony of Mr Westall	This was the cost of waggon carriage. For carriage of articles that could be carried by coach the expense was 1*d*. per lb. See also Aris's *Birmingham Gazette*, April 2, 1821, p. 1; June 11, 1821, p. 1
	London—Oxford	60	3*s*. to 3*s*. 6*d*.	·6 to ·7*d*.	Brit. Mus. 8235. ee. 4 (1), 'Oxford and Didcot Railway Bill,' p. 3	
1835	Cirencester—London	98	4*s*. to 4*s*. 6*d*.	$\frac{1}{2}$*d*.	*Great Western Railway Bill. Evidence before Lords Committee*, pp. 489, 490	
	Tiverton—London	160	5*s*. 0*d*.	$\frac{3}{8}$*d*.	Ibid., p. 29	On woollens
	London—Tiverton	160	4*s*. 0*d*.	$\frac{3}{10}$*d*.	Ibid.	On wool
	London—Reading	46	1*s*. 9*d*.	$\frac{1}{2}$*d*.	Ibid., pp. 396, 398	On tallow and woollens
	Henley—Reading	7	2*s*. 0*d*.	3$\frac{1}{2}$*d*.	Ibid., p. 414	On woollen cloths

APPENDIX 8

COST OF CARRIAGE BY INLAND NAVIGATIONS

In the following statistics, I have presented some available information regarding the expenses of conveyance by canal, and, where possible, have made a comparison of these with the costs of conveyance by the ordinary roads. It will be observed in the first table that the opening of the Grand Trunk Canal reduced the cost of carriage to an amount only one-third to one-fourth of that paid for land carriage. The following figures are taken from Baines, *History of Liverpool*, pp. 439–40, into which, perhaps, they were incorporated from Salt, *Statistics and Calculations*, p. 71:

	Canal carriage per ton			Land carriage per ton		
	£	s.	d.	£	s.	d.
From Liverpool to Etruria		13	4	2	10	0
„ „ Wolverhampton	1	5	0	5	0	0
„ „ Birmingham and Stourport	1	10	0	5	0	0
From Manchester to Lichfield	1	0	0	4	0	0
„ „ Derby	1	10	0	3	0	0
„ „ Leicester	1	10	0	6	0	0
„ „ Newark	2	0	0	5	6	8
„ „ Nottingham	2	0	0	4	0	0
„ „ Wolverhampton	1	5	0	4	13	4
„ „ Birmingham	1	10	0	4	0	0
From Liverpool or Manchester to Shardlow ...	1	10	0	3	0	0

The freight rates on the navigations connecting Manchester with the Trent and Severn, and with the Birmingham Canal, in 1781, were:

for perishable goods, 3*d.* per ton per mile.
for non-perishable goods, 2½*d.* „ „

The freight from Shardlow to Gainsborough (on Trent Navigation) was 10*s.* per ton (v. *Manchester Collectanea*, in Chetham Society Publications, LXVIII, p. 136).

The following table of freight rates, taken from Allnutt, *Rivers and Canals West of London*, p. 3, shows us what the rates of carriage were along the Thames, as compared with the rates for land carriage in the same sections:

Price of Carriage on rhe River Thames Navigation.

	Water carriage per ton			Land carriage per ton		
	£	s.	d.	£	s.	d.
From London to Windsor or Maidenhead ...		9	0	1	10	0
„ „ Marlow or Henley		12	0	2	5	0
„ „ Reading or Caversham ...		15	0	2	10	0
„ „ Wallingford or Bensington ...		18	0	2	18	0
„ „ Abingdon or Oxford ...	1	2	0	3	5	0
„ „ Faringdon or Lechlade ...	1	8	0	4	0	0

Therefore, according to Allnutt, the cost of land carriage along this route was from three to three and one-third times as much as water carriage along the same route. But Mavor gives slightly different figures for practically the same year, and these we subjoin.

Mavor, *Agriculture of Berkshire*, 1808, p. 531, gives us the following Table showing Prices of Carriage on the Thames and Isis Navigation, from Lechlade to London and back, downward and upward; also Cost of Land Carriage to and from the Several Places undermentioned and London; and also the Time generally taken in navigating a Barge from such Places to London downward (with aid of stream) and upward (by horse-towing):

Between London and	Prices of water carriage		Price of land carriage each way	General time of passage	
	Down	Up		Down	Up
	per ton	per ton	per ton	days	days
	s. d.	*s. d.*	*£ s. d.*		
Staines	5 0	7 0	1 0 0	1	1½
Windsor	6 0	8 0	1 6 8	1	2
Maidenhead	7 6	9 6	1 10 0	1¼	2¼
Marlow	9 0	11 6	2 0 0	1½	3
Henley	10 6	13 6	2 5 0	1¾	3¼
Reading	12 0	16 0	2 10 0	2	4
Wallingford	14 0	18 0	3 0 0	3	5
Abingdon	16 6	23 0	3 10 0	3¼	5¼
Oxford	19 0	25 0	4 0 0	3¼	6
Lechlade	30 0	40 0	5 0 0	5	8

From this table, if we omit the last item, we learn that the cost of water carriage down-stream was roughly one-fourth, and the cost up-stream one-third, of the cost of carriage by land; or, in other words, the cost of land carriage was three to four times that of water carriage by the Thames Navigation.

Allnutt, p. 5. Prices of Carriage on the River Wey Navigation:

	Water carriage per ton	Land carriage per ton
	£ s. d.	*£ s. d.*
From London to Weybridge	8 0	1 5 0
,, ,, Guildford	12 0	1 15 0
,, ,, Godalming	13 6	2 0 0

Therefore, land carriage cost three times as much as water carriage.

Ibid., p. 6. Prices of Carriage on the Basingstoke Canal:

	Water carriage per ton	Land carriage per ton
	£ s. d.	*£ s. d.*
From London to Basingstoke	18 0	1 15 0

Therefore, land carriage cost twice as much as water carriage. (The cost of land carriage about 1786 seems to have been slightly higher than at this time (1810), for it was then £2 per ton. See Brit. Mus. B. 263 (6), 'Basingstoke Canal Navigation.')

Allnutt, *Rivers and Canals West of London*, p. 8:

Prices of Carriage on River Kennet Navigation:

	Water carriage per ton	Land carriage per ton
	£ s. d.	£ s. d.
From London to Reading (heavy goods, e.g., coal)	15 0	2 0 0
(other goods, e.g., groceries)	18 0	2 5 0
From London to Newbury (heavy goods) ...	18 0	3 10 0
(other goods) ...	1 5 0	3 15 0

Therefore, land carriage cost on the average three times as much as water carriage.

Ibid., p. 9. Prices of Carriage on the Wilts and Berks Canal:

	Water carriage per ton	Land carriage per ton
	£ s. d.	£ s. d.
From London to Abingdon	1 2 0	3 5 0
,, ,, Wantage	1 6 0	4 0 0
,, ,, Swindon	1 13 0	4 15 0
,, ,, Semington	2 5 0	5 10 0
,, ,, Bath	2 12 6	6 3 0
,, ,, Bristol	2 15 0	7 10 0

Therefore, the cost of land carriage was about three times the cost of water carriage.

Ibid., p. 11. Prices of Carriage on the Kennet and Avon Canal:

	Water carriage per ton	Land carriage per ton
	£ s. d.	£ s. d.
From London to Newbury (heavy goods, e.g., coal, timber)	18 0	3 10 0
(groceries, valuables, etc.)	1 5 0	3 15 0
From London to Devizes (heavy goods) ...	1 16 0	5 10 0
(groceries, etc.) ...	2 0 0	6 0 0
From London to Bath (heavy goods) ...	2 7 0	7 0 0
(groceries, etc.) ...	2 10 0	7 10 0
From London to Bristol (heavy goods) ...	2 9 6	8 0 0
(groceries, etc.) ...	2 13 0	8 10 0

Therefore, land carriage cost three times as much as water carriage.

From the tables given above, it would seem to be an almost universal rule that on the rivers and canals west of London the cost of water carriage was only about one-third of that by land.

On page 20 of Allnutt's work, he has shown the average ton-mile cost for the conveyance of goods on the various navigations. This table is next appended. A study of it will show that the price of carriage on river navigations was much less than on canal navigations. Where or how he obtained the data for this table, he does not say, and I have been unable to discover; but it almost seems as if his object were partisan, namely, to show that river navigation was cheaper than canal navigation. I insert the table here for what it is worth; but I warn the reader that my study of the subject does not warrant the above inference. The reduction of the cost to a ton-mile basis is contrary to the way in which, according to my researches, the prices were stated or reckoned.

Average Price of Carriage (including tolls, etc.):

Names of Navigations	Valuable or perishable goods or goods liable to risk (per ton per mile)	Other goods coarse or heavy (per ton per mile)
	d.	*d.*
By Basingstoke, Kennet and Avon, Wilts and Berks, Thames and Severn, and Stroud canals	5–5¾	3¼–4
By other canals, viz., Grand Junction, Oxford, Fazeley, Birmingham, Staffordshire and Worcestershire, and Grand Trunk ...	5½–6	3¾–4¼
On rivers Thames, Isis, Wey, Kennet, Avon, Severn	2½–3	2–2½
Average price of carriage by river Mersey, Runcorn to Liverpool	3–3½	1¾–2¼
By river Trent, Gainsborough to Shardlow ...	2¾–3¼	2–2½
By river Severn, Stourport to Bristol ...	2½–3	2–2¼

Before leaving this table, let me say that if any one will take the trouble to work out the ton-mile figures for the cost of carriage of goods, on the above-mentioned canals, as given in the foregoing tables, he will not find the figures in this last table to be correct, but will find them too high. Furthermore, we have facts from other sources which confirm our opinion that Allnutt's ton-mile figures here given for canal navigations are altogether too high. In 'Collection of Prospectuses, Maps, etc., of Railways and Canals,' p. 19, we have the distances and tonnage rates between Birmingham and London, via the Birmingham, Warwick and Birmingham, Warwick and Napton, Oxford, and Grand Junction canals, showing that the tonnage rate along this route for "general merchandise" was 29*s.* 8*d.*, and the distance 146 miles. This, when reduced to the standard that Allnutt (unwisely) adopted, would give us the cost of carriage along these canals as 2½*d.* per ton per mile, which is practically the same figure as he has given for the cost of carriage on the great rivers. On the whole, therefore, we are compelled to reject his inaccurate comparison, as given in this table. It must not be assumed that in making the above computation on a ton-mile basis we are giving any countenance to that basis of comparison of rates; we have used it simply tentatively, to prove the erroneous nature of what Allnutt has brought forward. The fact is that water rates were not quoted on that basis.

In the 'Proceedings of the Great Western Railway Company,' in 1833, in the evidence of Mr Stone (tea dealer) of Bristol (p. 11), it was shown that, since 1827, the cost of land carriage from London to Bristol was 4s. per hundredweight, and the cost of water carriage from London to Bristol was 2s. 6d. per hundredweight.

On page 12 (ibid.), in the evidence of Mr Shepherd (grocer) of Bath, it was shown that the cost of land carriage, London to Bath, was 6s. per hundredweight, and the cost of water carriage, London to Bath, was 2s. 6d. per hundredweight.

On page 19 (ibid.), in the evidence of Mr Harris (grocer) of Reading, it was shown that the cost of carriage by land, London to Reading, was 30s. per ton, and the cost of carriage by water, London to Reading, was 15s. per ton.

In the Reports of the Commissioners on the Thames Navigation, in 1811, comparative freight rates on canals and rivers are given; but they were issued by those who would be favourable to the Thames. They are given here because they show Allnutt's figures of ton-mile rates, on the canals, for valuable and perishable goods, to be altogether too high. The rates they quoted were as follows:

I. *By Canals.*

London—Birmingham (143 miles),
    Freight on valuable goods =55s. per ton, or 4½d. per ton per mile.
       ,,   heavy   ,, =35s.   ,,   3d.   ,,   ,,
Birmingham—Manchester (116 miles),
    Freight on valuable goods =40s. per ton, or 4½d. per ton per mile.
       ,,   heavy   ,, =26s.   ,,   3d.   ,,   ,,
Basingstoke—Weybridge (41 miles by canal and river),
    Freight on valuable goods =15s. 4d. per ton, or 4½d. per ton per mile.
       ,,   heavy   ,. =13s.   ,,   4¼d. ,,   ,,

The mean of the rates on the three canals shows that valuable goods were carried for 4½d. per ton per mile, and heavy goods were carried for 3¾d. per ton per mile.

II. *By River.*

Reading—London (78 miles),
    Freight on valuable goods =18s. per ton, or 2¾d. per ton per mile.
       ,,   heavy   ,, =13s.   ,,   2d.   ,,   ,,
London—Abingdon (108 miles),
    Freight on valuable goods =26s. per ton, or 2¾d. per ton per mile.
       ,,   heavy   ,, =20s.   ,,   2¼d. ,,   ,,

Taking the average of these rates, we would judge that, on the Thames river, valuable goods were carried for 2¾d. per ton per mile, and heavy goods were carried for 2⅛d. per ton per mile.

It will be noted from the above that the average rate here given on valuable goods carried by the canals was 4½d. per ton per mile, while Allnutt gives 5d. to 6d.; which seems to indicate that Allnutt's figures are very much overstated. Even the above figures given by the Thames Commissioners must not be accepted as authoritative, since they were given, doubtless for a purpose, by a body of men who wanted to present the Thames Navigation in as favourable a light as possible.

In the evidence of Mr Westall, a linen draper of Birmingham, before the Committee on the London and Birmingham Railway Bill, we learn that from London to Birmingham the rate on light goods carried by coach was 1d. per lb.; on heavy goods carried by waggon it was 5s. per cwt.; and on the canal, linen goods and mercery paid 2s. 9d. per cwt., while extra heavy goods paid 2s. 6d. per cwt. (v. *Great Western Railway. Evidence on the London and Birmingham Railway Bill*, pp. 32–33). Goods that came by coach from London were delivered in fifteen to sixteen hours from the time of leaving London. Goods that came by waggon generally took four days. Goods that came by canal took five to six days. From these statements it would appear that the cost by canal was just half of the cost by waggon. This is in accordance with information culled from other sources. For instance, that the cost of land carriage of Birmingham iron manufactures from Birmingham to London was 5s. per cwt., is confirmed by Brit. Mus. 214. i. 4 (120), 'Report of Committee on Oxford Canal,' p. 1; and the freight cost of the same goods by fly-boat on the canal from Birmingham to London, at a somewhat later period, was 45s. to 50s. per ton, or (say) 2s. 6d. per cwt. (*Remarks upon Pamphlet by "Investigator" on the Proposed Birmingham and London Railway*, p. 13). It is evident, therefore, from this that the expense of carriage by canal was only one-half of that by land.

Again, from the 'Report of the Committee on the Oxford Canal' [Brit. Mus. 214. i. 4 (120)], p. 3, we have the following statistics:

	£	s.	d.
From Birmingham to Oxford, by canal, 160 miles, cost of carriage per ton was	1	6	8
From Oxford to London, by Thames, cost per ton was ...	1	4	0
∴ total charge by canal and river from Birmingham to London was	2	10	8
But total charge by road from Birmingham to London was	5	0	0

Therefore the expense of water carriage was about one-half of that by road.

The great difference in cost between canal carriage and land carriage is also brought out in Phillips, *Plan for a Navigable Canal*, p. 21. Here he says that near the Staffordshire Canal the cost of road carriage was 8–9s. per ton for 10 miles, while the cost of water carriage by the canal was a half-crown per ton for 10 miles. This would indicate that canal carriage there was only about one-fourth the cost of carriage by road.

In the elaboration of this subject we find much diversity of statement as to the cost of carriage, and this is but natural, since the various waterways were very much unlike one another and also because different classes of goods could only be moved at different costs of transportation [see, for instance, *Communications to the Board of Agriculture*, i, p. 179; *Observations on the Comparative Merits of Navigations and Railroads*, p. 40; Gooch, *Agriculture of Cambridge*, p. 28]. But from what we have here presented, we may broadly generalize by saying that the cost of canal conveyance was from one-fourth to one-half of the cost of carriage by road.

APPENDIX 9

RAILWAY AND CANAL AMALGAMATIONS UP TO 1866

From Brit. Doc. 1872 (364), xiii, 1 (Part 2), pp. 755–6, we take the following table as to the canals and navigations acquired by railway companies by amalgamation, purchase, or lease:

Year when acquired	Canals and navigations acquired	Terms
	Bristol and Exeter Railway	
1864	Grand Western Canal	Sale authorized
1866	Bridgewater and Taunton Canal	Transfer authorized
	Great Eastern Railway	
1846	Stowmarket Navigation	Leased for 42 years from January 1846
1846	Lowestoft Navigation	Leased in perpetuity
	Great Northern Railway	
1846	Fossdyke and Witham Navigation	Authority to lease
1846	Grantham Canal	Authority to purchase
1847	Louth Navigation	Authority to purchase
1846	Nottingham Canal	Authority to purchase
	Great Western Railway	
1846	Stratford-upon-Avon Canal	Purchased
1852	Kennet and Avon Canal	} Transfer
	Kennet Navigation	
	Lancashire and Yorkshire Railway	
1846	Manchester, Bolton and Bury Canal	Vested
	Manchester, Sheffield and Lincolnshire Railway	
1846	Chesterfield and Gainsborough Canal	} Vested in perpetuity for an annuity
1846	Macclesfield Canal	
1846	Peak Forest Canal	
	London and North Western Railway	
1846–7	Shropshire Union Canals, viz.	Authorized in 1847 to be leased in perpetuity to the London and North Western Railway Company
	Shrewsbury Canal	
	Montgomeryshire Canal	
	Birmingham and Liverpool Junction Canal	
	Ellesmere and Chester	
	Shropshire Canal	Vested in Shropshire Union Co. 1857 and in the lease to the L. and N. W. Ry. Co.
1864	Lancaster Canal	Transfer by lease in perpetuity, 1864
1847	Huddersfield and Manchester Canal	Vested
1864	St Helen's Canal	Vested
1852	Cromford and High Peak Canal	Vested jointly in the Midland and L. & N. W. Ry. Cos.
	Midland Railway	
1846	Ashby Canal	Authorized to purchase
1846	Oakham Canal	Authorized to purchase
	Midland Great Western Railway	
1845	Royal Canal	Purchased

Year when acquired	Canals and navigations acquired	Terms
	Monmouthshire Railway	
1845	Monmouthshire Canal Navigations	Vested
1865	Brecon and Abergavenny Canal	Purchased
	Newport Pagnell Railway	
1863	Newport Pagnell Canal	Authority to purchase
	North British Railway	
1848	Edinburgh and Glasgow Union Canal	Vested
	North Eastern Railway	
	Hull and Leven Canal	
1847	Leven Canal	} Authority to purchase
	Pocklington Canal	
	North Staffordshire Railway	
1846	Trent and Mersey Canal	Vested
1864	Newcastle-under-Lyme Canal	Leased in perpetuity
	Somerset Central Railway	
1852	Glastonbury Canal	Transferred
	South Yorkshire Railway	
		Leased by Act of 1864,
1847	Stainforth and Keadby Navigation	together with the rail-
1847	Don Navigation	ways, for 999 years, to
1848	Sheffield Canal	the Manchester, Shef-
1850	Dearne and Don Navigation	field and Lincolnshire
		Railway
	Tenbury and Bewdley Railway	
1860	Leominster Canal (part of)	Purchased

APPENDIX 10

EFFECT OF RAILWAY COMPETITION ON CANAL CHARGES

IN the following tables, I have brought together such information of a statistical character as could be found, in the hope that it might make more definite the results that accrued from the competition of railways with the previously existing canals. It will be seen that the variations which are found in these rates are too wide to base any general statement upon them, and have it reflect with mathematical precision the reductions which were made. The only conclusion which we may reach from the facts as presented is that the minimum reduction was about one-sixth of the former rates and the maximum reduction was about six-sevenths of those rates. These limits, however, are of little practical value; and to generalize somewhat further and still be within the truth, we may say that, on the whole, the minimum reduction of rates was from one-third to one-half of the former rates.

The following table shows a statement of the reduction which took place in the rates on the Grand Junction and Leicester lines of canal since the introduction of railways in that section (1836). The Grand Junction Canal forms the main trunk of canal communication between London and the North. It extends from Paddington to Braunston, where it runs into the Oxford Canal, which communicates by other canals with Liverpool, Manchester and Birmingham. Five miles short of its entrance to the Oxford Canal it is joined by the Leicester lines. The lines here included were amalgamated with the Grand Junction Canal.

TABLE A. *Tonnage rates on undermentioned lines of canal.*

Canals	Rates authorized under their Acts, and which they did charge		Reduced since 1836 to	
	s.	*d.*	*s.*	*d.*
Grand Junction, 97 miles,				
On sundries	16	3¾	2	0¼
On coal	9	1	2	0¼
Grand Union, 24 miles,				
On sundries	6	0		5½
On coal	2	11		5½
Union, 19 miles,				
On sundries	4	9		5¼
On coal	2	1		5½
Leicester, 16 miles,				
On sundries	2	6		4
On coal	1	2		4
Loughborough, 10 miles,				
On sundries	2	6		4
On coal	1	2		4
Erewash, 11 miles,				
On sundries	1	0		4
On coal	1	0		4

Here the reduction was very great; so great as to make the competitive rates only one-third to one-eighth of the former canal rates (v. Brit. Doc. 1846 (275), XIII, 93, 'Minutes of Evidence,' p. 48).

From the Report of the Royal Commission of 1867, Brit. Doc. 1867 [3844], XXXVIII, 1, p. lxv, we take the following information:

TABLE B. *Rates per ton, Bristol to London.*

Articles	Rates charged by Carriers						Rates charged by Railway 1866	
	1820		1830		1840			
	s.	*d.*	*s.*	*d.*	*s.*	*d.*	*s.*	*d.*
Drapery	66	0	60	6	47	6	40	0
Hops	68	0	63	0	49	6	40	0
Oil	60	6	47	0	42	0	20	0
Tobacco	66	0	66	0	47	6	26	8

TABLE C. *Rates per ton, Birmingham to London.*

Articles	Rates by Canal		Rates charged by Railway 1866
	1836	1842	
	s. *d.*	*s.* *d.*	*s.* *d.*
Undamageable iron	25 0	—	15 0
Hardware	60 0	40 0	27 6
Sugar	40 0	37 6	21 8
Tallow	35 0	30 0	21 8
Drapery	70 0	45 0	40 0
Glass	70 0	—	27 6

This road was opened in 1837–8, and therefore the difference between the rates of 1836 and 1842 would show the effect of the railway. The canal rates of 1842 were only from two-thirds to six-sevenths of those of 1836; that is, there had been a reduction of from one-seventh to one-third of the former canal rates.

From Manchester to London, bales and grain were carried:

in 1833, by quick vans, at £20 per ton.
  „ 1834, by canal   „ £4   „
  „ 1840, by railway   „ £3. 4s. 8d. per ton.⎱ This was before the railway
  „ 1866, by railway   „ £1. 15s.   „   ⎰ company acted as carrier.

The railway rate of 1840 was, therefore, only three-fourths of the former canal rate of 1834.

Tables D, E, F, G are taken from Brit. Doc. 1881 (374), XIII, 1, 'Report of the Select Committee on Railway Rates and Fares,' Appendix No. 59.

TABLE D. *Tonnage rates, London to Birmingham. By canal in 1836; by railway and canal in 1842; by railways in 1866 and 1880.*

Articles	Rates by Canal, Collected and Delivered		Rates by Railways, Collected and Delivered		Rates by Railways, Collected and Delivered
	Before Railway opened	After Railway opened			
	1836	1842	1842	1866	1880
	s. *d.*	*s.* *d.*	*s.* *d.*	*s.* *d.*	*s.* *d.*
Undamageable iron	25 0	—	—	15 0	15 0
Damageable iron	27 6	—	—	17 6	17 6
Hardware ...	60 0	35 0	40 0	27 6	—
Nails	40 0	32 6	32 6	21 8	—
Raw Sugar ...	40 0	37 6	37 6	21 8	20 0
Lump Sugar ...	50 0	37 6	40 0	27 6	—
Tallow	35 0	35 0	30 0	21 8	—
Tea	50 0	37 6	40 0	32 6	34 2
Drapery ...	70 0	—	45 0	40 0	40 0
Spelter	—	18 0	20 0	17 6	15 0
Glass	70 0	—	—	27 6	—

During the stoppage of the canal by frost, etc., before the opening of the railway, goods had to be sent by road waggon at these charges:

Glass,   140*s*. per ton.
Other goods, 120*s*.   ,,

From the above figures for the canal rates of 1836 and 1842, it appears that, on account of the railway competition, the canal rates of 1842 were only from two-thirds to three-fourths of the canal rates of 1836.

TABLE E. *Tonnage rates, London to Manchester. By quick vans in 1833–4; by canal in 1834; by railways in 1840, 1866 and 1880.*

Articles	Quick Vans	Quick Vans	Canal	When Railways opened, before Ry. Cos. were carriers	Rates by Railways, the Railway Cos. being carriers	Rates by Railways, the Railway Cos. being carriers
	1833	1834	1834	1840	1866	1880
		£ s. d.	s.	s. d.	s. d.	s. d.
Sugar, raw ...	—	—	—	61 8	28 4	4-ton lots 22 6
						less lots 31 8
Sugar for refiners	—	—	—	—	20 0	20 0
Tallow ...	—	—	—	61 8	25 0	piping 27 6
Lead	—	—	—	61 8	28 4	31 8
Bales, Packs, and Trusses	£20	18 13 4	80	64 8	35 0	40 0
Hardware ...	—	—	—	64 8	40 0	43 4
Silk	—	—	100	—	insured 87 6	
					smalls 3 2	
					uninsured 62 6	62 6
					smalls 2 2	
Glass	—	—	100	69 8	40 0	40 0
Furniture ...	—	—	140	—	O. R. 70 0	O. R. 70 0
Luggage ...	—	—	—	79 8	55 0	55 0
Wines and Spirits	—	—	—	69 8	40 0	40 0
Hides	—	—	—	69 8	28 4	30 0

From the above table, nothing very definite can be learned as to the effect of railways in reducing the rates formerly charged by canals; for in a comparison of the rates of 1834 and 1840 there are only two articles for which the rates are given in both of these years, namely, "Bales, Packs, and Trusses," and "Glass." The table is more valuable in showing the reduction of rates between 1840 and 1866, after the railway companies became carriers.

The following table is much more valuable in showing the effect of railways in reducing the rates charged by the canals (q.v.).

From this table, the rates by canal, before and after the railway was opened, are easily compared, without any disturbing elements; and it is apparent that the reduced rates after the railway was opened were only from one-half to two-thirds of those in effect before the railway was opened; or, in other words, there was a reduction of one-third to one-half of the former cost of carriage.

TABLE F. *Tonnage rates, Birmingham to Manchester. By canal in 1836; by railways and canal in 1842; by railways in 1866 and 1880.*

Articles	Rates by Canal, Collected and Delivered		Rates by Railways, Collected and Delivered		Rates by Railways, Collected and Delivered
	Before Railways opened	After Railways opened			
	1836	1842	1842	1866	1880
	s. d.	s. d.	s. d.	s. d. s. d.	s. d.
Undamageable iron	22 6	—	—	10 0 to 11 6	12 0
Damageable iron	25 0	—	—	11 6 to 13 0	13 6
				s. d.	
Hardware ...	40 0	20 0	25 0	20 0	21 8
Nails	30 0	17 6	20 0	16 8	17 6
Iron wire ...	—	17 6	20 0	16 8	17 6
Parcels and bales	30 0	20 0	25 0	22 6	24 2
Flint Glass ...	40 0	—	—	O. R. 22 6	A. 20 10

TABLE G. *Tonnage rates, South Staffordshire to Liverpool. By canal in 1831, and by railways in 1866 and 1880.*

Articles	Rates by Canal, Collected and Delivered	Rates by Railways, Collected and Delivered	Rates by Railways, Collected only (but delivered alongside ship in 10-ton lots)
	1831	1866	1880
	s. d.	s. d. s. d.	s. d. s. d.
Undamageable iron	18 0	10 0 to 11 6	10 0 to 11 6
Damageable iron	20 0	11 0 to 13 0	12 6 to 14 0
Hardware ...	40 0	15 10 to 18 4	20 0 to 23 4
Nails	27 6	14 2 to 16 8	15 0 to 17 6
Glass	40 0	O. R. 22s. 6d.	17 6 to 18 4 O. R.
Timber	13 4	10 0 to 12 6	11 8 to 12 6
Grain	13 4	10 0 to 12 6	11 8 to 12 9
		(including collection in Liverpool)	

These figures do not show the immediate effect of the railways, but only the ultimate reduction of rates which they brought about, a reduction which amounted to one-sixth to one-half of the former canal rates.

APPENDIX 11

STATISTICS SHOWING EXTENT TO WHICH FREIGHT RATES WERE RAISED THROUGH AMALGAMATIONS OF CANALS AND RAILWAYS

In the following tables we give some authoritative information as to the extent to which railway and canal rates were raised, through the amalgamation of the canals with the railways. From the nature of the case, it is impossible to generalize, since each instance had no connexion with any other but was arranged solely on its own merits.

About the middle of the century, pig iron was brought in large amounts to Runcorn (chiefly from Scotland), because that was a great depot and distributing centre. From there it was sent along the Bridgewater Canal to Leigh, thence along the Leeds and Liverpool Canal and the Lancaster Canal into the country northward. The following table contrasts the tolls that were charged before and after the formation of the working arrangement between the railways and canals in that section.

Illustration of the Operation of the Advanced Toll on Pig Iron from Runcorn to the following places:

To	Total Railway charges with which the Canal from Runcorn must compete		Canal Tolls							
			Previous to the Lease			After the Lease			
	From Fleetwood	From Poulton	Lancaster Canal, 1d. per ton per mile	Leeds and Liverpool Canal, ½d. per ton per mile	Total Toll formerly	Lancaster Canal, 1d. per ton per mile	Leeds and Liverpool Canal, 1¼d. per ton per mile	Total Toll now
	s. d.	s. d.	miles	d.	miles	s. d.	s. d.	miles	d.	miles	s. d.	s. d.
Wigan	4 0	—	—	—	8	4	4	—	—	8	1 0	1 0
Blackburn	5 6	5 0	11	11	19	9½	1 8½	11	11	19	2 4½	3 8½
Accrington	—	5 6	11	11	25	1 0½	1 11½	11	11	25	3 1½	4 0½
Burnley	7 0	5 6	11	11	35	1 5½	2 4½	11	11	35	4 4½	5 8½
Marsden	—	5 6	11	11	40	1 8	2 7	11	11	40	5 0	5 11
Colne	—	5 6	11	11	45	1 10½	2 9½	11	11	45	5 7½	6 6½

It will be observed from these figures that the total tolls charged after the leasing of the canals by the railways were twice or three times as much as before the lease was effected. Brit. Doc. 1852–3 (246), xxxviii, 175, 'Evidence of Mr Loch.'

In the case of the Bolton and Bury Canal, great changes were made in the rate of tolls after its amalgamation with the Lancashire and Yorkshire Railway, as indicated by the following schedule, from which we see that in some cases the railway freight rate was slightly more than the canal tolls, and in other cases slightly less; so that, on the whole, we may say that the average railway freight mileage rate was probably about the same as the average canal tolls. In other words, independent carriers sending goods by the canal would pay as much in tolls only as the railway would charge for the whole service, including haulage. Brit. Doc. 1852–3 (246), xxxviii, 175, 'Evidence of Mr Loch.'

Bolton and Bury Canal.

Comparison of the Relative Charges per ton per mile made by the Lancashire and Yorkshire Railway Co., for freight on their railway, and for Toll on their Canal.

		Railway Freights				Canal Tolls		
		per ton	Distance	Mileage		per ton	Distance	Mileage
		s. d.	miles	*d.*	*s. d.*	*s. d.*	miles	*d.*
Manchester—Bolton	Iron	2 0	10	2·4	Toll 1 9½ Lockage 0¾ Wharfage 0½	1 10¾	10¾	2·12
					If from Liverpool or Runcorn direct	1 10¾	10¾	2·12
Liverpool—Bury	Timber Dye woods	6 8	34	2·35	*s. d.* Toll 2 1 Lockage 0¾ Wharfage 0½	2 2¼	12½	2·1
Liverpool—Bolton	Timber	6 8	28	2·86	Lockage and wharfage as above	1 10¾	10¾	2·12
Fleetwood—Bolton	Iron	5 0	40	1·5	" "	1 10¾	10¾	2·12
	Grain	6 4	40	1·9	" "	1 10¾	10¾	2·12
Fleetwood—Bury	Iron	6 0	46	1·56	" "	2 2¼	12½	2·1
	Grain	8 0	46	2·09	" "	2 2¼	12½	2·1

The following comparison is made of the Charges on the Huddersfield Canal, before and after amalgamation with the London and North Western Railway:

Huddersfield Canal.

Charges before Amalgamation	Charges subsequent to Amalgamation
Toll of 1*d.* per ton per mile	Toll, 1*d.* per ton per mile Tunnel dues, 1*s.* 6*d.* per boat each way Light dues, equal to 10 to 15 tons Wharfage, 1*d.* per ton

Illustration.

	£ s. d.	£ s. d.
Formerly, a boat with a cargo of 20 tons, and returning light, would pay for 20 miles		1 13 4
Subsequently, Toll	1 13 4	
Tunnel dues, 1*s.* 6*d.* each way ...	3 0	
Light dues, say for 15 tons ...	1 5 0	
Wharfage, on 20 tons	1 8	3 3 0
	Difference =	1 9 8

Brit. Doc. 1852–3 (246), xxxviii, 175, 'Evidence of Mr Loch.' See also ibid., p. 84, showing how, after a prolonged contest, the Bridgewater Canal Trustees were forced to put up their rates in obedience to the demand of the railways.

Two other cases are here quoted from the 'Report of the Committee of 1872 on Railway Amalgamations,' Brit. Doc. 1872 (364), XIII, 1, 'Minutes of Evidence,' p. 332:

Between Leeds and Manchester, there were three navigations, namely, the Aire and Calder, the Calder and Hebble, and the Rochdale Canal. Manchester packs were being conveyed along these waterways at a rate with which the railways could not compete; and in order to destroy this competition the London and North Western, the Lancashire and Yorkshire, the North Midland, and the Manchester, Sheffield and Lincolnshire Railways jointly obtained a lease of the Rochdale Canal for a term of years, and raised the tonnage upon Manchester packs to a rate prohibiting their conveyance upon the water any longer.

Another instance is that of the London and North Western Railway Company in dealing with the food supply from Liverpool and from Gloucester to the mining districts of Staffordshire. Distance was all in favour of Gloucester, and the rate from that port to the Staffordshire collieries was originally 7s. a ton. The London and North Western, having obtained practical control over the old Birmingham Canal, by which the food supplies were conveyed from the terminus of the Birmingham and Worcester Canal at Birmingham to the collieries, an average distance of ten miles, raised the rate on that canal so as to increase the total rate from 7s. to 10s. per ton, and by so doing turned the supply of corn for that district from Gloucester to Liverpool, in order that this supply should be conveyed over seventy-five miles of their railway. By 1865 the bankers, merchants, etc., of Liverpool were loud in their complaints against the excessive rates charged by "that leviathan monopolist," the London and North Western. Brit. Mus. C. T. 809 (7), 'Rates of Carriage to and from Liverpool,' pp. 1–10.

We have elsewhere noted the extortionate prices charged for carriage by the navigations connecting Liverpool and Manchester, before the opening of the railway there in 1830; and the strong protests of those who were the projectors of the railway. With reference to this subject, there is an interesting remark by Mr Francis R. Conder, C.E., in a paper read before the Manchester Statistical Society, on Nov. 30, 1882. He says: "The statement might well be regarded as incredible, were it not supported by indisputable evidence, that fifty years after the opening of the Liverpool and Manchester Railway, it costs more to convey a bale of cotton from the one city to the other than it did in 1829." Brit. Doc. 1883 (252), XIII, 1, Appendix, p. 239. Within half a century, one monopoly was displaced by a more progressive but equally exacting one.

APPENDIX 12

ILLUSTRATIONS OF THE WAY IN WHICH CANALS SOMETIMES MAINTAINED COMPETITION AGAINST THE RAILWAYS

It may serve to exemplify more fully how any competition between the canals and the railways was possible, if we take one or two illustrations:

About 1850, the Grand Junction Canal Company was the largest of the carriers by canal. In 1847 the canal companies generally were afraid that the carriers would be forced to leave the waterways. They had been driven off the London and North Western road and forced to give up some of their most important traffic. The Trent and Mersey Canal had allied its interests with, and was under the control of, the North Staffordshire Railway Company, and therefore the canal route from Birmingham to Liverpool and Manchester was practically closed to private carriers.

Under these circumstances, the Grand Junction Canal Company determined to fight for their right to carry between London and Birmingham. They entered into negotiations with other independent canals, asking them to share in starting a carrying establishment. They all refused; and the Grand Junction Company, before entering upon a carrying business, made agreements with almost all the other independent canals between the end of their line and Birmingham, so as to make sure what tolls these canals would charge them. Under these arrangements, the Grand Junction Company was able for years to keep the traffic on the canal and even to increase the *absolute* amount of it. The statistics of this trade we append below; Brit. Doc. 1852–3 (246), xxxviii, 175, 'Evidence of Mr Mellish,' pp. 14–15:

Amount of Trade on Grand Junction Canal.

Year	Through Trade	Local Trade	Total Trade
	tons	tons	tons
1833	186,029	522,228	708,257
1834	192,253	527,528	719,781
1835	192,859	631,786	824,645
1836	191,043	826,518	1,017,561
1837	216,706	890,251	1,106,957
1838	202,134	746,354	948,488
1839	231,953	712,169	944,122
1840	224,819	729,430	954,249
1841	235,511	851,954	1,087,465
1842	227,782	714,053	941,835
1843	239,116	749,386	988,502
1844	295,100	794,421	1,089,521
1845	294,257	847,616	1,141,873
1846	229,282	858,689	1,087,971
1847	253,141	910,325	1,163,466
1848	227,736	803,548	1,081,284
1849	206,390	771,865	978,255
1850	221,853	804,879	1,026,732
1851	219,886	879,988	1,099,874
1852	228,935	915,644	1,144,579

The foregoing statistics begin with 1833, the year the London and Birmingham Railway Act was passed, so that the figures for the first years were not affected by railway competition.

The total trade on the canal had increased, therefore, 25 per cent. in twenty years. To show how large this trade was, compare the following figures:

	Length of line	Tonnage
1852, London and North Western Railway	639 miles	3,898,622 tons.
„ Grand Junction Canal	135 „	1,144,579 „

That is, with a length of line about five times as great as that of the canal, the railway carried only three times the amount of freight that was carried by the canal. Of course, we must remember that the heavy and bulky freight that went by the canal paid a much lower carriage rate than the goods that were sent by rail, so that, ton for ton, the revenue on the railway was much greater than that on the freight carried by the canal. Consequently, from the standpoint of the operating revenue, the above comparison may mean very little.

Note, that much the larger part of the above traffic was local. We have said that the absolute amount of traffic on the canal increased, but, of course, the relative amount did not keep pace with that on the railway.

Another case which shows us the way in which, by the aid of the carriers, a canal was able to keep its traffic, at least for some time, from going over to the railway, was that of the Bridgewater Canal. When the Liverpool and Bury Railway was opened, which was another line between Liverpool and Manchester, it had to be satisfied with some part of the traffic to commence with. At that time, the Bridgewater Trustees were carrying about twice as much traffic between these two places as the Liverpool and Manchester Railway (v. table at end of this Appendix). The railway companies proposed to the Bridgewater Trustees that instead of continuing to carry what they could collect, and what they conceived themselves to be entitled to, they should be content with only half the traffic, and the other half should be divided between the two railway companies. The Trustees objected to this, but the railway companies insisted on the division; and at the same time they required the Trustees to exclude the private carriers from the canals, for it was felt that the success of the Trustees in collecting so large a traffic was due very much to the exertions and independent energy of the carriers. The Trustees declined both proposals: either to yield up the trade which was their own, or to exclude the carriers from their canals. (They had bought up the Mersey and Irwell Navigation, in 1844, as almost a bankrupt concern.) The railway companies persevered in their demands, and as a result the rates between Liverpool and Manchester were reduced from an average of 7*s.* and 9*s.* a ton, to 2*s.* 6*d.* a ton, for six months or so. The Trustees, to avert the railway companies' intention, made arrangements with the private carriers that they would carry them through safely and that they would bear their losses from the beginning to the close of the contest. In return for this, they required the carriers to act almost as their agents and to charge the freight rates that the Trustees might direct. It answered the purpose; the carriers were thus able to pass through the contest and aid the trade on the canal: and the final result was that the proportion of traffic on the canals was as large as, if not larger than, it had been previously.

The railway companies again applied for a division of traffic, and it was agreed to, with the stipulation that the Trustees should pay over to the railway companies 5*s.* per ton on the excess which the Trustees might carry above their one-half. This went on for about nine months, but it was a losing game for the Trustees, and they put an end to it about the close of the year 1850, after which there was no division of traffic, but a tariff of rates for the three parties.

Under previous arrangement, the canal charged 8*s.* 4*d.* a ton on manufactured goods from Manchester to Liverpool, while the railway charged 10*s.* for the same service, the difference being regarded as an equivalent for the faster carriage on the railway. But, later, the railway company forced the canals, after long-continued resistance, to put their rates up to 10*s.* Brit. Doc. 1852–3 (246), xxxviii, 175, 'Minutes of Evidence,' pp. 23, 84.

The tonnage of freight carried on these navigations, during these critical years, is given in the following table. It shows what an important factor the private carriers were in the maintenance of the traffic, as compared with the amount carried by the owners of the navigations:

Statement of Traffic by Water between Liverpool and Manchester, 1839–52 inclusive, and by Railway for the years 1838–48 inclusive.

Year	Bridgewater Canal			Mersey and Irwell Navigation			Gross Total Tonnage			Carried by Railway
	Carried by Owners	Carried by Carriers	Total Tonnage	Carried by Owners	Carried by Carriers	Total Tonnage	Carried by Owners	Carried by Carriers	Total Tonnage	
1839	30,526	162,600	193,126	96,488	29,280	125,768	127,014	191,880	318,894	Average 164,625
1840	26,862	169,631	196,493	90,455	51,352	141,807	117,317	220,983	338,300	
1841	26,263	154,013	180,276	74,429	69,155	143,584	100,692	223,168	323,860	
1842	19,217	169,154	188,371	54,113	85,385	139,448	73,330	254,489	327,819	
1843	21,207	235,788	256,995	54,947	101,063	156,010	76,154	336,851	413,005	
1844	19,311	266,599	285,910	36,424	107,408	143,832	55,735	374,007	429,742	190,914
1845	23,930	241,747	265,677	42,344	156,689	199,033	66,274	398,436	464,710	200,239
1846	29,134	254,233	283,307	50,219	164,442	214,661	79,353	418,675	498,028	217,416
1847	18,196	223,763	241,959	41,937	129,885	171,822	60,133	353,648	413,781	191,144
1848	16,334	262,064	278,398	28,689	122,850	151,539	45,023	384,914	429,937	181,968
1849	22,756	271,839	294,505	48,050	135,990	184,040	70,815	407,829	478,644	—
1850	24,025	230,655	254,680	29,867	129,036	158,903	53,892	359,691	413,583	—
1851	29,777	172,109	201,886	9,150	138,682	147,832	38,927	310,791	349,718	—
1852	27,993	201,768	229,761	7,701	131,442	139,143	35,694	333,210	368,904	—

APPENDIX 13

STATISTICAL VIEW OF HIGHWAY AND CANAL LEGISLATION

THE accompanying tabular view of the Road Acts and Canal Acts must not be understood to be reduced to the most careful mathematical exactness of absolute accuracy; but within the limits of accuracy which are at all possible in the application of statistics to the subject in hand, we venture to assert that no defects will be found, and that the presentation here given will show concisely the relative importance which the roads and canals assumed at the different periods and in the different sections of England. It has been the endeavour to group the counties by natural divisions, according to the great industrial characteristics which have been predominant in each group, and not according to any artificial geographical arrangement.

What, then, are the limits within which we may expect accuracy? In the first place, a road or a canal which extended into two or more counties has been noted under each county; so that if one road were built through three counties, it would be made to appear as three roads. But since this has been done consistently through the whole time between 1700 and 1830, the relative accuracy of our statistics will not be affected.

We must not suppose, however, that all these Acts represent actual road construction immediately after the passing of the Acts. Sometimes roads authorized to be made at a certain time were not made until years afterwards; and this would seem to vitiate any conclusions we might draw; but when we remember that this dilatoriness in constructing roads after they were sanctioned would not be much different at one period from another, we can easily see that our results are still quite comparable at all the periods during this epoch.

Again, *all* the Acts here enumerated were not for construction of new roads. A great many terms are used in the statutes in describing the purposes of the Acts, such as "building," "constructing," "amending," "repairing," "widening," "altering," etc., the roads; and in many other cases the Acts were passed for continuing the provisions of former Acts. While, therefore, our figures do not give us exact information as to new construction, they give us a very accurate guide as to the relative importance which the roads assumed at the different periods and in different sections of the kingdom. What we have said in this connexion regarding the roads is not so pertinent concerning the canals, for in connexion with them there was comparatively little legislation that was not followed by actual construction.

Another reservation we must make as to the roads, namely, that these Acts do not include the general road or turnpike Acts which were intended to apply to all the roads alike. As we have seen, these general Acts were scarce during the last half of the eighteenth century, because legislation was passed for each road separately according as the claim of each was presented. The number of these general Acts was so small in comparison with the number of separate road Acts (there being only four of any great consequence), that we may safely neglect them, as being insignificant for statistical purposes.

Now, what do our statistics, as thus defined, show in regard to the development of the means of communication? That there was a great increase in the attention given to road improvement beginning about 1750, is evident from the average road Acts per decade in the period given, the number per decade from 1751–90 being five

Divisions of England	Total Road Acts 1701–50	1751–70	1771–90	1791–1810	1811–30	Total Road Acts for the periods 1701–50	1751–90	1791–1830	Average Road Acts per decade 1701–50	1751–90	1791–1830	Percentage Increase of Road Acts 1751–90	1791–1830	Percentage of the whole area of England	Percentage of Road Acts for the periods 1701–50	1751–90	1791–1830
1. Northern Counties—Northumberland, Cumberland, Westmorland, Durham	16	39	39	62 / 4	76 / 1	16	78	138 / 5	3·2	19·5	34·5 / 1·2	509	77	10·5	3·8	4·8	5·6
2. Yorkshire, Lancashire, Cheshire	72	110 / 7	114 / 11	211 / 48	268 / 13	72	224 / 18	479 / 61	14·4	56· / 4·5	119·7 / 15·2	289	114	17·4	17·2	13·7	19·6
3. North Midlands—Derby, Stafford, Nottingham, Shropshire, Warwick, Leicester and Rutland, Northampton	55	189 / 12	190 / 22	237 / 63	277 / 26	55	379 / 34	514 / 89	11·	94·7 / 8·5	128·5 / 22·2	761	36	14·3	13·2	23·2	21·1
4. West Midlands—Hereford, Monmouth, Worcester	20	36 / 3	45 / 4	49 / 15	52 / 7	20	81 / 7	101 / 22	4·	20· / 1·7	25· / 5·5	400	25	4·2	4·8	5·	4·1
5. South Midlands—Berkshire, Oxford, Buckingham, Bedford, Hertford	89	89 / 1	78 / 1	98 / 14	111 / 8	89	167 / 2	209 / 22	17·8	41·7 / ·5	52·2 / 5·5	134	25	6·6	21·3	10·2	8·6
6. Eastern Counties—Lincoln, Huntingdon, Cambridge, Norfolk, Suffolk, Essex	44	75 / 1	55 / 1	95 / 13	108 / 7	44	130 / 2	203 / 20	8·8	32·5 / ·5	50·7 / 5·	269	56	17·6	10·5	8·	8·3
7. South-eastern Counties—Middlesex, Kent, Surrey, Sussex, Hampshire	82	162 / 0	122 / 3	195 / 17	220 / 24	82	284 / 3	415 / 41	16·4	71· / ·7	103·7 / 10·2	333	46	11·2	19·6	17·4	17·
8. South-western Counties: (a) Gloucester, Wilts, Somerset	40	112 / 0	78 / 2	114 / 32	134 / 17	40	190 / 2	248 / 49	8·	47·5 / ·5	62· / 12·2	494	31	8·4	9·6	11·6	10·2
(b) Dorset, Devon, Cornwall	0	58 / 0	42 / 2	50 / 9	83 / 5	0	100 / 2	133 / 14	0	25· / ·5	33·2 / 3·5	388	33	9·8	0	6·1	5·5
Total Road Acts / Total Canal Acts	418	870 / 24	763 / 46	1111 / 215	1329 / 108	418	1633 / 70	2440 / 823	83·6	408·2 / 17·5	610· / 80·7	388	49				

N.B. Canal Acts are lower figures in each case.

times, and from 1791–1830 between seven and eight times, greater than those of 1701–50. The immediate change at the decade 1751–60 is very marked from the number of Acts of that period as compared with those of the period 1741–50, namely 140 in the latter decade and 403 in the former, which is almost 200 per cent. increase. This change would seem to point very strongly to the belief that the Industrial Revolution was already in progress in the decade from 1751–60. The percentage of increase of the road Acts in 1751–90 over the preceding fifty years is markedly characteristic of England's progress.

Another feature of the table which will be at once discerned, is the way in which certain sections increased their road Acts, as a sign of the industrial advance in these localities. Perhaps this is best brought out by considering the average road Acts per decade in the three periods given. Two divisions are very prominent in this respect, namely, the North Midland counties in one group, and the counties of Lancaster, York and Chester in the other. These were the great manufacturing sections, which were much in need of improved means of carriage and communication. But the causes for these changes we have dwelt upon fully in the chapter dealing with the roads after 1750.

It is needless for us to follow out in detail all the information obtainable from such a view of the legislation; but one other fact deserves to be mentioned, that is, the enormous change in the number of canal Acts in the period 1791–1800 over any previous period. In that period there were almost exactly six times as many Acts as in the preceding decade; and the figures we have here given for twenty-year periods show the change at a glance. Had we the opportunity to give the figures for the year 1793, and compare them with the corresponding figures for any previous year, we should at once discover that the thirty-third year of the reign of George III was the year of the "canal mania." There were exactly three times as many Acts passed that year as in the preceding. With the opening of the railway era in 1830, we come to a time when there were very few, almost no, Acts passed for the construction of canals, and the break at this year was very abrupt.

APPENDIX 14

PICKFORD ET AL. *v.* THE GRAND JUNCTION RAILWAY CO.

An important chapter in the history of railways is that which deals with their relation to the carriers, and shows us the way in which the latter were driven off the lines of railway and their trade came into the hands of the railway companies. We must not give the impression, however, that *all* the firms of carriers were driven off the rails, nor that those which were ousted from their trade had their business overthrown and their connexions despoiled immediately. Some lines dealt more liberally with the carriers, and even encouraged their trade as a means of increasing the revenues of their roads; while others were opposed to the carriers from the first and did all they could to take the trade away from them. We have elsewhere examined this subject in general, and therefore do not need to revert to it here; but, instead, we wish to present the history of a particular case, which throws much light upon the general subject, a case which was unique in the history of railway traffic development, and which made the carrying trade the topic of vigorous and sometimes acrimonious discussion. The case of Pickford et al. *v.* The Grand

Junction Railway Co. is all the more important from the fact that at the present time Pickfords are probably the chief survivors of the early carriers who began their work on the canals and have had a continuous activity as forwarders of goods ever since.

For some years after the establishment of the steam railway as a common carrier, it was thought by many that the public interest could best be safeguarded by allowing competition on the railway lines; and, with this object in view, the private carriers were in most cases admitted to these lines. As we have already seen, there were several different systems under which the carriers were allowed to work. On the London and Birmingham Railway all carriers were admitted under certain regulations; but the Grand Junction Railway Co., while they allowed private carriers engaged in the London trade upon their line, retained to themselves the conveyance of all Birmingham and Lancashire goods.

The presence of the carriers upon their line was irksome to the Grand Junction Railway Co., for the latter were anxious to secure a regulated monopoly; and, while seeming to give more freedom to the carriers and the public, the railway company, in 1839, invited the carriers to enter into arrangements with them, by which the Birmingham and Lancashire traffic was to be opened to competition among the carriers, and the charges to the public were to be reduced. But the condition was imposed that the carriers were "not to charge less than" the railway company. The rates were reduced to the public, subject to this reservation. It would seem as if this were but a cloak to cover up a deeper design. In 1838 the railway company found that, even with their monopoly of the carriage between Lancashire and Birmingham, their business as general carriers was very small; for the public had become accustomed to their former carriers, whose established connexions gave them facilities for safe and economical carriage. It would, therefore, be advantageous for the railway company to have the old carriers supplanted and their connexions appropriated. Hence, while the carrier was, apparently, freely admitted upon the line, in reality he was admitted as a mere servant of the railway company. By the agreement, Pickfords got but a small percentage of the traffic receipts from the goods that were turned over to the railway company. Then, when the railway rates were reduced, ostensibly to benefit the public, the returns of the Pickfords became so small as to be unremunerative. The railway company did not stop here, however, but compelled Pickfords, under an ingenious arrangement, to contribute, out of their small percentage, a certain amount in payment of railway services at terminals (J. Moss, *Railways*, p. 386; *Railway Times*, iv, p. 186).

In 1840 a crisis was finally reached in the relation of the carrier to the railway company. By their Acts of Parliament the Grand Junction Railway Co. were authorized to make reasonable charges for the carriage of goods, and to fix what they regarded as proper charges for carrying small parcels, not exceeding 500 lbs. each. The railway company became carriers of goods for hire between Manchester and London, using for that purpose their own line and the lines of the Liverpool and Manchester and London and Birmingham Railway Cos. They published a list of charges, which divided the "rates by merchandize trains" into seven classes, from 16s. to 60s. a ton; and then followed "boxes, bales, hampers, or other packages," when they contained parcels, etc., under 112 lbs. weight each, directed, consigned, or intended for different persons, or for more than one person, on which the rate was made 1d. per lb. weight. On Nov. 24, 1840, Pickford & Co. packed several parcels (consisting of teas, books, and hardware, which had been delivered to them by various persons to be carried from Birmingham to Manchester) in a hamper, the gross weight of which, including the parcels, was 8 cwt. 3 qrs., although each parcel separately was less than 112 lbs. weight, and would, if delivered separately, have been a small parcel and thus have fallen under the title "smalls" according to the sevenfold classification above-named. This hamper was tendered to the Grand Junction Railway Co., and they were asked to carry it to its destination, for which service Pickfords offered to pay all that the railway company could legally charge, namely, 60s. per ton or a total of £1. 6s. 6d. The railway company's agent at Birmingham refused to receive the hamper unless the senders allowed him to open it, so that the number of parcels might be known, and each parcel might be charged and paid for separately at the rate fixed in the railway company's list, or unless they would pay the railway company 1d. per lb. upon the total weight of 8 cwt. 3 qrs., which would have amounted to £4. 1s. 8d. Pickfords refused to pay the latter, and the railway company refused to carry the hamper (*Railway and Canal Cases*, III, pp. 193–5. A similar case is given in ibid., III, pp. 197–8). This case was tried in 1841 and was decided in favour of Pickford & Co. (*Railway and Canal Cases*, II, p. 592 et seq.).

It will present the situation more clearly if we give briefly the two sides of the case, without attempting to weigh the merits of either; and if the reader will remember that there were other cases similar to this (e.g., Parker v. Great Western Railway Co., as given in *Railway and Canal Cases*, III, pp. 563–87), he will understand that the arguments in this case were applicable in the others. The evidence and affidavits in favour of the Grand Junction Railway Co. are found in *Railway Times*, IV (1841), pp. 208–9, 236–8, 289–92; V (1842), pp. 739–41; VI, pp. 176, 206; VII, pp. 217–18; and those in favour of Pickford & Co. are given in *Railway Times*, IV, pp. 293–6, 297–8, 366–7; V, pp. 739–41; VI, pp. 113, 152, 198–9, 238–9; VII, p. 328; and in *Railway and Canal Cases*, III, pp. 203–4, 538, 551–5.

The central difficulty, as we noted above, turned upon the carrying of "smalls" in hampers. It was said that Pickford & Co. had sent small parcels for different persons packed together in hampers, for the carriage of which they had paid the railway a certain rate per ton, the same as for goods, and afterwards, in distributing these parcels, charged each person the full amount of carriage that would have been paid had each parcel been carried separately. This was represented as having been done in order that the carriers might pocket the difference and thus swell their profits from the carrying trade. Pickford & Co. denied this allegation, and the court did not find any truth in it. On the other hand, the court decided that it was illegal for a railway company to charge for a hamper of small parcels for delivery to one consignee, the same as if they had had the trouble of collecting and delivering each separate article.

The statement was made that Pickford & Co.'s charges were not uniform to all persons under like circumstances, and thus there was introduced upon the railway a system which the public had found very objectionable on the canals. Not only were they accused of discrimination between customers, but it was also said that they commonly made insufficient, and, in some cases, untrue declarations of the description or quantity of the goods they delivered to the railway company for carriage, and thus deprived the railway company of their just and normal charges for carrying. The Grand Junction Railway Co. also believed it true that, because they did not charge anything for packages returning empty, Pickford & Co. declared as empty certain packages which were actually discovered to contain goods that Pickfords knew to be liable to charge like all other goods. To remedy these injuries that were believed to exist, and to benefit both the public and themselves, the Grand Junction Railway Co. determined to put an end to certain special agreements between them and the common carriers, and so some of the carriers continued their business and others ceased to carry on the Grand Junction Railway.

The Grand Junction Railway Co. had acquired the right to carry goods on the lines of the London and Birmingham Railway Co. and the Liverpool and Manchester Railway Co. In Liverpool and Manchester the Grand Junction Railway Co. had no arrangements for collecting and distributing their traffic and so employed the Liverpool and Manchester Railway Co. as their forwarding and delivering agent in these cities. They were in the same position with regard to London, and for some time they had employed the old carrying firm of Chaplin and Horne to be their agents in the metropolis, to unload and deliver in London all goods brought thither by the Grand Junction Railway Co., and to collect and load in London all goods that could be sent by that railway company. (The arrangements by which Chaplin and Horne acted as agents of the Grand Junction Railway Co. in London are given in *Railway and Canal Cases*, III, pp. 199–201; see also the advertisement of Chaplin and Horne in *Railway Times*, VII, p. 1447.) For this work Chaplin and Horne got 10*s.* per ton. The Grand Junction Railway Co. had been accustomed to charge 65*s.* per ton for goods sent from Manchester to London, and to pay 10*s.* per ton to Chaplin and Horne for their work. Pickford & Co. wanted the railway company to give them the same favourable rate as had been given Chaplin and Horne; and they tendered the railway company 55*s.* per ton for the carriage of the goods, they doing the work of distributing their own goods in London. But the railway company said 65*s.* per ton was their charge to all persons for carrying goods, and thus Pickford & Co. were refused any concessions. The court held that it was unreasonable for the Grand Junction Railway Co. to discriminate between the two carrying firms, and their decision was that the railway company should make their rates for carriage to all parties, under like circumstances, the same (*Railway and Canal Cases*, III, pp. 203–4).

In addition to the foregoing, another of the great points in dispute was the desirability of having the common carriers on the railway line. We have elsewhere shown that practically all the early railway acts contemplated the competition of the carriers on the line and made provision therefor according to the varying circumstances. Of course, subsequent experience had clearly shown that this right could not be exercised without great danger to the public, and therefore some deemed it wise either to regulate this competition, or else to have the railway companies cooperate with the carriers so that the latter, with their well-equipped establishments, might be valuable adjuncts of the railways in extending their traffic. The question, therefore, was, as to whether it was the interest of the public and the railways to overthrow or to encourage the carriers upon the lines. It was evident that if the railways could get all the carriers driven off their lines and could obtain a monopoly of the traffic, the public would have little security as to economy of charge and efficiency of management in connexion with the operation of the railways; and concerning this matter Pickford & Co., while opposing the Grand Junction Railway Co. in their monopolistic policy, appealed to the public to support them in their efforts to prevent this monopoly (see their letter addressed to "The Merchants and Trade of Liverpool," as given in *Railway Times*, VI, p. 152). The railway company, in their turn, pointed out that it would be more economical to eliminate all middlemen and their profits, as thereby rates would be lower for the public, and also referred with much satisfaction to the fact that their rates were much lower than the rates charged by the carriers on their line (v. *Railway Times*, VI, p. 206). Consequently, they urged that the public interests would be best served by giving them, rather than the common carriers, all the traffic. Pickford & Co., on the other hand, reminded the public that it was their competition that caused the railway company to put down the rates, and that if the public allowed them to be driven off the line, the Grand Junction Railway Co. would then be in a position to unduly increase the rates and recoup themselves for their present competitive loss.

The foregoing were the chief points in the controversy, but there were many minor elements which also entered into it (see digest of the essential points of the dispute in *Railway Times*, v, pp. 739–41). As we have already noted, the decision of the court (rendered July 7, 1842) and of the Vice-Chancellor was in favour of Pickford & Co., and the Grand Junction Railway Co. were required to live up to the law in the application of rates and to desist from discrimination against Pickford & Co. (*Railway and Canal Cases*, iii, pp. 203–4).

As a matter of fact, however, the Grand Junction Railway Co. did not obey the decision of the court, but continued their existing policy. A long and desultory correspondence was kept up between these two parties, ostensibly, from the railway company's point of view, to arrive at some satisfactory arrangement as to the legal principles to be observed, but, really, it would seem, to prevent the carrier from getting any hold upon traffic to be carried on the railway. Pickford & Co., by letter, appealed to the railway company, and urged the latter to deal fairly with their shareholders and the carriers by obedience to the law (v. letter in *Railway Times*, vi, p. 113); but the railway company refused to accept the court's decision. Pickfords then addressed the traders of Liverpool, and, doubtless, those of other places also, showing the determination of the Grand Junction Railway Co. to secure a monopoly of the carrying trade, and earnestly soliciting the support of the merchants in their efforts to prevent this (*Railway Times*, vi, p. 152). Their claims were upheld by some of the most influential of those who were closely in touch with traffic affairs, and were also supported by the experience of some roads which had found it desirable to change their former plan of exclusion of the carriers (*Railway Times*, vi, p. 152); but the Grand Junction Railway Co. continued their system (although slightly altered) with singular pertinacity and in almost entire disregard of the court, notwithstanding the statement of their Secretary that they were living up to the court's decree (v. Letter of Mark Huish, in *Railway Times*, vi, p. 206).

During the two years which followed the first decision, Pickford & Co. had been unable to get the railway company to grant them reasonable charges for carriage in accordance with the law. The Grand Junction Railway Co. took an appeal against the decision of the Vice-Chancellor, and reopened the case before the Lord Chancellor (*Railway and Canal Cases*, iii, p. 538); but upon the evidence showing the unreasonable and discriminating way in which the railway company had treated the carriers on their line (ibid., pp. 551–5), the Lord Chancellor upheld the decision of the Vice-Chancellor against the railway company.

It is almost impossible to ascertain exactly the attitude taken by the Grand Junction Railway Co. in regard to this decision, but it seems fairly certain that they practically ignored the decree of the court. We arrive at this conclusion from the action that was taken by the merchants of Liverpool in 1849, when they presented to the Railway Commissioners a memorial with reference to the carriage of parcels by railway. After citing the decision of the court that the sending of hampers packed with small parcels was legally and morally justifiable, they referred to the oppressive regulations issued by the railway companies calculated to put a stop to that privilege altogether, and then asked that an investigation be made into this course of procedure and the remedy be applied, so that shippers might be protected and that railways might be prevented from securing a monopoly of the carrying trade (v. 'Memorial' given in *Railway Times*, xii, p. 624. Memorials of like nature were presented from the merchants of Birmingham, Leeds, etc., as shown in *Herepath's Railway and Commercial Journal*, xi, p. 585). These complaints against the railway companies became loud and persistent; and in the same year (1849) a deputation, consisting of the most influential carriers, appeared before the Railway Commission, and presented petitions from Liverpool, Birmingham, Leeds, Edinburgh,

Sheffield, Newcastle, Bristol, and other large places, praying that the railway companies might be prevented from resorting to illegal and improper means in order to defeat fair competition (*Herepath's Railway and Commercial Journal*, xi, p. 599). With all this testimony, the conclusion is almost inevitable that the Grand Junction Railway Co. paid little attention to the verdict of the court; but continued the policy which was considered as most effective for driving the private carriers off their line.

On the whole subject of the relations of the railways to the carriers, see Nash, *Railway Carrying and Carriers' Law* (1846), Pt. ii, Chap. xi, which takes up the two great cases, Pickfords *v.* Grand Junction Railway Co., and Parker *v.* Great Western Railway Co. Hodges, *The Law relating to Railways and Railway Companies* (1847), Chap. ii, also treats the whole question fully from the legal side, and goes exhaustively into the two great cases. See also the public discussion of it in the *Railway Chronicle*, 1844, pp. 110–11, 184–5, 159–60, 184–5, and ibid., 1845, pp. 173 and 879, in which the railway side is taken; as it is also in Brit. Mus. 8235. b. 57 (1), 'The Carriers' Case considered in Reference to Railways' (1841), a small pamphlet written in a very biased vein. Both sides are presented in *The Times*, Mar. 5, 1844, p. 5; June 10, 1844, p. 6; Sept. 27, 1844, p. 6; April 21, 1845, p. 6; Mar. 21, 1846, p. 5; July 24, 1846, p. 4; Aug. 3, 1846, p. 3. Refer also to Whitehead, *Railway Management*, pp. 6–8, and Boyle, 'Hope for the Canals, pp. 5–6, 14–18.

BIBLIOGRAPHY

It is necessary to say that the following references do not comprise more than a majority of the works that have been consulted, but they are those from which material has been chiefly obtained for the discussion of the subject in hand. Occasionally others are mentioned in the text, but it has been essential to limit the voluminous bibliography to include only the most important sources of information.

OFFICIAL PUBLICATIONS

British Documents to 1911, including Journals of the House of Commons and House of Lords, Statutes of the Realm, Reports of Committees and Commissions, etc.

Rotuli Parliamentorum and Rotuli Hundredorum.

Hansard's Parliamentary Debates.

Mirror of Parliament.

Calendar of State Papers, Domestic.

Calendar of Treasury Papers.

Rymer's Foedera, Conventiones, Literae, et cujuscunque generis Acta Publica inter reges Angliae, etc.

HAYNES, S. A Collection of State Papers relating to Affairs in the Reigns of King Henry VIII, King Edward VI, Queen Mary, and Queen Elizabeth, from the year 1542 to 1570...left by William Cecil, Lord Burghley.

Acts of the Privy Council of England, 1429–1601.

RUSHWORTH, JOHN. Historical Collections of Private Passages of State, etc., 1618–49.

FIRTH, C. H., and TAIT, R. S. Acts and Ordinances of the Interregnum, 1642–1660. 3 vols. London, 1911.

Railway and Canal Cases. Vols. I–VII. London, 1835–54.

GENERAL WORKS

A., M. The Imperial Railway of Great Britain. London, 1865.

ACWORTH, WILLIAM M. The Railways of England. London, 1889.

ADAMS, WILLIAM BRIDGES. Practical Remarks on Railways and Permanent Way; as adapted to the Various Requirements of Transit. London, 1854.

—— Roads and Rails and their Sequences, Physical and Moral. London, 1862.

—— English Pleasure Carriages; their Origin, History, Varieties, Material, Construction, etc. London, 1837.

Agriculture, Board of. Communications to the Board of Agriculture; on Subjects relative to the Husbandry and Internal Improvement of the Country, 1797–1811. 7 vols. London, 1797–1811. Ditto for year 1819. London, 1819.

ALEXANDER, JOHN. The Two Stephensons. Pioneers of the Railway System. London, [1903].

ALLEN, JOHN.   History of the Borough of Liskeard and its Vicinity.   London, 1856.

ALLEN, LAKE.   The History of Portsmouth; containing a full and enlarged Account of its Ancient and Present State, etc.   London, 1817.

[ALLEN, THOMAS.]   A History of the County of Lincoln, from the Earliest Period to the Present Time; by the Author of the Histories of London, Yorkshire, Lambeth, Surrey, Essex, etc.   Assisted by Several Gentlemen residing in the County.   2 vols.   London, 1834.

ALLNUTT, ZACHARIAH.   Considerations on the Best Mode of Improving the Present imperfect State of the Navigation of the River Thames from Richmond to Staines, showing the Advantages to the Public, the Navigator, and the Owners and Occupiers of Houses, Mills, etc....in preference to the making any Canal. Maps, etc.   Henley, 1805.

—— Useful and Correct Accounts of the Navigation of the Rivers and Canals West of London.   Map.   2nd ed.   Henley, [ca. 1810].

ANDERSON, JAMES, LL.D.   Recreations in Agriculture, Natural History, Arts, and Miscellaneous Literature.   6 vols.   London, 1799–1802.

ANDERSON, JOHN.   History of Edinburgh.   Edinburgh, 1856.

ANDERSON, T.   An Historical and Chronological Deduction of the Origin of Commerce.   4 vols.   London, 1764.

Anonymous.   A Cursory View of a Proposed Canal: from Kendal to the Duke of Bridgewater's Canal, leading to the great Manufacturing Town of Manchester, etc.   (Birmingham Free Reference Library, No. 90,319.)   [N.P.]   [1769.]

—— Case in Support of the Bill now depending in Parliament for removing the Bar between the Birmingham, and the Birmingham and Worcester Canals. (1815.)   (Birmingham Free Reference Library, No. 87,368.)

—— Observation on the General Comparative Merits of Inland Communication by Navigations or Rail-Roads, with particular Reference to those projected or existing between Bath, Bristol, and London: In a Letter to Charles Dundas, Esq., M.P., Chairman of the Kennet and Avon Canal Company. London, 1825.

—— Statement by a Subscriber to the London and Birmingham Railway, to the Proprietors and Occupiers of Estates on the Line of the proposed...Railway. Dated, London, 27 Jan., 1831.   (Birmingham Free Reference Library, No. 88,482.)

—— Observations on Railways particularly on the Proposed London and Birmingham Railway.   London, 1831.

—— Railways; their Uses and Management.   London, 1842.

—— An Account of the Proceedings of the Great Western Railway Company, with Extracts from the Evidence given in Support of the Bill, before the Committee of the House of Commons, in the Session of 1834.   London, 1834.

—— Copy of the Wearmouth Bridge Acts of Parliament, viz., 32 Geo. III. c. 90, for building the Bridge, and 54 Geo. III. c. 117, to authorize the Disposal of Certain Securities on the Tolls by way of Lottery....   Bishopwearmouth, 1830.

—— Two Reports of the Commissioners of the Thames Navigation, on the Objects and Consequences of the Several Projected Canals, which interfere with the Interests of that River; and on the present sufficient and still improving State of its Navigation.   Oxford, 1811.

—— The Queen's Famous Progress, or Her Majesty's Royal Journey to the Bath, and Happy Return.   London, 1702.

—— Observations on a General Iron Railway: Showing its great Superiority over all the Present Methods of Conveyance, and claiming the particular attention of Merchants, Manufacturers, Farmers, and, indeed, every class of

Society. 2nd ed. (Author was Thomas Gray.) London, 1821. Brit. Museum 1029. e. 19.

Anonymous. The Parish of Ashburton in the Fifteenth and Sixteenth Centuries; as it appears from Extracts from the Churchwardens' Accounts, A.D. 1479–1580. With Notes and Comments. London, 1870.

—— A Compleat History of Somersetshire. Sherborne, 1742.

—— The History of Cheshire: Containing King's Vale-Royal entire, together with considerable Extracts from Sir Peter Leycester's Antiquities of Cheshire; and the Observations of Later Writers, particularly Pennant, Grose, etc. 2 vols. Chester, 1778.

—— Railway Competition. A Letter to George Carr Glyn, Esq., M.P., Chairman of the London and North Western Railway. London, 1849.

—— The History of Guildford, the County-Town of Surrey. Containing its Ancient and Present State, civil and ecclesiastical, Collected from Public Records and other Authorities, etc. Guildford, 1801.

—— The History of Inland Navigations, particularly those of the Duke of Bridgewater in Lancashire and Cheshire. Folding Map of the Bridgewater Canal. London, 1766. 2nd ed., 1799.

—— The Monthly Chronicle of North-Country Lore and Legend. 5 vols. Newcastle-on-Tyne, 1887–91.

—— Salopian Shreds and Patches (reprinted from "Eddowes's Shrewsbury Journal"). 10 vols. Shrewsbury, 1875–92.

ANSTICE, ROBERT. Remarks on the Comparative Advantages of Wheel Carriages of different Structure and Draught. Bridgewater and London, 1790.

ARCHER, MARK. William Hedley, the Inventor of Railway Locomotion on the Present Principle. 3rd ed. Newcastle-upon-Tyne, [1885].

ASHTON, JOHN. The Dawn of the XIXth Century in England. 2 vols. London, 1886.

ASTLE, THOMAS. The Will of Henry VII. London, 1775.

ASTON, JOSEPH. A Picture of Manchester. Manchester, [1816].

ATKINSON, D. H. Ralph Thoresby, the Topographer; His Town and Times. 2 vols. Leeds, 1885–87.

ATKINSON, THOMAS DINHAM, and CLARK, JOHN WILLIS. Cambridge Described and Illustrated, being a short History of the Town and University. London, 1897.

ATKYNS, Sir ROBERT. The Ancient and Present State of Glocestershire. London, 1712. (Reprinted in 1768.)

Avon, River. A Reply to the Answer to the Objections usually raised against the Embankment of the River Avon within the Port of Bristol...in a Letter to George Gibbs, Esq.,..... Bristol, 1791.

—— Orders in Council, His Majesty's Commission, and Certificate of the Commissioners, in the Year 1636, relative to the Navigation of the River Avon, in the Counties of Gloucester, Worcester and Warwick. (Birmingham Free Reference Library, No. 90,318.) Tewkesbury, 1826.

AXON, ERNEST (editor). Manchester Quarter Sessions. Notes of Proceedings, Vol. I, 1616–23. Edited from the Manuscript in the Reference Library, Manchester. [Manchester], 1901.

AXON, WILLIAM EDWARD ARMITAGE. The Annals of Manchester: A Chronological Record from the Earliest Times to the end of 1885. Manchester and London, 1886.

BACON, NATHANIELL. The Annalls of Ipswche. The Lawes, Customes and Governmt of the same. Collected out of Ye Records Bookes and Writings of that Towne. Edited by William H. Richardson, M.A.; with a Memoir by Sterling Westhorp, Mayor of Ipswich, 1884–85. Ipswich, 1884.

BADESLADE, THOMAS. The History of the Ancient and Present State of the Navigation of the Port of King's-Lyn, and of Cambridge, and the rest of the Trading-
Towns in those Parts: And of the Navigable Rivers that have their Course
through the Great-Level of the Fens, called Bedford Level. Also The History
of the Ancient and Present State of Draining in that Level....From Authentick
Records, and Ancient Manuscripts; and from Observations and Surveys....
London, 1725.

—— Reasons humbly offered to the Consideration of the Publick; shewing how
the Works now executing by Virtue of an Act of Parliament to Recover and
Preserve the Navigation of the River Dee, will destroy the Navigation; and
occasion the Drowning of all the Low Lands adjacent to the said River, etc.
Chester, [1735].

—— The New Cut Canal, intended for Improving the Navigation of the City of
Chester, with the Low Lands adjacent to the River Dee, compared with the
Welland, alias Spalding River, now silted up, and Deeping-Fens adjacent, now
Drowned, etc. Chester, [1736].

—— The Ancient and Present State of the Navigation of the Towns of Lyn,
Wisbeach, Spalding, and Boston; of the Rivers that pass through those Places....
2nd ed. London, 1751. [This is, in substance, the same as Badeslade's
"Account of the Present State, etc." with additions from MSS. in possession
of the Corporation of Adventurers.]

BAIGENT, FRANCIS JOSEPH, and MILLARD, JAMES ELWIN. A History of the Ancient
Town and Manor of Basingstoke in the County of Southampton. Basingstoke
and London, 1889.

BAILEY, JOHN, and CULLEY, GEORGE. General View of the Agriculture of the
County of Northumberland.... London, 1794.

—— General View of the Agriculture of the County of Cumberland. With
Observations on the Means of Improvement. Drawn up for the Consideration
of the Board of Agriculture and Internal Improvement. London, 1794.

BAILEY, THOMAS. Annals of Nottinghamshire. History of the County of Nottingham, including the Borough. 4 vols. in 2. London and Nottingham, [1852–55].

BAINES, EDWARD. The History of the County Palatine and Duchy of Lancaster.
2 vols. London and Manchester, 1868.

—— The History of the County Palatine and Duchy of Lancaster. New edition,
revised and edited by John Harland; continued and completed by Brooke
Herford. 2 vols. London, 1868.

BAINES, F. E. On the Track of the Mail Coach; being a Volume of Reminiscences,
personal and otherwise. London, 1895. (Author was sometime Inspector-
General of Mails, and author of "Forty Years at the Post Office.")

BAINES, THOMAS, and FAIRBAIRN, WILLIAM. Lancashire and Cheshire, Past and
Present. London, 1867.

BAINES, THOMAS. History of the Commerce and Town of Liverpool, and of the
Rise of Manufacturing Industry in the Adjoining Counties. London and
Liverpool, 1852.

BAIRD, THOMAS. General View of the Agriculture of the County of Middlesex, With
Observations on the Means of its Improvement.... London, 1793.

BARRETT, C. R. B. Somersetshire: Highways, Byways, and Waterways. London,
1894.

BARTLETT, WILLIAM A. The History and Antiquities of the Parish of Wimbledon,
Surrey. London and Wimbledon, 1865.

BATEMAN, JOSEPH. The General Turnpike Road Act, 3 Geo. IV. cap. 126; with
an Appendix of Forms...to which are added an Index and Notes. London, 1822.

BATEMAN, JOSEPH. A Second Supplement to the General Turnpike Road Acts for 1827. Containing the Amended Acts 5 Geo. IV. c. 69 and 8 Geo. IV. c. 24, etc. London, 1827.

—— The General Highway Act, 5 & 6 W. IV. c. 50, with Notes and an Index. London, 1835.

BATESON, MARY. Records of the Borough of Leicester, being a series of Extracts from the Archives of the Corporation of Leicester. 3 vols. London, 1899–1905.

BAYLEY, THOMAS BUTTERWORTH. Observations on the General Highway and Turnpike Acts passed in the Seventh Year of His Present Majesty, etc. London, 1773.

BAYNE, A. D. A Comprehensive History of Norwich (including municipal, political, religious, and commercial History). London and Norwich, 1869.

BEARCROFT, PHILIP, D.D. An Historical Account of Thomas Sutton, Esq.; and of his Foundation in Charter-House. London, 1737.

BECKMANN, JOHN. A History of Inventions and Discoveries. 3 vols. London, 1797.

BENNETT, JAMES. The History of Tewkesbury. Tewkesbury, 1830.

BERGIER, NICOLAS. The General History of the Highways, in all Parts of the World, more particularly in Great Britain. London, 1712. (Good on Roman administration of the highways.)

BILLINGSLEY, JOHN. General View of the Agriculture of the County of Somerset, with Observations on the Means of its Improvement.... Bath and London, 1797.

BIRD, JAMES BARRY. The Laws Respecting Travellers and Travelling, comprising all the Cases and Statutes relative to that Subject, etc. London, 1801.

—— The Laws respecting Highways and Turnpike Roads, comprising the Common Law relating to Highways...the Statute Law relating to Highways and Turnpike Roads, etc. London, 1801.

BISHTON, J. General View of the Agriculture of the County of Salop.... Brentford, 1794.

BLACKLOCK, F. GAINSFORD. The Suppressed Benedictine Minster, and other Ancient and Modern Institutions, of the Borough of Leominster. Leominster, [1900].

BLACKNER, JOHN. The History of Nottingham, embracing its Antiquities, Trade, and Manufactures, from the Earliest Authentic Records to the Present Period. Nottingham, 1815.

BLEW, WILLIAM C. A. Brighton and Its Coaches. A History of the London and Brighton Road, with some Account of the Provincial Coaches that have run from Brighton. London, 1894.

BLEWITT, REGINALD JAMES. New Monmouthshire Railway from Newport to Nanty-glo and Blaenafon through Pontypool. [Newport, 1844.]

BLOMEFIELD, FRANCIS. An Essay towards a Topographical History of the County of Norfolk. 2nd ed. 11 vols. London, 1805–10.

BLOMFIELD, J[AS]. C[HARLES]. History of the Present Deanery of Bicester, Oxon. 8 vols. Oxford and London, 1882–94.

BOOTH, HENRY. An Account of the Liverpool and Manchester Railway, comprising a History of the Parliamentary Proceedings, preparatory to the Passing of the Act, a Description of the Railway, in an Excursion from Liverpool to Manchester....Liverpool, [1830].

BOURN, DANIEL. A Treatise upon Wheel Carriages; shewing their Present Defects: with a Plan and Description of a New Constructed Waggon: which will effectually preserve and improve the Public Roads, and be more useful, cheap, and handy to the Proprietor. London, 1763.

—— Some Brief Remarks upon Mr Jacob's Treatise on Wheel Carriages. London, 1773.

BOWES, I. George Stephenson and M. Ferdinand de Lesseps. The Men and their Work: the First Public Railway in the World: and the Suez and Panama Canals. London, 1893.

BOYLE, EDWARD, and WAGHORN, THOMAS. The Law relating to Traffic on Railways and Canals. 3 vols. London, 1901.

BOYLE, J. R. The Early History of the Town and Port of Hedon, in the East Riding of the County of York. Hull and York, 1895.

—— The County of Durham. London, 1892.

BOYLE, THOMAS. Hope for the Canals! showing the Evil of Amalgamations with Railways to Public and Private Interests, and the Means for the Complete and Permanent Restoration of Canal Property to a Position of Prosperity, upon its Present Basis of Original and Independent Enterprise. London, 1848.

BOYMAN, BOYMAN. Steam Navigation, its Rise and Progress. London, 1840.

BOYS, JOHN. General View of the Agriculture of the County of Kent.... Brentford, 1794.

BOYS, WILLIAM. Collections for an History of Sandwich in Kent. Canterbury, 1892.

BRAND, JOHN. The History and Antiquities of the Town and County of the Town of Newcastle upon Tyne. 2 vols. London, 1789.

BRASBRIDGE, JOSEPH. The Fruits of Experience; or Memoir of Joseph Brasbridge, written in his 80th year. London, 1824.

BRASSINGTON, W. SALT. Historic Worcestershire. London, 1895. (Pp. 21–80 deal with Roman, British and Saxon Roads.)

BRAY, WILLIAM. Sketch of a Tour into Derbyshire and Yorkshire. 2nd ed., 1788. (In Pinkerton's Voyages and Travels, Vol. II, pp. 336–464.)

BRAYLEY, EDWARD W. A Topographical History of Surrey. 5 vols. London, 1850.

BRENT, JOHN. Canterbury in the Olden Time, From the Municipal Archives and other Sources. Canterbury and London, 1860.

BREWER, J. S., and BULLEN, WILLIAM. Calendar of the Carew Manuscripts, preserved in the Archiepiscopal Library at Lambeth, 1515–1574. London, 1867–73.

BRINDLEY, JAMES. Queries proposed by the Committee of the Common-Council of the City of London, about the intended Canal from Monkey Island to Isleworth, answered. Also, An Estimate of the Expence of making a Navigable Canal from Monkey Island to Isleworth, upon the Lower Line laid down in the Plan, by the Way of Stanes, etc. London, 1770.

—— [Report] to the Committee of the Common-Council of the City of London [regarding the improvement of the river Thames]. [London], 1770.

BRISCOE, JOHN POTTER. History of the Trent Bridges at Nottingham. (A Paper read before the Royal Historical Society, Apr. 22, 1878.)

BROME, JAMES. Travels over England, Scotland, and Wales. Giving a True and Exact Description of the Chiefest Cities, Towns, etc. London, 1700.

BROWN, THOMAS. General View of the Agriculture of the County of Derby.... London, 1794.

BROWNE, Sir THOMAS (1605–82). Works: including his Life and Correspondence. Edited by Simon Wilkin. 4 vols. Norwich and London, 1836.

BUCHANAN, ROBERTSON (C.E.). A Practical Treatise on Propelling Vessels by Steam, etc. Glasgow, 1816.

BULL, FREDERICK WILLIAM. A History of Newport Pagnell. Kettering, 1900.

[BULL, HENRY.] A History, Military and Municipal, of the Ancient Borough of the Devizes, and subordinately of the entire Hundred of Potterne and Cannings in which it is included. London and Devizes, 1859.

BULLER, E. C. A Compendious History and Description of the North Union Railway. Preston, 1838.

BUNCE, JOHN THACKRAY. History of the Corporation of Birmingham, with a Sketch of the Earlier Government of the Town. 2 vols. Birmingham, 1878–85.

BUND, J. W. WILLIS (compiler and editor). Worcestershire County Records. Division I, Documents relating to Quarter Sessions. Calendar of the Quarter Sessions Papers. Vol. I, 1591–1643. Worcester, 1900.

BURN, RICHARD. The Justice of the Peace, and Parish Officer. 2 vols. London, 1755. (Burn was a J.P. for Westmorland County.)

—— A New and Accurate Description of the Present Great Roads and the Principal Cross Roads of England and Wales. London, 1756.

—— The History of the Poor Laws: with Observations. London, 1764. (Pp. 236–238 refer to the highways.)

BURTON, ROBERT. A New View, and Observations on the Ancient and Present State of London and Westminster (continued by an Able Hand). London, 1730.

BURTON, THOMAS. Diary of Thomas Burton, Esq., Member in the Parliaments of Oliver and Richard Cromwell, from 1656 to 1659....Edited and illustrated with notes historical and biographical by John Towill Rutt. 4 vols. London, 1828.

BURTON, WILLIAM. The Description of Leicestershire, Containing Matters of Antiquitye, Historye, Armorye and Genealogy. (MS. notes by Peter Le Neve.) London, 1622.

BUSCH, Dr WILHELM. England under the Tudors. Vol. I, King Henry VII (1485–1509). Translated under the supervision of Rev. A. H. Johnson, M.A., by Alice Todd. With Introduction and some comments by James Gairdner. London, 1895.

BUTTERWORTH, A. KAYE. A Treatise on the Law relating to Rates and Traffic on Railways and Canals, with special reference to The Railway and Canal Traffic Act, 1888, and the Practice of the Railway and Canal Commission. 2nd ed. London, 1889.

CALDERWOOD, Mrs. Letters and Journals of Mrs Calderwood of Polton, from England, Holland, and the Low Countries in 1756. Edited by Alexander Fergusson. Edinburgh, 1884.

[CANNING, RICHARD.] The Principal Charters Which have been Granted to the Corporation of Ipswich in Suffolk. London, 1754.

CAPPER, CHARLES HENRY. Observations on "Investigator's" Pamphlet relative to Railways. London, 1831.

CAPPER, ROBERT. Proposed Birmingham Ship Canal. Birmingham, 1886.

CARÖE, W. D., and GORDON, E. J. A. Sefton: A Descriptive and Historical Account, comprising the Collected Notes and Researches of the late Rev. Engelbert Horley; Together with the Records of the Mock Corporation. London and New York, 1893.

CARTER, N. H. Letters from Europe, comprising the Journal of a Tour through Ireland, England, Scotland, France, Italy, and Switzerland in the years 1825, '26, and '27. 2 vols. New York, 1827.

CARTWRIGHT, JAMES J[OEL]. Chapters in the History of Yorkshire; being a collection of Original Letters, Papers, and Public Documents, illustrating the State of that County in the Reigns of Elizabeth, James I, and Charles I. Wakefield, 1872.

—— (editor). The Memoirs of Sir John Reresby, of Thrybergh, Bart., M.P. for York, etc., 1634–1689, written by Himself. London, 1875.

CARY, JOHN. Cary's new Itinerary, or an Accurate Delineation of the Great Roads, both direct and cross, throughout England and Wales. London, 1798.

—— Inland Navigation; or Select Plans of the Several Navigable Canals throughout Great Britain, etc. London, 1795.

CARY, ROBERT. Memoirs of the Life of Robert Cary, Baron of Leppington, and Earl of Monmouth. Written by Himself, and now published from an Original MS. in the Custody of John Earl of Corke and Orrery,.... London, 1759.

CATHRALL, WILLIAM. The History of Oswestry, comprising the British, Saxon, Norman, and English Eras; the Topography of the Borough, etc. Oswestry, [1855].

CHALLENOR, BROMLEY. Selections from the Municipal Chronicles of the Borough of Abingdon from A.D. 1555 to A.D. 1897. Abingdon, 1898.

CHALMERS, GEORGE. Caledonia: or a Historical and Topographical Account of North Britain from the most ancient to the present Times. New edition. 8 vols. Paisley, 1887–1902.

CHAMBERLAIN, HENRY. A New and Compleat History and Survey of the Cities of London and Westminster, the Borough of Southwark, and Parts adjacent; from the Earliest Accounts to the Beginning of the Year 1770....By a Society of Gentlemen; Revised, Corrected and Improved, by Henry Chamberlain. London, 1770.

CHAMBERLAYNE, EDWARD. Angliae Notitia, or the Present State of England: Together with Divers Reflections upon the Antient State thereof. 1st, 2nd and 3rd editions. London, 1669.

CHAMBERS, A. H. Observations on the Formation, State and Condition of Turnpike Roads, and other Highways, with Suggestions for their Permanent Improvement on Scientific Principles.... London, 1820.

CHANTER, J. R., and WAINWRIGHT, THOS. Reprint of the Barnstaple Records. 2 vols. Barnstaple, 1900.

CHANTREAU, M. Voyage dans les trois Royaumes d'Angleterre, d'Écosse et d'Irlande, fait en 1788 et 1789. 3 vols. Paris, 1792.

CHAPMAN, WILLIAM. Report on the Proposed Navigation between the East and West Seas, so far as extends from Newcastle to Haydon-Bridge,.... Newcastle, 1795.

—— Observations on the Various Systems of Canal Navigation, with Inferences practical and mathematical; in which Mr Fulton's Plan of Wheel-Boats, and the Utility of Subterraneous and of small Canals, are particularly investigated. London, 1797.

CHATTAWAY, E. D. Railways: Their Capital and Dividends, with Statistics of their Working in Great Britain and Ireland. London, 1855–56.

CITIZEN, A (pseud.). A Short Historical Account of Bristol Bridge; with a Proposition for a new Stone Bridge, from Temple Side to the Opposite shore. As also An Account of some remarkable stone Bridges abroad....Likewise An Account of some Bridges of Note in England. Bristol, 1759.

—— A Reply to A most Partial Pamphlet, entitled A Letter from a By-Stander, to the Commissioners for Re-building the Bridge at Bristol. Bath, 1762.

CIVIS (pseud.). The Railway Question. Practical Suggestions for a Fundamental Reform of the Railway System, on a Principle combining National Benefits with the Permanent Interests of Shareholders. London, 1856.

CLARIDGE, JOHN. General View of the Agriculture in the County of Dorset.... London, 1793.

CLARK, ANDREW. The Life and Times of Anthony Wood, Antiquary, of Oxford, 1632–1695, described by himself. Collected from his Diaries and other Papers. 5 vols. Oxford, 1891–1900.

CLARK, ANDREW. Survey of the Antiquities of the City of Oxford, composed in 1661–66 by Anthony Wood; edited by Andrew Clark. 3 vols. Oxford, 1889, 1890, 1899.

CLARK, JOHN. General View of the Agriculture of the County of Hereford. London, 1794.

CLARKE, LEONARD W. The History of Birmingham. 7 vols. in MS. Birmingham, [N.D.].

CLELAND, JAMES. Enumeration of the Inhabitants of the City of Glasgow and County of Lanark. For the Government Census of M.DCCC.XXXI. with Population, and Statistical Tables relative to England and Scotland. 2nd ed. Glasgow, 1832.

CLIFFORD, FREDERICK. History of Private Bill Legislation. 2 vols. London, 1885–87.

CLUBBE, JOHN. History and Antiquities of the Ancient Villa of Wheatfield, in the County of Suffolk. (Contained in "Fugitive Pieces on Various Subjects," Brit. Mus. 12,352. ee. 21. London, 1761.)

CLUTTERBUCK, ROBERT. The History and Antiquities of the County of Hertford, compiled from the best authorities and Original Records. 2 vols. London, 1815.

CLUTTERBUCK, Rev. R. H. The Archives of the Corporation of Andover. Reprinted from the "Andover Advertiser," 1891 (?).

COBBETT, RICHARD STUTELY. Memorials of Twickenham. London, 1872.

COBBETT, WILLIAM, Jr. The Law of Turnpikes; or an Analytical Arrangement of, and Illustrative Commentaries on, all the General Acts relative to the Turnpike Roads of England: etc. London, 1824.

COBBETT, WILLIAM, M.P. Tour in Scotland; and in the Four Northern Counties of England in the Autumn of the Year 1832. London, 1833.

CODRINGTON, THOMAS. Roman Roads in Britain. London and Brighton, 1903.

COLBORNE, JOHN. Report on a View taken in Pursuance of an Order of the Bedford Level Corporation, in the Months of June and July last, of the Middle and South Levels, and their Outfalls to sea; with a Plan for the effectual Draining the said Levels. [Chester], 1777.

COLBORNE, THOMAS. The Monmouthshire Railways and the Monmouthshire Freighters. Newport, 1890.

[COLLINS, B. C.] A Treatise on Inland Navigation. Salisbury: Printed for B. C. Collins, 1788.

COLQUHOUN, PATRICK. A Treatise on the Commerce and Police of the River Thames.

COLVILLE, Mrs ARTHUR. Duchess Sarah: being the Social History of the Times of Sarah Jennings Duchess of Marlborough, with Glimpses of her Life and Anecdotes of her Contemporaries in the Seventeenth and Eighteenth Centuries. London, New York, etc., 1904.

CONYBEARE, Rev. EDWARD. A History of Cambridgeshire. London, 1897.

COOKE, GEO. ALEXANDER. A Topographical and Statistical Description of the County of Lancaster; containing an Account of its Situation, Extent, Towns, Rivers, Canals, etc. London, [N.D.].

—— Topographical and Statistical Description of the County of York. London, [N.D., but about 1812].

CORDIER, J. Considérations sur les chemins de fer. Paris, 1830.

[CORRY, JOHN.] The History of Liverpool, from the Earliest Authenticated Period down to the Present Time.... Liverpool, 1810.

CORRY, JOHN, and EVANS, Rev. JOHN. The History of Bristol, Civil and Ecclesiastical; including Biographical Notices of Eminent and Distinguished Natives. (Vol. I by John Corry and Vol. II by Rev. John Evans.) Bristol, 1816.

CORRY, JOHN. The History of Lancashire. 2 vols. London, 1825.

CORT, R. Rail-road Impositions Detected: or Facts and Arguments to prove that the Liverpool and Manchester Railway has not paid One per cent. Nett Profit; and that the Birmingham, Bristol, Southampton, Windsor, and other Railways are, and must forever be, only Bubble Speculations. London, 1834.

COTTERILL, CHARLES FORSTER. The Past, Present and Future Position of the London and North Western, and Great Western Railway Companies...with an Inquiry how far the Principle of Competing Lines, in Reference to Railway Undertakings, is sound and defensible. London, 1849.

COULSON, H. J. W., and FORBES, URQUHART A. The Law relating to Waters, Sea, Tidal and Inland, Including Rights and Duties of Riparian Owners, Canals, Fishery, Navigation, Ferries, Bridges, and Tolls and Rates thereon. 2nd edition. London, 1902.

COURT, M. H. A Digest of the Realities of the Great Western Railway. London, 1848.

COWPER, EDWARD ALFRED. The Steam Engine. (In "Heat and Its Mechanical Applications." A series of Lectures delivered at the Institution of Civil Engineers, session 1883–84, pp. 55–100.) London, 1885.

COWPER, J. M. Notes from the Records of Faversham, 1560–1600. (In Trans. Royal Historical Society, 1872, pp. 324–343.)

COX, Rev. J. CHAS. Three Centuries of Derbyshire Annals, as illustrated by the Records of the Quarter Sessions of the County of Derby, from Queen Elizabeth to Queen Victoria. 2 vols. London, 1890.

COXE, WILLIAM. A Historical Tour Through Monmouthshire. (First published in 1801.) Brecon, 1904.

CROKER, JOHN WILSON. The Croker Papers. The Correspondence and Diaries of the late Right Honourable John Wilson Croker, LL.D., F.R.S., Secretary to the Admiralty from 1809–1830. Edited by Louis J. Jennings, in 3 vols. London, 1885.

CROMWELL, THOMAS. History and Description of the Ancient Town and Borough of Colchester, in Essex. 2 vols. London, 1825.

CROSTON, JAMES. The History of the County Palatine and Duchy of Lancaster, by the late Edward Baines, with additions of the late John Harland and Rev. Brooke Herford. A New, Revised, and Enlarged Edition, edited by James Croston. 5 vols. Manchester and London, 1888.

CRUDEN, ROBERT PEIRCE. The History of the Town of Gravesend in the County of Kent and of the Port of London. London, 1843.

CRUTCHLEY, JOHN. General View of the Agriculture of the County of Rutland.... London, 1794.

CUMMING, ALEXANDER. Observations on the Effects which Carriage Wheels with Rims of different Shapes have on the Roads; respectfully offered to the Consideration of the Legislature. (Birmingham Free Reference Library, No. 210,445.) Pentonville, 1797.

—— A Supplement to the Observations on the Contrary Effects of Cylindrical and Conical Carriage Wheels. London, 1809. (Discusses mechanical features.)

CUMMING, T. G. Description of the Iron Bridges of Suspension now Erecting over the Strait of Menai, at Bangor, and over the River Conway, in North Wales.... Also some Account of the Different Bridges of Suspension in England and Scotland, etc. London, 1824.

—— Illustrations of the Origin and Progress of Rail and Tram Roads, and Steam Carriages, or Locomotive Engines; also Interesting Descriptive Particulars of the Formation...Extent, and Mode of Working some of the Principal Railways now in Use within the United Kingdom. Denbigh, 1824.

CUNDY, N[ICHOLAS] W. Inland Transit. The Practicability, Utility, and Benefit of Railroads; the Comparative Attraction and Speed of Steam Engines, on a Railroad, Navigation, and Turnpike Road.... 2nd ed. London, 1834.

—— Observations on Railways, addressed to the Nobility, Gentry, Clergy, Agriculturists, Merchants, Manufacturers, Ship Owners, and Traders, particularly to those situate on the Line and connected with the Grand Northern and Eastern Railroad. 2nd ed. Yarmouth, 1835.

CUNNINGHAM, W. The Growth of English Industry and Commerce. 3 vols. 4th ed. Cambridge, 1903–05. (5th ed., 1910–12.)

DANIELL, Rev. J. J. The History of Chippenham. Compiled from Researches by the Author, and from the Collections of the Late Rev. Canon Jackson, F.S.A. Chippenham, Bath, and London, 1894.

DAVIES, Rev. JOHN SILVESTER. A History of Southampton. Partly from the Manuscript of Dr Speed, in the Southampton Archives. Southampton and London, 1883.

DAVIES, Rev. ROWLAND. Journal of the Very Rev. Rowland Davies, LL.D., Dean of Ross (and afterwards Dean of Cork), from Mar. 8, 1688–89, to September 29, 1690. Edited by Richard Caulfield, B.A. Printed for the Camden Society, 1857.

DAVIS, HENRY GEORGE. The Memorials of the Hamlet of Knightsbridge. (Edited by Chas. Davis.) London, 1859.

DAVIS, RICHARD. General View of the Agriculture of the County of Oxford.... London, 1794.

DAVIS, THOMAS. General View of the Agriculture of the County of Wilts.... London, 1794.

DEACON, WILLIAM. Observations on Stage Waggons, Stage Coaches, Turnpike-Roads, Toll-Bars, Weighing Machines, &c. Occasioned by a Committee of the House of Commons being appointed to inquire into the Principles and Effects of Broad and Narrow Wheels. London, 1807.

DECHESNE, LAURENT. L'Évolution économique et sociale de l'industrie de la laine en Angleterre. Paris, 1900.

DEE, JOHN. The Private Diary of Dr John Dee, and the Catalogue of his Library of Manuscripts, from the Original MSS. in the Ashmolean Museum at Oxford, and Trinity College Library, Cambridge. Edited by James Orchard Halliwell, Esq., F.R.S. London, 1842.

DEERING, CHARLES, M.D. Nottinghamia Vetus et Nova, or an Historical Account of the Ancient and Present State of the Town of Nottingham. Gathered from the Remains of Antiquity and Collected from Authentic Manuscripts and Ancient as well as Modern Historians. Nottingham, 1751.

[DEFOE, DANIEL.] An Essay upon Projects (pp. 68–112). London, 1697. Brit. Mus. 1029. b. 24.

—— A Tour through the Whole Island of Great Britain. By a Gentleman. 4th edition. 4 vols. London, 1748.

—— A Tour Through the Whole Island of Great Britain. 4th edition, with very great Additions, Improvements, and Corrections; which bring it down to the year 1748. 4 vols. London, 1748.

DEHANY, WILLIAM KNIGHT. The General Turnpike Acts, 3 Geo. IV. c. 126 and 4 Geo. IV. c. 95, with the Reasons for Passing the Explanatory Act, and for the Alterations introduced into it, etc. London, 1823.

DEKKER, THOMAS. A Knight's Conjuring. Done in earnest: Discovered in jest. London, [1607].

DELANE, W. F. A. The Present General Laws for Regulating Highways in England: consisting of the 41 Geo. III. c. 109; 41 Geo. III. c. 28; and the Last General

Highway Act 5 & 6 W. IV. c. 50. Arranged Alphabetically as one Act. London, 1835.

DELAUNE, THOMAS. Angliae Metropolis: or, The Present State of London: with Memorials comprehending a Full and Succinct Account of the Ancient and Modern State thereof...continued by a careful hand. London, 1690.

[DENNE, SAMUEL.] The History and Antiquities of Rochester and its Environs, etc. Rochester, 1772.

DENT, ROBERT K. The Making of Birmingham: being a History of the Rise and Growth of the Midland Metropolis. Birmingham and London, 1894.

D'EWES, Sir SIMONDS. The Journals of all the Parliaments During the Reign of Queen Elizabeth, both of the House of Lords and House of Commons. London, 1682.

D'HAUSSEZ, Baron. Great Britain in 1833. 2 vols. Philadelphia, 1833. (French edition, 2 vols. Bruxelles, 1833.)

DIBDIN, THOMAS FROGNALL. A Bibliographical, Antiquarian and Picturesque Tour in the Northern Counties of England and in Scotland. 2 vols. London, 1838.

DICKSON, R. W. General View of the Agriculture of Lancashire....By R. W. Dickson. Revised and prepared for the Press by W. Stevenson. London, 1815.

DODD, RALPH. Report on the First Part of the Line of Inland Navigation from the East to the West Sea by way of Newcastle and Carlisle, as originally projected, and lately surveyed. With Map. Newcastle, 1795.

—— Report on the Proposed Canal Navigation, forming a Junction of the Rivers Thames and Medway: with General Estimates, etc. London, 1799.

—— A Short Historical Account of the Greater Part of the Principal Canals in the Known World; with some Reflections upon the General Utility of Canals. Newcastle-upon-Tyne, 1795.

—— Report on the Intended Grand Surrey Canal Navigation. London, 1800.

DODDS, JOHN. Railway Reform A Public Necessity, with Practical Suggestions. Belfast, 1868.

DONALDSON, JAMES. General View of the Agriculture of the County of Northampton.... Edinburgh, 1794.

DONKIN, BRYAN. A Paper read before the Institution of Civil Engineers, on the Construction of Carriage-way Pavements. London, 1824.

DRAKE, FRANCIS. Eboracum: or the History and Antiquities of the City of York. London, 1736.

DRIVER, ABRAHAM and WILLIAM. General View of the Agriculture of the County of Hants.... London, 1794.

DUGDALE, Sir WILLIAM. The Life, Diary, and Correspondence of Sir William Dugdale, edited by William Hamper. London, 1827.

DUNCUMB, JOHN. Collections towards the History and Antiquities of the County of Hereford. 3 vols. (Vol. III is by W. H. Cooke; and the continuations are by W. H. Cooke, in 1892, and M. G. Watkins, in 1897.) Hereford and London, 1804–82.

—— General View of the Agriculture of the County of Hereford.... London, 1805.

DUNDAS, CHARLES. Observations on the General Comparative Merits of Inland Communication by navigations or railroads. Bath, 1825.

DUNSFORD, MARTIN. Historical Memoirs of the Town and Parish of Tiverton in County of Devon. Collected from the Best Authorities. Exeter, 1790.

—— Miscellaneous Observations in the Course of Two Tours, through several Parts of the West of England. Tiverton, 1800.

DUPIN, (Baron) CHARLES. Mémoires sur la Marine et les ponts et chaussées de France et d'Angleterre. Paris, 1818.

DUPIN, (Baron) CHARLES. Voyages dans la Grande-Bretagne, entrepris relativement aux Services Publics de la Guerre, de la Marine, et des Ponts et Chaussées en 1816, 1817, 1818 et 1819. 3 Parties. [Part III deals with roads and canals.] Paris, 1820–24.

—— The Commercial Power of Great Britain; exhibiting a Complete View of the Public Works of this Country....Translated from the French. 2 vols. London, 1825. [This is a translation of part of Part III of "Voyages dans la Grande-Bretagne."]

DUTENS, J. Mémoires sur les travaux publics de l'Angleterre. Paris, 1819.

EARWAKER, JOHN PARSONS (editor). The Court Leet Records of the Manor of Manchester from the year 1552 to the year 1686, and from the year 1731 to the year 1846. 12 vols. Manchester, 1884–90.

EAST, ROBERT. Extracts from Records in the Possession of the Municipal Corporation of the Borough of Portsmouth and from Other Documents relating thereto. New and enlarged edition. Portsmouth, 1891.

ECKERSLEY, PETER. Railway Management. Observations on Two Letters to George Carr Glyn, Esq., M.P., by John Whitehead...and Mark Huish. London, 1848.

EDGEWORTH, RICHARD LOVELL. An Essay on the Construction of Roads and Carriages. 2nd edition. London, 1817.

EGERTON, FRANCIS HENRY. The Second Part of a Letter to the Parisians, and the French Nation, upon Inland Navigation, containing a Defence of the Public Character of His Grace, Francis Egerton, late Duke of Bridgewater, etc. By the Honorable Francis Henry Egerton....And including some Notices and Anecdotes concerning Mr James Brindley. [Paris, 1820.]

EGREMONT, JOHN. The Law relating to Highways, Turnpike-Roads, Public Bridges, and Navigable Rivers. Vol. I. (No other was published.) London, 1830.

ELSTOBB, WILLIAM, Jr. The Pernicious Consequences of Replacing Denver-Dam and Sluices, etc., considered; in a Letter to Mr John Leaford. Wherein his Arguments (concerning the same) contained in a Pamphlet entitled, Observations, etc., are fairly and candidly examined. Cambridge, 1745.

ENFIELD, WILLIAM. An Essay towards the History of Leverpool, drawn up from Papers left by the late Mr George Perry, and from other Materials since collected. Warrington, 1773.

ENGLEFIELD, Sir HENRY C. Observations on the Probable Consequences of the Demolition of London Bridge. London, 1821.

ESCOTT, THOS. HAY SWEET. England: Its People, Polity, and Pursuits. 2 vols. London, Paris, and New York, 1879.

ESPRIELLA, Don MANUEL ALVAREZ [pseudonym for Robert Southey]. Letters from England (translated from the Spanish). 1st American edition, Boston, 1807. 2nd American edition, New York, 1808.

[EVELYN, JOHN.] A Character of England. As it was lately presented in a Letter to a Noble Man of France. London, 1659. Brit. Mus. 292. a. 43.

FAIRBAIRN, HENRY. A Treatise on the Political Economy of Railroads; in which the New Mode of Locomotion is considered in its Influence upon the Affairs of Nations. London, 1836.

FAIRBAIRN, WILLIAM. Remarks on Canal Navigation, illustrative of the Advantages of the Use of Steam as a Moving Power on Canals. London, 1831.

FALKNER, JOHN MEADE. A History of Oxfordshire. London, 1899.

FAREY, JOHN, Sr. General View of the Agriculture and Minerals of Derbyshire. 3 vols. London, 1811.

FARRER, WILLIAM. The Court Rolls of the Honor of Clitheroe in the County of Lancaster, Vol. I. Manchester and Burnley, 1897.

FAUJAS DE SAINT-FOND, BARTHELÉMY. Travels in England, Scotland, and the Hebrides. Translated from the French. 2 vols. London, 1799.

FAULKNER, THOMAS. An Historical and Topographical Description of Chelsea and its Environs. 2 vols. Chelsea, 1829.

—— The History and Antiquities of the Parish of Hammersmith. London, 1839.

FAY, SAM. A Royal Road: being the History of the London and South Western Railway, from 1825 to the present Time. Kingston-on-Thames, 1883.

FERGUSON, RICHARD SAUL. Cumberland and Westmorland M.P.s from the Restoration to the Reform Bill of 1867. London, 1871.

—— A History of Cumberland. London, 1890.

—— A Boke of Recorde or Register, containing all the acts and doings in or concerning the Corporation within the Town of Kirkbiekendall beginning (1575), etc. Kendal and Carlisle, 1892.

FERGUSON, R. S., and NANSON, W. Some Municipal Records of the City of Carlisle. Carlisle and London, 1887.

FIENNES, CELIA. Through England on a Side Saddle in the time of William and Mary. With an introduction by Hon. Mrs Griffiths. London and New York, 1888.

FISHWICK, Lieut.-Col. HENRY. The History of the Parish of Preston in Amounderness in the County of Lancaster. Rochdale and London, 1900.

FLETCHER, WILLIAM. The History and Development of Steam Locomotion on Common Roads. London, 1891.

FOOT, PETER. General View of the Agriculture of the County of Middlesex.... London, 1794.

FOWLER, JOHN KERSLEY. Echoes of Old Country Life. London and New York, 1892.

—— Records of Old Times: Historical, Social, Political, Sporting and Agricultural. London, 1898.

FOX, JOHN. General View of the Agriculture of the County of Monmouth.... Brentford, 1794.

FRASER, ROBERT. General View of the County of Devon.... London, 1794.

—— General View of the County of Cornwall.... London, 1794.

FREEMAN, EDWARD A. Exeter (in Historic Towns Series). London, 1887.

FRY, JOSEPH STORRS. An Essay on the Construction of Wheel Carriages....With Suggestions relating to the Principles on which Tolls ought to be imposed, etc. London, 1820.

FULTON, ROBERT. A Treatise on the Improvement of Canal Navigation; exhibiting the numerous Advantages to be derived from small canals and boats from two to five feet wide, etc. London, 1796.

[GALT, WILLIAM.] Railway Reform; its Expediency and Practicability considered. 3rd ed. London, 1844.

GALT, WILLIAM. Railway Reform: its Importance and Practicability considered as affecting the Nation, the Shareholders and the Government. London, 1865.

GASQUET, ABBOT. Parish Life in Mediaeval England. London, [1906].

GASQUET, FRANCIS AIDAN. Henry VIII and the English Monasteries. London, 1906.

GATTY, ALFRED. Sheffield: Past and Present. London, 1873.

GAUSSEN, ALICE C. G. A Later Pepys. The Correspondence of Sir William Weller Pepys, Bart., Master in Chancery 1758–1825, with Mrs Chapone, Mrs Hartley, etc. Edited with an introduction and notes by Alice C. G. Gaussen. 2 vols. London and New York, 1904.

GENTLEMAN (pseud.). Proposals àt large for the Easy and Effectual Amendment of the Roads, by some further necessary Laws and Regulations concerning the Wheels of all Carriages; and the Methods or Rules of Travelling;...To which are added, Brief Remarks and Considerations on the foregoing Proposals; tending to prove the Necessity and Utility thereof. By a Gentleman. London, 1758.

GENTLEMAN (pseud.). A Tour from London to the Lakes: containing Natural, Economical, and Literary Observations. London, 1792.

—— A Tour in Monmouthshire and Part of Glamorganshire in July 1806. Halesworth, 1807. Brit. Mus. 578. b. 41 (1).

GEORGE, C. M. The National Waggon-Post; to travel at the Rate of twenty miles per hour, carrying One Thousand Ton Weight, all over the kingdom of England, with Passengers, Goods, and Stock. Paris, 1825.

GERSTNER, F. DE. Mémoire sur les grandes routes, les chemins de fer et les canaux de navigation; traduit de l'Allemand de M. F. de Gerstner...et précédé d'une introduction, par M. P. S. Girard. Paris, 1827.

GIBBS, A. E. The Corporation Records of St Albans. St Albans, 1890.

GILBERT, C. S. An Historical Survey of the County of Cornwall: to which is added a Complete Heraldry of the same. 2 vols. Plymouth and London, 1817–20.

GILBERT, DAVIES. The Parochial History of Cornwall, founded on the Manuscript Histories of Mr Hals and Mr Tonkin. 4 vols. London, 1838.

GILBEY, Sir WALTER. Early Carriages and Roads. London, 1903.

GODWIN, GEORGE, Jr. An Appeal to the Public on the Subject of Railways. London, 1837.

GONZALES, Don MANOEL. The Voyage of Don Manoel Gonzales (late Merchant) of the City of Lisbon in Portugal, to Great Britain (1731)....Translated from the Portuguese Manuscript. London, 1808.

GOOCH, Rev. WILLIAM. General View of the Agriculture of the County of Cambridge.... London, 1813.

GOODMAN, WILLIAM. The Social History of Great Britain during the Reigns of the Stuarts, beginning with the Seventeenth Century. New York, 1843.

GORDON, ALEXANDER. An Historical and Practical Treatise upon Elemental Locomotion, by Means of Steam Carriages on Common Roads: showing the Commercial, Political, and Moral Advantages, etc. London, 1832.

—— The Fitness of Turnpike Roads and Highways for the most Expeditious, Safe, Convenient, and Economical Internal Communication. London, 1835.

—— Observations addressed to Those Interested in either Rail-ways or Turnpike-Roads; showing the Comparative Expedition, Safety, Convenience, and Public and Private Economy of these two Kinds of Road for Internal Communication. London, 1837.

—— Observations on Railway Monopolies, and Remedial Measures. London, 1841.

GORDON, W. J. Our Home Railways: How they began and How they are worked. 2 vols. London, [1910].

GOSSON, STEPHEN. Pleasant Quippes for Upstart Newfangled Gentlewomen. London, 1595. Reprinted, 1847.

GRAHAME, THOMAS. Treatise on Internal Intercourse and Communication in Civilized States and particularly in Great Britain. London, 1834.

—— Essays and Letters on Subjects conducive to the Improvement and Extension of Inland Communication and Transport. Westminster, 1835.

—— A Letter to the Traders and Carriers on the Navigations connecting Liverpool and Manchester. 2nd ed. Glasgow, 1834.

GRANGER, JOSEPH. General View of the Agriculture of the County of Durham.... London, 1794.

GRAY, THOMAS. Observations on a General Iron Railway, or Land Steam-Conveyance; to supersede the Necessity of Horses in all Public Vehicles; showing its vast Superiority in every Respect, etc. 5th ed. London, 1825. (1st ed., 1821.)

[GRAY, W.] Chorographia, or a Survey of Newcastle upon Tine. London, 1649.

Great Western Railway. Extracts from the Minutes of Evidence on the London and Birmingham Railway Bill, together with Abstracts from Acts of Parliament, etc. (Birmingham Free Reference Library, No. 63,041.) London, 1838.

Great Western Railway Bill. Minutes of Evidence taken before the Lords Committees to whom the Bill was committed. 1835.

Great Western Railway. Speech of Counsel on the 30th May, 1848, before A Select Committee of the House of Commons, on behalf of the Head, Lower and Assistant Masters of Eton against the Great Western Railway Extension from Slough to Windsor. (Printed for private circulation only.)

[GREEN, JOHN RICHARD.] Oxford during the Last Century, being two Series of Papers published in the Oxford Chronicle and Berks and Bucks Gazette during the Year 1859. Oxford, 1859.

GREEN, JOHN RICHARD, and ROBERSON, Rev. GEORGE. Studies in Oxford History, chiefly in the Eighteenth Century. Edited by C. L. Stainer. Oxford, 1901.

GREEN, JOHN RICHARD. Oxford Studies. (Edited by Mrs J. R. Green and Miss K. Norgate.) London, 1901.

GREEN, VALENTINE. The History and Antiquities of the City and Suburbs of Worcester. Two volumes in one. London, 1796.

GRIFFITH, EDWARD. A Collection of Ancient Records, relating to the Borough of Huntingdon, with Observations illustrative of the History of Parliamentary Boroughs in General. London, 1727.

GRIGGS, Messrs. General View of the Agriculture of the County of Essex.... London, 1794.

GRINLING, CHARLES H. The History of the Great Northern Railway, 1845–1902. London, 1903.

GRUNDY, JOHN. Philosophical and Mathematical Reasons, Humbly offered to the Consideration of the Publick, To prove that the Present Works, executing at Chester, to recover and preserve the Navigation of the River Dee, must intirely destroy the same. With some Remarks on Mr Badeslade's Reasons, etc., thereon. Chester, [1735 or 1736].

GUILDING, Rev. JOHN MELVILLE (editor). Reading Records. Diary of the Corporation, 1431–1654. 4 vols. London, 1892–96.

GURNEY, GOLDSWORTHY. Observations on Steam Carriages on Turnpike Roads. With Returns of the Daily Practical Results of Working; the Cause of the Stoppage of the Carriage at Gloucester; and the consequent Official Report of the House of Commons.... London, 1832.

GUTHRIE, JAMES. The River Tyne: its History and Resources. (Guthrie was late Secretary to the R. Tyne Commission.) Newcastle-upon-Tyne, 1880.

H., W. The Infallible Guide to Travellers, or Direct Independants. Giving a most Exact Account of the Four Principal Roads of England, beginning at the Standard in Cornhill, and extending to the Sea-Shore, and branching to most of the Cities...with their true distance...according to Mr Oglesby's Dimensuration. By W. H. Gent. London, 1682.

HADLEY, GEORGE. A New and Complete History of the Town and County of the Town of Kingston-upon-Hull. Kingston-upon-Hull, 1788.

HALL, JAMES. A History of the Town and Parish of Nantwich,...in the County Palatine of Chester. Nantwich, 1883.

HALLIWELL, JAMES ORCHARD (editor). The Autobiography and Correspondence of Sir Simonds D'Ewes, during the reigns of James I and Charles I. 2 vols. London, 1845.

HANCOCK, WALTER. Narrative of Twelve Years' Experiments (1824–36) Demonstrative of the Practicability and Advantage of Employing Steam-Carriages on Common Roads.... London, 1838.

HANDYSIDE, GEORGE. Review of the Manchester, Sheffield and Lincolnshire Railway. Newcastle-upon-Tyne, 1863.

HANSHALL, J. H. The History of the County Palatine of Chester. Chester, 1817.

HARCOURT, LEVESON FRANCIS VERNON. Rivers and Canals. The Flow, Control, and Improvement of Rivers, and the Design, Construction and Development of Canals both for Navigation and Irrigation, etc. 2 vols. 2nd edition, rewritten and enlarged. Oxford, 1896.

HARDWICK, CHARLES. History of the Borough of Preston and its Environs, in the County of Lancaster. Preston and London, 1857.

HARDY and PAGE. Bedfordshire County Records. Notes and Extracts from the County Records comprised in the Quarter Sessions Rolls (1714–1832) and Sessions Minute Books (1651–1660). [N.P., N.D. (1907?).]

HARDY, WILLIAM JOHN (editor). A Calendar to the Records of the Borough of Doncaster. 4 vols. Doncaster, 1899–1903.

HARGROVE, WILLIAM. History and Description of the Ancient City of York; comprising all the most interesting Information, already published in Drake's Eboracum; enriched with much Entirely New Matter, from other Authentic Sources. 2 vols. York, 1818.

HARPER, CHARLES G. The Holyhead Road: the Mail Coach Road to Dublin. 2 vols. London, 1902.

—— The Exeter Road: the Story of the West of England Highway. London, 1899.

—— The Bath Road: History, Fashion and Frivolity on an old Highway. London, 1899.

—— The Cambridge, Ely, and King's Lynn Road: the great Fenland Highway. London, 1902.

—— The Norwich Road: an East Anglian Highway. London, 1901.

—— The Great North Road: the Old Mail Road to Scotland. 2 vols. London, [N.D.].

—— The Newmarket, Bury, Thetford, and Cromer Road: Sport and History on an East Anglian Turnpike. London, 1904.

—— The Brighton Road: Old Times and New on a Classic Highway. London, 1892.

—— The Dover Road: Annals of an Ancient Turnpike. London, 1895.

—— The Portsmouth Road, and Its Tributaries: To-day and in Days of Old. London, 1895.

HARRIS, GEORGE. Materials for a Domestic History of England. (In Trans. Royal Historical Society, 1873, pp. 142–57.)

HARRIS, STANLEY. Old Coaching Days. London, 1882.

—— The Coaching Age. London, 1885.

[Wm.] Harrison's Description of England in Shakspere's Youth. Being the Second and Third Books of his Description of Britaine and England. Edited from the first two Editions of Holinshed's Chronicle, A.D. 1577, 1587. By Frederick J. Furnivall. 3 vols. London, 1877–1908.

HASSALL, CHARLES. General View of the Agriculture of the County of Monmouth.... London, 1812.

HAWKINS, JOHN. Observations on the State of the Highways, and on the Laws for Amending and Keeping them in Repair, etc. London, 1763.

—— Tyburn's Worthies. London, 1722.

HAWKINS, WILLIAM. A Full True and Impartial Account of all the Robberies committed in City, Town, and Country, for several Years past by William Hawkins, in Company with Wilson, Wright, Butler, Fox, and others not yet Taken, etc. London, 1722. Brit. Mus. G. 19,418 (1).

HAWKSMOOR, NICHOLAS. A Short Historical Account of London-Bridge; with a Proposition for a New Stone-Bridge at Westminster....In a Letter to the Right Honourable the Members of Parliament for the City and Liberty of Westminster. London, 1736.

HEAD, Sir GEORGE. A Home Tour Through the Manufacturing Districts of England in the summer of 1835. New York, 1836.

HEALEY, CHARLES E. H. CHADWICK. The History of the Part of West Somerset comprising the Parishes of Luccombe, Selworthy, Stoke Pero, Porlock, Culbone and Oare. London, 1901.

HEARNE, THOMAS. Remarks and Collections. 6 vols. Edited by C. E. Doble, D. W. Rannie, and others. Oxford, 1885-1902.

HEARNSHAW, F. J. C., and D. M. (editors). Southampton Court Leet Records. Vol. I, Parts I and II, 1550-1602. Southampton, 1905-06.

HEDGES, JOHN KIRBY. The History of Wallingford, in the County of Berks, from the Invasion of Julius Caesar to the Present Time. 2 vols. London, 1881.

HELY, JAMES. Representation of the Benefits and Advantages of making the River Avon Navigable from Christ-Church to the City of New Sarum; &c. London, 1672.

HEMINGWAY, JOSEPH. History of the City of Chester, from its Foundation to the Present Time. 2 vols. Chester, 1831.

HEMMEON, J. C. The History of the British Post Office. London, 1912. (An excellent piece of work.)

HERVEY, JOHN, First Earl of Bristol. The Diary of John Hervey, First Earl of Bristol. With Extracts from his Book of Expenses, 1688-1742. With Appendices and Notes. Wells, 1894.

HERVEY, JOHN, Lord. Memoirs of the Reign of George the Second, from his Accession to the Death of Queen Caroline. Edited from the Original Manuscript at Ickworth, by the Right Hon. John Wilson Croker, LL.D., F.R.S. 2 vols. London, 1848.

HERVEY, JOHN. Letter Books of John Hervey, First Earl of Bristol...1651-1750. 3 vols. Wells, 1894.

HIBBARD, JOHN. Statements on the Great Utility of a Circular and other Inland, etc., Canal Navigation, and Drainage; with the great Interest locally arising therefrom to the Improvement of Agriculture, Cattle, Manufacture, Commerce, and Fisheries, of Great Britain and Ireland, etc. London, 1796.

HILLS, WALLACE HENRY. The History of East Grinstead. East Grinstead, 1906.

HITCHINS, FORTESCUE. The History of Cornwall, from the Earliest Records and Traditions to the Present Time. Edited by Mr Samuel Drew of St Austell. 2 vols. Helston, 1824.

HOBHOUSE, Right Rev. Bishop. Churchwardens' Accounts of Croscombe, Pilton, Yatton, Tintinhull, Morebath and St Michael's Bath, ranging from A.D. 1849 to 1560. [London], 1890.

HODGES, WILLIAM. The Law relating to Railways and Railway Companies. London, 1847.

HODGSON, JOHN. A History of Northumberland, in three Parts. (Pt III, Vol. I, 1820; Pt I by J. Hodgson Hinde, 1858.) 6 vols. Newcastle-upon-Tyne, 1820-58.

HÖFKEN, GUSTAF. Englands Zustände, Politik, und Machtentwickelung mit Beziehung auf Deutschland. Theile I, II. Leipzig, 1846.

HOLINSHED, RAPHAEL. Description of Britaine. London, 1585.

—— Chronicles of England, Scotland, and Ireland. 6 vols. London, 1807–08.

HOLLAND, HENRY. General View of the Agriculture of Cheshire.... London, 1808.

[HOLLAND, JOHN.] The Tour of the Don. A Series of Extempore Sketches made during a pedestrian Ramble along the Banks of that River, and its principal Tributaries. Originally published in the "Sheffield Mercury" during the Year 1836. 2 vols. London, 1837.

HOLLOWAY, WILLIAM. The History and Antiquities of the Ancient Town and Port of Rye, in the County of Sussex. London, 1847.

HOLMES, RICHARD. The Booke of Entries of the Pontefract Corporation, 1653–1726. Pontefract, 1882.

HOLT, JOHN. General View of the Agriculture of the County of Lancaster.... London, 1794.

HOMER, HENRY. An Inquiry into the means of Preserving and Improving the Publick Roads of this Kingdom. London, 1767.

HOPKIN, EVAN. An Abstract of the Particulars contained in a Perambulatory Survey of above Two Hundred Miles of Turnpike Road, through the Counties of Carmarthen, Brecknock, Monmouth, and Gloucester. Swansea, 1805.

HORSFIELD, THOMAS WALKER. The History, Antiquities, and Topography of the County of Sussex. 2 vols. Lewes and London, 1835.

HOUGHTON, JOHN. A Collection of Letters for the Improvement of Husbandry and Trade. 2 vols. London, 1681.

—— A Collection [of Letters] for the Improvement of Husbandry and Trade, etc. Revised, corrected, and published by...Richard Bradley, F.R.S. 4 vols. London, 1727.

HOWARD, Hon. Sir ROBERT. The Committee, a Comedy. London, 1665.

HUDDLESTON, LAWSON. Method of Conveying Boats or Barges from a higher to a lower Level and the contrary on Canals; by means of a Plunger, instead of loosing Water by Locks. (Nicholson's Journal, IV, p. 236, anno 1703.)

HUDSON, Rev. WILLIAM, and TINGEY, JOHN COTTINGHAM. The Records of the City of Norwich. 2 vols. (I have examined only Vol. I.) Norwich and London, 1906.

HUERNE DE POMMEUSE, M. Des Canaux Navigables considérés d'une manière générale, avec des recherches comparatives sur la navigation intérieure de la France et celle de l'Angleterre. Tome deuxième. Paris, 1822. (Vol. I unpublished.)

HULLS, JONATHAN. A Description and Draught of a new-invented Machine for carrying Vessels or Ships out of, or into any Harbour, Port, or River, against Wind and Tide, or in a calm. London, 1737.

HUMPHREYS, ARTHUR L. The Materials for the History of the Town of Wellington, County of Somerset. London and Wellington, 1889.

HUNTER, Rev. JOSEPH. South Yorkshire: The History and Topography of the Deanery of Doncaster. 2 vols. London, 1828, 1831.

HUNTER, JOSEPH. Hallamshire. The History and Topography of the Parish of Sheffield in the County of York....A new and enlarged edition, by the Rev. Alfred Gatty, D.D. London, 1869.

HUTCHINS, JOHN. The History and Antiquities of the County of Dorset, compiled from the best and most Ancient Historians, Inquisitiones post mortem, and other Valuable Records and Manuscripts in the Public Offices, and Libraries, and in Private Hands, etc. 2 vols. London, 1774.

HUTCHINSON, W. The History of the County of Cumberland, and some Places Adjacent. 2 vols. Carlisle, 1794.

HUTTON, W. An History of Birmingham. 3rd edition, 4 vols., with MS. Notes by Hamper. Birmingham, 1795.

HUTTON, W[ILLIAM]. The Scarborough Tour in 1803. London, 1804.

INVESTIGATOR (pseud.). Beware the Bubbles!!! Remarks on Proposed Railways, more particularly on that between Birmingham and London. 2nd ed. London, 1831.

IRVING, JOSEPH. The Annals of Our Time, 1837–71. London and New York, 1871.

JACKSON, Sir GEORGE. The Bath Archives. A further Selection from the Diaries and Letters of Sir George Jackson, K.C.H., from 1809 to 1816. Edited by Lady Jackson. 2 vols. London, 1873.

JACOB, JOSEPH. Observations on the Structure and Draught of Wheel Carriages. London, 1773.

JEAFFRESON, J. CORDY. The Life of Robert Stephenson, F.R.S. (with Descriptive Chapters on some of his most Important Professional Works, by Prof. Wm. Pole, F.R.S.). 2 vols. London, 1864.

JEANS, J. S. Jubilee Memorial of the Railway System. A History of the Stockton and Darlington Railway and a Record of its Results. London, 1875.

JEBOULT, EDWARD. A General Account of West Somerset, Description of the Valley of the Tone, and the History of the Town of Taunton. Taunton, 1873.

JESSOP, WILLIAM. Report to the Committee of the Subscribers to the Grand Junction Canal. Northampton, 1792.

[JESSOP, WILLIAM.] Estimate [by Wm. Jessop] of the Expence of making Navigable the River Ouse, from Lewes into Barcomb Mill-Pond, at the High Water of Neap Tides, for Vessels of 45 ft. in Length, 12 ft. in Breadth, and drawing 3 ft. 6 in. of Water; so as to carry 30 Tons. (It is accompanied by other papers in MS. regarding the Navigation and Drainage Works along this route near Lewes.)

JEWITT, LLEWELLYNN. The Wedgwoods: being a Life of Josiah Wedgwood; with Notices of his Works and their Productions, Memoirs,...and a History of the Early Potteries of Staffordshire. London, 1865.

JOHNSON, CHARLES. A General History of the Lives and Adventures of the Most Famous Highwaymen, Murderers, Street-Robbers, etc. London, 1734.

JOHNSON, CUTHBERT W. The Advantages of Railways to Agriculture. [Lewes, 1837.]

JORDAN, Rev. JOHN. A Parochial History of Enstone, in the County of Oxford. London and Oxford, 1857.

JOYCE, HERBERT. The History of the Post Office from its Establishment down to 1836. London, 1893.

JUSSERAND, J. J. Les Anglais au moyen âge: la vie nomade et les routes d'Angleterre au XIVe siècle. Paris, 1884. (Translated by L. T. Smith under the title "English Wayfaring Life in the Middle Ages." London, 1889. 4th ed., 1892.)

KEMP, THOMAS. A History of Warwick and its People. Warwick, 1905 (?).

KENNEDY, JOHN. The History of Steam Navigation. Liverpool, 1903.

KENT, NATHANIEL. General View of the Agriculture of the County of Norfolk.... London, 1794.

KING, AUSTIN J., and WATTS, B. H. The Municipal Records of Bath, 1189–1604. London, [1885].

KING, Lieut.-Col. COOPER. A History of Berkshire. London, 1887.

KITCHIN, THOMAS. The Traveller's Guide through England and Wales. London, 1783.

[LABELYE, CHARLES.] The Present State of Westminster Bridge. Containing a Description of the said Bridge, as it has been ordered into Execution by the Right Honourable, etc., the Commissioners appointed by Parliament, etc. 2nd edition. London, 1743.

LABELYE, CHARLES. The Result of a View of the Great Level of the Fens, taken at the Desire of His Grace the Duke of Bedford, etc., Governor, and the Gentlemen of the Corporation of the Fens, in July 1745. London, 1745.

LABELYE, CHARLES. An Abstract of Mr Charles Labelye's Report, Relating to the Improvement of the River Wear, and Port of Sunderland. Newcastle-upon-Tyne, 1748.

—— A Description of Westminster Bridge. To which are added, An Account of the Methods made use of in laying the Foundations of its Piers. And an Answer to the Chief Objections that have been made thereto. With an Appendix, containing Several Particulars relating to the said Bridge, or to the History of the Building thereof, etc. London, 1751.

LAMBERT, B. The History and Survey of London and its Environs. 4 vols. London, 1806.

LANGFORD, JOHN ALFRED. A Century of Birmingham Life: or a Chronicle of Local Events from 1741 to 1841. 2 vols. Birmingham and London, 1868.

—— Modern Birmingham and its Institutions: a Chronicle of Local Events, from 1841 to 1871. 2 vols. Birmingham (1873–77).

LANGLEY, J. BAXTER. The Dangers of the North British Railway Policy, or a Question for the Consideration of the Inhabitants of Newcastle and the Surrounding Towns. 2nd ed. Newcastle, [1861].

LARDNER, DIONYSIUS. Railway Economy: A Treatise on the New Art of Transport, its Management, Prospects and Relations,.... London, 1850.

—— Railway Economy: A Treatise on the New Art of Transport....With an Exposition of the Practical Results of the Railways in Operation in the United Kingdom, on the Continent, and in America. London, 1850.

LATIMER, JOHN. The Corporation of Bristol in the Sixteenth Century. 1531–1600 (from original sources). Bristol, 1903.

—— The Annals of Bristol in the Seventeenth Century. Bristol, 1900.

—— The Annals of Bristol in the Eighteenth Century. Bristol, 1893.

—— The Annals of Bristol in the Nineteenth Century. Bristol, 1887.

LAYSON, JOHN F. George Stephenson; The Locomotive and the Railway. London, 1881.

LEACH, EDMUND. A Treatise of Universal Inland Navigations, and the Use of All Sorts of Mines. A Work entirely New. Recommended to the Inhabitants of Great Britain and Ireland. Plainly demonstrating the Possibility of making any River and Stream of Running Water in the World Navigable, by Canals of a New Construction, without Locks and Dams, etc. London, [1790].

LEADER, JOHN DANIEL. Extracts from the Earliest Book of Accounts belonging to the Town Trustees of Sheffield, dating from 1566 to 1707, with Explanatory Notes. Sheffield, 1879.

—— The Records of the Burgery of Sheffield, commonly called the Town Trust. London, 1897.

LEADER, ROBERT EADON. History of the Company of Cutlers in Hallamshire in the County of York. 2 vols. Sheffield, 1905.

—— Sheffield in the Eighteenth Century. Sheffield, 1901.

LEATHAM, ISAAC. General View of the Agriculture of the East Riding of Yorkshire.... London, 1794.

LECKY, WILLIAM EDWARD HARTPOLE. A History of England in the Eighteenth Century. 8 vols. London, 1878–90.

LEE, LANCELOT J., and VENABLES, R. G. Shropshire County Records. Issued in fourteen numbers.

LELAND, JOHN. The Itinerary of John Leland, the Antiquary. 9 vols. 2nd edition; collected and improved from the Original Manuscript. With the addition also of a General Index. Oxford, 1745.

[LE SAGE, G. L.] Brit. Mus. 10,848. aa. 12. Remarques sur l'état present d'Angleterre, faites par un voyageur inconnu [G. L. Le Sage] dans les années 1713 et 1714. Amsterdam, 1715.

LESLIE, JAMES. Description of an Inclined Plane for conveying Boats from One Level to another on the Monkland Canal, at Blackhill, near Glasgow. Edinburgh, 1852.

LEYLAND, JOHN. The Yorkshire Coast, and the Cleveland Hills and Dales. London, 1892.

LINDSAY, W. S. History of Merchant Shipping and Ancient Commerce. 4 vols. London, 1874–76.

LIPSCOMB, GEORGE. A Journey into Cornwall, through the Counties of Southampton, Wilts, Dorset, Somerset and Devon. Warwick, 1799.

LITTLETON, E. Proposal for Maintaining and Repairing the Highways. London, 1692.

LLOYD, JOHN, Jr. Papers relating to the History and Navigation of the Rivers Wye and Lug. Hereford, [1873].

LLOYD, SAMUEL. A National Canal between the Four Rivers a National Necessity. London, 1888.

LOFTIE, W. J. A History of London. 2 vols. London, 1883.

—— Old and New London. 5 vols. London, [N.D.]. Vols. I and II are by Walter Thornbury.

London and Birmingham Railway Bill. Extracts from the Minutes of Evidence given before the Committee of the Lords on this Bill. London, 1832.

LOVEDAY, JOHN EDWARD TAYLOR. Diary of a Tour in 1732 through Parts of England, Wales, Ireland and Scotland. Edinburgh, 1890.

LOWE, ROBERT. General View of the Agriculture of the County of Nottingham.... London, 1794.

MACADAM, JOHN LOUDON. Remarks on the Present System of Road-making, with Observations deduced from Practice and Experience with a view to the Revision of the Existing Laws and the Introduction of Improvements in the Method of Making, Repairing, and Preserving Roads, and defending the Road Trusts from Misapplication of Funds. Bristol, 1816; 2nd edition, Bristol, 1819; 3rd and 4th editions, London, 1821; 5th edition, London, 1822.

—— Observations on the Management of Trusts for the care of Turnpike Roads, as regards the Repair of the Road, the expenditure of the Revenue, and the appointment and Quality of Executive Officers, etc. London, 1825.

MACE, THOMAS. Profit, Conveniency, and Pleasure, to the Whole Nation. Being a short Rational Discourse, lately presented to His Majesty concerning the Highways of England: Their Badness, the Causes thereof, the Reasons of those Causes, the impossibility of ever having them well-mended according to the old way of mending But may...according to This New Way.... London, 1675.

MACERONI, FRANCIS. A Few Facts concerning Elementary Locomotion. 2nd ed. London, 1834.

—— Maceroni versus Mechanics' Magazine. London, 1834.

—— Memoirs of the Life and Adventures of Colonel Maceroni.... 2 vols. London, 1838.

MACGREGOR, J. Observations on the River Tyne, with a View to the Improvement of the Navigation;.... Newcastle, 1832.

MACKENZIE, E. A Descriptive and Historical Account of the Town and County of Newcastle-upon-Tyne, including the Borough of Gateshead. 2 vols. in one. Newcastle-upon-Tyne, 1827.

[MACKY, J.] A Journey through England. In Familiar Letters from a Gentleman Here, to His Friend Abroad. 3 vols. London, 1714–29.

MACLAREN, C. Railways compared with Canals and Common Roads, and their Uses and Advantages explained. Edinburgh, 1825. (Contained in the "Pamphleteer," Vol. xxvi. London, 1826.)

MACPHERSON, DAVID. Annals of Commerce, Manufactures, Fishery and Navigation. 4 vols. London, 1805 (?).

MACRITCHIE, Rev. WILLIAM. Diary of a Tour Through Great Britain in 1795. (With introduction and notes by David MacRitchie.) London, 1897.

MACTURK, G. G. A History of the Hull Railways. Hull, [1879].

MAITLAND, WILLIAM. The History and Survey of London from its Foundation to the Present Time. 2 vols. London, 1760.

MALCOLM, WILLIAM, JAMES and JACOB. General View of the Agriculture of the County of Surrey.... London, 1794.

—— General View of the Agriculture of the County of Buckingham.... London, 1794.

MALDEN, H. ELLIOT. A History of Surrey. London, 1900.

MALET, Captain [HAROLD ESDAILE]. Annals of the Road or Notes on Mail and Stage Coaching in Great Britain. To which are added Essays on the Road, by Nimrod. London, 1876.

MANNING, OWEN. The History and Antiquities of the County of Surrey. 3 vols. London, 1804.

MANTOUX, PAUL. The Industrial Revolution in the 18th Century. London, 1928.

MARKHAM, CHRISTOPHER ALEXANDER. The Liber Custumarum: the Book of the Ancient Usages and Customs of the Town of Northampton, from the Earliest Record to 1448. Northampton, 1895.

MARKHAM, CHRISTOPHER A., and COX, Rev. JOHN CHARLES. The Records of the Borough of Northampton. 2 vols. Northampton, 1898.

MARSH, A. E. W. A History of the Borough and Town of Calne, and some Account of the Villages, etc., in its Vicinity. Calne and London, 1904.

MARSHALL, WILLIAM. The Rural Economy of Norfolk: comprising the Management of Landed Estates, and the Present Practice of Husbandry in that County. 2 vols. London, 1787.

—— The Rural Economy of Yorkshire. Comprising the Management of Landed Estates, and the Present Practice of Husbandry in the Agricultural Districts of that County. 2 vols. London, 1788.

—— The Rural Economy of Glocestershire; including its Dairy, etc. 2 vols. Glocester, 1789.

—— The Rural Economy of the Midland Counties; including the Management of Live Stock &c. 2nd edition. 2 vols. London, 1796.

—— The Rural Economy of the West of England: including Devonshire, and parts of Somersetshire, Dorsetshire and Cornwall.... 2 vols. London, 1796.

—— The Rural Economy of the Southern Counties; comprising Kent, Surrey, Sussex; the Isle of Wight; the Chalk Hills of Wiltshire, Hampshire, etc. 2 vols. London, 1798.

—— Minutes, Experiments, Observations, and General Remarks on Agriculture in the Southern Counties; a new Edition. To which is prefixed a sketch of the Vale of London, and an Outline of its Rural Economy now first published. 2 vols. London, 1799.

—— A Review of the Reports to the Board of Agriculture; from the Northern Department of England. York, 1808.

—— A Review (and Complete Abstract) of the Reports to the Board of Agriculture; from the Midland Department of England:.... York, 1815.

—— A Review (and Complete Abstract) of the Reports to the Board of Agriculture; from the Southern and Peninsular Departments of England:.... York, 1817.

MARSHALL, WILLIAM. Railway Legislation and Railway Reform considered with Special Reference to Scottish Lines. 2nd edition. Edinburgh, 1853.

MARTIN, R. M. Railways—Past, Present, and Prospective. London, 1849.

MASON, ROBERT HINDRY. The History of Norfolk: from Original Records and other Authoritative Sources preserved in public and private collections. London, 1884.

MASSINGBERD, Rev. W. O. History of the Parish of Ormsby-cum-Ketsby, in the Hundred of Hill and County of Lincoln, compiled from Original Sources. Lincoln, [1893].

—— Court Rolls of the Manor of Ingoldmells in the County of Lincoln. London, 1902.

MATHER, WILLIAM. Of Repairing and Mending the Highways. London, 1696.

MATHEW, FRANCIS. Of the Opening of Rivers for Navigation, The Benefit exemplified by The Two Avons of Salisbury and Bristol. With a Mediterranean Passage by Water for Billanders of Thirty Tun between Bristol and London. With the Results. London, 1655.

—— A Mediterranean Passage by Water between the Two Sea Towns Lynn and Yarmouth, upon the Two Rivers The Little Owse, and Waveney. With farther Results. Producing the Passage from Yarmouth to York. London, 1656.

—— A Mediterranean Passage by Water from London to Bristol &c. And from Lynne to Yarmouth, and so consequently to the City of York: for the great Advancement of Trade and Traffique. London, 1670.

MATHEWS, J. Remarks on the Cause and Progress of the Scarcity and Dearness. London, 1797.

MAVOR, WILLIAM. General View of the Agriculture of Berkshire.... London, 1808.

MAXWELL, GEORGE. General View of the Agriculture of the County of Huntingdon.... London, 1793.

MAXWELL, Rt. Hon. Sir HERBERT. The Creevey Papers: A Selection from the correspondence and Diaries of the late Thomas Creevey, M.P. 2 vols. London, 1904.

MAY, GEORGE. A Descriptive History of the Town of Evesham, from the foundation of its Saxon Monastery. Evesham and London, 1845.

McDONNELL, G. Railway Management, with and without Railway Statistics. London, [1854?].

MEIGE, GUY. The Present State of Great Britain and Ireland. 11th edition, revised and completed by Solomon Bolton. London, 1757.

MERCATOR (pseud.). Remarks on the Tonnage Rates and Drawbacks of the Grand Junction Canal, with Observations on the proposed London and Birmingham Canal, by Mercator. London, 1836.

MERITON, GEORGE. A Guide to Surveyors of the Highways, shewing the Duty of Such Surveyors...collected and gathered out of Public Acts of Parliament, etc., now in force. London, 1694.

METCALFE, JOHN. The Life of John Metcalfe, commonly called Blind Jack of Knaresborough. With many entertaining anecdotes...and also a Succinct Account of his various Contracts for Making Roads, Erecting Bridges, and other Undertakings,.... York, 1795.

METEYARD, ELIZA. The Life of Josiah Wedgwood from his Private Correspondence and Family Papers...and other Original Sources. 2 vols. London, 1865–66.

MIDDLETON, JOHN. View of the Agriculture of Middlesex; with Observations on the Means of its Improvement.... London, 1798.

MISSON, HENRY DE VAUDROY. Mémoires et observations faites par un voyageur en Angleterre. [Paris], 1698. (This book was translated into English by John

Ozell and published in London, 1719, under the title "M. Misson's Memoirs and Observations in his Travels over England etc.")

MOLYNEUX, WILLIAM. Burton-on-Trent: its History, its Waters, and its Breweries. London, [1869].

MOMSEN, Dr P. Einiges über die öffentlichen Arbeiten in England,.... Kiel, 1838.

MONEY, WALTER. The History of the Ancient Town and Borough of Newbury in the County of Berks. Oxford and London, 1887.

MONK, JOHN. General View of the Agriculture of the County of Leicester.... London, 1794.

MOORE, HENRY CHARLES. Omnibuses and Cabs: Their Origin and History. London, 1902.

MOORE, THOMAS. The History of Devonshire, from the Earliest Period to the Present. London, 1829.

MORANT, PHILIP. The History and Antiquities of the County of Essex, compiled from the best and most ancient historians; from Domesday-Book, etc.; the whole digested, improved, perfected, and brought down to the present time. 2 vols. London, 1768. (Reprinted at Chelmsford, 1816.)

MOREAU, P. Description of the Railroad from Liverpool to Manchester. Together with a History of Railroads, and matters connected therewith, compiled by A. Notré,....Translated from the French by J. C. Stocker, Jr., Civil Engineer. Boston (Mass.), 1833.

MORITZ, Rev. CHARLES P. Travels, chiefly on foot, through Several Parts of England, in 1782, described in Letters to a Friend. Translated from the German by a Lady. (In Pinkerton's Voyages and Travels, Vol. II, pp. 489–573.) The German text is entitled "Reisen eines Deutschen in England im Jahr 1782," by Carl Philip Moriz. Berlin, 1785.

MORRIS, RUPERT HUGH. Chester in the Plantagenet and Tudor Reigns. [Chester, 1893.]

MORRISON, JAMES. Observations illustrative of the Defects of the English System of Railway Legislation, and of its injurious Operation on the Public Interests; with Suggestions for its Improvement. London, 1846.

MORTON, JOHN. The Natural History of Northamptonshire; with Some Account of the Antiquities. To which is Annexed A Transcript of Doomsday-Book, so far as it relates to That County. London, 1712.

MORYSON, FYNES. An Itinerary written by Fynes Moryson Gent. First in the Latine Tongue, and then translated by him into English. Containing His Ten Yeeres Travell through the Twelve Dominions of Germany, Bohmerland, Sweitzerland, Netherland, Denmarke, Poland, Italy, Turkey, France, England, Scotland, and Ireland. London, 1617.

MOULE, H[ENRY] J[OSEPH]. Descriptive Catalogue of the Charters, Minute Books and Other Documents of the Borough of Weymouth and Melcombe Regis, A.D. 1252–1800. Weymouth, 1883.

MUDGE, RICHARD Z. Observations on Railways, with Reference to Utility, Profit, and the Obvious Necessity for a National System. London, 1837.

MUIR, RAMSEY. A History of Liverpool. London, 1907.

MURRAY, ADAM. General View of the Agriculture of the County of Warwick.... London, 1813.

NASH, CHARLES. Railway Carrying and Carriers' Law; the Liabilities and Non-Liabilities of Railways, Carriers, and others,...the Dispute between Railways and Carriers; Extracts from Judgments; with the Law, and Cases illustrating the Reasoning on those Subjects; etc. London, 1846.

NELSON, W. The Office and Authority of a Justice of Peace: Collected out of all the Books, whether of Common or Statute Law, hitherto written on that subject, etc. 6th edition. London, 1718.

NEWBALL, JOHN. A Concern for Trade, and the Various Consequences; relating to the Encrease and Decrease....Also Remarks on the Bill, intitl'd A Bill to amend and render more effectual the Laws in Being for the Preservation of the publick Roads, &c. Stamford [1745–6 (?)].

Newcastle. An Account of the Great Flood in the River Tyne, on Saturday Morning, Dec. 30, 1815. To which is added, A Narrative of the Great Flood in the Rivers Tyne, Tease, and Wear, &c. on the 16th and 17th Nov., 1771. Newcastle, 1816.

—— Copy of the Evidence taken before a Committee of the House of Commons on the Newcastle and Carlisle Railway Bill. Taken from the shorthand notes of Mr Gurney. To which is added the Report of Mr Leather, on the Projected Line of Railway. Newcastle-upon-Tyne, 1829.

[NICHOLS, JOHN.] Illustrations of the Manners and Expences of Antient Times in England, in the Fifteenth, Sixteenth, and Seventeenth Centuries, deduced from the Accompts of Church-wardens, and other Authentic Documents, collected from Various Parts of the kingdom, with explanatory Notes. London, 1797.

NICHOLS, JOHN. The Progresses and Public Processions of Queen Elizabeth...now first printed from Original Manuscripts of the Times; or collected from scarce Pamphlets, etc. 3 vols. London, 1788–1805.

NICHOLSON, CORNELIUS. The Annals of Kendal: being a Historical and Descriptive Account of Kendal and the Neighbourhood. London and Kendal, 1861.

NICHOLSON, JOSEPH, and BURN, RICHARD. The History and Antiquities of the Counties of Westmorland and Cumberland. 2 vols. London, 1777.

NORDEN, JOHN. Notes on London and Westminster. 1592. [Brit. Mus. 010347. m. 16.]

—— Speculi Britanniae Pars. A Description of Hartfordshire. 1598. (Reprinted) London, 1903.

—— The Surveyors Dialogue. London, 1607.

—— Speculi Britanniae Pars Altera: or a Delineation of Northamptonshire (in 1610). London, 1720.

—— Speculi Britanniae Pars: An Historical and Chorographical Description of the County of Essex. London, 1840.

—— An Intended Guyde for English Travailers. London, 1625.

—— Speculum Britanniae: An Historical and Chorographical Description of Middlesex and Hartfordshire. London, 1723.

—— Speculi Britanniae Pars: A Description of Hartfordshire, by John Norden, 1598. Reprinted with portrait and biography of John Norden by W. B. Gerish. Ware and London, 1903.

NORTH, ROGER. The Life of the Right Honourable Francis North, Baron of Guilford, Lord Keeper of the Great Seal, under King Charles II and King James II.... London, 1742.

Northumberland County History Committee. A History of Northumberland. 7 vols. Vols. I and II, by Edward Bateson, London, 1893–95. Vol. III, by Allen B. Hinds, London, 1896. Vols. IV–VII, by John C. Hodgson, London, 1897–1904.

O'BRIEN, W. The Prize Essay on Canals and Canal Conveyance, for which a Premium of £100 was awarded by the Canal Association. London, 1858.

OBSERVER, An (pseud.). Brief Remarks on the Proposed Regent's Canal. London, 1812.

OGILBY, JOHN. [Brit. Mus. G. 4697.] Britannia Depicta: or, Ogilby Improved. Being an Actual Survey of all the Direct and Principal Cross Roads of England

and Wales....Engraved by Emanuel Bowen, Geographer....Compiled from the Best Authorities, by John Owen, Gent.

OGILBY, JOHN. Britannia, Volume the First: or, an Illustration of the Kingdom of England and Dominion of Wales: by a Geographical and Historical Description of the Principal Roads thereof. Actually Admeasured and Delineated.... London, 1675.

—— An Actual Survey of all the Principal Roads of England and Wales; described by One Hundred Maps from Copper Plates,...now improved, very much corrected, and made portable by John Senex. 2 vols. London, 1719.

OGILVIE, A. M. Ralph Allen's Bye, Way, and Cross-Road Posts (from Documents in the Home Office Papers in the Record Office). London, 1897.

OLIVER, Rev. GEORGE. The History of the City of Exeter. Exeter and London, 1861.

ORMEROD, GEORGE. The History of the County Palatine and City of Chester. (2nd edition, revised and enlarged, by Thomas Helsby.) 3 vols. London, 1882.

[OWEN, HUGH.] Some Account of the Ancient and Present State of Shrewsbury. Shrewsbury, 1808.

OWEN, HUGH, and BLAKEWAY, JOHN BRICKDALE. A History of Shrewsbury. 2 vols. London, 1825.

OWEN, WILLIAM. New Book of Roads; or a Description of the Roads of Great Britain. London, 1805.

P., E. M. S. The Two James's and the Two Stephensons; or, the Earliest History of Passenger Transit on Railways. London, 1861.

P., F. A Letter to a Friend, containing Observations on the Comparative Merits of Canals and Railways, occasioned by the Reports of the Committee of the Liverpool and Manchester Railway. London, 1832.

P., J. For Mending the Roads of England: its proposed. Signed J. P. Brit. Mus. T. 1816 (13). [1715 (?).]

P., X. A. A Descant upon Railroads. London, 1842.

PAGAN, WILLIAM. Road Reform: A Plan for abolishing Turnpike Tolls and Statute Labour Assessments, and for providing the Funds necessary for maintaining the Public Roads by an Annual Rate to be levied on Horses. Edinburgh, 1845.

PAGE, WILLIAM (editor). The Victoria History of the Counties of England. (We omit the complete list of these volumes, but many of them have excellent material.)

PALMER, HENRY R. Report on the Improvement of the Rivers Mersey and Irwell, between Liverpool and Manchester, describing the Means of adapting them for the Navigation of Sea-going Vessels. London, 1840.

—— Description of a Railway on a New Principle; with Observations on those hitherto constructed. London, 1824.

PALMER, J. E. British Canals. Problems and Possibilities. London, 1910.

PALMER, JOHN. Papers relative to the Agreement made by Government with Mr Palmer, for the Reform and Improvement of the Posts. London, 1797.

—— Mr Palmer's Case Explained. By C. Bonnor. London, 1797.

—— Debates in Both Houses of Parliament in the Months of May and June 1808, relative to the Agreement made by Government with Mr Palmer for the Reform and Improvement of the Post Office and its Revenue. London, 1809. Brit. Mus. 8244. cc. 37.

PARK, THOMAS (compiler). The Harleian Miscellany: A Collection of Scarce, Curious, and Entertaining Pamphlets and Tracts, as well in Manuscript as in Print: Selected from the Library of Edward Harley, Second Earl of Oxford, etc. 10 vols. London, 1808–13.

PARKES, W. Claim of the Subscribers to the Birmingham and Liverpool Railroad. London, 1824.

PARKINSON, R. General View of the Agriculture of the County of Huntingdon....
London, 1813.

PARNELL, Rt. Hon. Sir HENRY. A Treatise on Roads. London, 1834.

PARSLOE, JOSEPH. Our Railways. Sketches Historical and Descriptive, with
Practical Information as to Fares and Rates, etc., and a Chapter on Railway
Reform. London, 1878.

PARSONS, EDWARD. The Civil, Ecclesiastical, Literary, Commercial and Miscellaneous History of Leeds, Halifax, Huddersfield, Bradford, Wakefield,
Dewsbury, Otley, and the Manufacturing District of Yorkshire. 2 vols. Leeds,
1834.

[PATCHING or PATCHEN, RESTA.] Four Topographical Letters, written in July 1755,
upon a Journey through Bedfordshire, Northamptonshire, Leicestershire,
Nottinghamshire, Derbyshire, Warwickshire, etc. from a Gentleman in London
to his Brother and Sister in Town, etc. Newcastle-upon-Tyne, 1757.

PATERSON, JAMES. A Practical Treatise on the Making and Upholding of Public
Roads,...And a Dissertation on the Utility of Broad Wheels, and other
Improvements. Montrose, 1819.

PEARCE, WILLIAM. General View of the Agriculture of Berkshire.... London, 1794.

PEASE, Sir ALFRED E. The Diaries of Edward Pease, the Father of English Railways.
London, 1907.

PEGGE, SAMUEL. Curialia Miscellanea, or Anecdotes of Old Times; Regal, Noble,
Gentilitial, and Miscellaneous. London, 1818.

PEMBERTON, T. EDGAR. James Watt, of Soho and Heathfield, Annals of Industry
and Genius. Birmingham, 1905.

PENNANT, THOMAS. The Journey from Chester to London. London, 1782.

—— A Tour from Downing to Alston Moor (1773). London, 1801.

—— A Tour from Alston Moor to Harrogate (1773). London, 1804.

—— A Journey from London to the Isle of Wight (1787). London, 1801.

PEPYS, SAMUEL. Diary and Correspondence of Samuel Pepys, Esq., F.R.S., from
his Manuscript Cypher in the Pepysian Library, with a Life and Notes by
Richard Lord Braybrooke. Deciphered, with additional notes, by Rev. Mynors
Bright. 6 vols. London, 1875–79.

PERCY, H. A. [Brit. Mus. 684. e. 27.] Northumberland Household Book (edited
by T. Percy, Bishop of Dromore). The more correct and full title is, "The
Regulations and Establishment of the Household of H. A. Percy...at his
Castles of Wresill and Lekinfield in Yorkshire [at 1512 and after]." 1770.

PHELPS, Rev. WILLIAM. The History and Antiquities of Somersetshire. 2 vols.
London, 1839.

PHIL'ANGLUS (pseud.). The Contrast: also Further Considerations on the Concern
for Trade. With Proposals how to amend and render more effectual the Laws
in Being for the Preservation of the Public Roads, and to preserve Trade. In
Two Letters to a Member. London, [1746]. Br. Mus. 1139. 1. 11 (2).

PHILIPS, FRANCIS. Analysis of the Defective State of Turnpike Roads and Turnpike
Securities, with Suggestions for their Improvement. London, 1834.

PHILLIPS, J. A General History of Inland Navigation, Foreign and Domestic;
containing a Complete Account of the Canals already executed in England;
with Considerations on those projected. 4th ed. London, 1803.

PHILLIPS, JOHN. A Treatise on Inland Navigation; illustrated with a Whole-Sheet
Plan, delineating the Course of an Intended Navigable Canal from London to
Norwich and Lynn, through the Counties of Essex, Suffolk and Norfolk...
With Two Other Plans, to prove the Practicability of executing the Whole
with Success:.... London, 1785.

PHILLIPS, Sir RICHARD. A Personal Tour through the United Kingdom; describing living objects and contemporaneous interests. Nos. 1 and 2. London, 1828.

PHILLIPS, ROBERT. A Dissertation concerning the Present State of the High Roads of England, especially of those near London. Wherein is proposed a new Method of Repairing and Maintaining Them. London, 1737.

PICTON, J. ALLANSON. Memorials of Liverpool, Historical and Topographical, including a History of the Dock Estate. 2nd ed. 2 vols. London and Liverpool, 1875.

—— Liverpool Municipal Records. 2 vols. London and Liverpool, 1883, 1886.

PILKINGTON, JAMES. A View of the Present State of Derbyshire; with an Account of its most remarkable Antiquities. 2 vols. Derby, 1789.

PITT, W. General View of the Agriculture of the County of Stafford.... London, 1794.

PITT, WILLIAM. General View of the Agriculture of the County of Leicester.... London, 1809.

—— General View of the Agriculture of the County of Northampton.... London, 1809.

—— General View of the Agriculture of the County of Worcester.... London, 1813.

PLYMLEY, JOSEPH. General View of the Agriculture of Shropshire.... London, 1808.

POMEROY, WILLIAM THOMAS. General View of the Agriculture of the County of Worcester.... London, 1794.

POOLE, BENJAMIN. Coventry: its History and Antiquities. London and Coventry, 1870.

POOLE, BRAITHWAITE. The Commerce of Liverpool. London and Liverpool, 1854.

PRATT, EDWIN A. British Canals: Is their Resuscitation Practicable? London, 1906.

—— A History of Inland Transport and Communication in England. New York, 1912.

PRICE, A[UBREY] C[HARLES]. Leeds and its Neighbourhood: an Illustration of English History. Oxford, 1909.

PRIEST, Rev. ST JOHN. General View of the Agriculture of Buckinghamshire.... London, 1813.

PRIESTLEY, JOSEPH. Historical Account of the Navigable Rivers, Canals, and Railways throughout Great Britain (to refer to the Map of Inland Navigation). London, 1831.

PRINGLE, ANDREW. General View of the Agriculture of the County of Westmorland.... Edinburgh, 1794.

PROCTER, THOMAS. A Profitable Worke to this whole Kingdome. Concerning the mending of all High-ways as also for Waters and Iron Workes. London, 1610.

PROGRESS, PETER (pseud.). The Rail, its Origin and Progress. London, 1847.

—— The Locomotive, or the Steam Engine applied to Railways, Common Roads, and Water. And an Account of the Atmospheric Railway. London, 1848.

PROPRIETOR, A (pseud.). The Advantages of Railways with Locomotive Engines, especially the London and Greenwich Railway or Viaduct...to the Public, the Proprietors of Property along and near the Line of Road, and the shareholders, explained. 2nd ed. London, 1833.

PROVIS, W. A. Suggestions for Improving the Canal Communication between Birmingham, Wolverhampton, Shropshire, Cheshire, North Wales, and Manchester, by means of a New Canal from Middlewich to Altringham. London, 1837.

PUBLICUS (pseud.). Observations on Bridge Building, and the Several Plans offered for a New Bridge. In a Letter addressed to the Gentlemen of the Committee, appointed by the Common Council of the City of London, for putting in Execution a Scheme of Building a New Bridge across the Thames, at or near Black Friars. London, 1760.

[QUINCEY, THOMAS.] A Short Tour in the Midland Counties of England; performed in the summer of 1772. Together with an Account of a similar Excursion undertaken September 1774 [by Thomas Quincey]. London, 1775.

RADCLIFFE, Sir GEORGE. The Life and Original Correspondence of Sir George Radcliffe,....By Thomas Dunham Whitaker. London, 1810.

Railways. A Few General Observations on the Principal Railways executed, in progress, and projected, in the Midland Counties and North of England, with the Author's Opinion upon them 'as Investments. London, 1838.

Railways as they Really Are: or Facts for the Serious Consideration of Railway Proprietors. No. I. London, Brighton and South Coast Railway. London, 1847. No. II. The Dover, or South-Eastern Company. London, 1848. No. VII. The Lancashire and Yorkshire Railway. London, 1847.

RAUMER, FREDERICK VON. England in 1841: being a series of letters written to friends in Germany. Translated from the German by H. Evans Lloyd. 2 vols. London, 1842.

RENDEL, G[RACE] DAPHNE. Newcastle-on-Tyne: its Municipal Origin and Growth. London, 1898.

RENNIE, BROUN, and SHIRREFF, Messrs. General View of the Agriculture of the West Riding of Yorkshire.... London, 1794.

RENNIE, JOHN. Report of a Survey of the River Thames, between Reading and Isleworth, and of Several Lines of Canals projected to be made between those Places; with Observations on their Comparative Eligibility. London, 1794.

—— Report and Estimate of the Grand Southern Canal, proposed to be made between Tunbridge and Portsmouth: by means of which and the River Medway, an Inland Navigation will be opened between the River Thames and Portsmouth. [With large Plan of the Canal.] London, 1810.

—— Autobiography of Sir John Rennie, F.R.S. Comprising the History of his Professional Life, together with Reminiscences dating from the Commencement of the Century to the Present Time. London, 1875.

RICHARDSON, M. A. The Borderer's Table Book. 8 vols. Newcastle-upon-Tyne, 1846.

—— Reprints of Rare Tracts, Imprints of Antient Manuscripts, &c., chiefly illustrative of the History of the Northern Counties. Vol. III, Historical. Newcastle, 1849.

ROBERSON, GEORGE, and GREEN, JOHN RICHARD. Oxford during the last Century. Oxford, [N.D.].

ROBERTS, GEORGE. The Social History of the People of the Southern Counties of England in past Centuries. London, 1856.

ROBERTS, SAMUEL. Autobiography and Select Remains of the late Samuel Roberts (of Sheffield). London, 1849.

ROBERTSON, ARCHIBALD. A Topographical Survey of the Great Road from London to Bath and Bristol, etc., in 2 Parts. London, 1792.

ROBERTSON, Rev. DAVID. Diary of Francis Evans, Secretary to Bishop Lloyd, 1699–1706. Printed for the Worcestershire Historical Society, by James Parker & Co. Oxford, 1903.

ROGERS, JAMES E. THOROLD. A History of Agriculture and Prices in England. 6 vols. London, 1886–87.

[RUDDER, S.] The History of the Ancient Town of Cirencester, in two Parts: Pt I— The Ancient State; Pt II—the Modern and Present State. Cirencester, 1814.

RUDGE, THOMAS. General View of the Agriculture of the County of Gloucester.... London, 1807.

RUSHWORTH, JOHN. Historical Collections of Private Passages of State, Weighty Matters in Law, Remarkable Proceedings in Five Parliaments (1618–1629). 8 vols. London, 1721.

RYDOCK, WILLIAM. The Life and Notorious Practices of William Rydock, alias Wreathocke. Who was condemned for Felony in Robbing the Reverend Dr Lancaster on the Highway. June 11, 1735. London, 1736.

RYE, W. B. Collections for the History of Rochester. 3 vols. Rochester, 1817–87.

RYE, WILLIAM BRENCHLEY. England as Seen by Foreigners in the Days of Elizabeth and James the First....Extracts from the Travels of Foreign Princes and Others. London, 1865.

—— England as seen by Foreigners in the days of Elizabeth and James I. Comprising translations of the Journals of the two Dukes of Wirtemberg in 1592 and 1610. London, 1865.

RYMER, JAMES. The Practice of Navigation upon a New Plan. Bath, 1778.

RYMER, THOMA. Foedera, Conventiones, Literae, et Cujuscunque Generis Acta Publica, inter Reges Angliae et Alios quosvis Imperatores, Reges, Pontifices, Principes, vel Communitates, ab ineunte Saeculo Duodecimo, viz., ab anno 1101, ad nostra usque Tempora, Habita aut Tractata, etc. Editio secunda. Tomi I–XX. Londini, 1726–1735.

S., E. The Discoverie of the Knights of the Poste: Or the Knightes of the Post, or comon common baylers newly Descried.... London, 1597.

S., R. (pseud.). [Br. Mus. 1028. a. 3.] Avona; or a Transient View of the Benefit of making Rivers of this Kingdom Navigable. Occasioned by observing the Scituation of the City of Salisbury, upon the Avon, and the Consequence of opening that River to that City. Communicated by Letter to a Friend at London. By R. S. London, 1675.

SALIS, HENRY RODOLPHE DE. A Chronology of Inland Navigation in Great Britain. London, 1897.

SALT, SAMUEL. Statistics and Calculations essentially necessary to Persons connected with Railways or Canals, containing a Variety of Information not to be found elsewhere. Manchester, 1845.

SALZMANN, L. F. The History of the Parish of Hailsham, the Abbey of Otham and the Priory of Michelham. Lewes, 1901.

SANDARS, JOSEPH. A Letter on the Subject of the Projected Rail Road between Liverpool and Manchester, pointing out the Necessity for its Adoption and the Manifest Advantages it offers to the Public, with an Exposure of the exorbitant and unjust Charges of the Water Carriers. 3rd edition. Liverpool, 1824.

SAUSSURE, CESAR DE. A Foreign View of England in the Reigns of George I and George II. Translated and edited by Madame Van Muyden. London, 1902.

SCOTCHER, RICHARD. The Origin of the River Wey Navigation. Being an Account of the Canalization of the River, from a Manuscript written by Richard Scotcher in 1657. Now first published. Edited with an Introduction by Thomas J. Lacy. Guildford, 1895.

SCOTT, BENJAMIN. The Progress of Locomotion; being Two Lectures on the Advances made in Artificial Locomotion in Great Britain. London, 1854.

SCOTT, JOHN. Digests of the General Highway and Turnpike Laws, with the Schedule of Forms,...and Remarks. Also an Appendix on the Construction and Preservation of Roads. London, 1778.

—— Berwick-upon-Tweed. The History of the Town and Guild. London, 1888.

SCRIVENOR, HARRY. The Railways of the United Kingdom Statistically considered in relation to their Extent, Capital, Amalgamation, etc. From authentic documents. London, 1849.

SEKON, G. A. (pseud. for G. A. Nokes). A History of the Great Western Railway, being the Story of the Broad Gauge. London, 1895.

SEYER, SAMUEL. Memoirs Historical and Topographical of Bristol and its Neighbourhood. 2 vols. Bristol, 1821, 1823.

SHAEN, SAMUEL, Jr. A Review of Railways and Railway Legislation at Home and Abroad. London, 1847.

SHAPLEIGH, JOHN. Highways: a Treatise shewing the Hardships, and Inconveniences of Presenting, or indicting Parishes, Towns, etc., for not repairing the Highways. And offering several material Additions, and Amendments to the Laws now in being, for the better, and more effectual Repairing of the same. London, 1749.

SHAREHOLDER, A. (pseud.). Railroads. Statements and Reflections thereon: particularly with Reference to the Proposed Railroad without a Tunnel, and the competition for the Line between London and Brighton. By a Shareholder, at the Request of Other Shareholders. London, 1836.

SHAW, GEORGE T. and ISABELLA. Liverpool's First Directory. A Reprint of the Names and Addresses from Gore's Directory for 1766....Also a History of the Liverpool Directories from 1766 to 1907. Liverpool, 1907.

SHAW, Rev. S. A Tour to the West of England in 1788 (in Pinkerton's Voyages and Travels, Vol. II, pp. 172–335). London, 1789.

SHAW, SIMEON. History of the Staffordshire Potteries; and the Rise and Progress of the Manufacture of Pottery and Porcelain. First published in Hanley, 1829, and Reissued at London, 1900.

SHEARDOWN, WILLIAM. The Great North Road and the Great Northern Railway; or, Roads and Rails. Doncaster, [1863].

SHELFORD, LEONARD. The General Highway Act of the 5 & 6 W. IV. c. 50. With Notes Explaining the Alterations in the Law of Highways, etc. London, 1835.

SIMOND, LOUIS. Journal of a Tour and Residence in Great Britain, during the years 1810 and 1811, by a French Traveller. 2 vols. New York, 1815. (2nd ed., corrected and enlarged: to which is added an Appendix on France, written in December 1815, and October 1816. 2 vols. Edinburgh, 1817.)

SIMPSON, ROBERT. A Collection of Fragments Illustrative of the History and Antiquities of Derby, compiled from authentic sources. 2 vols. Derby, 1826.

SINCLAIR, Rev. J. S. Memoirs of Sir John Sinclair, Bart. 2 vols. London, 1837.

SKEY, ROBERT S. Report to the Committee of the Birmingham and Liverpool Junction Canal, on the Present State of the Competition between the Canal Carriers using that Line and the Grand Junction Railway Company, with Suggestions for a more economical and effective working of the Canal. Westminster, 1841.

SLEIGH, JOHN. A History of the Ancient Parish of Leek, in Staffordshire; including Horton, Cheddleton and Ipstones. 2nd ed. London, [1883].

SLUGG, J. T. Reminiscences of Manchester Fifty Years Ago. Manchester and London, 1881.

SMEATON, JOHN. Begin, From a Survey of the River Calder, from Wakefield to Brooksmouth, and from thence to Salter-Hebble Bridge, near Halifax, taken in the Months of October and November 1757, by John Smeaton, it appears as follows:— Halifax, 1757.

—— Reports of the late John Smeaton, F.R.S., made on Various Occasions, in the Course of his Employment as a Civil Engineer. 3 vols. London, 1812.

SMILES, SAMUEL. Lives of the Engineers. 4 vols. London, 1861–62.

—— Industrial Biography. Iron Workers and Tool Makers. London, 1879.

SMITH, ARTHUR. The Bubble of the Age; or the Fallacies of Railway Investments, Railway Accounts, and Railway Dividends. 3rd ed. London, 1848.

SMITH, WORTHINGTON G. Dunstable: Its History and Surroundings. London, 1904.

SORBIÈRE, SAMUEL. A Voyage to England, containing many Things relating to the State of Learning, Religion, and other Curiosities of that Kingdom (English translation). London, 1709.

STEUART, HENRY. Account of a Plan for the better supplying the City of Edinburgh with Coal; together with an Examination of the Merits of the Two Principal Lines pointed out for the Intended Canal between Edinburgh and Glasgow. Edinburgh, 1800.

STEVENSON, DAVID. Fifty Years on the London and North Western Railway, and other Memoranda in the Life of David Stevenson. Edited by Leopold Turner. London, 1891.

STEVENSON, W. HENRY. Records of the Borough of Nottingham; being a series of extracts from the Archives of the Corporation of Nottingham, 5 vols., 1155–1702. (Vol. v edited by W. T. Baker.) London and Nottingham, 1882–90.

—— Calendar of the Records of the Corporation of Gloucester. Gloucester, 1893.

STEVENSON, WILLIAM. General View of the Agriculture of the County of Dorset.... London, 1812.

—— General View of the Agriculture of the County of Surrey.... London, 1818.

STEWART, CHARLES S. Sketches of Society in Great Britain and Ireland. 2nd ed. 2 vols. Philadelphia, 1835.

STOCKS, JOHN EDWARD, and BRAGG, W. B. Market Harborough Parish Records to A.D. 1530. London, 1890.

STONE, THOMAS. General View of the Agriculture of the County of Huntingdon.... London, 1793.

—— General View of the Agriculture of the County of Bedford.... London, 1794.

—— General View of the Agriculture of the County of Lincoln.... London, 1794.

STOW, JOHN. The Abridgement of the English Chronicle, First Collected by M. John Stow, and after him augmented with very many memorable Antiquities, and continued...unto the beginning of the Yeare 1618. By E. H. Gentleman. London, 1618.

—— Annales, or a Generall Chronicle of England. Begun by John Stow: Continued and augmented...unto the end of this present yeere, 1631. By Edmund Howes, Gent. London, 1631.

—— A Survey of London, written in the year 1598. New Edition edited by William J. Thoms, F.S.A. London, 1842. (Another edition, London, 1876.)

—— A Summarie of the Chronicles of England to 1590. London, [N.D.].

STRETTON, CLEMENT E. Early Tramroads and Railways in Leicestershire. Burton-on-Trent, 1900.

—— History of Various English Railways (The Stretton Collection for the Chicago Exhibition, 1893).

—— A Few Notes on Early Railway History. London, 1884.

—— History of the South Staffordshire Railway. Wolverhampton, 1885.

—— The Stour Valley Railway and Great Western. Shrewsbury, 1886.

—— History of the Manchester and Birmingham Railway. Manchester, 1886.

—— The Grand Junction Railway. [N.P.], 1887.

—— The History of the Birmingham and Derby Junction Railway. Birmingham, 1892.

STRETTON, CLEMENT E. The History of the Birmingham, Wolverhampton, and Stour Valley Railway. Printed for Chicago Exhibition, 1893.
—— The History of the Railways of Birmingham. Birmingham, 1897 (?).
—— The History of the London and Birmingham Railway. Leeds, 1901.
—— The History of the Amalgamation and the Formation of the London and North Western Railway Company. Leeds, 1901.
—— History of the London and North Western Railway. Leeds, 1902.
—— The History of the Birmingham and Gloucester, and Bristol and Gloucester Railways. Leeds, 1902.
—— The Stone Roads, Canals, Edge-Rail-Ways, Outram-Ways, and Electric Rail-Ways in the County of Leicester. Leicester, [1907].
STRICKLAND, H. E. General View of the Agriculture of the East Riding of York-shire.... York, 1812.
SUBSCRIBER (pseud.). Remarks upon the Pamphlet by "Investigator," on the Proposed Railway between Birmingham and London. By a Subscriber to the Railway. London, 1831.
SUMMERS, JEREMIAH WILLIAM. The History and Antiquities of Sunderland, Bishop-wearmouth, etc., from the earliest Authentic Records down to the Present Time. Vol. I. Sunderland, 1858.
SURTEES, ROBERT. The History and Antiquities of the County Palatine of Durham: compiled from original records preserved in public repositories and private collections. 4 vols. London, 1816–40.
Surtees Society Publications. Vols. LXV and LXXVII, for the years 1875 and 1883, entitled "Yorkshire Diaries and Autobiographies in the Seventeenth and Eighteenth Centuries." Durham, London, etc., 1877 and 1886.
SUTCLIFFE, JOHN. Report on the Line of Navigation from Hexham to Haydon-Bridge,....And A Report on the Line from Newcastle to Haydon-Bridge.... Newcastle, [1797].
—— A Treatise on Canals and Reservoirs, and the best Mode of Designing and Executing them; with Observations on the Rochdale, Leeds and Liverpool, and Huddersfield Canals, &c. Rochdale, 1816.
SWINDEN, HENRY. The History and Antiquities of the Ancient Burgh of Great Yarmouth in the County of Norfolk. Norwich, 1772.
SYDENHAM, JOHN. The History of the Town and County of Poole; collected and arranged from Ancient Records and other authentic documents, etc. Poole, 1839.
SYDNEY, WILLIAM CONNOR. England and the English in the Eighteenth Century. 2 vols. London, 1891.
—— The Early Days of the Nineteenth Century, in England. 2 vols. London, 1898.
SYKES, JOHN. Local Records: or Historical Register of Remarkable Events which have occurred in Northumberland and Durham, Newcastle-upon-Tyne, and Berwick-upon-Tweed. 2 vols. Newcastle, 1833. (Reprinted and a third volume added by T. Fordyce, Newcastle, 1866–67.)
SYLVESTER, CHARLES. Report on Rail Roads, and Locomotive Engines, addressed to the Chairman of the Committee of the Liverpool and Manchester projected Railroad. 2nd ed. Liverpool, 1825.
SYMPSON, E. MANSEL. Lincoln: A Historical and Topographical Account of the City. London, 1906.
TAIT, JAMES. Mediaeval Manchester and the Beginnings of Lancashire. Manchester, 1904.
TATE, GEORGE. The History of the Borough, Castle and Barony of Alnwick. 2 vols. London, 1866–69.

TATHAM, WILLIAM. Remarks on Inland Canals, the small System of interior Navigation, Various Uses of the Inclined Plane, etc. London, 1798.

—— The Political Economy of Inland Navigation, Irrigation and Drainage.... London, 1799.

TAYLOR, JAMES. A Brief Account of the Rise and Early Progress of Steam Navigation: intended to demonstrate that it originated in the Suggestions and Experiments of the late Mr James Taylor of Cumnock, in connection with the late Mr Miller of Dalswinton. Ayr, 1844.

TAYLOR, JOHN ("Water Poet"). The Old, Old, Very Old Man; or The Age and long Life of Thomas Par, the Son of John Parr of Winnington in the Parish of Alberbury, in the County of Salopp (or Shropshire) who was Borne in the Raigne of King Edward the 4th being aged 152 yeares and odd Monethes etc. London, 1635.

—— The Coaches Overthrow. Or A Joviall Exaltation of Divers Tradesmen, and Others, for the Suppression of Troublesome Hackney Coaches. London, [1635?].

—— All the Workes of John Taylor the Water Poet being 63 in number Collected into one Volume by the Author. London, 1630.

—— The Carriers' Cosmography: or A Brief Relation of the Inns, Ordinaries, Hostelries, and other Lodgings in and near London; where the Carriers, Waggons, Foot-posts and Higglers do usually come from any parts, towns, shires,...with nomination of what days of the week they do come to London, and on what days they return; etc. London, 1637.

—— John Taylor's Last Voyage, and Adventure, Performed from the twentieth of July last 1641 to the tenth of September following. In which time he past, with a Scullers Boate from the Citie of London, to the Cities and Townes of Oxford, Gloucester, Shrewesbury, Bristoll, Bathe, Monmouth, and Hereford, etc. London, 1641.

TAYLOR, JOSEPH. A Journey to Edenborough in Scotland (in 1705). Printed from the MS. with Notes, by William Cowan. Edinburgh, 1903.

TEISSERENC, EDMOND. Études sur les voies de communication perfectionnées et sur les lois économiques de la production du transport. Paris, 1847.

[TELFORD, THOMAS, and DOUGLASS, L.] An Account of the Improvements of the Port of London, and more particularly of the Intended Iron Bridge, consisting of one Arch, of Six Hundred Feet Span. London, 1801.

TELFORD, THOMAS, and NICHOLLS, Capt. Ship Canal for the Junction of the English and Bristol Channels. Reports by these two men. London, [1824].

TELFORD, THOMAS. Life of Thomas Telford, Civil Engineer, written by himself; containing a Descriptive Narrative of his Professional Labours: with a Folio Atlas of Copper Plates. Edited by John Rickman, one of his Executors.... London, 1838.

[THOMAS, JOHN.] The Local Register and Chronological Account of Occurrences and Facts connected with the Town and Neighbourhood of Sheffield. Sheffield, 1830.

THOMAS, WILLIAM. Observations on Canals and Railways, illustrative of the... Advantages to be derived from an Iron Rail-Way, adapted to Common Carriages, between Newcastle, Hexham and Carlisle;...also second edition, Report of Robert Dodd...on a Proposed Navigable Canal between Newcastle and Hexham. Newcastle-upon-Tyne, 1825.

THOMPSON, PISHEY. The History and Antiquities of Boston, and the Villages of Skirbeck, Fishtoft, etc. Boston (Eng. and Mass.) and London, 1856.

[THORPE, J.] A Collection of Statutes concerning Rochester Bridge. London, 1733.

THRUPP, GEORGE A. The History of Coaches. London, 1877.

TIMMINS, SAMUEL. William Murdock. (In Birmingham Miscellaneous Pamphlets, Vol. H. 2. Birmingham Free Central Library, No. 145,025.)

TOMBS, R. C. The King's Post. Being a volume of historical facts relating to the Posts, Mail Coaches, Coach Roads, and Railway Mail Services of and connected with the Ancient City of Bristol from 1580 to the Present time. Bristol, 1905.

—— The Bristol Royal Mail, Post, Telegraph, and Telephone. Bristol, [N.D.].

TOULMIN, JOSHUA. The History of the Town of Taunton, in the County of Somerset. Taunton, 1791.

TREDGOLD, THOMAS. A Practical Treatise on Railroads and Carriages, showing the Principles of estimating their strength, Proportions, Expense, and Annual Produce, etc. London, 1825.

TREVITHICK, FRANCIS. Life of Richard Trevithick, with an Account of his Inventions. 2 vols. London, 1872.

TRISTRAM, WILLIAM OUTRAM. Coaching Days and Coaching Ways. London, 1888.

TRUSLER, JOHN. Description of the Road from London to Bath. London, 1797.

—— A Delineation of the Roads of the West and South of England. Bath, 1800.

TUKE, Mr, Jr. General View of the Agriculture of the North Riding of Yorkshire.... London. 1794.

TURNER, GEORGE. General View of the Agriculture of the County of Gloucester.... London, 1794.

TURNER, WILLIAM H[ENRY]. Selections from the Records of the City of Oxford, with Extracts from Other Documents illustrating the Municipal History, 1509–1583. Oxford and London, 1880.

TWINING, Rev. THOMAS. Recreations and Studies of a Country Clergyman of the Eighteenth Century. London, 1882.

VALLANCE, JOHN. Considerations on the Expedience of Sinking Capital in Railways. London, 1825.

VANCOUVER, CHARLES. General View of the Agriculture of the County of Cambridge.... London, 1794.

—— General View of the Agriculture of the County of Essex.... London, 1795.

—— General View of the Agriculture of the County of Devon.... London, 1808.

—— General View of the Agriculture of Hampshire.... London, 1813.

VANDERSTEGEN, WILLIAM. The Present State of the Thames Considered; and a Comparative View of Canal and River Navigation. London, 1794.

VATKE, THEODOR. Culturbilder aus Alt-England. Berlin, 1887.

VOWELL, JOHN, alias HOKER, JOHN. The Antique Description and Account of the City of Exeter: in three Parts. Exon, 1765.

WAGNER, CARL A. Über die wirthschaftliche Lage der Binnenschiffahrtsunternehmungen in Grosbritannien und Irland. 2 Pts. Berlin, 1901. (Reprinted from Archiv für Eisenbahnwesen.)

WAKE, BERNARD JOHN. Turnpike Roads: Lenders of Money on Mortgage of Tolls, &c. cannot, under the Present Acts, have any Legal Security:—A Reply, in Support of this Doctrine to William Knight Dehany, Esq....the avowed Draughtsman of the Recent Turnpike Road Acts, and who has attempted to refute it, in Answer to a Former Publication by the Author of this. London, 1823.

WALFORD, CORNELIUS. Fairs, Past and Present: A Chapter in the History of Commerce. London, 1883.

WALFORD, EDWARD. Old and New London. 6 vols. London, [N.D.].

WALKER, D. General View of the Agriculture of the County of Hertford.... London, 1795.

WALKER, JAMES, and RASTRICK, JOHN URPETH. Liverpool and Manchester Railway. Report to the Directors on the Comparative Merits of Loco-motive & Fixed Engines, as a Moving Power. Liverpool, 1829.

WALKER, JAMES SCOTT. An Accurate Description of the Liverpool and Manchester Railway, and the Branch Railways to St Helen's, Warrington, Wigan and Bolton; with an Account of the Opening of the Railway, and of the Melancholy Accident which occurred to the late Rt. Hon. William Huskisson, etc. 3rd edition. Liverpool, 1832.

WALLIS, JOHN. The Natural History and Antiquities of Northumberland: and so much of the County of Durham as lies between the Rivers Tyne and Tweed; commonly called North Bishoprick. 2 vols. London, 1769.

WALPOLE, HORACE. Paul Hentzner's Travels in England, during the Reign of Queen Elizabeth (1598), translated by Horace, late Earl of Orford, etc. London, 1797.

—— Journal of the Reign of King George the Third, from the year 1771 to 1783. Now first published from the Original Manuscripts. Edited, with notes, by Dr Doran. 2 vols. London, 1859.

—— The Letters of Horace Walpole...edited by Mrs Paget Toynbee. 16 vols. London, 1903.

WARD, JOHN. The Borough of Stoke-upon-Trent, in the Commencement of the Reign of Her Most Gracious Majesty Queen Victoria, etc. London, 1843.

WARD, THOMAS HUMPHRY. The Reign of Queen Victoria: A Survey of Fifty Years of Progress. 2 vols. London, 1887.

WARDELL, JAMES. The Municipal History of the Borough of Leeds, in the County of York, including numerous extracts from the Court Books of the Corporation, and an Appendix, containing copies and translations of Charters, and other Documents relating to the Borough. London and Leeds, 1846.

WARNER, Rev. RICHARD. A Walk through some of the Western Counties of England. Bath, 1800.

—— The History of Bath. Bath, 1801.

—— A Tour through the Northern Counties of England and the Borders of Scotland. 2 vols. Bath, 1802.

—— A Tour through Cornwall in the Autumn of 1808. Bath, 1809.

WATKINS, JOHN. An Essay towards a History of Bideford, in the County of Devon. Exeter, 1792.

WEBB, SIDNEY and BEATRICE. English Local Government: The Story of the King's Highway. London: Longmans, Green & Co., 1913.

WEDEL, LUPOLD VON. Journey through England and Scotland, made by Lupold von Wedel in the years 1584 and 1585. Translated from the Original Manuscript by Dr Gottfried von Bülow, Supt. of the Royal Archives in Stettin. (In Trans. Royal Historical Society, N.S., Vol. IX, 1895, pp. 223-70.)

WEDGE, JOHN. General View of the Agriculture of the County of Warwick.... London, 1794.

WEDGE, THOMAS. General View of the Agriculture of the County Palatine of Chester.... London, 1794.

WELFORD, RICHARD. A History of the Parish of Gosforth, in the County of Northumberland. Newcastle-upon-Tyne (1879).

—— History of Newcastle and Gateshead in the Fourteenth and Fifteenth Centuries. London and Newcastle-upon-Tyne, [N.D.; probably about 1884].

WELLBELOVED, ROBERT. A Treatise on the Law relating to Highways; comprehending Turnpike Roads, Public Bridges, and Public Foot-paths, etc. London, 1829.

[WENDEBORN, F. A.] Beyträge Zur Kentniss Grosbritanniens vom Jahr 1779. Aus der Handschrift eines Ungenanten [F. A. Wendeborn] herausgegeben von Georg Forster, Professor am Carolino in Cassel. Lemgo, 1780.

WENDEBORN, FRED. AUG. A View of England towards the Close of the Eighteenth Century. Translated from the Original German, by the Author himself. 2 vols. London, 1791.

WENTWORTH, THOMAS, Earl of Strafford. The Earl of Strafford's Letters and Dispatches, with an Essay towards his Life by Sir George Radcliffe....By W. Knowler. 2 vols. London, 1739.

—— The Wentworth Papers (1705–39). Selected from the Private and Family Correspondence of Thomas Wentworth, Lord Raby, created, in 1711, Earl of Strafford, of Stainborough, co. York. With a Memoir and Notes by James J. Cartwright, M.A. London, 1883.

WEST, WILLIAM. The History, Topography and Directory of Warwickshire, etc. Birmingham, 1830.

WESTALL, GEORGE. Inland Cruising on the Rivers and Canals of England and Wales. London, 1908.

WHEELER, JAMES. Manchester, its Political, Social, and Commercial History, Ancient and Modern. Manchester and London, 1836.

WHEELER, WILLIAM HENRY. A History of the Fens of South Lincolnshire, being a Description of the Rivers Witham and Welland and their Estuary, etc. 2nd edition. Boston and London, 1896.

WHISHAW, FRANCIS. Analysis of Railways: consisting of a Series of Reports on the Twelve Hundred Miles of Projected Railways in England and Wales, now before Parliament; together with those which have been abandoned for the present Session, etc. London, 1837.

WHITAKER, THOMAS DUNHAM. The History and Antiquities of the Deanery of Craven, in the County of York. 2nd edition. London, 1812.

—— Loidis and Elmete; or an attempt to illustrate the District described in those Words by Bede; and supposed to embrace the Lower Portions of Aredale and Wharfdale, together with the Entire Vale of Calder, in the County of York. Leeds, 1816.

—— The History and Antiquities of the Deanery of Craven, in the County of York, 3rd edition with many Additions and Corrections, edited by A. W. Morant. London, 1878.

WHITE, J. Some Account of the Proposed Improvements of the Western Part of London, by the Formation of the Regent's Park.... 2nd edition. London, 1815.

WHITEHEAD, JOHN. Railway Management, Letter to George Carr Glyn, Esq., M.P., Chairman of the London and North Western Railway Company. 2nd edition. London, 1848.

WHITMORE, W. W. Letter to Lord John Russell on Railways. London, 1847.

WHITWORTH, RICHARD. The Advantages of Inland Navigation; or some Observations offered to the Public to shew that an Inland Navigation may be easily effected between the three great ports of Bristol, Liverpool, and Hull; together with a Plan for executing the same. London, 1766.

WHITWORTH, ROBERT. A Report and Survey of the Canal proposed to be made on one Level from Waltham-Abbey to Moorfields. Also of a line which may be continued from Mary-bone to the said proposed Canal, etc. With an Address on the Importance and great Utility of Canals in General. London, 1778.

WILD, Rev. J. Tetney, Lincolnshire. A History. (Printed for the Author by Albert Gait.) Grimsby, 1901.

WILKINSON, W. A. The Toll Question on Railways exemplified in the case of the Croydon and Greenwich Companies. London, 1841.

WILLIAMSON, Captain JAMES. The Clyde Passenger Steamer: Its Rise and Progress during the Nineteenth Century. Glasgow, 1904.

WILLMORE, FREDERIC WILLIAM. A History of Walsall and its Neighbourhood. Walsall and London, 1887.

—— Records of Rushall, County Stafford, with a Transcript of the Old Parish Register and Extracts from the Churchwardens' Accounts. Walsall, 1892.

WILLMORE, GRAHAM, WOLLASTON, F. L., and HODGES, W. Reports of Cases argued and determined in the Court of Queen's Bench and upon Writs of Error from that Court to the Exchequer Chamber, and in the Bail Court. London, 1840.

WILSON, HENRY. Hints to Road Speculators, together with the Influence Railroads will have upon Society, in promoting Agriculture, Commerce, and Manufactures. London, 1845.

WILSON, THOMAS. The Railway System and its Author, Thomas Gray, now of Exeter. A Letter to the Right Hon. Sir Robert Peel, Bart. London, 1845.

WODDERSPOON, JOHN. Memorial of the Ancient Town of Ipswich, in the county of Suffolk. Ipswich and London, 1850.

WOOD, NICHOLAS. A Practical Treatise on Railroads, and Interior Communication in General; with Original Experiments and Tables of the Comparative Value of Canals and Rail-Roads. London, 1825.

WOODCROFT, BENNET. A Sketch of the Origin and Progress of Steam Navigation from Authentic Documents. London, 1848.

WOODRUFF, CHAS. EVELEIGH. A History of the Town and Port of Fordwich, with a Transcription of the XVth Century Copy of the Custumal. Canterbury, [1895].

WOODWARD, B. B., WILKS, T. C., and LOCKHART, CHARLES. A General History of Hampshire, or the County of Southampton, Including the Isle of Wight. 3 vols. London, [1861–69].

WOOLRYCH, HUMPHREY W. A Treatise on the Law of Ways, including Highways, Turnpike Roads and Tolls, Private Rights of Way, Bridges, and Ferries. London, 1829.

—— A Treatise on the Law of Waters and of Sewers; including the Law relating to Rights in the Sea, and Rights in Rivers, Canals, Dock Companies, Fisheries, Mills, etc. London, 1830.

WORDSWORTH, CHRISTOPHER. The Fifteenth Century Cartulary of St Nicholas Hospital, Salisbury, with Other Records. Salisbury, 1902.

WORGAN, G. B. General View of the Agriculture of the County of Cornwall.... London, 1811.

WORTH, RICHARD NICHOLLS. History of Plymouth, from the Earliest Period to the Present Time. Plymouth, 1890.

—— Calendar of the Plymouth Municipal Records. Plymouth, 1893.

WYLIE, JAMES HAMILTON. History of England under Henry the Fourth. 4 vols. London, 1884–98.

YONGE, WALTER. Diary of Walter Yonge, edited by George Roberts for the Camden Society. London, 1848.

York and North Midland Railway Reports of the Committee of Investigation. York, 1849.

YOUNG, ARTHUR. The Farmer's Tour Through the East of England. London, 1771.

—— A Six Weeks' Tour Through the Southern Counties of England and Wales. London, 1768.

—— A Six Months' Tour Through the North of England. London, 1771.

YOUNG, ARTHUR. The Farmer's Letters to the People of England. London, 1768.

—— General View of the Agriculture of the County of Lincoln; Drawn up for the Consideration of the Board of Agriculture and Internal Improvement, by the Secretary to the Board. London, 1799.

—— General View of the Agriculture of Hertfordshire; Drawn up for the Consideration of the Board of Agriculture and Internal Improvement, by the Secretary to the Board. London, 1804.

—— General View of the Agriculture of the County of Suffolk; Drawn up for the Consideration of the Board of Agriculture and Internal Improvement, by the Secretary to the Board. 3rd edition. London, 1804.

—— The Farmer's Letters to the People of England...to which is added Sylvae: or, Occasional Tracts on Husbandry and Rural Œconomics. London, 1767.

—— General View of the Agriculture of the County of Sussex.... London, 1793.

—— General View of the Agriculture of the County of Suffolk.... London, 1794.

—— General View of the Agriculture of the County of Norfolk.... London, 1804.

—— General View of the Agriculture of the County of Essex.... 2 vols. London, 1807.

—— View of the Agriculture of Oxfordshire.... London, 1809.

YOUNG, CHARLES FREDERICK T. (C.E.) The Economy of Steam Power on Common Roads, in Relation to Agriculturists, Railway Companies, Mine and Coal Owners,...with its History and Practice in Great Britain....And its Progress in the United States, by Alex. L. Holly, C.E., and J. K. Fisher. London, 1860.

YOUNG, Rev. GEORGE. A History of Whitby, and Streoneshalh Abbey; with a Statistical Survey of the Vicinity to the Distance of Twenty-five Miles. 2 vols. Whitby, 1817.

BRITISH MUSEUM MANUSCRIPTS

No. 5,489, pp. 108–21. Papers concerning the River Stour in County of Kent. 1774–75.

No. 5,865, p. 183. Petition for leave to bring in a Bill into Parliament to recover the Navigation of River Cam. 1701–02.

No. 5,866, p. 222 b. Memoranda regarding Coaches.

No. 5,957, pp. 1–48. Tour in England, July 20 to Oct. 19, 1735.

No. 6,668, pp. 900 14. Itinerary from Winster, co. Derby, to Ormskirk, co. Lancaster.

No. 6,707. Reynolds' Derbyshire Collection.

No. 6,767. Notes of a Tour in Nottinghamshire and Derbyshire, etc., in 1766.

No. 11,052, pp. 80–136. Papers regarding River Wye Navigation. [1649–51.]

No. 11,302. (Scudamore Papers, Vol. XII.) Papers and Accounts on the County of Hereford. Bridges, Iron Works, and River Wye.

No. 11,394, pp. 28–44. Papers on the Dee Navigation, 1733.

No. 11,571. Journey through Surrey and Sussex by John Burton, 1752.

No. 12,496, pp. 263–91. Orders and Directions, Together with a Commission for the better Administration of Justice, and more perfect Information of His Majestie, etc. (Printed.) London, 1630.

No. 12,497, p. 360. The Presentment of the Jury touching Mordon (Wandle) River (co. Surrey).

No. 14,256–259. Dr Richard Pococke's Journey Round Scotland to the Orkneys, 1760. 4 vols.

No. 14,260–261. Travels of Dr Pococke in England in 1764. 2 vols.

No. 14,823. Journal of Tours by Craven Ord in the Counties of Norfolk and Suffolk, 1781–97.

No. 15,776. Journals of Rev. Jeremiah Milles, and of his Travels in different Parts of England and Wales, 1735–43.

No. 15,800. Plut. CLXXXII. B. Dr Richard Pococke's Journey into England from Dublin, by way of the Isle of Man, also his Journey into Cornwall and Devonshire, the Original Manuscripts, 1750.

No. 16,179. Extracts from the Registers and other Documents of the City of Chester from the time of Edward II to the year 1701.

No. 21,550. Tour Through the Counties of Suffolk, Norfolk, Cambridge, and Essex, Sept. 10 to Dec. 19, 1777, by Captain (Francis) Grose.

No. 21,567, p. 2. A Survey of the Rivers of Wye and Lugg, in reference to portation and fishing.

No. 22,926. Some Observations made in a Journey begun June 7, and finished July 9, 1742.

No. 22,999–23,001. Dr Richard Pococke's Travels in England. Original MSS. Vols. I–III.

No. 23,087. Tours of Mr Vertue through the Counties of Essex, Suffolk, Norfolk, Hants, Wilts, Warwick, Gloucester and Oxford, 1739–40.

No. 23,089. Journals of Tours in Sussex and Surrey, 1747.

No. 23,749. A few Notes about the River Nene.

No. 24,809. Papers relating to the City of London, the Thames, and Middlesex.

No. 28,566. Iter Boreale: anno salutis 1639, &......inter Anglos & Scotos, etc. Journal with the King's Army, 1639, by John Aston.

No. 32,442, pp. 1–76. Sketch of a Tour from London to the Lakes, made in the Summer of the Year 1799.

No. 32,443, pp. 240–64. Fragment of a Description of a Tour from London through Oxford to Southampton, 1829.

No. 34,754. A Relation of a Short Survey of 26 Counties, briefly describing the Citties and their Scytuations, and the Corporate Towns and Castles Herein. Observed in a Seven Weekes Journey begun at the City of Norwich, (1634)....... By a Captaine, a Lieutenn*t*, and an Ancient, All three of the Military Company at Norwich. Also A Relation of a Short Survey of the Westerne Counties, in which is briefely described the Citties, Corporations, Castles, and some other Remarkables in them. Observed in a seven Weekes Journey begun at Norwich, ...Aug. 4, 1635. By the same Lieutennant, that with the Captaine, and Ancient of the Military Company in Norwich Made a Journey into the North the yeere before.

No. 35,686, pp. 27, 53–54, 61 ff., 68–69, 71–165. Papers on the River Ouze Navigation. 1793–94.

BRITISH MUSEUM ADDITIONAL MANUSCRIPTS

Add. 5,842, pp. 244–71. A Tour through Part of England in 1735 (by John Whaley). Various other memoranda, e.g., pp. 103, 359, etc.

Add. 6,693, p. 305. The State of the Rivers of Wye and Lugg in Herefordshire. (Printed.)

Add. 11,395. Memorandum Book and Diary of Thomas Warton, 1780–84.

Add. 15,662, p. 225. License granted by Edward III (21 Ed. III, 1347) to New-castle-upon-Tyne to levy customs for repair of the Bridge across the river in that town.

Add. 17,398. The Original Itinerary of Grose the Antiquary. Journeys in England in 1775 and 1777.

Add. 18,003. A Journal of the Proceedings of the Trustees appointed for putting in Execution an Act of Parliament passed in the Twelfth Year of the Reign of Queen Anne, Intituled An Act for the Speedy and Effectual Preserving the Navigation of the River of Thames....

Add. 18,047. Official Returns of the Public Revenue, from 1699 to 1745.

Add. 18,054. A Particular State of the Receipts and Issues of the Public Revenue, from 1688 to 1702.

Add. 19,200. The History and Antiquities of the Ancient Villa of Wheatfield in the County of Suffolk (by the Rev. William Myers). London, 1758.

Add. 19,942. Dr J[eremiah] Milles's Sketches of Horse Panniers, Pack Saddles, etc. (ca. 1755).

Add. 24,466. Tours in Nottinghamshire and Derbyshire, 1801–02.

Add. 24,933. An Account of the Antient Wooden and present Stoned Bridge at Rochester and of the Lands proper and Contributary thereto.

Add. 27,587. Letter-Book of A. Jelfe, relating to Westminster Bridge, 1734–44 (showing accounts also).

Add. 27,951. Journals of Visits to England by an Irish Clergyman in 1761 and 1772.

Add. 28,570, p. 37 et seq. Tour to the West, 1791.

Add. 28,648, pp. 2–28. Rochester Bridge.

Add. 28,649. The Description of Devonshire by Sir William Pole, of Shute, Kt. Transcribed by J. Prince.

Add. 28,793. Journal of Rev. John Skinner during his Tour in Cornwall, etc. 1797, 1798.

Add. 28,802. Llewelyn Meyrick's Journal of a Tour through Part of England and North Wales in the summer of 1821.

Add. 30,172. Journal of Tours in the Midland and Western Counties of England and Wales 1794, and in Devonshire in 1803.

Add. 30,173. Journey made by William Phillipps from Broadway (County Wor-cester) to Manchester and Liverpool, and back, 1792.

Add. 30,302. Remarks upon Wayside Chapels with Observations on the Archi-tecture and Present State of the Chantry on Wakefield Bridge, by John Chessell Buckler and Charles Alban Buckler, architects. London and Oxford, 1870.

Add. 30,929. Diary of Charles Danvers, of Tours in the West of England and in the Lake District, 1795–1812.

Add. 31,337. Journal of a Tour from Brighton to Weymouth in 1816.

Add. 31,857. William Cobbett's Eastern Tour, 1830.

Add. 32,442, pp. 1–76. Sketch of a Tour from London to the Lakes made in the Summer of the Year 1799.

Add. 33,576, p. 63. Suit concerning the Lock near Waltham Abbey on the River Lea, 1683.

Add. 33,640. Sketches in Wales, Derbyshire, etc.; 1803. By Rev. J[ohn] Skinner.

Add. 33,683–685. Journal of a Northern Tour (by Rev. John Skinner) from Camer-ton, to London, to Peterborough, and Lincoln, to York and Furham, and thence to Edinburgh and Linlithgow in the year 1825. 3 vols.

Add. 34,105, pp. 188–95. On the Medway Navigation, 1627.

Add. 34,218, pp. 37–58. Statements, Arguments and Correspondence concerning the Navigation and Overflowing of the River Medway between Maidstone and Yalding. 1600.

Add. 34,727. West Papers, Vol. I, pp. 14, 16.

Add. 35,154; 35,142–147 (6 vols.); 27,825–830—Place Manuscripts—Biography of Francis Place, and articles on Social and Industrial Conditions.

Add. 35,649, pp. 222 and 224. Papers on the Tolls charged on several Navigations.

Add. 35,679, pp. 316–23. Papers showing troubles with a certain lock-keeper on River Cam. 1759–60.

Add. 35,689, pp. 21–22. Papers concerning the River Cam Navigation. [1811.]

Add. 35,691, pp. 67–89, 94, 116. Papers relating to the Turnpike from Arrington to Biggleswade. Pp. 169–74, 275, 315, Papers regarding the Navigation of the Rivers Cam and Ouze.

Add. 36,663, p. 558, gives 'A List of Coaches passing through and coming into Newcastle-under-Lyme, Nov. 1830 (by Mr T. P. Platt).'

Add. 36,767, pp. 1–4, give an Order of Privy Council in regard to the Conservation of River Thames. Page 146, a paper entitled, 'The State of the Cause concerning the Milles and Causey at Chester.' 27 Feb., 1607.

Add. 36,914. River Weaver Navigation and Cheshire Salt Works, 1699–1720.

BRITISH MUSEUM, EGERTON MANUSCRIPTS

Eger. 784. Diary of Mr (William) Whiteway of Dorchester, co. Dorset, from Nov. 1618 to March 1634.

Bibl. Eg. 926. Original Letters from the Rev. S. Denne, Rector of Wilmington, Kent, and others, on Kentish Topography and Antiquities, 1771–84.

Eger. 2235, pp. 84–93. Tours of the Rev. Joseph Spence to Kimbolton and Ormesby in 1743, and to Kirkby and the Peak in 1752.

BRITISH MUSEUM, STOWE MANUSCRIPTS

Stowe 747, p. 84. Letter from (Sir) A. Copley, dated Saturday, Jan. 1, 1697, to Thomas Kirke, Leedes, regarding a Bill to make river Dun navigable.

Stowe 818, pp. 83–87. Memoranda on making Derwent river Navigable.

Stowe 877. Sampson Erdeswicke's Staffordshire. MS. additions by E. Vernon. Pp. 13–22 contain 'A Scheme or Proposal for making a Navigable Communication between the Rivers of Trent and Severn, in the County of Stafford. By Dr Thomas Congreve of Wolverhampton.' Printed in London, 1717.

BRITISH MUSEUM, LANSDOWN MANUSCRIPTS

Lansd. 32. Papers regarding the River Lee Navigation.

Lansd. 38, pp. 84–85. An Answer to the Complaint of the Inhabitants of Enfield against the carrying of grain from Ware to London by the River of Ley (Lea), Oct. 29, 1583. P. 88, A Petition of the Inhabitants of Ware to Lord Burghley, for a Commission to hinder damage done to the River Lea, 1583. Pp. 91–95, The Defects of the Water of the River Lea from Waltham Holy Cross to the mills beneath Stratford at Bow, 1583.

Lansd. 41, p. 44. Orders issued for the Conservation of the River Thames, Sept. 1584; pp. 169–76, Ordinances, Laws and Decrees of Oct. 1, 6 Elizabeth, of the Commissioners of Sewers of Lincolnshire.

Lansd. 49, pp. 74–75. An Order for the Repair of the Bridge at Walmesford. [1586]

Lansd. 55, pp. 109, 114. Complaint to Lord Burghley as to the impassability of Christmas Lane in Suffolk. Jan. 8, 1587.

Lansd. 60, p. 96. The Inhabitants of Enfield complain to the Queen that they are impoverished by the Navigation of the River Lea, and ask redress or relief. 1589.

Lansd. 76, No. 55, pp. 125–28. Order of the Star Chamber concerning the Right of Navigation on River Lea, June 20, 1594.

Lansd. 84, pp. 55–57. Papers concerning the exactions made by the people of King's Lynn on goods carried to or from Cambridge. 1597.

Lansd. 142. Several Papers concerning the Watermen of the River Thames.

Lansd. 166, pp. 84–93. Papers regarding the repairing of Berwick Bridge. [1607–13]

Lansd. 328. A Volume containing extracts from Ancient Records concerning the Repairing of bridges in various Counties of England.

Lansd. 561. Trial in Court of King's Bench, 1754, regarding the Ancient Highway from Richmond, through Richmond Park, to Croydon.

Lansd. 688. Tour in the Midland Counties, 1710.

Lansd. 722, pp. 29–40. Sir William Dugdale's Journal of his Itinerary to the Fens of Ely, begun from London, May 19, 1657.

Lansd. 896, pp. 162–67. Papers on the Navigation of Beverley Beck.

BRITISH MUSEUM, HARLEY MANUSCRIPTS

Harl. 368, p. 7 ff. A Paper giving Reasons [in favor of Bristol] against the Continuance of the Port of Gloucester. May 25, 1584.

Harl. 2003. This whole Volume deals with the River Dee and its Mills and Causey, etc.

Harl. 2022, p. 66. Papers regarding River Dee.

Harl. 2046, p. 1. Indenture made April 28, anno 12 Eliz., for repairing a highway 2½ miles long adjoining to Chester.

Harl. 2057, p. 116. An Order of the Assembly at Chester, July 17, 1612, for cleansing etc. Horne Lane.

Harl. 2077, p. 21. Regarding the maintenance of Huntington Lane, near the City of Chester, in repair.

Harl. 2081. Law Papers Concerning Dee Mills. Ca. 1607.

Harl. 2082. Cheshire Collections of Manuscripts on Chester and Dee Mills, etc.

Harl. 2084. This Volume deals almost wholly with the River Dee and the Mills and Causey at Chester.

Harl. 2150, p. 182. Indenture under which Thomas Bennett of Chester was to keep the Streets of that city in good repair. Dated Oct. 8, 12 Eliz.

Harl. 2263, p. 323. A License for inclosing a certain horse way through private ground. Dated Dec. 23, 1708.

Harl. 2264, p. 272. Docquet of a License for the enclosure of a certain common. 24 Jan., 1710.

Harl. 6166, p. 229 ff. The defaults and common Nuisance of Bridges and of Causeys and of Ways at every end of the Bridges within the Hundreds of Tandridge and Ryegate and Divers Other Places within the said County [Surrey] to be inquired of by the Justices of Peace of the same County at their General Sessions to be held at Croyden the Tuesday next after Twelfth day in the 25th Year of our Sovereign Lord King Henry VIII.

Harl. 6211. An Account of my Travels from Venice through Germany into England [temp. Charles I], Vol. II, pp. 132–206. [By John Lawson (?).]

Harl. 6494, p. 129 ff. Journey into the West of England, 1637.

BRITISH MUSEUM, SLOANE MANUSCRIPTS

Sloane 1156. Relation du voyage de Monsieur de la Villeauxcleres en Angleterre faict en l'année 1624.

Sloane 1731 b. Journal du voyage de mon frère Jean de Cardonnel en Irlande etc. [including West England]. 1649–50.

Sloane 1899. Dr Thomas Browne's Journal of a Tour in Kent. 1698.

Sloane 1900, pp. 36–60. Dr Edward Browne's Memorandum Book, 1662; Diary of a Journey in England, begun Sept. 8, 1662.

Sloane 1983 A & B. Memoranda made in a Journey from London to Oxford.

Sloane 3323, pp. 267–69. Papers concerning the River Dee Navigation.

WYATT MANUSCRIPTS

Nos. 93,189 and 93,190, 2 vols., in Birmingham Free Reference Library. Deal with a variety of subjects.

BRITISH MUSEUM, PAMPHLETS AND TRACTS

012314. e. 88. Coach and Sedan Pleasantly Disputing for Place and Precedence, the Brewers-Cart being Moderator. London, [1636].

08226. aaa. 29. Reasons humbly offered to the Consideration of Parliament, for the suppressing such of the Stage-Coaches and Caravans now travelling upon the Roads of England, as are unnecessary, and Regulating such as shall be thought fit to be continued. [1700?]

08235. f. 18. Seasonable Considerations on a Navigable Canal Intended to be cut from the River Trent, at Wilden Ferry, in the County of Derby, to the River Mersey, in the County of Chester. [1766]

08235. f. 77. Observations on the General Comparative Merits of Inland Communication by Navigations or Railroads, with particular Reference to those projected or existing between Bath, Bristol, and London: in a Letter to Charles Dundas, Esq., M.P., Chairman of the Kennet and Avon Canal Company. London, 1825.

08235. h. 12. Reports and Pamphlets on the Subject of Norwich a Port, from the Year 1818 to the passing of the Norwich and Lowestoft Navigation Act in 1827. Norwich, [1818–27].

21. h. 5 (2). By the Maior. Orders set down by the right Honorable, Sir John Watts, Knight, Lord Maior of this Citty of London with the Co[n]sent of the Aldermen his brethren concerning the rates of Carriages with Cartes within this Cittie and Borough of Southwarke. At the Guild Hall the xxv day of November, 1606. London, [1606].

21. h. 5 (36). Ad Session Oier' & Terminer & Gaolae Domini Regis de Newgate Deliberation' tent' pro Civitat' London apud Justice-Hall in le Old Baily in Paroch' S. Sepulchri in Ward' de Faringdon extra London praed', die Mercurii, scilicet decimo septimo die Junii, Anno Regni Regis Caroli Secundi nunc Angliae &c. vicesimo, etc. [London, 1666 or 1667?]

21. h. 5 (52). Ad General' Quarterial' Session' Pacis Dom' Regis tent' pro Civitat' London...die Mercurii, scilicet sexto decimo die Octobris, Anno Regni Dom' nostri Caroli Secundi...vicesimo quarto, etc. [London, 1672]

101. i. 59. Plan for Raising Three Hundred Thousand Pounds, for the Purpose of compleating the Bridge at Black-Friars, and Redeeming the Toll thereon; Embanking the North Side of the River Thames...; Redeeming the Antient Toll upon London Bridge.... London, 1767.

102. k. 52. An Act of Common Council for the better Regulation of Hackney-Coaches. London, 1683.

191. e. 9. A New Description of England and Wales, with the Adjacent Islands. Wherein are contained Diverse useful Observations and Discoveries, in respect to Natural History, Antiquities, Customs...with a Particular Account of the Products, Trade, and Manufactures...to which is added a new and correct Set of Maps of each County, their Roads and Distances;...by Hermann Moll, Geographer. London, 1724.

213. i. 1 (44). A Bill for Enlarging the Term and Powers granted by an Act passed in the Twentieth Year of the Reign of His present Majesty [Geo. II] for repairing the High-Road leading from...Stockton-upon-Tees, to Darlington, and from thence through Winston to Barnard Castle, etc.

213. i. 1 (45). A Bill for Continuing, and making more Effectual, Two Acts of Parliament, one passed in the Tenth Year of the Reign of Her late Majesty Queen Anne, and the other in the Eighth Year of the Reign of His late Majesty King George the First, for Repairing the Highway between a certain Place called Kilburn Bridge in the County of Middlesex, and Sparrows Herne in the County of Hertford.

213. i. 1 (86). The Case of the Deputy Postmasters [as to keeping horses for supplying those riding post]. [London, 1748 (?)]

213. i. 1 (87). The Case of the Innkeepers and Keepers of Livery Stables, and also of the several Owners and Proprietors of Inns and Livery-Stables, and of Lands lett therewith. [London, 1748]

213. i. 1 (91). The Case of the Land-Owners interested in the Banks on each Side of the River Ouze, in the County of Norfolk, between Stowbridge and the Port of Lynn.

213. i. 2 (60). A State of the Road from Keighley to Kendall; and of the Expence the Country will be put to, by paying Interest and Salaries, in case the Bill for a Turnpike-Road should pass; with an estimate of what may be yearly raised by the Laws now in being.

213. i. 2 (61). The Reasons published against the Bill for a Turnpike between Keighley and Kirkby Kendal, Answered.

213. i. 2 (62). Reasons against the Bill for an intended Turnpike between Keighley, in the County of York, and Kendall, in the County of Westmorland.

213. i. 2 (76). A Bill for Confirming an Agreement entered into between the Company of Proprietors of the Undertaking for recovering and preserving the Navigation of the River Dee, and Sir John Glynne, Baronet, Lord of the Manor of Hawarden, and several Freeholders and Occupiers of Land within the said Manor; and for Explaining and Amending Three several Acts of Parliament, of the Sixth, Fourteenth, and Seventeenth Years of His present Majesty's Reign, for recovering and preserving the Navigation of the said River Dee.

213. i. 3 (100). Considerations about the Method of Preserving the Public Roads.

213. i. 3 (101). Reasons against a Bill for Permitting only Carriages with broad Wheels, and those drawn by two Horses, to pass on Turnpike Roads, with regard to the Countries within Twenty-five or Thirty Miles of London. [About 1754–55.]

213. i. 5 (94). A State of the Case, relating to the several Navigations made, or proposed to be made in or adjoining to the North-Eastern Parts of the County of Chester.

213. i. 5 (94). Remarks upon the Observations on the intended Navigation from Witton Bridge, to the Towns of Knutsford, Macclesfield, Stockport, and Manchester. [Ca. 1766]

213. i. 5 (95). A State of Facts tending to shew the Utility of the Proposed Canal from Witton near Northwich to Knutsford and Macclesfield, and by Stockport to Manchester. [Ca. 1766]

213. i. 5 (96). Navigation. Some Observations relative to Navigation: Humbly submitted to the Consideration of the Legislature. [Ca. 1766]

213. i. 5 (97). Observations on the Intended Navigation from Witton Bridge to Knutsford; and from thence through Macclesfield and Stockport to the Town of Manchester. [Ca. 1766]

214. i. 4 (103). Observations on the Effects of the Intended Oxford Canal Navigation, with respect to the Diminution or Increase of Seamen. [1769]

214. i. 4 (104). Begin, As a Bill is now depending in Parliament before the Right Honourable the House of Commons, for making a Canal, from the City of Coventry, to the City of Oxford, the following Queries are humbly submitted to the Consideration of both Houses of Parliament, before the said Bill passes into a Law. [1769]

214. i. 4 (119). A Letter (dated Mar. 4, 1769) from Yarmouth to the Representatives in Parliament for Yarmouth and Lynn, on the Subject of the Canal Navigation from Coventry to Oxford, now depending in Parliament.

214. i. 4 (120). Report from the Committee appointed by the Chamber of London to consider the Bill for extending the Coventry Canal to Oxford; with Facts and Observations subjoined respecting the Coal and Coasting Trade. [1769?]

214. i. 4 (124). The Case of the Petitioner, George Perrott, Esquire, on the Bill for Making a Canal from the Coventry Canal-Navigation, at or near Coventry, to the City of Oxford. [1769?]

215. i. 1 (105). Thames Navigation. (Reply to a printed Bill lately handed about, entitled, Some few of the many Objections that occur to the Bill now depending in Parliament for making a Navigable Cut from Sunning...to Monkey Island, &c.) [1771?]

215. i. 4 (117). Case and Reasons for Disusing Weighing Engines on the Turnpike Roads. 1774.

290. c. 30. A New and Compleat Survey of London. By a Citizen and Native of London. 2 vols. London, 1742.

856. m. 1 (26). An Act for making the River Stour Navigable from the Town of Maningtree in the County of Essex, to the Town of Sudbury in the County of Suffolk. (Act 4 & 5 Anne, c. 15, Public.)

856. m. 1 (84). A Bill to make the River Darwent, in the County of Derby, Navigable (Act 6 Geo. I—Public—c. 27).

856. m. 1 (66). The Case of the Carriers and Waggoners who carry Goods to hire. [1720?]

857. b. 7 (39 or 99). Reasons against the Navigable Scheme.

857. b. 7 (109). An Answer to the Reasons against the Navigable Scheme.

857. b. 9 (2). A Particular State of the Receipts and Issues of the Publick Revenue Taxes and Loanes during the Reigne of his late Majesty King William. That is to say, From the 5th day of November 1688 from which Day the Parliament appointed the said Accounts should comence to the 25th day of March 1702 being the ffirst Determinaċon of the Accompts since the Demise of his said late Majestye which happened on the 8th day of March preceeding. The same reduced to one Generall Accompt or State for the whole time above menċoned. [This is all in MS.]

857. b. 9 (72). Reasons Humbly Offered by John Daniel and William Blackburn, Esquires, for themselves, and on Behalf of Charles Duckenfield, Thomas Butterworth, and John Reddish, Esquires, and others, Gentlemen and Freeholders of the County of Chester, against a Bill for Repealing an Act made in the Seventh Year of His Majesty's Reign, for making the River Weaver Navigable from Frodsham Bridge to Winsford Bridge in the County of Chester. [1726?]

857. b. 9 (73). Reasons Humbly Offered Against allowing the County of Chester any part of the Tonnage Duty for making the River Weaver Navigable; and that the same may be made Navigable on the Easiest Terms.

857. b. 9 (74). Reasons Humbly Offered Against a Bill passed the Honourable House of Commons and now Depending before Your Lordships, Entitled, A Bill for Repealing an Act passed in the Seventh Year of His present Majesty's Reign, for making the River Weaver Navigable, from Frodsham Bridge to Winsford Bridge in the County of Chester, and for the more speedy and effectual carrying on and perfecting the Navigation of the said River, from Frodsham Bridge to Northwych in the said County.

857. b. 9 (75). Proposals Humbly Offered for Making the River Weever Navigable from Frodsham Bridge, to Northwich in the County of Chester.

857. b. 9 (76). Reasons Humbly Offered for passing the Bill for making the River Weaver Navigable from Frodsham Bridge to Northwich in the County of Chester: With Remarks upon the Proposals from Liverpool.

857. b. 9 (77). The Case of the Barge-Masters and others, Navigating on the Rivers of Isis and Thames, from Oxford to London; showing the Hardships they labour under, by the exorbitant Sums they pay for passing through the several Locks, Wears, Bucks, Gates and for the Use of Boats belonging to the same, and going over Towing-Paths on the Banks of the said Rivers.

857. b. 9 (78). Reasons Humbly Offered by the Trustees of Richard Vernon, Esq., deceased, against the Bill for Repealing an Act made in the Seventh Year of His Majesty's Reign, For making the River Weaver Navigable from Frodsham Bridge to Winsford Bridge, in the County of Chester.

857. c. 1 (28). The Case of the Cheesemongers, in and about the Cities of London and Westminster, relating to the Bill to recover and preserve the Navigation of the River Dee, in the County of Chester. [1732]

857. c. 1 (37). The Case of the Inhabitants of the County and City of Chester, Petitioners for the Bill to Recover and Preserve the Navigation of the River Dee; In Answer to the Petition of the Cheesemongers in and about the City of London against the said Bill. [1732]

857. c. 3 (69). Reasons against building a Bridge over the Thames at Westminster.

857. c. 4 (32). An Act for making Navigable the River or Brook, called Worsley Brook, from Worsley Mill, in the Township of Worsley, in the County Palatine of Lancaster, to the River Irwell in the said County. (Act 10 Geo. II—Public —c. 9.)

857. c. 13 (54). The Case of the Undertakers for making Navigable the Rivers Aire and Calder, in the County of York, and of their Lessees. [31 Geo. II?]

858. b. 3 (41). Reasons for extending the Navigation of the River Calder from Wakefield to Halifax. (This would be about 1757 or 1758, for Smeaton's Survey was made 1757.)

858. b. 4 (36). An Act for Explaining and Amending an Act passed in the Sixth Year of His present Majesty's Reign, intitled, An Act to Recover and Preserve the Navigation of the River Dee, in the County Palatine of Chester; and another Act passed in the Fourteenth Year of his present Majesty's Reign, intituled, An Act for Incorporating the Undertakers of the Navigation of the River Dee; and for Repealing the Tonnage-Rates payable to the said Undertakers; and for granting to them other Tonage or Keelage-Rates in lieu thereof; and for other Purposes therein mentioned. (Act 17 Geo. II, c. 28, Public.)

517. b. 31. The Statute 7 Geo. IV, cap. cxlii, for Consolidating the Trusts of the Turnpike Roads, in the Neighbourhood of the Metropolis, north of the Thames.... London, 1826.

517. k. 16 (3). England's Wants: or several Proposals probably beneficial for England, Humbly offered to the Consideration of all Good Patriots in both Houses of Parliament. By a true lover of his Country [Edward Chamberlayne]. London, 1667.

567. e. 7. Beyträge Zur Kenntniss vorzüglich des Innern von England und seiner Einwohner. Aus den Briefen eines Freundes gezogen von der Herausgeber. 4 Bde. Leipzig, 1791.

577. b. 6–10. A Description of England and Wales. Containing a particular Account of each County...and the Lives of the illustrious Men each County has produced, &c. 10 vols. London, 1769.

578. k. 30. Observations made during a Tour through Parts of England, Scotland, and Wales, in a Series of Letters [by Richard Joseph Sulivan, F.R.S.]. London, 1780.

579. c. 41 (3). A Description of the Ancient and Present State of the Town and Abbey of Bury St Edmund's, in the County of Suffolk. 3rd edition. Bury St Edmunds, 1782.

579. c. 42 (4). A Description of Manchester: giving a Historical Account of those Limits in which the Town was formerly included....By a Native of the Town. Manchester, 1783.

712. a. 4. The Traveller's Companion. Containing Variety of Useful yet Pleasant Matters relating to Commerce.... London, 1702. Pp. 54–58 give an outline of the Post Roads (at that time) from London, with their several stages and branches.

712. g. 15. A Few General Observations on the Principal Railways Executed, in Progress, and Projected, in the Midland Counties and North of England, with the Author's Opinion upon them as Investments. Maps. London, 1838.

712. g. 16 (17). A Treatise of Wool and Cattel. In a Letter written to a Friend, Occasioned upon a Discourse concerning the great Abatements of Rents, and the Low Value of Lands, &c. London, 1677.

712. g. 16 (20). The Trade of England Revived: and the Abuses thereof Rectified, in Relation to Wool and Woollen-Cloth, Silk and Silk-Weavers, Hawkers, Bankrupts, Stage-Coaches, Shop-Keepers, Companies, Markets, Linnen-Cloth. Also What Statutes in force may be injurious to Trade and Tradesmen, with several Proposals. London, 1681.

725. c. 40. News from the Fens, or An Answer to a Pamphlet entituled, Navigation Prejudiced by the Fen-Drainers....Wherein is set forth the Vanitie and Falshood of that Discourse, and it is Proved, That Navigation is meliorated by the Fen-Drainers, &c. London, 1654.

796. c. 36. A Brief Director for those that would send their Letters to any Parts of England, Scotland or Ireland. Or a List of all the Carriers, Waggoners, Coaches...that come to London, from the most parts and places, by Land and Sea. [1710?]

816. l. 4 (21). Robinson, Mayor. Commune Concilium tentum in Camera Guihaldae Civitatis London. decimo die Octobris, Anno Domini 1663. Annoque Regni Domini nostri Caroli Secundi, nunc Regis Angliae, etc. Decimo quinto. [London, 1663]

816. m. 7 (131). A Proposal [for regulating Cars, Carts, etc., in London].

816. m. 8 (4). The Case of the Town and Port of King's-Lynn in Norfolk, as to their Navigation.

816. m. 8 (5).   The Case of the Corporation of the Great Level of the Fenns; relating to a Bill depending in Parliament, for the better Preservation of the Navigation of the Port of King's-Lynn;....

816. m. 8 (6).   The State of the Adventurers Case, in Answer to a Petition exhibited against them by the Inhabitants of the Soake of Peterburgh.

816. m. 8 (11).   A Short Demonstration, That Navigation to Bedford, is for the Benefit of Bedfordshire.

816. m. 8 (38).   The Case of the Citizens of Chester in Answer to several Petitions from Leverpool, Parkgate, and the Cheesemongers; and also to Printed Reasons [by Thomas Badeslade] against the Act to Recover and Preserve the Navigation of the River Dee. 1735.

816. m. 8 (39).   Reasons for making the River Dunn in the West Riding of the County of York navigable, and the great advantages which will accrue to the Nation in general by it.

816. m. 8 (49).   The Case of the Barge-Masters and others, Navigating on the Rivers of Isis and Thames, from Oxford to London, shewing the Hardships they labour under, by the exorbitant sums they pay for passing through the several Locks, etc. [1720?]

816. m. 8 (50).   Reasons for making Navigable the Rivers of Stower and Salwerp, and the Rivulets and Brooks running into the same, in the Counties of Worcester and Stafford. [1720?]

816. m. 8 (51).   An Answer, as well to a Paper, intituled Reasons wherefore the making Navigable of the Rivers of Stower and Salwerp in the County of Worcester, will be of great advantage to the County of Salop, and especially to the Towns of Shrewsbury, Bridge-North, Wenlocke, Wellington, and Towns adjoyning to the River of Severn. As also to another Paper, intituled, An Answer to some partiall pretences, called, Reasons dispersed by some Shropshire Coal Masters. [1720?]

816. m. 8 (52).   The Case of making the River Avon, in the County of Somerset and Gloucester, Navigable, from Bristol to Bath....

816. m. 8 (53).   The Case for Making the Rivers Aire and Calder, in the County of York, Navigable to Leeds and Wakefield.

816. m. 8 (54).   Reasons against the Bill for making the Rivers Ayre and Calder, in the West Riding of the County of York, Navigable.

816. m. 8 (55).   Reasons humbly offered to the Consideration of the Parliament, for the making Navigable the River Derwent, from the Town of Derby to River Trent.

816. m. 8 (56).   The Case of the Navigation of the River Wye, in the County of Surry. [London (?), 1670.]

816. m. 8 (57).   A Reply to a Paper Intituled: An Answer to the Pretended Case Printed concerning the Navigation of the River Wye, in the County of Surrey, by shewing the true state thereof. [1670?]

816. m. 8 (58).   The Proposals for making the River Chelmer navigable, from Malden to Chelmsford, are as follow.   Also, Objections against these Proposals.

816. m. 9 (13).   An Abstract of the Forfeitures and Penalties set and imposed on Offences done contrary to the Act of Parliament for Paving and Cleansing the Streets.

816. m. 12 (79).   An Act of Common Council for the Government of Cars, Carts, Carrooms, Carters, and Carmen; and for the Prevention of Frauds in the Buying and Selling of Coals.

816. m. 12 (151).   To the Honourable the Commons of England in Parliament Assembled.   The Humble Petition of a great Number of the Licensed Hackney Coachmen.

816. m. 12 (152). The Case of the Antient Hackney-Coachmen....Humbly Presented to the Honourable House of Commons.

816. m. 12 (153). The Case of John Nicholson, Walter Storey, (and others) in Behalf of themselves and the First 400 Ancient Hackney-Coachmen, and their Widows.

816. m. 12 (154). The Case of the Hackney Coachmen.

816. m. 12 (155). The Hackney Coachmen's Case, Humbly presented to the Honourable House of Commons; with a Proposal to Raise for Her Majesty £200,000 per annum.

816. m. 12 (156). The Case of Thomas Blunt...and the Rest of the Eight Hundred Licens'd Hackney Coach-men....

816. m. 12 (157). To the Honourable the House of Commons, The Humble Petition of Charles Sewell, Thomas Holland, and George Garrett, on behalf of themselves, and the rest of the Eight Hundred Licensed Hackney Coach-men, and their Widows.

816. m. 12 (158). To this Honourable House, The Case of the Coachmen; of divers Coachmakers, Harnessmakers...and other Traders depending upon them,... in relation to the Bill for encreasing the Number of Hackney Coaches.

816. m. 12 (159). The Case of Divers Tradesmen, Creditors of the Hackney Coachmen, in London and Westminster, and Stagemen to several Places of England.

816. m. 12 (161*). The Humble Proposals of James Lord Mordington, and Martin Laycock, Esq.; for the Farming of the Hackney-Coaches.

816. m. 12 (162). Stage-Coaches Vindicated: or Certain Animadversions and Reflections upon several Papers writ by J. C. of the Inner Temple, Gent., against Stage Coaches.

816. m. 12 (163). A Copy of a Printed Letter from J. C. to a Post-Master in the Country, with Directions about the Management of his Designe for Putting down Stage Coaches.

816. m. 14 (26). Reasons Humbly Offered to the Honourable House of Commons, why the Waggoners ought not to be obliged to any certain Weight.

816. m. 14 (27). The Case of Richard Fielder, in Relation to the Petition of the Waggoners.

816. m. 14 (28). The Case of John Littlehales, against the pretended Petition of the Waggoners travelling the Northern Roads of England, etc.

883. h. 16. The Laws of Sewers; or the Office and Authority of Commissioners of Sewers. 2nd edition. London, 1732.

982. b. 22. A Description of the River Thames, etc., with the City of London's Jurisdiction and Conservacy thereof proved, both in point of Right and Usage, by Prescription, Charters, etc....to which is added...Observations...also of the Water Carriage on the River Thames.... London, 1758.

1028. h. 24. The Methods proposed for making River Dunn Navigable, and the Objections to it answered. With an Account of the Petitioner's Behaviour to the Landowners. To which is annexed, a Mapp of the River, and the Reasons lately Printed for making it Navigable, with the Advantages of it. London, 1723.

1130. c. 43 (2). Statement of the General Laws respecting Highways, and Turnpike Roads; including the Substance of the New Acts. London, [1825?].

1138. b. 11. The Ancient Trades Decayed, Repaired Again: Wherein are declared the several Abuses that have utterly impaired all the Ancient Trades in the Kingdom;....Written by a Countrey Tradesman. London, 1678.

1246. l. 16 (1). Act 7 Geo. I, Stat. 1, c. 15. An Act for making the Rivers Mercy and Irwell navigable from Liverpool to Manchester, in the County Palatine of Lancaster.

1246. 1. 16 (3). An Act for making a Navigable Cut, or Canal, from the River Trent, at or near Wilden Ferry in the County of Derby, to the River Mersey, at or near Runcorn Gap.

1302. g. 8 (3). Canal between the Eastern and the Western Seas. Newcastle-upon-Tyne, 1817. [A series of reports of various meetings, regarding better communication between Newcastle and Carlisle, taken from local newspapers of that year.]

1396. e. 22 (4). Ruminations on Railways. No. I, Railway Speculation; No. II, The Railway Board of Trade. London, 1845.

1396. g. 13. Railways; their Uses and Management. London, 1842.

1396. g. 21. Railways and the Board of Trade. 3rd edition. London, 1845.

1396. g. 49 (13). The Amalgamation of Railways considered as affecting the Internal Commerce of the Country. London, 1846.

1851. b. 2 (15). Begin: Pilkington, Mayor. [A Proclamation of the London Common Council.] [London, 1691]

1865. c. 17 (28). The Case of many Coachmen in London and Westminster, and within the Weekly Bills of Mortality, Licensed according to the Act for Licensing Hackney-Coaches, but yet turned out by the present Commissioners. [London, 1670?]

1879. c. 4 (28). The Case of the Waggoners of England, Humbly presented to the Consideration of Parliament. [1700?]

1890. c. 9. Plans, Prospectus, Reports, and Minutes of Evidence, in reference to the London and Birmingham Railway. London, 1832–65.

1890. e. 4 (57). Proposals for Raising by Subscription the Sum of £400,000 in Shares of £100 for the Purpose of building and maintaining a Bridge over the River Thames, from the South side of the said River, at or near...Horse Shoe Alley,...to the Bottom of Queen Street, Cheapside, in the city of London. London, 1810.

2064. a. A General History of the County of Norfolk, intended to convey all the Information of a Norfolk Tour, with the more extended Details.... 2 vols. Norwich, 1829.

2099 (5). An Essay to shew the Advantages that will follow the Progressive Formation of Railways throughout the Kingdom. By E. P. London, 1836.

2390. Tracts on the Proposed Stamford Junction Navigation. Stamford, [1810–11].

6376. b. 28 (2). Bye Laws made by the Trustees of the River Lee Navigation; and Penal Clauses in the Acts of Parliament, passed for improving, extending, and preserving the Navigation of the River Lee; from the Town of Hertford to the River Thames. Hertford, 1827.

6426. de. 13. An Act for better Regulating the Poor; Maintaining a Nightly Watch; Lighting, Paving, and Cleansing the Streets, Rows, and Passages; Providing Fire-Engines and Firemen, and Regulating the Hackney-Coachmen, Chairmen, Carmen, and Porters, within the City of Chester. London, 1772.

6485. c. 11. A Brief Account of Wilkinson and Hetherington, two Notorious Highwaymen, who were executed at Morpeth...1821. Newcastle-upon-Tyne, 1821.

8223. e. 9 (12). The Case of the Unlicensed Hackney-Chairmen, usually Imploy'd within the Cities of London and Westminster and the Suburbs and Liberties thereof.

8223. e. 10 (70). Prospectus of the Kentish Railway Company.

8223. e. 10 (101). Prospectus of the Norfolk, Suffolk, and Essex Rail-Road Company. 1825.

8223. e. 10 (116). Prospectus of the Patent Steam Carriage Company, for England and Wales. [1825]

8223. e. 10 (147). Prospectus of the Surrey, Sussex, and Hants Railroad Company. London, 1825.

8223. e. 10 (148). Prospectus of the Surrey, Sussex, Hants, Wilts, and Somerset Rail-Road Company. Lothbury, 1825.

8223. e. 10 (149). Taunton Great Western Rail-Road Company. [Resolutions drawn up] At a Meeting of Land-Owners and Others, resident in Taunton and its Neighbourhood the Expediency of forming a Rail-Road from...Taunton to Bristol, and from Taunton to Exeter, with a Branch to Tiverton. Taunton, 1825.

8223. e. 10 (151). Proposed Tontine Bridge across the Swale, at or near King's Ferry, Isle of Sheppy.

8229. bbb. 60. Considerations on the Probable Commerce and Revenue that may arise on the Proposed Canal between Newcastle and Maryport. Carlisle, 1807.

8235. a. 71. The Railways of England; containing an Account of their Origin, Progress, and Present State...together with a Map. London, 1839.

8235. aaa. 5. History of the Darlington and Barnard Castle Railway: with Notices of the Stockton and Darlington, Clarence, West Hartlepool, and other Railways and Companies in the District. By an Inhabitant of Barnard Castle. London, 1877.

8235. b. 57 (1). The Carriers' Case Considered in Reference to Railways. London, 1841.

8235. c. 72. Railway Management; or how to make Ten Per Cent. London, 1860.

8235. cc. 41 (1). An Authentic Description of the Kennet and Avon Canal. To which are added, Observations upon the Present State of the Inland Navigation of the South-Western Counties of England; and of the Counties of Monmouth, Glamorgan, and Brecon, in South Wales. London, 1811.

8235. d. 27. Railways and Shareholders; with Glances at Railway Transactions—Shareholders' Powers—Accounts and Audits—Railway Meetings—Defective Legislation, etc. By an Edinbro' Reviewer. 2nd edition. London, 1849.

8235. ee. 4 (1). Oxford and Didcot Railway Bill. Copy of the Evidence taken before a Committee of the House of Commons. Oxford, [1843].

8235. ee. 12 (1). Reasons in favour of a Direct Line of Railroad from London to Manchester. London, 1846.

8235. h. 44. Remarks relating to a Canal intended to be made from the City of Chester, to join the Navigation from the Trent to the Mersey, at or near Middlewich. Chester, 1770.

8245. bb. 14. A Letter to the Inhabitants of Hertford [as to the desirability of Turnpike Trusts working together to secure the best results]. [1771?]

8775. b. 49 (1). Reasons against the Bill now depending in Parliament, for the Scouring out and Deepening of the River Nene.... Cambridge, 1754.

8775. c. 66. Extracts from the Book of Minutes of the Commissioners, and from Reports of Engineers; with other matters relating to the Wear Navigation Act. Sunderland, 1819.

8775. f. 20. Considerations on the Idea of Uniting the Rivers Thames and Severn through Cirencester, with some Observations on other intended Canals. London 1782.

8776. a. 17. A Short Narrative of the Proceedings of the Gentlemen, concerned in obtaining the Act, for building a Bridge at Westminster.... London, 1738.

8776. a. 45. Reflections on the General Utility of Inland Navigation to the

Commercial and Landed Interests of England; with Observations, by Publicola, on the Intended Canal from Birmingham to Worcester, etc. London, [N.D.].

8776. aaa. 33. Facts and Arguments respecting the Great Utility of an Extensive Plan of Inland Navigation in Ireland. Dublin, 1800.

8776. b. 40. Considerations on the Proposed Cut from the Medway to the Thames, ...and its probable Effects on the Navigation of the Medway. London, 1827.

8776. c. 14. Thames Navigation. Observations upon the Evidence adduced before the Committee of the House of Commons, upon the late Application to Parliament for a Bill for making a Navigable Canal from the River Kennet...to join the Basingstoke Canal, etc. Maidenhead, 1825.

8776. c. 21. Proposals at Large for the easy and effectual Amendment of the Roads, by some further Necessary Laws and Regulations, concerning the Wheels of all Carriages;....By a Gentleman. London, 1753.

8776. ee. 17 (6). A Treatise on Inland Navigation. Salisbury: Printed by B. C. Collins, 1788. [Probably Collins was also the author.]

10,347. e. 13. Tables for the Calculation of Lock Dues, payable upon the Calder and Hebble Navigation by virtue of an Act of Parl. (5 Geo. IV). Halifax: Printed for the use only of the Company of Proprietors of the Calder and Hebble Navigation, 1825.

10,348. a. 5. Meine Fussreise durch die drey brittischen Königreiche. Voran einige Nachrichten von dem Feldzuge in Champagne. Von einem französischen Offizier. Riga, 1797.

10,348. ccc. 56. North of England and Scotland in 1704 [A Journal published from the Original MS. of an unknown author]. Edinburgh, 1818.

10,349. a. 1. A Brief Description of England and Wales; containing a Particular Account of each County.... London, [1780?].

10,349. bb. 17. A New and Accurate Description of the Present Great Roads and the Principal Cross Roads of England and Wales commencing at London, and continued to the farthest Parts of the Kingdom, with the several Branches.... London, 1756.

10,349. g. 11. Narrative of the Journey of an Irish Gentleman through England in the Year 1752. Edited (by Henry Huth) from a Contemporary Manuscript, with a few illustrative Notes. London, 1869.

10,815. c. 35. Will of the Duke of Bridgewater. London, 1836.

B. 263. (4). Facts and Reasons Tending to shew, that the Proposed Canal, from the Trent to the Mersey, ought not to terminate at Northwich and Burton; and to prove, That this Plan hath been well digested and hath not wanted public Notoriety.

B. 263. (5) and (7). An (engraved) Plan of the Intended Navigable Canal from Basingstoke to the River Wey; and a List of Landowners through whose Grounds the Basingstoke Canal is intended to pass.

B. 263. (6). Basingstoke Canal Navigation. Supposed Annual Carriage on the said Canal, &c. [1786 or 1787]

B. 503. (5) and (6). Reports of the Engineers appointed by the Commissioners of the Navigation of the Rivers Thames and Isis to survey the State of the said Navigation from Lechlade to Days Lock.... Printed 1791.

B. 503. (7). A Report of the Committee of Commissioners of the Navigation of the Thames and Isis, appointed to Survey the Rivers from Lechlade to Whitchurch, 1791. Printed at Oxford, 1791.

B. 504. (2). An Address to the Public on the New Intended Canal from Stourbridge to Worcester; with the Case of the Staffordshire and Worcestershire Canal Company. 1786.

B. 504. (3). Reflections on the General Utility of Inland Navigation to the Commercial and Landed Interests of England; with Observations on the Intended Canal from Birmingham to Worcester, and some Strictures upon the Opposition given to it by the Proprietors of the Staffordshire and Worcestershire Canal. It is signed " Publicola."

B. 504. (4). A View of the Advantages of Inland Navigations: with a Plan of a Navigable Canal, intended for a Communication between the Ports of Liverpool and Hull. London, 1765. [The author was probably R. Bentley, partner of Josiah Wedgwood.]

C. 32. d. 8. A Direction for the English Traviller, by which he shal be inabled to Coast about all England and Wales &c. [London, 1643] It contains ' A Brief Director for those That would send their Letters to any parts of England, Scotland, or Ireland. Or, a List of all the Carriers, Waggoners, Coaches, Posts, Ships, Barks, Hoys, and Passage Boats that come to London, from the most parts and places, by Land and Sea.'

C.T. 217. Letter [of W. W. Whitmore] to Lord John Russell on Railways. London, 1847.

C.T. 309. (7). Report of the Proceedings of a Public Meeting of the Inhabitants of...Liverpool...[to consider the Rates of Carriage to and from Liverpool].

E. 856. (4). An Act of Common-Councell made the eleventh day of September, in the Yeare of our Lord 1655. For the better avoiding and prevention of Annoyances within the City of London, and Liberties of the same. [London], 1655.

E. 927. (4). The Devil's Cabinet Broke Open: or a New Discovery of the Highway Thieves. Being a Seasonable Advice of a Gentleman lately converted from them, to Gentlemen and Travellers to avoyd their Villanies.... London, 1657.

E. 1063. (59). An Ordinance for Better Amending and Keeping in Repair the Common Highwaies within this Nation. London (Mar. 31), 1654.

E. 1064. (5). An Ordinance of Explanation of a Former Ordinance, Entituled, An Ordinance for Better Amending and Keeping in Repair the Common Highwaies within this Nation. May 16, 1654.

E. 1064. (18). An Ordinance for the Regulation of Hackney Coachmen in London and the Places Adjacent.

E. 1064. (38). An Ordinance for the Giving Libertie for the Carrying of Millstones, Stone, Timber, etc. Sept. 2, 1654.

G. 6463. (228). Articles set down by the Right Worshipfull Thomas Randolph Esquier, Master and Comptroller generall of all her Majesties Posts, and straightly by him commanded to be kept by the Postes from London, to the Northerne borders against Scotland.... London, 1588.

G. 6463. (232). Orders set downe and allowed by the Lordes of her Majesties Privie Counsell, and appoynted to be put in print for the Postes betweene London and the borders of Scotland. [At Westminster, Jan. 14, 1583] [1583]

K. 6. 58 (c). Basingstoke Canal Navigation.

Maps 46. b. 23. A Book of the Names of all Parishes, Market Towns, Villages, Hamlets, and smallest Places, in England and Wales.... London, 1657.

Maps 46. b. 26. A Book of the Names of all Parishes, Market Towns, Villages, Hamlets, and smallest Places, in England and Wales. London, 1677.

Maps 88. d. 13. Documents and Plans relating to Canals of England.

T. 100* (14). The Case of the River Derwent, in respect of Navigation, and of the Bill now in Parliament concerning the same. [1705 (?)]

T. 1157. (2). The Fingerpost; or Direct Road from John-o'-Groat's to the Land's End: being a Discussion of the Railway Question. London, [1825?].

T. 1157. (4). Highways Improved. A Letter to a Member of Parliament on the Expediency of Appointing County or District Surveyors of Highways. London, 1825.

T. 1371. (18). The Probable Effects of the Proposed Railway from Birmingham to London Considered. London, 1831.

T. 1860. (3). A Journey to England. With some Account of the Manners and Customs of that Nation. London, 1700.

SERIAL PUBLICATIONS

Annual Register.

Antiquary, The.

Archaeologia: or Miscellaneous Tracts relating to Antiquity. Published by the London Society of Antiquaries. London, 1770–present time.

Archaeologia Aeliana: or Miscellaneous Tracts relating to Antiquity. Published by the Society of Antiquaries, of Newcastle-upon-Tyne, New Series.

British Almanac and Companion. London, 1829–46.

Camden Society. Annals of the First Four Years of the Reign of Queen Elizabeth by Sir John Hayward, Knt. Edited from Harl. MS. 6021, by John Bruce. London, 1840.

Camden Society. Wills and Inventories from the Registers of the Commissary of Bury St Edmunds and the Archdeacon of Sudbury. Edited by Samuel Tymms. London, 1850.

Camden Society. Letters and Papers of the Verney Family down to the End of the Year 1639. Printed from the Original MSS. in the possession of Sir Harry Verney, Bart. Edited by John Bruce. London, 1853.

Chetham Society Publications. Vol. XLVI contains The Portmote or Court Leet Records of the Borough or Town and Royal Manor of Salford, from 1597 to 1669 inclusive. Transcribed and edited by J. G. de T. Mandley. 1902.

Vol. LXV contains the Continuation of the Court Leet Records of the Manor of Manchester, A.D. 1586–1602. Compiled and edited by John Harland, 1865.

Chetham Society Publications. Vols. LXVIII and LXXII, Collectanea relating to Manchester and its Neighbourhood at Various Periods. Compiled, arranged, and edited by John Harland, F.S.A. Printed for the Society, 1866 and 1867.

The Commercial and Agricultural Magazine, 1799–1802, continued as The Agricultural Magazine, 1802–

Edinburgh Review. Vol. XXXII (1819), pp. 477–87; Vol. CXIX (1864), pp. 340–68.

Gentleman's Magazine.

The Grand Magazine of Universal Intelligence, and Monthly Chronicle of Our Own Times. 3 vols. London, 1758–60.

Highland Society of Scotland. Prize Essays and Transactions of the Highland Society of Scotland. Vol. VI (Edinburgh, 1824), pp. 1–146, deals with Railroads in a series of Essays edited by Robert Stephenson.

Local Notes and Queries. 7 vols. (Birmingham Free Reference Library, No. 144,953.)

The London Magazine; Or, Gentleman's Monthly Intelligencer. 27 vols. London, 1732–58.

The London Magazine. Vol. I, N.S., 1825.

Memoirs and Proceedings of the Literary and Philosophical Society of Manchester. 2nd series, Vol. III (1819), and 4th series, Vol. VIII (1894).

Monthly Magazine.

Oxford Historical Society. Collectanea. 4 vols. Oxford, 1885–1905.

Quarterly Review, The. Vols. xxxii and xcvii. London, 1825 and 1855.

Quatrième Congrès International de Navigation Intérieure, tenu à Manchester en 1890. Rapports des délégués Français sur les travaux du Congrès. Paris, 1892.

Railway Chronicle, 1844–49. London, 1844–49.

Railway and Commercial Journal, Herepath's. Vols. ix–xi. London, 1847–49.

Railway Magazine, The, and Annals of Science, by John Herepath: containing copious accounts of all Railways at Home and Abroad, N.S., Vols. i–vi. London, 1836–39.

Railway Times. Vols. i–xii (1838–49). London, 1838–49.

Repertory of Arts and Manufactures, The. 16 vols. London, 1794–1802; continued as The Repertory of Arts, Manufactures, and Agriculture, consisting of Original Communications, etc. 2nd series, 46 vols. London, 1802–25.

Royal Statistical Society, Journal of the. London, 1839–present time.

Transactions of the Cumberland and Westmorland Antiquarian and Archaeological Society, N.S., Vols. i–iv. Kendal, 1901–04.

Transactions of the Lancashire and Cheshire Antiquarian Society. Manchester, 1883–present time.

NEWSPAPERS

The Bath Chronicle. Jan. 11, 1787–Jan. 8, 1789.

Birmingham Advertiser, 1833. Birmingham, 1833.

Aris's Birmingham Gazette. (Birmingham), 1740–1833.

The Birmingham Journal, 1825–34. Birmingham, 1825–34.

Bonner and Middleton's Bristol Journal. (Bristol), Jan. 14, 1775–Dec. 23, 1786.

The Bristol Gazette and Public Advertiser. (Bristol), Aug. 7, 1777–May, 15, 1788.

The Bristol Liberal. Jan. 7, 1832.

The Bristol Mercury. June 30, 1832–Oct. 5, 1833.

The Constitutional Chronicle. (Bristol), 1780–82 (irregular numbers only).

Felix Farley's Bristol Journal. (Bristol), Nov. 2, 1776–May 17, 1788; Nov. 17, 1832–Dec. 13, 1834.

Sarah Farley's Bristol Journal. (Bristol), April 27, 1782–August 3, 1782.

The Chronicle of Bristol. Aug. 1, 1829–Jan. 1, 1830.

The Cambridge Chronicle and Journal, from Nov. 2, 1810–Nov. 3, 1815.

Glocester Journal, 1789–91. Gloucester, 1789–91 (irregular numbers only).

Hampshire Advertiser and Salisbury Guardian. Southampton, 1832–34.

The Hampshire Telegraph and Sussex Chronicle. 1814–19.

Hereford Journal. April 20, 1803–Dec. 4, 1805 (irregular numbers only).

Hull Advertiser. 1834–35.

Leeds Intelligencer, 1758–1836. Leeds, 1758–1836.

Billinge's Liverpool Advertiser, 1823–28; continued as The Liverpool Times, 1829–32. Liverpool, 1823–32.

The City Mercury: or, Advertisements concerning Trade. London, Nov. 4, 1675; July 4, 1692; March 13, 1693; March 20, 1693; March 27, 1693; June 11, 1694; Dec. 10, 1694.

The County Chronicle and Weekly Advertiser, for Essex, Herts, Kent, Surrey, Middlesex, Berks, etc. London, 1818–36.

Domestic Intelligence, or News both from City and Country. London, Oct. 24, 1679– ; continued as The Protestant (Domestick) Intelligence, or News both from City and Country, from Jan. 16, 1680, on. London, 1679–82.

The Economist, Weekly Commercial Times, and Bankers' Gazette, 1844–47. London, 1844–47.
The London Evening Post. (London), Jan. 20–22, 1732–36.
The Gazetteer and New Daily Advertiser. (London), Feb. 27, 1787–April 4, 1788.
The General Advertiser. (London), Feb. 26, 1785–May 21, 1787.
The Morning Chronicle and London Advertiser. Jan. 8, 1785–April 1, 1786.
The Morning Chronicle. (London), April 1, 1791–April 11, 1792.
The Morning Post and Daily Advertiser. (London), July 20, 1791.
The Oracle. (London), March 22, 1792–Sept. 22, 1792.
The Public Advertiser. (London), April 8, 1786–Oct. 22, 1793.
The St James Chronicle. (London), Dec. 7, 1786–Aug. 29, 1793.
The London Times. March 3, 1794–Nov. 8, 1847.
The Manchester Advertiser. (Manchester), Aug. 30, 1825 to end of 1826.
Wheeler's Manchester Chronicle, 1831–32. Manchester, 1831–32.
Manchester Courier, 1825–27. Manchester, 1825–27.
Manchester Gazette, 1824–31. Manchester, 1824–31.
Manchester Guardian, 1825–46. Manchester, 1825–46.
The Newcastle Courant. (Newcastle), April 16, 1774–Dec. 4, 1794; 1821–32.
Oxford Gazette and Reading Mercury, 1767–71. Reading, 1767–71.
Jackson's Oxford Journal. July 29, 1780–Nov. 2, 1793.
The Reading Mercury and Oxford Gazette. Reading, 1792–96.
Sheffield Iris, 1828–37. Sheffield, 1828–37.
Ward's Sheffield Public Advertiser. (Sheffield), 1760–65.
The Shrewsbury Chronicle or Wood's British Gazette. (Shrewsbury), Feb. 20, 1773–April 15, 1775.
The York Herald, County and General Advertiser. York, 1812–14.

GUIDE BOOKS

Bath. The New Bath Guide, or useful Pocket Companion for the Year 1784.
The New Birmingham Directory, and Gentleman and Tradesman's Compleat Memorandum Book. (Printed by and for M. Swinney.) Birmingham, 1774 (?).
The Birmingham Directory; or Merchant and Tradesman's Useful Companion. Birmingham, 1777.
Chapman's Birmingham Directory. Birmingham, 1808.
Brit. Mus. 579. c. 41 (1). Brighton New Guide; or a Description of Brighthelmston and the Adjacent Country. London, 1800.
The Tradesman's and Traveller's Pocket Companion; or the Bath and Bristol Guide.... Bath, [1750?)].
Matthews, William. Matthews New Bristol Directory, for the Year 1793–94, etc. Bristol, [1793–94].
Bristol. The New History, Survey and Description of the City and Suburbs of Bristol, or Complete Guide, and informing and useful Companion for the Residents and Visitants of this ancient, extensive and increasing City, etc. Bristol, 1794.
A Description of the University, Town, and County of Cambridge...Directions concerning the Posts, Roads, Stage Coaches, Waggons, etc., to and from Cambridge. Cambridge, 1796.
The Chester Guide. Chester, 1795.
Battle's Hull Directory for the Year 1791. Hull, [1791].
A Complete Guide to All Persons who have any Trade or Concern with the City of London, and Parts adjacent; containing the Names of all streets....An

Account of all the Stage-Coaches, Carriers, &c. (for the years 1740, 1772, 1774, 1783). London, 1740, 1772, 1774, 1783.

(W.) Holden's Annual List of Coaches, Waggons, Carts, Vessels, &c., from London, to all parts of England, Wales, Scotland, and Parts of Ireland, including the villages near the Metropolis. London, 1802, 1809.

Brit. Mus. 579. c. 43 (3). The Ancient and Modern History of Portesmouth, Portsea, Gosport and their Environs. Gosport, [N.D.]

Southampton. The Southampton Guide for 1774 and 1797.

The York Guide (1796): Containing a Description of the Cathedral and other Public Buildings...to which is added...Times of the Posts coming in and going out; Mail and other Coaches, etc. 2nd edition. York, 1796.

MAPS

Ogilby, John. Itinerarium Angliae, or, A Book of Roads, wherein are contained the Principal Road-Ways of His Majesty's Kingdom of England and Dominion of Wales. London: Printed by the Author at his House in White-Fryers, M.DC.LXXV.

Brit. Mus. Maps 86. d. 10 (33). A New Map containing all the Cities, Market Townes...Roads, etc. London, 1696.

England Exactly Described, or a Guide to Travellers: in a Compleat Sett of Mapps of all the County's of England. London, 1715.

Brit. Mus. B. 264. (3). A Set of Fifty New and Correct Maps of England and Wales, &c. with the Great Roads and Principal Cross-Roads, &c. Shewing the Computed Miles from Town to Town. A Work long wanted, and very useful for all Gentlemen that travel to any Part of England. All, except two, composed and done by Hermann Moll, Geographer. London, 1724.

Badeslade, Thomas. Chorographia Britanniae, or a set of Maps of all the Counties in England and Wales. (Comprises a Map for showing the great roads from London and a Map of all the cross roads.) London, 1742.

Bowles's New Traveller's Guide, through the Principal Direct and Cross Roads of England and Wales. London, [ca. 1750].

Armstrong, Mostyn John. An Actual Survey of the Great Post-Roads between London and Edinburgh. London, 1776.

Paterson, Daniel. A New and Accurate Description of all the Direct and Principal Cross Roads in Great Britain. 4th edition. London, 1778. Fifth edition, corrected, and improved, with Additions, London, 1781. Fifteenth edition of same, London, 1811.

Kitchin, Thomas. The Roads in England and Wales. Engraved from the latest Surveys, by Thomas Kitchin, Hydrographer to His Majesty. 1783.

Brit. Mus. 1190. 7. A Map of England, shewing the Lines of all the Navigable Canals, with those which have been proposed.... 1795.

A Collection of Prospectuses, Maps, and other Documents relating to Early Railways, Canals, etc., 1798–1846. Birmingham Free Central Library, Iron Room, Box No. III. (Some fine material.)

Enouy, Joseph. A New Map of England and Wales, compiled from Actual Surveys of the Counties...with the Turnpike Roads according to the New Admeasurements, Navigable Rivers and Canals. London, 1801.

Brit. Mus. Maps 30. a. 47. Map of England and Wales with Part of Scotland [showing clearly the navigable canals and rivers]. 1802.

Smith's New Map of England and Wales, with Part of Scotland, including the Turnpike and principal Cross Roads, the Course of the Rivers and Navigable Canals, &c. London, 1806. Ibid., for the year 1818, and for 1827.

Owens, William. New Book of Roads, or a Description of the Roads of Great Britain. New edition. London, 1814.

Mogg, Edward. A Survey of the Highroads of England and Wales, with Part of Scotland...exhibiting at one View...the various Branches of Roads and Towns to which they lead, together with the actual Distance of the same from the Main Road, Rivers, Navigable Canals, Railways...&c. London, 1814–15. Brit. Mus. Maps 47. c. 14.

Cary, John. Cary's New Itinerary: or an Accurate Delineation of the Great Roads, both Direct and Cross, throughout England and Wales: with many of the principal roads, in Scotland, from an Actual Admeasurement. 7th edition, with Improvements. London, 1817.

(Huerne de Pommeuse.) Des Canaux Navigables. Atlas de la Navigation Intérieure de l'Angleterre et de la France. 1819. Paris, 1822.

Cheffins's Official Maps of the London and Birmingham, and the Grand Junction Railways. London, 1839.

Brit. Mus. Maps 92. d. 29. Wyld's Railroad Map of England and Wales [showing also those in progress and those projected, and outlining the high roads]. London, 1840.

A Map of England, Wales, and Scotland, describing all the Direct and Principal Cross Roads in Great Britain, and showing the Railroads, Great Rivers and Navigable Canals. London, 1841.

Bradshaw's Railway Companion, containing the Times of Departure, Fares, etc., of the Railways of England, etc. Manchester and London, 1842.

Cary's Reduction of his Large Map of England and Wales, with Part of Scotland; comprehending the whole of the Turnpike Roads, with the Great Rivers and the Course of the different Navigable Canals. London, [1850?].

Collins' Railway Map of England and Wales and Part of Scotland, shewing the Inland Navigation by means of Rivers and Canals, together with the Rail and Principal Turnpike Roads, from the most recent surveys. London, [1852?].

England and Wales showing the Railroads, completed, proposed, and in Progress. London, [1853].

Bett's New Map of England and Wales compiled from the latest Parliamentary Documents, showing the Roads, and Railroads. London, [ca. 1853].

Stanford's Railway and Road Map of England and Wales. London, 1862.

Brit. Mus. Maps 29. a. 37. A Pocket-Guide to the English Traveller: Being a Compleat Survey and Admeasurement of all the Principal Roads and most considerable Cross-Roads in England and Wales. In One Hundred Copper Plates. London, 1719.

Brit. Mus. 1175. (106). A New Map of England and Wales, comprehending the whole of the Turnpike Roads, with the Great Rivers and Navigable Canals. London, 1804. Very complete in its details.

Brit. Mus. C. 32. d. 8. A New Map of the Kingdome of England and Principality of Wales, taken out of I. S. (i.e., John Speed). London, 1673. It is contained in 'A Direction for the English Traviller.' London, 1643.

Brit. Mus. 1205. (9). A New Travelling Map of England, Wales and Scotland; Drawn from all the Surveys which have hitherto been made of particular Counties, describing the Direct and principal Cross Roads, Cities, Boroughs, Market Towns and Villages, to which is added the actual distance from one

Market Town to another, and exact admeasurement of each place from London. By Edward Mogg. London, 1810. Excellent map.

Brit. Mus. 577. e. 2. The Traveller's Guide: or, a Most Exact Description of the Roads of England. Being Mr Ogilby's Actual Survey, and Mensuration by the Wheel, of the Great Roads from London...together with the Cross Roads. London, 1699. This is simply a pocket edition of Ogilby's 'Itinerarium Angliae.'

Brit. Mus. 1175. (45). A New Map of England and Wales, describing all the Turnpike and principal Bye Roads, the Great Rivers and Navigable Canals. By Robert Rowe. London, 1819. Great detail.

Brit. Mus. 1205. (10). Wallis's New Travelling Map of England and Wales, with Part of Scotland, on which are delineated the Direct and Principal Cross Roads,... Also the Course of the Great Rivers.... London, 1815. Clear representation of the roads because avoiding a mass of other details.

Brit. Mus. 1175. (17). The Post Roads through England and Wales; by T. Jefferys, Geographer to His Majesty. [London, 1792] Omits all other roads except post roads.

Brit. Mus. K. 5. 68. Bowles's New Pocket Map of England and Wales, Revised and corrected from the best Authorities; with the Addition of New Roads, and other Improvements. By D. Paterson. London, 1773.

Brit. Mus. K. 191. g. 13. Nouveau Théâtre de la Grande Bretagne, Vol. iv. After the first two plates, there is 'A New Map of South Britain or England and Wales. Containing all the Cities and Market Towns with the Roads from Town to Town....' London, 1720. Does not show London as the centre of a series of great roads to all parts of the kingdom; but shows many smaller places than London which were greater road centres than it.

Brit. Mus. K. 5. 52. A New Map of England. Containing the Adjacent Parts of Scotland, Ireland, France, Flanders and Holland. Shewing the true Scituation and Distance of London from Edinburgh, Dublin,....With a Description of the Post Roads, and their several Branches from Town to Town,....By R. Greene. London, [1690]. Does not show such a vast network of roads as some other maps of this time.

Brit. Mus. K. 5. 60. The South Part of Great Britain, called England and Wales. Containing all yᵉ Cities, Market Towns, Boroughs,...with yᵉ Names of ye Rivers,...Great or Post Roads, and principal Cross Roads &c. By Hermann Moll, Geographer. London, 1710.

Brit. Mus. K. 5. 61. A New and Accurate Map of England and Wales. Describing in a more distinct and correct manner than any other Map Extant, all the Cities, Boroughs, & Market Towns &c. also all the direct and principal Cross Roads, with the post Towns and computed distances between Town and Town....By E. Bowen, Geographer. London, 1734. Roads clearly delineated.

Brit. Mus. K. 5. 62. Le Royaume d'Angleterre. Divisé en Comtez et Baronies. Dressé sur les dernières Observations par et chez le Sʳ le Rouge Ingénieur Géographe du Roi. Paris, 1745. Gives the great roads only.

Brit. Mus. 1205. (4). The Traveller's Guide or Ogilby's Roads Epitomized. A Sett of Tables in which are described all the grand Roads & several of the Cross Roads of England and Wales....By I. V. Kircher. London, [1706]. Shows a vast network of roads at that time, as does also

Brit. Mus. Maps 86. d. 10 (33). A New Map Containing all the Citties, Market Townes, Rivers, Bridges, & other considerable Places in England and Wales. Wherein are delineated yᵉ Roads from Towne to Towne,.... London, [1696].

Brit. Mus. K. 5. 88. The Traveller's Companion or the Post Roads of England and Wales; with Distances in Measured Miles. By the late John Rocque, Chorographer to the King. London, 1771.

Brit. Mus. K. 5. 84. The Roads of England according to Mr Ogilby's Survey. London, [1712]. Circular map of the great roads.

Brit. Mus. Maps 89. e. 3 (50). A New Mapp of the Kingdome of England, representing the Princedome of Wales, and other Provinces, Cities, Market Towns, with the Roads from Town to Town. Amsterdam, [1590]. This volume gives also a complete series of county maps.

Brit. Mus. 1220. 4. Smith's New Map of the Railways and Canals of England, Wales and Scotland. Containing also the principal Roads, and the distances of the places from London. London, 1838. Gives dates of Canal Acts.

N.B. The foregoing are only a few of the large number of maps consulted in order to secure the greatest degree of accuracy in this work.

INDEX

9 780367 175962